EXPLORING CREATION WITH

CHEMISTRY

3rd EDITION

Kristy Plourde, M.S.

Technical Editorial Contributors:
Mary Manson McManamy, Ph.D.
Chad Snyder, Ph.D.
Brian Meier, M.A.
Doug Simpson, M.S.
Daniel Martin, M.S.

apologia

Exploring Creation with Chemistry
3rd Edition

Published by
Apologia Educational Ministries, Inc.
P.O. Box 896844
Charlotte, NC 28289-6844
www.apologia.com

ISBN: 978-1-946506-58-0

Exploring Creation with Chemistry, 3rd Edition, is revised and updated from a previous edition authored by Jay L. Wile.

Cover and Book Design: Doug Powell

Printed by Bradford & Bigelow, Newburyport, MA
December 2022

20 19 18 17 16 15 14 13 12 11

INSTRUCTIONAL SUPPORT

Did you know that in addition to publishing award-winning curriculum, Apologia also offers instructional support? We believe in helping students achieve their full potential, whatever their learning style. When you choose an Apologia curriculum, you are not just selecting a textbook. Every course has been designed with the student's needs in mind. Please note that not all options are available for every course.

INDEPENDENT LEARNERS

Apologia Textbooks and Notebooks are written to students in a conversational tone so that young people can easily navigate through the curriculum on their own. Apologia curriculum helps students methodically learn, self-check, and master difficult concepts before moving on.

AUDITORY LEARNERS

Apologia Audio Books are the complete text of each course read aloud. Sometimes students learn best when they can see and hear what they're studying. Students can follow along with the audio while reading.

VISUAL LEARNERS

Apologia Video Instruction enhances the student's education with on-location video footage, lectures, animated diagrams, and video presentations of all experiments.

SELF-PACED LEARNERS

Apologia Self-Paced Courses are designed as the best of all worlds. Students have online access to the entire textbook with audio and instructional videos included within the online text. Tests and grading are included.

SOCIAL LEARNERS

Apologia Live Online Classes provide an environment where students can interact in real time with their classmates and a professional instructor in a structured virtual classroom.

At Apologia, we believe in homeschooling. We are here not only to support your endeavors but also to help you and your student thrive! Find out more at apologia.com.

apologia

STUDENT NOTES

EXPLORING CREATION WITH

CHEMISTRY

3rd EDITION

You are about to embark upon an amazing journey! In this text, you will be introduced to the fascinating subject of chemistry. You will learn about the matter that makes up God's creation and how it changes. Although this course will be hard work, you will learn some truly incredible things. We hope that you develop an even deeper appreciation for the wonderful creation that God has given us!

ABOUT THIS BOOK

This text contains 16 modules. Each module should take you about 2 weeks to complete, as long as you devote at least an hour of every school day to studying chemistry. At this pace, you will complete the course in 32 weeks. Since most school years are longer than 32 weeks, there is some built-in flex time. You should not rush through a module just to make sure that you complete it in 2 weeks. Set that as a goal, but be flexible. Some of the modules might be harder for you than others, so take more time on them to make sure you understand the material.

To help guide your study, there are several student exercises that you should complete:

- Answer the "On Your Own" questions as you read the text. This will cement in your mind the concepts you are trying to learn. Answers are provided at the end of each module. Once you have answered an "On Your Own" question, turn to the end of the module and check your work. If you did not get the correct answer, study the answer key to learn why.
- Answer the review questions after you have completed the entire module. They will help you recall the important concepts from the reading.
- Solve the practice problems after completing the module. They allow you to review the important quantitative skills from the module.

The separate Solutions and Tests Manual includes answers to the review questions and practice problems, as well as tests and test answers. It also includes sample calculations for several of the experiments in this textbook.

You should memorize any information that is centered in the text, such as definitions and key facts, which are printed in blue type. Words that appear in blue type in the text are important terms that you should know for the test. You should also memorize all the numbered equations provided in the text, unless otherwise specified.

There are also several items at the back of the book that you will find useful in your studies. The glossary defines many of the terms used in the course, and the index tells you where topics can be found in the course. Appendix A lists all the tables and equations from the text. Appendix B contains extra problems for each module. If you are having difficulty with a certain type of problem, you can get more practice by solving the related problems in appendix B. The solutions for those problems are provided in the Solutions and Tests Manual. Appendix C contains a complete list of all of the supplies you need to perform the experiments in this course.

EXPERIMENTS

The experiments in this course are designed to be done as you are reading the text. We recommend that you keep a notebook of these experiments or purchase the student notebook that goes with this text. A notebook serves two purposes. First, as you write about the experiment in the notebook, you will be forced to think through all of the chemistry concepts that you explored in the experiment, which will help cement them into your mind. Second, some colleges may ask for some evidence that your chemistry course had a laboratory component. The notebook will not only provide such evidence but will also show colleges the quality of your chemistry instruction.

We recommend that you perform the experiments in the following way:

- When you come to an experiment in the text, read through it in its entirety to gain a quick understanding of what you are to do.
- Once you have read the experiment, start a new page in your laboratory notebook. Record a purpose statement and a hypothesis for the experiment. While conducting the procedure, write down all of the data—any observations or measurements you make during the experiment. For example, if you see a solution begin to bubble when you mix 2 chemicals, that is part of the experiment's data and should be written down. If you measure the mass of something, it is part of the experiment's data and should be written down. In addition, use this page to do any calculations that you are asked to do as a part of the experiment.
- When you have finished the experiment and any necessary calculations, write a brief report in your notebook, right after the page where you wrote the data and calculations. You should not write a step-by-step procedure. Instead, write a brief summary that will allow someone who has never read the text to understand what you did and what you learned.
- PLEASE OBSERVE COMMONSENSE SAFETY PRECAUTIONS! The experiments in this course are no more dangerous than most normal household activity. Remember, however, that the vast majority of accidents happen in the home. Chemicals used in the experiments should never be ingested, and hot beakers and flames should be regarded with care. OSHA recommends wearing some sort of eye protection such as safety glasses or goggles while performing all experiments.

If you are interested in performing more than the experiments contained in this course, you can purchase a MicroChem kit and its accompanying laboratory manual. This kit contains a total of 17 experiments, 14 of which are referenced in this course. The graphic on the right will alert you to an opportunity to perform a MicroChem experiment.

Instructions next to the graphic will indicate which experiment from the MicroChem kit you can do at that time. Please note that these experiments are *optional*. You will have a solid high school experience in chemistry without doing these extra experiments. However, they do enhance this course if you choose to do them. If you look up "MicroChem experiments" in the index, you will find the page numbers that contain references to these experiments.

LABORATORY EQUIPMENT

Although the chemicals used in the experiments are common substances that you probably have around your home, there are specialized materials (such as beakers and test tubes) that make the experiments much easier to perform. You can purchase a laboratory kit that contains these materials at apologia.com. Please consult Apologia's website for the current price of the kit.

Laboratory Equipment Set Available for *Exploring Creation with Chemistry*, 3rd Edition

Item	# of Experiments*	Item	# of Experiments*
Safety goggles	30	250 mL beaker	11
Mass scale (0–500 grams)	8	100 mL beaker	11
Blue and red litmus paper	1	Watch glass	2
Thermometer (–10–110 °C)	6	2 eyedroppers	4
6 test tubes	6	Stirring rod	7
Test tube cleaning brush	4	50 mL glass graduated cylinder	8
Funnel	2	50 mL plastic graduated cylinder	1
Alcohol burner with stand	6	Round pieces of filter paper	2

*The "# of Experiments" column tells how many experiments require the item or its household substitute.

If you already have items that meet these specifications, or if you do not wish to do *every* experiment in the course, you need not order the kit. These items make it *easier* for you to do the experiments, but you can use common household substitutes for each of them. Once you use a household item for a chemistry experiment, it should not be placed back into everyday use.

Be sure to clean your equipment after each experiment.

- If you use the burner, you may store it with the fuel inside it as long as you have a storage space where it won't spill. If there is a chance of spilling, then pour the fuel back into its container for safe storage. The metal frame and wire gauze used over the burner may be discolored, but this is considered normal.

- If you use the mass scale, then make sure there are no residual chemicals on it. Do not submerge the body of the scale into water. Wash the plastic cup that sits on top of the scale with soap and water.
- Throw paper products that need to be discarded into your trash can. Wrap all leftover solids in a paper towel and discard them in the trash.
- Pour all liquids left over at the end of our experiments down the drain with plenty of water. Let the water run for at least a minute to ensure that all chemicals have been washed down the drain.
- Hand wash glassware with soap and water. Rinse it at least 3 times to ensure that it is not contaminated and will be ready for the next time you use it.
- When you use a chemical such as lye or ammonia, wipe the table surface first with a paper towel to remove any remaining chemical(s). Then clean the table surface using soap and water and dry it with a paper towel. Clean and dry it at least 3 times. Discard the paper towels in your trash can.
- Wipe chemical bottles with a paper towel(s) to remove any remnants of the chemical on the outside of the bottle. Discard the paper towel in your trash can.
- Sweep or vacuum the floor.
- Wash your hands with soap and water to remove any final chemical residues.
- Store your equipment and supplies in a dry and safe place. You should consider locking them up to keep them away from younger children.

LEARNING AIDS

A special website for this course contains links to web-based information about chemistry. If you are having trouble with a particular topic, you can go to the website and look at the links for that module. Most likely, you will find help there. Finally, if you are enjoying a particular module in the course and would like to learn more about it, there are links which will lead you to advanced material related to that module.

Please go to apologia.com and click on the Book Extras link to find instructions for accessing our Book Extras material. For this course, you will need the following password:

Howgreatareyourworks!

TABLE OF CONTENTS

WELCOME

When you look around, do you ever wonder where everything came from and how it was made? Have you ever contemplated why a tree is hard, a sponge is soft, and a breeze is invisible?

> By faith we understand that the universe was formed at God's command, so that what is seen was not made out of what was visible.
> —Hebrews 11:3, NIV

Welcome to the world of chemistry! This year you are going to take a journey that allows you to explore God's creation with the eyes of a scientist. Studying God's creation at the molecular and atomic level can enable you to understand how wonderfully everything fits together in this world God has created for us.

first reactions

What do you see when you look at the periodic table of elements? At first glance, most students see so many scientific symbols that they become intimidated. But the periodic table is the essence of chemistry. The years of discovery and knowledge summarized in it make it unlike any other common science tool.

From the galaxies in the universe to the tiniest microscopic cell, you experience the elements wherever you look!

> How many are your works, LORD! In wisdom you made them all; the earth is full of your creatures.
> —Psalm 104:24, NIV

Science is the endeavor of explaining the truth of the world around us, and God is the source of both creation and truth. You will discover that proper application of scientific principles will help you uncover how the world around you operates. Since science and faith both search for truth, they complement each other. The more you know about your world, the more you will wonder at the complex beauty of God's creation.

> He has made everything beautiful in its time. He has also set eternity in the human heart; yet no one can fathom what God has done from beginning to end.
> —Ecclesiastes 3:11, NIV

God gave humans dominion over the earth, so we can understand many things about it. This textbook is not just a compilation of facts and figures for you to memorize. This textbook is designed to take you on a remarkable journey that involves facts about chemistry, figures to help you understand the facts, and truth from your Creator. We at Apologia pray that this text will enable you to say:

> "How great are your works, LORD, how profound your thoughts!"
> —Psalm 92:5, NIV

MEASUREMENT, UNITS, AND THE SCIENTIFIC METHOD

Our understanding of life has changed more in the past 2 centuries than all the previously recorded span of human history. The earth's population has increased more than 5-fold since 1800, and our life expectancy has nearly doubled because of our ability to synthesize medicines, control diseases, and increase crop yields. Many goods are now made of polymers (plastics) and ceramics instead of wood and metal because of our ability to manipulate and manufacture materials with properties unlike any found in nature. In one way or another, all of these changes involve chemistry.

first reactions

When God told Noah to build an ark 300-cubits long, Noah had to know how long a cubit was and what the 300 meant. If Noah had built an ark 300-inches long, he would have failed his mission. In science and engineering, a common language is also needed as foundation to further learning. Some of the mathematical language of chemistry will be introduced in this module. Like Noah, you will learn some basic terms of measurement and units, handling of numbers, and performing calculations.

What is chemistry? Quite simply, chemistry is the study of matter. Of course, this definition doesn't do us much good unless we know what matter is. So, to understand what chemistry is, we first need to define matter.

Matter—Anything that has mass and takes up space.

If matter is defined in this way, almost *everything* around us is matter. Your family car has a lot of mass. That's why it's so heavy. It also takes up a lot of space in the driveway or the garage. Your car must be made of matter. The food you eat isn't as heavy as a car, but it still has some mass. It also takes up space. So food must be made up of matter as well. Indeed, almost everything you see around you is made up of matter because nearly everything has mass and takes up space. There is one thing that has no mass and takes up no space. It's all around you right now. Can you think of what it might be?

You might think that the answer is air. However, that's not the right answer. Perform Experiments 1.1 and 1.2 to see what we mean.

EXPERIMENT 1.1

PURPOSE: To determine if air has mass.

MATERIALS:

- Meterstick (A yardstick will work as well, but a 12-inch ruler is not long enough.)
- Two 8-inch or larger balloons
- 2 pieces of string long enough to tie the balloons to the meterstick
- Tape
- Safety goggles

QUESTION: Does air have mass?

HYPOTHESIS: Pick one: Either air has mass or air does not have mass.

PROCEDURE:

1. Without blowing them up, tie the balloons to the strings. Be sure to make the knots loose so that you can untie one of the balloons later in the experiment.
2. Tie the other end of each string to an end of the meterstick. Try to attach the strings as close to the ends of the meterstick as possible.
3. Once the strings have been tied to the meterstick, tape them down so that they cannot move.
4. Go into your bathroom and pull back the shower curtain so that a large portion of the curtain rod is bare. Balance the meterstick (with the balloons attached) on the bare part of the shower curtain rod. You should be able to balance it very well. If you don't have a shower curtain rod or you are having trouble using yours, you can use any surface that is adequate for delicate balancing like the upper part of a chair.
5. Once you have the meterstick balanced, stand back and look at it. The meterstick balances now because the total mass on one side equals the total mass on the other side. To knock it off balance, you would need to move the meterstick or add more mass to one side. You will do the latter.
6. Have someone else hold the meterstick so that it does not move. For this experiment to work properly, the meterstick must stay stationary.
7. While the meterstick is held stationary, remove one of the balloons from its string (do not untie the string from the meterstick), and blow up the balloon.
8. Tie the balloon closed so that the air does not escape, then reattach it to its string.
9. Have the person holding the meterstick let go. If the meterstick was not moved while you were blowing up the balloon, it will tilt toward the side with the inflated balloon as soon as the person lets it go. This is because you added air to the balloon. Since air has mass, it knocks the meterstick off balance. So air does have mass!
10. Clean up and return everything to the proper place.

CONCLUSION: What did you think? Write something about what you observed related to the fact that air has mass.

think about this

Air is typically used as a metaphor for nothingness. It is, however, very complex. Aristotle (384–322 BC) is generally given credit for being the first to state that air has weight, although many did not believe him. Would it surprise you to know that some 1,400 years *before* Aristotle, it was known that air had weight? It's true. The Bible tells us that God "gave to the wind its weight and apportioned the waters by measure" (Job 28:25). Think about this verse in the context of the timeline of scientific knowledge. Job may have lived anywhere from 2300 to 1700 BC. God created everything visible and invisible, including air. The Bible is His word, and we can trust it to be true. We explore science to understand what God already knows about His creation.

EXPERIMENT 1.2

PURPOSE: To determine if air takes up space.

MATERIALS:

- Tall glass
- Paper towel
- Sink full of water
- Safety goggles

QUESTION: Does air take up space?

HYPOTHESIS: Pick one: Either air takes up space or air does not take up space.

PROCEDURE:

1. Fill the sink with water until the water level is high enough to submerge the entire glass.
2. Make sure the inside of the glass is dry.
3. Wad up the paper towel and shove it down into the bottom of the glass.
4. Turn the glass upside down and be sure that the paper towel does not fall out of the glass.
5. Submerge the glass upside down in the water, being careful not to tip the glass at any time.
6. Wait a few seconds and remove the glass, still being careful not to tilt it.
7. Pull the paper towel out of the glass. You will find that the paper towel is completely dry. Even though the glass was submerged in water, the paper towel never got wet. Why? When you tipped the glass upside down, there was air inside the glass. When you submerged it in the water, the air could not escape the glass, so the glass was still full of air. Since air takes up space, there was no room for water to enter the glass, so the paper towel stayed dry.
8. Repeat the experiment, but this time be sure to tip the glass while it is underwater. You will see large bubbles rise to the surface of the water. When you pull the glass out, you will find that it has water in it and that the paper towel is wet. This is because tilting the glass allowed the air trapped inside it to escape. Once the air

escaped, there was room for the water to come into the glass.

9. Clean up and return everything to the proper place.

CONCLUSION: What did you think? Write something about what you observed related to the fact that air takes up space.

Now that you see that air does have mass and does take up space, have you figured out the correct answer to our original question? What very common thing that is surrounding you right now has no mass and takes up no space? The answer is light. As far as scientists can tell, light does not have any mass and takes up no space. Light is not considered matter. Instead, it is pure energy. Everything else that you see around you is considered matter. Chemistry, then, is the study of nearly everything! As you can imagine, studying nearly everything can be a very daunting task. However, chemists have found that even though there are many forms of matter, they all behave according to a few fundamental laws. If we can clearly understand these laws, then we can clearly understand the nature of the matter that exists in God's creation.

Before we start trying to understand these laws, we must step back and ask a more fundamental question: *How* do we study matter? The first thing we have to be able to do in order to study matter is to measure it. If we want to study an object, we first must learn things like how big it is, how heavy it is, and how old it is. To learn these things, we have to make some measurements. The rest of this module explains how scientists measure things and what those measurements mean.

UNITS OF MEASUREMENT

Let's suppose you are making curtains for a friend's windows. You ask him to measure the window and give you the dimensions so that you can make the curtains the right size. Your friend tells you that his windows are 50 by 60, so that's how big you make the curtains. When you go over to his house, it turns out that your curtains are more than twice as big as his windows! Your friend tells you that he's certain he measured the windows correctly, and you tell your friend that you are certain you measured the curtains correctly.

think about this

"Who has measured the waters in the hollow of his hand and marked off the heavens with a span, enclosed the dust of the earth in a measure and weighed the mountains in scales and the hills in a balance?" Isaiah 40:12

Can you determine what the author is trying to do here? He is trying to describe the greatness of God by using measurements that would have been understood by anyone living at this time. How would scientists of this day try to describe the greatness of God?

How can this be? The answer is quite simple. Your friend measured the windows in centimeters. You, on the other hand, measured the curtains in inches. The problem was not caused by measuring incorrectly. Instead, the problem was the result of measuring with different units.

When we are making measurements, the units we use are just as important as the numbers that we get. If your friend had told you that his windows were 50 centimeters by 60 centimeters, there would have been no problem. You would have known exactly how big to make the curtains. Since he failed to do this, the numbers that he gave you (50 by 60) were useless. A failure to indicate the units involved in measurements can lead to serious problems. For example, the Mars Climate Orbiter, a NASA (National Aeronautics and Space Administration) spacecraft built for the exploration of Mars, vanished during an attempt to put it into orbit around the planet. In an investigation that followed, NASA determined that a mix-up in units had caused the disaster. One team of engineers had used metric units in its calculations, while another team had used English units in executing an engine burn. The teams did not indicate the units they were using, and as a result, we lost a spacecraft worth several billion dollars.

Scientists should never simply report numbers. They must always include units so that everyone knows exactly what the numbers mean. That will be the rule in this chemistry course.

If you answer a question or solve a problem and do not list units with the numbers, your answer will be considered incorrect.

FIGURE 1.1
Two Consequences of Not Using Units Properly
Window illustration by David Weiss. Mars Climate Orbiter image courtesy of NASA/JPL/Caltech

These curtains are too long for this window because the window was measured in centimeters, but the curtains were made assuming the measurements were in inches.

The Mars Climate Orbiter did not successfully make it into orbit because 2 engineering teams involved used different units in their designs.

Since scientists use units in all of their measurements, it is convenient to define a standard set of units that will be used by everyone. This system of standard units is called the metric system. If you do not fully understand the metric system, don't worry. By the end of this module, you will be an expert at using it. If you do fully understand the metric system, you can skim this section as a review.

THE METRIC SYSTEM

We need to measure many different things when studying matter. First, we must determine how much matter exists in the object that we want to study. We know that there is a lot more matter in a car than there is in a feather because a car is heavier. To study an object precisely, we need to know exactly how much matter is in it. To accomplish this, we need to measure mass and the amount of space the object takes up. So how do we measure the object's mass? In the metric system, the unit for mass is the gram. If an object has a mass of 10 grams, we know that it has 10 times the matter in an object with a mass of 1 gram. To give you an idea of the size of a gram, the average mass of a housefly is about 1 gram. A gram is a rather small unit. Most of the things that we will measure will have masses of 10 to 10,000 grams. For example, this book has a mass of about 2,300 grams.

Now that we know what the metric unit for mass is, we need to know a little more about the concept of mass. Some people might think of mass as weight. That's not exactly true. Mass and weight are 2 different measurements. Mass measures how much matter exists in an object. Weight, on the other hand, measures how hard gravity pulls on that object.

For example, if I were to use my bathroom scale and weigh myself, I would find that I weigh 150 pounds. However, if I were to take that scale to the moon and weigh myself, I would find that I weigh only 25 pounds. Does that mean I'm thinner on the moon than I am at home? Of course not. It means that on the moon, gravity is not as strong as it is in my house on Earth.

On the other hand, if I were to measure my mass at home, I would find it to be 68,000 grams. If I were to measure my mass on the moon, it would still be 68,000 grams. That's the difference between mass and weight. Since weight is a measure of how hard gravity pulls, an object weighs different amounts depending on the gravity that is present. Because mass measures how much matter is in an object, it does not depend on the gravity present.

Unfortunately, there are many other unit systems in use today besides the metric system. In fact, the metric system is probably not the system with which you are most familiar. You are probably most familiar with the English system. The English unit for mass is (believe it or not) called the slug. Although we will not use the slug often, it is important to understand what it means, especially when you study physics. The English system uses pounds as a measurement of how much material an object has. However, pounds are not a measure of mass; they are a measure of weight. The metric unit for weight is called the Newton.

There is more to measurements than mass, however. We might also want to measure how big an object is. For this, we must use the metric system's unit for distance, the meter. You are probably familiar with a yardstick. A meter is slightly longer than a yardstick. The English unit for distance is the foot. There are many other units of distance like inches, yards, and miles, but we'll talk about those a little later.

We also need to be able to measure how much space an object occupies. This measurement is commonly called volume and is measured in the metric system with the unit called the liter. The main unit for measuring volume in the English system is the gallon. To give you an idea of the size of a liter, it takes just under 4 liters to make a gallon.

Finally, we have to be able to measure the passage of time. When studying matter, we will see that it has the ability to change. The shape, size, and chemical properties of certain substances change over time. It is important to be able to measure time so that we can determine how quickly the changes take place. In both the English and metric systems, time is measured in seconds. That is a good thing, isn't it?

Since it is very important for you to be able to recognize which units correspond to which measurements, Table 1.1 summarizes what you have just read. The letters in parentheses are the commonly used abbreviations for the units listed. You should memorize this table.

TABLE 1.1
Physical Quantities and Their Base Units

Physical Quantity	Base Metric Unit	Base English Unit
Mass	gram (g)	slug (sl)
Distance	meter (m)	foot (ft)
Volume	liter (L)	gallon (gal)
Time	second (s)	second (s)

MANIPULATING UNITS

Let's suppose we asked you to measure the width of your home's kitchen using the English system. What unit would you use? Most likely, you would express your measurement in feet. However, suppose instead we asked you to measure the diameter of a penny. Would you still use the foot as your measurement unit? Probably not. Since you know the English system already, you would probably recognize that inches are also a unit for distance; since a penny is relatively small, you would use inches instead of feet. In the same way, if we asked you to measure the distance between 2 cities, you would probably express your measurement in terms of miles, not feet. This is why we used the term *base English unit* in table 1.1. Even though the English system's normal unit for distance is the foot, there are alternative units for length when measuring very short or very long distances. The same holds true for other English units. For example, volume can be measured in cups, pints, and ounces. We choose the unit based on the object to be measured.

The metric system also has alternative units for measuring small things compared to measuring big things. These alternative units are constructed by placing a prefix in front of the metric base unit. You will soon see the metric system is easier to use and understand than the English system because the prefixes always have the same relationship to the base unit, regardless of what measurement is used.

To use a larger or smaller scale in the metric system, simply add a prefix to the base unit. For example, in the metric system, the prefix *centi-* means one-hundredth, or 0.01. So, if we wanted to measure the length of a sewing needle in the metric system, we would probably express the measurement with the *centi*meter unit. Since a centimeter is one-hundredth of a meter, it can be used to measure relatively small things. On the other hand, the prefix *kilo-* means 1,000. If we want to measure the distance between 2 states, we would probably use the *kilo*meter. Since each kilometer is 1,000 times longer than the meter, it can be used to measure long things.

The beauty of the metric system is that these prefixes mean the same thing *regardless of the physical quantity being measured!* If we were measuring something with a very large mass (such as a car or boat), then we would use the kilogram unit. One kilogram is the same as 1,000 grams. In the same way, if we were measuring something that had a large volume, we might use the kiloliter, which is 1,000 liters. Adding a *kilo-* prefix to a unit multiplies the scale of the unit by 1,000—that is, 1,000 times larger than the base unit scale.

Compare this incredibly logical system of units to the chaotic English system. If we want to measure something short, we use the inch unit, which is equal to one-twelfth of a foot. On the other hand, if we want to measure something with small volume, we might use the quart unit, which is equal to one-fourth of a gallon. In the English system, every alternative unit has a different relationship to the base unit, and we must remember all of those crazy numbers. We have to remember that there are 12 inches in a foot, 3 feet in a yard, and 5,280 feet in a mile, while at the same time remembering that for volume there are 8 ounces in a cup, 2 cups in a pint, 2 pints in a quart, and 4 quarts in a gallon. That's a lot to memorize! Thankfully, the majority of science operates around the metric system.

In the metric system, all we have to remember is what the prefix means. Since the *centi-* prefix means one-hundredth, then we know that 1 centimeter is one-hundredth of a meter, 1 centiliter is one-hundredth of a liter, and 1 centigram is one-hundredth a gram. Since the *kilo-* prefix means 1,000, we know that there are 1,000 meters in a kilometer, 1,000 grams in a kilogram, and 1,000 liters in a kiloliter. Doesn't that make a lot more sense?

Another advantage to the metric system is that the prefixes are all based on a factor of 10. Table 1.2 summarizes the most commonly used prefixes and their numerical meanings. The prefixes in boldface type are the ones that we will use over and over again. Memorize those 3 prefixes and their meanings before you take the test for this module. The commonly used abbreviations for these prefixes are listed in parentheses.

TABLE 1.2
Common Prefixes Used in the Metric System

Prefix	Numerical Meaning
micro (μ)	0.000001
milli (m)	**0.001**
centi (c)	**0.01**
deci (d)	0.1
deca (D)	10
hecta (H)	100
kilo (k)	**1,000**
mega (M)	1,000,000

Remember that each of these prefixes, when added to a base unit, makes an alternative scale for measurement. If you wanted to measure the length of something small, you would have all sorts of options for which unit to use. If you wanted to measure the length of someone's foot, you could use the decimeter. Since the decimeter is one-tenth of a meter, it measures things that are only slightly smaller than a meter. On the other hand, if you wanted to measure the length of a sewing needle, you could use the centimeter because a sewing needle is significantly smaller than a meter. Or if you want to measure the thickness of a piece of paper, you might use the millimeter since it is one-thousandth of a meter, which is a *really* small unit.

You can see that the metric system is much more logical and versatile than the English system. That is, in part, why scientists and most countries in the world use it as their main system of measurement. With the exception of the United States, almost every country in the world uses the metric system as its standard system of units. Since scientists in the United States frequently work with scientists from other countries around the world, they must use and understand the metric system. Throughout all of the modules of this chemistry course, the English system of measurement will be presented only for illustration purposes. Since scientists must thoroughly understand the metric system, it will be the main system of units that we will use.

CONVERTING BETWEEN UNITS

Now that you understand what prefixes are and how they are used in the metric system, you must become familiar with converting between units within the metric system. In other words, if you measure the length of an object in centimeters, you should also be able to convert your answer to any other distance unit. For example, if you measure the length of a sewing needle in centimeters, you should be able to convert that length to millimeters, decimeters, meters, etc. Accomplishing this task is relatively simple as long as you remember the skills you learned in multiplying fractions. Suppose we asked you to complete the following problem:

$$\frac{7}{64} \times \frac{64}{13} =$$

There are 2 ways to figure out the answer. The first way would be to multiply the numerators and the denominators together and then simplify the fraction. If you did it that way, it would look something like this:

$$\frac{7}{64} \times \frac{64}{13} = \frac{448}{832} = \frac{7}{13}$$

You could get the answer much more quickly, however, if you remember that when multiplying fractions, common factors in the numerator and the denominator cancel each other out. The 64 in the numerator cancels with the 64 in the denominator, and the only factors left are the 7 in the numerator and the 13 in the denominator. In this way, you reach the final answer in one less step:

$$\frac{7}{\cancel{64}} \times \frac{\cancel{64}}{13} = \frac{7}{13}$$

Another skill in multiplying fractions you will need is illustrated in this next equation:

$$\frac{7}{13} \times \frac{3}{3} = \frac{21}{39}$$

Notice what happens when the original fraction is multiplied by a value equal to 1. The answer is equal to the original fraction, but it looks different. Multiplying a fraction by a value of 1 does not change the value of the fraction.

We will use these fraction concepts in converting between units. Suppose we measure the length of a pencil to be 15.1 centimeters, but the person who wants to know the length of the pencil would like us to tell him the answer in meters. How would we convert between centimeters and meters? First, we would need to know the relationship between centimeters and meters. According to table 1.2, *centi-* means 0.01. So 1 centimeter is the same thing as 0.01 meters. In mathematical form, we would say:

$$1 \text{ centimeter} = 0.01 \text{ meter}$$

Look what happens when we divide both sides by 1 centimeter:

$$\frac{1 \text{ centimeter}}{1 \text{ centimeter}} = \frac{0.01 \text{ meter}}{1 \text{ centimeter}}$$

The fraction on the right equals 1! Do you see how that works? If 1 centimeter equals 0.01 meters, then the numerator and the denominator on the right side are saying the same thing, which is why that fraction equals 1. So if we use this fraction and multiply it by a measurement, then we will not change the value of the measurement. We call this fraction a conversion factor.

Now that we know how centimeters and meters relate to one another, we can convert from one to another. First, we write down the measurement that we know:

$$15.1 \text{ centimeters}$$

We then realize that any number can be expressed as a fraction by putting it over the number 1. So we can rewrite our measurement as:

$$\frac{15.1 \text{ centimeters}}{1}$$

Now we can convert that measurement into meters by multiplying it with the conversion factor we determined above. We have to do it the right way so that the original measurement unit cancels out when multiplied by the conversion factor. Here's how we do it:

$$\frac{15.1 \text{ \sout{centimeters}}}{1} \times \frac{0.01 \text{ meter}}{1 \text{ \sout{centimeter}}} = 0.151 \text{ meter}$$

Given Unit   Conversion Factor   Wanted Unit

So 15.1 centimeters is the same as 0.151 meters. This conversion method, called the factor-label method, works for 2 reasons. First, since 0.01 meters is the same as 1 centimeter, multiplying our measurement by 0.01 meters over 1 centimeter is the same as multiplying by 1. Since nothing changes when we multiply by 1, we haven't altered the value of our measurement at all. Second, by putting the 1 centimeter in the denominator of the second fraction, we allow the centimeters unit to cancel (just like the 64 canceled in the previous discussion). Once the centimeters unit has canceled, the only thing left is meters, so we know that our measurement is now in meters.

This is how we will do all of our factor-label setups. We will first write the measurement we are given in fraction form by putting it over 1. We will then put the unit we do not want in the denominator of the conversion factor and put the unit we do want in the numerator. Finally, the numerical meaning of any prefixes needs to go on the opposite side of the conversion factor to get the fraction to equal 1. We will see many examples of this method, so don't be concerned if you are a little confused right now.

Here is a list of steps for the factor-label method:

1. Create a fraction out of the given measurement by placing it over 1.
2. Place the original measurement unit in the denominator of the conversion factor.
3. Place the wanted unit in the numerator of the conversion factor.
4. Place the numerical meaning of any prefixes on the opposite side of the conversion factor.
5. Multiply the given measurement fraction by the conversion factor.

It may seem odd that words can be treated exactly the same as numbers, but measuring units have just that property. Whenever a measurement is used in any mathematical equation, the units for that measurement must be included in the equation. Those units are then treated the same way numbers are treated. We will come back to this point in an upcoming section of this module.

We will be using the factor-label method for many other types of problems throughout this course, so it is very, very important for you to learn it. Also, since we will be using it so often, we should start abbreviating things so that they will be easier to write down. We will use the abbreviations for the base units listed in table 1.1, along with the prefix abbreviations listed in table 1.2. For example, kilograms will be abbreviated as kg, while milliliters will be abbreviated as mL.

Since the factor-label method is so important in our studies of chemistry, let's see how it works in Example 1.1.

EXAMPLE 1.1

A student measures the mass of a rock to be 14,351 grams. What is the rock's mass in kilograms?

First, we use the definition of *kilo-* to determine the relationship between grams and kilograms:

$$1 \text{ kg} = 1{,}000 \text{ g}$$

Then we put our given measurement in fraction form:

$$\frac{14{,}351 \text{ g}}{1}$$

Then we multiply our measurement by a fraction that contains the relationship noted above, making sure to put the 1,000 g in the denominator so that the unit of grams will cancel out:

$$\frac{14{,}351 \cancel{\text{g}}}{1} \times \frac{1 \text{ kg}}{1{,}000 \cancel{\text{g}}} = 14.351 \text{ kg}$$

Given Unit — Conversion Factor — Wanted Unit

So, 14,351 grams is the same as **14.351 kilograms**.

Because we will use it over and over again, you must master this powerful technique. Also, you will see toward the end of this module that the factor-label method can become extremely complex; therefore, it is very important that you take the time now to answer "On Your Own" questions 1.1–1.3. Once you have solved the problems on your own, check your answers using the solutions provided at the end of the module.

ON YOUR OWN

1.1 A student measures the mass of a bag of sand to be 9,321 g. What is the bag's mass in kg?

1.2 If a glass contains 0.465 L of water, what is the volume of water in mL?

1.3 On a professional basketball court, the distance from the 3-point line to the basket is 724.0 cm. What is this distance in meters?

CONVERTING BETWEEN UNIT SYSTEMS

As you may have guessed, the factor-label method can also be used to convert *between systems* of units as well as within systems of units. If a measurement is done in the English system, the factor-label method can be used to convert that measurement to the metric system, or vice versa. To be able to do this, we must learn the relationships between metric and English units, as shown in Table 1.3. Although these relationships are important, we will not use them very often, so you don't need to memorize them.

TABLE 1.3
Relationships between English and Metric Units

Measurement	English/Metric Relationship
Distance	1.00 inch = 2.54 cm
Mass	1.00 slug = 14.59 kg
Volume	1.00 gallon = 3.78 L

We can use this information to form conversion factors for the factor-label method the same way we did for the metric system conversions. Study Example 1.2 to see how this works.

EXAMPLE 1.2

The length of a tabletop is measured to be 37.8 inches. What is that length in cm?

To solve this problem, we first put the given measurement in its fraction form:

$$\frac{37.8\ \text{in}}{1}$$

We then multiply this fraction by the conversion relationship so that the inches unit cancels:

$$\frac{37.8\ \cancel{\text{in}}}{1} \times \frac{2.54\ \text{cm}}{1\ \cancel{\text{in}}} = 96.012\ \text{cm}$$

Given Unit — Conversion Factor — Wanted Unit

So a measurement of 37.8 inches is equivalent to **96.012 cm**.

Give yourself a little more practice with the factor-label method by answering "On Your Own" questions 1.4–1.5.

ON YOUR OWN

1.4 How many kilograms are in 8.465 slugs?

1.5 If an object occupies a volume of 6.1236 liters, how many gallons does it occupy?

MORE COMPLEX UNIT CONVERSIONS AND PROBLEM SOLVING

Now that we have seen some simple applications of the factor-label method, let's look at more complex problems. For example, suppose we measure the volume of a liquid to be 4,523 centiliters but would like to convert this measurement into kiloliters. This is a more complicated problem because the relationship between cL and kL is not direct. In all of the previous examples, we knew the relationship between the unit we had and the unit to which we wanted to convert. In this problem, we need to add an extra step to convert from one to the other.

We would say there is an indirect relationship between the 2 units. We know how many cL are in a L and how many L are in a kL, so we can use these 2 relationships to do a 2-step conversion. First, we can convert centiliters into liters:

$$\frac{4{,}523\ \cancel{\text{cL}}}{1} \times \frac{0.01\ \text{L}}{1\ \cancel{\text{cL}}} = 45.23\ \text{L}$$

Given Unit — Conversion Factor — Wanted Unit

Then we can convert liters into kiloliters:

$$\frac{45.23\ \cancel{\text{L}}}{1} \times \frac{1\ \text{kL}}{1{,}000\ \cancel{\text{L}}} = 0.04523\ \text{kL}$$

Given Unit — Conversion Factor — Wanted Unit

We are forced to do this 2-step process because of the indirect relationship between the 2 prefix units in the metric system. However, we can always convert between 2 prefix units if we first convert to the base unit. To speed up this kind of conversion, we can combine these 2 steps into 1 line:

$$\frac{4{,}523\ \cancel{\text{cL}}}{1} \times \frac{0.01\cancel{\text{L}}}{1\ \cancel{\text{cL}}} \times \frac{1\ \text{kL}}{1{,}000\ \cancel{\text{L}}} = 0.04523\ \text{kL}$$

Given Unit — Conversion Factors — Wanted Unit

You will be seeing mathematical equations like this one as we move through the subject of chemistry, so it is important for you to understand what's going on. The first fraction in the equation above represents the measurement that we were given. Since there isn't a relationship between the unit we were given and the unit to which we will convert, we first convert the given unit to the base unit. This is accomplished with the second fraction in the equation. When the first fraction is multiplied by the second fraction, the cL unit cancels and is replaced by the L unit. The third fraction then cancels the L unit and replaces it with the kL unit, which is the unit we want. This gives us our final answer.

We solved all of the previous problems using the same steps.

1. The problem gave a measurement with a unit and asked for a different unit.
2. We planned and laid out conversion factors. Think of the sequence of unit conversions you will need to get from the initial unit to the final unit for the answer. For each change of unit in your plan, you will need a conversion factor.
3. Finally, we set up the factor-label method using the steps we learned earlier.

Try to use the above approach in Example 1.3.

EXAMPLE 1.3

The mass of an object is measured to be 0.030 kg. What is the object's mass in mg?

Given Unit: kg
Wanted Unit: mg

Plan Conversion Factors: Because the relationship between milligrams and kilograms is indirect, we will plan a 2-step conversion. First, we know that:

$$1 \text{ kg} = 1{,}000 \text{ g}$$

We also know that:

$$1 \text{ mg} = 0.001 \text{ g}$$

So we need to multiply our original measurement in fraction form with both of these relationships. We must do it in such a way as to cancel out the kg and replace it with g, and then cancel out the g and replace it with mg.

Set Up Problem:

$$\frac{0.030 \cancel{\text{kg}}}{1} \times \frac{1{,}000 \cancel{\text{g}}}{1 \cancel{\text{kg}}} \times \frac{1 \text{ mg}}{0.001 \cancel{\text{g}}} = 30{,}000 \text{ mg}$$

Given Unit   Conversion Factors   Wanted Unit

The object's mass is 0.030 kg, which is the same as **30,000 mg**.

ON YOUR OWN

1.6 A balloon is blown up so that its volume is 1,500 mL. What is its volume in kL?

1.7 If the length of a race car track is 2.0 km, what is it in cm?

1.8 How many mg are in 0.01 Mg?

The factor-label method is one of the most important tools you can learn for the study of chemistry (and physics, for that matter). Therefore, you must become a veritable expert at it. Try your hand at "On Your Own" questions 1.6–1.8 so that you can get some more practice.

DERIVED UNITS

We mentioned previously that units can be used in mathematical expressions in the same way that numbers can be used. Just as there are rules for adding, subtracting, multiplying, and dividing numbers, there are also rules governing those operations when using units. You will have to become very adept at using units in mathematical expressions, so let's discuss those rules now.

Adding and Subtracting Units: When adding and subtracting units, the most important thing to remember is that the units we are adding or subtracting must be identical. For example, we cannot add grams and liters. The result would not make sense physically. Since gram is a unit of mass and liter is a unit of volume, there is no way we can add or subtract them. We also cannot add or subtract kilograms and grams. Even though both units measure mass, we cannot add or subtract them unless the units are *identical.* If we did want to add or subtract them, we would have to convert the kilograms into grams or convert the grams into kilograms. It doesn't matter which way we go, as long as the units we add or subtract are identical.

Once we have identical units, we can add and subtract them using the rules of algebra. Since 2x + 3x = 5x, we know that 2 cm + 3 cm = 5 cm. In the same way, 3.1 g – 2.7 g = 0.4 g. When adding or subtracting units, we add or subtract the numbers they are associated with and then simply carry the unit along into the answer.

Multiplying and Dividing Units: When multiplying and dividing units, it doesn't matter whether or not the units are identical. Unlike addition and subtraction, we can multiply or divide any unit by any other unit. In algebra:

$$3x \times 4y = 12xy$$

When multiplying units:

$$3 \text{ kg} \times 4 \text{ mL} = 12 \text{ kg} \times \text{mL}$$

Similarly, in algebra:

$$6x \div 2y = 3\,\frac{x}{y}$$

When dividing units:

$$6 \text{ g} \div 2 \text{ mL} = 3\,\frac{\text{g}}{\text{mL}}$$

So when multiplying or dividing units, you multiply or divide the numbers and then do exactly the same thing to the units.

Let's use the rules we've just learned to explore a few other things about units. First, let's see what happens when we multiply measurements that have the same units. Suppose we wanted to measure the surface area of a rectangular table. From geometry, we know that the area of a rectangle is the length times the width. So, let's suppose we measure the length of a table to be 1.1 meters and the width to be 2.0 meters. Its area would be:

$$1.1 \times 2.0 = 2.2$$

What would the units be? In algebra, we would say that:

$$1.1x \text{ x } 2.0x = 2.2x^2$$

Therefore:

$$1.1 \text{ m x } 2.0 \text{ m} = 2.2 \text{ m}^2$$

This tells us that m^2 (square meter) is a unit for area.

Let's take this one step further. Suppose we measure the length, width, and height of a small box to be 1.2 cm, 3.1 cm, and 1.4 cm, respectively. What would the volume of the box be? From geometry, we know that volume is length times width times height, so the volume would be:

$$1.2 \text{ cm x } 3.1 \text{ cm x } 1.4 \text{ cm} = 5.208 \text{ cm}^3$$

Therefore, cm^3 (usually called cubic centimeters or cc's) is a unit for volume. If you've ever listened to doctors or nurses talking about how much liquid to put in a syringe when administering a shot, they usually use cc's as the unit. When a doctor tells a nurse, "Give the patient 4 cc's of penicillin," he is telling the nurse to inject a 4-cm^3 volume of penicillin into the patient.

Wait a minute. Wasn't the metric unit for volume the liter? Yes, but another metric unit for volume is the cm^3. In addition, m^3 (cubic meters) and km^3 (cubic kilometers) are possible units for volume. This is a very important point. Often, several different units exist for the same measurement. The units you use will depend, to a large extent, on what information you are given in the first place. We'll see more about this fact later.

Units like cm^3 are called derived units because they are derived from math calculations with basic units that make up the metric system. Many of the units you will use in chemistry are derived units. We will discuss one very important physical quantity with derived units in an upcoming section of this module, but first you need to understand exactly how to use derived units in mathematical equations.

Let's suppose we want to take the volume that we previously determined for the box and convert it from cubic centimeters to cubic meters. You might think the conversion would look something like this:

$$\frac{5.208 \text{ cm}^3}{1} \text{ x } \frac{0.01 \text{ m}}{1 \text{ cm}}$$

This conversion might look correct, but there is a major problem with it. Remember what the factor-label method is designed to accomplish. In the end, the old units are supposed to cancel out, leaving the new units in their place. The way this conversion is set up, the old units *do not cancel!* When we multiply these 2 fractions together, the cm in the denominator does not cancel out the cm^3 in the numerator. When multiplying fractions, the numerator and denominator must be *identical* for them both to cancel. The cm in the denominator above must be replaced with cm^3.

How is this done? It's quite simple. Do you remember the math equation for volume? It is length times width times height. You have to provide a conversion factor for each unit of the length, width, and height.

$$\frac{5.208 \text{ cm}^3}{1} \text{ x } \frac{0.01 \text{ m}}{1 \text{ cm}} \text{ x } \frac{0.01 \text{ m}}{1 \text{ cm}} \text{ x } \frac{0.01 \text{ m}}{1 \text{ cm}}$$

We can shorten this setup by using an exponent:

$$\frac{5.208\ \text{cm}^3}{1} \times \left(\frac{0.01\ \text{m}}{1\ \text{cm}}\right)^3$$

Inside the parentheses, the m becomes m^3, the cm becomes cm^3, the 0.01 becomes 0.000001, and the 1 stays as 1:

$$\frac{5.208\ \cancel{\text{cm}^3}}{1} \times \frac{0.000001\ \text{m}^3}{1\ \cancel{\text{cm}^3}} = 0.000005208\ \text{m}^3$$

Given Unit Conversion Factor Wanted Unit

Now since both the numerator and denominator have a unit of cm^3, that unit cancels and is replaced with the m^3. So a volume of 5.208 cm^3 is equivalent to a volume of 0.000005208 m^3.

Since cubic meters, cubic centimeters, and the like are measurements of volume, you might have already guessed that there must be a relationship between these units and the other volume units we discussed earlier. In fact, 1 cm^3 is the same thing as 1 mL. This is a very important relationship and is something you will have to know before you can finish this module. Commit it to memory now:

1 cubic centimeter is the same as 1 milliliter (1 cm^3 = 1 mL).

Let's combine this fact with the mathematics we just learned and perform a very complicated unit conversion. If you can understand Example 1.4 and successfully complete "On Your Own" questions 1.9–1.11, then you have mastered unit conversion. If things are still a bit shaky for you, don't worry. There are plenty of practice problems at the end of this module to help you practice your unit conversion skills.

EXAMPLE 1.4

The length, width, and height of a small box are measured to be 1.1 in, 3.2 in, and 4.6 in, respectively. What is the box's volume in liters?

Given Unit: in (length)
Wanted Unit: L (volume)

Plan Conversion Factors: First, determine volume. To solve this problem, we first use the geometric equation for the volume of a box:

$$V = l \times w \times h$$

Set Up Problem:

$$V = 1.1\ \text{in} \times 3.2\ \text{in} \times 4.6\ \text{in} = 16.192\ \text{in}^3$$

Plan Conversion Factors: Second, convert to liters. Now that we know the volume, we just have to convert from in^3 to L. This is a little more difficult than it sounds, however. Since there is no direct relationship between in^3 and L, we must go through a series of conversions to get to the desired unit. First, we can convert our unit from the English system to the metric system by using the following relationship:

$$1 \text{ in} = 2.54 \text{ cm}$$

To do this, we will have to cube the fraction we multiply by so that we have cm^3 and in^3.

Set Up Problem:

$$\frac{16.192 \text{ in}^3}{1} \times \left(\frac{2.54 \text{ cm}}{1 \text{ in}}\right)^3 = \frac{16.192 \cancel{\text{in}^3}}{1} \times \left(\frac{16.387 \text{ cm}^3}{1 \cancel{\text{in}^3}}\right) = 265.338 \text{ cm}^3$$

Now that we have the metric volume unit, we can use the fact that a cm^3 is the same as a mL:

$$265.338 \text{ cm}^3 = 265.338 \text{ mL}$$

Now we can convert from mL to L:

Set Up Problem:

$$\frac{265.338 \cancel{\text{mL}}}{1} \times \frac{0.001 \text{ L}}{1 \cancel{\text{mL}}} = 0.265338 \text{ L}$$

The volume, then, is **0.265338 L**.

Alternative approach: First, convert inches to centimeters, then get the volume in cubic centimeters, then convert to milliliters, and finally convert to liters.

Set Up Problem:

We convert initial inch units to cm:

$$\frac{1.1 \cancel{\text{in}}}{1} \times \frac{2.54 \text{ cm}}{1 \cancel{\text{in}}} = 2.794 \text{ cm} \qquad \frac{3.2 \cancel{\text{in}}}{1} \times \frac{2.54 \text{ cm}}{1 \cancel{\text{in}}} = 8.128 \text{ cm}$$

$$\frac{4.6 \cancel{\text{in}}}{1} \times \frac{2.54 \text{ cm}}{1 \cancel{\text{in}}} = 11.684 \text{ cm}$$

We calculate volume in cm^3:

$$V = 2.794 \text{ cm} \times 8.128 \text{ cm} \times 11.684 \text{ cm} = 265.338 \text{ cm}^3$$

Finally, we convert cm^3 to L (remember that cm^3 is equal to mL):

$$\frac{265.338 \cancel{\text{mL}}}{1} \times \frac{0.001 \text{ L}}{1 \cancel{\text{mL}}} = \mathbf{0.265338 \text{ L}}$$

You will see that this is the same answer as above. We simply changed the order of how we did it.

MAKING MEASUREMENTS

Now that we've learned about measurement units, we need to spend a little time learning how to make measurements. After all, being able to manipulate units in mathematical equations isn't going to help us unless we can make measurements with those units to begin with. To learn how to make measurements properly, we have to know how to use measuring instruments. Let's start with a simple measuring instrument: the ruler.

ON YOUR OWN

1.9 Should you be impressed if someone says she can hold her breath for 0.00555 hours? You must convert this to seconds to answer. (HINT: 1 hour = 60 minutes and 1 minute = 60 seconds.)

1.10 How many cm^3 are in 0.0091 kL?

1.11 The area of a room is 32 m^2. What is the area of the room in mm^2?

Suppose we wanted to measure the length of this purple line with an English ruler. We would make the measurement something like this:

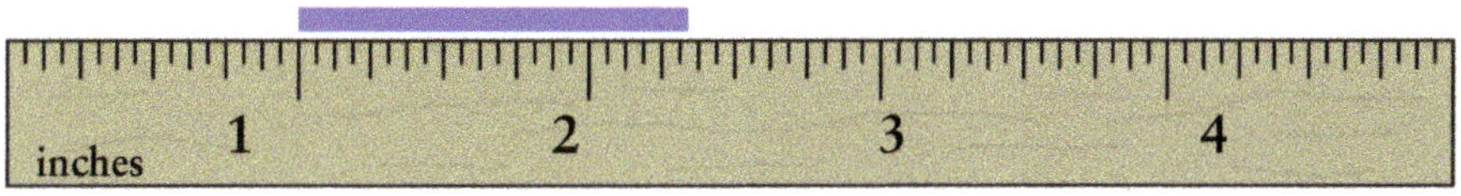

Illustration by Megan Whitaker

First, notice that we did not start the measurement at the beginning of the ruler. Instead, we lined up the ribbon with the first inch mark because it is slightly more accurate. It is very difficult to line up the edge of a ruler with the edge of the object you are measuring. This is especially true when the ruler is old and the edges are worn. So, the first rule for measuring with a ruler is to start at 1, not 0.

How would you read this measurement? First, you need to see what the scale on the ruler is. If you count the number of dashes between 1 inch and 2 inches, you will find that there are 15 of them. Every dash is worth one-sixteenth of an inch because 15 dashes break up the area between 1 inch and 2 inches into 16 equal regions.

Now that we know the scale is marked off in sixteenths of an inch, we can see that the ribbon is a little bigger than 1 5/16 of an inch. Is that the best we can do? Of course not! Because the edge of the ribbon falls between 5/16 (10/32) and 6/16 (12/32) of an inch, we can estimate that it is approximately 11/32. The proper length of the ribbon is 1 11/32 inches. Generally, chemists do not like fractions in their final measurements, so we will convert 11/32 into its decimal form to get a measurement of 1.34375 inches. Later on we will see that this measurement has far too many digits in it, but for now we will assume that it is okay.

Let's measure that same ribbon with a metric ruler:

Illustration by Megan Whitaker

Now what measurement do we get? There are 9 small dashes between each cm mark; therefore, the scale of this ruler is one-tenth of a cm, or 0.1 cm. This is typical of metric rulers. They are almost always marked off in tenths since all the prefixes in the metric system are multiples of 10. If you think about it, 1 mm = 0.1 cm, so you could also say that each small dash is 1 mm. The ribbon is between 3.4 cm and 3.5 cm long. (The printing of your specific page could alter your measurement slightly.) Using our method of approximating between the dashes, we would say that the ribbon is 3.41 cm long.

How do we deal with the part of the line between the dashes? Here is a rule you will need to follow with any scientific instrument. You will always estimate one digit beyond what the instrument is marked. We can say 3.41 cm is the length of the line. Our instrument is marked off in 0.1 cm, so we will always report the measurement to the nearest 0.01 cm. What if you measure 3.43 cm and someone else measures 3.41 cm? There is nothing wrong with this because the last digit of any measurement is an estimate.

Whenever you are using a measuring device marked with a scale, be sure to use it the way we have used the rulers here. First, determine what the dashes on the scale mean. Then, try to estimate between the dashes if the object you are measuring does not exactly line up with a dash. That gives you as accurate a measurement as possible. You should always strive to read the scale to *the next decimal place* if possible. For example, on a metric ruler, the scale is marked off in 0.1 cm, so you should read the ruler to 0.01 cm, as discussed above.

One physical quantity that chemists measure frequently is volume because they spend a great deal of time mixing liquids. When chemists measure volume, one of the most useful tools is the graduated cylinder. This device looks a lot like a rain gauge. It is a hollow cylinder with markings on it. These markings, called graduations, measure the volume of liquid that is poured into the cylinder.

Many of your chemistry experiments in this textbook will use a graduated cylinder (or a suitable substitute) to measure volume; so you need to know how to do this. When liquid is poured into a cylinder, the liquid tends to creep up the edges of the cylinder. This is because there are attractive forces between the liquid and the cylinder. Therefore, liquid poured into a graduated cylinder does not have a flat surface. Instead, it looks something like Figure 1.2:

FIGURE 1.2
The Meniscus
Illustration by Megan Whitaker

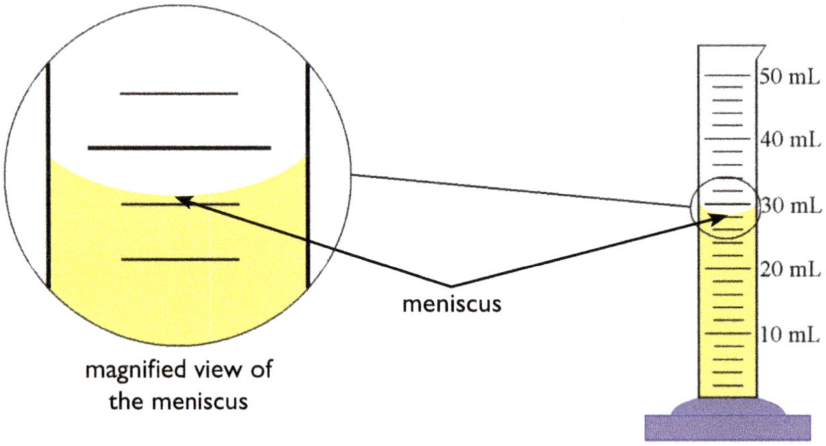

The curved surface of the yellow liquid is called the meniscus (muh nis' kus). To determine the volume of the liquid in any graduated cylinder, you must read the level of the liquid from the bottom of the meniscus. On the graduated cylinder in Figure 1.2, there are 4 dashes between each marking of 10 mL, splitting the distance between the 10 mL marks into 5 divisions. This means that each dash must be worth 2 mL. If you look at the bottom of the meniscus in the figure, you will see that it is between 28 and 30 mL. Is the volume 29 mL then? No, not quite.

In the 2 examples we considered before, it was hard to guess how far between the dashes the object's edge was because the dashes were very close together. In this example, the dashes are farther apart, so we can be a bit more precise in our final answer. For the volume to be 29 mL, the bottom of the meniscus would have to be exactly halfway between 28 and 30. Clearly, the meniscus is much closer to 28 than to 30. So the volume is really between 28 and 29, and probably a little closer to 28. We would estimate that the proper reading is 28.3 mL. It could be as low as 28.1 or as high as 28.5, so 28.3 is a good compromise. Remember, you need to try to read the scale to the next decimal place. Since the scale is marked off in increments of 2 mL, you must try to read the answer to 0.1 mL. Experiment 1.3 will give you some practice at this kind of estimation.

ACCURACY, PRECISION, AND SIGNIFICANT FIGURES

Now that we've learned a bit about taking measurements, we need to discuss when measurements are good and when they are not. In chemistry, there is always some uncertainty in the value of a measurement. We can describe the measurement in 2 ways—its accuracy and its precision. Even though these words are used interchangeably in daily life, there's an important distinction between them.

Accuracy—An indication of how close a measurement is to the true value.

Precision—An indication of the scale on the measuring device that was used.

In other words, the more correct a measurement is, the more accurate it is. The smaller the scale on the measuring instrument, the more precise the measurement.

So let's look again at the ribbon measurement as an example. Suppose we used a ruler with a scale marked off in 0.1 cm and measured a length of 3.45 cm. What does that number mean? It means that the length of the ribbon, as far as we could tell, was somewhere between 3.445 and 3.455 cm long. We could not determine the length of the ribbon any better than that because our ruler was not *precise* enough to do any better.

On the other hand, if we found a ruler with a scale marked in increments of 0.01 cm, we could get a more precise measurement. For example, we could get a measurement of 3.448 cm because estimating between the dashes gives us one more decimal place. Since the ruler is marked off in hundredths, we can get a measurement out to the thousandths place. That extra digit in the thousandths place nails down the length better. It would be impossible to obtain so precise a measurement from the ruler we used in the example above, so the new ruler provides a more precise measurement. These examples show that the precision of our measurements depends completely on the measuring devices we use. The smaller the scale on the instrument, the more precise our measurement will be.

However, suppose we used the second ruler improperly. Maybe we read the scale incorrectly or didn't line up the ribbon to the ruler very well and got a measurement of 3.118 cm. Even though this measurement is more *precise* than the one we made with the first ruler, it is significantly less *accurate* because it is way off of the correct value of 3.448 cm. The *accuracy* of our measurement depends on how carefully and correctly we use the measuring device. In other words, a measurement's *precision* depends upon the instrument, whereas a measurement's *accuracy* depends upon the person doing the measurement.

Since a measurement's precision depends on the instrument used, the only way to improve precision is to get a better, more precise instrument. However, there are other ways to improve accuracy. First, make sure you understand the proper methods of using each instrument. Second, practice making measurements, which will help your skill and your accuracy.

The most practical way to improve your accuracy in measurement is to make your measurement several times and average the results. This averages out all the little differences that can occur between measurements, even while using the same instrument. An even better way of assuring accuracy is to have several *different* people make the measurements and average all their answers together. The more individual measurements are made, the more accurate the average of them will be.

Let's illustrate this whole idea of accuracy and precision in another way. Suppose you were throwing darts at a dart board. Figure 1.3 shows 3 possible outcomes that we will discuss.

FIGURE 1.3
Accuracy and Precision
Illustration by Megan Whitaker

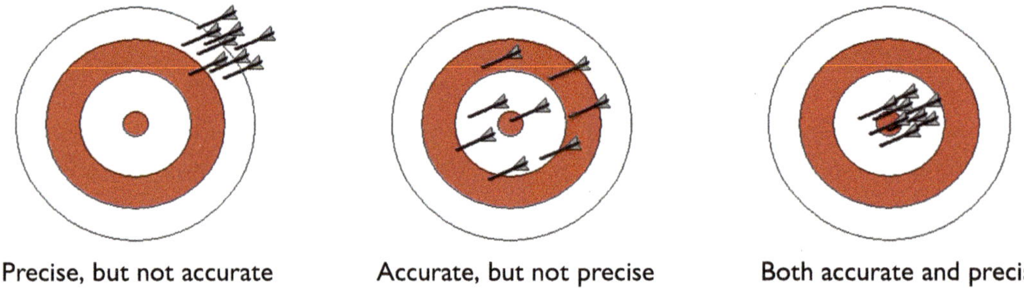

The first target on the left has all the darts clumped together, but they are way off the bull's-eye. If we made several measurements with a precise device but used it wrongly every time or the device had a flaw, it would give us numerous measurements that were very close to one another but far from the true value. This example shows a lot of precision, but not much accuracy.

The middle target has the darts surrounding the bull's-eye, but they are far from one another. If we used a measuring device that is not very precise but used it correctly, it would give us similar results. This may be because we had a hard time estimating between the marks on the scale of the instrument. This example shows accuracy, but not much precision.

The last target on the right has all darts clumped together, and they are on the bull's-eye. This is an example of measurements that are both accurate and precise. They are accurate because they average out to the correct value of the bull's-eye, and they are precise because they are very close to one another, indicating that a precise measuring device was used. This example is what every chemist desires when making measurements.

Since accuracy and precision are very important, we need to know how to evaluate the accuracy and precision of our measurements. The way to determine the accuracy of a measurement is to compare it to the correct value. If you have no idea what the correct value is, determining your measurement's accuracy is difficult. It is not impossible, but we will not spend time in this course on this topic.

Determining the precision of a measurement is quite easy. To determine the precision of a measurement and an instrument, you merely need to look at its significant figures. But what is a significant figure? It is a digit that is read from an instrument. That purple line was 3.41 cm long. All of those digits came from the ruler, so that measurement has 3 significant figures. The volume of liquid in the graduated cylinder was 28.3 mL. Again, all of those digits came from the instrument, so they are significant figures. A significant figure is a measured digit!

If 2 instruments measure the same thing, the one which gives a significant figure in the smallest decimal place is the more precise instrument. For example, 2.545 cm is more precise than 2.54 cm because 2.545 cm has more significant figures. There are 4 rules you need to remember to determine the number of significant figures:

1. All nonzero digits are significant.
2. All zeros in front of the first 1–9 digit are not significant.
3. All zeros between 2 significant figures are significant.
4. All zeros at the end of a number *and* to the right of the decimal point are significant.

TABLE 1.4
Examples of Significant Figure Rules

Number	Example	Number of Significant Figures
Nonzero digits	9,341	4
Zeros in front of first 1–9 digit	0.000564	3
Zero between 2 significant figures	120.043	6
Zero at end of number *and* to right of decimal point	510.0 510	4 2

Counting significant figures is very important in science and in our ability to understand measurements, so you must have a firm grasp on this concept. Read through Example 1.5 and follow the logic. After that, try "On Your Own" question 1.12.

EXAMPLE 1.5

Count the significant figures in each of the following numbers:

a. 3.234 b. 6.016 c. 105.340 d. 0.00450010 e. 2,330

(a) Since every digit is nonzero, every digit is a significant figure (first rule). This means there are **4 significant figures** in this number.

(b) Three of the digits are nonzeros and therefore are significant figures (first rule). The 0 is also significant because it is between 2 significant figures (third rule). So there are **4 significant figures** in the number.

(c) Four of the digits are nonzeros and therefore are significant figures (first rule). The 0 between the 1 and 5 is significant because it's between 2 significant figures (third rule). The last 0 is also significant because it is at the end of the number and to the right of the decimal point (fourth rule). The answer is there are **6 significant figures** in the number.

(d) The 3 nonzero digits are all significant figures, as are the zeros between the 5 and 1. The 0 at the end is also a significant figure. However, the first 3 zeros are not significant because they are not between 2 significant figures and they are not at the end of the number to the right of the decimal. They are important, but they are not measured. There are **6 significant figures** in the number.

(e) The 3 nonzero digits are significant figures. The zero is not significant because it is before, not at the end of, the number to the right of the decimal. There are **3 significant figures** in the number.

ON YOUR OWN

1.12 How many significant figures are in the following measurements?

a. 3.0220 cm
b. 0.0060 m
c. 1.00450 L
d. 61.054 kg
e. 300,000 mm

As we have seen, the precision of our instrument determines the number of significant figures we can report in a measurement. In our graduated cylinder example, we decided that we could reasonably approximate our measurement to somewhere between 28 and 29 mL. If we can do that, then we can report our answer to the nearest tenth of a mL, giving us a 3-significant-figure answer of 28.3 mL. If, instead, our graduated cylinder had been marked off in tenths of a mL, we could probably approximate the measurement to somewhere between 28.2 and 28.3 mL, allowing us to have a 4-significant-figure answer like 28.26 mL.

In the example which used an English ruler, we said that our final answer, 1.34375 inches, had too many digits in it. Now hopefully you can see why. According to this number, our ruler was precise enough to measure distance of one-hundred-thousandth of an inch! That's far too much precision. Most English rulers are, at best, precise to 0.01 inches. Therefore, the proper length of the ribbon that you should report is 1.34 in.

Suppose we had another ribbon to measure:

Illustration by Megan Whitaker

How would we report its measurement? Would we say that this ribbon is 3 cm long? Actually, that measurement is not quite right. The ribbon does seem to end right on the 4 cm line, so there is no need to do any approximations here. Why, then, is 3 cm a wrong answer for the length of the ribbon?

The problem with reporting the length of the ribbon as 3 cm is that we are not being as precise as we can be. Since the ruler's scale is marked off in 0.1 cm, we can safely report our answers to the hundredths of a cm. If the ribbon's edge had fallen between 2 dashes, we could have approximated as we did above. Therefore, the precision of the ruler is to the hundredths place. Therefore, if the object's edge falls right on one of the dashes, do not throw away precision. Report this length as 3.00 cm. This tells someone who reads the measurement that the ribbon's length was measured to a precision in the hundredths of a cm.

If you report the length as 3 cm, that means the ribbon could be as short as 2.5 cm or as long as 3.4 cm. Both of those measurements round to 3. But the ruler we used was much more precise. It determined the length of the ribbon to be 3.00 cm. So the ribbon is somewhere between 2.995 and 3.004 cm long. Keeping all of the significant figures that you can is a very important part of doing chemistry experiments. You will get some practice at this in Experiment 1.3.

Reporting the precision of a measurement is just as important as reporting the number itself. Why? Let's suppose that the No-Weight Cookie Co. just produced a diet cookie that they claim has only 5 calories per cookie. To confirm this claim, researchers did several careful experiments and found that, in fact, there were 5.4 calories per cookie. Does this result mean that the No-Weight Cookie Co. lied about the number of calories in its cookies? No. When the company reported that there are 5 calories per cookie, the precision of their claim indicated that there could be anywhere from 4.5 to 5.4 calories per cookie. The researchers' finding was more precise than the company's claim; nevertheless, the company's claim was accurate.

SCIENTIFIC NOTATION

Since reporting the precision of a measurement is so important, we need a notation system that allows us to do this no matter what number is involved. As numbers get very large, it becomes more difficult to report their precision properly. For example, suppose we measured the distance between 2 cities as 100.0 km. According to our rules of precision, reporting 100.0 km as the distance means that our measuring device was marked off in units of 1 km, and we estimated between the marks to come up with 100.0 km. However, suppose our measurement wasn't that precise. Suppose the instrument we used could determine the distance only to within 10 km? How could we write down a distance of 100 km and indicate that the precision was only to within 10 km?

The answer to this question lies in the technique of scientific notation. In scientific notation, we write numbers so that no matter how large or how small they are, they always include a decimal point. Remember, a number can be represented in many different ways. The number 4, for example, could be written as 2×2 or 4×1 or simply 4. Each one of these is an appropriate representation of the number 4. In scientific notation, we always represent a number as a something times a power of 10. For example, 50 could be written in scientific notation as 5×10. The number 150 could be written as 1.5×100 or 15×10.

Do you see why this helps us in writing down the precision of our original measurement? Instead of writing the distance as 100 km, we could write it as 1×100 or 1.0×100 or 1.00×100. How does this help? According to our rules of significant figures, the 0 in 1.0 is significant because it is at the end of the number and to the right of the decimal. By writing down our measurement as 1.0×100, we indicate that the 0 was measured and that the measurement is precise to within 10 km. There is no way to do that with normal decimal notation because neither of the zeros in 100 is significant. Scientific notation, then, gives us a way to *make* zeros significant if they need to be. If our measurement of 100 km was precise to within 1 km, we could indicate that by reporting the measurement as 1.00 x 100 km. Since both zeros in 1.00 are significant, this tells us that both zeros were measured, so our precision is within 1 km.

Since numbers that we deal with in chemistry can be very big or very small, we use one piece of mathematical shorthand in scientific notation. Recall from algebra that 100 is the same as 10^2. We will use this shorthand to make the numbers easier to write down. Scientific notation always has a number with a decimal point right after the first digit times a 10 raised to some power.

One other advantage of using scientific notation is that it simplifies the job of recording very large or very small numbers, making mistakes in computations less likely. For example, there are roughly 20,000,000,000,000,000,000,000 particles in each breath of air that we take. Numbers like that are very common in chemistry. In scientific notation, the number would be 2×10^{22}. That's much easier to write down!

How did do we know to raise the 10 to the 22nd power? To get the decimal point right after the 2, we would have to move it to the left 22 digits, which is equivalent to multiplying by 10^{22}. When putting a large number into scientific notation, all we need to do is count the number of spaces the decimal point needs to move and then raise the 10 to that power.

$$20{,}000{,}000{,}000{,}000{,}000{,}000{,}000.0 = 2 \times 10^{22}$$

Chemistry also deals with very small numbers. For example, one of the things we will discuss in great detail in several upcoming modules is a particle called a proton. The proton has a mass of about 0.00000000000000000000000000167 kg. This number is very cumbersome to write down, so we will use scientific notation to make our job a little easier. In scientific notation, the proton's mass is 1.67×10^{-27} kg. Why raise the 10 to the –27th power? When numbers are raised to the negative power, they are smaller than 1. So when we multiply a number by 10 raised to a negative power, we shift the decimal place the other way. To get the decimal point to be right after the 1, we have to move it 27 places. Since we moved it to the right 27 places, we multiply it by 10^{-27}.

$$0.00000000000000000000000000167 = 1.67 \times 10^{-27}$$

You will need to follow 2 basic rules for scientific notation in this course.

1. Place only 1 digit (not a 0) in front of the decimal point.
2. Only significant figures go in front of the multiplication sign.

See how this works by following Example 1.6, and then make sure you understand this technique by completing "On Your Own" questions 1.13–1.14.

EXAMPLE 1.6

Convert the following numbers into scientific notation:

a. 20,300 b. 3,151,367 c. 234,000 d. 0.000002340 e. 0.000875

(a) The decimal place must be moved to the left by 4 digits to get it next to the 2. Since we are dealing with a big number, we have to multiply by a 10 raised to a positive power. Therefore, the answer is **2.03×10^4**. Since the last 2 zeros to the right of the 3 are *not* significant as the number is written, we must drop them in our answer. All zeros are significant in 2.0300×10^4.

(b) The decimal place must be moved to the left 6 places and the number is big, so the answer is **3.151367×10^6**.

(c) The decimal place must be moved to the left 5 places, and since it is a big number, the answer is **2.34×10^5**. The last 3 zeros were dropped because as written, they are not significant.

(d) The decimal must be moved 6 places to the right. Since this is a small number, we are dealing with a negative exponent, so the answer is **2.340×10^{-6}**. In this case, the final 0 cannot be dropped because, based on our rules of significant figures, a 0 at the end of a number and to the right of the decimal point is significant.

(e) The decimal point must be moved 4 places to the right. Since it is a small number, the answer is **8.75×10^{-4}**.

Convert the following numbers from scientific notation back into decimal form.

(a) 3.45×10^{-5} (b) 2.3410×10^7 (c) 1.89×10^{-9} (d) 3.0×10

(a) Since the 10 is raised to a negative power, the decimal point must be moved to make it small. The power of –5 tells us that we move it 5 spaces, so the answer is **0.0000345**.

(b Since the power of 10 is positive, we must move the decimal point to make the number bigger. The power of 7 tells us we must move it 7 places, so the answer is **23,410,000**. Note that we cannot indicate that the 0 after the 1 is significant in this notation. It is clearly significant in the original number, so it is impossible to properly represent the precision of this number in decimal form.

(c) We must move the decimal point 9 places and make the number smaller, so the answer is **0.00000000189**.

(d) Since the exponent is not listed, we assume it's 1. That means that we move the decimal point 1 place so that the number gets bigger, so the answer is **30**. Once again, there is no way to indicate that the 0 is significant, as it is in the original number.

ON YOUR OWN

1.13 Convert the following numbers from decimal form to scientific notation.

a. 26,089,000 c. 0.00009870
b. 12,000,000,003 d. 0.980

1.14 Convert the following numbers from scientific notation to decimal form.

a. 3.456×10^{14} c. 3.45×10^{-5}
b. 1.2341×10^{3} d. 3.10×10^{-1}

USING SIGNIFICANT FIGURES IN MATHEMATICAL PROBLEMS

Now that we have the ability to write down any measurement with its proper precision, there is only one more topic on significant figures that we need to discuss. We need to know how to use our concepts of significant figures when we work mathematical problems. Suppose we had 2 measurements and wanted to add them together. Since each measurement has its own precision, the final answer would also have a certain precision. How do we know the precision of our answer?

For example, suppose we measured the total length of a knife to be 25.46 cm. Later, someone else measured the length of the knife handle with a less precise ruler and got 7.8 cm. If we wanted to determine the length of the knife's blade, either we could measure it, or we could say that the blade's length was the total length of the knife minus the length of the handle, or 25.46 cm – 7.8 cm. If we do the subtraction, we get 17.66 cm. This answer is too precise because the knife handle was measured with a less precise ruler. The answer is limited to the least precise instrument, so the proper answer is 17.7 cm.

To add, subtract, multiply, or divide measurements, we will use 2 rules about using significant figures in mathematical equations. You will be using these rules over and over again throughout this course, so you will be expected to know them:

1. *Adding and Subtracting with Significant Figures:* When adding and subtracting measurements, round your answer so that it has the same precision as the least precise measurement in the calculation.

2. *Multiplying and Dividing with Significant Figures:* When multiplying and dividing measurements, round the answer so that it has the same number of significant figures as the measurement with the fewest significant figures.

Example 1.7 shows how these rules work.

EXAMPLE 1.7

A student measures the mass of a jar that is filled with sand and finds it to be 546.2075 kg. A note on the jar says, "When empty, this jar has a mass of 87.61 kg." What is the mass of the sand in the jar?

Since 546.2075 kg is the mass of both the jar and the sand, and since 87.61 kg is the mass of the jar alone, the mass of the sand must be the difference between the 2:

$$\begin{array}{r} 546.2075 \text{ kg} \\ -\ 87.61 \quad \text{kg} \\ \hline 458.5975 \text{ kg} \end{array}$$

However, since the precision of the jar's mass only goes out to the hundredths place, that's the best we can do in our final answer. Therefore, the mass of the sand is **458.60 kg**. Note that this number has more significant figures than 87.61. That doesn't matter because in addition and subtraction, we do not count significant figures; we look only at precision.

A woman runs 3.012 miles in 0.430 hours. What is her average speed?

We can find her average speed by dividing the distance traveled by the time:

$$\text{Speed} = 3.012 \text{ miles} \div 0.430 \text{ hour} = 7.004651163$$

The 3.012 miles has 4 significant figures, while 0.430 hours has 3. Therefore, our final answer must have 3 significant figures, making it **7.00 miles/hour**

Now that we have learned these rules, you will be expected to use them *in all further mathematical operations!* Whether you are working an "On Your Own" problem, a practice problem, a test problem, or an experiment, you will use these rules. In the examples and answers for all previous problems, these rules have not been followed, but they will be from now on. By the time you finish the next couple of modules, keeping track of significant figures and precision should be second nature to you.

There is one more point that you must understand about significant figures before you get some practice using the rules. When making unit conversions, you might be tempted to round everything to 1 significant figure because of the conversion relationships. For example, when you convert 121 g into kg, you use the following equation:

$$\frac{121 \cancel{\text{g}}}{1} \times \frac{1 \text{ kg}}{1{,}000 \cancel{\text{g}}}$$

Note that the 1 kg, the 1 in the denominator of the first fraction, and 1,000 g all look like they have only 1 significant figure. You might be tempted to round your answer to 1 significant figure. However, that would not be correct. The reason is simple: These 3 numbers all come from definitions. They are infinitely precise. The 1 kg is really 1.000... kg, and the 1,000 g is really 1,000.000... g because *exactly* 1 kilogram is defined to be *exactly* 1,000 g. In the same way, the 1 on the bottom of the first fraction is really 1.000... because it is an integer. The *only* number in this equation that has a limited number of significant figures is the 121 g (it is a measurement), so the answer is 0.121 kg.

In general, then, the prefixes used in the metric system as well as the integers used in fractions are infinitely precise and have an infinite number of significant figures. As a result, we ignore them when determining the significant figures in a problem. This is a very important rule:

The definitions of the prefixes in the metric system and the integers used in fractions are not considered when determining the significant figures in the answer.

Get some practice making measurements, using them in mathematical equations, and keeping track of significant figures by performing Experiment 1.3.

EXPERIMENT 1.3

PURPOSE: To compare conversions to measurements.

MATERIALS:
- Book (not oversized)
- Metric and English rulers
- Safety goggles

QUESTION: How do measurements compare to conversions?

HYPOTHESIS: Write a hypothesis about how close you expect your conversions to be to measurements.

PROCEDURE:
1. Lay the book on a table and measure its length in inches. Read the ruler as shown in the measurement section above, estimating any answer that falls between the markings on the scale. Next, convert the fraction to a decimal (as we did in the measurement section above) and round it to the hundredths place because that's the precision of an English ruler.
2. Measure the width of the book in the same way.
3. Now that you have the length and width measured, multiply them together to get the surface area of the book. Since you are multiplying inches by inches, your area unit should be in^2. Remember to count the significant figures in each of the measurements and round your final answer so that it has the same number of significant figures as the measurement with the least number of significant figures.
4. Use the relationship given in table 1.3 to convert the length measurement into cm. Do the same thing to the width measurement, making sure to keep the proper number of significant figures. Note that the relationship between inches and centimeters is exact. The 2.54 cm should not be taken into account when considering the significant figures because 1 inch is exactly 2.54 cm.
5. Use the metric ruler to measure the length and width of the book in centimeters. Once again, do it as shown in the measurements section above. If the scale of the ruler is marked off in 0.1 cm, then your length and width measurements should be written to the hundredths of a centimeter. Compare these answers to the length and width you calculated by converting from inches. They should be nearly the same. If they are different by only a small percentage, there is no problem. However, if they differ by more than a small percentage, recheck your measurements and your conversions.
6. Multiply the length and width measurements you took with the metric ruler to calculate the surface area of the book in cm^2. Use the relationship between inches and centimeters to convert your answer into in^2. Remember, since you are using a derived unit, the conversion is more complicated. You might want to review Example 1.4.

7. Now compare the converted value for the surface area to the one you calculated in step 3 using your English measurements. Once again, they should be equal or close to equal. If not, you have either measured wrongly or made a mistake in your conversion.
8. Clean up and return everything to the proper place.

CONCLUSION: Write something about how well you made your measurements.

MEASURING TEMPERATURE

In chemistry, we will be measuring and making calculations with temperature, so we need to know what units are used in the measurement. The temperature unit that you are probably most familiar with is Fahrenheit (abbreviated as °F). We hear this unit used often by weather reporters talking about tomorrow's weather and how warm or cold it will be. Although this is a very common temperature unit, it is not used by chemists. Instead, chemists use one of 2 other temperature units: Celsius (sel' see us; abbreviated °C) or Kelvin (kel' vuhn; abbreviated K). First we need to see how these units are defined, and then we will see how they relate to Fahrenheit and why chemists use them.

When we measure temperature, we are measuring how much the liquid within the thermometer is expanding. Therefore, we must find a way to relate that measurement to a unit which means temperature. We do this in the following way:

1. Immerse a thermometer in a mixture of ice and water.
2. Make a mark where we see the liquid in the thermometer and assign a value to that mark. In the Celsius temperature scale, we call it exactly 0°C. In the Fahrenheit scale, we give that mark a value of exactly 32°F.
3. Immerse the thermometer in a pot of boiling water.
4. Make a mark where we see the liquid in the thermometer and assign it a value of exactly 100°Celsius or exactly 212°Fahrenheit.
5. Divide the distance between the 2 marks into equal divisions. Then we have a temperature scale.

This method for defining a temperature scale is illustrated in Figure 1.4.

FIGURE 1.4
Making a Celsius Thermometer
Illustration by Megan Whitaker

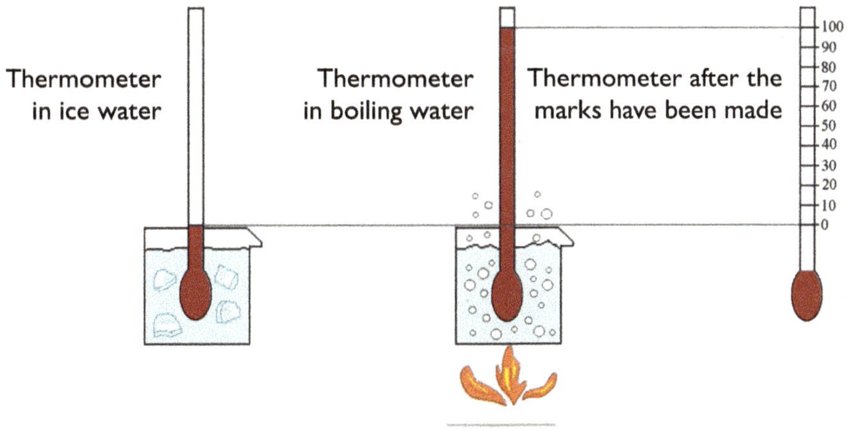

This process of using certain physical measurements to define the scale of a measuring device is called calibration. This particular calibration makes use of a surprising fact in chemistry:

If ice and water are thoroughly mixed, the temperature of the mixture will stay the same (0.0°C or 32.0°F), regardless of the amount of ice or water present.

This might sound rather surprising, but it is true. Even though you might think that a little water with a lot of ice is colder than a lot of water with a little ice, they are actually the same temperature! Equally surprising is this fact:

Boiling water is always at the same temperature (100.0°C or 212.0°F at standard atmospheric pressure) whether it is boiling rapidly or hardly boiling at all.

Now that we know how the Celsius temperature unit is defined, we can learn how it relates to the Fahrenheit unit. It makes sense that Fahrenheit and Celsius relate to one another since they both measure the same thing: temperature. They are related by a very simple equation:

$$°C = \frac{5}{9}(°F - 32) \qquad \text{Equation 1.1}$$

In this equation, °C represents the temperature in degrees Celsius, and °F stands for the temperature in degrees Fahrenheit. So if you must use a Fahrenheit thermometer in your experiments, you can use this equation to convert your measurements into Celsius. Please note one very important thing about this equation: The 5, 9, and 32 are all exact. Therefore, you need not consider their significant figures. They have infinite precision and an infinite number of significant figures. The only significant figures you must consider are those of the original measurement. Using this equation, we will report our answer using the multiplication rule even though there is a subtraction in the equation. Example 1.8 will help show you how this is done.

EXAMPLE 1.8

A student uses a Fahrenheit thermometer to do a chemistry experiment but then must convert his answer to Celsius. If the temperature reading was 50.0°F, what is the temperature in Celsius?

To solve this one, we simply use equation 1.1:

$$°C = \frac{5}{9}(50.0 - 32)$$

$$°C = 10.0$$

There are 3 significant figures in the original measurement. Since the other numbers in this equation are exact, the answer must have 3 significant figures. Therefore, the answer is **10.0°C**.

We usually say that room temperature is about 25°C. What is this temperature in Fahrenheit?

To solve this one, we must first rearrange equation 1.1 using algebra. Once we do this, we get the following equation:

$$°F = \frac{9}{5}(°C) + 32$$

$$°F = \frac{9}{5}(25) + 32$$

$$°F = 77$$

The presence of only 2 significant figures in the original number means only 2 significant figures in the end; therefore, the temperature is **77°F**.

Try "On Your Own" questions 1.15–1.16 to see whether or not you fully understand this type of conversion.

ON YOUR OWN

1.15 Normal body temperature is 98.6°F. What is this temperature in Celsius?

1.16 Rubbing alcohol boils at 180.5°C. What is the boiling temperature of rubbing alcohol in Fahrenheit?

What about the other unit mentioned earlier? The Kelvin temperature unit is a special unit that we will use quite a bit in later modules. It is special because we can never reach a temperature of 0 Kelvin or lower, for reasons we will see in a later module. This fact makes the Kelvin temperature scale different from most others. After all, anything colder than ice water has a negative temperature in Celsius units. This means that temperatures less than 0 are quite common in the Celsius scale. Although not quite as common, it is possible to reach temperatures less than 0 in the Fahrenheit scale as well. It is impossible for anything in nature to reach 0 Kelvin or below. Since we can never get to or go below 0 Kelvin, the Kelvin temperature scale is often called an absolute temperature scale.

Once we have a temperature in units of degrees Celsius, converting it to Kelvin is simple. All we do is add 273.15 to the measurement. In mathematical terms, we would use this equation:

$$K = °C + 273.15 \qquad \text{Equation 1.2}$$

K is the temperature in units of Kelvin, and °C is the temperature in units of Celsius. In this equation, the 273.15 is *not* exact. Its precision plays a role. Note that since this equation involves adding, we use the rules of addition and subtraction when determining the significant figures involved. Those rules are *different* from the ones for multiplication and division, so be aware of that. Example 1.9 will show you how to use this equation.

EXAMPLE 1.9

What is the boiling temperature of water in Kelvin?

This conversion is a snap. We just realize that water boils at 100.0°C. If we put that temperature into equation 1.2, we get this:

$$K = 100.0 + 273.15 = 373.15$$

The original temperature goes out to the tenths place, while 273.15 goes out to the hundredths place. The rules for significant figures in adding tell us that the answer must have the same precision as the least precise number in the equation. Therefore, our final answer is **373.2 K**.

The lowest temperature that has ever been recorded in the United States of America is -80.0°F. What is this temperature in Kelvin?

Since the only way we can get to Kelvin is by adding 273.15 to the temperature in Celsius, we must first convert °F into °C:

$$°C = \frac{5}{9}(-80.0 - 32)$$

$$°C = -62.2$$

Now that we have the answer in °C, we can easily convert to Kelvin:

$$K = -62.2 + 273.15 = 210.95$$

Our final answer is **211.0 K**. Our rules for adding tell us that the precision must be kept to the tenths place because that is the same precision as the least precise number in the equation. That's why our final answer goes out to the tenths place. So you see, even very, very cold temperatures in the Celsius and Fahrenheit temperature scales are still rather large numbers in the Kelvin temperature scale!

ON YOUR OWN

1.17 What is the Fahrenheit equivalent of 0.00 Kelvin? (Use 3 significant figures for this measurement.)

Now cement your knowledge of temperature conversions with "On Your Own" question 1.17.

THE NATURE OF A SCIENTIFIC LAW

One way to approach chemistry or any other science is to look around you and try to think of logical explanations for what you see. You would certainly observe, for instance, that different substances have different forms and appearances. For example, some substances are gases, some are liquids, and some are solids. Some are hard and shiny, but others are soft and dull. You would also observe that different substances behave differently. Iron rusts, but gold does not; copper conducts electricity, but sulfur doesn't. How can you explain these and a vast number of other observations?

God made our world far too complex for us to understand by looking and thinking alone. We need to ask specific questions and conduct experiments to find answers. Scientists develop laws through experimentation and observation. After experimenting on or observing some facet of nature, they formulate a hypothesis to explain their observations. A hypothesis is no more than an educated guess that attempts to explain some aspect of the world around us. For example, when early scientists observed rotting meat, they always saw maggots crawling around on it. This led them to form the hypothesis that maggots are created from rotting meat.

Once a hypothesis has been formulated, scientists test it with more rigorous experiments. For example, after forming the hypothesis that maggots are created from rotting meat, early scientists did experiments to make sure that the maggots were not coming from something else. They would put rotting meat on a shelf high in the air to make sure that no maggot could crawl up to it. Even when the rotting meat was put high in the air, maggots still appeared on it. To early scientists, such experiments confirmed their hypothesis. Although there was no way for maggots to crawl up to the rotting meat, they did indeed appear on it. Many similar experiments convinced early scientists that their original hypothesis was correct even though it wasn't.

Once a hypothesis is confirmed by more rigorous experimentation, it is considered a theory. After numerous experiments, a scientific law might also be constructed. A scientific law is different from a scientific theory in that rather than explaining why or how something happens, it describes what is happening. This usually involves mathematics. Both theories and laws are formed after many experiments and observations. Theories and laws can change if contradictions are ever found in future experiments.

science and creation

We study science to learn more about creation and, ultimately, the Creator. God is the one who holds it all in His hands and uses it for His glory. Early scientists who were experimenting with rotting meat and maggots called their theory the theory of spontaneous generation. As the centuries passed, many more experiments were done to test the theory. Those experiments seemed to support the idea that life, such as maggots, could be created from nonlife, such as rotting meat. All of the experiments done to confirm this theory, however, were flawed. For example, Francesco Redi, an Italian physician, showed that if the rotting meat was completely isolated from the outside world, no maggots would appear; however, microscopic organisms did. French scientist Louis Pasteur eventually performed careful experiments that overturned the theory of spontaneous generation. His work showed that even microscopic organisms could not arise from nonlife but came to the meat by dust particles that blew in the wind.

The point of this story is to illustrate that when you read about scientific results, you must keep in mind that scientific theories can never be absolutely proven. There is always a chance that a new experiment might give results that can't be explained by a present theory. All a theory can do is provide the best explanation available at the time. If new experiments uncover results that present theories can't explain, the theories will have to be modified or perhaps even replaced. Science experiments are a good way to try to understand the nature of God's creation, but they are not absolute truth. Remember that.

salt and light

Bishop Robert Grosseteste (1175–1253) was an English statesman, philosopher, theologian, and scientist. He was one of the first scientists to establish a framework for what would later become the scientific method. Besides being a great scientist, Grosseteste was a strong Christian. He taught that the purpose of inquiry was not to come up with great inventions, but instead to learn the reasons behind the facts. In other words, he wanted to explain why things happened the way they did. That's the essence of science. He said that "just as the light of the sun irradiates the organ of vision and things visible, enabling the former to see and the latter to be seen, so too the irradiation of a spiritual light brings the mind into relation with that which is intelligible" (Stevenson 1899, 52).

Stained glass window by William Morris (to the designs of Burne-Jones) [Photo: public domain]

EXPERIMENTATION AND THE SCIENTIFIC METHOD

As we go about studying chemistry, your experiments will help you understand the concepts being presented in this course. You should use the scientific method to document the results of your experiments. The scientific method includes techniques for investigating, acquiring new knowledge, or correcting and integrating previous knowledge.

1. **Purpose:** What is your goal? What do you want to answer? Example: I would like to know what phase water is at when it is at room temperature.
2. **Hypothesis/Prediction:** What do you think will happen in this experiment? Example: I believe that water is liquid at room temperature.
3. **Experiment:** Test the hypothesis with systematic observations, measurements, and laboratory techniques.
4. **Analysis:** Determine what the results of the experiment show and decide on the next actions to take.
5. **Conclusion:** Describe what the results of the experiment show and the next actions to take.

As we do experiments, we will use this experimental method to document our findings.

Now that you have read this module and answered the "On Your Own" questions, it is time for you to shore up your new skills and knowledge with the practice problems and review questions at the end of this module. As you go through them, check your answers with the solutions provided and be sure you understand any mistakes you made. If you need more practice problems, you can find them in appendix B. Once you are confident in your abilities, take the test. If you do not score at least 70% on the test, then you should probably review this module before you proceed to the next one.

SUMMARY OF KEY EQUATIONS AND TABLES IN MODULE 1

Equation 1.1:	$°C = \frac{5}{9}(°F - 32)$	Converting °F to °C
Equation 1.2:	$K = °C + 273.15$	Converting °C to K

Table 1.1: Physical Quantities and Their Base Units

Table 1.2: Common Prefixes Used in the Metric System

Table 1.3: Relationships between English and Metric Units

Table 1.4: Examples of Significant Figure Rules

ANSWERS TO THE "ON YOUR OWN" QUESTIONS

1.1 $\frac{9{,}321\ \cancel{g}}{1} \times \frac{1\ kg}{1{,}000\ \cancel{g}} = 9.321\ kg$

1.2 $\frac{0.465\ \cancel{L}}{1} \times \frac{1{,}000\ mL}{1\ \cancel{L}} = 465\ mL$

1.3 $\frac{724.0\ \cancel{cm}}{1} \times \frac{0.01\ m}{1\ \cancel{cm}} = 7.240\ m$

1.4 $\frac{8.465\ \cancel{sl}}{1} \times \frac{14.59\ kg}{1\ \cancel{sl}} = 123.5\ kg$

1.5 $\frac{6.1236\ \cancel{L}}{1} \times \frac{1\ gal}{3.78\ \cancel{L}} = 1.62\ gal$

1.6 $\frac{1{,}500\ \cancel{mL}}{1} \times \frac{0.001\ \cancel{L}}{1\ \cancel{mL}} \times \frac{1\ kL}{1{,}000\ \cancel{L}} = 0.0015\ kL$

1.7 $\frac{2\ \cancel{km}}{1} \times \frac{1{,}000\ \cancel{m}}{1\ \cancel{km}} \times \frac{1\ cm}{0.01\ \cancel{m}} = 200{,}000\ cm$

1.8 $\frac{0.01\ \cancel{Mg}}{1} \times \frac{1{,}000{,}000\ \cancel{g}}{1\ \cancel{Mg}} \times \frac{1\ mg}{0.001\ \cancel{g}} = 10{,}000{,}000\ mg$

1.9 $\frac{0.00555\ \cancel{hr}}{1} \times \frac{60\ \cancel{min}}{1\ \cancel{hr}} \times \frac{60\ sec}{1\ \cancel{min}} = 19.98\ sec$

That is not a long time to hold one's breath. So we would not be impressed.

1.10 $\frac{0.0091\ \cancel{kL}}{1} \times \frac{1{,}000\ \cancel{L}}{1\ \cancel{kL}} \times \frac{1\ mL}{0.001\ \cancel{L}} = 9{,}100\ mL = 9{,}100\ cm^3$ (mL and cm^3 equivalent)

1.11 The relationship between m and mm is easy:

$$1\ mm = 0.001\ m$$

To set up the conversion, we start with:

$$\frac{32\text{ m}^2}{1} \times \frac{1\text{ mm}}{0.001\text{ m}}$$

This expression *does not cancel m²*. There is an m² on the top of the first fraction and only an m on the bottom of the second fraction. To cancel m² (which we must do to get the answer), we have to square the conversion fraction:

$$\frac{32\text{ m}^2}{1} \times \left(\frac{1\text{ mm}}{0.001\text{ m}}\right)^2$$

Then we get:

$$\frac{32\text{ }\cancel{\text{m}^2}}{1} \times \frac{1\text{ mm}^2}{0.000001\text{ }\cancel{\text{m}^2}} = 32{,}000{,}000\text{ mm}^2$$

1.12 (a) All 3 nonzero digits are significant figures, as are both zeros. One 0 is between 2 significant figures, and the other is at the end of the number to the right of the decimal. There are 5 significant figures.

(b) The first 3 zeros are not significant because they are not between 2 significant figures. The 6 is a significant figure, as is the last 0 because it is at the end of the number to the right of the decimal. So there are 2 significant figures.

(c) All digits are significant figures here. The first 2 zeros are between significant figures, and the last 0 is at the end of the number to the right of the decimal. Therefore, there are 6 significant figures.

(d) All digits are significant figures. The 0 is between 2 significant figures. There are 5 significant figures.

(e) All the zeros are not significant in this number. There is 1 significant figure.

1.13 (a) 2.6089×10^7 (b) $1.2000000003 \times 10^{10}$
(c) 9.870×10^{-5} (d) 9.80×10^{-1}

1.14 (a) 345,600,000,000,000 (b) 1,234.1
(c) 0.0000345 (d) 0.310

1.15 To solve this one, we simply need to use equation 1.1:

$$°\text{C} = \frac{5}{9}(°\text{F} - 32)$$

$$°\text{C} = \frac{5}{9}(98.6 - 32)$$

$$°\text{C} = 37.0$$

Our measurement starts with 3 significant figures, and the other numbers in the equation are exact, so we must end up with 3 significant figures. The answer is 37.0°C.

1.16 This one requires that we use algebra to rearrange equation 1.1 so that we can solve for °F:

$$°F = \frac{9}{5}(°C) + 32$$

$$°F = \frac{9}{5}(180.5) + 32$$

$$°F = 356.9$$

Since 180.5 has 4 significant figures and everything else in the equation is exact, our answer is 356.9°F.

1.17 The only way we can convert to Fahrenheit is if we have a temperature in Celsius. Before we can get the answer, we must first convert 0.00 K to degrees Celsius by rearranging equation 1.2:

$$°C = K - 273.15$$

$$°C = 0.00 - 273.15 = -273.15$$

Since we are subtracting, we look at precision. The original measurement goes out to the hundredths place, as does 273.15. Therefore, our answer should go out to the hundredths place. Now we can convert to Fahrenheit:

$$°F = \frac{9}{5}(°C) + 32$$

$$°F = \frac{9}{5}(-273.15) + 32$$

$$°F = -459.67$$

Since 273.15 is the only number in the equation that is not exact, the answer must have the same number of significant figures. The answer is –459.67°F.

STUDY GUIDE FOR MODULE 1

REVIEW QUESTIONS

1. Which of the following contains no matter?
   a. A baseball
   b. A balloon full of air
   c. Heat
   d. A light ray

2. List the base metric units used to measure length, mass, time, and volume.

3. In the metric system, what does the prefix *milli-* mean?

4. All conversion factors, when in the form of a fraction, must equal _________.

5. Which has more liquid: a glass holding 0.05 kL or a glass holding 12,000 mL?

6. How long is the bar in the picture below?

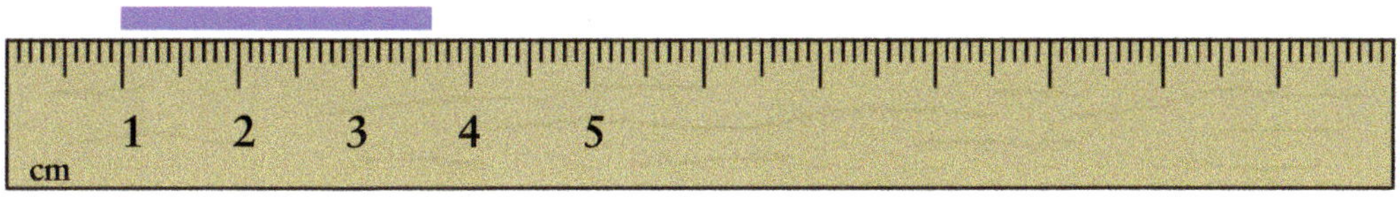

Illustration by Megan Whitaker

7. Two students measure the mass of an object that is known to be 50.0 grams. The first student measures the mass to be 49.8123 grams. The second measures the mass to be 50.1 grams. Which student was more precise? Which student was more accurate?

8. Explain what a significant figure is.

9. How many significant figures are in the following numbers?
   a. 120350 b. 10.020 c. 0.000000012 d. 7.20×10^2

10. A student measures the mass of object A to be 50.3 grams and measures the mass of object B to be 200.24 grams. She then reports the combined mass to be 251 grams. Is this student correct? Why or why not?

11. What would be the units on the following calculations? You do not have to do the math since this question only wants to know the units.
    a. 8 cm + 2 cm =
    b. 4 g ÷ 2 mL =

12. Which answer for question 11 will be a derived unit?

13. What are the 2 basic rules for using scientific notation?

14. Which is colder: 50.0 grams of water at 0.00°C or 50.0 g of water at 32.00°F?

PRACTICE PROBLEMS

Be sure to use the proper number of significant figures in all of your answers!

1. Convert 2.4 mL into L.

2. Convert 69.00 km into m.

3. Convert 0.091 kg into cg.

4. If an object has a volume of 69.2 mL, how many kL of space does it occupy?

5. A box is measured to be 23 cm by 45 cm by 38 cm. What is its volume in cubic meters?

6. A nurse injects 71.0 cc of medicine into a patient. How many liters is that?

7. Convert the following decimal numbers into scientific notation:
   a. 12.45000
   b. 3,040,000
   c. 6,100.500
   d. 0.001234

8. Convert the following numbers back into decimal:
   a. 6×10^{9}
   b. 3.0450×10^{-3}
   c. 1.56×10^{21}
   d. 4.50000×10^{-7}

9. Convert 85.6°C into Fahrenheit.

10. The temperature of the moon during its day is 396 K. What is that in Celsius? In Fahrenheit?

11. The average low temperature of International Falls, MN, in January is –7.0°F. What is that in °C?

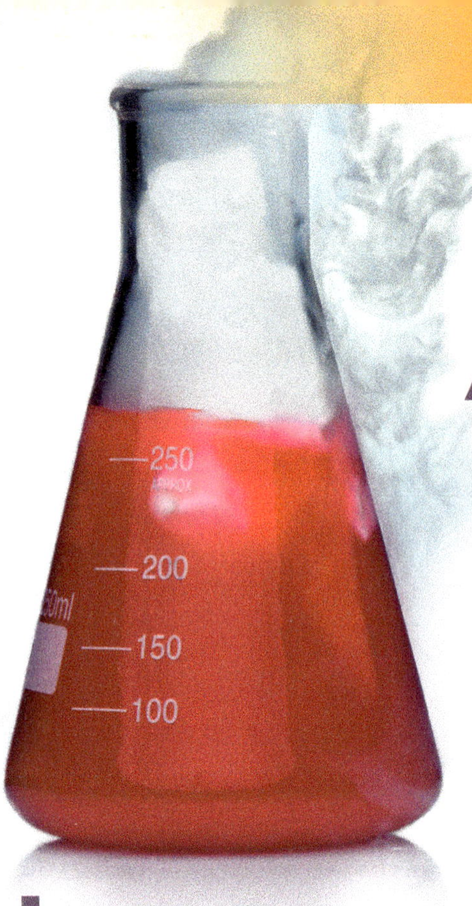

ATOMS AND MOLECULES

In the first module, we discussed matter. In this module, we will explore some of the most fundamental aspects of matter. We need to know what matter is composed of, why matter exists in such a wide variety of substances, and how matter is able to change from one substance to another.

Pretend you were given a radio and asked to tell what it is made of. What would you do? Most likely, you would begin to take it apart. You would find a way to remove the back cover and look inside. Of course, you would find a lot of things inside the radio, so you would probably have to take it apart piece by piece until you had broken it down into all of its individual components. You could then learn what each component was and make a list of everything that makes up the radio. By taking the radio apart, you would be able to learn about its composition.

first reactions

In the story of creation, Adam named things created by God. We continue to exercise that ability in science. Chemistry uses a Periodic Table of Elements to display the properties of these basic elements. In this module we will look at matter, the elements, their physical and chemical properties and begin to discuss interaction of these basic building blocks. A working understanding of the Periodic Table is a necessary foundation for the rest of chemistry. Matthew 7 tells us the wise man built his house upon a rock. Your future study of chemistry rests on a good understanding of the properties, mixtures and compounds you will begin to learn about here.

FIGURE 2.1
An Old Radio
© iStockphoto.com

Taking a radio apart tells you more than just what it's made of, however. If you wanted to learn how that radio works, you could learn how each individual component works. That knowledge would help you immensely in understanding how the entire radio works and help you gain some understanding about how other electronic devices, such as televisions, work. To truly understand the nature of something, we have to break it down into all of its parts. Once we have done that, we can learn about each individual component, and that will allow us to understand whatever thing we are investigating.

This has long been the chemist's method for trying to understand the nature of matter. Chemists believe that if you can break matter down into its fundamental components and learn everything there is to learn about those components, you will then know everything there is to know about matter. In this module, we will begin to explore how chemists over the ages have pursued this method and what they have learned as a result.

EARLY ATTEMPTS TO UNDERSTAND MATTER

During the earliest years of scientific investigation, ancient Greek philosophers thought that the matter inside a substance was continuous. By this they meant that you could divide any substance in half over and over again. You would eventually have to use tweezers and a very powerful magnifying glass, but no matter how small the substance became, you could still divide it in half again. In other words, they thought that substances were composed of long, unbroken blobs of matter. This concept was known as the continuous theory of matter.

About 4 centuries before Jesus Christ walked this earth, the Greek philosopher Democritus was walking toward the shore of a beach and was suddenly struck with an amazing thought. He noticed that from a distance, the sand along the shore looked like it was one long, continuous blob of yellow. As he walked closer to the beach, however, he could see that the sand was not at all continuous. Instead, it was made up of hundreds of thousands of individual particles called grains.

FIGURE 2.2
Distant and Close Views of the Beach
Wikimedia Commons, public domain

Democritus then looked out at the ocean and wondered to himself whether or not water was made exactly the same way. From the shore, the water seemed to be a huge mass of unbroken matter. But suppose he could magnify the water. Would he eventually see grains of water? Could it possibly be that the continuous theory of matter was based on an illusion? Perhaps matter was not really continuous. Maybe it just appeared that way because he could not magnify it enough to see the individual particles that make it up. Democritus wrote a great deal about this idea, which is referred to as the discontinuous theory of matter. Instead of being continuous, he argued, matter is composed of tiny individual particles. Our eyes simply cannot see these particles, so the apparent continuity of matter is an illusion. Democritus's musings represent man's first attempt to think in terms of atoms.

Since no experiments could be imagined to test the validity of either the continuous or the discontinuous theory of matter, scientists could not come to any firm conclusions about which theory was more scientifically sound. Each theory had its fervent supporters and its ardent critics. Unfortunately, neither side had any empirical evidence (facts gained by experiment and observation) to support its theory, so scientific progress on this point stalled for many centuries.

THE LAW OF MASS CONSERVATION

Progress in understanding the nature of matter picked up in the late 1700s. During that time, Antoine Lavoisier (lah vwah zyay'), considered by some to be the founder of chemistry, developed a theory that stunned the scientific world. This theory is now considered a scientific law and is called the law of mass conservation.

The law of mass conservation—Matter cannot be created or destroyed; it can only change forms.

When you burn wood in the fireplace, the matter in the wood is not destroyed. Instead, it changes from wood into other substances. Today, we know that those substances are carbon dioxide, water, and ash.

This may not sound very surprising. However, to the chemistry community in Lavoisier's day, this was a startling revelation. You see, early chemists were guided by what they observed. Since the human eye cannot observe everything in nature, this led them to some very wrong conclusions.

For example, early chemists would watch wood burning in the fire and would see the wood slowly disappear. They were unable to see the carbon dioxide and water vapor forming from the wood because those substances are colorless gases. As a result, they simply thought that the matter in the wood was disappearing—that matter was being destroyed. Because their experiments were not accurate, early chemists developed scientific laws that were quite wrong.

Today we have developed methods that allow us to detect the presence of colorless gases, so we can measure the mass of the carbon dioxide and water vapor produced when wood burns. We can also determine that when wood burns, the oxygen in the air around the wood is used up. This is why a fire cannot burn unless there is a plentiful supply of oxygen. If we measure the mass of the wood and the oxygen before a fire starts and we trap all of the carbon dioxide and water vapor produced by the fire after it starts, we will find out that the mass of the wood and oxygen used by the fire is exactly equal to the mass of the ash, water vapor, and carbon dioxide produced by the fire. Experiment 2.1 allows you to see that even when matter goes through substantial change, the total mass is not affected.

EXPERIMENT 2.1

PURPOSE: To investigate the conservation of mass.

MATERIALS:

- 100-mL beaker (A juice glass can be used instead.)
- 250-mL beaker (A juice glass can be used instead.)
- Watch glass (A small saucer can be used instead. It must cover the mouth of the 100-mL beaker or juice glass listed above.)
- Teaspoon
- Lye (This is commonly sold with the drain cleaners in hardware stores and supermarkets. Make sure that your bottle is labeled 100% lye. If you cannot find lye, you can order it online. It may be listed under its chemical name, sodium hydroxide.) ***Always use this chemical in a well-ventilated area. This experiment can cause irritating fumes.***
- White vinegar
- Several leaves of red (often called purple) cabbage
- Water
- Small pot for boiling water
- Measuring cup
- Stove
- Mass scale
- Stirring rod (A spoon will work.)
- Safety goggles
- Rubber cleaning gloves

QUESTION: Is mass conserved during a chemical reaction?

HYPOTHESIS: Propose an answer to the question.

PROCEDURE:

1. Add 2 cups of water to the pot.
2. Place the cabbage leaves into the water.
3. Place the pot on the stove and turn on the burner. Let the water boil for 5 minutes.
4. While you are waiting for the water to boil, use the scale to measure the mass of your beaker.
5. Add approximately 60 mL of vinegar to the beaker. The beaker should have volume levels marked, so you can just fill it to the 60 mL mark. You needn't add exactly 60 mL. Anywhere from 50 to 60 will be fine. If you are not using a beaker, add about

¼ cup. When chemists make measurements like these, where they don't need to get an exact amount, they call them qualitative measurements. On the other hand, when chemists make measurements as exactly as possible, they call them quantitative measurements. Remember these terms because we will use them again.

6. Put on the gloves.
7. Measure 1 teaspoon of lye and put it on the watch glass. You can take the gloves off, but ***whenever you are touching anything that has lye on it (as you will do later), put the gloves back on because lye is caustic and can cause chemical burns.***
8. Once the water in the pot has been boiling for 5 minutes, take the pot off the stove and allow it to cool for a few minutes.
9. Carefully pour about 50 mL of the solution in the pot into the 250-mL beaker. If some cabbage leaves get into the beaker, that is fine.
10. Allow the solution in the 250-mL beaker to cool.
11. Once the solution is cool enough that you can comfortably pick up the beaker in your hand, pour about 10 mL of the solution into the 100-mL beaker. The vinegar in the beaker should turn pink. This is due to chemicals (anthocyanins) that come from the cabbage leaves. As you will learn in Module 9, anthocyanins turn pink in the presence of acids such as vinegar.
12. Place the stirring rod in the beaker so that it stands in the beaker.
13. Place the watch glass on top of the beaker so that it covers the mouth of the beaker.
14. Place the beaker (with the stirring rod in it and the watch glass covering it) on the scale, and measure the mass of the total assembly. Your setup should look like the picture below:

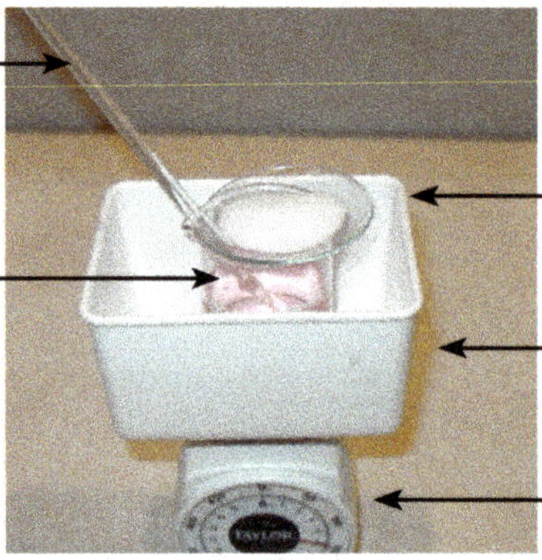

15. After you have measured the total mass of the experimental setup, dump the lye into the vinegar.
16. Stir the solution with the stirring rod. Do not lift the stirring rod out of the solution because you would risk losing some of the solution.
17. You should quickly observe a color change because the lye (a base) neutralizes the vinegar (an acid). Since anthocyanins are pink in the presence of acids and the pink color is replaced by a greenish-yellow color, you know that the vinegar is being changed. You will learn a lot more about this process in Module 9. For now, just realize that you are observing a chemical reaction, where the vinegar and lye are forming 2 completely different chemicals.
18. Notice also that most of the lye seems to have disappeared. Of course, it didn't really disappear. It simply reacted with the vinegar and changed into a chemical that is now

dissolved in the water.

19. Rest the stirring rod against the side of the beaker and place the watch glass back on the top of the beaker so that (aside from the lye and the color change), the set-up looks like it did in step 14.
20. Carefully touch the sides of the beaker. The beaker should be noticeably warmer than when you put it on the scale. This is further evidence that a chemical reaction occurred because energy in the form of heat was released.
21. Read the mass of the entire assembly again.
22. Clean up and return everything to the proper place.

CONCLUSION: The mass in step 21 should be the same as what you read in step 14. This demonstrates the law of mass conservation. Although vinegar and lye were changed into 2 different substances, although heat energy was released, and although the lye seemed to disappear, the mass of the entire assembly did not change.

Once chemists accepted the law of mass conservation, they started using it to analyze many of the changes they observed in matter. Example 2.1 illustrates one way that the law of mass conservation can be used in analyzing how matter changes.

EXAMPLE 2.1

A chemist notices that when given enough energy, a certain white powder changes into 2 different substances: tin and oxygen. A chemist watches 151 grams of powder undergo this change. The chemist easily collects the tin that was formed and measures its mass to be 119 grams. Unfortunately, he could not collect the oxygen that was formed and therefore could not determine its mass. Use the law of mass conservation to determine how much oxygen was made.

The law of mass conservation tells us that we cannot create or destroy matter. If the chemist started with 151 grams of matter, no matter what change occurred, he must end up with 151 grams of matter. Since we know that only 2 substances—tin and oxygen—were formed in the course of the experiment, all 151 grams must be accounted for between the tin and the oxygen. Mathematically, we would say:

$$\text{Mass of powder} = \text{Mass of tin} + \text{Mass of oxygen}$$

Rearranging the equation gives us:

$$\text{Mass of oxygen} = \text{Mass of powder} - \text{Mass of tin}$$

Plug in our numbers:

$$\text{Mass of oxygen} = 151 \text{ grams} - 119 \text{ grams} = 32 \text{ grams}$$

Notice the significant figures here. Although both 151 and 119 have 3 significant figures, the answer can have only 2 because we are subtracting and thus do not count significant figures. Instead, we look at the decimal place. Since both numbers have their last significant figure in the ones place, the answer must also have its last significant figure in the ones place. Therefore, **32 grams of oxygen** were formed when 151 grams of the white powder underwent the observed change.

ON YOUR OWN

2.1 A compound is decomposed into its constituent elements of carbon, nitrogen, and hydrogen. If, in the decomposition, 12.0 grams of carbon, 14.0 grams of nitrogen, and 3.0 grams of hydrogen were formed, what was the mass of the compound before it was decomposed?

Make sure that you understand how to use the law of mass conservation by answering "On Your Own" question 2.1

ELEMENTS: THE BASIC BUILDING BLOCKS OF MATTER

Once chemists began using the law of mass conservation, they noticed something quite interesting: Some substances can undergo change in such a way as to make many other substances, each less massive than the original. For example, they noticed that a black powder called iron sulfide can undergo a change that forms iron and sulfur. Starting with 88 grams of iron sulfide would make 56 grams of iron and 32 grams of sulfur. Similarly, 85 grams of a white powder known as sodium nitrate could undergo a change that would produce 23 grams of sodium, 48 grams of oxygen, and 14 grams of nitrogen. In both of these cases, the change that occurred seemed to divide the matter that was in the powder into smaller bundles that each became a new substance. Chemists called this type of change decomposition.

On the other hand, chemists noticed that there were some substances that would not undergo decomposition in any way. For example, once they had made sulfur from iron sulfide, there was nothing they could do to force the sulfur to decompose into several less massive substances. No matter what they tried to do, the sulfur would not change into different substances that were less massive than the original sample of sulfur. It seemed that the matter in the sulfur could not be made into smaller substances. This told chemists that there was something special about sulfur.

Remember the example of the radio used earlier? If you began to take apart a radio, you would eventually get to the point where you had a lot of parts (screws, wires, and dials, for example) that you could not disassemble any further. When you got to that point, you would say that you had broken the radio down into all of its components.

When chemists saw that iron sulfide could be decomposed into iron and sulfur but that the iron and sulfur could not be decomposed any further, they assumed that this meant iron and sulfur were the components of iron sulfide. These components could not be decomposed any further, so they represented the simplest form of the matter within the iron sulfide. Chemists called these components elements.

Element—Any substance that cannot be broken down (decomposed) into simpler components.

Chemists soon discovered that there were many, many substances that could not be decomposed into smaller substances; there were many, many elements. Since elements were considered the components of all forms of matter, chemists wanted to catalog them carefully. They compiled the most important information about each element into a table that was called the periodic table of elements. As time went on, this table became the chemist's most important tool. It remains so even to this day. The modern form of the periodic table of elements is shown on the next page.

The periodic table of elements can also be found in the end pages of this book. Feel free to photocopy it and keep it handy so you can easily look at it as you read the rest of this and other modules. In addition, you will use it heavily while doing all problems and tests. You might want to put it in a clear plastic sheet protector and use it as a bookmark so you will always have it handy while you read. The periodic table of elements contains a huge amount of information, so we will be referring to it over and over again. It will help you quite a bit to look at the table every time we refer to it so that you will see how to use it.

The periodic table of elements is often called simply the periodic table. Notice that it has 118 boxes. Each box represents an element, and there are currently 118 elements. Believe it or not, only 92 of these elements (those with boxes numbered 1–92) appear naturally in creation. The 26 other elements (numbered 93–118) have been manufactured by scientists in a lab. We say there are currently 118 elements because sometimes a scientist will figure out how to manufacture a new one. For example, element 118 was manufactured in 2002 by experiments conducted at Dubna, Russia, at the Flerov Laboratory of Nuclear Reactions by workers from the Joint Institute for Nuclear Research in Russia and the Lawrence Livermore National Laboratory in the United States. The researchers collided a Californium (Cf) atom with a calcium (Ca) atom.

Think for a minute about what this means: Since elements are the building blocks of matter, everything you see around you is composed of elements which are listed on the table. If you were to break down anything in nature into its individual components of matter, then each of those components would be one of the elements listed on the table.

Each box in the table contains 1–3 letters that make up the chemical abbreviation for each element. For example, box 6 has a C in it. This is the chemical abbreviation for the element carbon. In the same way, box 20 contains a Ca. This is the chemical abbreviation for the element calcium. If an abbreviation contains more than one letter, the first letter is always capitalized and the others are not capitalized.

The abbreviations for some elements are temporary until they get their assigned names and symbols from the International Union of Pure and Applied Chemistry (IUPAC). You'll notice the letter U often in higher atomic numbers. The IUPAC temporarily names new elements by assigning each digit a Latin root. For example, the element roentgenium, atomic number 111, was temporarily called Uuu—*un* for one—so Uuu is 1–1–1. Then they add the suffix *–ium*. So element 111 was unununium. Can you guess the temporary name for element 115? It would be Uup—ununpentium, or 1–1–5!

In 1937, the first artificially created element, technetium, was made. In 1940, neptunium was made, followed quickly by element 94, plutonium. Between 1940 and 1961, elements 95–103 were created. Elements 104–106 were created by 1974, and it wasn't until 1981 that element 107 was confirmed. It took 15 years to create elements

THE PERIODIC TABLE OF ELEMENTS

1A	2A	3B	4B	5B	6B	7B	8B	8B	8B	1B	2B	3A	4A	5A	6A	7A	8A
1 **H** 1.01																	2 **He** 4.00
3 **Li** 6.94	4 **Be** 9.01											5 **B** 10.8	6 **C** 12.0	7 **N** 14.0	8 **O** 16.0	9 **F** 19.0	10 **Ne** 20.2
11 **Na** 23.0	12 **Mg** 24.3											13 **Al** 27.0	14 **Si** 28.1	15 **P** 31.0	16 **S** 32.1	17 **Cl** 35.5	18 **Ar** 39.9
19 **K** 39.1	20 **Ca** 40.1	21 **Sc** 45.0	22 **Ti** 47.9	23 **V** 50.9	24 **Cr** 52.0	25 **Mn** 54.9	26 **Fe** 55.8	27 **Co** 58.9	28 **Ni** 58.7	29 **Cu** 63.5	30 **Zn** 65.4	31 **Ga** 69.7	32 **Ge** 72.6	33 **As** 74.9	34 **Se** 79.0	35 **Br** 79.9	36 **Kr** 83.8
37 **Rb** 85.5	38 **Sr** 87.6	39 **Y** 88.9	40 **Zr** 91.2	41 **Nb** 92.9	42 **Mo** 96.0	43 **Tc** (98)	44 **Ru** 101.1	45 **Rh** 102.9	46 **Pd** 106.4	47 **Ag** 107.9	48 **Cd** 112.4	49 **In** 114.8	50 **Sn** 118.7	51 **Sb** 121.8	52 **Te** 127.6	53 **I** 126.9	54 **Xe** 131.3
55 **Cs** 132.9	56 **Ba** 137.3	57 **La** 138.9	72 **Hf** 178.5	73 **Ta** 181.0	74 **W** 183.8	75 **Re** 186.2	76 **Os** 190.2	77 **Ir** 192.2	78 **Pt** 195.1	79 **Au** 197.0	80 **Hg** 200.6	81 **Tl** 204.4	82 **Pb** 207.2	83 **Bi** 209.0	84 **Po** (209)	85 **At** (210)	86 **Rn** (222)
87 **Fr** (223)	88 **Ra** (226.0)	89 **Ac** (227)	104 **Rf** (265)	105 **Db** (268)	106 **Sg** (271)	107 **Bh** (270)	108 **Hs** (277)	109 **Mt** (276)	110 **Ds** (281)	111 **Rg** (280)	112 **Cn** (285)	113 **Nh** (284)	114 **Fl** (289)	115 **Mc** (288)	116 **Lv** (293)	117 **Ts** (294)	118 **Og** (294)

58 **Ce** 140.1	59 **Pr** 140.9	60 **Nd** 144.2	61 **Pm** (145)	62 **Sm** 150.4	63 **Eu** 152.0	64 **Gd** 157.2	65 **Tb** 158.9	66 **Dy** 162.5	67 **Ho** 164.9	68 **Er** 167.3	69 **Tm** 168.9	70 **Yb** 173.0	71 **Lu** 175.0
90 **Th** 232.0	91 **Pa** 231.0	92 **U** 238.0	93 **Np** (237)	94 **Pu** (244)	95 **Am** (243)	96 **Cm** (247)	97 **Bk** (247)	98 **Cf** (251)	99 **Es** (252)	100 **Fm** (257)	101 **Md** (258)	102 **No** (259)	103 **Lr** (262)

- Nonmetals
- Standard metals
- Transition metals
- Inner transition metals

108–112, and then another 9 years to synthesize 113–118. So now the heaviest element, 119, waits to be confirmed, and if it is, the periodic table will get a new row!

While this is generally the rule followed for the periodic table, you will soon see that there are exceptions. What would you expect the abbreviation for sodium to be? Probably S or So, right? That would be consistent with the nomenclature described above, but it's not right. The chemical abbreviation for sodium is Na. Where did that come from?

Sometimes chemists do not use the English name of the element to determine its abbreviation; they use its Latin name. The Latin name for sodium is *natrium*, so its abbreviation is Na. In addition, chemists sometimes abbreviate elements with just the first letter of the Latin name. For example, the abbreviation for potassium is K because the Latin name for potassium is *kalium*. This system has come about because chemists all over the world discovered elements at different times in history. Since chemists in some countries used Latin as the scientific language, they used Latin to name the elements. In other countries, chemists used English. Other elements are simply named after the person who discovered them or where the discovery occurred.

As time goes on, you will begin to learn the names that go with the abbreviations on the table. For now, you will be expected to memorize a few of them so that you can understand some of the things you read in this module. By the end of this module, you will be expected to know the first 30 elements on the table, H through Zn, plus Br and I. You need not know the numbers that appear with them on the periodic table. You need only know that H is the abbreviation for hydrogen, He is the abbreviation for helium, etc. Table 2.3 at the end of this module lists all element names and their corresponding abbreviations so that you can begin to learn them.

The next thing you might notice is that each box on the table contains 2 numbers. These numbers tell us some very important things about each element. We will not get to the meaning of those numbers for a couple of modules, so for now you can simply ignore them. Also, you might think that the table is structured in a funny way. The boxes are not simply piled on top of each other or arranged in nice, even rows. We will see in a later module that the table is laid out this way to group the elements that act alike in the same column.

Finally, notice the heavy, jagged line that starts at the element boron (B) and moves diagonally down the table to astatine (At). This very important line separates the elements into 2 classes: metals and nonmetals. Elements that are called metals are usually malleable (can be easily bent and shaped), have luster (are shiny), and are able to conduct electricity. On the other hand, nonmetals are typically brittle (break easily when you try to bend or shape them), lack luster, and do not conduct electricity.

Elements that appear on the table to the left of the jagged line (Zn, Ca, and Pt, for example) are classified as metals. Elements that appear on the table to the right of the jagged line (B, N, and Br, for example) are nonmetals. As usual, this rule has a few exceptions. The most important exception (and one you will be expected to remember) is hydrogen.

Even though hydrogen is left of the jagged line, it is always considered a nonmetal.

You will have to wait for a while before you have enough information to understand why hydrogen is an exception to this rule. For now, you just need to memorize the fact that it is.

The other exception to this rule is a class of elements known as metalloids. These elements have some metal properties and some nonmetal properties. For example, silicon (Si) is brittle, but it has luster. In addition, it conducts electricity under certain conditions but not under other conditions. For now, you do not have to consider metalloids. As far as you are concerned, there are only 2 classes of elements—metals and nonmetals—and these classes are separated by the heavy jagged line on the table.

The classification of elements into metals and nonmetals is important because we will use it again and again to identify the nature of different substances. It is important, therefore, that you can recognize whether an element is a metal or a nonmetal by looking at the table. In addition, it is important for you to know the 3 properties that describe metals and nonmetals. Answer "On Your Own" questions 2.2–2.4 to make sure you understand this classification scheme.

ON YOUR OWN

2.2 Which of the following elements is (are) metal(s)? S, Ca, At, Tl, Hs, Xe, H

2.3 If you want to make a wire for an electrical system and you have only the following elements to work with, which should you choose? Ag, P, Br

2.4 Which of the following would you expect to be brittle? Co, Cs, C

COMPOUNDS

Since chemists classify as elements all substances that cannot be decomposed into simpler substances, we also need to classify the substances that can be decomposed into simpler substances. We call these compounds.

Compounds—Substances that can be decomposed into elements by chemical means.

Based on these definitions, we can say that matter comes in 2 forms: elements and compounds. Elements are the basic building blocks of all substances. Compounds are formed when elements group together to form a new substance. Alternatively, we could say that elements are formed when compounds are broken down into their components. Either way of looking at these 2 forms of matter is correct.

In 1794, French chemist Joseph Proust (proost) demonstrated a fact about compounds that later became known as the law of definite proportions.

The law of definite proportions—The proportion of elements in any compound is always the same.

Proust discovered that when several elements come together to form a compound, they always come together in the same proportion. For example, water is made when the element hydrogen (H) and the element oxygen (O) react together. Whenever this happens, it always requires 8 grams of oxygen for every 1 gram of hydrogen. If the reaction started with 10 grams of hydrogen, it would need 80 grams of oxygen. When the reaction was done, 90 grams of water would be formed, in agreement with the law of mass conservation. No matter how much hydrogen the reaction started with, the mass of oxygen had to be 8 times greater, or some hydrogen would be left over.

You can think about the law of definite proportions in the same way that you think about recipes. If a cake recipe says that the cake serves 8 people, you could make a cake to serve 40 people by using 5 times the ingredients called for in the recipe. The law of definite proportions tells us that there is a single recipe that governs how much of each element is needed to make a particular compound. If you don't add elements together in exactly the proportion the recipe calls for, then there will be some leftover ingredients at the end. Figure 2.3 tries to illustrate this idea.

FIGURE 2.3
The Law of Definite Proportions

A chemist decides that she wants to make table salt. She has already determined that table salt is a compound composed of 2 elements: sodium (Na) and chlorine (Cl). What she doesn't know is what proportion of each element she must use. To determine this, she does a few experiments. First, she reacts 10.0 grams of sodium with 10.0 grams of chlorine. This is a very dangerous reaction because sodium (a solid metal) is very reactive and chlorine (a gas) is poisonous. When the reaction is over, she finds that she has 16.5 grams of table salt and 3.5 grams of sodium left over.

10.0 g sodium and 10.0 g chlorine makes 16.5 g table salt and 3.5 g leftover sodium.
Sodium: Dennis s.k. Collection/Dnn87 (CC BY-SA) . Illustration of chlorine by David Weiss. Salt: © Crestock.com (nenovbrothers).

First, we see that this experiment confirms the law of mass conservation. Since the chemist started out with 20.0 grams (10.0 g + 10.0 g) of matter, she also had to end up with 20.0 grams (16.5 g + 3.5 g) of matter. Since the sodium and chlorine were not added in the correct proportion, the substance that she ended up with was not pure table salt. Instead, only 16.5 grams of table salt could be made. The rest of the 20.0 grams, then, is accounted for by sodium that could not react because too much was added in the first place.

Second, we see that the *chlorine* must have run out. There must have been plenty of sodium left over. Since there was no leftover chlorine, that must mean that the chlorine was completely used up in the reaction. This little fact allows us to determine the proper proportions (the recipe) of sodium and chlorine a chemist must have to make table salt. Since 3.5 grams of sodium were left over, then that mass of sodium was not used in the reaction. To find out how much sodium was used, we simply subtract the leftover amount from the amount we started with:

$$\text{Amount of sodium used} = \text{Starting amount} - \text{Amount left over}$$

$$\text{Amount of sodium used} = 10.0\text{ g} - 3.5\text{ g} = 6.5\text{ g}$$

We see that only 6.5 grams of sodium were used when reacted with 10.0 grams of chlorine. The proper proportion in which to add sodium and chlorine is 6.5 grams of sodium for every 10.0 grams of chlorine.

To test this recipe, our chemist then reacts 10.0 grams of chlorine with 6.5 grams of sodium. According to the law of definite proportions, these 2 elements should now react to form table salt with no leftovers since she is adding the elements together in the proper proportions:

6.5 g sodium and 10.0 g chlorine makes 16.5 g table salt and no leftover sodium.

So she did, indeed, determine the proper recipe for making table salt: 10.0 grams chlorine plus 6.5 grams sodium makes 16.5 grams table salt. To convince herself that the law of definite proportions really works, she tripled the masses of sodium and chlorine and tried again:

19.5 g sodium and 30.0 g chlorine makes 49.5 g table salt.

Once again, the experiment ended with no leftovers because the elements were added together in the proper proportion.

As a final experiment, the chemist adds 6.5 grams of sodium to 15.0 grams of chlorine. In this case, the law of definite proportions shows that she used too much chlorine because 6.5 grams of sodium needs only 10.0 grams of chlorine. There should, therefore, be 5.0 grams of chlorine left over in the end.

6.5 g sodium and 15.0 g chlorine makes 16.5 g table salt and 5.0 g leftover chlorine.

The law of definite proportions, coupled with the law of mass conservation, can be used to solve problems like Example 2.2. After studying the example, answer "On Your Own" questions 2.5–2.6.

EXAMPLE 2.2

A chemist reacts 15.0 grams of calcium (Ca) with 15.0 grams of oxygen (O). This reaction makes 21.0 grams of a compound known as lime. Along with the lime, there is also some leftover oxygen. If the chemist wants to make 55 grams of lime and have no leftover oxygen or calcium, how much of each element should he use?

To solve this problem, we first need to determine the recipe for making lime. To do that, we need to figure out how much oxygen was used in the reaction. Since there was no leftover calcium, we know that all 15.0 grams of calcium were used. How much leftover oxygen was there? We can find out by using the law of mass conservation:

$$\text{Total mass before reaction} = \text{Total mass after reaction}$$

$$15.0\text{ g} + 15.0\text{ g} = 21.0\text{ g} + \text{Mass of leftover oxygen}$$

$$\text{Mass of leftover oxygen} = 30.0\text{ g} - 21.0\text{ g} = 9.0\text{ g}$$

Since 9.0 grams of oxygen were left over:

$$\text{Mass of oxygen used} = \text{Starting mass} - \text{Mass left over}$$

$$\text{Mass of oxygen used} = 15.0\text{ g} - 9.0\text{ g} = 6.0\text{ g}$$

So the proper recipe to use in making lime is this: 15.0 grams of calcium plus 6.0 grams of oxygen makes 21.0 grams of lime. The problem asks how much of each element is needed to make 55 grams of lime. We obviously need to increase the recipe, but by how much? We have to find what number, when multiplied by 21.0, gives us 55. That will tell us by what factor we need to increase our recipe:

$$21.0\text{ g} \cdot x = 55\text{ g}$$

$$x = \frac{55\ \cancel{\text{g}}}{21.0\ \cancel{\text{g}}} = 2.6$$

Notice that in the expression for x, the unit of grams cancels. As a result, x has no units. Numbers like this are called dimensionless quantities.

Now we know that to make 55 grams of lime, we must increase our recipe by a factor of 2.6. If the original recipe was 15.0 grams calcium + 6.0 grams oxygen, then the new recipe is:

$$\text{Mass of calcium} = 15.0\text{ g} \times 2.6 = 39\text{ g}$$

$$\text{Mass of oxygen} = 6.0\text{ g} \times 2.6 = 16\text{ g}$$

Therefore, to make 55 grams of lime, the chemist needs to react **39 grams of calcium with 16 grams of oxygen**.

ON YOUR OWN

2.5 When 24.0 grams of the element carbon (C) react with 8.08 grams of the element hydrogen (H), natural gas (the same stuff that is burned in a gas stove or a gas furnace) is produced. There is no leftover carbon or hydrogen when these quantities are used. How much carbon and hydrogen would you need to make 128.4 grams of natural gas?

2.6 A chemist decomposes 60.0 grams of a purple powder into 32.8 grams of the element chlorine (Cl) and an unmeasured amount of the element cobalt (Co). The chemist uses the results of this experiment to help him determine the recipe for making this purple powder from its elements. He then decides to make 2.00 kg of it. How much cobalt and how much chlorine gas will he need?

THE LAW OF MULTIPLE PROPORTIONS

One more law that deals with the nature of compounds is the law of multiple proportions.

The law of multiple proportions—If 2 elements combine to form different compounds, the ratio of masses of the *second* element that react with a fixed mass of the *first* element will be a simple, whole–number ratio.

This law shows that sometimes there are different ways of combining elements. If this is the case, there is a simple relationship between the recipes for the 2 compounds. Although the law sounds a bit hard to understand when you read it, an example will make it very clear.

When 12.0 grams of carbon (C) and 32.0 grams of oxygen (O) combine, they make 44.0 grams of carbon dioxide, a gas that humans exhale. On the other hand, under the right conditions, 12.0 grams of carbon can combine with just 16.0 grams of oxygen to make 28.0 grams of carbon monoxide, a completely different compound. To illustrate just how different these compounds are, carbon monoxide is poisonous to humans. When 12.0 grams of carbon combine with 32.0 grams of oxygen, they form a compound that humans exhale each time they breathe. However, when carbon and oxygen combine in a different proportion (12.0 grams and 16.0 grams), they make a compound that is lethal to humans.

The law of multiple proportions indicates that there is a simple relationship between the amount of oxygen in the recipe for carbon dioxide and the amount of oxygen in the recipe for carbon monoxide. It first says that we must use a *fixed* amount of the first element (carbon). We have already done that; in each recipe, the amount of carbon is the same (12.0 grams). Then we determine the ratio of the mass of oxygen used to make the first compound and the mass of oxygen used to make the second compound. It took 32.0 grams of oxygen to make carbon dioxide but only 16.0 grams of oxygen to make carbon monoxide. The ratio of the mass of oxygen in carbon dioxide to the mass of oxygen in carbon monoxide, then, is 32.0:16.0, or 2:1. That's a simple, whole–number ratio.

Let's look at one more example just to make sure you understand this law. When 36.0 grams of carbon react with 12.0 grams of hydrogen, they produce 48.0 grams of methane, which is also called natural gas. On the other hand, when 36.0 grams of carbon react with 8.0 grams of hydrogen, they produce 44.0 grams of propane, which is the gas typically burned by outdoor gas grills. Once again, 2 elements (carbon and hydrogen) are reacting in different proportions to make different gases. We have a fixed amount of carbon (36.0 grams). If we then take the ratio of the masses of hydrogen that react with this fixed mass of carbon, we get 12.0:8.0, or 3:2. Once again, it's a simple, whole–number ratio. That's what the law of multiple proportions says should happen.

What's the big deal about the law of multiple proportions? It was the main evidence for an atomic theory proposed by John Dalton in the early 1800s. Dalton used his theory to predict the law of multiple proportions, and other scientists then did experiments to demonstrate that the law was correct. This provided excellent evidence for the validity of Dalton's theory.

DALTON'S ATOMIC THEORY

The law of definite proportions, combined with the law of mass conservation, led John Dalton, an English chemist, to propose a theory he thought would help chemists better understand the nature of matter. With this theory, Dalton thought that he could explain *why* these laws were always obeyed. He called it his atomic theory, and it forms the basis for our modern–day interpretation of the nature of matter. Dalton's theory included 4 vital assumptions:

1. All elements are composed of small, indivisible particles called atoms.

2. All atoms of the same element have exactly the same properties.

3. Atoms of different elements have different properties.

4. Compounds are formed when atoms are joined together. Since atoms are indivisible, they can join together only in simple, whole–number ratios.

salt and light

John Dalton (1766-1844) was a Christian, chemist, physicist, and meteorologist. He is best known for the development of atomic theory, including the idea that atomic weights of elements were proportional to each other.

Portrait by Charles Turner (Public Domain)

Remember, these 4 ideas were assumptions. He could not offer definitive proof that they were true. However, he said that if you assume that they are true, you can explain why the laws of mass conservation and definite proportions are always obeyed. Also, he predicted the law of multiple proportions (as you see in statement 4), which was later demonstrated to be true.

How do these assumptions explain the laws of mass conservation and definite proportions? Consider the law of mass conservation. According to Dalton's assumptions, all compounds and elements that undergo change are simply rearranging their atoms. Since those atoms are indivisible, they obviously cannot be destroyed, so the total number of atoms in the entire system must stay the same. If the total number of atoms stays the same, the total amount of matter stays the same, and therefore, the mass stays the same.

Now let's consider the law of definite proportions. Dalton's theory assumes that atoms can join together to make compounds only in simple, whole–number ratios. In other words, a compound might have 1 carbon atom and 2 oxygen atoms, while another compound might have 1 carbon atom and 1 oxygen atom. A compound cannot have 1 carbon atom and 1.5 oxygen atoms because atoms are indivisible. It can have either 1 atom or 2 atoms, never 1.5 atoms. If atoms combine in this way, their masses would always combine in the same proportion, which is exactly what the law of definite proportions says.

Of course, the real evidence for Dalton's theory came with the fact that he predicted the law of multiple proportions. Since Dalton believed that atoms combine in simple, whole–number ratios, he predicted that those ratios should be seen in the recipes for different compounds. When scientists studied this, they found out that he was right. This helped Dalton's atomic theory gain acceptance in the scientific community. After all, it is nice for a theory to be able to explain facts that are already known. However, if a theory *predicts* facts that have not already been determined, it adds a lot of weight to the theory.

As time has gone on, chemists have seen that a great many of the facts that we know about nature can be explained quite nicely if we assume that Dalton's atoms do indeed exist. Atoms are far too small to be seen, so we can never really *prove* that they exist. However, if we assume they exist, we can explain a wealth of facts that we have accumulated over the years. Chemists today simply accept the idea that atoms exist, even though their existence can never be proven.

If we could see these atoms, what would they look like? Figure 2.4 gives you an idea of what a group of atoms might look like if we were able to actually see them.

FIGURE 2.4
A Scanning Tunneling Electron Microscope Image of the Surface of a Nickel Foil
Image courtesy of the IBM research division

Something important about this figure is that you are *not* looking at a picture of atoms. For us to see an object (or take a picture of it), visible light must bounce off the object and hit our eyes (or the camera's film). That cannot happen with atoms because atoms are too small for visible light to bounce off them. You are not seeing a picture here. You are seeing a *computer image* based on some data.

In this figure, a scanning tunneling electron microscope was used to examine the surface of a nickel foil. In this kind of microscope, an electrified probe is placed very close to the surface of the metal, and the flow of electricity from the probe to the metal is monitored by a computer. Based on a theory called quantum mechanical tunneling, the computer then draws what the theory says the surface of the metal must look like for the electricity to flow the way that it did. If the theory is correct (and if the computer used the theory correctly), this is a silhouette of the nickel atoms on the surface of the foil. If the theory is incorrect, then who knows what this image represents? This image is certainly not proof that atoms exist. However, it does give more evidence for their existence.

The image in Figure 2.4 is pretty much how Dalton pictured atoms. Today we are fairly sure that atoms have a more detailed structure than what is shown in the figure. The scanning tunneling electron microscope does not have enough resolution to show that detailed structure. However, as you will learn in Module 3, other experiments have been done to probe the detailed structure of the atom, and today we have a theory that helps us picture the detailed structure of an atom. Before we can get to that, we need to learn a little more about matter and how it changes.

One final note about Dalton's atomic theory: Although chemists do accept the fact that atoms exist, we have also come to realize that 2 of Dalton's assumptions are not quite right. For example, the first assumption contains a small mistake. Under the right circumstances, atoms can, indeed, be split apart. Atoms are not truly indivisible. For the purposes of this course, however, we can assume that Dalton was right on this point because we won't be splitting atoms. His second assumption is not quite right either. Certain atoms within an element can be heavier than other atoms in that same element. These atoms are called isotopes and will be discussed in detail in Module 3.

MOLECULES: THE BASIC BUILDING BLOCKS OF COMPOUNDS

We have already said that elements are simply vast collections of identical atoms. If we could see a sample of the element helium (He), the gas put in balloons to make them float, at the atomic level, we would see a bunch of identical particles jumbled together much like grains of sand on a beach. Democritus (way back in the fourth century B.C.) was right. Matter is discontinuous: It contains little "grains" that we call atoms.

What about compounds, however? If we looked at the water in the ocean on the atomic level, would we see a bunch of atoms jumbled together? No, not exactly. Water is a compound that is formed when 2 hydrogen (H) atoms join with 1 oxygen (O) atom. If we looked at water on an atomic level, we would not see just a bunch of hydrogen and oxygen atoms jumbled together; we would instead see little groups of atoms that each contained 2 hydrogen atoms and 1 oxygen atom. Those *groups* would be jumbled up together like grains of sand. The groups that are formed when atoms join together to form compounds are called molecules. In other words, while all elements are composed of identical atoms, all compounds are composed of identical molecules. Figure 2.5 shows the difference between a mixture of atoms and a compound made of molecules.

FIGURE 2.5
Atoms and Molecules
Illustrations by David Weiss

A bunch of oxygen atoms (red) and hydrogen atoms (yellow) grouped together. This is not a compound. It is just a mixture of atoms.

The same oxygen and hydrogen atoms grouped together to make water molecules. This is what water would look like on the atomic scale.

The basic building blocks of matter, then, are atoms.

If a substance consists of identical atoms, then it is called an element.

On the other hand, when atoms join together, they form molecules.

If a substance consists of identical molecules, then it is called a compound.

ABBREVIATING AND CLASSIFYING COMPOUNDS

We've already seen that each element has its own name and abbreviation. What about compounds? Each compound needs to have a name, and it would be nice to be able to abbreviate those names. How do we do this? We name elements based on the atoms that make them up. For example, the element fluorine (F) is a substance made up of individual fluorine atoms. An element's name is the same as the atoms that make it up.

Compounds, then, should be named after the molecules that make them up. So how do we name a molecule? To learn how to name molecules, we first have to learn how to abbreviate them. Since molecules are made up of atoms, we can use the abbreviations on the periodic table to help us abbreviate molecules as well. For example, a water molecule is formed when 2 hydrogen (H) atoms join with 1 oxygen (O) atom. The abbreviation for a water molecule, then, is H_2O.

How do we get this abbreviation? The letters in the abbreviation come from the atoms which make up the molecule. Since water includes hydrogen atoms and oxygen atoms, the abbreviations for these 2 atoms, H and O, must be in the abbreviation for the molecule. In addition to the letters in the molecule's abbreviation, there are also numbers. These numbers, written as subscripts, tell us how many of each atom is in the molecule. The subscript 2 after the H tells us that water contains 2 hydrogen atoms. Since there is 1 oxygen atom in a water molecule, why isn't there a 1 subscript after the oxygen? That's because chemists love to leave off ones. In chemistry, if a number should exist but is not written down, then it is assumed to be 1. So to abbreviate a molecule, we simply use the abbreviation for each atom that makes up the molecule and follow it with a subscript that indicates the number of those atoms that are in the molecule. If the number is 1, we don't write it down.

Before we do some examples and problems, 2 terms must be introduced. Chemists do not use the term *abbreviation* when dealing with atoms and molecules. Instead, they use the terms *chemical symbol* and *chemical formula*. A chemical symbol is the abbreviation for an atom. So H is the chemical symbol for hydrogen. A chemical formula simply tells how many of each atom make up a particular molecule. Therefore, H_2O is the chemical formula for a water molecule. Sometimes chemists drop the word *chemical* and simply use the terms *symbol* and *formula*. You will have to get used to these terms, so we will begin using them now. Example 2.3 shows how these formulas work, and "On Your Own" questions 2.7–2.8 will give you some practice with them.

EXAMPLE 2.3

What is the chemical formula for a molecule that contains 1 atom of sodium, 1 atom of nitrogen, and 3 atoms of oxygen?

The chemical symbols for sodium, nitrogen, and oxygen are Na, N, and O, respectively. Since there is only 1 Na atom and 1 N atom, we needn't put number subscripts after them. However, there are 3 O atoms, so a subscript of 3 must follow the O:

$$NaNO_3$$

How many atoms are in the molecule whose formula is Na_2HPO_4?

The formula indicates that there are 2 Na atoms, 1 H atom, 1 P atom, and 4 O atoms. How do we know that H is one atom's symbol, P is another atom's symbol, and O is another's? Remember, each atom's symbol only contains 1 capital letter. Therefore, every time you see a capital letter, you know you are dealing with a new atom. The total number of atoms, then, is simply the sum of all the individual atoms:

$$\text{Total number of atoms} = 2 + 1 + 1 + 4 = 8$$

So there are a total of **8 atoms** in this molecule.

ON YOUR OWN

2.7 Give the formula for each of the following molecules:

a. A molecule that has 1 atom of potassium and 1 atom of bromine

b. A molecule that is made up of 5 carbon atoms, 10 hydrogen atoms, and 5 oxygen atoms

c. A molecule with 2 lithium atoms, 1 sulfur atom, and 4 oxygen atoms

2.8 How many of each atom exist in the following molecules?

a. $RbCl$ b. $KMnO_4$

c. Na_3PO_4 d. $HC_2H_3O_2$

CLASSIFYING MATTER AS IONIC OR COVALENT

Once you have determined a compound's chemical formula, you are well on your way to figuring out its name. There is one other thing you must do to name a compound properly: You must classify it. Just as elements are classified into 2 groups—metals and nonmetals, compounds are classified based on their properties as well. There are many, many different classification schemes for compounds, and you will learn several of them as this course continues. For now, we will look at the broadest, simplest classification scheme.

Compounds can be classified as either ionic or covalent. You will learn later about subclasses within the covalent class, but this is a good start for now. A compound is considered ionic if, when dissolved in water, it conducts electricity. Pure water cannot conduct electricity, but when an ionic compound is dissolved in water, it can.

"Now wait a minute," you might be saying. "Aren't we always told not to use water around electricity? If water cannot conduct electricity, then why are we concerned about it?" As we said earlier, *pure* water cannot conduct electricity. However, the water that comes out of your sink tap is not pure water. It has several ionic compounds dissolved in it. The most notable ionic compound in water is sodium fluoride, which helps protect your teeth against decay. Almost every city in the United States adds sodium fluoride (or something equivalent) to its drinking water for precisely that reason. Even rain is not pure water. It also has ionic compounds dissolved in it. About the only way to get pure water is to buy or make distilled water. Distilled water is about 99% pure water. We will talk later in this module about what pure water is.

If a compound does not allow water to conduct electricity after it is dissolved, it is a covalent compound. How can we tell whether a compound is ionic or covalent? One way would be to dissolve it in water and see if it conducts electricity. That's what you will do in Experiment 2.2.

If you purchased the MicroChem kit mentioned in the Student Notes, you can perform Experiment 3 in that kit in addition to Experiment 2.2.

EXPERIMENT 2.2

PURPOSE: To determine the electrical conductivity of compounds dissolved in water.

SUPPLIES

- Distilled water (Available at grocery stores. Half a gallon is plenty.)
- Baking soda
- Sugar
- 9-volt battery ***(Do not use an electrical outlet in place of the battery. The electrical energy in a wall socket will hurt you and can kill you!)***
- 2 pieces of wire (preferably insulated), each of which is at least 15 cm long
- Scissors or wire cutters to strip insulation from wire (if it is insulated)
- Tape (preferably black electrical tape)
- Optional: Stores like Radio Shack sell a cap that will fit on 9V batteries. The cap has wires that you can use to replace the wires, scissors or wire cutters, and tape listed above.
- ½ teaspoon
- 100-mL beaker or small glass
- Safety goggles

HYPOTHESIS: Ionic compounds will conduct electricity in water; covalent compounds will not conduct electricity in water.

PROCEDURE:

1. Rinse out your beaker or glass with tap water to get rid of any detergent residues left in it from washing. Then rinse it twice with distilled water to get out any impurities left behind from the tap water.
2. Add 80 mL of distilled water to the beaker. Treat this volume as a qualitative measurement.
3. If your wires are insulated, strip the insulation off of each wire end so that you have about 2 cm of bare wire at all ends. When that is done, attach one wire's end to one post of the battery and the other wire's end to the other post. The best way to do this is to wad up the bare wire into a ball and shove the ball inside the battery post.

Then, while pressing firmly, tape down the insulated part of the wire at the top of the post. The main concern is that the bare wire must be touching the battery post firmly so that it makes good electrical contact.

4. Once each wire is attached to its own battery post, immerse the other ends of the wires into the water contained in your glass or beaker. Make sure the ends do not touch each other. In the end, your experiment should look something like the drawing below:

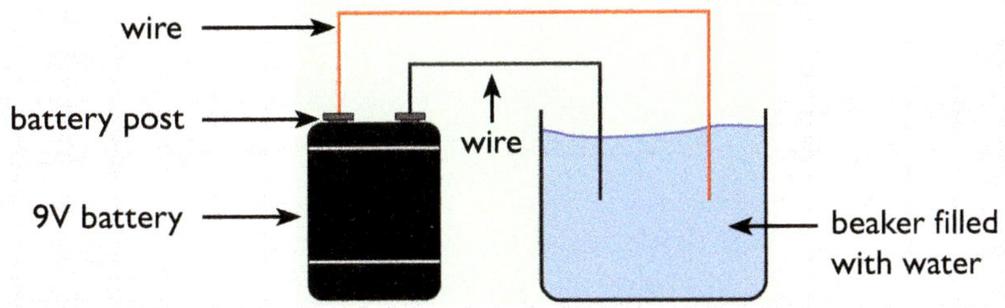

5. Watch the bare ends of the wire closely for a little while. Nothing exciting happens, right?
6. Now remove the wires from the water, add half a teaspoon of baking soda to the water, and mix it around a bit to get it to dissolve. The water should be cloudy at first, but the undissolved baking soda will eventually settle, and the solution will be relatively clear again.
7. Immerse the ends of the wire as you did before. Now what's happening? If the other ends of your wires are firmly touching their individual battery posts, you should see bubbles forming on the wire ends that are immersed in the solution.
8. Why is the bubbling happening? You'll learn in physics that when electricity is conducted, particles called electrons (ee lek' trons) move from one end of the battery to the other. This motion contains a lot of kinetic energy. The way we have set the experiment up, the electrons must travel through one wire, through the water, and through the other wire to reach the other end of the battery. When this motion takes place, some of that energy can be given to individual water molecules. Under the conditions of this experiment, the kinetic energy generated by the electricity is enough to cause the water to decompose into hydrogen and oxygen, its constituent elements. Since hydrogen and oxygen are gases, they form bubbles at the ends of the wires. For this energy to exist to begin with, electricity must be conducted. Since pure water cannot conduct electricity, nothing happened when you immersed the wires in the pure water. However, when you dissolved baking soda in the water, it became able to conduct electricity. This means that baking soda must be an ionic compound.
9. Rinse the glass or beaker with tap water and distilled water as you did before.
10. Repeat the experiment with table sugar. You will find that no bubbles form in this case. This must mean that no electricity can be conducted, so table sugar must be a covalent compound.
11. Clean up and return everything to the proper place.

CONCLUSION: What did this experiment show you about ionic and covalent compounds? Write something about what you have learned.

In this experiment, we used the decomposition of water as an indicator for electrical conductivity. If bubbles appeared, that told us water was being decomposed, which meant that electricity was being conducted. This is another example of measuring something indirectly, as we discussed in Module 1.

To classify compounds as ionic or covalent, you could go through an experiment like this for each compound you want to classify. Fortunately, there is an easier way. To classify compounds, all you have to do is look at the chemical formula and follow 2 rules:

1. **If a compound contains at least 1 metal atom and at least 1 nonmetal atom, the compound is ionic.**
2. **If a compound is made up solely of nonmetal atoms, the compound is covalent.**

Following these rules, you can now see why our experiment turned out the way it did. The chemical formula for table sugar is $C_{12}H_{22}O_{11}$. The atoms C, H, and O are all nonmetals, so table sugar is a covalent compound. It will not conduct electricity when dissolved in water. On the other hand, the chemical formula for baking soda is $NaHCO_3$. The atom Na is a metal, while H, C, and O are nonmetals; therefore, baking soda is an ionic compound and will conduct electricity when dissolved in water.

We've already told you that there are exceptions to nearly every rule in chemistry. Although all ionic compounds conduct electricity when dissolved in water, there are some compounds that are not ionic but will nevertheless conduct electricity when mixed with water. HCl is an excellent example. It is clearly a covalent compound because both hydrogen and chlorine are nonmetals. When mixed with water, however, HCl *will* conduct electricity because it chemically reacts with the water to produce ions. You will learn about this later. At this point in your chemistry education, you can ignore such exceptions. As far as you are concerned, if a compound conducts electricity when dissolved in water, it is an ionic compound.

ON YOUR OWN

2.9 Classify the following compounds as ionic or covalent:

a. SiF_4 b. $CaBr_2$
c. $LiC_2H_3O_2$ d. $C_6H_{12}O_6$

In the same way, although the vast majority of compounds made up solely of nonmetal atoms are covalent, there are some exceptions. For example, ammonium hydroxide (NH_4OH) is made up solely of nonmetal atoms. However, it is an ionic compound. You will learn why this is the case later. Once again, at this point in your chemistry education, you can ignore such exceptions and just follow the rules as discussed in this module. Make sure you understand these rules by classifying the molecules in "On Your Own" question 2.9.

NAMING COMPOUNDS

Now that you can classify and interpret a molecule's chemical formula, you are finally ready to learn how to name compounds. The first thing to remember is that we use a different system to name ionic compounds than we use to name covalent compounds. Since the ionic–compound naming system is a little easier, we will begin there. To name an ionic compound:

1. **Start with the name of the first atom in the molecule.**
2. **Replace the name of the next atom in the molecule with its *–ide* name.**
3. **Putting those 2 names together gives the compound's name.**

This is why sodium chloride is the name for NaCl. The symbol Na means sodium, and Cl stands for chlorine. According to the rules, we replace *chlorine* with *chloride*. Putting these names together, we get sodium chloride. In the same way, we would call K_2O potassium oxide. Pretty easy, huh? Later we will see that this gets more complicated when an ionic compound includes more than 2 different kinds of atoms. For now, we will stick with ionic compounds that have just 2 different kinds of atoms in them, so naming ionic compounds will be pretty simple at first.

You need to memorize Table 2.1 for the test at the end of this module.

TABLE 2.1
The *–ide* Atom Names of Nonmetals

Atom	*–ide* Name	Atom	*–ide* Name
H	hydride	P	phosphide
C	carbide	S	sulfide
N	nitride	Cl	chloride
O	oxide	Br	bromide
F	fluoride	I	iodide

Naming covalent compounds is a little trickier because covalent compounds are more complicated than ionic compounds. When metal atoms and nonmetal atoms form ionic compounds, they can form only one type of molecule. For example, when sodium (Na) and chlorine (Cl) get together, the only molecule they can form is NaCl. They cannot form $NaCl_2$ or Na_2Cl or anything like that. Therefore, when we say the name *sodium chloride*, we know it means only one possible compound: NaCl.

Compare this to the situation with covalent molecules. When carbon (a nonmetal) joins with oxygen (a nonmetal), multiple compounds can be formed: CO or CO_2. In this case, if we used the ionic compound naming system, we would call both of these molecules carbon oxide. Clearly, we can't have the same name for 2 different molecules. To fix this, we add prefixes in front of the name of each atom in the compound. This way, the number of atoms in each molecule is explicitly stated in its name. The prefixes we use in this naming system are summarized in Table 2.2, which you need to memorize.

TABLE 2.2
Prefixes for Naming Covalent Compounds

Prefix	Meaning	Prefix	Meaning
mono	1	hexa	6
di	2	hepta	7
tri	3	octa	8
tetra	4	nona	9
penta	5	deca	10

These prefixes are inserted before each name in the compound to indicate how many of each type of atom are in the molecule. Once again, this general rule has one exception. If the prefix used on the first atom is *mono–*, we drop it. For example, CO_2 has the name *carbon dioxide*. Since there is only 1 carbon atom in the molecule, we should use the *mono–* prefix. However, since carbon is the first atom, we drop it. There are 2 oxygen atoms, so we must use the *di–* prefix in front of oxygen. Just like naming ionic compounds, we change the last atom's name to an *–ide* ending.

We can drop the *mono–* prefix only if it is on the first atom in the molecule. Therefore, CO is named carbon monoxide. Since there is only 1 oxygen atom, we need to use the *mono–* prefix, and since oxygen isn't the first atom in the molecule, we cannot drop it. Finally, we can never drop any prefix unless it is *mono–*, so the molecule C_2H_6 is named dicarbon hexahydride. See if you understand both of these naming systems by studying Example 2.4 and then answering "On Your Own" questions 2.10–2.11.

EXAMPLE 2.4

Name the compound whose chemical formula is Al_2O_3.

To name a compound, we must first determine whether it is ionic or covalent. Since aluminum (Al) lies on the left of the jagged line, it is a metal. This means we are working with an ionic compound. We therefore simply change *oxygen* to *oxide* and put the names together: **aluminum oxide**.

What is the name of the molecule PH_3?

Since P and H are both nonmetals, this is a covalent compound. We have to use the prefixes in the name. The prefix for phosphorus (P) is *mono–*, but since it is the first atom, we can drop it. The prefix for hydrogen is *tri–*, and we must add the *–ide* ending to hydrogen: **phosphorus trihydride**.

ON YOUR OWN

2.10 Name the following compounds:

a. CF_4 b. Al_2S_3 c. P_2O_3 d. CaO

2.11 Give the chemical formulas for the following covalent compounds:

a. sulfur hexafluoride
b. pentacarbon decahydride

This lesson in naming compounds might lead you to believe that the process of giving names to compounds is easy. Don't be fooled into thinking this. We have merely scratched the surface of naming compounds. Notice that the only compounds we have named contain only 2 different types of atoms. That's because compounds with more than 2 types of atoms are more difficult to name. In addition, some ionic molecules break the rules, and we have to develop a new naming system for them. Finally, there is a class of covalent compounds called organic compounds that has a whole new naming scheme. We will revisit some of these tougher topics later on, so don't dismiss as easy the very difficult process of naming compounds.

One other complication that crops up when we start naming compounds is the problem of common names. Think for a minute about why we don't call water by its real chemical name: dihydrogen monoxide. This is because many compounds (like water) were named long before chemists knew about atoms and molecules and decided upon a standard naming system. These compounds were named without following any of the rules we just studied. Since old habits die hard, many compounds are still called by their common names rather than their proper chemical names. Ammonia (NH_3) is another example. The proper name for ammonia is nitrogen trihydride, but chemists called it ammonia long before our naming system was developed. Common names like these are here to stay, so we just have to get used to recognizing these names and associating them with the proper compound.

CLASSIFYING MATTER

Now that we have named compounds, we can finish this module with classification of matter. If we can properly classify the various types of matter we see in God's creation, it will be easier for us to determine how that matter is changing. First, every bit of matter that we see can be classified into 2 large groups: mixtures and pure substances.

Pure substance—A substance that contains only one element or compound.

Mixture—A substance that contains different compounds and/or elements.

These definitions may sound a bit confusing at first, but don't worry; you'll soon get the hang of them. Examples of pure substances include sulfur, iron, sugar, and table salt. In each of these substances, only 1 element or compound exists. The first 2 examples are elements since they are made up of atoms. In a sample of sulfur, only 1 type of atom exists: sulfur (S) atoms. In a sample of iron, only 1 type of atom exists: iron (Fe) atoms. The last 2 examples are compounds since they are made up of molecules. Sugar is made up of only 1 type of molecule: $C_{12}H_{22}O_{11}$. Likewise, table salt is composed entirely of NaCl molecules. You should recognize that sugar is a covalent compound (made up solely of nonmetals), whereas table salt is ionic (made up of a metal and a nonmetal).

Earlier, you used distilled water in one of your experiments. We noted that the distilled water was 99% pure. What that means is that 99% of all molecules in distilled water are H_2O molecules. If 100% of all the molecules in distilled water were H_2O molecules, then distilled water could be labeled a pure substance. This is what we mean when we say that pure substances contain only 1 element or compound.

Mixtures, on the other hand, contain more than 1 compound and/or element. If we were to dissolve table salt in pure water, the result would be a mixture of 2 compounds: H_2O and NaCl. This is what we mean when we say that mixtures have different compounds and/or elements. The H_2O molecules and the NaCl molecules in a saltwater mixture do not combine to make new molecules; they just occupy the same general area. That's what a mixture is. The important thing to realize about mixtures is that individual elements or compounds retain their individual properties while in a mixture. For example, pure water is tasteless. Table salt, on the other hand, has a distinct, sharp taste. A mixture of saltwater, then, tastes sharp like salt, but not as sharp as pure salt. Each compound retains its taste, and the tasteless water dilutes the sharp taste of the salt in the mixture.

This leads to another way we can distinguish between pure substances and mixtures: The individual components in a mixture can be physically separated based on their individual properties. For example, consider Figure 2.6.

FIGURE 2.6
Mixtures Can Be Separated Based on the Properties of Their Components

Photos by Flanders Video Productions

Sulfur (the yellow powder) and iron (the black powder) are mixed together.

A magnet (the white bar) is brought near the mixture.

The magnet pulls the iron out of the mixture and leaves the sulfur behind.

Sulfur and iron are mixed together and then heated so that they form the compound iron sulfide (the black powder).

A magnet (the white bar) is brought near the iron sulfide.

The magnet doesn't pick up anything because the iron no longer has its individual properties.

In the top portion of the figure, we are working with a *mixture* of sulfur and iron. Since this is a mixture, the sulfur retains its properties (its yellow color and nonmagnetism, for example), and the iron retains its properties (its black color and magnetism, for example). These properties can be used to separate the 2 elements. If a magnet is brought near the mixture, the iron will cling to the magnet, but the sulfur will not. We could separate the mixture back into its components (sulfur and iron) by using the magnet to pull out all of the iron. In the end, we would have 2 pure substances, iron and sulfur. A mixture, then, can be separated using the individual properties of its components.

Compare this to the lower portion of the figure, where the same 2 elements (sulfur and iron) are heated together so that they form the *compound* iron sulfide. Although iron sulfide contains both sulfur and iron, it is *not* a mixture; it is a pure substance. We can see that in several ways. For example, notice that there is no yellow in iron sulfide: it is black. The sulfur does not retain its characteristic yellow color. Notice also that when the magnet is brought near the iron sulfide, nothing clings to it. That's because the iron does not retain its ability to respond to a magnet. Both the sulfur and the iron have changed or given up their individual properties to become a compound. This is a pure substance. Its elements cannot be separated based on their individual properties because they do not retain their individual properties when they form a compound.

The only way we could get the iron and sulfur out of iron sulfide is to *decompose* the iron sulfide back into a mixture of iron and sulfur and then separate the 2 substances. Therefore, mixtures can be *physically separated* into their individual components, but pure substances must be decomposed before their individual components can be physically separated. Experiment 2.3 gives you an opportunity to find another way of separating the components in a mixture.

EXPERIMENT 2.3

PURPOSE: To separate a mixture of sand and salt.

SUPPLIES:

- 2 beakers - 100 mL and 250 mL (or 2 glasses—1 large and 1 small)
- Sand (Kitty litter is an acceptable substitute, but don't use the kind that clumps.)
- Table salt
- Funnel
- Water
- Filter paper (You can cut a circle out of the bottom of a coffee-maker filter.)
- Stirring rod (or small spoon)
- Teaspoon
- Heat source such as a stove or alcohol burner
- Safety goggles

QUESTION: Can mixtures be separated?

HYPOTHESIS: Write down a hypothesis related to separating mixtures into elements and compounds

PROCEDURE:

1. Pour a teaspoon of sand and a teaspoon of salt into the 100-mL beaker.
2. Use your stirring rod to mix them well. At this point, you have a mixture of sand and salt in your beaker. The molecules of sand (SiO_2) do not combine with the molecules of salt (NaCl) in any way. These 2 different types of molecules simply occupy the same general space now. In this experiment, you will learn how to separate this mixture back into sand and salt.
3. One way you could accomplish this separation would be to use fine tweezers and a powerful magnifying glass. If you looked at the mixture through the magnifying glass, you could see little grains of salt and little grains of sand. You could then use the tweezers to slowly pick the salt out of the sand. With enough patience and time, you could separate the mixture in that way. Unfortunately, this method is quite tedious and would take an enormous amount of time. To speed up the procedure, you will use a different method. Remember that in a mixture, the individual molecules still retain their own unique properties. One property of salt is that it dissolves in water. However, sand does not. We will use this property to separate them.

4. Add about 25 mL (1/8 cup) of water (a qualitative measurement) to the mixture and use your stirring rod to mix it in well. As the water mixes with the sand and salt, the salt will dissolve in water and the sand will not. Now you have an ugly mixture of sand, salt, and water in your beaker.
5. To filter your mixture so that the saltwater will separate from the sand, first fold a circle of filter paper as shown below. In the last step of opening the triangle to make a cone, it's important to note that the triangle has 4 layers (from folding it in half 2 times). To get a cone that works as a filter, 3 layers need to be on one side and 1 layer on the other.

Illustration by Megan Whitaker

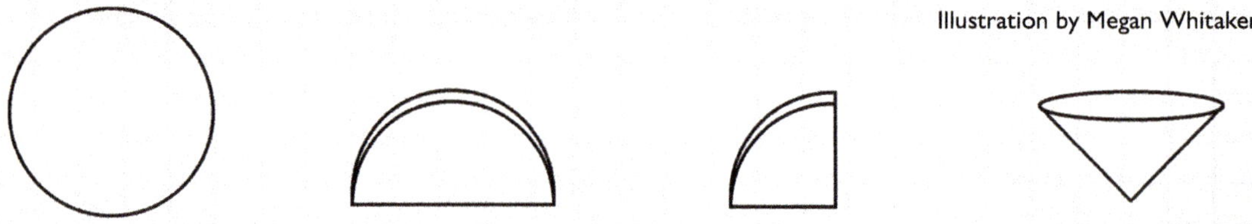

Start with a large piece of circular filter paper. Fold it in half. Fold it in half again so it makes a triangle. Open the triangle to make a cone.

6. Place the resulting cone of filter paper in the funnel. To get the filter paper to retain its cone shape and stick to the funnel, wet it with some water.
7. Hold the funnel above the empty 250-mL beaker so that anything falling through the funnel will land in the beaker.
8. Pour the mixture of water, salt, and sand into the filter paper. Make sure that the water level never rises above the top of the filter-paper cone. In addition, for best results, try to pour off just the liquid first and leave the majority of the sand in your beaker until the very end. As the liquid filters through the paper, you will see it fall into the 250-mL beaker. As long as you do not allow the liquid level to rise above the top of the filter-paper cone, the liquid in the beaker will be clear. Continue this process until everything in the beaker (including the sand) has been poured into the funnel.
9. Once all of the liquid has filtered through the filter paper (this may take a while!), rinse the 100-mL beaker with 10 mL (1 tablespoon) of water and dump it all into the filter paper. Allow all of the liquid to drain through the filter paper.
10. What you have now is pure sand on the filter paper and a mixture of salt and water in your beaker. To get rid of the water, simply start heating it so that it will boil. ***Warning: If you are using glasses instead of beakers for this experiment, do not heat the glass. Instead, pour the salt water into a pot and boil it there. Regular glasses tend to break when they are heated to 100°C!***
11. Allow all of the water to boil away. What's left behind? Pure table salt is left as a residue in the beaker.
12. Clean up and return everything to the proper place.

CONCLUSION: You have separated your mixture of sand and salt back into its components. The sand is on the filter paper, and the salt is in the 250-mL beaker.

If you purchased the MicroChem kit mentioned in the Student Notes, you can learn another method used to separate mixtures by performing Experiment 1.

This is the idea behind the concept of mixtures. Since the individual molecules within a mixture retain their unique properties, we can use those properties to separate them. Since salt dissolves in water and sand does not, we used that particular property as a means of separation in the experiment.

Mixtures and pure substances, then, are the 2 broad classifications of matter. However, they can both be broken down into 2 subclassifications as well. Pure substances can be either elements or compounds. Therefore, the classifications you learned before are really subclasses of the pure substance classification. Mixtures can be further classified as either heterogeneous (het' uh roh jee' nee us) or homogeneous (ho' moh jee' nee us).

Homogeneous mixture—A mixture with a composition that is always the same no matter what part of the sample is observed.

Heterogeneous mixture—A mixture with a composition that is different depending on what part of the sample is observed.

These definitions make sense when you think about the prefixes *homo–* and *hetero–*. *Homo–* means the same, and *hetero–* means different. Therefore, homogeneous solutions are the same throughout the sample, whereas heterogeneous solutions are different in different parts of the sample.

The best way to explain the difference between heterogeneous and homogeneous mixtures is by example. In Experiment 2.3, you made up a mixture of sand and salt. When you first dumped the sand and the salt together, it was a heterogeneous mixture. If you looked at one portion of the mixture, you might see mostly sand, whereas another portion of the mixture might be mostly salt. As you stirred the mixture, it started to become homogeneous because the sand and salt were mixing evenly throughout the sample. If you stirred the sand and salt enough, you would end up with a mixture in which the salt was evenly spread throughout the sand. That would be a homogeneous mixture.

Examples of heterogeneous mixtures include Italian salad dressing, a bowl of cereal and milk, and a cake with icing on it. In each case, these mixtures have a different composition depending on where you observe the sample. When you take Italian dressing out of the refrigerator, the seeds, herbs, and spices are mostly at the bottom of the bottle. The seeds, herbs, and spices are more concentrated in one part of the sample than the other. A bowl of cereal and milk typically has mostly cereal on the top and milk at the bottom because the cereal tends to float on top of the milk. Finally, an iced cake is a mixture of cake and icing, but the icing is completely on the top, while the cake is underneath.

Examples of homogeneous mixtures include saltwater, Coke, and steel. Each of these mixtures has the same composition throughout. A glass of saltwater has the same amount of salt and water at the top of the glass as at the bottom. Coke is a mixture of several ingredients, but there is no difference in taste or composition from the top of the Coke can to the bottom. Finally, steel is a mixture of iron (Fe) and carbon (C). Iron gives steel its strength, and carbon helps the steel stay rigid. The amounts of carbon and iron are the same no matter what sample of steel is observed, so it is also a homogeneous mixture. Figure 2.7 sums up these classifications of matter.

FIGURE 2.7
A Classification Scheme for Matter

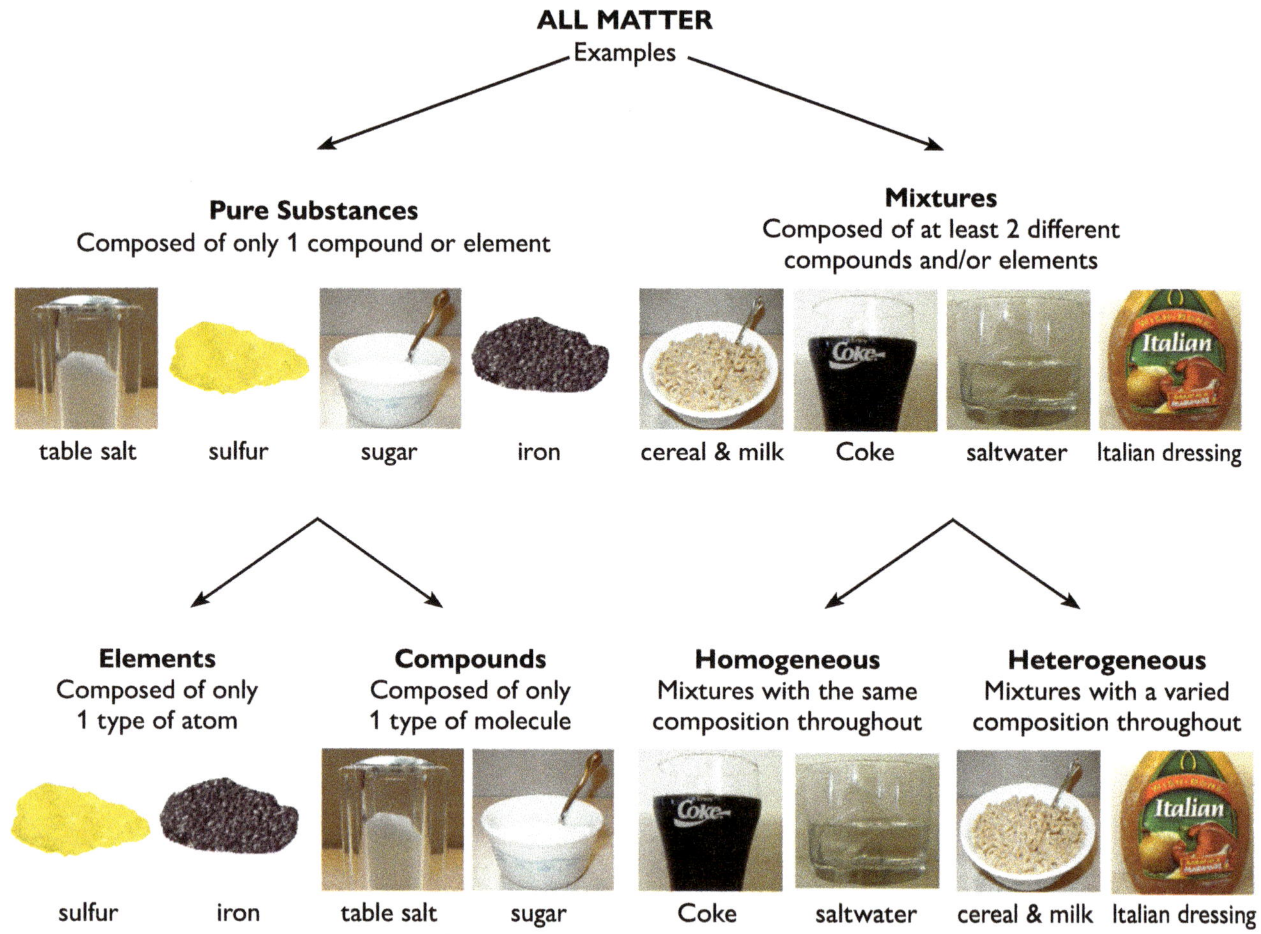

A couple of interesting facts related to the concept of mixtures are worth discussing. First, if you buy a carton of milk, the label says that the milk is homogenized. Now that you know this classification scheme, you can understand what that word means. When milk is taken from a cow, it is a heterogeneous mixture. The cream in the milk tends to rise to the top, and the water tends to fall to the bottom. The milk is thick and creamy if it comes off the top of the bottle, whereas it is thin and watery if it comes off the bottom of the bottle. Before you buy the milk, it is treated so that the cream stays evenly distributed throughout the milk. Therefore, the formerly heterogeneous mixture is turned into a homogeneous mixture. This is what the term *homogenized* means.

Second, did you know that air is a mixture? Most people think that since animals breathe in oxygen, the air around them must be pure oxygen. Nothing could be farther from the truth. In fact, air is a mixture of several gases, including nitrogen, argon, water vapor, carbon dioxide, and oxygen. Oxygen actually makes up less than ¼ of the air that we breathe. Assuming there is no humidity, the air we breathe is 78% nitrogen, 21% oxygen, 0.9 % argon, 0.03% carbon dioxide, and 0.07% other gases. Figure 2.8 shows this relationship in pie graph form.

Figure 2.8
The Composition of Dry Air

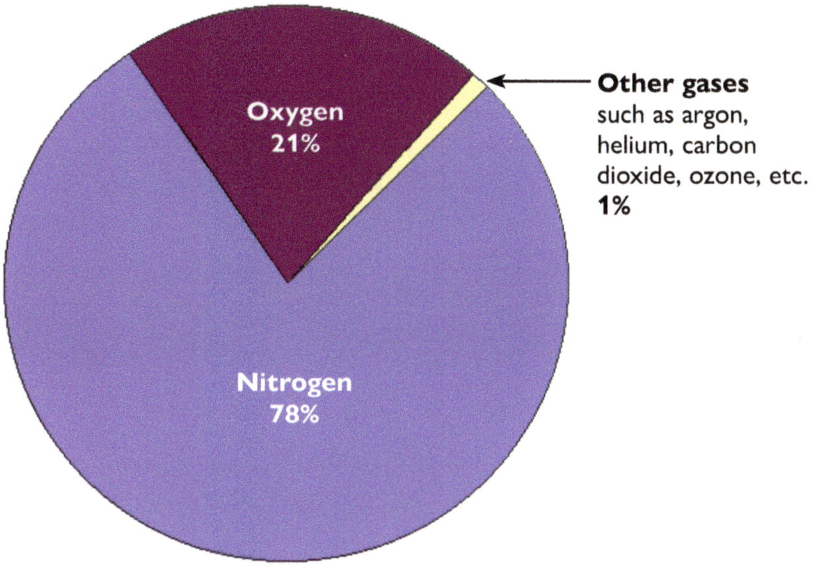

This mixture of gases is ideal for human life. Even though we need oxygen to live, we can have too much oxygen. Breathing elevated levels of oxygen for prolonged periods of time can have serious consequences, such as lung damage, chest pains, and even blindness! In addition, the frequency of natural forest fires increases significantly when the amount of oxygen in the air increases. For every 1% increase in the amount of oxygen in the air, the chance of a natural forest fire occurring goes up by 70%! Therefore, we need oxygen in our air, but not too much of it.

The air in our atmosphere does not have too much oxygen in it because the oxygen has been diluted with nitrogen. Nitrogen is the perfect gas to accomplish this dilution because it does not react with our bodies in any way but is important in the biological world through the nitrogen cycle. We breathe in nitrogen with every breath, but then we breathe it right out again because it does not interact with our bodies. Almost all other gases—carbon monoxide, sulfur dioxide, nitrogen dioxide, etc.—are poisonous

science and creation

Doesn't it seem like an amazing coincidence that the air on this planet just happens to have oxygen, but not too much oxygen? It has the perfect amount of oxygen for human life. The other gas that makes up the vast majority of the rest of our planet's air just happens to be one of the few gases that does not affect our bodies in any negative way. Clearly, such a perfect mixture of gases would never have come about by chance. God made the air around us perfect for sustaining human life. As we go on in this course, you will see that there are thousands and thousands of "coincidences" like this one that clearly show that this world has a Creator. This is why we say that real science points unerringly to God.

ON YOUR OWN

2.12 Classify the following as elements, compounds, homogeneous mixtures, or heterogeneous mixtures:

a. mercury
b. carbon monoxide
c. chocolate
d. apple
e. tea (no ice)
f. cesium
g. ocean water
h. barium chloride
i. chocolate chip cookie

to human life, so it is very good that those gases do not make up any significant fraction of the air we breathe.

See if you understand our matter classification scheme by answering "On Your Own" question 2.12.

In the end, then, we can see that Democritus's musings on the nature of matter were correct. Chemists have shown throughout the years that matter does not come in continuous sheets; it comes in little packages known as atoms. When identical atoms are jumbled together, the substance that results is called an element. When atoms join together, they make molecules. When these molecules are collected together, they form substances called compounds. Even though he didn't know any of these details, Democritus's ideas were essentially right. Not bad for someone who lived more than 2,000 years ago, is it?

SUMMARY OF TABLES IN MODULE 2

TABLE 2.3
Element Names and Symbols

The list is compiled in the order in which they appear on the periodic table.
The elements in bold are the ones you need to memorize for the test.

Name	Symbol	Name	Symbol	Name	Symbol
Hydrogen	H	Niobium	Nb	Thallium	Tl
Helium	He	Molybdenum	Mo	Lead	Pb
Lithium	Li	Technetium	Tc	Bismuth	Bi
Beryllium	Be	Ruthenium	Ru	Polonium	Po
Boron	B	Rhodium	Rh	Astatine	At
Carbon	C	Palladium	Pd	Radon	Rn
Nitrogen	N	Silver	Ag	Francium	Fr
Oxygen	O	Cadmium	Cd	Radium	Ra
Fluorine	F	Indium	In	Actinium	Ac
Neon	Ne	Tin	Sn	Thorium	Th
Sodium	Na	Antimony	Sb	Protactinium	Pa
Magnesium	Mg	Tellurium	Te	Uranium	U
Aluminum	Al	**Iodine**	I	Neptunium	Np
Silicon	Si	Xenon	Xe	Plutonium	Pu
Phosphorus	P	Cesium	Cs	Americium	Am
Sulfur	S	Barium	Ba	Curium	Cm
Chlorine	Cl	Lanthanum	La	Berkelium	Bk
Argon	Ar	Cerium	Ce	Californium	Cf
Potassium	K	Praseodymium	Pr	Einsteinium	Es
Calcium	Ca	Neodymium	Nd	Fermium	Fm
Scandium	Sc	Promethium	Pm	Mendelevium	Md
Titanium	Ti	Samarium	Sm	Nobelium	No
Vanadium	V	Europium	Eu	Lawrencium	Lr
Chromium	Cr	Gadolinium	Gd	Rutherfordium	Rf
Manganese	Mn	Terbium	Tb	Dubnium	Db
Iron	Fe	Dysprosium	Dy	Seaborgium	Sg
Cobalt	Co	Holmium	Ho	Bohrium	Bh
Nickel	Ni	Erbium	Er	Hassium	Hs
Copper	Cu	Thulium	Tm	Meitnerium	Mt
Zinc	Zn	Ytterbium	Yb	Darmstadium	Ds
Gallium	Ga	Lutetium	Lu	Roentgenium	Rg
Germanium	Ge	Hafnium	Hf	Copernicium	Cn
Arsenic	As	Tantalum	Ta	Nihonium	Nh
Selenium	Se	Tungsten	W	Flerovium	Fl
Bromine	Br	Rhenium	Re	Moscovium	Mc
Krypton	Kr	Osmium	Os	Livermorium	Lv
Rubidium	Rb	Iridium	Ir	Tennessine	Ts
Strontium	Sr	Platinum	Pt	Oganesson	Og
Yttrium	Y	Gold	Au		
Zirconium	Zr	Mercury	Hg		

ANSWERS TO THE "ON YOUR OWN" QUESTIONS

2.1 Since the compound contains only carbon, nitrogen, and hydrogen, the sum of those individual masses must add up to the mass of the compound:

$$\text{Mass of compound} = \text{Mass of nitrogen} + \text{Mass of carbon} + \text{Mass of hydrogen}$$

$$\text{Mass of compound} = 14.0 \text{ g} + 12.0 \text{ g} + 3.0 \text{ g} = 29.0 \text{ g}$$

By the law of mass conservation, the mass of the compound before the decomposition must have been 29.0 g.

2.2 Since S, At, and Xe are to the right of the jagged line, they are nonmetals. In addition, H is always a nonmetal because it is an exception to the rule. Ca, Tl, and Hs are metals.

2.3 Electrical wires have to be able to conduct electricity and should bend relatively easily. You should choose a metal to work with. Metals lie to the left of the jagged line, and the only one that does is Ag.

2.4 Nonmetals are brittle, so they all crumble into pieces instead of bending. The only nonmetal on the list is C.

2.5 According to the law of mass conservation, 24.0 grams of carbon + 8.08 grams of hydrogen must make 32.1 grams of natural gas since there are no leftover amounts of carbon or hydrogen. To make 128.4 grams of natural gas, all we have to do is determine the factor by which we need to increase our recipe.

$$(32.1 \text{ g})x = 128.4 \text{ g}$$

$$x = \frac{128.4 \cancel{\text{g}}}{32.1 \cancel{\text{g}}} = 4.00$$

If we need to increase our recipe by a factor of 4.00, then all we need to do is multiply all of our ingredient amounts by 4.00:

$$\text{Mass of carbon} = 24.0 \text{ g x } 4.00 = 96.0 \text{ g}$$

$$\text{Mass of hydrogen} = 8.08 \text{ g x } 4.00 = 32.3 \text{ g}$$

So to make 128.4 grams of natural gas, we must add 96.0 grams of carbon to 32.3 grams of hydrogen. *Note: You might notice that these 2 numbers do not add up to 128.4 grams as they should by the law of mass conservation. Instead, they add to 128.3 g. This is a result of the rounding that we must do to keep track of significant figures; it is considered normal.*

2.6 Before we can answer this question, we need to determine how much cobalt was in the original sample. That will then tell us the recipe. For this, we use the law of mass conservation. Since we started with 60.0 grams, we must end with 60.0 grams as well. If 32.8 grams of mass is accounted for by the chlorine, then the rest must have gone to the cobalt.

$$\text{Mass of cobalt} = 60.0\text{ g} - 32.8\text{ g} = 27.2\text{ g}$$

So the recipe for making 60.0 grams of the purple powder is to react 27.2 grams of cobalt with 32.8 grams of chlorine. But the chemist wants to make 2.00 kg of the stuff. First we have to convert that to grams so all of our units are the same. The amount 2.00 kg is the same as 2.00×10^3 g. Now we need to determine what to multiply our recipe by so it will make 2.00×10^3 g:

$$(60.0\text{ g})x = 2.00 \times 10^3\text{ g}$$

$$x = \frac{2.00 \times 10^3\ \cancel{\text{g}}}{60.0\ \cancel{\text{g}}} = 33.3$$

Now we just multiply our list of ingredients by this number, and we'll have the answer:

$$\text{Mass of cobalt} = 27.2\text{ g} \times 33.3 = 906\text{ g}$$

$$\text{Mass of chlorine} = 32.8\text{ g} \times 33.3 = 1090\text{ g}$$

To make 2.00 kg of the purple powder, react 906 grams of cobalt with 1090 grams of chlorine.

2.7 (a) The chemical symbols for potassium and bromine are K and Br. Since there is only 1 atom of each, then there is no need to put any numbers in the formula. Therefore, KBr is the chemical formula.

(b) The symbols for carbon, hydrogen, and oxygen are C, H, and O. Since there are 5 carbon atoms and 5 oxygen atoms, there must be a 5 subscript after the C and O. Since there are 10 hydrogen atoms, a 10 must follow the H. So $C_5H_{10}O_5$ is the proper chemical formula.

(c) The proper formula is Li_2SO_4.

2.8 (a) This molecule is made up of 1 rubidium atom and 1 chlorine atom.

(b) Since there is no number after the K (potassium), we assume there is 1 potassium atom. The same assumption holds true for the Mn, so there is 1 manganese atom. The 4 after the oxygen means there are 4 oxygen atoms.

(c) In this molecule, there are 3 sodium atoms, 1 phosphorus atom, and 4 oxygen atoms.

(d) This molecule is made up of 2 carbon atoms, 4 hydrogen atoms, and 2 oxygen atoms.

2.9 (a) Since Si and F are on the right of the jagged line, this molecule is comprised solely of nonmetals. This makes it covalent.

(b) Because Ca (calcium) is on the left of the jagged line, it is a metal. Boron (Br), on the right of the jagged line, is a nonmetal. This means the compound is ionic.

(c) This is also ionic because Li is a metal.

(d) All atoms in this molecule are nonmetals, so the compound is covalent.

2.10 (a) This molecule is covalent, so we need to use prefixes. The prefix for carbon (C) is *mono*–, but we drop it because it is first. The prefix for fluorine (F) is *tetra*–, so we have carbon tetrafluoride.

(b) Al is a metal, so this is an ionic compound. We use no prefixes: aluminum sulfide.

(c) diphosphorus trioxide

(d) calcium oxide

2.11 (a) SF_6 (b) C_5H_{10}

2.12 (a) Mercury is on the periodic table, so it is an element.

(b) Carbon monoxide is the name for the molecule CO, and molecules form a compound.

(c) Chocolate contains sugar, cocoa, and other ingredients that are mixed together so the composition is the same throughout. Chocolate would be a homogenous mixture.

(d) If you cut open an apple, you can see that there is a skin on the outside and pulp and seeds on the inside. Since the composition is different inside and outside, an apple is a heterogeneous mixture.

(e) Tea has items dissolved in water. The composition is the same throughout, so it is a homogeneous mixture.

(f) Cesium is on the periodic table, so it is an element.

(g) Ocean water has more than just salt and water in it. There are all sorts of items in the ocean water. Also, the composition of ocean water is not the same throughout. This makes ocean water a heterogeneous mixture.

(h) Barium chloride is the name for the molecule $BaCl_2$, so it is a compound.

(i) A chocolate chip cookie contains several different ingredients, and the composition varies throughout, so it is a heterogeneous mixture.

STUDY GUIDE FOR MODULE 2

REVIEW QUESTIONS

1. What is the difference between the continuous theory of matter and the discontinuous theory of matter?

2. What 2 laws were instrumental in the development of Dalton's atomic theory? What law did Dalton predict using his theory?

3. Describe how the law of mass conservation can be used to explain the result of Experiment 2.1.

4. List the 4 assumptions of Dalton's atomic theory. Which of the assumptions are wrong? Why?

5. a. What is the difference between an atom and an element?
   b. What is the difference between a molecule and a compound?
   c. What is the difference between an atom and a molecule?
   d. What is the difference between an element and a compound?

6. What are the physical characteristics that distinguish metals from nonmetals?

7. How can you determine whether an atom is a metal or a nonmetal from the periodic table? Are there any exceptions?

8. How can you experimentally determine whether a compound is ionic or covalent?

9. How can you determine from the periodic table whether a compound is ionic or covalent?

10. Which law governs why carbon and oxygen can form 2 different molecules: CO and CO_2?

11. Why do chemists use 2 different naming systems for compounds?

12. If a substance can be physically separated into its components, is it a pure substance or a mixture?

13. If a substance can be given a chemical name, such as calcium bromide, is it a pure substance or a mixture?

14. What element makes up the majority of the air we inhale?

15. Describe the difference between quantitative and qualitative measurements.

16. A change that seems to break a molecule apart is called a ________________.

17. A student writes this formula for sodium chloride: NACL. What is wrong with this formula?

18. What are the elements that lie along the jagged line called?

19. Describe a dimensionless quantity.

20. Why doesn't distilled water conduct electricity?

PRACTICE PROBLEMS

1. A chemist uses an experiment similar to Experiment 2.2 to decompose 150.0 grams of water (H_2O). If he finds that he makes 16.7 grams of hydrogen from the decomposition, how much oxygen must he have produced?

2. To burn a piece of wood, oxygen gas has to be added. Ash, carbon dioxide, and water are products of the burning process. If 3.8 kg of oxygen gas is used to burn a piece of wood, then what was the mass of the wood if a chemist collects 3.12 kg water, 925 g of ash, and 15.14 kg of carbon dioxide?

3. Identify the following as metal or nonmetal: Br, Sc, P, Ra.

4. In an experiment to make hydrogen peroxide, a chemist mixes 40.0 grams of hydrogen with 40.0 grams of oxygen. She finds that 42.6 grams of hydrogen peroxide are produced. In addition, a lot of hydrogen is left over at the end of the experiment. How many grams of hydrogen should have been added to the 40.0 grams of oxygen so that there would be no leftovers at the end of the experiment?

5. In making calcium nitride, 100.0 grams of calcium plus 100.0 grams of nitrogen makes 152.5 grams of product along with some leftover nitrogen. How much nitrogen and calcium should be added together to make 1.0 kg of calcium nitride without any leftovers?

6. What is the chemical formula of a molecule that is made up of 12 carbon atoms, 24 hydrogen atoms, 10 oxygen atoms, and 2 sulfur atoms?

7. How many total atoms are in 1 molecule of $C_3H_6C_{12}O$?

8. Which of the following molecules are covalent?
   a. C_2H_6 b. PH_3 c. BeS d. Li_2O

9. Name all of the compounds in problem 8.

10. What are the chemical formulas for the following compounds?
    a. tetranitrogen hexahydride   b. dihydrogen monoxide

11. Classify the following as either a mixture or a pure substance:
    a. Soil   b. Silver   c. Nitric acid, HNO_3 d. Lemonade

12. Reclassify everything in problem 11 as an element, compound, homogeneous mixture, or heterogeneous mixture.

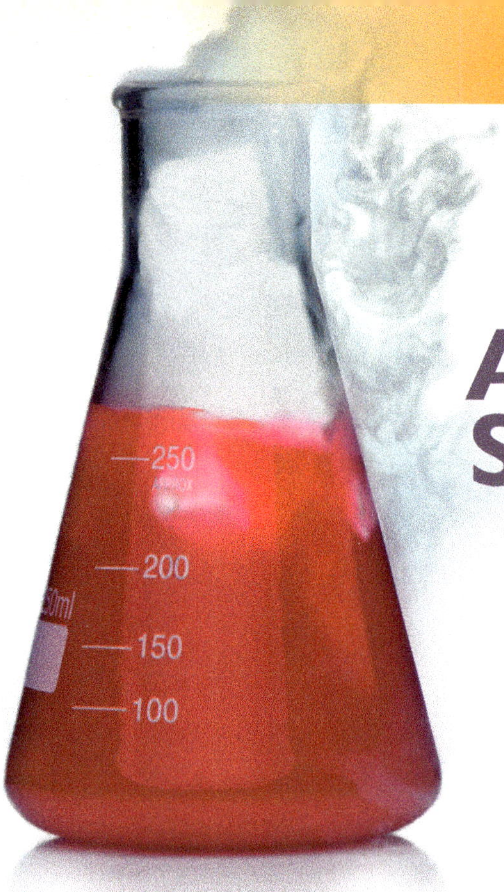

ATOMIC STRUCTURE

In the last module, we began working with atoms. We have used them only in chemical formulas so far, and you might have wondered what they look like. Unfortunately, it's not possible for us to actually see atoms. But for a moment, let's suppose we could. If we could see an atom, what would it look like? That's what we're going to learn about in this module.

You might wonder how in the world we can determine what atoms look like if we cannot see them. Remember, chemists often use things that we can see to learn about things we cannot see. For example, if we were to ask you if it is windy outside, then how would you answer us? You would probably go to the window and look outside. But you cannot see the wind, so how do you know it is windy? You look for evidence of the wind, don't you? Maybe you see a flag flapping, tree branches bending, or paper blowing along the ground. All of these evidences point to the fact that it is windy outside. We can also use indirect observations to help us learn about atoms. Observing how matter responds to certain situations can help us learn more about the structure of the atom.

first reactions

In this module, we begin to explore the limits of our knowledge of the smallest of things, the inside of the atom. We will take a closer look at the construction of atoms, how to identify them; their electrons, protons and neutrons, electromagnetic spectrum properties and a few formulas we use to describe them. Names like isotope, Bohr, and quantum mechanical model will be introduced as we begin to look inside atoms. Always remember the words in Isaiah 40; God's knowledge is beyond scrutiny. As scientists learn more and more of the world around us, we explore deeper into how things are created. We can never reach the limit of how God created out of nothing, nor the love that chose to create this world we explore.

HISTORICAL OVERVIEW

Ever since Democritus first theorized about atoms some 2,400 years ago, scientists have wondered what they might look like. Democritus thought that each type of atom had its own unique shape: some were triangles, some were pentagons, and some were squares. These shapes, he thought, fitted together like jigsaw puzzle pieces to form molecules. As time went on, chemists began to do experiments that started to unlock the mysteries of the atom.

One of the first experiments that shed some light on the question of the atom's structure was performed by an English chemist named William Crookes (krooks). In the second half of the 1800s, Crookes studied how certain gases behaved when exposed to electricity. He filled a glass tube (which we often call a Crookes tube) with a tiny amount of gas and connected the tube to a battery, allowing electricity to pass through it. Crookes noticed that when the battery was hooked up, the end of the tube would glow a faint greenish–yellow color, as we see in Figure 3.1.

FIGURE 3.1
A Crookes Tube
Illustration by David Weiss.

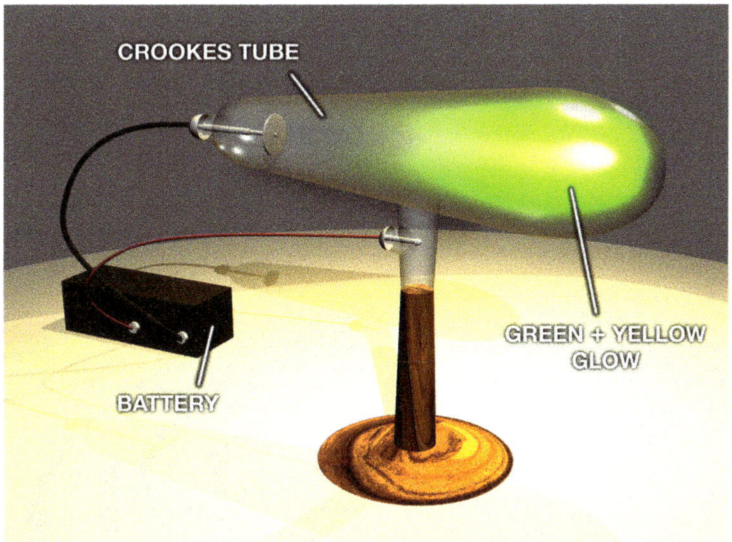

When the battery was not hooked up, there would be no glow. He concluded from this experiment that the battery was producing particles in the gas. These particles, he theorized, traveled through the gas and hit the end of the tube, causing the glow that he saw. Since these particles always seemed to travel from the post on the battery called the cathode (kath' ohd), Crookes called them cathode rays. As a result, a Crookes tube is sometimes referred to as a **cathode ray tube**.

To try to confirm his theory that these cathode rays were particles, he put an obstacle between the cathode and the end of the tube. He reasoned that if these cathode rays were, in fact, particles, they would be blocked by the obstacle and would not hit the end of the tube. Since they could not hit the end of the tube, they would not cause a glow. This is, indeed, just what he saw, as illustrated in Figure 3.2.

FIGURE 3.2
A Crookes Tube with an Obstacle
Illustrations by David Weiss

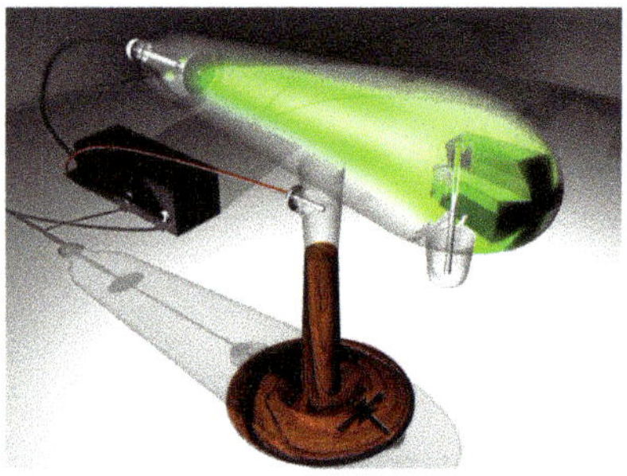

The Experiment
An obstacle was placed between the cathode and the wall of the tube. A shadow of the obstacle appeared at the end of the tube.

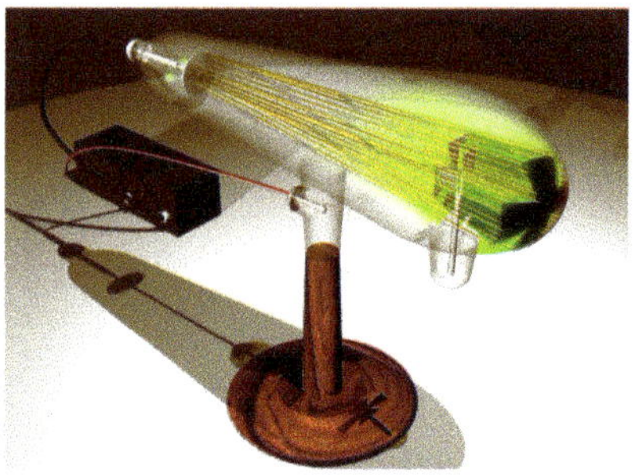

The Interpretation
The results seemed to indicate that particles were coming from the cathode. If they struck the obstacle, they could not make it to the wall, and that explains the shadow.

About 20 years after Crookes did his experiments, another English chemist, J. J. Thomson, revisited them to try to determine more about the nature of these cathode rays. Thomson's work ended up demonstrating that one of John Dalton's main assumptions about atoms was wrong. In the last module, we discussed Dalton's theory that assumed that atoms were indivisible. He reasoned that since atoms were the smallest unit of matter, they could not be broken down into smaller pieces. However, Thomson showed that regardless of what type of gas he put in the Crookes tube, cathode rays were always formed. He further showed that these particles were precisely the same, whether he filled the Crookes tube with helium, argon, nitrogen, or any other gas. This led Thomson to believe that these cathode rays were a part of every atom and that his experiments were breaking them away from the atoms.

Thomson's idea that cathode rays were small parts of the atom was further boosted by his discovery that these rays were electrically charged. He determined this by seeing how they reacted in the presence of magnetic and electric fields. To see why this fact was critical in tearing down Dalton's assumption that atoms are indivisible, you first have to understand a few things about electrical charge.

FIGURE 3.3
Crooke's Tube with Magnet
This diagram shows the path cathode rays take in the presence of a magnet. Notice how a magnet can exert a force on moving charges.
Illustration by David Weiss

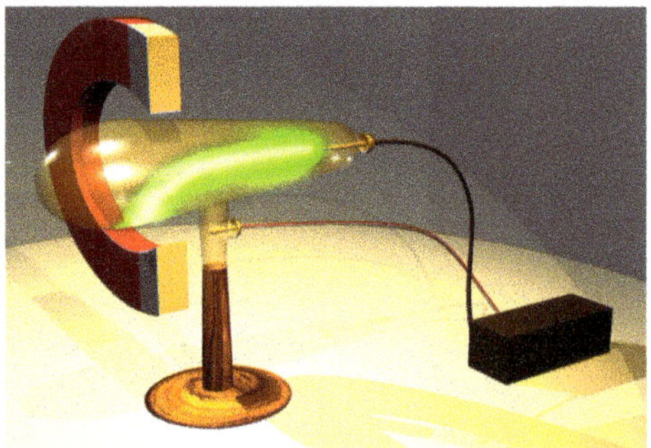

ELECTRICAL CHARGE

Everyone knows that there is something called electricity. We know that it can be used to operate lights, televisions, motors, etc. Most people do not understand what electricity is, however. In fact, most scientists don't have a really good understanding of electricity. That's because understanding electricity begins with understanding the concept of electrical charge. Unfortunately, scientists do not have a clear understanding of what electrical charge is or why it exists. We know how electrical charges behave and how they generate electricity, but we really don't know much about what they are or where they come from. Don't worry if you don't understand everything in the discussion that follows; we will point out the concepts that are important for you to understand at the end.

There are 2 types of electrical charge: positive and negative. We don't know what makes something negative or positive in charge, but we know that these charges exist. We also know that the vast majority of matter on earth does not have any overall electrical charge. We call these things electrically neutral. To learn a little more about what it means to be electrically positive or negative, perform Experiment 3.1.

EXPERIMENT 3.1

PURPOSE: To investigate electrical charge.

MATERIALS:

- Comb
- Aluminum foil
- Cellophane (Scotch) tape
- Safety goggles

QUESTION: How do electrical charges behave?

HYPOTHESIS: Write down your hypothesis about what the electrical charges will do in the presence of each other.

PROCEDURE:

1. Take 2 pieces of cellophane tape, each approximately 5 inches long, and attach them to the top of a table. Attach them so that about 1/2 inch of tape hangs off the table.
2. Once the tape is firmly attached to the table, grab the ends of the tape that hang over the table edge and quickly pull both pieces so that they detach from the table.
3. Hold the tape so that each piece hangs down.
4. Turn the pieces of tape so that their sticky sides face each other, and slowly bring them together. As you move the sticky sides toward each other, you will see them begin to repel each other. The pieces of tape will float away from each other as you try to move them together.
5. Cut some aluminum foil into several very small pieces and lay those pieces in a small pile on the table.

6. Pass the comb over the foil pieces. Bring the comb as close as you can to the foil without actually touching it. Nothing should happen.
7. Comb your hair with the comb. For this experiment to work, your hair must be reasonably clean. If it is too greasy, the next step of the experiment will not work.
8. After combing your hair, pass the comb over the aluminum pieces as you did before. This time, the aluminum pieces should move on their own toward the comb. If this didn't happen, your hair was probably a bit too greasy. Wash your hair, let it dry completely, then try again.
9. Clean up and return everything to the proper place.

CONCLUSION: How do we explain what happened in the experiment? Write down some thoughts and observations.

How do we explain what happened in the experiment? Let's start with what happened in the first part. To explain the results, you must understand that when you pulled the tape off the table, the sticky substance pulled some electrical charges off the table. Since each piece was pulling off the same type of charge, each piece of tape ended up having the same type of electrical charge. As those like charges were brought together, they repelled each other. Therefore, we come to our first rule of electrical charge:

Like charges repel one another.

What about the second part of the experiment? In that part, you started with a comb and some aluminum foil, neither of which was electrically charged. In fact, both the comb and the aluminum had electrical charges in them, but there were as many positive electrical charges as negative ones. Just as $-1 + 1 = 0$, the negative and positive charges cancelled each other out to give no charge at all.

However, when you combed your hair, the comb picked up some stray negative electrical charges from your hair. This caused the comb to have more negative charges than positive charges, so it became negatively charged. When the comb was then placed near the aluminum foil, the negative charge of the comb repelled all of the negative charges in the aluminum foil. The negative charges in the aluminum foil got as far away from the comb as possible. But when that happened, the parts of the foil near the comb had more positive charges than negative ones, and those positive charges were attracted to the comb. Therefore, the foil pieces began to move toward the comb. Therefore, we come to the second rule of electrical charge:

Opposite charges attract each other.

The other important thing to remember about electrical charge was mentioned in the previous discussion:

Every substance on earth has electrical charges. However, if the number of positive charges equals the negative charges, the substance is neutral with no overall charge ($^-1 + 1 = 0$). If there is more of one charge than the other, the substance takes on that electrical charge ($^-2 + 1 = {}^-1$).

Therefore, the comb started with no overall electrical charge because it had an equal number of positive and negative charges in it. However, when it picked up a few extra negative charges from your hair, it became negatively charged because there were more negative charges than positive charges in it.

If you don't understand everything you've just read, don't worry. The concepts of electrical charge are difficult to comprehend. Even the most brilliant chemist in the world still doesn't fully understand them. What you have to be able to understand are the 3 blue statements above. Those are the foundational concepts of electrical charge that we will use throughout this module.

ELECTRICAL CHARGE AND ATOMIC STRUCTURE

We now have enough information under our belts to understand why Crookes's and Thomson's experiments destroyed Dalton's postulate that atoms are indivisible. When the Crookes tube was filled with gas, the gas was always electrically neutral. The gas had as many negative charges as positive charges. However, when the battery was hooked up to the Crookes tube, negatively charged particles were produced. Thomson reasoned that those negatively charged particles must have been taken from the atoms of gas, much like the comb you used in the previous experiment pulled negative charges from your hair.

If negative charges could be pulled from an electrically neutral atom, then the atom must be composed of smaller components. Some of those components must be negatively charged, and the other components must be positively charged. Since the gas atoms started out neutral, there must be an equal number of positively and negatively charged components in atoms. Crookes and Thomson showed that these charged components could be separated, demonstrating that atoms could be broken down into smaller components. This destroyed Dalton's assumption that atoms are indivisible.

Thomson later did a famous experiment that measured the ratio of the charge and mass of these negatively charged particles. He showed that no matter what atoms they were pulled from, these particles always had the same ratio of charge to mass. He considered these particles to be something common to all atoms, and he called them **electrons** (ee lek' trahnz). As we have already said, since electrons are pulled from electrically neutral atoms, there must also be positive components in the atom. Thomson's student, Ernest Rutherford, discovered those components and called them **protons** (pro' tahns). In 1932, another English scientist, James Chadwick, discovered that there was an additional component to the atom. This component was electrically neutral (it had no charge), so Chadwick called it the **neutron** (new' trahn).

By 1932, scientists had determined that atoms were made up of 3 components: protons, electrons, and neutrons. What makes one atom different from another atom is the number of protons, neutrons, and electrons each contains. For example, look at the periodic table again. The first box on the periodic table represents the atom hydrogen. A hydrogen atom is composed of 1 proton, 1 electron, and 0 neutrons. The second box on the table represents the atom helium, which contains 2 protons, 2 electrons, and 2 neutrons. Hydrogen and helium have totally different properties. For example, hydrogen is explosive when placed in contact with a flame or even a spark. On the other hand, helium will never burn, no matter how hot a flame it comes into contact with. This difference is the result of the different number of protons, neutrons, and electrons that make up the atoms.

DETERMINING THE NUMBER OF PROTONS AND ELECTRONS IN AN ATOM

You might wonder how we know that a hydrogen atom has 1 proton and a helium atom has 2. This is what the top number in each box of the periodic table tells us. We call that number the atomic number, and it tells how many protons are in an atom.

An atom's atomic number tells how many protons it contains.

In addition, since all atoms are electrically neutral, we know they must contain an equal number of protons (positive charges) and electrons (negative charges).

All atoms have an equal number of electrons and protons.

Therefore, if you look at the periodic table, you will see that sulfur (S) has an atomic number of 16. This tells you that a sulfur atom has 16 protons and 16 electrons in it. Make sure you understand this idea by answering "On Your Own" question 3.1.

ON YOUR OWN

3.1 How many protons and electrons are in the following atoms?

a. molybdenum (Mo)
b. beryllium (Be)
c. hafnium (Hf)

DETERMINING THE NUMBER OF NEUTRONS IN AN ATOM

As you can see, when you are armed with the periodic table, it is very simple to determine the number of protons and electrons in an atom. Determining the number of neutrons in an atom is a little more difficult, however. To understand why this is the case, we must talk more about the periodic table. We said earlier that each box in the periodic table represents a single type of atom, but that's not really true. In fact, each box on the periodic table represents several different atoms, but all of those atoms are part of the same element. For example, the sixth box on the table represents carbon. There are 3 different types of carbon atoms. Each has the same number of protons and electrons (6), but they have different numbers of neutrons. When atoms have the same number of protons and electrons but different numbers of neutrons, we call them isotopes (eye' suh tohps).

Isotopes—Atoms with the same number of protons but different numbers of neutrons.

The most common type of carbon atom is composed of 6 protons, 6 electrons, and 6 neutrons. There are 2 other types of atoms which belong to the element we call carbon. The second type of carbon atom still has 6 protons and 6 electrons, but it has 7 neutrons.

The third type of carbon atoms has 6 protons, 6 electrons, and 8 neutrons. All 3 of these types of atoms are made up of the same number of protons and electrons but different numbers of neutrons. Therefore, they are all isotopes.

What is the difference between these carbon isotopes? Not much. The one with 8 neutrons is a little heavier than the one with 7 neutrons. Similarly, the one with 7 neutrons is a bit heavier than the one with 6 neutrons. Other than that, their chemical natures are precisely the same. Any reaction that one of the isotopes can participate in, they all can participate in. In fact, if you were to have a lump of carbon (a black solid) in your hand, all 3 isotopes would be present. The isotope with 6 neutrons would make up about 98.9% of all atoms in the sample, the isotope with 7 neutrons would make up about 1.1%, and the isotope with 8 neutrons would make up less than 1.0×10^{-12}%. These isotopes would appear identical in almost every way. They would all participate exactly the same in any chemical reaction. You could not find a way to separate them from one another chemically. This is a very important point:

Isotopes behave identically in their chemistry;
the main difference between them is their mass.

Every element is composed of several isotopes. This means that every box on the periodic table actually represents several different isotopes. Since that is the case, it is impossible to look at the periodic table and determine how many neutrons are in a given atom. To determine the number of neutrons in an atom, we must be given a little more information. This information is provided by the mass number.

Mass number—The total number of neutrons and protons in an atom.

This mass number is not always provided when discussing elements because often it is not necessary to consider the number of neutrons in an atom. However, if the mass number is given, then it is easy to determine how many neutrons are in that particular atom.

Remember, the most common isotope of carbon has 6 protons and 6 neutrons. Its mass number is 6 + 6 = 12. We can include this mass number in 2 ways. When writing out the full name of the element (carbon), we can add a hyphen followed by the mass number and call this isotope carbon–12. On the other hand, if we just want to write the symbol for carbon (C), then the mass number must appear as a superscript *preceding* the symbol. The symbol for this isotope would be ^{12}C. What about the other isotopes of carbon? The one with 6 protons and 7 neutrons would have a mass number of 6 + 7 = 13. Therefore, we could call it carbon–13 or ^{13}C. Similarly, the isotope with 6 protons and 8 neutrons would be called carbon–14 or ^{14}C.

FIGURE 3.4
Atomic Symbol (Magnesium)
Illustration by David Weiss

Before we go any further, let's make a point to clear up some confusion you might have. Some students interpret the term *isotope* incorrectly, thinking that there is one normal version of an atom and that all the others are isotopes. For example, they think that carbon–12 is the normal version of carbon and that carbon–13 and carbon–14 are isotopes. That's *not* the proper use of the term. *All 3 types of carbon are isotopes*. Look back at the definition. Carbon–12, carbon–13, and carbon–14 all have 6 protons, but they have different numbers of neutrons. They are *all* isotopes of carbon. Think about it this way: The term *isotope* is *relational*. It refers to how atoms relate to one another. Consider 3 men who have the same father. We would say that these men are brothers. The term *brother* is relational. It tells how the men relate to one another. There is not one normal man, while the other 2 men are the brothers! They are *all* brothers. In the same way, if several atoms all have the same number of protons (the same atomic number) but different numbers of neutrons, they are *all* isotopes.

Now that you have learned about isotopes, you have what you need to determine the makeup of any atom on the periodic table, provided you have enough information. Study Example 3.1 and then tackle "On Your Own" questions 3.2–3.3 to be certain you understand how this is done.

EXAMPLE 3.1

What are the name and symbol of an atom that is made up of 10 protons, 10 electrons, and 11 neutrons?

Since the atom has 10 protons, its atomic number is 10. The periodic table shows that neon (Ne) is the element with that atomic number. To give the full name, we must also include its mass number. Since it has 10 protons and 11 neutrons, its mass number is 10 + 11 = 21. Therefore, the name of the atom is **neon–21**, and its symbol is **^{21}Ne**.

How many protons, electrons, and neutrons make up a ^{129}I atom?

To determine the number of protons and electrons, all we have to do is find I (iodine) on the periodic table. We see that its atomic number is 53. This means it has **53 protons and 53 electrons**. In addition, we are given its mass number, which represents the total number of protons and neutrons. If the atom has 53 protons, and its total number of protons + neutrons is 129, then the number of neutrons it has is 129 - 53 = **76 neutrons**.

ISOTOPES AND NUCLEAR BOMBS

Before we leave this discussion of protons, electrons, neutrons, and isotopes, we should point out one more thing. If you pay attention to the news at all, you have probably heard about how the United States and other nuclear powers in the world are trying to keep nuclear technology away from certain countries and are concerned about those countries that are developing nuclear technology. Have you ever wondered exactly what nuclear technology is? Most people would say that it refers to the ability to make a nuclear bomb. In fact, that's not true at all! Any person who has studied nuclear physics at the college level has all of the knowledge needed to build a nuclear bomb! The trick is not in making the nuclear bomb; almost anyone (with a little training) can do that. The real trick to making a nuclear bomb is in making the bomb's fuel.

ON YOUR OWN

3.2 Give the number of protons, electrons, and neutrons that make up the following atoms:
a. ^{22}Ne b. Fluorine-19 c. ^{235}U

3.3 An atom has 33 protons, 33 electrons, and 41 neutrons. What are its full name and symbol?

Uranium is the usual fuel used to make nuclear bombs. However, the element uranium has 3 naturally occurring isotopes: ^{238}U, ^{235}U, and ^{234}U. A sample of natural uranium is about 99.3% ^{238}U, about 0.7% ^{235}U, and less than 0.01% ^{234}U. Only the ^{235}U isotope can be used to make a bomb. For a nuclear bomb to work, the uranium used in the bomb must have a significantly larger percentage of ^{235}U than natural uranium has. In other words, to be able to produce a nuclear bomb, the amount of ^{235}U in the uranium used must be artificially increased. The way to do this, of course, is to get rid of some of the ^{238}U in the uranium. This process, called isotopic (eye' suh top' ik) enrichment, is very tricky. After all, the only real difference between ^{235}U and ^{238}U is the mass. Chemically, they behave in exactly the same way. This makes it nearly impossible to separate them for isotopic enrichment.

Of course, nearly impossible and impossible are 2 different things. During World War II, scientists in a group called the Manhattan Project were able to figure out a way to partially separate ^{238}U from ^{235}U. The process is incredibly inefficient, but it is able to take natural uranium and enrich the ^{235}U content from 0.7% to a little more than 7%, which is enough to fuel a nuclear bomb. Ultimately, making a nuclear bomb is simple; however, getting the fuel to power it is nearly impossible, and the technique for making this fuel is difficult and not readily shared.

ATOMIC STRUCTURE IN MORE DETAIL

Now that you know how to determine the number of protons, neutrons, and electrons in an atom, you might ask how these particles arrange themselves in an atom. Are they jumbled together like a bunch of marbles in a sack? Do they arrange themselves in groups? Do they form any neat arrangements or designs? These are the same questions that chemists were asking themselves in the early 1900s.

After electrons were first discovered (and before neutrons were discovered), chemists thought that an atom looked a lot like a dish of plum pudding. This concept, called the plum pudding model, was developed by British physicist, J.J. Thomson. It is illustrated in Figure 3.5.

FIGURE 3.5
The Plum Pudding Model of the Atom
Illustration by David Weiss

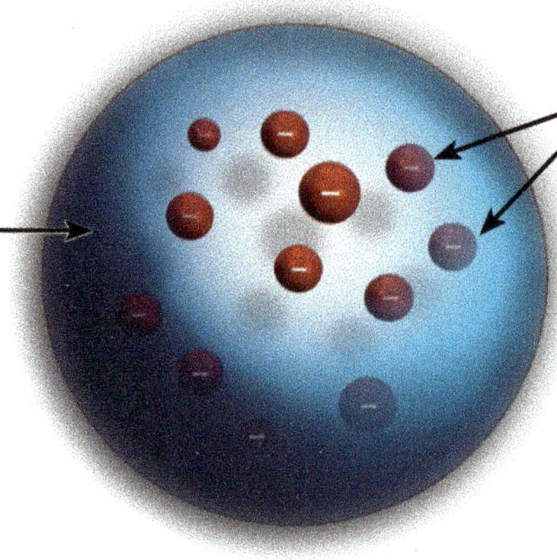

In essence, chemists thought that the atom was made up of a pudding–like substance which had positive electrical charge. Electrons, they thought, were suspended in this "pudding" like plums, making the entire atom electrically neutral. Remember, it's impossible to see atoms; so chemists just had to imagine what an atom might look like. That's why we call this concept a model of the atom. A model is simply a constructed image of something we cannot see with our eyes.

Since we can't see atoms, how do we know whether or not the plum pudding model is correct? The best way to determine that is to compare it to experimental data. If the model is inconsistent with the experimental data, then the model is obviously incorrect. However, if the model is consistent with the data, then we don't necessarily know that the model is right, but at least we know that it might not be wrong. The trick is in designing an experiment that will produce data that really test all aspects of the model.

Ernest Rutherford (Thomson's student) designed the perfect experiment to test the validity of the plum pudding model. In earlier experiments, Rutherford had identified a type of radiation called alpha particles. He determined that these alpha particles have positive electrical charges and are emitted by certain isotopes that were called radioactive. You will learn a great deal more about radioactive isotopes if you take advanced chemistry. For now, just remember that alpha particles are positive particles emitted by certain radioactive isotopes.

Rutherford decided that what he needed to do was to shoot some of the alpha particles at a thin metal foil. He reasoned that if the plum pudding model of the atom were correct, the alpha particles should pass through the foil without changing their direction much. After all, the positive alpha particles would be attracted by the negative "plums" (the electrons) but repelled by the positive "pudding." Since the plums were randomly distributed around the pudding, the alpha particles would be attracted and repelled in many different directions, and the result would be that they would travel straight through the atom.

If you have trouble understanding why the alpha particles should travel straight through a plum pudding atom, think about it this way. Suppose you were trying to move a large boulder by tying many ropes to it and having several people pull on the ropes while others pushed on the boulder. If everyone pulled and pushed in different, random directions, the boulder wouldn't go anywhere. To move the boulder, everyone would need to pull in one direction and push in roughly the same direction. The same principle applies here. When traveling through a plum pudding atom, the alpha particles would be pulled and pushed in different, random directions. This would result in the alpha particles not being moved from their original path.

We hope that you can see the ingenuity behind Rutherford's experiment. The data provided would be an excellent test of the model in question. If the alpha particles did not behave as Rutherford expected, then clearly the plum pudding model was wrong. However, if they did behave as expected, then that would be one more bit of evidence in favor of the plum pudding model. In the end, the experiment looked something like Figure 3.6.

FIGURE 3.6
Rutherford's Experiment
Illustration by David Weiss

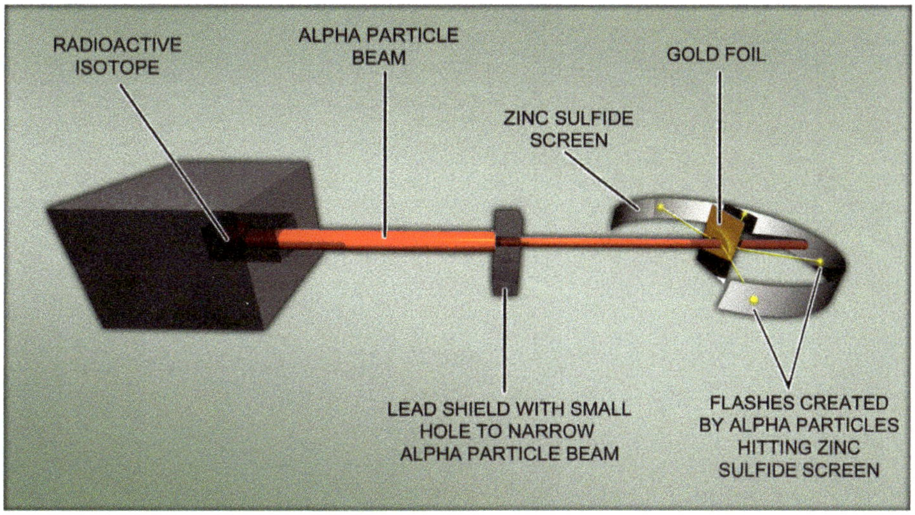

In this experiment, Rutherford put a radioactive isotope in a thick lead box with a single hole in it. The radioactive isotope inside the box was emitting alpha particles (a type of nuclear radiation). Most of the alpha particles were stopped inside the box, but any that were emitted in the direction of the hole escaped. This made a beam of alpha particles that could be pointed toward a target. The beam was narrowed by another piece of lead with a hole in it. In the end, Rutherford built an alpha-particle shotgun out of the lead box that directed a thin beam of alpha particles to hit a thin piece of gold foil.

Rutherford surrounded the target with a zinc sulfide screen. When an alpha particle hits zinc sulfide, a bright green glow is emitted. Rutherford could see where the alpha particles went after they collided with the gold foil by simply looking at where the zinc sulfide screen was glowing. Wherever it was glowing very brightly, many alpha particles were hitting it. Wherever it was glowing faintly, only a few alpha particles were hitting it.

Rutherford ran his experiment and, to his astonishment, the zinc sulfide screen glowed everywhere! The glow was the brightest directly in front of the target, but alpha particles were hitting the zinc sulfide screen at all other positions as well. Rutherford noticed that the farther away from the front of the target, the dimmer the glow was. Nevertheless, alpha particles were colliding with the target and bouncing in every conceivable direction.

This, of course, spelled the end of the plum pudding model of the atom. After all, there was no way to imagine how the alpha particles could possibly bounce off of the target and change direction if the plum pudding model of the atom were correct. Rutherford tried his experiment with several different types of targets to make sure that his conclusion applied to all atoms. In the end, the results were the same, regardless of what target was used.

Of course, it wasn't enough for Rutherford to simply determine that the plum pudding model of the atom was wrong. Now that he had all this experimental data, he needed to try to develop a new model of the atom that was consistent with all of the information he had up to that point. Rutherford developed such a model and called it the planetary model of the atom, but it is usually referred to today as the Rutherford model of the atom. This model is illustrated in Figure 3.7.

FIGURE 3.7
The Rutherford Model of the Atom
Illustration by David Weiss

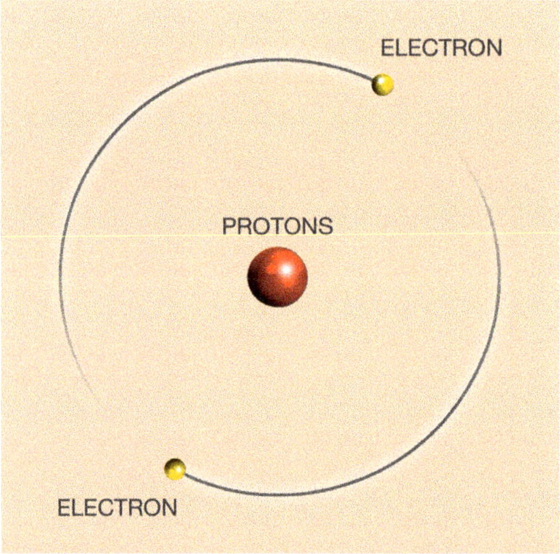

In Rutherford's model, the protons were clustered together at the center of the atom, called the nucleus (new' klee us), while the electrons orbited around the nucleus in the same way that the planets in our solar system orbit around the sun. This similarity between his view of the atom and our solar system is why Rutherford called it the planetary model of the atom. Notice that Rutherford's model does not contain neutrons because neutrons had not been discovered yet.

Realize that this idea of what the atom looks like was not something Rutherford just dreamed up. It was the only feasible model that explained the data he got from his experiment. Remember, Rutherford's experiment showed that when positive particles were shot at atoms, most of those particles passed through the atoms undisturbed. However, some were deflected away from their original path. Sometimes the deflection was enormous, and the alpha particle would bounce backward after hitting an atom.

Rutherford's model is consistent with all of this data. To help you understand why, look at Figure 3.8.

FIGURE 3.8
Why Rutherford's Model Is Consistent with His Data
Illustration by David Weiss

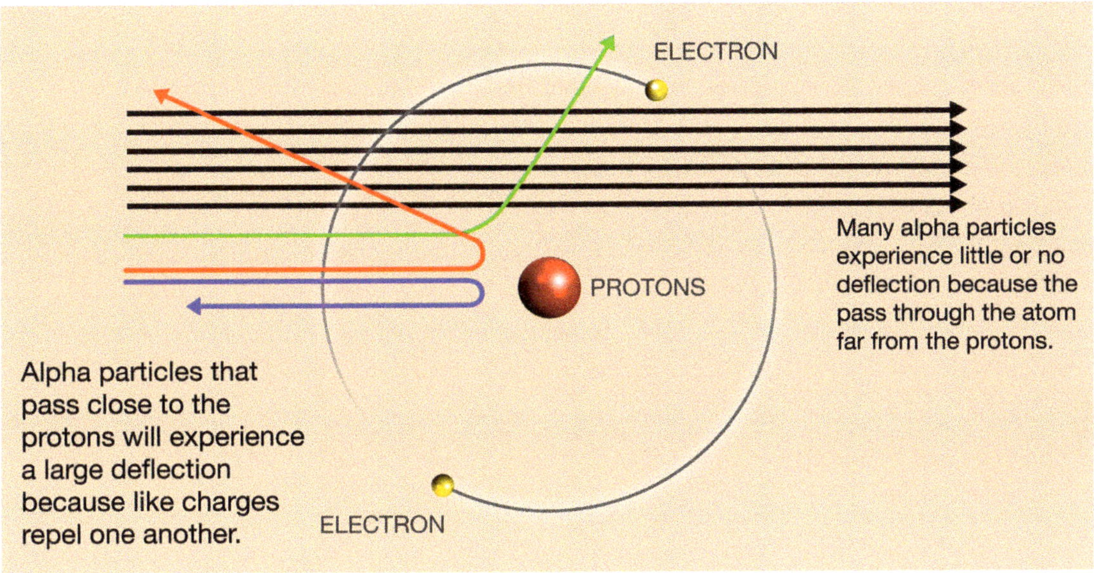

In the figure, the arrows represent alpha particles that are passing through the atom. Unlike the plum pudding model, this model has the positive electrical charge concentrated at the center of the atom. The negative charges (the electrons) are orbiting around the protons. Since the electrons are not concentrated in one spot, they do not affect the path of the alpha particles greatly. The protons cause most of the deflection of the alpha particles. The vast majority of alpha particles that pass through the atom, then, will not be deflected because they will pass through the atom far from the protons. These alpha particles, represented by the straight arrows, will continue to travel straight, hitting the zinc sulfide screen in front of the target.

However, if an alpha particle passes through the atom close to the protons, it will be repelled by the protons' positive charge and will be deflected from its original path. The closer it passes to the protons, the more its path will bend. If it is headed directly toward the proton, it will be repelled so much that it will turn around and head in the opposite direction or appear to bounce back. These deflected alpha particles are represented by the curved arrows.

The vast majority of alpha particles are not deflected because they do not pass very close to the protons and they strike the zinc sulfide screen directly in front of the target, as Rutherford saw in his experiment. Rutherford's observation that the number of alpha particles that were deflected is very small also makes sense in terms of this model because the number of particles that pass close to the protons is quite small. In the end, then, Rutherford's model was consistent with all of the data produced in his experiment.

The one thing Rutherford's model was not consistent with, however, was a current theory in physics. Physicists during Rutherford's time had noticed that whenever electrically charged particles moved in circles, they emitted light. When they emitted this light, they would lose energy and slow down. According to this theory, the Rutherford model of the atom would never be stable. This theory would predict that the electrons moving in a circle would continue to emit light until they lost all of their energy. It would also predict that as the electrons lost energy, they would spiral toward the protons in the atom until they collided with them.

Even though Rutherford's model was not consistent with a current theory in physics, it was the only model of the atom that could be devised that was consistent with his experimental data. It became the accepted model of the atom until it was later modified by Niels Bohr (boor). To understand Bohr's model of the atom and what experiments led up to it, you must first learn a little about light.

think about this

We associate light with illumination. At first, you might think of the physical properties of light that you experience every day. However, light is used to describe many things, including God, our testimony, and righteousness, to name a few. For example, Psalm 18:28 says, "For it is you who light my lamp; the LORD my God lightens my darkness." In this verse, light is used as a metaphor to describe God as He directs our path. Without God illuminating our path, we will stumble and fall into darkness. Also, Matthew 5:16 says, "In the same way, let your light shine before others, so that they may see your good works and give glory to your Father who is in heaven." Here the apostle Matthew uses light to describe how Christians are to contrast with a dark and sinful world.

THE NATURE OF LIGHT

In Module 1, you learned that since light neither has mass nor takes up space, it is not matter. However, you haven't yet learned what light is and what its properties are. That's the purpose of this discussion.

The first thing you must realize about the nature of light is that scientists don't completely understand it. There are currently 2 models for light: the particle model and the wave model. The particle model states that light is made up of small particles called photons. Any beam of light that you see is made up of billions of individual particles (photons) that are all traveling in the same direction. The wave model, on the other hand, says that a beam of light is a wave that travels much like a wave on the ocean.

To make things even more confusing, sometimes light behaves like a particle, and sometimes it behaves like a wave. Scientists really aren't sure they understand this phenomenon, but they accept it. Currently, theory states that light acts like both a particle and a wave. The situation that light is in determines whether it will behave as a particle or a wave. Scientists call this theory the particle/wave duality theory, and it is used frequently in describing light.

Particle/wave duality theory—The theory that light sometimes behaves as a particle and sometimes behaves as a wave.

If you don't understand the particle/wave duality theory, don't worry. Scientists don't understand how light can be both a particle and a wave, but for now we are forced to believe it because that's what the experimental data indicate.

The wave model of light is by far the most useful way to picture light, so we will start there. Consider the wave in Figure 3.9.

FIGURE 3.9
A Wave
Illustration by David Weiss

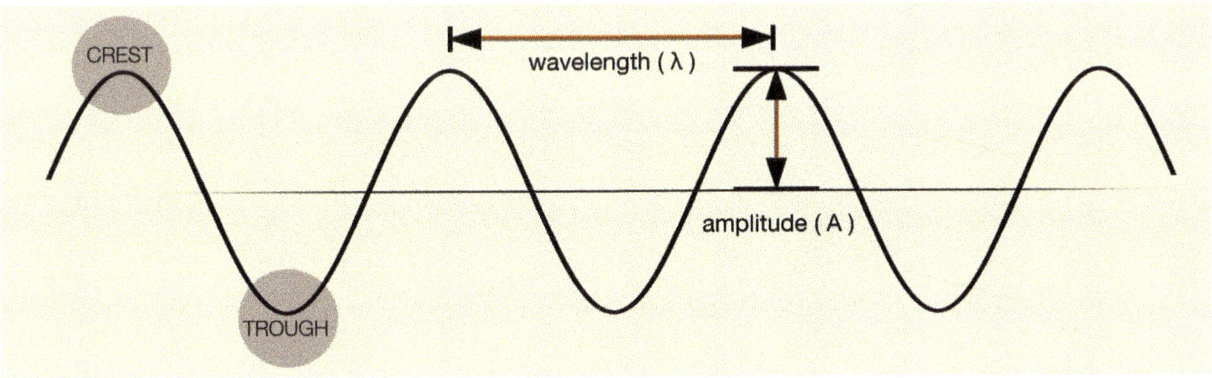

This is one way to picture light. It moves through space just like a wave moves through water. The high points of the wave are called crests, and the low points are called troughs. If we were to describe the above wave, we would need to specify 2 things. First, we have to say how far apart the waves are. We use the term wavelength to accomplish this.

Wavelength—The distance between the crests (or troughs) of a wave.

Since wavelength is a measurement of distance, its units will be meters, centimeters, etc. The wavelength of a wave is usually given the symbol λ (the lower–case Greek letter *lambda*), as illustrated in the figure.

To describe the wave completely, we not only have to tell how far apart the waves are; we also need to tell you how big the waves are. This is referred to as the amplitude of the wave.

Amplitude—A measure of the height of the crests or the depths of the troughs on a wave.

Amplitude is usually given the symbol A, as illustrated above. Amplitude is not a distance measurement, however. The units for amplitude vary from wave to wave, and we will not get into those distinctions here. If you take physics, you will learn about the units of amplitude. For now, just remember that amplitude means how big the wave is. The larger the amplitude, the bigger the wave.

How does all of this apply to light? The wavelength of light corresponds to its *color*. Perhaps you have seen what happens when sunlight hits a prism. Once it passes through the prism, the apparently white light turns into a rainbow of colors. This is because through laws you will learn in physics, a prism breaks light into its different wavelengths. Each color you see represents a range of wavelengths.

Think about the last time you saw a rainbow. What colors did you see? Red was at the top of the rainbow; orange was next; yellow followed orange; and then there were green, blue, indigo, and violet. These are the basic colors of light, and they always appear in a rainbow in that order. Why?

FIGURE 3.10

The Visible Spectrum

Rainbow photo ©GoodShoot/SuperStock, Inc. / Visible spectrum illustration by Megan Whitaker

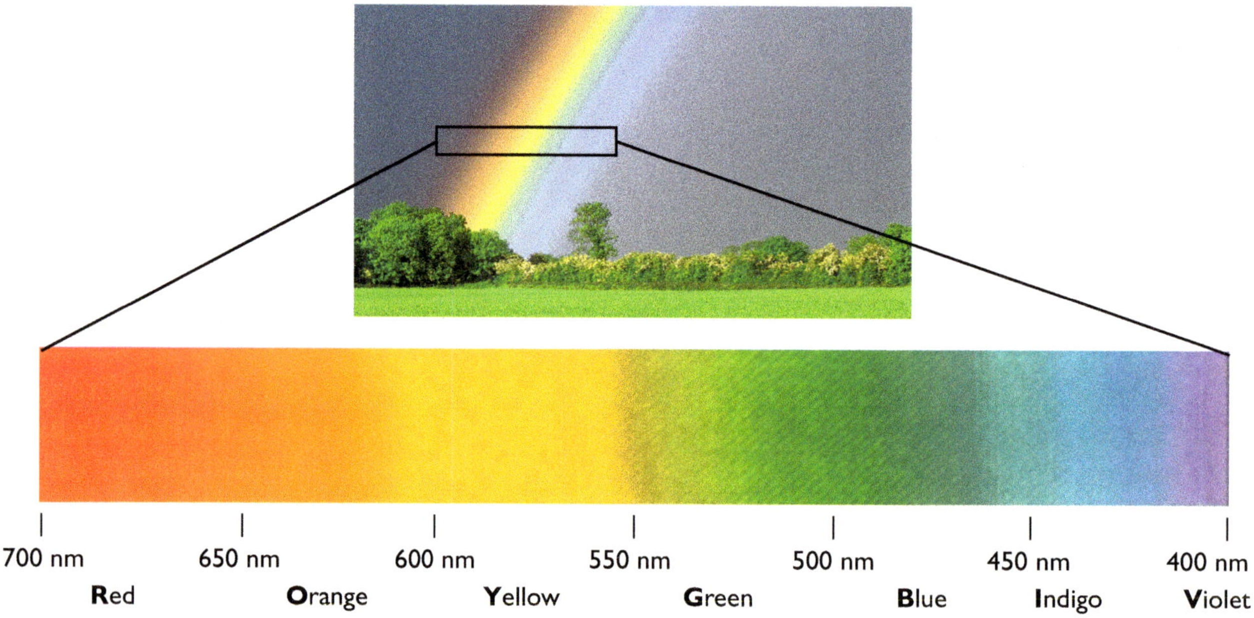

For example, red light is made up of waves whose wavelengths are between 650 and 700 nanometers long. The prefix *nano-* means 10^{-9}, so red light's wavelengths are very small. However, of all the colors, red has the longest wavelength. Figure 3.10 summarizes each basic color and the approximate range of wavelengths that make up that color. This range of light wavelengths is typically called the visible spectrum. You will learn why it is called this a bit later.

think about this

In Genesis 9:11-17, God promised Noah that the rainbow would symbolize His promise to never again destroy the earth with a flood. A rainbow is formed by little water droplets suspended in the air after a rainfall. These water droplets act like little prisms, separating light into different wavelengths. What an amazing God we serve who can take billions of tiny little prisms and create an eternal promise for us!

As you will learn in physics, the wavelength of light determines how the light passes through a prism. Since wavelength determines color, the colors of light in a rainbow always appear in the same place because they always pass through the water droplets in the air in exactly the same way.

Although you needn't memorize the wavelengths that each color corresponds to, you *do* need to memorize the *relative* size of the wavelengths in question—that is, you need to know that red light has the largest wavelengths, orange light has smaller wavelengths, and so on. This is easy if you think about the colors as a single name. The first letter of each color is listed in the figure in boldface type to emphasize that when you put them together in the proper order, you come up with a man's name: ROY G. BIV. If you think of ROY G. BIV every time you think of the colors in the visible spectrum, you will always know that red light has the longest wavelength and violet light has the smallest, and you will also know the order of all colors between them.

Now that we know what the wavelength of light means, what does the amplitude of light mean? Simply put, the amplitude of a light wave tells how bright the light wave is. A dim light bulb produces light waves with a small amplitude, while a bright light bulb produces light with a large amplitude. If we had 2 light bulbs of the same color but one was dimmer than the other, we could say that the light bulbs were both producing light of the same wavelength but of different amplitudes. Similarly, if we had 2 light bulbs of different color but the same brightness, we could say that the light bulbs produced light waves of equal amplitude but of different wavelengths.

There's a bit more to describing a light wave than just its amplitude and wavelength; the speed of the light wave is also important. After all, we know that light travels from place to place. If we turn on a light bulb, the light rushes from the light bulb and fills up the room. Therefore, light does travel. But how fast does it travel? Have you ever watched a thunderstorm and seen a lightning flash but heard the thunder crash a few moments later? This happens because the light generated by the lightning flash travels more quickly than the sound it makes. The light reaches your eyes before the sound reaches your ears. For every 5 seconds between the time you see the lightning flash and the time you hear the thunderclap, the lightning is striking 1 mile away from you.

Light, then, travels fast—much faster than sound. In fact, light travels in air with a speed of 3.0×10^8 meters per second (m/s). In units you are more familiar with, that translates to 9.8×10^8 feet per second (ft/s). That's really moving! To put it another way, light travels 190,000 miles every second. One thing that's very interesting about this number is that it does not change with the wavelength or amplitude of the light wave. Regardless of the color or the brightness of light, its speed in air is always 3.0×10^8 m/s. Since this value doesn't change, we call it a physical constant.

Physical constant—A measurable quantity in nature that does not change.

You will be introduced to several physical constants in this course, and the speed of light is the first one. Each physical constant has its own symbol. The symbol for the speed of light is c. When we say that $c = 3.0 \times 10^8$ m/s, we are giving you the speed of light. You do not need to memorize this number. We will always provide you with any physical constants you need to solve the assigned problems.

Now that we know about the speed of light, we can finally speak of another very important means of describing a wave: frequency.

Frequency—The number of wave crests (or troughs) that pass a given point each second.

To best understand frequency, suppose you are standing in the ocean. If you look out at the waves, the distance between each crest is the wavelength. On the other hand, if you stand still and count the number of wave crests that hit you over a certain interval of time, that is frequency.

Think for a moment about the unit associated with frequency. Suppose you stand still in the water at the beach and count that 2 ocean wave crests hit you every second. You would report your measurement as 2 wave crests per second. However, since everyone at the beach can see that you are counting waves, you might just say "2 per second." What does *per second* mean? The unit meters per second is mathematically expressed as meters/second. *Per second* just means 1/second. This means that the standard unit for frequency is simply 1/second. This unit is usually called the Hertz (abbreviated Hz), in honor of Heinrich Rudolf Hertz, a German physicist who greatly expanded our knowledge of light and its relationship to electricity.

The frequency, wavelength, and speed of light are all related to each other by a simple equation:

$$f = \frac{c}{\lambda} \quad \text{Equation 3.1}$$

In this equation, f stands for frequency, c is the speed of light, and λ stands for wavelength. Notice how the units work out. Speed is in m/s, while wavelength is in meters. When we divide the speed by the wavelength, meters cancel, and we are left with 1/s, which is also called Hertz. This is a very important equation, and you should memorize it. See how we use it in Example 3.2.

salt and light

Photo: Public Domain

Imagine holding a world record that no one has been able to surpass for nearly 100 years. **Edward William Morley (1838-1923)** has a record in science that no one has been able to surpass. He was able to determine the atomic weight of oxygen to more precision using wet chemistry techniques than anyone else has ever accomplished. Today, atomic masses are determined by physical means, using sophisticated instruments with computers. However, Morley was able to figure the atomic mass of oxygen using techniques involving solutions, reactions, and typical glassware found in a chemistry laboratory. His work set the standard for laboratory precision that can still be followed today.

Morley, who was homeschooled until the age of 19, wrote these words while a professor at Western Reserve University: "I believe that Jesus Christ shall come with the clouds of heaven to judge the world in righteousness and that those who have believed in Him shall inherit eternal life through the grace of God" (Graves 1996, 155).

EXAMPLE 3.2

A violet light wave has a wavelength of 405 nm. What is its frequency? ($c = 3.0 \times 10^8 \frac{m}{s}$, and the prefix *nano–* means 10^{-9}).

To solve this problem, we first must recognize that we use equation 3.1:

$$f = \frac{c}{\lambda}$$

To find the frequency, we just need to divide the speed, which is given, by the wavelength, which is also given. However, before we can use the equation, we must make sure our units work out. The wavelength is in nm, while the speed is in m/s. For the equation to work, our units must agree. Either m needs to change to nm or vice versa. We choose to change nm into m. According to what's given:

$$1 \text{ nm} = 10^{-9} \text{ m}$$

$$\frac{405 \cancel{\text{nm}}}{1} \times \frac{10^{-9}\text{m}}{1 \cancel{\text{nm}}} = 4.05 \times 10^{-7} \text{ m}$$

Try to put that equation into your calculator to see if you can get the right answer. If your calculator is a scientific calculator, you can put the numbers into the calculator in scientific notation. However, most calculators want you to input scientific notation in a very specific way. If you don't do it correctly, your answers will be off by a factor of 10 or more.

Now that our units agree, we can put the numbers into the equation:

$$f = \frac{3.0 \times 10^8 \cancel{\frac{m}{s}}}{4.05 \times 10^{-7} \cancel{m}} = 7.4 \times 10^{14} \frac{1}{s}$$

Remember, since the speed of light as given has only 2 significant figures, we can only report our answer to 2 significant figures as well. The frequency is **7.4×10^{14} Hz.** What does the answer mean? It means that when you look at violet light, 7.4×10^{14} crests of the light wave hit you every second! Once again, make sure you can put this equation into your calculator and get the right result.

A light wave has a frequency of 4.70×10^{14} Hz. What are its wavelength and color according to the chart in Figure 3.10? ($c = 3.0 \times 10^8$ m/s; *nano–* means 10^{-9})

Once again, we use equation 3.1:

$$f = \frac{c}{\lambda}$$

But now we have to rearrange it so that we are solving for wavelength:

$$\lambda = \frac{c}{f}$$

Now we can plug in the numbers:

$$\lambda = \frac{3.0 \times 10^{8} \frac{m}{\cancel{s}}}{4.70 \times 10^{14} \frac{1}{\cancel{s}}} = 6.4 \times 10^{-7}\,m$$

The chart in Figure 3.10 lists wavelength in nm. If we are to refer to the chart, we must convert to nm:

$$\frac{6.4 \times 10^{-7}\,\cancel{m}}{1} \times \frac{1\,nm}{10^{-9}\,\cancel{m}} = 640\,nm$$

The wavelength is **640 nm**. According to Figure 3.10, then, this light is **orange light**.

ON YOUR OWN

($c = 3.0 \times 10^{8} \frac{m}{s}$; *nano-* means 10^{-9})

3.4 What color is light that has a frequency of 6.4×10^{14} Hz?

3.5 If red light has wavelengths from 7.00×10^{2} nm to 6.50×10^{2} nm, what is the frequency range for red light?

Make sure you understand how to use Equation 3.1 by answering "On Your Own" questions 3.4–3.5.

THE ELECTROMAGNETIC SPECTRUM

Now that you've had some experience using equation 3.1, go back and just look at it for a moment. Notice how the equation shows that frequency and wavelength are related. As wavelength gets larger, Equation 3.1 tells us that we will be dividing by a large number. What does that tell us about frequency? It tells us that for large wavelengths, frequency is small. Conversely, if wavelength is small, frequency is large. When 2 quantities behave like this, we say that they are inversely related to one another. When studying light, this inverse relationship between frequency and wavelength is very important to remember.

When wavelength is large, frequency is small.

When wavelength is small, frequency is large.

This inverse relationship between frequency and wavelength should make sense if you think about it. Imagine yourself once again standing in the ocean. As you look at the waves coming toward you, if their crests are far apart (they have a large wavelength), they will not hit you very often (they will have a small frequency). However, if the wave crests are close together (small wavelength), they will hit you very often (large frequency).

You may not realize that the light we see with our eyes (the visible spectrum) is only a small part of the light that comes to us from the sun. The sun bathes our planet with light of many, many different wavelengths and frequencies. Our eyes perceive only a small fraction of the total amount of this light. Figure 3.11 is a more complete representation of all light that comes from the sun, which we call the electromagnetic spectrum.

FIGURE 3.11
The Electromagnetic Spectrum
Illustrations from MasterClips collection

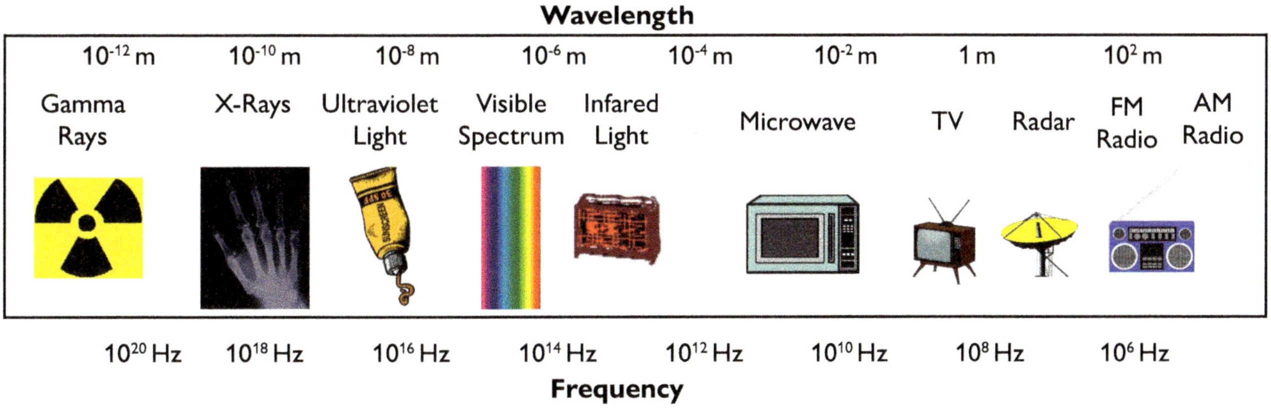

If you examine Figure 3.11, you will first realize just how small a part of the electromagnetic spectrum we actually see with our eyes. The rainbow under the words *visible spectrum* represents that part of the electromagnetic spectrum. Everything else shown in Figure 3.11, from wavelengths of 10^{-14} m all the way to wavelengths of 10^{4} m, is still light. Our eyes are sensitive only to the light contained in the visible spectrum, which means that is the only light we can see. However, as you can tell by the figure, we find uses for much of the light we cannot see.

For example, the signals that are sent out from a radio station to our radio antennae are, in fact, waves of light. We cannot see these light waves, but they are there nevertheless. Since they are there, we can use them to carry radio signals (FM and AM radio signals, for example). The only difference between these light waves and the ones we see with our eyes is the wavelength. Wavelengths of light used to carry radio signals are longer than those of visible light. In addition, television antennae also use light with wavelengths longer than those of visible light. What may be even more surprising to you is that microwave ovens use light to cook food. Microwaves also have longer wavelengths than those of visible light.

Notice that while we often use several of the light waves with wavelengths *longer* than visible light, we do not frequently use any of the light waves with wavelengths *shorter* than visible light. There is a very good reason for this. Light, of course, has energy. Its energy is kinetic since the light is moving. The energy of a light wave is directly proportional to its frequency. This is an important enough fact that you need to memorize it.

As a light wave's frequency increases, its energy increases.

As its frequency decreases, its energy decreases.

Since frequency and wavelength are inversely related to each other, we could also say this:

As a light wave's wavelength increases, its energy decreases.

As its wavelength decreases, its energy increases.

Based on these 2 facts, then, light waves with wavelengths shorter than those of visible light have higher frequencies and therefore higher energy. This is a very important consideration because when light strikes something (your eye, your skin, etc.), it deposits its energy into what it strikes. When visible light strikes your eye, it deposits its energy there. As you'll see in the next experiment, your eye uses that energy to transmit signals to the brain, and that's what causes you to see. Light with wavelengths shorter than visible light has enough energy to kill living tissue. When ultraviolet light (the first thing in Figure 3.11 with wavelengths shorter than those of visible light) strikes your skin, it can kill some of your cells. If your skin is exposed to too much ultraviolet light, a large number of cells will die, and you will get a burn on your skin, which we call a sunburn. You can avoid a sunburn by covering your skin with sunscreen, a lotion containing chemicals that absorb the ultraviolet light before it hits your skin.

Because of the nature of ultraviolet light, God has built into this wonderful planet a very efficient system that filters out the majority of these rays. You may have heard some talk about the ozone layer. We will go into this in detail in the next module, but for now we'll say the ozone layer is part of the filtering system that God has designed to shield us from this destructive type of light. Although it is incredibly efficient, the ozone layer does allow some ultraviolet light through, and that is why you can get sunburned if you stay out in the sun too long without protection.

Gamma rays and X–rays are more energetic than ultraviolet light, so they are even more dangerous to living tissue. Nevertheless, we still use both of these forms of light occasionally in medical procedures. X–rays, of course, are used to examine bones and other aspects of our internal anatomy for diagnostic purposes. Although these X–rays will kill some living tissue, that risk is worth the benefit of being able to diagnose internal problems without surgery. Therefore, as long as you do not get X–rays frequently, your risk associated with the nature of the light is low, and the benefit you get from the diagnosis of internal problems is high. Gamma rays are also used in some medical applications. For example, certain types of cancer can be treated by exposing the cancerous site to gamma rays. Even though this leads to the death of healthy tissue, it can also lead to the death of cancerous tissue, so the risk of gamma ray exposure can be worth the medical benefit.

Before we leave this section, let's talk about one point of terminology. Notice that Figure 3.11 is titled "The Electromagnetic Spectrum." That's because we know that light is an electromagnetic phenomenon. It is produced by the interaction of electrically charged particles. Light is often called electromagnetic (ee lek' troh mag net' ik) radiation. Although that sounds like a scary term, it just means light, but it includes all wavelengths of light, not just the visible wavelengths.

THE RELATIONSHIP BETWEEN FREQUENCY AND ENERGY

You need to know one more equation that deals with light. This equation relates the frequency of a light wave to its energy. As already mentioned, as a light wave's frequency increases, its energy does as well. The mathematical equation that governs this relationship is:

$$E = h \times f \qquad \text{Equation 3.2}$$

In this equation, E is the energy of the light wave and f is the frequency. The symbol h refers to another physical constant, called Planck's (Plahnks) constant. Named after its discoverer, German physicist Max Planck, this constant allows us to relate energy and frequency. Its value is $6.63 \times 10^{-34} \frac{J}{Hz}$. You needn't memorize the number; it will be given in any problem for which it is needed. Instead of learning the value, look at the units. Notice that if we put those units into equation (3.2), then we will get the following:

$$E = \frac{J}{\cancel{Hz}} \times \cancel{Hz}$$

When this happens, the Hertz unit cancels, leaving Joules, an energy unit.

Remember, Hertz means 1/second. With that in mind, let's look at the unit for Planck's constant once again. If we divide the unit Joule by 1/second, we get this:

$$\frac{J}{Hz} = \frac{J}{\frac{1}{s}} = J \div \frac{1}{s} = J \times s$$

"Joule–second" is another way of expressing the unit for Planck's constant.

Now that you understand the units for Planck's constant, see how we use it in Example 3.3.

EXAMPLE 3.3

A light wave has a frequency of 2.3×10^{16} Hz. What is its energy?

$$h = 6.63 \times 10^{-34} \frac{J}{Hz};\ c = 3.0 \times 10^{8} \frac{m}{s}$$

This is easy. We simply plug our numbers into equation 3.2:

$$E = h \times f$$

$$E = 6.63 \times 10^{-34} \frac{J}{\cancel{Hz}} \times 2.3 \times 10^{16} \cancel{Hz}$$

$$E = 1.5 \times 10^{-17} J$$

We can keep only 2 significant figures because our frequency had only 2. This might seem like a small amount of energy, but it corresponds to the energy of ultraviolet light, so it is enough to kill living tissue!

A light wave has a wavelength of 1.21 x 10^{-12} m. What is its energy?

This problem is a little harder because the only equation we can use to calculate the energy of light is equation 3.2, and it uses *frequency*, not *wavelength*. To solve this, then, we must first turn the wavelength we've been given into frequency. We can do this with equation 3.1:

$$f = \frac{c}{\lambda}$$

$$f = \frac{3.0 \times 10^{8} \frac{\cancel{m}}{s}}{1.21 \times 10^{-12} \cancel{m}} = 2.5 \times 10^{20} \frac{1}{s} = 2.5 \times 10^{20}\ Hz$$

Now that we have the frequency, we can use equation 3.2:

$$E = h \times f$$

$$E = 6.63 \times 10^{-34} \frac{J}{\cancel{Hz}} \times 2.5 \times 10^{20} \cancel{Hz}$$

$$E = 1.7 \times 10^{-13}\ J$$

The energy is **1.7 x 10^{-13} J.**

ON YOUR OWN

($h = 6.63 \times 10^{-34} \frac{J}{Hz}$; $c = 3.0 \times 10^{8} \frac{m}{s}$; *nano-* means 10^{-9})

3.6 If a light wave has energy of 3.4 x 10^{-14} J, what is its frequency?

3.7 If a visible light wave has energy of 3.3 x 10^{-19} J, what color is it? (You can use Figure 3.10 if you need to.)

Make sure you understand how to use equation 3.2 by answering "On Your Own" questions 3.6–3.7.

HOW THE EYE DETECTS COLOR

Before we go on to discussing the Bohr model of the atom, perform Experiment 3.2. Hopefully, it will give you some insight into how humans perceive color.

EXPERIMENT 3.2

PURPOSE: To determine how our eyes detect color.

MATERIALS:

- 2 plain white sheets of paper (There cannot be lines on them.)
- A bright red marker (A crayon will also work, but a marker is better.)

QUESTION: How does the eye detect color?

HYPOTHESIS: Write down your hypothesis about what will happen to what we see if we stare at one color for 60 seconds and then close our eyes.

PROCEDURE:

1. Take one of the sheets of paper and make a thick cross on it with the red marker. The cross should be about 6 inches long, and the 2 legs that make it up should be about 3/4 inch thick.
2. Color the entire cross so that you have a large, solid bright red cross in the middle of a white sheet of paper.
3. Put the clean sheet of white paper underneath the sheet with the cross on it. Make sure the cross faces you so that you can see it.
4. Stare at the cross for a full 60 seconds. You can blink if you need to, but do not take your eyes off the cross.
5. After a full 60 seconds of staring at the cross, quickly pull the top sheet of paper out of the way so that you can see only the clean sheet of paper on the bottom. What happens? For about 90% of students, a blue-green cross will appear for a few seconds on the blank sheet of paper. After a few moments, it should vanish. This optical illusion will not work for some people, especially if they have a tendency toward color blindness.
6. Clean up and return everything to the proper place.

CONCLUSION: Write something about what happened in the experiment.

What happened in the experiment? To see light, your eyes are equipped with cells called rods and cones. These cells transmit electrical signals to the brain whenever they are hit by certain energies of light. The brain receives the electrical transmissions and uses them to form an image in your mind. The rods are sensitive mostly to low levels of light and are not very sensitive to color. However, the cones respond to certain specific energies of light. Some cones are sensitive only to low–energy visible light (red), while others are sensitive to medium–energy visible light (green), and still others are sensitive to the highest–energy visible light (blue). When colored light hits these cells, they will send signals to the brain only if the light that they are sensitive to is hitting them. If blue light hits your eyes, the cones that respond to blue light will send signals to your brain, but the cones that respond to green and red light will not. As a result, your brain interprets the signals coming from your eyes and puts a blue image in your head. Similarly, if red light hits your eyes, the cones that respond to red light will send signals to your brain, but the cones that respond to green and blue will not. As a result, your brain constructs a red image.

If a mixture of colors hits your eyes, the cones send signals to your brain in proportion to the amount of light to which they are sensitive. For example, suppose you are looking at a purple ball. When that purple hits your eyes, the cones that respond to red light send some signals to your brain, and the cones that respond to blue light send some signals to your brain. Your brain then constructs a purple image of the ball in response. If the purple color of the ball is on the blue side of purple, then the cones that respond to blue light send more signals to the brain than the cones that respond to red light. On the other hand, if the purple is more red than blue, the cones that respond to red light send more signals to the brain than those that respond to blue light. All of the colors that you see are the result of the brain receiving signals from 3 types of cone cells and adding those 3 primary colors (red, green, and blue) with the weight indicated by the amount of signal coming from each type of cell.

What happened in the experiment? While you were looking at the red cross, your cones that respond to low–energy (red) light were sending signals to the brain, but your other cones weren't doing anything. Cones get tired pretty quickly, and when they have sent the same signal to the brain for a period of several seconds, they eventually just shut off. The brain, sensing that no more signals are coming from the cones, assumes that they have shut off simply because they are tired, and it holds the same image in your mind until new signals come along. As you were staring at the cross, your low–energy cones eventually shut off. Since no more signals were coming from the low–energy cones and since no signals had ever come from the medium and high–energy cones, the brain was receiving no more signals. It therefore assumed you were still looking at the cross and continued to hold the image in your mind.

When you yanked the top sheet away, white light began to hit your eyes where only red light had hit them before. Since white light contains all energies, your medium– and high–energy cones began to receive light and transmit signals to your brain. Your low–energy cones were still shut off, so they didn't send any signals, even though they should have. The brain started receiving new signals, but only from the high– and medium–energy cones. It constructed an image of green (medium–energy) and blue (high–energy) light. Eventually, your low–energy cones realized that they had to start transmitting again, and once they did, the brain realized that the eyes were seeing all energies of light and so formed a white image in your mind.

The way we perceive color, then, is based on the energy of the light that hits our eyes. Isn't it marvelous how well designed the eye is to handle such a complex operation? That should tell you something about how marvelous its Designer is!

THE BOHR MODEL OF THE ATOM

During our discussion of Rutherford's model of the atom, we mentioned that it was inconsistent with electromagnetic theory. Nevertheless, since that model seemed to be the only way to explain Rutherford's experimental data, it became the accepted model of the day. As chemists sought to understand more about atomic structure, they did more experiments.

One of the experiments that shed some light on atomic structure involved adding energy to atoms by heating them or passing electricity through them. When this was done, the atoms emitted light, some of which was in the visible spectrum. The interesting thing was that each type of atom seemed to emit its own unique color or colors of visible light. For example, when mercury atoms were heated, a pale blue glow would occur. On the other hand, when electricity was passed through neon atoms, a bright pink light was emitted.

As chemists investigated these experiments further, the mystery deepened. They learned that the colors which were being emitted during these experiments were the result of a handful of individual wavelengths of visible light being emitted by the atom. For example, when electricity is passed through hydrogen gas, the atoms emit a purple light. This light is composed of exactly 4 wavelengths: 410 nm (violet), 434 nm (indigo), 486 nm (blue), and 656 nm (red–orange). These wavelengths mix together to form a purple color.

What makes these results interesting is the fact that atoms seem to emit *individual* wavelengths of light. This is very unusual. Most colors that we see are the result of a *range* of wavelengths, not *individual* ones. For example, if we were to make a red light, we would probably paint a normal light bulb red. The paint would keep all of the wavelengths except the red ones from getting through. The light coming from this painted bulb would not be made up of just one individual wavelength. Instead, all wavelengths from 700 to 650 nm would be there. The fact that atoms emit only a handful of individual wavelengths indicates that something unique is at work.

Another interesting fact is that each element produces its own *unique* set of wavelengths when heated or electrified. While hydrogen emits visible light with wavelengths of 410 nm, 434 nm, 486 nm, and 656 nm, tantalum (Ta) emits visible light of only a single wavelength: 535 nm. This little fact is quite useful. After all, if you need to determine the atoms in a substance, one way to do it is to heat the substance up and determine the wavelengths of light that are emitted. Since each element in the substance will emit its own unique wavelengths of light, you should be able to match the elements with their respective wavelengths and eventually determine the elemental composition of the substance.

This is the main way that astronomers determine the elemental makeup of distant stars. Stars are bright because the substances that make them up are very hot and they therefore emit light. They emit so many different wavelengths of light that, to our eyes, the light appears white. However, a scientific instrument known as a spectrometer (spek trahm' uht er) can analyze the light and determine all of the individual wavelengths that make it up. Then, matching the elements to their expected individual wavelengths can determine the elemental composition of the star. This method of analysis, known as spectroscopy (spek trahs' kuh pee), can be applied to chemicals as easily as it is applied to stars. It is one of the most useful tools a chemist has to analyze unknown substances.

The ability of atoms to emit individual wavelengths of light also helped Niels Bohr develop his own model of the atom. Bohr took the planetary model developed by Rutherford and added his own little twist. Rutherford assumed that all of the electrons orbited around the protons in a single orbit. Bohr suggested, instead, that there were several possible orbits that the electrons could be in. Furthermore, he suggested, the electrons could jump from one orbit to another. In addition, by the time Bohr had come up with his model of the atom, the neutron had been discovered by Chadwick. Bohr put the neutrons and protons together in the center of the atom and called it the nucleus. This model is shown in Figure 3.12.

FIGURE 3.12
The Bohr Model of the Atom
Illustration by David Weiss

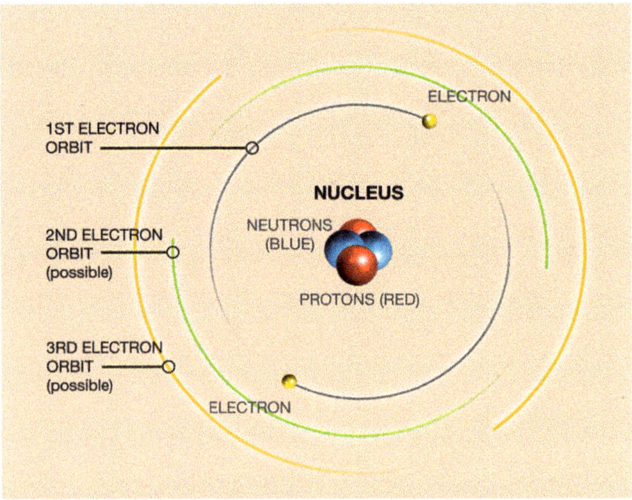

One very important advance that Bohr made in developing his model was his assumption that the electrons could only jump into and out of specific orbits. So, if an electron decided to jump out of orbit 1, it could not go just anywhere. It had to go into another allowed orbit, such as orbit 2. And even though it could be in orbit 1 and jump to orbit 2, *it could not be anywhere in between.* This type of assumption is called a quantum (kwahnt' uhm) assumption, and at the time it was considered rather odd. After all, if the electron could go to either orbit, why couldn't it go somewhere in between? Bohr had no answer to this question (indeed, even modern chemists have only a rough answer), but it was a necessary assumption in his model.

One of the first questions you might ask is this: How do the electrons jump from one orbit to another? The orbits that are farther away from the nucleus require more energy. If an electron wants to jump from an orbit close to the nucleus (orbit 1) to an orbit far away from the nucleus (orbit 3), it can do so by absorbing some energy. This is what happens when you heat a substance or pass electricity through it. When an electron moves from an orbit close to the nucleus to an orbit far from the nucleus, we say that the electron has been excited.

If, on the other hand, an electron would like to move from an orbit far away from the nucleus to an orbit close to the nucleus, the electron must somehow get rid of its extra energy. In other words, while in the orbit far from the nucleus, the electron has more energy than it would in an orbit close to the nucleus. To get to that closer orbit, the electron must find a way to get rid of that excess energy. It does this by emitting light. When an electron moves from an orbit far from the nucleus to an orbit close to the nucleus, we say it has de–excited.

When you heat a substance and its atoms emit light, you are witnessing a 2–step process. Electrons in the atoms first gain energy (because of the heat) and move into orbits that are far away from the nucleus. Then, after they sit in the far orbits for a short time, they decide they would rather be back in their original orbit, so they emit light to get rid of the excess energy and go back to their original orbit.

This is what makes the flames of a fire. When wood is burned, the chemicals in the wood are first changed into carbon dioxide gas and water vapor via a combustion reaction, as we have discussed before. However, heat is also produced in the reaction. This heat excites the electrons in the gases surrounding the fire, and they move to orbits farther away from their nucleus. However, after being there for a while, they decide to de–excite back to their original orbit. To do this, they must emit light. This is where the pleasant yellow–orange glow of the flames comes from.

You might think this is similar with colors you might see on a natural gas stove, but natural gas burns as a blue color naturally. The orange colors you might see come from incomplete combustion. You can examine another similar phenomenon to wood if you burn some wrapping paper. The dyes in most wrapping papers contain elements that are different from other papers. So while most papers burn with the same yellow–orange glow that wood burns with, wrapping paper typically burns with green, blue, or red flames because of the elements that make up the dyes in it.

Do you remember that Bohr assumed that an atom's electrons could only be in certain orbits? Since each orbit has a specific energy associated with it, this means that an electron can have only certain energies. If it is in orbit 1, it has a certain energy (E_1). If it is in orbit 2, it has another, higher energy (E_2). But since the electron can never be between orbit 1 and orbit 2, it can also never have any of the energies between E_1 and E_2. This assumption makes no sense at all because common sense tells us that any object should be able to attain any energy it wants, provided there is some way of transferring the necessary energy. Bohr said there were only certain energies that the electrons in the atom could have. This, another quantum assumption in Bohr's model, is the one we now see as critical to understanding atomic structure.

Bohr did not just dream up his model for the atom. In fact, for the hydrogen atom, he had detailed mathematical descriptions of the electron moving in its various orbits. He could use his mathematics to determine the energy of each electron orbit. He could determine how much energy the electron needed to jump from one orbit to another. Once he knew the energy necessary for the electron to jump from one orbit to the next, he could then determine the wavelength of light that the electrons must emit to de–excite. He found that the wavelengths he calculated were *exactly* equal to the wavelengths that chemists observed for de–exciting hydrogen atoms! This indicated that Bohr was on the right track, and his model for the atom quickly gained acceptance.

In fact, even though today's accepted model of the atom is slightly different from Bohr's model, the detailed mathematics that he developed are still taught in college chemistry today. These mathematics, which are a bit too detailed to tackle in this course, provide good insight into the physical mechanisms that hold the atom together. If you end up taking advanced chemistry, you will learn the mathematics of the Bohr model of the atom.

Additionally, the stylized pictures of atoms you often see are based on the Bohr model of the atom. Although science has moved past Bohr's model, the scientific community recognizes his achievements to this day.

THE QUANTUM MECHANICAL MODEL OF THE ATOM

Today's model of the atom, called the quantum mechanical model, is a bit more complicated than Bohr's model, but it still retains many of Bohr's concepts. For example, chemists still believe that electrons orbit around the nucleus and that there are many, many different orbits that electrons can continually jump between. In addition, we still say that the electrons need energy to go into orbits that are far away from the nucleus and that they need to release energy in the form of light to get back to orbits close to the nucleus. Finally, Bohr's quantum assumption about energy is still followed. We assume that there are only certain energies that an electron in an atom can have. An electron might have an energy of E_1 in one orbit and an energy of E_2 in the next higher orbit, but the electron can have no energy between E_1 and E_2.

Where the quantum mechanical model mainly differs from the Bohr model is the type of orbits that the electrons can occupy. In the Bohr model, electrons orbit the nucleus in circular orbits. In the quantum mechanical model, this is not the case. The way we theorize that electrons orbit the nucleus today is a little more complex. We assume that electrons do not orbit in fixed circles but instead orbit in clouds we call orbitals (or' buh tuls). When orbiting in a fixed circle, the electron is always the same distance from the nucleus. In today's model of the atom, electrons can, at different times, be at different distances from the nucleus but still be in the same orbital. However, if you were to watch the electron for quite a while, you would see that it stayed within a certain boundary.

If you are having trouble understanding this, think about it this way. Suppose you had a dog in your backyard and you drove a stake in the center of the yard and tied a rope to it. When the dog was tied to the rope, he could be anywhere in the backyard within a circle determined by the length of the rope. Although the dog could be anywhere inside the circle, he could not go outside the circle. This is like an electron in an orbital. The orbital has a general shape, and the electron can be anywhere within that shape. On the other hand, suppose the rope got wet during the winter and froze while it was stretched out. Then, because the rope was rigid and would always stay stretched out, the dog could not move anywhere except on the edge of the circle. This is like an electron orbiting in the Bohr model. It cannot be anywhere except on the edge of the circular orbit. That's the difference between electron orbits (the Bohr model) and electron orbitals (the quantum mechanical model).

The other difference between the Bohr model and the quantum mechanical model is the shape of the orbitals. In the Bohr model, all electrons had to orbit in circles. In the quantum mechanical model, there are several differently shaped orbitals that electrons can use to orbit around the nucleus. The shapes of these orbitals can become very complex, so we will limit our discussion to the first 3 types. The majority of atoms in nature use only these orbitals, so that will be enough for an introduction into this rather esoteric concept.

The simplest type of orbital an electron can occupy is called the s orbital. As illustrated in Figure 3.13, this orbital is spherical, with the nucleus at the center. An electron that occupies an s orbital can be anywhere inside the sphere but cannot venture outside of the sphere.

FIGURE 3.13
s Orbitals
Illustration by Megan Whitaker

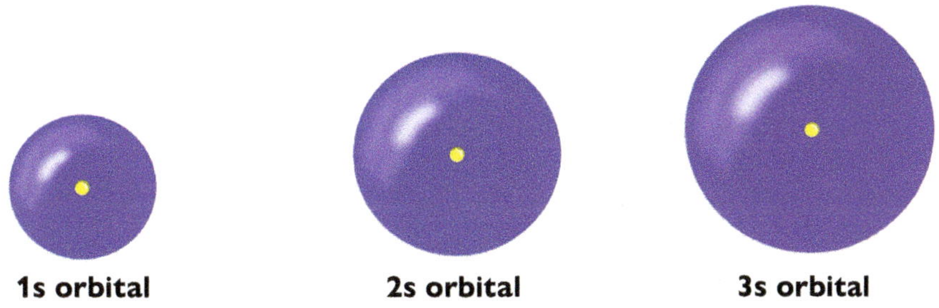

Notice in Figure 3.13 that there are 3 s orbitals, each of differing size. This is because in the quantum mechanical atom, just like in the Bohr atom, electrons can orbit far away from the nucleus or close to the nucleus depending on their energy. In the Bohr atom, the orbital number (1, 2, 3, etc.) determined how far they were from the nucleus. In the same way, the quantum mechanical model of the atom allows electrons to orbit far from the nucleus or close to it. If an electron orbits far away from the nucleus in a spherical pattern, it will occupy a large s orbital. If it orbits close to the nucleus in a spherical pattern, it will occupy a small s orbital. We distinguish small s orbitals from large ones with the number that appears next to the orbital letter. A 1s orbital is always smaller than a 2s orbital, which is always smaller than a 3s orbital.

Just like in the Bohr model, the farther away from the nucleus the electron is, the more energy it must have. The electrons in a 1s orbital have less energy than the electrons in a 2s orbital. Since the size of the orbital determines the energy of the electrons, the number that appears next to the orbital letter is often called the energy level of the atom.

The next orbital shape we will consider is the p orbital. This orbital is dumbbell–shaped, as illustrated in Figure 3.14. The nucleus is in the center of the dumbbell. Just as there are different sizes of s orbitals that correspond to different energies of electrons, there are also different sizes of p orbitals that correspond to different electron energies. One interesting thing about p orbitals is that there are none of them on the first energy level of the atom. This means there is no such thing as a 1p orbital. The first p orbital is the 2p orbital.

FIGURE 3.14
p Orbitals
3D Illustration by Michael Dailey

Another interesting thing about p orbitals is that for every energy level (or size of the orbital), there are 3 different p orbitals. They are all shaped the same, but they are oriented differently in space. One p orbital is oriented horizontally (as drawn in Figure 3.14), while the second one is oriented vertically. The third is oriented 90 degrees from both of the others.

In other words, each of the 3 p orbitals is oriented along a different axis in 3–dimensional space. Figure 3.15 illustrates how the 3 p orbitals all exist together in an atom.

FIGURE 3.15
Three Different p Orbitals in the Same Energy Level of an Atom
3D Illustration by Michael Dailey

So, as the figure shows, there can be 3 different p orbitals for each energy level. There are three 2p orbitals, three 3p orbitals that are all bigger than the 2p orbitals, three 4p orbitals that are all bigger than the 3p orbitals, and so on. All of this is a little obscure and may be hard to picture in your mind. If that is the case, don't be too concerned. At the end of this discussion, we will summarize what you need to know about the quantum mechanical model.

Finally, the last type of orbital we will use in this course is the d orbital. The shapes of the d orbitals are very complex, so we will not attempt to illustrate them. What you need to know about d orbitals is that there are 5 different d orbitals for each energy level. Also, the lowest–energy (smallest) d orbitals are located on the third energy level.

One important thing to note about the different shaped orbitals is that within the same atomic energy level, the different shaped orbitals require different amounts of energy. It is rather easy for an electron to whirl around the nucleus in a sphere, so s orbitals are the lowest–energy of all orbitals. On the other hand, a dumbbell is a little more complex, so p orbitals require more energetic electrons than s orbitals. Finally, since d orbitals have even more complex shapes, they require more energy than either s or p orbitals.

One more important thing to know about orbitals is that each individual orbital can hold only 2 electrons. After 2 electrons, the orbital is full, and any new electrons must find another orbital to fill. Since there are 3 different p orbitals on an energy level, 6 electrons could fit into all 3 p orbitals on that energy level (2 per orbital). In the same way, since there are 5 different d orbitals on an energy level, up to 10 electrons can go into the d orbitals of a given energy level.

Think a bit about all of these orbitals and how electrons use them. If an electron is orbiting as close as it can to the nucleus, we say that it is in the first energy level. On the first energy level, only 1 type of orbital exists: the s orbital. If an electron moves as close as it possibly can to the nucleus, then it must do so in a spherically shaped way (1s orbital). However, if the electron goes up to the next energy level (2), it is a little farther away from the nucleus and has the option of moving around the nucleus in a spherical way (2s orbital) or in 1 of any 3 dumbbell–shaped orbitals (2p orbitals). If the electron goes away from the nucleus even farther into the third atomic energy level, then it has the choice of being spherically shaped (3s orbital), 1 of 3 dumbbell–shaped orbitals (3p orbitals), or 1 of 5 more complex orbitals (3d orbitals).

This is the modern–day view of the atom. The atom has a nucleus of tightly packed protons and neutrons. Its electrons whirl around that nucleus in several different energy levels (1, 2, 3 . . .). Depending on the energy level, the electrons have several different shapes of orbitals (s, p, d, and other orbitals) in which they can travel around the nucleus. Each specific energy level (1, 2, 3 . . .) has its own energy requirements for the electrons. In addition, each orbital shape has additional energy requirements for the electrons. Based on the energy an electron has, it will go into the energy level and orbital whose energy requirements it best meets. Each orbital has a maximum capacity of 2 electrons. This is the summary of things that you must understand about the quantum mechanical model of the atom.

BUILDING ATOMS IN THE QUANTUM MECHANICAL MODEL (ELECTRON CONFIGURATIONS)

If all of the previous discussion confused you a bit, don't worry too much. In this section, we will try to apply what we have learned about the quantum mechanical model of the atom to describe what an atom might look like. After you have seen this application, you might gain a better understanding of what you read in the last section.

Before we start describing what atoms look like, there is one more fact of nature that you must learn:

All forms of matter try to stay in their lowest possible energy state.

This is a fact of nature that we will use over and over again. Think about it this way: Matter is basically lazy. It doesn't want to work any harder than it needs to. If at all possible, matter will always try to stay in its lowest–energy configuration.

For example, remember when we were discussing the light that was emitted from excited atoms? We said that when an atom is heated, its electrons absorb some of that kinetic energy and use it to travel to an orbital farther away from the nucleus. The electrons then emit light to go back down to their original orbital. Why do they do that? Once they jump to the higher–energy orbital, why don't they just stay there? They don't stay there because once they are in a higher–energy orbital, they are not in their lowest possible energy state. Therefore, the electrons release energy in the form of light to get back to their lowest energy state.

Since the lowest possible energy state is so important to matter, we give it a special name. We call it the ground state of the substance.

Ground state—The lowest possible energy state for a given substance.

Be sure you understand that the ground state is different for every substance on earth. For example, the ground state of a hydrogen atom is different from the ground state of a helium atom, as we will soon see.

Now that we know this very important fact about nature, we are finally ready to start building up atoms. First, let's describe the simplest atom in nature: the hydrogen-1 (^1H) atom. By looking at the periodic table, we can tell how many protons, neutrons, and electrons are in this atom. On the table, hydrogen's atomic number is 1. This means there are 1 proton and 1 electron in a hydrogen atom. Also, since the mass number of this hydrogen atom is 1, then there are 1 - 1 = 0 neutrons in the atom. We already know that a hydrogen atom has 1 proton and no neutrons in its nucleus. What about the electron? Where is it?

Since all matter in nature tries to get to its lowest energy state, the electron will go into the lowest energy orbital available. Energy level 1 is the lowest energy level, and energy level 1 has only an s orbital in it, so the electron goes into the 1s orbital. This means that the electron is orbiting relatively close to the nucleus and the orbit has the shape of a sphere. Therefore, the overall picture of a hydrogen atom in its ground state is 1 electron orbiting close to 1 proton, and the electron's orbital is spherical.

Now let's move to another simple atom, the helium–4 (^4He) atom. After looking at the table, we see that it has 2 protons and 2 electrons. In addition, since its mass number is 4, it has 4 - 2 = 2 neutrons. The 2 electrons will try to go into the lowest energy orbits available. Since energy level 1 is the lowest energy level and since it has only an s orbital in it, then both electrons will go into the 1s orbital. Our picture for the helium–4 atom in its ground state is 2 electrons orbiting close to a nucleus tightly packed with 2 protons and 2 neutrons. The electrons whirl about this nucleus in a spherical orbit.

Continuing, we can consider a lithium–6 (^6Li) atom. After looking at the table and dealing with the mass number, we see that it has 3 protons, 3 electrons, and 3 neutrons. The 3 electrons will try to go into the lowest energy orbits available. Since energy level 1 is the lowest energy level and it has only an s orbital, 2 electrons will go into the 1s orbital. However, any single orbital can hold only 2 electrons. The third electron in the atom cannot fit into the 1s orbital. It must, therefore, find the next lowest energy orbital to go into. Since there is no more room on the first energy level, it goes into energy level 2. At this point, it can either go into a 2s orbital or a 2p orbital. Since s orbitals are lower in energy than p orbitals, it will go into the 2s orbital. Our picture for the lithium–6 atom is 2 electrons orbiting close to and 1 electron orbiting farther away from the nucleus tightly packed with 3 protons and 3 neutrons. All 3 electrons whirl about this nucleus in spherical orbitals.

Do you see how the thinking goes here? Let's try a much more complicated atom, and we will then introduce an abbreviated notation that will make things a little easier to describe. Consider a ^{22}Ne atom . According to our rules, it has 10 protons, 10 electrons, and 12 neutrons. Where do the electrons go?

The electrons will look for the lowest–energy orbitals available. The first 2 will occupy the 1s orbital since energy level 1 is the lowest energy level available. Since only 1 s orbital is available on energy level 1, the rest of the electrons will have to go into higher energy levels. On energy level 2, both s and p orbitals are available. Since the s orbitals are lower in energy, the next 2 electrons will fill that orbital. So far, we have placed 4 electrons in their orbitals, and we have 6 to go. On energy level 2, there are 3 p orbitals that can also be filled. Each p orbital can hold 2 electrons, so the 6 remaining electrons can go into the 2p orbitals.

Our picture of the ^{22}Ne atom is a tightly-packed nucleus of 10 protons and 12 neutrons with 10 electrons whirling around it. Two of the electrons are orbiting the nucleus closely in a spherical pattern. Two more electrons are whirling around the nucleus a bit farther away, also in a spherical pattern. However, 6 more electrons are whirling about in dumbbell-shaped clouds. This is the way a modern-day chemist would describe a ^{22}Ne atom. Figure 3.16 illustrates what this would look like.

FIGURE 3.16
The Quantum Mechanical View of a ^{22}Ne Atom
3D Illustration by Michael Dailey

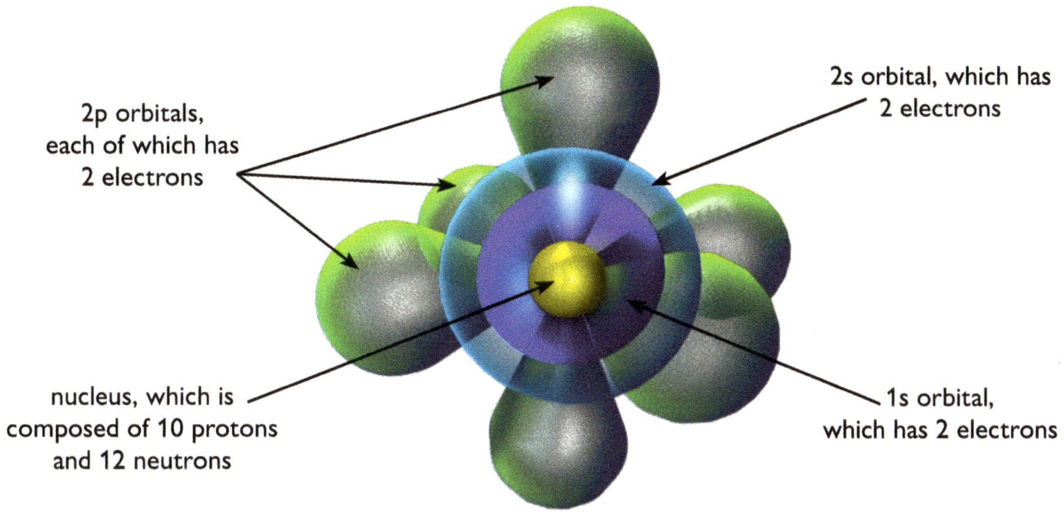

Hopefully you now have a reasonably good grasp of the quantum mechanical model of the atom. If you understand how we came up with the description of the ^{22}Ne atom, then you are all set. If you don't quite understand it, then you need to reread the material or seek out another source because you can't go further in this course without understanding this important concept.

Since the quantum mechanical model of the atom is complicated, we will simplify it a bit with some notation. We will learn in the next module that all of the chemical behavior of an atom is governed solely by the number of electrons it has. As a result, the nucleus of an atom is relatively unimportant to a chemist. When giving the quantum mechanical description of an atom, chemists do not bother to mention the nucleus. Also, rather than saying in words where all of the electrons go, we develop an abbreviation to help us. The abbreviation for the neon atom we just discussed is this:

$$1s^2\ 2s^2\ 2p^6$$

This says all a chemist needs to know about the structure of a neon atom. The numbers to the left of the letters represent the energy level, the letters represent the shape of the orbital, and the superscript numbers tell how many electrons are in the orbital. The abbreviation above tells that there are 2 electrons in the 1s orbital, 2 electrons in the 2s orbital, and 6 electrons in the 2p orbitals. This type of abbreviation is called an electron configuration, and you are going to become *very* familiar with this kind of notation in this and the next module.

Before we do a few sample electron configurations to show you how easy this is, you need to be introduced to one more interesting fact in chemistry. Remember how we decided which electrons in an atom went into which orbitals? We did it by using energy arguments. We said that the first energy level should fill up before electrons go into the second energy level. In the same way, the second energy level should fill all of its orbitals before electrons go into the third energy level. This is all true. However, at the third energy level, things get a little messy. The d orbitals in the third energy level take a lot of energy. In fact, the 3d orbitals require more energetic electrons than the 4s orbitals. If we had enough electrons to fill up the 3s and 3p orbitals, we would then start filling the 4s orbitals before we went back to fill up the 3d orbitals. Messy, isn't it?

As atoms get more and more electrons, the order of orbitals gets even more complicated. To be able to determine the proper electron configuration of any atom, then, we need to have some way of determining the order in which orbitals fill up. Believe it or not, we *already have that* because we have the periodic table. You see, the periodic table gets its weird shape *because* of electron configurations. By looking at the table, we can easily determine an atom's electron configuration, which means we can avoid all of the messy energy arguments we have been making. These energy arguments, of course, still form the *reason* that electron orbitals fill up in the way that they do. However, we needn't go through those arguments every time we want to determine an atom's electron configuration. All we have to do is look at the table.

Of course, to do this, you have to see why the table is arranged the way it is. Examine Figure 3.17 to understand how the periodic table helps you with electron configurations.

FIGURE 3.17

Periodic Table of Elements and Orbitals

s block

	1A	2A
1	1 1s	
2	3 2s	4 2s
3	11 3s	12 3s
4	19 4s	20 4s
5	37 5s	38 5s
6	55 6s	56 6s
7	87 7s	88 7s

d block

3B	4B	5B	6B	7B	8B			1B	2B
21 3d	22 3d	23 3d	24 3d	25 3d	26 3d	27 3d	28 3d	29 3d	30 3d
39 4d	40 4d	41 4d	42 4d	43 4d	44 4d	45 4d	46 4d	47 4d	48 4d
57 5d	72 5d	73 5d	74 5d	75 5d	76 5d	77 5d	78 5d	79 5d	80 5d
89 6d	104 6d	105 6d	106 6d	107 6d	108 6d	109 6d	110 6d	111 6d	112 6d

p block

3A	4A	5A	6A	7A	2A
					2 1s
5 2p	6 2p	7 2p	8 2p	9 2p	10 2p
13 3p	14 3p	15 3p	16 3p	17 3p	18 3p
31 4p	32 4p	33 4p	34 4p	35 4p	36 4p
49 5p	50 5p	51 5p	52 5p	53 5p	54 5p
81 6p	82 6p	83 6p	84 6p	85 6p	86 6p
113 7p	114 7p	115 7p	116 7p	117 7p	118 7p

f block

58 4f	59 4f	60 4f	61 4f	62 4f	63 4f	64 4f	65 4f	66 4f	67 4f	68 4f	69 4f	70 4f	71 4f
90 5f	91 5f	92 5f	93 5f	94 5f	95 5f	96 5f	97 5f	98 5f	99 5f	100 5f	101 5f	102 5f	103 5f

Looking at Figure 3.17, you should first see that for the purposes of this discussion, we pretend that helium sits right next to hydrogen on the table. The next thing you should see in the figure is that the columns have been separated. We call the first 2 columns the s orbital block because the last electrons in the atoms of that block end up in an s orbital. Since each s orbital can have only 2 electrons in it, the s orbital block can be only 2 columns wide. The next group represents those atoms whose last electrons end up in d orbitals. Since there are 5 d orbitals and each can have 2 electrons, there are 10 elements in each row that have their last electrons in d orbitals. Finally, the last block is the p orbital block. These elements have their last electrons in a p orbital. Since the 3 p orbitals can each hold 2 electrons, the p orbital block is 6 columns wide. The 2 lowest rows on the periodic table (the lanthanides and actinides) are not going to appear on your test. However, if you are interested, their last electrons go into f orbitals, which are even more complex than d orbitals.

Since the columns represent the electron orbitals in the atoms , the rows of the table represent the energy level that those electrons are in. For example, notice that Ne is in row 2. This means that the last electron in Ne goes in the second energy level. The fact that Ne is in a column that is a part of the p orbital block tells us that the last electron is in a p orbital. Neon's last electron goes in the 2p orbital. We figured this out earlier with energy arguments, but figuring it out this way is easier.

To use the arrangement of the periodic table to determine electron configurations, you just have to look at the periodic table and find the atom whose electron configuration you wish to determine. Then, starting with hydrogen, walk through the table, assigning electrons in each orbital one box at a time. When you get to the box that represents the atom you are interested in, you're done. The only trick you have to remember is that in the d orbital block, the row that the elements are on is one number higher than the energy level that the electrons are in. So, when filling up d orbitals, subtract 1 from the row number to get the energy level. If this all sounds a bit confusing, Example 3.4 should clear things up a bit.

EXAMPLE 3.4

Write the electron configuration for a carbon atom.

First, we look at the periodic table and find out that carbon (C) is element 6. To get to this atom, we start walking through the table from the beginning. The first 2 elements are H and He. Their boxes are in row 1 and the s orbital group. The first 2 electrons go into the 1s orbital. Going on through the table, we get to Li and Be. They are in row 2 and the s orbital group, so we put 2 electrons in the 2s orbital. Continuing on, we hit B and then C, where we want to stop. This means that we went through 2 boxes in the row 2, p orbital block. So there are 2 electrons in the 2p set of orbitals. Since we've hit carbon, we are done. So the electron configuration is:

$$1s^2\ 2s^2\ 2p^2$$

Remember, the superscripts in the electron configuration represent the number of electrons in each orbital. The sum of the superscripts must equal the number of electrons in the atom. Carbon has 6 electrons, and our superscripts add up to 6. Our answer, therefore, checks out.

Write the electron configuration for a Cl atom.

To get to Cl (17), we have to go through several boxes. We must walk through all of row 1, which has 2 boxes in the s orbital block. That gives us $1s^2$. We must also go through all of row 2, which has 2 boxes in the s orbital block and 6 boxes in the p orbital block. This gives us $2s^2 2p^6$. We also must go through the s orbital block in row 3. This gives us $3s^2$. Finally, to get to Cl, we go through 5 boxes in the row 3, p orbital block. This gives us $3p^5$. The electron configuration, then, is:

$$1s^2\ 2s^2\ 2p^6\ 3s^2\ 3p^5$$

The superscripts add up to the number of electrons in a Cl atom (17), so our answer checks out.

Give the electron configuration of a Ag atom.

To get to Ag, we have to walk through several boxes on the table. We must walk through all of row 1, which has 2 boxes in the s orbital block. That gives us $1s^2$. We must also go through all of row 2, which has 2 boxes in the s orbital block and 6 boxes in the p orbital block. This gives us $2s^2 2p^6$. Row 3 has 2 boxes in the s orbital group and 6 in the p orbital box, so that gives us $3s^2 3p^6$. Row 4 has 2 boxes in the s orbital block, 10 in the d orbital block, and 6 in the p orbital block. Remembering that for d orbitals, we must subtract 1 from the row number, this gives us $4s^2 3d^{10} 4p^6$. In row 5, we must go through the 2 boxes in the s orbital block, so that gives us $5s^2$. In row 5, we also have to go through 9 boxes to get to Ag. Once again, we subtract 1 from the row number with d orbitals, so this gives us $4d^9$. In the end, then, our electron configuration is:

$$1s^2\ 2s^2\ 2p^6\ 3s^2\ 3p^6\ 4s^2\ 3d^{10}\ 4p^6\ 5s^2\ 4d^9$$

Our superscripts add up to 47, so our answer checks out.

This is all you do to get the electron configuration of an atom. With the periodic table as a guide, it is not a hard task at all. However, let's give you a couple of words of advice:

1. Some chemistry books teach a different method to get electron configurations. Their method requires memorization and is not the way real chemists do electron configurations. That's why we don't teach it their way. Don't get confused if you look at other books, however. Their method might be different, but the answers should be the same.

2. Remember that these electron configurations are not just a worthless exercise. They have very, very deep meaning to chemists. They not only tell the structure of a particular atom, but, as we will see in the next module, they also tell a great deal about the chemistry of the atom.

See if you can do electron configurations in "On Your Own" questions 3.8–3.10. Feel free to use the periodic table, but do not use Figure 3.17. You should be able to do electron configurations from a normal table, not one specially enhanced for electron configurations.

ON YOUR OWN

3.8 Write the electron configuration for silicon (Si).

3.9 Write the electron configuration for selenium (Se).

3.10 What is the electron configuration of silver (Ag)?

ABBREVIATED ELECTRON CONFIGURATIONS

Before we leave electron configurations, let's give you a handy abbreviation that will keep you from writing a lot when you give your electron configurations. The last column in the table, column 8A, is very important. We will see why it is so important in the next module. For now, just realize that most chemists gauge an atom by how far it is away from this column in the table. One way you can abbreviate electron configurations is to look for the nearest atom in 8A that has a *lower* atomic number than the atom you are interested in. Then, you simply add any extra electrons on top of that atom's configuration. See how we do this in Example 3.5.

EXAMPLE 3.5

Write the abbreviated electron configuration for Sr.

To abbreviate an electron configuration, we just find the nearest 8A element that has a lower atomic number than the atom we are interested in. In this case, Kr fits the bill. We can then say that the only difference between Sr and Kr is that there are 2 more boxes in the row 5, s orbital block. We can write the electron configuration as:

$$[\text{Kr}]\ 5s^2$$

Putting Kr in square brackets tells a chemist that you are abbreviating the electron configuration in terms of the 8A element Kr.

Write the abbreviated electron configuration for Se.

The nearest 8A element that has a lower atomic number than Se is Ar. Don't be fooled here. Kr is the closest 8A element, but it does not have a *lower* atomic number. The only differences between Se and Ar are that there are 2 boxes in the row 4, s orbital group, 10 boxes in the row 4, d orbital group, and 4 boxes in the row 4, p orbital group. The abbreviated electron configuration for Se is:

$$[\text{Ar}]\ 4s^2\ 3d^{10}\ 4p^4$$

ON YOUR OWN

3.11 What is the abbreviated electron configuration for Br?

3.12 What is the abbreviated electron configuration for In?

Answer "On Your Own" questions 3.11–3.12 to cement the idea of abbreviated electron configurations in your mind.

As we said before, you will soon learn that the meaning of electron configurations is very deep, so your ability to predict them is a necessity.

THE AMAZING DESIGN OF ATOMS

Before we leave this module, there are a couple of interesting things to tell you about atoms. First, the atom is made up mostly of empty space. Since matter is made up entirely of atoms, this also means that matter is mostly empty space. This may sound crazy. After all, when you hit your head against the wall, the wall certainly doesn't feel like empty space. Nevertheless, the statement is true.

Nuclear chemistry experiments have demonstrated that the nucleus of a hydrogen atom has a radius of about 1×10^{-15} meters. On the other hand, the mathematics of the quantum mechanical model of the atom tell us that the radius of the 1s orbital in that same atom (the closest orbital to the nucleus) is about 5×10^{-11} meters. On the atomic scale, the electrons are *very far away* from the nucleus. If you don't understand the implications of these numbers, think about it this way: If the nucleus of the hydrogen atom were 1 inch in diameter, the electron would be orbiting it about 3/4 of a mile away! Between the nucleus and the electron there would be absolutely nothing. No air, no matter. Nothing. Just empty space. Therefore, the vast majority of an atom is simply empty space.

The most interesting aspect of an atom is not that it is made up of mostly empty space. The most interesting aspect of the atom is how well designed it is. You see, for all atoms except ^{1}H to exist, they must have neutrons. The neutron is just slightly more massive than the proton. It's a good thing, too. Nuclear chemistry experiments indicate that the neutron must be more massive than the proton, or it would not exist. Calculations indicate that if the mass difference between the neutron and the proton were changed by as little as 0.2%, all neutrons would spontaneously decay into protons, and the only atom in the universe would be ^{1}H. If that were the case, no life could exist in the universe! (See the Apologia book extras website mentioned in the Student Notes for more information.)

The mass of the electron is also quite crucial for the existence of life. Remember, the electrons whirl around the nucleus in the orbitals that we just discussed. The *size* of those orbitals depends on the strength of the electrical attraction between the protons and electrons, as well as the mass of the electron. If the electron's mass were larger, the size of the orbitals would be smaller. If the electron's mass were smaller, the size of the orbitals would be larger. Why does that matter?

If the electron's mass were larger, the electron orbitals would be smaller. As a result, atoms would hold on to their electrons more tightly, and it would be difficult to form ionic compounds. Many of the chemical processes that occur in living organisms depend on the existence of ions from ionic compounds. Without these ionic compounds, the chemical processes that depend on them would cease, and living organisms would die.

On the other hand, most of the compounds that make up living organisms are covalent compounds. Atoms in covalent compounds share their electrons. If the electron mass were smaller, the electron orbitals would be larger and, as a result, atoms would hold onto their electrons rather loosely. This would make it very easy to form ionic compounds but very hard to form covalent compounds. If that were the case, living organisms could not even form, much less exist!

In the end, then, the electron mass is just right to produce electron orbitals that are neither too big nor too small. The balance between ionic and covalent molecules is perfect for the formation and existence of life, due to the mass of the electron. Estimates indicate that a variation of as little as 2% in the mass of the electron would make it impossible for life as we know it to exist! (Corey 2002)

What is really amazing about the perfect charge balance between the proton and the electron is that these 2 particles are incredibly different. The proton is about 2,000 times heavier than the electron. In addition, the proton seems to be made up of 3 smaller particles called quarks, while as far as we can tell, the electron is not made up of any smaller particles. Nevertheless, although these 2 particles are quite different, they have *perfectly* balanced charges!

The most important thing to realize is that these atoms are the *simplest* building blocks of matter. Even these very simple building blocks have delicately balanced parts that all work together to give the atom its properties. If one of these parts were even slightly altered, the delicate balance would be destroyed, and the chemistry necessary for life would be forever lost. How is it, then, that these 2 incredibly different particles are able to carry *exactly* the right amount of charge to make a stable atom? How is it that the neutron has *precisely* the right amount of extra mass to make the atom stable? How is it that the electron has *precisely* the right mass to make the existence of living organisms possible? Without these 3 incredible "coincidences," life could never exist in the universe. Nevertheless, we are here.

As noted before, there are many such incredible "coincidences" in nature. There are so many that it might seem impossible for anyone who really understands physics or chemistry to believe that all this we are studying just came about by chance! The universe is too perfectly designed for that. From the smallest atom to the largest galaxy, incredible design features abound in our universe. These design features are the fingerprints of God, making all who objectively study the universe aware of His existence.

SUMMARY OF KEY EQUATIONS IN MODULE 3

Equation 3.1: $f = \frac{c}{\lambda}$ Frequency and wavelength relationship

Equation 3.2: $E = h \times f$ Frequency and energy relationship

ANSWERS TO THE "ON YOUR OWN" QUESTIONS

3.1 (a) On the periodic table, Mo has an atomic number of 42.
So a Mo atom has 42 protons and 42 electrons.

(b) On the periodic table, Be has an atomic number of 4.
So a Be atom has 4 protons and 4 electrons.

(c) On the periodic table, Hf has an atomic number of 72.
So an Hf atom has 72 protons and 72 electrons.

3.2 (a) On the periodic table, Ne has an atomic number of 10. This means it has 10 protons and 10 electrons. Its mass number, according to the problem, is 22. If it has 22 total protons + neutrons, and it has 10 protons, then it has 22 - 10 = 12 neutrons.

(b) On the periodic table, fluorine (F) has an atomic number of 9. This means it has 9 protons and 9 electrons. Its mass number, according to the problem, is 19. If it has 19 total protons + neutrons, and it has 9 protons, then it has 19 - 9 = 10 neutrons.

(c) On the periodic table, U has an atomic number of 92. This means it has 92 protons and 92 electrons. Its mass number, according to the problem, is 235. If it has 235 total protons + neutrons, and it has 92 protons, then it has 235 - 92 = 143 neutrons.

3.3 First, since it has 33 protons, we know its atomic number is 33. The element on the table which has an atomic number of 33 is arsenic (As). The problem states that this atom also has 41 neutrons; its mass number is 33 + 41 = 74. Therefore, the atom is arsenic–74, which has a symbol of ^{74}As.

3.4 To solve this, we use equation 3.1:

$$f = \frac{c}{\lambda}$$

But we have to rearrange it so that we are solving for wavelength:

$$\lambda = \frac{c}{f}$$

Now we can plug in the numbers:

$$\lambda = \frac{3.0 \times 10^{8} \frac{m}{\cancel{s}}}{6.4 \times 10^{14} \frac{1}{\cancel{s}}} = 4.7 \times 10^{-7} m$$

Figure 3.10 lists wavelength in nm. If we refer to the chart, we must convert to nm:

$$\frac{4.7 \times 10^{-7} \cancel{m}}{1} \times \frac{1 \text{ nm}}{1 \times 10^{-9} \cancel{m}} = 4.7 \times 10^{2} \text{ nm}$$

According to Figure 3.10, this light is blue light.

3.5 To solve this problem, we need to simply convert the wavelength range into a frequency range. To do this, we just convert the 2 wavelengths in the range into frequencies:

$$\frac{7.00 \times 10^{2} \cancel{nm}}{1} \times \frac{1 \times 10^{-9} \text{ m}}{1 \cancel{nm}} = 7.00 \times 10^{-7} \text{ m}$$

Now that our units agree, we can put the numbers into the equation:

$$f = \frac{3.0 \times 10^{8} \frac{\cancel{m}}{s}}{7.00 \times 10^{-7} \cancel{m}} = 4.3 \times 10^{14} \frac{1}{s}$$

Now we convert the other part of the range:

$$\frac{6.5 \times 10^{2} \cancel{nm}}{1} \times \frac{10^{-9} \text{ m}}{1 \cancel{nm}} = 6.50 \times 10^{-7} \text{ m}$$

Now that our units agree, we can put the numbers into the equation:

$$f = \frac{3.0 \times 10^{8} \frac{\cancel{m}}{s}}{6.50 \times 10^{-7} \cancel{m}} = 4.6 \times 10^{14} \frac{1}{s}$$

So the frequency range is between 4.3×10^{14} Hz and 4.6×10^{14} Hz.

3.6 This problem is a direct application of equation 3.2. Since we know both E and h, we simply have to use algebra and rearrange the equation to solve for f:

$$f = \frac{E}{h}$$

$$f = \frac{3.4 \times 10^{-14} \cancel{J}}{6.63 \times 10^{-34} \frac{\cancel{J}}{Hz}} = 5.1 \times 10^{19} \text{ Hz}$$

Figure 3.11 shows that this is the frequency of X–rays. If you have ever had a medical X–ray, light of approximately this energy was projected onto you to produce the picture that the doctor or dentist examined.

3.7 In this problem, we are given the energy and asked to determine the color. The only thing that can tell us color is wavelength. Equation 3.2 only relates energy and frequency. So we must first determine the frequency, and then we can use equation 3.1 to figure out the wavelength.

To determine the frequency:

$$f = \frac{E}{h}$$

$$f = \frac{3.3 \times 10^{-19}\,\cancel{J}}{6.63 \times 10^{-34}\,\frac{\cancel{J}}{Hz}} = 5.0 \times 10^{14}\ Hz$$

Now that we have the frequency, we can use algebra to rearrange equation 3.1 so that we can determine the wavelength:

$$\lambda = \frac{c}{f}$$

$$\lambda = \frac{3.0 \times 10^{8}\ \frac{m}{s}}{5.0 \times 10^{14}\ Hz} = \frac{3.0 \times 10^{8}\frac{m}{\cancel{s}}}{5.0 \times 10^{14}\frac{m}{\cancel{s}}} = 6.0 \times 10^{-7}\ m$$

To use Figure 3.10, we must convert this answer to nm because that's the unit used in the figure. Remember, 1 nm = 10^{-9} m:

$$\frac{6.0 \times 10^{-7}\,\cancel{m}}{1} \times \frac{1\ nm}{1 \times 10^{-9}\,\cancel{m}} = 6.0 \times 10^{2}\ nm$$

According to Figure 3.10, light with this wavelength is yellow light.

3.8 To get to Si (14), we have to go through several boxes. We must walk through all of row 1, which has 2 boxes in the s orbital block. That gives us $1s^2$. We must also go through all of row 2, which has 2 boxes in the s orbital block and 6 boxes in the p orbital block. This gives us $2s^2 2p^6$. We also must go through the s orbital block in row 3. This gives us $3s^2$. Finally, to get to P, we go through 2 boxes in the row 3, p orbital block. This gives us $3p^2$. The electron configuration, then, is:

$$1s^2\ 2s^2\ 2p^6\ 3s^2\ 3p^2$$

The superscripts add up to the number of electrons in a Si atom (14), so our answer checks out.

3.9 To get to Se, we have to walk through several boxes on the table. We must walk through all of row 1, which has 2 boxes in the s orbital block. That gives us $1s^2$. We must also go through all of row 2, which has 2 boxes in the s orbital block and 6 boxes in the p orbital block. This gives us $2s^2 2p^6$. Row 3 has 2 boxes in the s orbital group and 6 in the p orbital box, so that gives us $3s^2 3p^6$. Row 4 has 2 boxes in the s orbital block and 10 in the d orbital block that we must go through. Remembering that for d orbitals we must subtract 1 from the row number, this gives us $4s^2 3d^{10}$. Finally, to get to Se, we must walk through 4 boxes in the row 4, p orbital block. That gives us $4p^4$. Therefore, the final electron configuration is:

$$1s^2\ 2s^2\ 2p^6\ 3s^2\ 3p^6\ 4s^2\ 3d^{10}\ 4p^4$$

Our superscripts add up to 34, so our answer checks out.

3.10 To get to Ag, we have to walk through several boxes on the table. We must walk through all of row 1, which has 2 boxes in the s orbital block. That gives us $1s^2$. We must also go through all of row 2, which has 2 boxes in the s orbital block and 6 boxes in the p orbital block. This gives us $2s^2 2p^6$. Row 3 has 2 boxes in the s orbital group and 6 in the p orbital box, so that gives us $3s^2 3p^6$. Row 4 has 2 boxes in the s orbital block, 10 in the d orbital block, and 6 in the p orbital block. Remembering that for d orbitals we must subtract 1 from the row number, this gives us $4s^2 3d^{10} 4p^6$. In row 5, we must go through the 2 boxes in the s orbital block, so that gives us $5s^2$. In row 5, we also have to go through 9 d orbital boxes to get to Ag. Once again, we subtract 1 from the row number with d orbitals, so this gives us $4d^9$. Our electron configuration is:

$$1s^2\ 2s^2\ 2p^6\ 3s^2\ 3p^6\ 4s^2\ 3d^{10}\ 4p^6\ 5s^2\ 4d^9$$

Our superscripts add up to 47, so our answer checks out.

3.11 To abbreviate an electron configuration, we just find the nearest 8A element that has a lower atomic number than the atom we are interested in. In this case, Ar fits the bill. We can then say the only differences between Br and Ar are that there are 2 more boxes in the row 4, s orbital block, 10 boxes in the row 4, d orbital block, and 5 boxes in the row 4, p orbital block. We can write the electron configuration as:

$$[Ar]\ 4s^2\ 3d^{10}\ 4p^5$$

3.12 The nearest 8A element that has a lower atomic number than In is Kr. The only differences between In and Kr are that there are 2 boxes in the row 5, s orbital group, 10 boxes in the row 5, d orbital group, and 1 box in the row 5, p orbital group. The abbreviated electron configuration for In is:

$$[Kr]\ 5s^2\ 4d^{10}\ 5p^1$$

STUDY GUIDE FOR MODULE 3

REVIEW QUESTIONS

1. What was Rutherford's main contribution to the study of atomic structure?

2. What do we call the experimental apparatus that William Crookes used in his experiments, and what did he discover with it?

3. If 2 electrically charged particles repel one another, what can we conclude about their charges?

4. If a substance has 9 positive charges and 11 negative charges, is it positively charged, negatively charged, or neutral?

5. Give the location within the atom for a proton, electron, and a neutron.

6. Why is it so hard to separate one isotope from another?

7. What were the differences between the plum pudding model of the atom and the planetary model of the atom?

8. What is meant by saying that wavelength and frequency are inversely proportional to each other?

9. If you have an orange light bulb and a violet one, which emits waves with the largest wavelength? Which emits light of higher frequency? Which emits the higher energy light? (Do *not* look at the figures in the book. Just remember the name you were told to remember!)

10. Two light bulbs emit light of the same color. However, one is much brighter than the other. What can you say about the wavelengths and amplitudes of the waves being emitted by each bulb?

11. If an atom absorbs energy, what happens to its electrons? If the atom is emitting light, what are its electrons doing?

12. What is the concept of ground state and why is it so important in chemistry?

13. The 3 fundamental particles that make up the atom are the proton, neutron, and electron. Order them in terms of decreasing mass.

14. Define and give one example of a physical constant.

15. Describe what an orbital is.

PRACTICE PROBLEMS

$$(c = 3.0 \times 10^{8}\ \frac{m}{s};\ \textit{nano}\text{– means } 10^{-9};\ h = 6.63 \times 10^{-34}\ \frac{J}{Hz})$$

1. Give the number of protons, electrons, and neutrons in the following atoms:

   a. ^{90}Zr b. ^{22}Ne c. ^{58}Ni d. ^{222}Rn

2. Which of the following atoms are isotopes?

   ^{22}Na, ^{22}Ne, ^{23}Na, ^{22}Mg, ^{24}Na

3. What is the symbol of the atom made up of 6 protons, 8 neutrons, and 6 electrons?

4. If light has a frequency of 2.4×10^{14} Hz, what is its wavelength?

5. What is the energy of a light wave with a frequency of 9.8×10^{20} Hz?

6. If a light wave has a wavelength of 712 nm, what is its energy?

7. One electron is in a 2s orbital, while another is in a 3p orbital. Which has the higher energy? What shape is each electron's orbital?

8. Give the full electron configurations of the following atoms:

   a. V b. N c. Cs

9. Give the abbreviated electron configurations for the following atoms:

   a. Ti b. S c. Rb

10. What is wrong with the following electron configurations?

    a. $1s^2\ 2s^2\ 2p^6\ 3s^2\ 3p^7\ 4s^2\ 4d^9$

    b. $1s^2\ 2s^2\ 2p^6\ 3s^2\ 3p^6\ 3d^{10}\ 4s^2\ 4p^5$

MOLECULAR STRUCTURE

In the last module, we learned a little about what atoms would look like if we were able to actually see them. We learned that atomic structure is very complex and depends mostly on the way that the atom's electrons are arranged in their orbitals. Now that we have this information, we can undertake the task of determining what molecules would look like if we could actually see them.

first reactions

Although it may seem large, the periodic table isn't that big when you consider that practically everything visible and invisible in our universe is created from the elements it displays. That is because the elements are capable of combining in so many different ways to form molecules. The periodic properties will help you determine how the elements can combine.

Since you learned that atomic structure is very complex, you might think that molecular structure is complex as well. Although this is true to some extent, you will be happy to know that in terms of what you will learn in this course, molecular structure is simple in comparison to atomic structure. Indeed, determining the structure of a molecule is much like putting together a jigsaw puzzle. Like most things you learn in this course, there is a fair amount of background material you must understand before we tackle the actual subject of molecular structure.

ELECTRON CONFIGURATIONS AND THE PERIODIC TABLE

In the last module, you became efficient at determining an atom's electron configuration by looking at the periodic table. Now you need to learn a little about what a chemist can learn by looking at an electron configuration. To do this, let's take a look at the electron configuration of bromine:

$$1s^2\ 2s^2\ 2p^6\ 3s^2\ 3p^6\ 4s^2\ 3d^{10}\ 4p^5$$

You have already learned what all of this means in terms of atomic structure, but what does it mean in terms of the chemical behavior of the compound?

Think for a minute about the electrons in a bromine atom. There are 35 of them, and they are all whirling around the nucleus. The electron configuration shows that they are whirling about in 3 different orbital shapes (because s, p, and d orbitals are all present) and at 4 different distances from the nucleus (since energy levels 1–4 are present). When a bromine atom wants to join with another atom to make a molecule, what part of the bromine atom will play the most important role in this joining? Clearly, the electrons will. After all, both the bromine atom and whatever atom it decides to join with have electrons whirling outside the nucleus. When these atoms move close to one another, their electrons will be the first things to come into contact with each other.

When joining to form a molecule, atoms never get close enough to each other for their nuclei to interact. As a result, when forming molecules, the only important part of the atom is its electrons. The nucleus of the atom plays no role whatsoever in making the molecule. In fact, not even *all* of the electrons are important. The only electrons that significantly affect the chemistry of an atom are those farthest away from the nucleus. Those are the electrons that will begin to interact with another atom's electrons.

If you look back at the electron configuration for bromine, which electrons are farthest away from the nucleus? In bromine's electron configuration, energy levels 1–4 are present. According to what we learned in the last module, the larger the energy level's number, the farther it is from the nucleus. Therefore, the electrons in energy level 4 are the ones that are farthest from the nucleus. These are the ones that affect the chemistry of bromine. Although the electrons in energy levels 1–3 are present, they do little to affect the way in which bromine behaves chemically.

Since the electrons farthest away from the nucleus are so important, we give them a name: valence (vay' lents) electrons.

Valence electrons—The electrons that exist farthest from an atom's nucleus. They are the electrons with the highest energy level number.

Since bromine has 2 electrons in the 4s orbital and 5 electrons in the 4p orbitals, it has 7 valence electrons. Those 7 electrons are responsible for almost all of bromine's chemical behavior. This is a very important point; you must remember it:

An atom's chemistry or chemical behavior is determined mostly by the number of valence electrons it has.

One thing we learn from the electron configuration of an atom is how many valence electrons it has. This fact will help us determine the atom's chemical behavior.

Now let's look at a few other atoms and their electron configurations:

F: $1s^2\ 2s^2\ 2p^5$
Cl: $1s^2\ 2s^2\ 2p^6\ 3s^2\ 3p^5$
I: $1s^2\ 2s^2\ 2p^6\ 3s^2\ 3p^6\ 4s^2\ 3d^{10}\ 4p^6\ 5s^2\ 4d^{10}\ 5p^5$

How many valence electrons does each of the above atoms have? If you count the number of electrons in each atom's highest energy level, you will see that they all have 7 valence electrons. Notice also that these atoms are all in the same column of the periodic table. This is no accident. The periodic table is arranged so that all atoms in a given column have the same number of valence electrons and thus have essentially the same chemical behavior. This is another important point worth remembering:

Atoms in the same column of the periodic table have the same number of valence electrons and have very similar chemistry or chemical behavior.

Since the columns of the periodic table contain atoms of similar chemistry, they are often called groups or families.

Certain groups of elements are rather important in chemistry. The first of these is called the noble gases. They are the atoms in column 8A of the table, and they are important because they have the ideal electron configurations in nature:

He: $1s^2$
Ne: $1s^2\ 2s^2\ 2p^6$
Ar: $1s^2\ 2s^2\ 2p^6\ 3s^2\ 3p^6$
Kr: $1s^2\ 2s^2\ 2p^6\ 3s^2\ 3p^6\ 4s^2\ 3d^{10}\ 4p^6$
Xe: $1s^2\ 2s^2\ 2p^6\ 3s^2\ 3p^6\ 4s^2\ 3d^{10}\ 4p^6\ 5s^2\ 4d^{10}\ 5p^6$

In each case, the s and p orbitals in their highest energy level are full. This is a very low–energy situation, and since all matter strives to get to the lowest energy possible, we consider such an electron configuration to be ideal.

Since an atom's electron configuration determines its chemistry and these atoms have ideal electron configurations, they have no reason to change and thus they rarely undergo chemical reactions. Substances like the noble gases, which rarely take part in chemical reactions, are called inert substances.

Since the electron configurations of the noble gases are ideal, all other atoms strive to attain them. This is the guiding principle behind chemical behavior. Chemical reactions are really just the result of atoms trying to rearrange their electrons so that they can get an ideal electron configuration. Since (with the exception of helium) all noble gases have 8 valence electrons, we could say that atoms strive to get 8 valence electrons. This very important principle is called the octet rule, and we will use it over and over again in this course.

Octet rule—Most atoms strive to attain 8 valence electrons.

As usual in chemistry, there are exceptions to this rule. Hydrogen, lithium, and beryllium strive to have 2 valence electrons (so they will be like helium). You need to remember that all atoms except for a few strive for 8 valence electrons by filling in their outer s and p orbitals. The exceptions strive for 2 because their electrons occupy only the 1s orbital.

You need to know one more thing about the periodic table and electron configurations. Since the number of valence electrons is so important to the chemistry of an atom, the periodic table has been designed so that chemists can get this information very quickly. Rather than determining the electron configuration of an atom each time you need to figure out how many valence electrons it has, just look at the number of the column that the atom resides in. For example, bromine has 7 valence electrons. What column is it in? 7A. The noble gases (with the exception of helium) have 8 valence electrons. What group are they in? 8A.

Any atom in columns 1A–8A has the same number of valence electrons as the column number. This shortcut does not work for the elements in columns 1B–8B, but that's okay. The chemistry of the elements in columns 1B–8B is more difficult to understand, so we will not be dealing with them in this module. Therefore, all you need to do is remember this important fact:

For atoms in groups 1A–8A, the number of the column in which the atom is located on the periodic table equals the number of valence electrons the atom has.

LEWIS STRUCTURES

Since the valence electrons play such a hefty role in the chemistry of an atom, chemists like to use a type of notation that emphasizes them. This notation, called Lewis structures, was developed by Gilbert N. Lewis, a professor of chemistry at the University of California at Berkeley, in the early 1900s. It has become a central concept in chemistry because it forms the basis of our understanding of how molecules form.

A Lewis structure consists of an atom's symbol surrounded by dots. Each dot represents a valence electron in the atom. For example, the atom lithium is in group 1A of the periodic table, so it has 1 valence electron. The Lewis structure for lithium looks like this:

Li·

The single dot to the right of the symbol represents lithium's 1 valence electron. The rest of the atoms in the second row of the periodic table have Lewis structures that look like this:

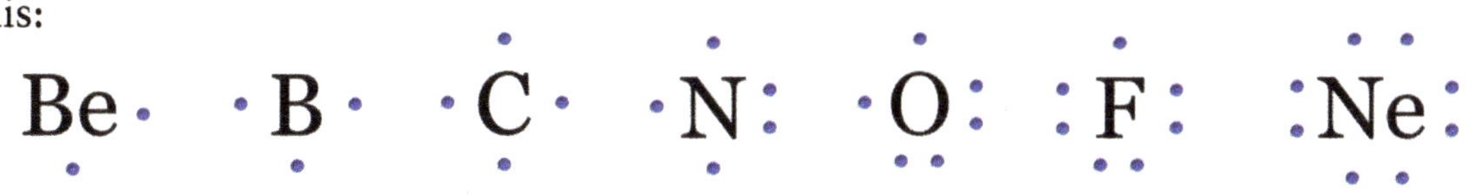

Beryllium (Be) has 2 dots because it is in group 2A and has 2 valence electrons. Boron has 3 valence electrons, carbon has 4, nitrogen 5, oxygen 6, fluorine 7, and neon 8. Realize that the dots do not represent the total number of electrons the atom has. That number is indicated by the atomic number on the table. The dots represent the number of *valence electrons* the atom has, which is indicated by the column number on the table.

Notice how the dots are put in a Lewis structure. The first dot is placed on the right side of the elemental symbol. If the atom has 2 valence electrons, the second dot is placed on the bottom of the symbol. The third valence electron's dot is placed to the left of the symbol, and the fourth is placed on top. If the element has more than 4 valence electrons, the remainder must be paired with the dots that are already there. Notice also that the ideal Lewis structure (shown in Ne because it has the ideal electron configuration) has a pair of dots on each side and at the top and bottom of the elemental symbol.

Since Lewis structures use dots to represent the valence electrons in an atom, they are sometimes called electron–dot diagrams. This name fails to credit Lewis for his seminal work in molecular structure, so most chemists do not like this terminology. Do not be confused if you read about electron–dot diagrams in other books. They are the same as Lewis structures. Make sure you understand how to determine an atom's Lewis structure by studying Example 4.1 and answering "On Your Own" question 4.1.

EXAMPLE 4.1

Draw the Lewis structure for the following atoms:

a. Na   b. Ga   c. Te

(a) Since Na is in group 1A, it has 1 valence electron. In a Lewis structure, a single electron is written on the right side of the symbol:

(b) Since Ga is in group 3A, it has 3 valence electrons. We start on the right side of the symbol and put a single dot on the right, then the bottom, then the left of the symbol:

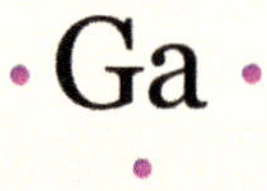

(c) Since Te is in group 6A, it has 6 valence electrons. The first 4 are put on the sides, bottom, and top of the symbol by themselves. After that, the last 2 must be paired with others. We pair the electrons in the same order that we put them down singly; so the last 2 dots will pair with the dots on the right and bottom of the symbol:

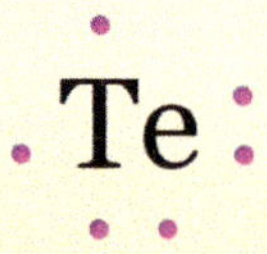

ON YOUR OWN

4.1 Draw the Lewis structures for the following atoms:
a. Ca   b. Si   c. At

LEWIS STRUCTURES FOR IONIC COMPOUNDS

Now we can learn how Lewis structures help us in determining molecular structure. We will start with ionic compounds because they are the easiest to understand. Consider table salt, NaCl. Why do the elements Na and Cl come together and make a molecule? Also, why is there 1 Na atom for every 1 Cl atom, as the formula indicates? Why aren't there 2 Na atoms for every 3 Cl atoms so that the formula is Na_2Cl_3? The answers to these questions are readily available by looking at the Lewis structures of these 2 atoms:

First, look at chlorine's Lewis structure. It is very close to the ideal valence electron configuration. All it needs to do is get 1 more electron, and then it will have 8 valence electrons. Sodium just happens to have 1 valence electron (1 dot in its Lewis structure).

Chlorine could reach the ideal electron configuration if it could just take the 1 valence electron from sodium and make that electron its own. Look at the Lewis structures that result when that happens:

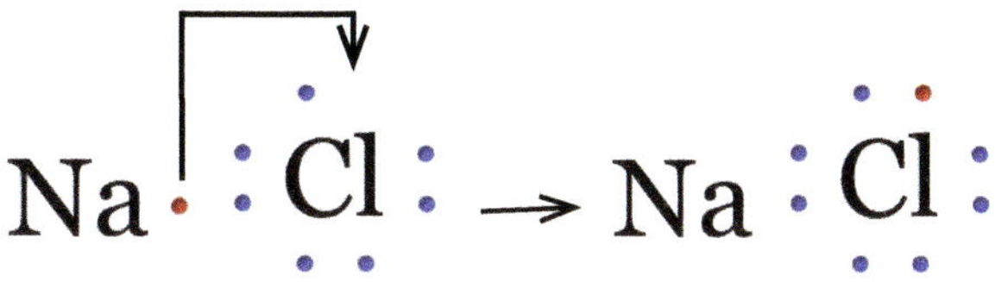

Once chlorine takes sodium's electron (the red dot) away, it has the ideal electron configuration because it has 8 valence electrons. This is how nonmetal atoms try to get an ideal electron configuration when they are a part of an ionic compound. They take electrons from other atoms until they have 8 valence electrons. This is a fact we must remember:

In ionic compounds, nonmetals try to gain electrons so their Lewis structures reach the ideal electron configuration with 8 dots around it.

Once a nonmetal's Lewis structure contains 8 dots, it has the ideal electron configuration and will be stable. Hydrogen, of course, is an exception to this rule. When hydrogen is a part of an ionic compound, it gains an electron to get 2 dots in its Lewis structure, not 8.

What about sodium? Once it loses its electron, it also has the ideal electron configuration. This may be hard to see, but if you go back to the full electron configuration, you'll understand what this means. Sodium's complete electron configuration is:

$$1s^2\ 2s^2\ 2p^6\ 3s^1$$

If the chlorine takes an electron away, sodium has only 10 electrons, not 11. Therefore, its new configuration is:

$$1s^2\ 2s^2\ 2p^6$$

This is the same as the electron configuration for neon, which is an ideal electron configuration.

So we see that metals behave differently than nonmetals. Whereas nonmetals tend to gain electrons until their Lewis structures have 8 dots around them, metals tend to give away their electrons until their Lewis structures have no dots around them. This is a fact you must remember:

In ionic compounds, metals try to lose electrons so their Lewis structures reach the ideal electron configuration with no dots around them.

When a metal's Lewis structure has no dots, it has the ideal electron configuration and will be stable. This is the fundamental difference between metals and nonmetals: Nonmetals try to gain electrons, while metals try to lose electrons. So when a single Na atom gives up its electron and donates it to a single Cl atom, both atoms end up with ideal electron configurations. This is what makes the chemical formula of table salt NaCl. Each Na atom has exactly enough electrons to satisfy each Cl atom's desires. One Na atom will always react with 1 Cl atom, and they will always form NaCl. This kind of consideration governs the formation of all molecules. Atoms will join in such a way as to make sure that they all end up with ideal electron configurations.

One thing has to be made clear, however. The last 2 Lewis structures drawn were *wrong and missing something*. You see, when a sodium atom loses an electron, it is no longer a sodium atom. Remember, all atoms have the same number of protons and electrons. A sodium atom, according to the periodic table, has 11 protons and 11 electrons. But when sodium gives up an electron, it ends up with 11 protons and only 10 electrons. It has more positive charges (protons) than negative charges (electrons), which gives it an overall positive charge. Since it has 1 more proton than it has electrons, we say it has a 1+ charge. Remember that chemists love to drop ones. They do not report ones in chemical formulas or chemical equations. In the same way, they often do not report one with electrical charges. Therefore, rather than listing the charge as 1+, it is often simply listed as +.

In the same way, a chlorine atom has 17 protons and 17 electrons. However, when the chlorine atom gets the electron from the sodium atom, it ends up with 17 protons and 18 electrons. This gives it an overall negative charge since it has more negative charges (electrons) than positive charges (protons). Since it has 1 more electron than it has protons, we say it has a 1- charge , which is usually reported as simply -. The correct Lewis structures in this situation are:

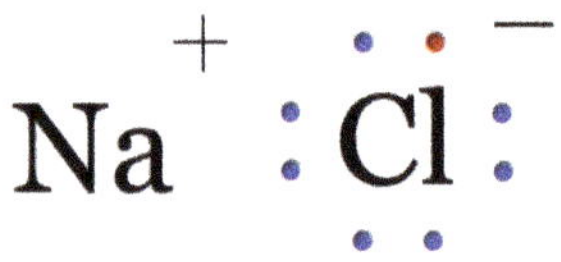

These Lewis structures emphasize the fact that each element in the molecule is electrically charged. When an atom gives up or gains 1 or more electrons to become electrically charged, we no longer refer to it as an atom. We call it an ion.

Ion—An atom that has gained or lost electrons, becoming electrically charged.

There are 2 kinds of ions: cations (cat' ions) and anions (an' ions):

Cation—A positively charged ion.

Anion—A negatively charged ion.

Therefore, we would say that a molecule of table salt is composed of 1 sodium cation and 1 chloride anion.

Notice the names that were just used. Positively charged cations keep the same name as the atoms from which they came. Negatively charged anions, on the other hand, add the *–ide* suffix to the name of the atoms from which that they came. This is an important fact to remember:

Positive cations have the same name as the atom from which they came.

Negative anions add an –ide suffix to the name of the atom from which they came.

This is where ionic compounds get the names we learned about in Module 2. To name an ionic compound, we simply name the ions (cations and anions) that make it up. Since table salt is made up of a sodium cation and a chloride anion, we call it sodium chloride.

Now let's consider a more difficult example: the ionic compound that is formed between magnesium and fluorine. The Lewis structures for these 2 atoms are:

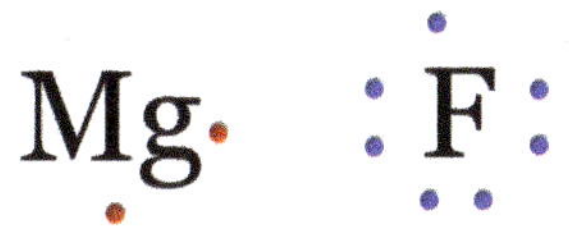

Since the fluorine atom has 7 valence electrons (7 dots), it needs to gain 1 more electron to attain the ideal electron configuration. Therefore, the fluorine takes 1 electron away from the magnesium:

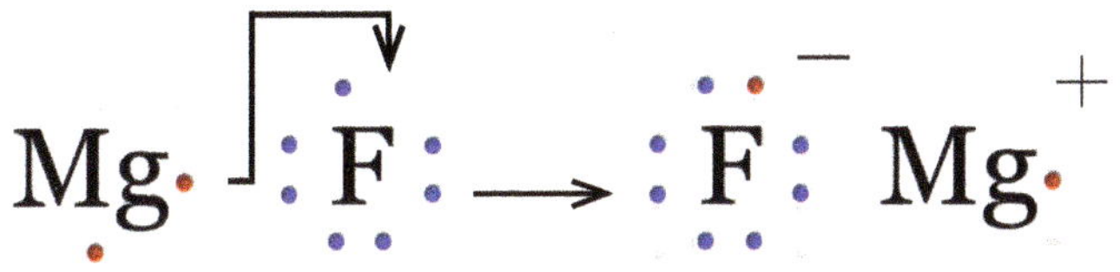

Now the fluorine has 8 valence electrons (8 dots). Since this is 1 more electron than it normally has, we assign it a - charge. Since the magnesium now has 1 less electron than it normally has, we assign it a + charge. Remember, electrons are negative, so the atoms that gain electrons (nonmetals) will develop negative charges, and the atoms that lose electrons (metals) will develop positive charges.

Are we finished with this molecule? The answer is no. Even though the fluorine has an ideal electron configuration, the magnesium still does not. Magnesium is a metal (on the left side of the jagged line) and will have its ideal electron configuration only when it has no dots in its Lewis structure. Since it still has 1 dot in the Lewis structure above, it does not have an ideal electron configuration. This means we still have work to do!

To get no dots in magnesium's Lewis structure, we need to give that last electron away. The problem is that the fluorine has all the electrons it wants. It will not take magnesium's last electron. However, there are, most likely, other fluorine atoms around. *Another* fluorine atom (one that hasn't gained any electrons yet) could take magnesium's last electron:

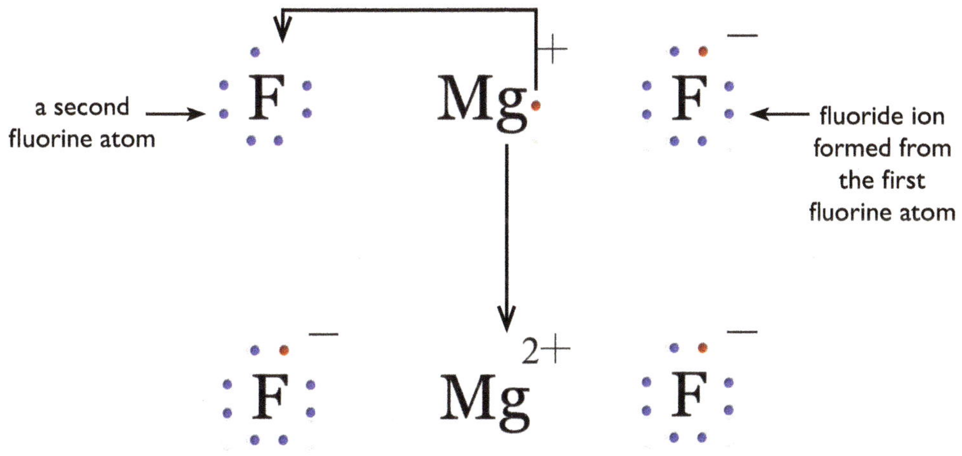

Now all the atoms have their ideal electron configurations. Notice that each individual fluorine atom gained 1 electron, so each individual fluoride anion has a 1- charge. On the other hand, a single magnesium lost 2 electrons, so it has a 2+ charge. We cannot drop the 2 like we drop ones, so we must list the charge as 2+. Since it took 2 fluorine atoms to gain the electrons and 1 magnesium atom to lose them so that every atom had an ideal configuration, the chemical formula for this compound must be MgF_2.

This is how ionic compounds get their chemical formulas. The atoms start exchanging electrons until all of them attain an ideal electron configuration. Once they have all achieved this, they are attracted to one another because of the electrical charges that they have gained. In magnesium fluoride, the magnesium cation will always be close to the 2 fluoride anions because the positive charge of the magnesium is attracted to the negative charges of the fluorides. This electrical attraction is what holds ionic compounds together.

Now that you have seen how ionic compounds get their chemical formulas, it's time to show you a simple shortcut that will allow you to determine the chemical formula of an ionic compound. The first step in this shortcut is to determine what charge an atom needs to attain an ideal electron configuration. For example, in the discussion above, we found that fluorine needed to gain 1 electron to attain an ideal configuration, so its ion's charge was 1-. Magnesium, on the other hand, needed to lose 2 electrons to attain an ideal configuration, so its ion developed a 2+ charge.

Since the number of valence electrons an atom has determines how many electrons it must gain or lose to attain an ideal configuration, all atoms in a given column on the periodic table tend to develop the same charge. In other words, we already determined that since fluorine has 7 valence electrons, it needs 1 more electron to attain its ideal configuration. The fluoride ion develops a charge of 1-. All of the other atoms in group 7A also have 7 valence electrons, so their ions will all develop a charge of 1-. In the same way, since Mg develops a 2+ charge to gain an ideal electron configuration, all of the atoms in its column (group 2A) develop a 2+ charge. Table 4.1 gives us a look at groups 1A–8A.

TABLE 4.1
Common Ions and Their Charges for Groups 1A–8A

Metals Lose Valence Electrons			Nonmetals Gain Valence Electrons				Noble Gases
1A	**2A**	**3A**	**4A**	**5A**	**6A**	**7A**	**8A**
Li^+							He
Na^+	Mg^{2+}	Al^{3+}	C^{4-}	N^{3-}	O^{2-}	F^-	Ne
K^+	Ca^{2+}		Si^{4-}	P^{3-}	S^{2-}	Cl^-	Ar
Rb^+	Sr^{2+}			As^{3-}	Se^{2-}	Br^-	Kr
Cs^+	Ba^{2+}					I^-	Xe

- Atoms in group 1A have 1 valence electron. To become like a noble gas, they could gain 7 electrons or lose 1 electron. They will lose that 1 valence electron, which will make them cations with a charge of 1+ in ionic compounds.
- Atoms in group 2A, as we have found out, will become cations and develop a 2+ charge in ionic compounds.

- Group 3A is a little more difficult. Boron is a nonmetal and is therefore different from the other elements in this column. We will look at boron in a moment as it is an exception to the rule. All others in this column will become cations and develop a 3+ charge because they want to lose 3 valence electrons to become like a noble gas.
- Group 4A is also difficult because C and Si are nonmetals, while the rest of the atoms in the column are metals. As a result, C and Si will *gain* electrons to get to their ideal configuration, while Ge, Sn, and Pb will lose electrons. C and Si need to get 8, so they need to gain 4 electrons, giving them a 4- charge in ionic compounds. Ge, Sn, and Pb want to be positive ions, but they don't follow the rules very well in how many electrons they lose. We will discuss them in a moment.
- Group 5A has mostly nonmetals, which want to gain 3 electrons to get to their ideal configuration. They become anions and develop a 3- charge in ionic compounds. All the metals in Group 5A are exceptions to the rule.
- Group 6A also has mostly nonmetals, which will gain 2 electrons to become anions and develop a 2- charge in ionic compounds. Once again, don't be concerned about the metals.
- Group 7A, as we have seen, has nonmetals, which become anions and develop a 1- charge in ionic compounds.

Using these rules, you need to be able to determine the charge that any atom in groups 1A–8A (minus the exceptions) develops in an ionic compound. At first this may seem like a huge memorization task, but it isn't. Remember, you will always be able to consult the periodic table. The column number on the table tells how many valence electrons the atom has. All you have to do is remember that nonmetals (right of the jagged line) will gain electrons and metals (left of the jagged line) will lose electrons to become like a noble gas. This determines the charge. So really, you needn't memorize any of the rules that were just listed. Just remember the principles discussed in this paragraph and, with the aid of the periodic table, you can determine the charges on your own.

Once you have determined the charges, determining the formula is a snap. Combine the ions in the lowest possible ratio to achieve a neutral compound.

- **If the charges have the same numerical value, then the subscript for each ion is 1 and can therefore be ignored.**
- **If the charges have different numerical values, drop the + and - signs, switch the numbers, and use them as subscripts.**
- **The chemical formula ratio for ionic compounds must always be the lowest possible ratio.**

This is all you do to determine an ionic molecule's chemical formula. If it sounds a little confusing now, don't worry. Example 4.2 will clear things up.

EXAMPLE 4.2

What is the chemical formula for the ionic compound formed by aluminum and sulfur? What is its name?

Step 1: Identify the ions. To solve this problem, we first need to determine the charges that Al and S develop in an ionic compound. Al is in group 3A, so it's a cation and wants to be 3+. S is in group 6A, so it wants to be an anion with a charge of 2-.

Step 2: Balance the charges. The charges do not have the same numerical values, so we drop the + and - signs, switch the numbers, and make them subscripts.

Step 3: Write the formula for the cation first, using subscripts from the charge balance. The 3 gets switched to the S, and the 2 goes to the Al:

$$Al_2S_3$$

Step 4: Name the compound, cation first. Al is the metal, so its ion name is aluminum. S is the nonmetal, so its anion name is sulfide. Therefore, the compound's name is **aluminum sulfide**.

What is the chemical formula of barium oxide?

Step 1: Identify the ions. According to the name given, this ionic compound is composed of a barium cation, which comes from the metal barium, and an oxide anion, which comes from the nonmetal oxygen. Barium is in group 2A, so it wants a 2+ charge. Oxygen is in group 6A, so it wants a 2- charge.

Steps 2 and 3: Balance the charges and write the formula cation first, using subscripts from the charge balance. The numerical values for the charges are the same, so we ignore them:

$$BaO$$

Notice that we did not have to be told that this is an ionic compound. Because it includes a metal and a nonmetal, we know that it is ionic and that we are using ionic rules to determine the formula.

What is the chemical formula for the compound that forms when Na and N are reacted together? What is its name?

Step 1: Identify the ions. The metal Na is in Group 1A, so it develops a charge of 1+. The nonmetal N is in group 5A, so it develops a charge of 3-.

Steps 2 and 3: Balance the charges and write the formula cation first, using subscripts from the charge balance. Ignoring the charges and switching the numbers gives us:

$$Na_3N$$

Step 4: Name the compound, cation first. The ion that comes from sodium is named sodium since it is positive. Since the ion that comes from nitrogen is negative, it is called nitride. So the name is **sodium nitride**.

Once again, we did not have to be told that this is an ionic compound. We know that because it contains a metal and a nonmetal and because Na loses an electron and N gains electrons to become ions.

This shortcut allows you to get the chemical formula of an ionic compound very quickly, but don't let that distract you from what is really going on here. Remember that this method is just a shortcut to skip writing out the Lewis structures and moving the dots around. In nature, the chemical formulas of these compounds are determined the way we showed you with Lewis structures. This method is simply a shortcut that saves time. Make sure you understand it by answering "On Your Own" questions 4.2–4.3.

ON YOUR OWN

4.2 Give the chemical formulas for the following compounds:

a. beryllium chloride
b. potassium oxide
c. aluminum phosphide
d. magnesium nitride

4.3 What is the chemical formula for the compound that forms when cesium and nitrogen are reacted together? What is its name?

HANDLING THE EXCEPTIONS IN IONIC COMPOUNDS

You might wonder about all the exceptions on the periodic table. In the list of rules on determining charge, we mentioned that Ge, Sn, and Pb are all exceptions. Why are they exceptions? Well, let's take a look at the Lewis structures, using Sn as an example.

$$\text{Sn: } 1s^2\ 2s^2\ 2p^6\ 3s^2\ 3p^6\ 4s^2\ 3d^{10}\ 4p^6\ 5s^2\ 4d^{10}\ 5p^2$$

Sn has 4 valence electrons; this is why it is in column 4A of the periodic table. Also, it has 10 electrons in the 4d orbital in addition to the full 4s and 4p orbitals it needs for a full octet. Losing d electrons will allow the metal to become more than one possible ion.

When we write out the name of the ionic compound involving these metals, we need to be very clear about which ion is present. For example, tin chloride is an ambiguous name because the tin could have any number of charges. How do we keep the multiple ion possibilities straight?

We are going to put a roman numeral in the name that represents the charge of the metal. So tin (IV) represents Sn^{4+} and tin (II) represents Sn^{2+}. We will do this for all metals. Instead of tin chloride, we should write tin (IV) chloride or tin (II) chloride; the roman numeral matches the charge on the metal ion.

This roman numeral helps us determine the chemical formula of the ionic compound. With tin (II) chloride the II represents Sn^{2+}, and chloride is Cl^{1-}. Combining the ions gives a formula of $SnCl_2$. In the same way, tin (IV) chloride represents $SnCl_4$ because tin (IV) is Sn^{4+} and chloride is Cl^{1-}.

When dealing with ions of atoms which we call exceptions to the rules, you will always be given a roman numeral that signifies the positive charge on the metal. With that information, you can use the same skill you just learned to determine the chemical formulas of those ionic compounds as well.

TABLE 4.2
Common Metals That Form More Than One Positive Ion

Element	Possible Ions	Name of Ions
Chromium	Cr^{2+}, Cr^{3+}	Chromium (II), Chromium (III)
Iron	Fe^{2+}, Fe^{3+}	Iron (II), Iron (III)
Copper	Cu^{+}, Cu^{2+}	Copper (I), Copper (II)
Gold	Au^{+}, Au^{2+}	Gold (I), Gold (II)
Tin	Sn^{2+}, Sn^{4+}	Tin (II), Tin (IV)
Lead	Pb^{2+}, Pb^{4+}	Lead (II), Lead (IV)

Knowing that some metals may have multiple charges, you can also use the chemical formula to determine the name of the compound. You do not need to memorize Table 4.2. Example 4.3 gives you some practice in naming ionic compounds with metals that have multiple charges.

EXAMPLE 4.3

What is the chemical name for the ionic compound $CuCl_2$?

Step 1: Identify the cation and anion and their charges. To solve this problem, we first need to determine the charges that Cu and Cl develop in an ionic compound. Cl is in group 7A, so it wants to be 1-. Cu is in group 1B, so it can possibly have multiple charges, but since Cl is 1- and there are 2 atoms of Cl for every Cu atom, then Cu must be 2+.

Step 2: Name the metal cation by its elemental name followed by the roman numeral. Copper (II) will be the name of the cation.

Step 3: Name the anion by its *–ide* name. Cl^{1-} is chloride.

Therefore, the compound's name is **copper (II) chloride.**

What is the chemical name for the ionic compound PbO_2?

Step 1: Identify the cation and anion and their charges. O is in group 6A, so it wants to be 2-. Pb is in group 4A, so it can possibly have multiple charges, but since O is 2- and there are 2 O atoms for every Pb atom, then Pb must have a charge of 4+.

Step 2: Name the metal cation by its element name followed by the roman numeral. Pb^{4+} is lead (IV).

Step 3: Name the anion by its *–ide* name: O^{2-} is oxide.

Therefore, the compound's name is **lead (IV) oxide.**

Make sure you understand this by answering "On Your Own" questions 4.4–4.5.

ON YOUR OWN

4.4 Give the chemical formulas of the following compounds:
a. tin (IV) fluoride
b. chromium (III) chloride
c. copper (II) oxide

4.5 Give the names of the following compounds:
a. FeF_3 b. NiI c. Mn_2O_3

IONIZATION ENERGY AND PERIODIC PROPERTIES

It is important to remember that although most atoms want to become charged to attain the ideal electron configuration, they never start out that way. When an atom is formed (we have only theories about *how* this happens), it is always formed as an atom. It always has just as many electrons as it has protons.

However, depending on how much an atom wishes to attain an ideal configuration, it may gain or lose electrons rather quickly. When this happens, we call the process ionization (eye' uh nuh zay' shun).

Ionization—The process by which an atom turns into an ion by gaining or losing electrons.

Other atoms may remain as atoms for a long time simply because they do not have a strong enough desire for an ideal electron configuration. We measure an atom's desire to become a positive ion by measuring its ionization energy.

Ionization energy—The amount of energy needed to take an electron away from an atom.

Each atom holds onto its valence electrons with a different strength. Some hold them tightly, while others hold them rather loosely. A lot of energy is required to take an electron away from an atom that holds its valence electrons tightly. On the other hand, if an atom holds on to its valence electrons loosely, not much energy is required to remove one.

But how do we know whether or not an atom holds on to its electrons tightly or loosely? As you might expect, it depends on the atom's electron configuration and can be determined from the periodic table. For example, consider the elements in group 1A. These atoms all have 1 valence electron. To get to an ideal electron configuration, all they have to do is lose that 1 electron. The ionization energy for these atoms is rather low. In other words, these atoms will be so much better off by the loss of one electron that they give that electron up willingly.

On the other hand, consider the atoms in group 7A. These atoms have 7 valence electrons and need only 1 more to attain an ideal electron configuration. How willingly do you think that these atoms will *give up* 1 of their electrons? Not at all. Since these atoms need to *gain* an electron to attain an ideal configuration, they are not very likely to give up an electron, taking them farther away from an ideal configuration.

The elements on the left side of the periodic table have a rather low ionization energy (they give up their electrons easily), while the atoms on the right side of the periodic table have high ionization energies (they give up their electrons grudgingly). This is a fact worth remembering.

In general, the ionization energy of atoms increases
from the left of the periodic table to the right of the periodic table.

In other words, we could say that since Mg lies left of P on the periodic table, Mg has a lower ionization energy than P.

Any atomic property that varies regularly across the periodic table is called a periodic property.

Periodic property—A characteristic of atoms that varies regularly across the periodic table.

Periodic properties are nice because they allow us to make predictions about the relative behavior of atoms by simply looking at the table. For example, suppose a chemist is working on an experiment and realizes that a source of electrons is needed to make the experiment a success. Looking at his inventory, the chemist sees that he has carbon and lithium. Looking at the periodic table, he sees that lithium lies to the left of carbon on the table. This tells him that lithium gives up its electron much more easily than carbon, so he knows to choose lithium for the experiment.

Not only does ionization energy change from left to right on the periodic table, but it also changes from the top to the bottom. Atoms near the top of the periodic table hold on to their electrons more tightly than those on the bottom on the table.

In general, the ionization energy of atoms increases
from the bottom of the periodic table to the top of the periodic table.

If 2 atoms lie in the same column of the periodic table, their relative ionization energy can be determined by their vertical position on the table. For example, O and Se are both in the same column of the table, so neither lies to the left or right of the other. However, since O is above Se on the periodic table, O has a higher ionization energy than Se.

Why does ionization energy increase from the bottom to the top of the periodic table? Look at the difference between O and Se. The valence electrons in oxygen are on energy level 2 (determined by the electron configuration), while selenium's valence electrons are in energy level 4. This means that selenium's valence electrons are farther away from the nucleus than oxygen's. As a result, the protons in the nucleus cannot attract selenium's valence electrons as strongly. This means that selenium holds on to its valence electrons more loosely, making its ionization energy lower.

The way in which ionization energy changes relative to an atom's position in the table is summarized in Figure 4.1. Such periodic trends are important and worth remembering.

FIGURE 4.1
The Periodic Behavior of Ionization Energy

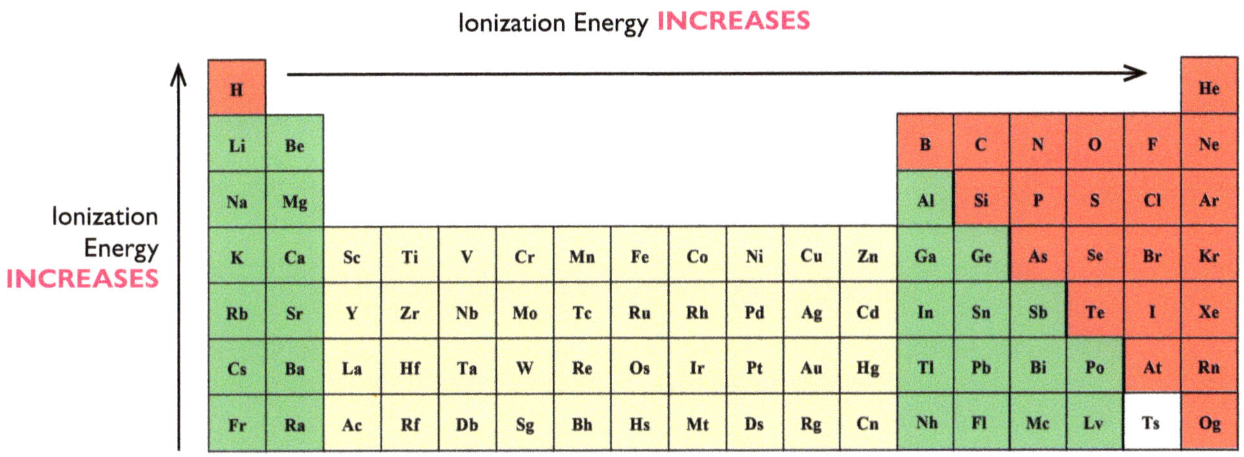

Make sure you understand how to use this periodic property by studying Example 4.4 and then answering "On Your Own" questions 4.6–4.7.

EXAMPLE 4.4

A chemist needs to find an element that readily gives up an electron. Which of the following would suit her needs best: Ca, As, or Br?

Since all of these atoms lie in the same row of the periodic table, they do not lie above or below each other. The fact that ionization energy decreases from the top to the bottom of the table is useless here. However, Ca lies to the left of As, which lies to the left of Br. Since ionization energy increases from left to right on the periodic table, Ca has the lowest ionization energy. **Ca gives up its electrons most readily and is the best choice for the experiment.**

Which of the following holds onto its electrons the tightest: Al, In, or B?

All of these elements lie in the same column, so the fact that ionization energy increases from left to right on the table is of no use. However, B lies above Al, which lies above In. Since ionization energy decreases from the top to the bottom of the table, B has the highest ionization energy. This means that **B holds on to its electrons the tightest.**

Which element has the highest ionization energy?

Helium has the highest ionization energy. This is because all noble gases do not want to add or lose any electrons.

ON YOUR OWN

4.6 Order the following atoms in terms of increasing ionization energy: At, Ir, Pb, Cs.

4.7 Which atom most readily gives up its electrons: Ca, Be, Ba, Ra?

ELECTRONEGATIVITY: ANOTHER PERIODIC PROPERTY

The way in which we have defined ionization energy deals only with positive ions. After all, we defined ionization energy as the energy needed to take an electron away from an atom. When electrons are taken away from atoms, the atoms become positive ions. What about negative ions? Is there a way of determining how much an atom desires to gain electrons? Of course there is. The property that determines an atom's desire to gain electrons is called electronegativity.

Electronegativity—A measure of how strongly an atom attracts extra electrons to itself.

As the electronegativity of an atom increases, its desire to become a negative ion increases.

Let's determine how electronegativity varies across the periodic table. First, consider atoms in group 1A. Do they desire extra electrons? No, because they need to lose an electron to attain the ideal electron configuration. On the other hand, atoms in group 7A have a strong desire to gain an extra electron because that will give them the ideal electron configuration. Therefore, we can conclude:

In general, the electronegativity of atoms increases from left to right on the periodic table.

What about an atom's vertical position on the table? How does that affect its electronegativity? As already noted, atoms that are higher on the periodic table tend to have their valence electrons closer to the nucleus as compared to atoms that appear near the bottom of the table. The attraction of an atom for electrons should decrease from top to bottom on the table because the farther the valence electrons are from the nucleus, the less attraction the protons can exert on them. Noble gases (group 8A) do not follow these electronegativity periodic properties because they have the ideal electron configuration and do not want to gain electrons.

In general, the electronegativity of atoms increases from bottom to top on the periodic table.

Figure 4.2 summarizes the way in which both ionization energy and electronegativity vary across the periodic table.

FIGURE 4.2
The Periodic Behavior of Ionization Energy and Electronegativity Except Electronegativity of Noble Gases

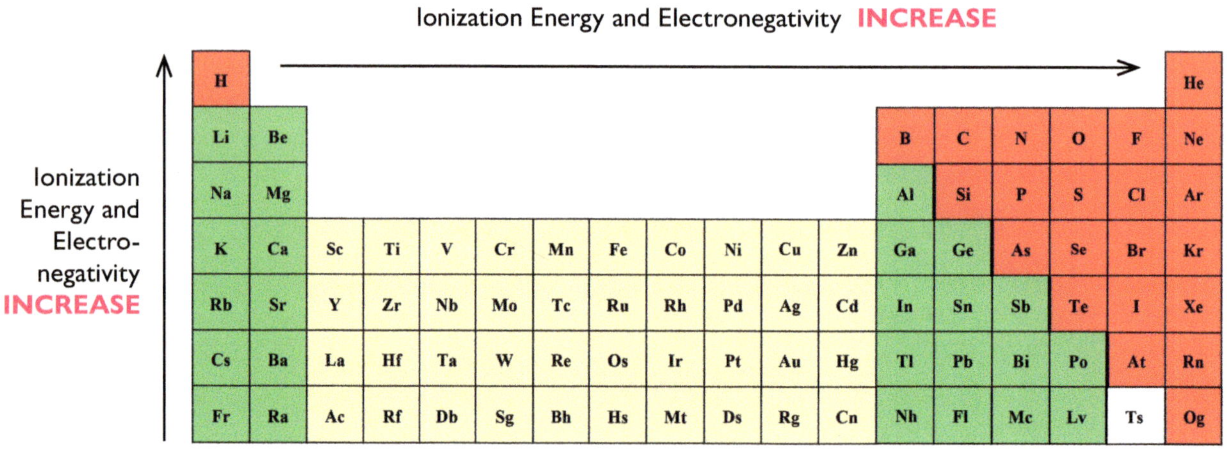

Use the same reasoning you learned in the previous section to answer "On Your Own" questions 4.8–4.9.

ON YOUR OWN

4.8 Which element desires electrons most: Ca, As, K, Se?

4.9 Order the following in terms of increasing desire for electrons: Br, Cl, I, At, F.

ATOMIC RADIUS: ANOTHER PERIODIC PROPERTY

Before we move on to the Lewis structures of covalent compounds, let's discuss one more periodic property: the atomic radius. Even though atoms have rather complex shapes due to the shapes of the electron orbitals, to some extent, we can approximate atoms as tiny spheres. If we do that, then the radius of an atom is an indicator of its size. The larger the atom's radius, the larger the atom.

If you think about it, you should quickly realize that (once again) the electron configuration of an atom is what determines its size. After all, the protons and neutrons in an atom are tightly packed at the center, but the electrons whirl around the nucleus far from the center. The electrons are the outer part of an atom. Since electron configuration determines the atom's size, it is not surprising that atomic radius varies according to the placement of the atom on the periodic table.

Think for a moment about how the electron configuration changes as you move *down* the table. Remember that atoms that are higher on the periodic table tend to have their valence electrons closer to the nucleus as compared to atoms that appear near the bottom of the table. Therefore, it should be pretty obvious that the lower an atom is on the periodic table, the bigger the atom is because its valence electrons are farther from the nucleus.

In general, the atomic radius increases from top to bottom on the periodic table.

What about an atom's horizontal position on the periodic table? How does that affect its atomic radius? As you move across the periodic table, the electrons are not filling up higher energy levels. Instead, they are filling orbitals in the same energy level or below. For example, look at row 4 on the periodic table. As you travel across row 4, the electrons first fill up the 4s orbital, then the 3d orbitals, then the 4p orbitals. The energy level of the orbitals does not increase. As a result, you might think that the size of the atom doesn't vary as you travel across the periodic table. However, if you are thinking that, you are forgetting something.

Remember, as you travel from left to right on the periodic table, the *atomic number* of the atom is increasing. What does that tell you? It tells you that the number of protons in an atom increases as you travel across the table. The total positive charge in the nucleus increases. What does that mean? It means that the nucleus can exert more electrical force on the electrons, *pulling them closer.* As a result, when you travel from left to right on the periodic table, the size of the atoms actually *decreases*!

In general, the atomic radius increases from right to left on the periodic table.

Figure 4.3 summarizes the way in which atomic radius varies across the periodic table.

FIGURE 4.3
The Periodic Behavior of the Atomic Radius

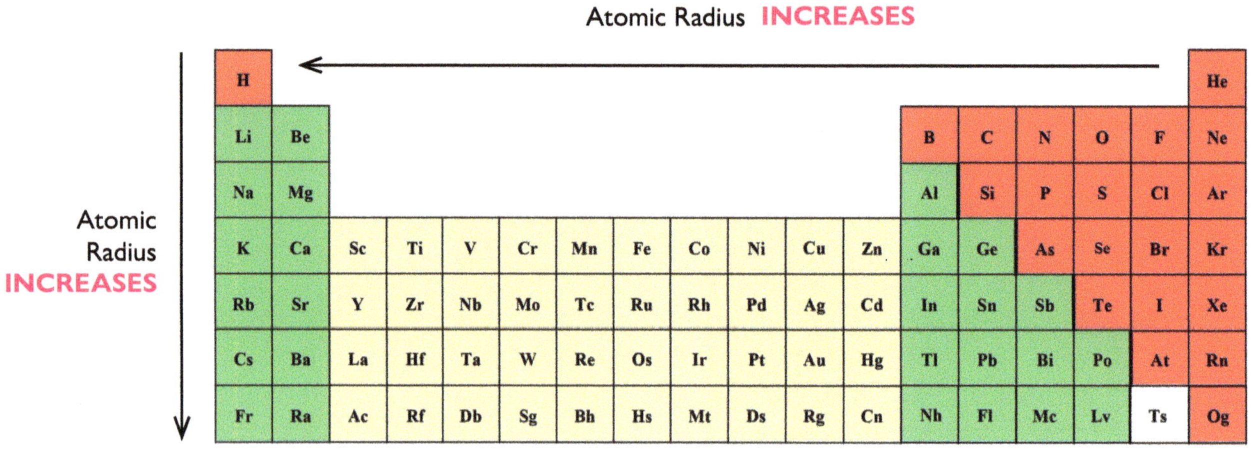

Use the information you've just learned to answer "On Your Own" questions 4.10–4.11.

ON YOUR OWN

4.10 Which element is the largest: Ca, As, K, Se?

4.11 Order the following in terms of increasing atomic radius: Br, Cl, I, At, F.

LEWIS STRUCTURES OF COVALENT COMPOUNDS

Now that we have learned a great deal about the nature of ionic compounds, let's turn our attention to covalent compounds. As you learned in Module 2, the difference between ionic and covalent compounds revolves around the type of elements that make them up. Ionic compounds contain both metals and nonmetals, while covalent compounds contain only nonmetals. Now that you understand Lewis structures and how atoms desire to attain an ideal electron configuration, you can finally learn the *fundamental* difference between these 2 types of compounds.

If the main driving force for the formation of molecules is the desire for each atom to attain a perfect electron configuration, there seems to be a major problem in explaining the formation of covalent compounds. After all, we already learned that covalent compounds contain only nonmetals. We also learned that nonmetals always gain electrons. They never give up electrons. How do nonmetals combine to get ideal electron configurations if none of them is willing to give up electrons so that others can gain them?

Think about what happens when your friend has something that you want but is unwilling to give it to you. What do you usually do? Hopefully, you end up *sharing* the item. The same goes for atoms. If 2 nonmetals come together and neither will give up its electrons, the 2 atoms will *share* electrons to help them both attain ideal electron configurations. The best way to understand this is by looking at some Lewis structures.

We mentioned in Module 2 that elements are composed of individual atoms, whereas compounds are composed of individual molecules. This rule is not completely correct.

Nearly every rule in chemistry has its exceptions. There are elements in nature that are not composed of individual atoms. These elements are called homonuclear (ho' moh new' klee er) diatomics (dye uh tom' iks) and are composed of individual molecules, not individual atoms. As their name implies, homonuclear diatomic molecules are made up of 2 atoms (diatomic) which are the same type (homonuclear).

Nitrogen, oxygen, chlorine, fluorine, bromine, iodine, astatine, and hydrogen are all elements that are made up of homonuclear diatomic molecules rather than individual atoms.

The chemical formulas for these elements are N_2, O_2, Cl_2, F_2, Br_2, I_2, At_2, and H_2. These exceptions are so important that you will have to memorize them. Notice that 5 of these 8 homonuclear diatomics are in group 7A on the periodic table. That should help you memorize them.

Please understand that the concept of a homonuclear diatomic applies only to an *element*, not to an atom within a molecule. For example, oxygen is a homonuclear diatomic. The element oxygen exists as O_2, not as O. However, when oxygen is a part of a molecule, it can exist as 1 oxygen atom, 2 oxygen atoms, 3 oxygen atoms, etc., depending on the molecule. For example, water (H_2O) molecules contain only 1 oxygen atom each. The fact that oxygen is a homonuclear diatomic is irrelevant when discussing the number of oxygen atoms in a water molecule. In the same way, nitric acid (HNO_3) contains 3 oxygen atoms. Once again, since the oxygen atoms are a part of a molecule, the fact that oxygen is a homonuclear diatomic is irrelevant. The *only time* to consider the fact that oxygen is a homonuclear diatomic is when we are dealing with the *element* oxygen.

Since elements H, N, O, F, Cl, Br, I, and At all exist as homonuclear diatomic molecules, no single atoms exist as a stable form of matter. The element fluorine exists as the molecule F_2. When we see the Lewis structure of the fluorine atom, we can see why:

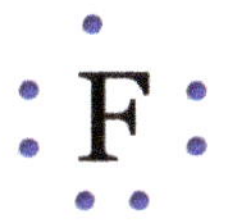

To attain an ideal electron configuration, fluorine needs to gain 1 electron. If a metal atom were around, it could take that electron away from the metal and become an F^- (fluoride) ion. However, suppose no metal were around. What could the fluorine atom do?

Wherever there is 1 fluorine atom, there are probably a few more. If another fluorine atom were to enter into the picture, a possibility would begin to emerge:

We made the second fluorine's electrons red so you would be able to tell one fluorine from another. Notice that each atom has a space for 1 electron, and each atom also has 1 unpaired electron. Suppose each atom's unpaired electron were to fit into the space that the other atom has:

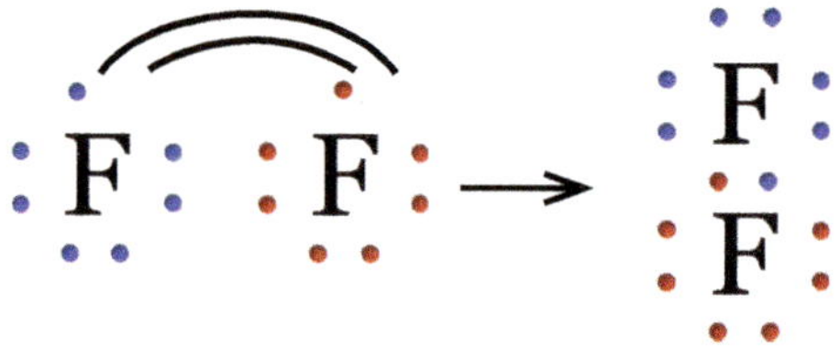

Make sure you understand what happened here. The electrons were *not* transferred from one atom to the other. Instead, the atoms were brought close enough so that each atom's unpaired electron was able to sit next to the other atom's unpaired electron. To make it more readable, the top *F* (the one with the blue electron dots) was turned over so that both atomic symbols were right side up.

Now look at what has happened. Although each atom still only has 7 valence electrons, each electron has another electron to pair up with. Because the 2 fluorine atoms share 1 of each other's electrons and all the electrons seem to be paired up, they think they each have 8 valence electrons. When all of an atom's electrons are paired up, the atom thinks that it has 8 electrons.

If you don't quite understand why sharing electrons like this makes an atom think it has an ideal electron configuration, think of it this way. Each fluorine atom is sharing an electron with the other. This means that, for certain moments in time, the electron pair pictured between the 2 atomic symbols above is whirling around the first fluorine atom. For those moments, the first fluorine atom has 8 valence electrons. During other moments, the shared electron pair is whirling around the second fluorine atom, making it have 8 valence electrons. Even though neither of the fluorine atoms always has 8 valence electrons, each of them has 8 valence electrons *sometimes*.

Since both fluorine atoms gain a little in this arrangement, they stay together, sharing the pair of electrons that are pictured between the 2 atomic symbols in the Lewis structure above. This shared electron pair is called a covalent bond.

Covalent bond—A shared pair of valence electrons that holds atoms together in covalent compounds.

This covalent bond is the fundamental thing that separates an ionic compound from a covalent one.

The atoms in covalent compounds share electrons to form molecules, while the atoms in ionic compounds give and take electrons to form ions.

Notice the difference here. The atoms in ionic molecules stay together because the positively charged ions are attracted to the negatively charged ions. In covalent compounds, there is a physical bond (a pair of shared electrons) which holds the atoms together.

Since the covalent bond created by shared electron pairs holds the atoms in a molecule together, we often emphasize this in the Lewis structure by replacing the pair of dots with a dash. This illustrates that there is something physically attaching one atom to the other. It also helps to distinguish the electron pairs that are not being shared from the ones that are being shared. The best way to represent the Lewis structure for a fluorine molecule is:

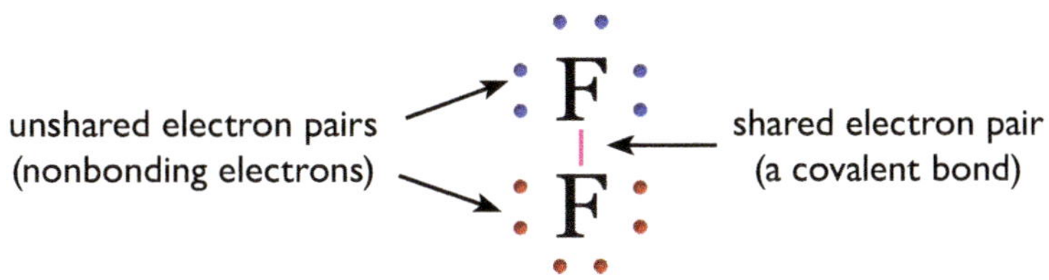

Now do you see why determining molecular structure is a lot like putting together a jigsaw puzzle? To determine what a molecule of any covalent compound looks like, all we have to do is find out how we can fit the atoms together in such a way as to give every atom its ideal electron configuration.

Let's look at another covalent compound so that we can better see how to determine the Lewis structure of molecules. Consider the compound carbon tetrachloride. According to the nomenclature rules you learned in the last module, this compound is composed of 1 carbon atom and 4 chlorine atoms. To determine its Lewis structure, we need to look at the Lewis structures for these atoms:

This is roughly equivalent to laying out all of the puzzle pieces before starting the puzzle. It helps us see what we have to work with.

The first thing we have to realize about working with these Lewis structures is that the atoms given in the chemical formula are all we have to work with. We cannot use any more or any less than the atoms listed in the chemical formula. No matter what, we must use exactly 1 C atom and 4 Cl atoms to make this molecule.

Secondly, the dots in the Lewis structures for each of these atoms are all the dots that we have to work with. We cannot use any more or any less than the dots drawn. Since carbon has 4 valence electrons (4 dots) and the 4 chlorine atoms each have 7 valence electrons (for a total of 28 dots), we have exactly 32 dots to work with. No more, no less. We must construct a molecule in which each atom has 8 dots around it, using only these 32 dots. How do we go about doing this?

When putting together a jigsaw puzzle, experts recommend constructing the frame of the puzzle first because the frame pieces are generally the easiest to find. In putting together Lewis structures, it is best to start at the *center* and work our way out. Generally, we put the atom with the *most unpaired electrons* in the center. We then try to attach the remaining atoms to this 1 central atom. If we happen to have more than 1 of the atom with the most unpaired electrons, we generally hook them all together and put the whole chain in the center.

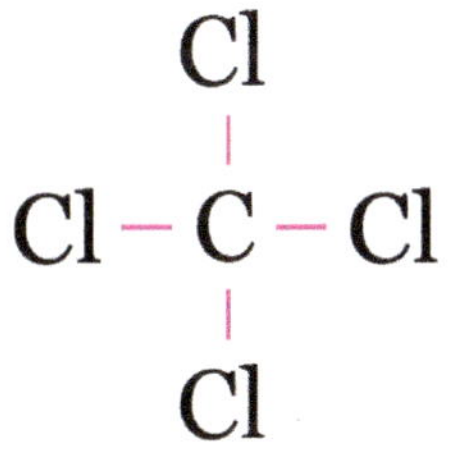

It does not matter where we put the Cl atoms around the central atom. Each line represents 2 shared electrons, so we have 8 electrons total in our Lewis structures. Where do the other 24 electrons go? Don't forget the ultimate goal of the atoms in a covalent molecule. The atoms want 8 (H wants 2) valence electrons. Let's now give the outside atoms their octet. This means placing 6 dots around each Cl atom.

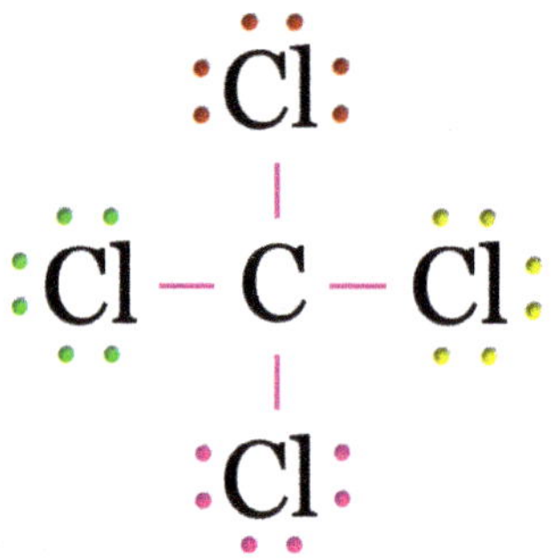

Finally, we must make sure the central atom has its octet. In this example, the C atom does have an octet with the 4 bonds that have 2 valence electrons each. This, then, is the final electron configuration for CCl_4.

We suggest using the following steps to construct Lewis structures.

1. **Count the valence electrons in the entire molecule by adding up the valence electrons of the individual atoms.**
2. **Put the atom with most unpaired electrons in the center. If there is more than 1 of those atoms, put them all in the center, bonded together.**
3. **Arrange the other atoms around the central atom and bond them with a single covalent bond.**
4. **Fill in the octets of the outside atoms.**
5. **Make sure the central atom has its octet.**

We need to warn you about an issue with constructing Lewis structures. When an atom enters into a covalent molecule, the number of valence electrons it started with is no longer important. The important issue is that the atom wants 8 valence electrons and the molecule can place the valence electrons anywhere in the molecule to ensure that all atoms have access to their 8 valence electrons. This point will become more evident in the next section when we discuss more complicated Lewis structures. In the meantime, read through Example 4.5 and try "On Your Own" question 4.12.

EXAMPLE 4.5

What is the Lewis structure for a water molecule?

Water has the chemical formula H_2O. This means that a water molecule has 2 hydrogen atoms and 1 oxygen atom. To determine the molecule's Lewis structure, we first have to look at the Lewis structures for the individual atoms:

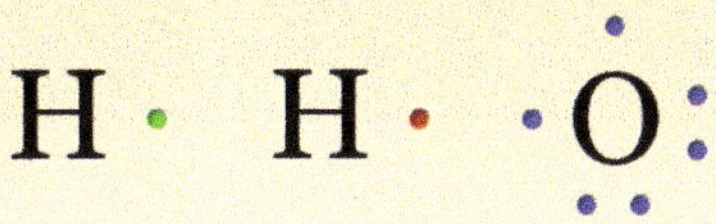

Remember that unlike most atoms, hydrogen needs only 2 valence electrons. Therefore, we must use the dots pictured above and arrange them to give the hydrogen atoms 2 valence electrons and the oxygen atom 8. We start by putting the atom with the most unpaired electrons in the middle. That would be oxygen:

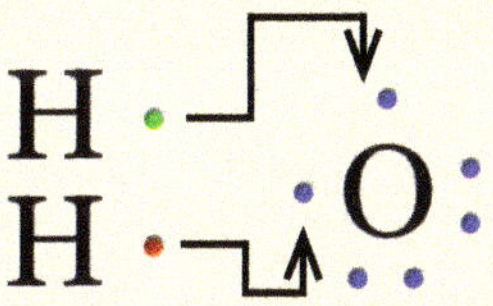

Now we have to attach the H atoms to the O atom. This isn't hard, since each hydrogen atom has 1 space open for an electron, and the oxygen atom has 2 spaces open:

This gives us 8 electrons around the oxygen atom and 2 around each hydrogen atom. We have a good Lewis structure. Now all we need to do is replace the shared electron pairs with dashes to indicate that they are covalent bonds.

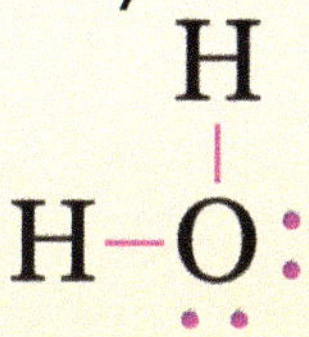

This is the final Lewis structure for water.

Please realize that the actual position of the H atoms around the O atom is irrelevant. Because molecules are free to rotate in 3-dimensional space, any of the following Lewis structures would be acceptable for water:

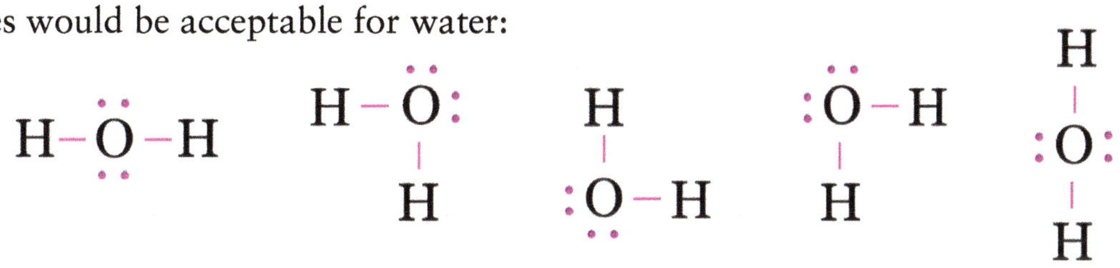

Your Lewis structure does not need to be *identical* to the ones drawn here. It just needs to have the same central atom, the same atoms bonded to that central atom, and the same number of nonbonding electron pairs for each atom. For example, in the case of water, as long as your Lewis structure has oxygen at the center with 2 nonbonding electron pairs, and as long as it has 2 hydrogen atoms bonded to the oxygen atom with 1 shared electron pair each, your Lewis structure is just as valid.

ON YOUR OWN

4.12 Draw the Lewis structures for the following molecules:

a. NH_3 b. $SiBr_4$ c. H_2S d. ICl

MORE COMPLICATED LEWIS STRUCTURES

If Lewis structures were always this simple, anyone could do them. However, there are much more complicated molecules in nature, requiring more complicated Lewis structures. For example, consider the very important molecule oxygen. Oxygen is a homonuclear diatomic, making its chemical formula O_2, which means that we have 2 oxygen atoms to work with:

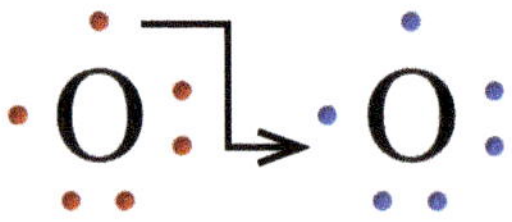

Each O atom has 2 unpaired electrons, so we could start by pairing 1 of the unpaired electrons on the first O atom with 1 of the unpaired electrons on the second O atom:

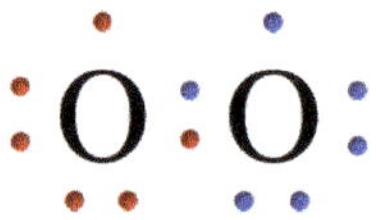

Now we have a problem. Each O atom has 7 electrons around it, but they each need 8. There are no more electrons, so what can we do? We just need to remember that a dot counts for an atom if it is next to the atom's symbol or *between the atom's symbol and another atom's symbol.* So, we just need to take the unpaired electron from the first O atom and move it between the 2:

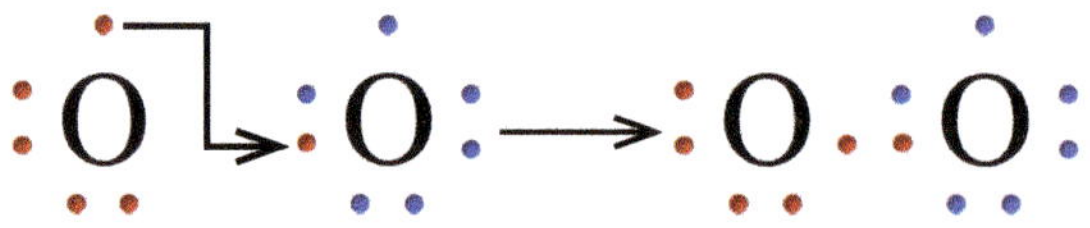

How does this help? Count the number of electrons that belong to the second O atom now:

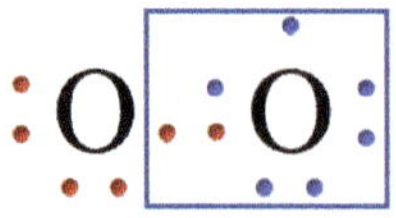

The dots inside the box all belong to the second O atom because they are all right next to the atomic symbol or between the 2 atomic symbols. If you count, there are 8 of them. This gives the second O atom its ideal electron configuration! What about the first O atom, however?

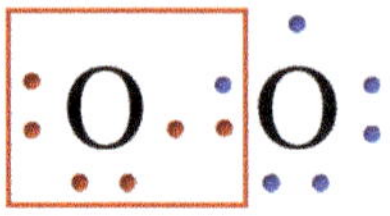

The first O atom still has only 7 dots. This is easy to fix because there is an unpaired electron on the second O atom. If we move it down between the 2 O atoms, it will count for both atoms:

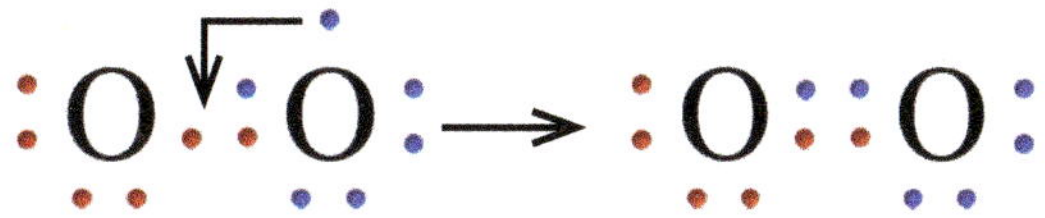

If you count the dots now, each O atom has 8 electrons either next to it or between it and the other O atom. Each atom has its ideal electron configuration. All we have to do is to replace each shared electron pair with a dash to indicate that it is a covalent bond. Since there are 2 shared electron pairs, we must use 2 dashes.

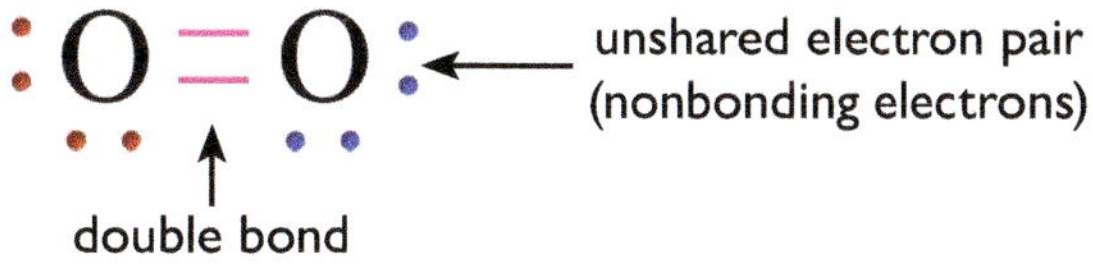

Since 2 shared electron pairs make up this bond, we call it a double bond. As you might expect, a double bond is much stronger than a single bond. Comparing F_2 (which we saw earlier is linked by a single bond) to O_2, we could say that it would be more difficult to break down an O_2 molecule than an F_2 molecule because of the stronger bond in O_2.

Be sure you realize what went on when we constructed the double bond in our Lewis structure above. We moved electrons from one place in the Lewis structure to another place. This is perfectly legal, as long as we do not get rid of any electrons. Also, notice that when we move an electron from 1 side of an atom to a spot between the 2 atoms, the electron suddenly counts for both atoms. This will be a very useful technique in determining the Lewis structure of complex molecules.

Let's look at one more complex Lewis structure. Two principal components of the air we breathe are N_2 and O_2. We just learned about the Lewis structure for O_2, so now let's look at N_2. As the chemical formula indicates, N_2 is composed of 2 nitrogen atoms:

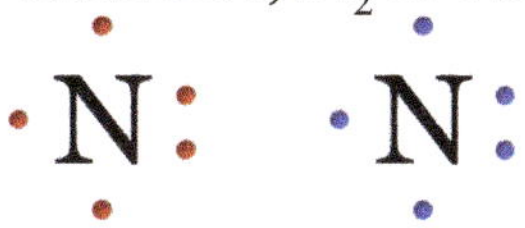

Each of these atoms has 3 unpaired electrons. The first thing we can do with them is pair up 1 of the unpaired electrons on the first N atom with 1 of the unpaired electrons on the second N atom:

This is a good start, but we are far from done. Each N atom has only 6 electrons now, but they both need 8. We can give the second N atom 8 electrons by taking the 2 unpaired electrons on the first N atom and putting them between the 2 N atoms. This allows the electron to count for both atoms.

Now the N atom on the right has 8 electrons, but the N atom on the left still has only 6. We can fix that by moving the 2 unpaired electrons on the second N atom between the 2 N atoms.

Once again, since the electrons between the 2 N atoms count for both of them, we now have 8 electrons with each atom. The only thing left to do is replace the shared electron pairs with dashes. Since there are 3 shared electron pairs, we must have 3 dashes:

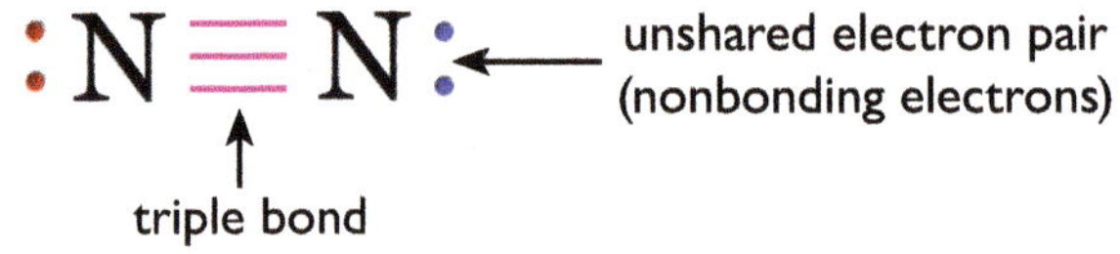

This type of bond is, naturally, called a triple bond. As you might expect, it is stronger than a double bond. N_2 would be more difficult to break apart than O_2 because the triple bond that holds N_2 together is stronger than the double bond that holds O_2 together.

The real trick to figuring out Lewis structures is just to keep trying. Start working on an idea and see if it pans out. If it doesn't, start over and try it again with a new idea. Don't be afraid to make some mistakes before you come up with the correct answer. Make sure you understand the way to make double and triple bonds in Lewis structures by studying Example 4.6 and answering "On Your Own" questions 4.13–4.14. The next module uses Lewis structures extensively, so it is important that you know how to work with them.

EXAMPLE 4.6

Determine the Lewis structure for carbon dioxide.

The name tells us that the chemical formula is CO_2, which means that we have 1 carbon atom and 2 oxygen atoms to work with, giving a total of 16 valence electrons in the molecule.

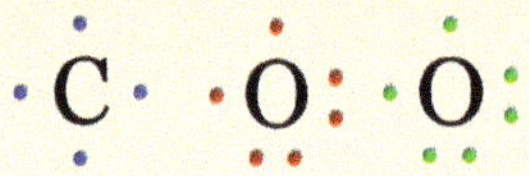

According to the procedure we are supposed to follow, we must put the C atom in the middle since it has the most unpaired electrons, and then we must try to attach the O atoms to it:

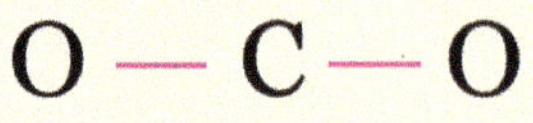

Fill in the octets on the outside atoms.

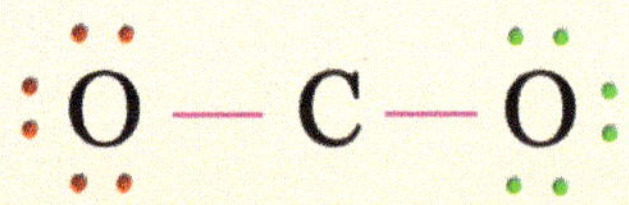

We have used up all 16 valence electrons. The carbon atom has 4 electrons. The oxygen atoms have 8, so we need to fix that carbon atom. But how do we do this when we have no more electrons to give carbon? Multiple bonds!

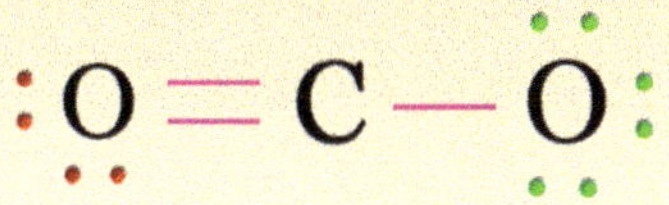

This gives carbon 6 electrons, so we will need 1 more multiple bond to give it 8. Should we double bond with the oxygen atom on the right or triple bond with the oxygen atom on the left?

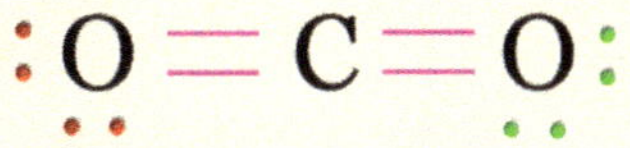

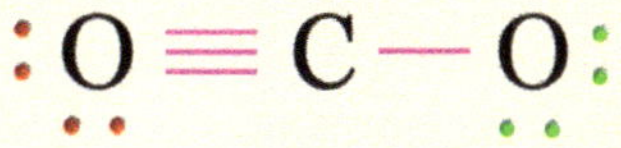

For this course, choose symmetry whenever possible. In college chemistry you will learn more about how the electronegativities of the outside atoms determine what kinds of bonds are formed. However, there is a rule you should follow with multiple bonds. Group 7A atoms do not double or triple bond due to their high electronegativities.

Now all atoms in the molecule have 8 electrons, so our Lewis structure is complete.

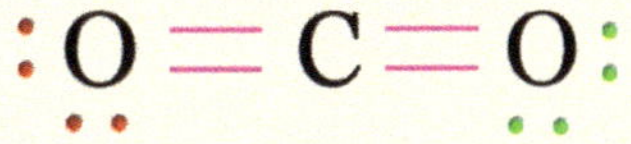

A carbon dioxide molecule, therefore, consists of 2 oxygen atoms, each double–bonded to a carbon atom.

What is the Lewis structure for HCN?

The formula tells us that we have 1 hydrogen atom, 1 carbon atom, and 1 nitrogen atom to build into our Lewis structure.

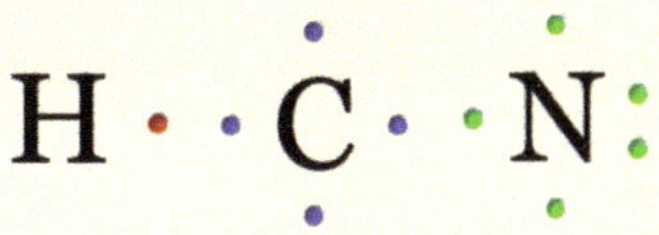

According to the procedure we are supposed to follow, we must put the carbon in the middle since it has the most unpaired electrons, and attach the others to it:

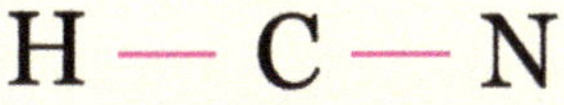

Fill in octets on outside atoms first. The H atom only wants 2 valence electrons, so once H is bonded, it is done. So the N atom needs 6 more valence electrons around it.

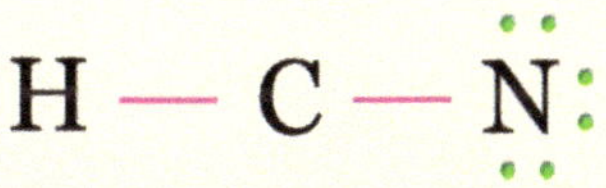

This uses all 10 of our valence electrons. We need to multiple bond with N to get C its 8 valence electrons.

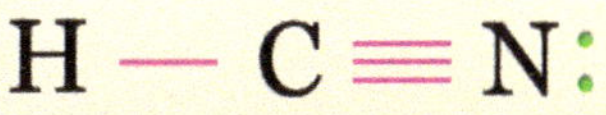

Sharing a double bond between C and N gets C access to only 6 valence electrons. This means we will need to share another bond between C and N, making a triple bond.

ON YOUR OWN

4.13 Draw the Lewis structure for CH_2O.

4.14 Draw the Lewis structure for C_2H_2.

AN APPLICATION OF LEWIS STRUCTURES

Lewis structures are not only helpful in determining what a molecule looks like; they are also helpful in understanding a molecule's chemical behavior. For example, as we mentioned in the last module, the surface of our planet is protected from the ultraviolet rays of the sun by the ozone layer. Your knowledge of Lewis structures will now allow you to understand how this system works.

Recall from the previous module that the sun is bathing the earth with all wavelengths of light. Our eyes use the visible light to see. We use light with longer wavelengths (infrared, microwave, etc.) for many diverse purposes, including microwave ovens, television, and radio. Light waves with wavelengths smaller than visible light (gamma rays, X–rays, ultraviolet rays), however, are quite dangerous to living organisms because they have enough energy to destroy any living tissue that they contact. If the surface of our planet were not protected from these light waves, nothing could survive. Any living organism would immediately be fried by these energetic light waves.

How is the earth protected from these light waves? A 3–tiered system blocks high–energy light that comes from the sun. First, the highest–energy light that comes from the sun consists of gamma rays. These light waves have very small wavelengths, indicating very large energy. As these light waves strike our atmosphere, they encounter an enormous amount of nitrogen. As you recall from our previous discussions, nitrogen's Lewis structure looks like this:

We told you before that this triple bond makes the nitrogen molecule very hard to break apart. These high–energy light rays have plenty of energy, so when they strike the nitrogen molecule, they are able to break it apart. This uses up all of the energy in the light wave. As mentioned in Module 1, light is pure energy. Therefore, when the energy gets used up, the light wave is gone!

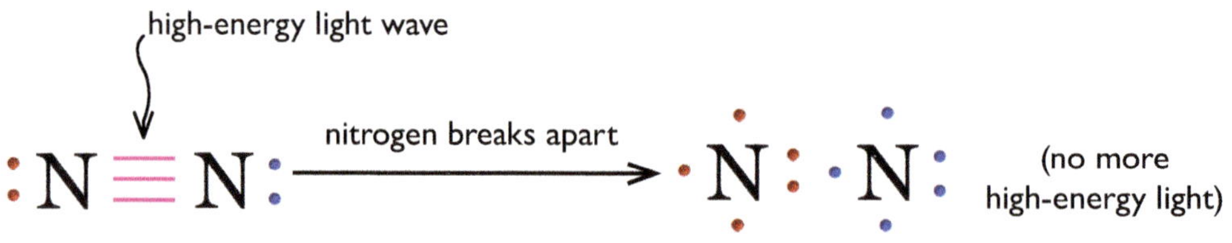

In this way, nitrogen protects the earth from harmful gamma rays.

But gamma rays are not the only high–energy light waves that come from the sun. X–rays also strike the earth's atmosphere daily. Although these light waves still have enough energy to destroy living tissue, they do not have enough energy to break apart nitrogen molecules. The nitrogen cannot block them. Fortunately there is also a lot of oxygen in the atmosphere. Oxygen, as we have learned, has the following Lewis structure:

The double bond indicates that the oxygen molecule is held together tightly, but not as tightly as the nitrogen molecule. Although X–rays cannot break nitrogen molecules apart, they can break oxygen molecules apart. The oxygen in our atmosphere blocks the X–rays that come from the sun.

There is still one more type of high–energy light that comes from the sun, however. Ultraviolet rays have enough energy to destroy living tissue, but not enough energy to break apart oxygen molecules. Oxygen cannot block ultraviolet rays. Instead, a substance known as ozone takes care of them. To see why ozone can block ultraviolet rays, we need to determine its Lewis structure.

Ozone is a strange substance. Its chemical formula is O_3. To determine its Lewis structure, we start by writing out the 3 O atoms that we have to work with:

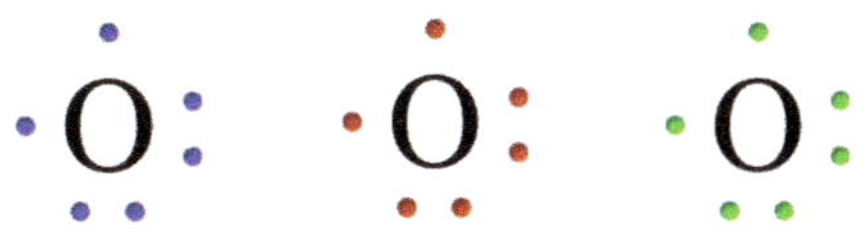

We have 18 valence electrons with these 3 atoms. But which one do we pick for the central atom? Well, it is not going to matter since the atoms are identical.

We will fill in the octets on the outside O atoms first.

We have used only 16 of our allotted 18 valence electrons. The 2 leftover electrons go on the central atom.

We still need 2 more electrons on the central atom. We will need to double bond with one of the oxygen atoms. But which one? It does not matter because all we would have to do is rotate the structure 180°, and it would be the same structure.

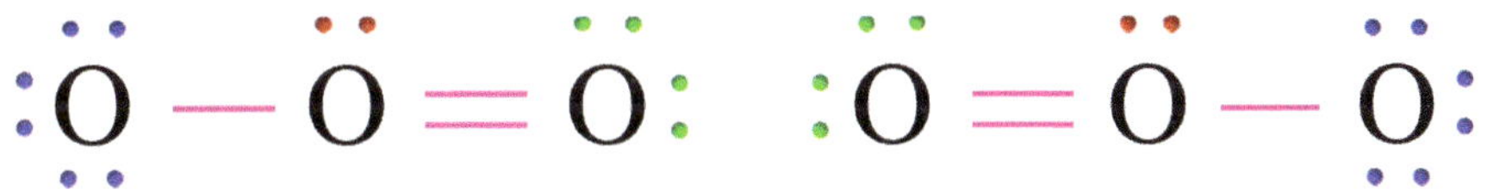

Now all the O atoms have 8 dots, so we have a valid Lewis structure. We could have also drawn the Lewis structure in a bent molecule. You will learn more about molecule geometry in the next module.

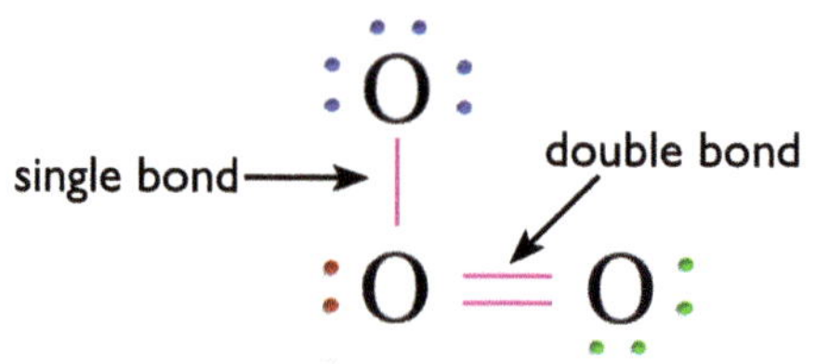

Now that we have the Lewis structure, we can see why ozone is uniquely designed to block ultraviolet light that comes from the sun. Ultraviolet light has less energy than gamma rays, so it cannot break apart a molecule with a double bond. However, ozone has a single bond, which is weaker than a double bond. Ultraviolet light has just enough energy to break this bond. Once again, when the bond is broken, the energy of the light is used up, so the light disappears.

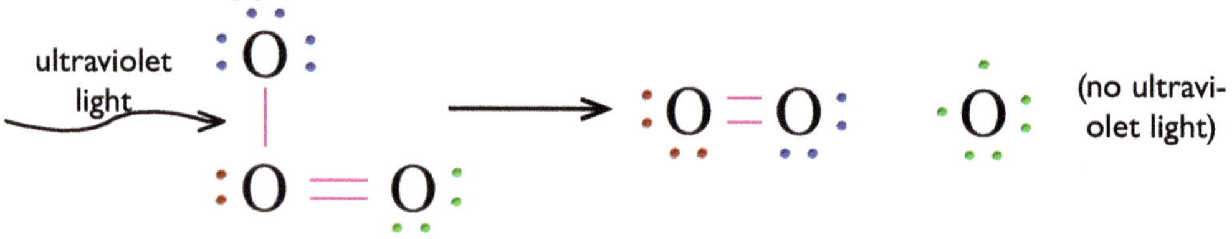

This is how ozone blocks ultraviolet light.

Do you see the marvelous amount of engineering that went into this system? Since many different types of high–energy light from the sun must be blocked, the Designer of this planet installed a 3–tiered protection system. The nitrogen in the atmosphere protects us from the most highly energetic light, the oxygen in the atmosphere protects us from the intermediately high–energy light, and the ozone in the atmosphere protects us from the moderately high–energy light. This complex system for protecting living organisms from harmful light rays is just one more piece of evidence that the earth was created, not formed by chance.

If you are impressed with this system so far, hang on! The most incredible aspect of this light–blocking system hasn't even been discussed yet. Although its ability to block ultraviolet light makes ozone necessary for life as we know

science and creation

Earth's atmosphere contains plenty of ozone, but the vast majority of it exists in a layer of the atmosphere (the ozone layer) that is 20 to 30 kilometers (12.4 to 18.6 miles) above sea level, where no living organism breathes! Think about all of this for a moment. Earth just *happens* to have all of the gases necessary to be protected from the sun's harmful rays, and they just *happen* to be in the right place. If nitrogen weren't in the atmosphere, life could never exist because of the gamma rays hitting the planet. If oxygen weren't present, living organisms would suffocate as well as die from exposure to X-rays. Finally, if ozone were not in the atmosphere, living organisms would die from exposure to ultraviolet rays. Having ozone in the atmosphere is not all that is necessary, however. If ozone were evenly distributed around the atmosphere, as are nitrogen and oxygen, living organisms would die from the poisonous effects of ozone. So we must have ozone, but it must be somewhere in the atmosphere where no living organisms breathe.

Once again, design features such as this one provide ample evidence that this earth did not appear by chance. The probability of a planet forming with just the right gases in just the right places in the atmosphere is simply too small to consider. Add to this the fact that the design features inherent in our atmosphere constitute only a minuscule fraction of the design features of the entire planet, and an unbiased scientist can come to only one conclusion: This planet is a system that has been designed and created by God.

it, ozone is also incredibly poisonous to living organisms. If living organisms breathe in too much ozone, they die. We must have ozone to protect us from the sun's ultraviolet rays, but we cannot breathe too much of it, or it would kill us. Seems like a contradiction, doesn't it? It would be, except the *Designer* of our planet is smarter than we are.

SUMMARY OF TABLES IN MODULE 4

Table 4.1: Positive and Negative Ions for Groups 1A–8A

Table 4.2: Common Metals That Form More Than One Positive Ion

ANSWERS TO THE "ON YOUR OWN" QUESTIONS

4.1 (a) Since Ca is in group 2A, it has 2 valence electrons:

(b) Since Si is in group 4A, it has 4 valence electrons:

(c) Since At is in group 7A, it has 7 valence electrons:

4.2 (a) Since beryllium is in group 2A, it has a 2+ charge. Chlorine is in group 7A, so chloride has a 1- charge. Ignoring the + and - signs and switching the numbers gives us $BeCl_2$.

(b) Since potassium is in group 1A, it has a 1+ charge. Oxygen is in group 6A, so oxide has a 2- charge. Ignoring the + and - signs and switching the numbers gives us K_2O.

(c) Since aluminum is in group 3A, it has a 3+ charge. Phosphorus is in group 5A, so phosphide has a 3- charge. The numerical values of the charge are identical, so we ignore them: AlP.

(d) Since magnesium is in group 2A, it has a 2+ charge. Nitrogen is in group 5A, so nitride has a 3- charge. Ignoring the + and - signs and switching the numbers gives us Mg_3N_2.

4.3 Since cesium is in group 1A, it has a charge of 1+. Nitrogen is in group 5A and has a 3- charge. Ignoring the + and - signs and switching the numbers gives us Cs_3N.

Cs is the metal, so its ion name is cesium. N is the nonmetal, so its anion name is nitride. Therefore, the compound's name is cesium nitride.

4.4 (a) In this problem, the roman numeral after tin tells that it has a 4+ charge. Fluorine is in group 7A, and fluoride has a 1- charge. Ignoring the + and - signs and switching the numbers gives us SnF_4.

(b) In this problem, the roman numeral after chromium tells that it has a 3+ charge. Chlorine is in group 7A, and chloride has a 1- charge. The numerical values of the charges are the same, so we ignore them, giving us $CrCl_3$.

(c) In this problem, the roman numeral after copper tells that it has a 2+ charge. Oxygen is in group 6A, and therefore oxide has a 2- charge. Ignoring the + and - signs and switching the numbers gives us CuO.

4.5 (a) The periodic table shows that iron (Fe) is a metal and can have multiple charges. However, the anion fluorine (F) is group 7A and always has a 1- charge. Since there are 3 F atoms for every Fe atom, Fe must have a charge of 3+. Fe is the metal cation, so its ion name is iron. Since iron can have multiple charges, it needs a roman numeral (III) to show the charge for this cation. F is the nonmetal anion, so its ion name is fluoride. Therefore, the compound's name is iron (III) fluoride.

(b) The periodic table shows that nickel (Ni) is a metal and can have multiple charges. However, the anion iodine (I) is group 7A and always has a 1- charge. Since there is 1 I atom for every Ni atom, then Ni must have a charge of 1+. Ni is the metal cation, so its ion name is nickel. Since nickel can have multiple charges, it needs a roman numeral (I) to show the charge for this cation. I is the nonmetal anion, so its ion name is iodide. Therefore, the compound's name is nickel (I) iodide.

(c) O is in group 6A, so it wants to be 2-. Manganese (Mn) is a metal, so it can possibly have multiple charges. Since O is 2- and there are 3 O atoms for every 2 Mn atoms, Mn must have a charge of 3+. Mn is the metal cation, so its ion name is manganese. Since manganese can have multiple charges, it needs a roman numeral (III) to show the charge for this cation. O is the nonmetal anion, so its ion name is oxide. Therefore, the compound's name is manganese (III) oxide.

4.6 These elements all lie in the same row. This means that the only difference in their position on the table is left to right. Since ionization energy increases from left to right on the table, the correct order is:

$$Cs < Ir < Pb < At$$

4.7 The only difference in position for these elements is vertical. Because ionization energy decreases from top to bottom on the table, Fr has the lowest ionization energy. This means that Ra gives up its electrons most readily.

4.8 All of these atoms are in the same row on the periodic table, so the only difference between their positions is left to right. Since Se lies farthest to the right, it has the highest electronegativity. This means that Se has the highest desire for electrons.

4.9 These elements all lie in the same column on the table, so their only difference in position is top to bottom. Because electronegativity decreases from top to bottom of the table, the order is:

$$At < I < Br < Cl < F$$

4.10 All of these elements are in the same row. Since the atomic radius decreases from left to right on the periodic table, the leftmost element will be largest. That means K is the largest.

4.11 All of these elements are in the same column. Since the atomic radius increases as you move down the table, the order is:

F < Cl < Br < I < At

4.12 (a) The chemical formula tells us that we have 1 N atom and 3 H atoms. That is a total of 8 valence electrons to work with.

Because N has the most unpaired electrons, it goes in the center and we attach the H atoms to it.

Remember, the ideal electron configuration for hydrogen is 2 valence electrons, while for nitrogen it is 8. We have only used 6 of our electrons, so the other 2 will go on the N atom. This will give all atoms access to their desired octets.

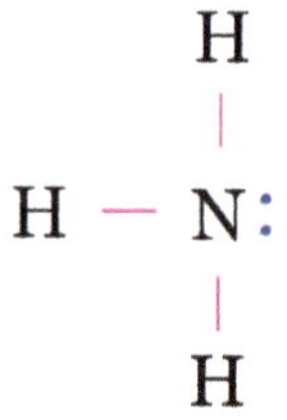

(b) The chemical formula tells us that we have 1 Si atom and 4 Br atoms to work with, making a total of 32 valence electrons.

Because Si has the most unpaired electrons, it goes in the center, and we attach the Br atoms to it.

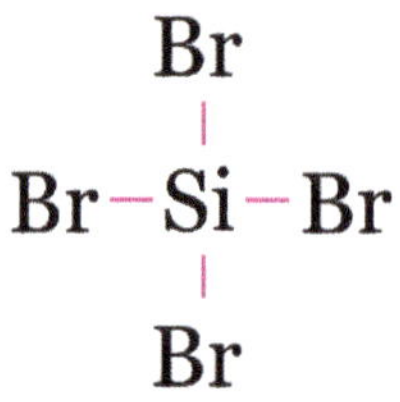

Fill in the octets on the outside atoms. The Lewis structure should look like this:

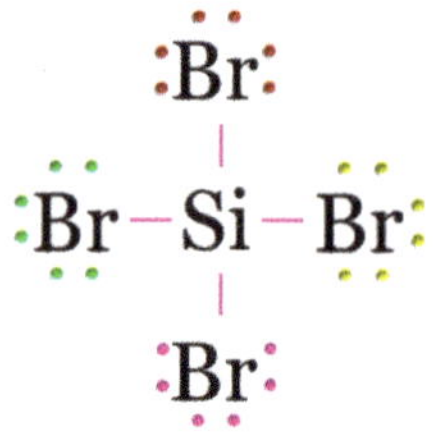

All atoms have 8 valence electrons now, and we have used up all 32 valence electrons. That structure is our final answer.

(c) The chemical formula tells us that we have 1 S atom and 2 H atoms, making a total of 8 valence electrons in this Lewis structure.

Because S has the most unpaired electrons, it goes in the center, and we attach the H atoms to it:

Placing these 3 atoms in a straight line will also work. We used up 4 valence electrons, and we do not have to put any more around the H atoms since they want only 2 electrons. What do we do with the 4 remaining electrons? We put them on the central atom because Si needs 4 more valence electrons. The final Lewis structure looks like this.

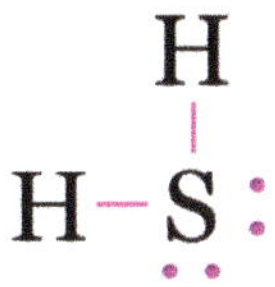

(d) The chemical formula tells us that we have 1 I atom and 1 Cl atom, making a total of 14 valence electrons.

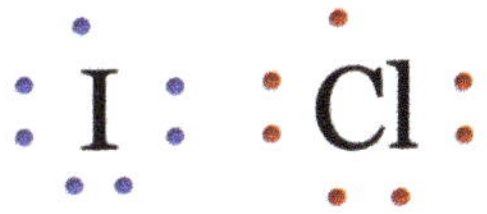

There is no need to pick a central atom because there are only 2 atoms in this molecule. Connecting the 2 atoms with a bond leaves 12 electrons to go around the 2 atoms. Since each atom needs 6 more electrons, the final structure should look like this.

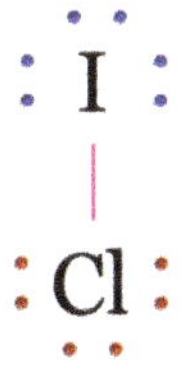

4.13 The chemical formula tells us that we have 1 C atom, 2 H atoms, and 1 O atom, making our Lewis structure have 12 valence electrons.

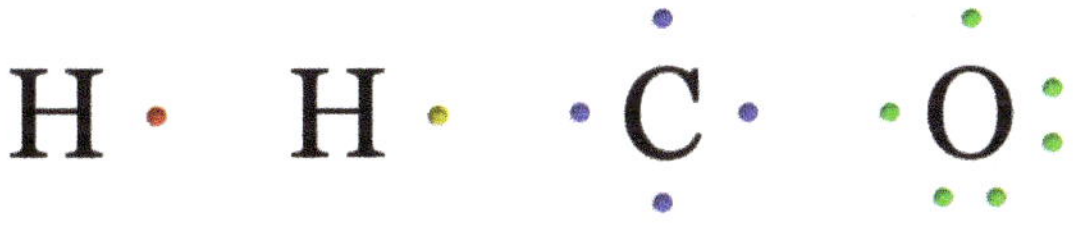

According to our procedure, we put the C atom in the middle (since it has the most unpaired electrons) and attach the other atoms to it.

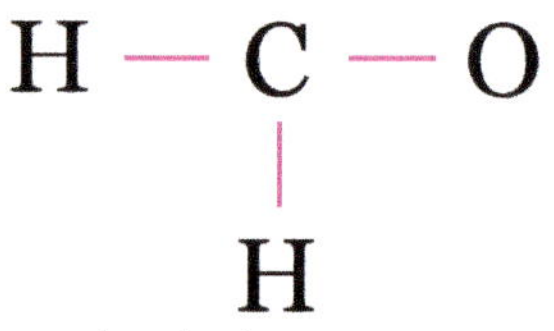

The H atoms already have their ideal electron configuration since they each have 2 electrons shared in the bond. We can take care of the O atom by giving it 6 valence electrons.

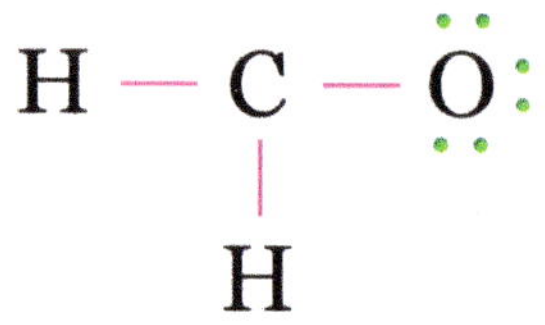

Now the oxygen atom has 8, but the carbon atom still has only 6. We have used all 12 of our allotted valence electrons. So how do we get the C atom another 2 electrons? We will create a double bond between C and O. This will give the following structure as the final answer:

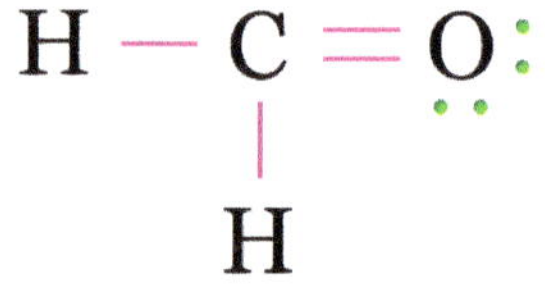

4.14 The chemical formula tells us that we have 2 C atoms and 2 H atoms. This means we will have 10 valence electrons in our Lewis structure.

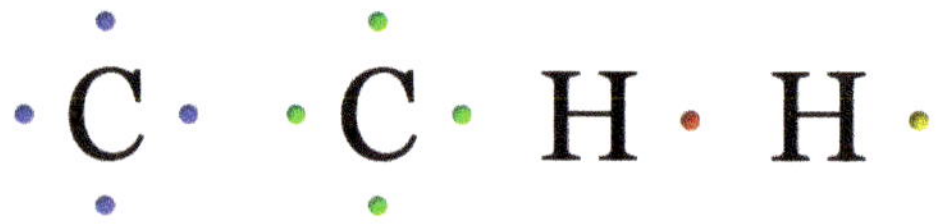

The procedure we have been following tells us that if we have 2 (or more) of the atom with the most unpaired electrons, we link them together and put them in the center.

We then bond the other atoms to them.

You might wonder how we decided to put 1 H atom on 1 C atom and the other H atom on the other C atom. Why aren't both H atoms on the same C atom? There is no way to know this when you are first starting to learn Lewis structures. This is simply the only way to get a good Lewis structure. You might have made several attempts before you came across this solution. One thing you might consider: Creation tends to like symmetry (balance). Putting an H atom on each C atom balances the molecule.

Now we can see that the H atoms have their ideal electron configuration, but both C atoms have only 6 electrons each. We have used 6 of our 10 total valence electrons, so we can divide the leftover electrons between the C atoms.

But this still leaves each C atom short of the 8 electrons it wants. Let's share a double bond between the 2 C atoms.

This gives the second C atom 8 electrons, but the first C atom still has only 6 We can fix that by making another bond between the 2 C atoms. This will give each atom access to the octet it wants.

NOTE: All Lewis structure illustrations in this module are by Conner Hughes.

STUDY GUIDE FOR MODULE 4

REVIEW QUESTIONS

1. What are valence electrons and why are they so important in chemistry?

2. How many valence electrons are in the following electron configuration?

$$1s^2\ 2s^2\ 2p^6\ 3s^2\ 3p^6\ 4s^2\ 3d^{10}\ 4p^6\ 5s^2\ 4d^{10}\ 5p^3$$

3. The chemistry of life is based on carbon. This means that all of the molecules which govern life–giving chemistry have carbon atoms as their principal component. Chemists have often speculated that there could be a life form with a chemistry based on silicon. Why?

4. What are the noble gases and why are they important in chemistry?

5. What is the octet rule? Give 2 exceptions to this rule.

6. What is the fundamental difference between metals and nonmetals?

7. Critique this statement: The number at the top of a column on the periodic table is the number of valence electrons that all elements have.

8. What is the difference between an atom and an ion?

9. What is ionization energy?

10. Name 3 periodic properties of atoms.

11. Describe how the periodic properties mentioned in question 10 vary as you move up a column and to the right along a row of the periodic table.

12. What is the fundamental difference between ionic compounds and covalent compounds?

13. What is wrong with the following formula: Mg_2O_2?

14. What gases are involved in earth's 3–tiered protection system against high–energy light that comes from the sun?

15. List the homonuclear diatomics.

PRACTICE PROBLEMS

1. Draw the Lewis structures for the following atoms:
   a. Ge b. Te c. Ba

2. Give the chemical formulas for the following molecules:
   a. aluminum sulfide b. iron (I) nitride
   c. magnesium oxide d. chromium(III) oxide

3. Which atom gives up its electrons most readily: Al, B, In, or Ga?

4. Order the following atoms in terms of increasing ionization energy: Sr, Sb, I.

5. Which atom has the greatest desire for extra electrons: N, Sb, or As?

6. Order the following in terms of increasing atomic radius: As, K, Br, Se.

7. Draw a Lewis structure for CH_4.

8. What is the Lewis structure for PCl_3?

9. What is the Lewis structure for FNO?

10. Based on their Lewis structures, which molecule is easiest to break apart:

$$C_2H_4, PN, \text{ or } H_2?$$

POLYATOMIC IONS AND MOLECULAR GEOMETRY

In the last module, we learned all about Lewis structures and how they relate to both ionic and covalent compounds. This module still deals with Lewis structures, but it applies them to new concepts. First, you will learn about more complicated ionic compounds. By design, the ionic compounds we have dealt with so far have been rather simple. The most important ionic compounds tend to be a little more complicated. To get a good understanding of chemistry, we must improve our knowledge of ionic compounds.

In addition to learning about more complicated ionic compounds, we will explore the 3–dimensional aspect of molecules. After all, the Lewis structures we learned in the last module are very informative, but they have a big drawback. They do not allow us to see what the molecules look like in 3–dimensional space. In other words, all Lewis structures are flat (2–dimensional) because they are drawn on a flat piece of paper. In reality, molecules are not flat; they are 3–dimensional. We will learn some concepts in this module that will help us visualize the 3–dimensional character of molecules.

The 3–dimensional nature of molecules will help us understand why certain liquids mix together easily (like milk and water), while other liquids do not (like gasoline and water). To learn all of these new concepts, you will have to be comfortable with the skills discussed in the last module. If you feel tentative about any of the material presented in

first reactions

As your understanding of basic chemistry grows, you are capable of entering a world where your mind can see molecules growing in 3-dimensional space. You can visualize how electrons whirling about the nucleus get shared by 2 atoms forming bonds, uniting them into structures you can describe and model. You can predict the outcome of reactions only because of the order God has created for our natural world. "For the LORD is God, and he created the heavens and earth and put everything in place. He made the world to be lived in, not to be a place of empty chaos. 'I am the LORD' he says, 'and there is no other'" (Isaiah 45:18, NLT).

the last module, it might be a good idea to go back and look at that material again or look at another resource that deals with such material before plunging into this module.

POLYATOMIC IONS

So far, you have learned that when a single atom gains or loses electrons, an ion forms. That is not the only way that an ion can form in nature. Several atoms can come together and gain or lose electrons as a group. Since these ions require more than 1 atom, we call them polyatomic (pahl' ee uh tom' ik) ions.

Polyatomic ions—Ions that are formed when a group of atoms gains or loses electrons.

Although you might not understand how this can happen, a quick look at the Lewis structures of a couple of polyatomic ions might help. For example, consider what might happen if a single hydrogen atom tried to form a molecule with a single oxygen atom. If we tried to construct the Lewis structure for such a molecule, we would start by drawing the Lewis structures of the individual atoms:

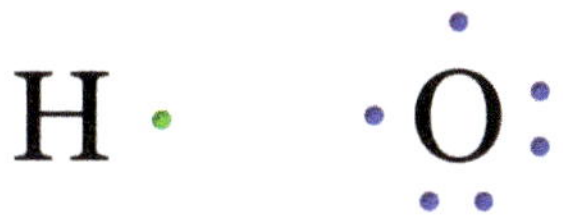

We would then link the 2 atoms together to see what we got:

In this Lewis structure, the H atom has its ideal electron configuration, but the O atom has only 7 electrons. We can play around with this Lewis structure as long as we like, but there is no way to move the electrons around so that both H and O get their ideal electron configurations. Since we cannot draw a good Lewis structure for the OH molecule, we have to assume that it doesn't exist.

Although the OH molecule doesn't exist, the OH^- *ion* does exist. After all, as the Lewis structure above shows, all that the O atom needs to attain its ideal electron configuration is 1 more electron. Suppose a sodium atom were willing to donate its electron to the oxygen in the OH:

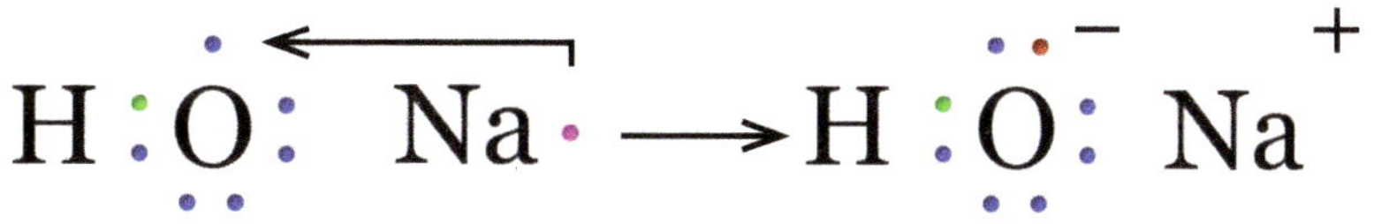

This situation is now ideal for all atoms concerned. The hydrogen atom has 2 electrons, the oxygen atom has 8 electrons, and the sodium atom has no electrons in its Lewis structure. This gives each atom its ideal electron configuration. But look what has happened. The oxygen and hydrogen atoms are linked together by shared electrons:

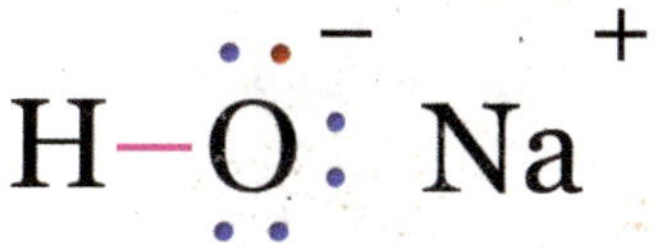

So although OH^- is an ion, it also contains a shared electron pair. These 2 characteristics make OH^- a polyatomic ion. Both the O and the H atoms share the negative charge brought about by the extra electron taken from the sodium atom. Since both the O and the H atoms share the negative charge, this is neither an oxide ion nor a hydride ion; it is something else. We call this ion a hydroxide ion. The resulting ionic compound, sodium hydroxide, has a chemical formula of NaOH. The O and the H are kept next to each other in the chemical formula to emphasize that they are hooked together by a pair of shared electrons.

Any ion that has more than 1 atom is a polyatomic ion. Another example of such a substance is the carbonate ion, which is composed of 1 carbon atom and 3 oxygen atoms:

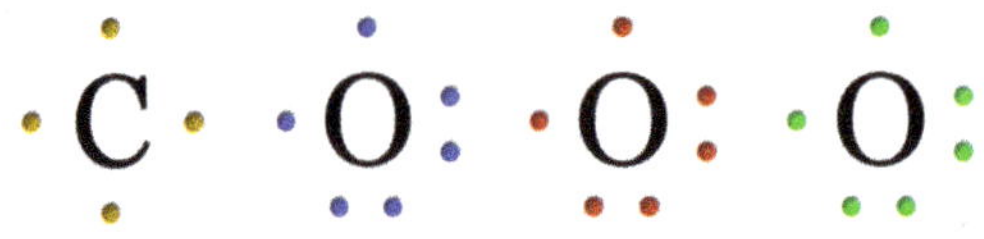

If we draw a Lewis structure according to the rules we followed in the previous module, we get:

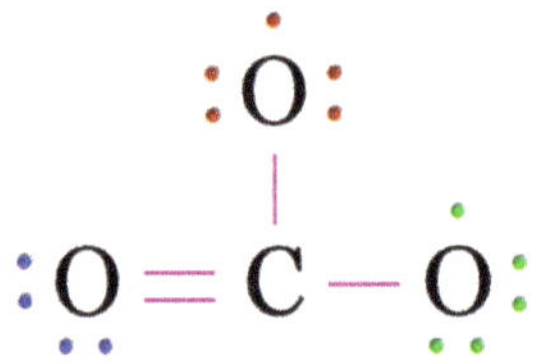

This gives the carbon atom and 1 of the oxygen atoms their ideal electron configurations, but 2 of the oxygen atoms are each 1 electron short. The only way that those oxygen atoms will come up with 8 electrons each is for each of them to get an extra electron. Suppose 2 potassium atoms were around and willing to donate their electrons:

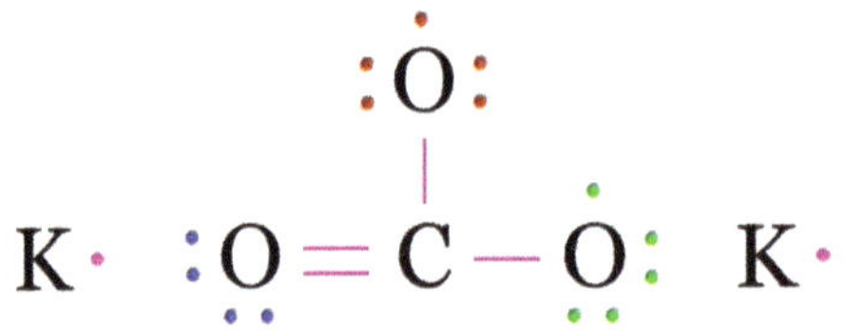

This would lead to:

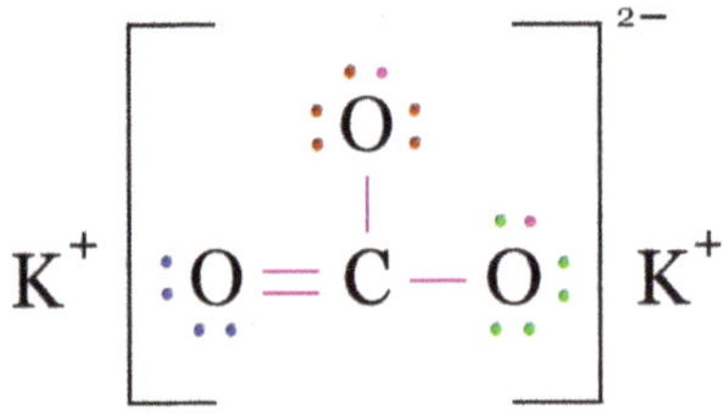

Since each potassium atom lost 1 electron, they each have a 1+ charge. The polyatomic ion, on the other hand, gained 2 electrons, so it has a 2- charge.

The resulting CO_3^{2-} ion is called the carbonate ion. The ionic compound that results from the above Lewis structure, K_2CO_3, is called potassium carbonate. Once again, the CO_3 is written together as a group to emphasize that these atoms are all linked together as a polyatomic ion.

Now that you know about polyatomic ions and why they exist, you need to become familiar with several of the most common ones. You need to know their chemical formulas, their charges, and how to determine the chemical formulas of the ionic compounds of which they are a part. The good news is that you do not have to learn their Lewis structures. You will never, in this course, be asked to construct a Lewis structure for a polyatomic ion because it involves adding (or subtracting) electrons like we did in the previous 2 Lewis structures for the hydroxide ion and the carbonate ion. We don't want you to do this because it breaks a rule we established in the last module. If you go on to advanced chemistry, you will be expected to draw polyatomic ion Lewis structures.

Although you do not need to know their Lewis structures, you do need to memorize the names, formulas, and charges of the polyatomic ions listed in Table 5.1.

TABLE 5.1
Important Polyatomic Ions

Ion Name	Formula	Ion Name	Formula
ammonium (uh moh' nee uhm)	NH_4^+	cyanide (sigh' uh nide)	CN^-
hydroxide (hye drox' ide)	OH^-	carbonate (kar' bun ate)	CO_3^{2-}
chlorate (klor' ate)	ClO_3^-	chromate (krohm' ate)	CrO_4^{2-}
chlorite (klor' ite)	ClO_2^-	dichromate (dye krohm' ate)	$Cr_2O_7^{2-}$
nitrate (nye' trate)	NO_3^-	sulfate (suhl' fate)	SO_4^{2-}
nitrite (nye' trite)	NO_2^-	sulfite (suhl' fite)	SO_3^{2-}
acetate (as' uh tate)	$C_2H_3O_2^-$	phosphate (fahs' fate)	PO_4^{3-}

This is a significant amount of memorization work, but it will get easier as time goes on because we will use these polyatomic ions quite a bit. Notice that even though there is not a lot of rhyme or reason behind these names, there is a little. When an ion ends with an oxygen atom, its name usually starts with the first atom in the ion and ends in *–ate* or *–ite*. In addition, for some starting atoms, the ion with more oxygen atoms has the *–ate* ending, while the one with fewer oxygen atoms has the *–ite* ending. These facts may help you in memorizing these ions.

Not only do you need to know the names, charges, and formulas of polyatomic ions, but you also have to be able to determine the chemical formulas of ionic compounds that contain polyatomic ions. This doesn't require learning anything new because the technique you used in the previous module for determining the chemical formula of ionic compounds still applies here. You just have to be careful about how you write the final result. Study Example 5.1 and then answer "On Your Own" questions 5.1–5.3 to make sure you understand how to do this.

EXAMPLE 5.1

What is the chemical formula for calcium nitrate?

Step 1: Identify the ions. According to the name, calcium nitrate is made up of the calcium ion and the nitrate ion. Calcium is in group 2A, so the calcium ion must have a charge of 2+. The nitrate ion is a polyatomic ion. According to the table, it has a charge of 1^-.

Step 2: Balance the charges. To determine the chemical formula using the shortcut, we ignore the charges and switch the numbers. The 1 goes with the calcium ion, and the 2 goes with the nitrate ion.

Step 3: Write the formula, cation first, using subscripts from the charge balance. We write the resulting formula like this:

$$Ca(NO_3)_2$$

Let's make sure we understand why the chemical formula is written this way. We said that the 2 went with the nitrate ion; there are 2 nitrate ions in the chemical formula. The only way we can show that the 2 goes with all of the atoms in the nitrate ion is to use parentheses. This tells us that the entire nitrate ion appears twice in the formula for calcium nitrate. Some compounds have parentheses in their chemical formulas.

What is the chemical formula of magnesium phosphate?
According to the name, this compound is composed of the magnesium ion and the phosphate ion. According to the periodic table, the magnesium ion will have a charge of 2+. According to the table, the phosphate ion has a charge of 3^-. To determine the chemical formula, we ignore the signs and switch the numbers. We have 3 magnesium ions and 2 phosphate ions. Using our parentheses notation, we get:

$$Mg_3(PO_4)_2$$

Once again, realize that the 2 outside of the parentheses goes with both atoms inside the parentheses. This means that the compound is composed of 3 magnesium atoms, 2 phosphorus atoms, and 8 oxygen atoms.

What is the chemical formula for copper (I) acetate?
According to the name, this substance is composed of the copper ion and the acetate ion. Since there is a roman numeral after the copper, we know that copper is an exception to our rules. The (I) tells us that in this compound, it has a charge of 1+. According to the table, the acetate ion has a charge of 1^-. Since the numerical values of the charges are the same, we ignore them. This gives us:

$$CuC_2H_3O_2$$

Notice that in this chemical formula, parentheses are not needed for the polyatomic ion because there is only 1 acetate ion.

ON YOUR OWN

5.1 Write the chemical names for the following compounds:
a. $Ba(NO_3)_2$ b. $NaClO_3$ c. $Ca(OH)_2$ d. LiCN

5.2 Give the chemical formula for each of the following compounds:
a. potassium sulfate b. ammonium carbonate c. aluminum acetate

5.3 When aluminum nitrite reacts with sodium carbonate, aluminum carbonate and sodium nitrite are produced.Write the chemical formulas for each compound.

MOLECULAR GEOMETRY: THE VSEPR THEORY

Now that we have spent a considerable amount of time on ionic compounds, we turn back to the subject of covalent compounds. In the last module, we learned how to use Lewis structures to help us picture what a molecule might look like. The problem is that this picture is still a bit wrong. As we saw earlier in this module, a 2–dimensional Lewis structure cannot fully represent a 3–dimensional molecule. Molecules not only have length and width, but they also have *depth*. Lewis structures do not allow us to picture the depth of a molecule. We must learn something else before we can fully picture what a molecule really looks like.

Consider the Lewis structure for CH_4:

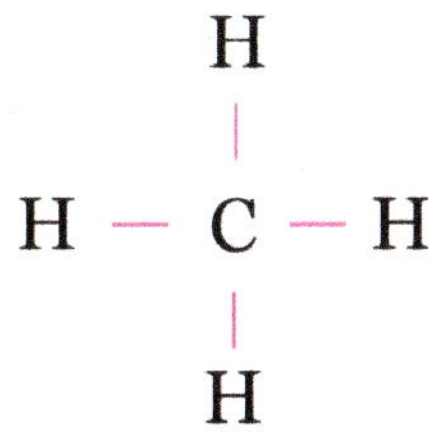

Remember that the dashes in this Lewis structure represent covalent bonds. Remember also that these covalent bonds are just electron pairs that are shared between 2 atoms. Since each of these covalent bonds is made up of electron pairs, we can easily predict something about their behavior: They will tend to repel one another because they both have the same type of electrical charge.

If these electron pairs repel each other, it stands to reason that they will try to move as far apart from each other as possible. How will they do that? If the molecule were to stay flat, then the electron pairs could not get any farther apart from each other than what is pictured in the Lewis structure above. In geometry terms, we could say that the largest angle between the chemical bonds would be 90 degrees. On the other hand, if the molecule took advantage of 3–dimensional space, the bonds could get farther apart from each other. Figure 5.1 illustrates this phenomenon.

Figure 5.1
CH_4 Puts the Maximum Distance between Its Bonds by Forming a Tetrahedron

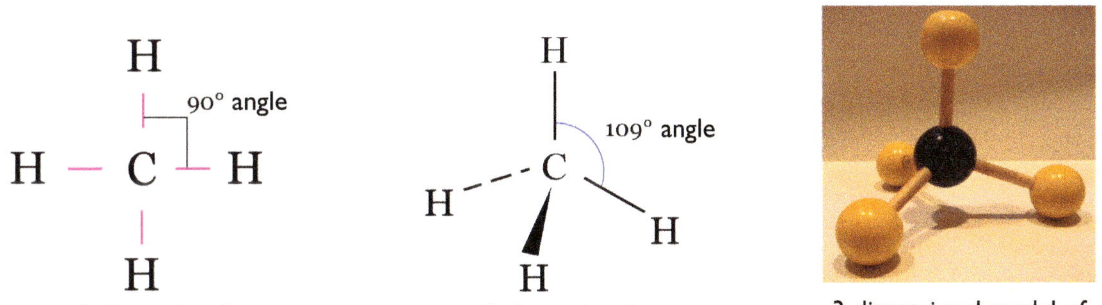

2-dimensional (flat) molecule

3-dimensional molecule

3-dimensional model of methane's tetrahedral shape

In a flat, 2–dimensional structure, the bonds between the C and H atoms cannot get any farther apart from one another than 90 degrees. However, if the molecule takes advantage of 3–dimensional space, the bonds can spread out as far as 109 degrees from each other. We try to picture this 3–dimensional structure by using solid, dashed, and triangular lines. The solid lines represent chemical bonds that are in the plane of the paper. The dashed line indicates a bond that goes behind the plane of the paper. Finally, the heavy triangular line represents a chemical bond that is out in front of the plane of the paper. The hydrogen atom attached to the dashed line sits behind the paper, while the hydrogen atom attached to the heavy triangular line is sitting in front of the paper. The other 2 hydrogen atoms are sitting on the paper. This shape, called a tetrahedron (teh truh he' drun), is one of the fundamental shapes that a molecule can attain. This shape is easiest to understand by looking at the photograph, where the black sphere represents the carbon atom and the yellow spheres represent the hydrogen atoms. The wooden sticks represent the bonds between the atoms.

The theory that allows us to predict that CH_4 has a tetrahedral shape is called VSEPR theory. VSEPR stands for Valence Shell Electron Pair Repulsion. The VSEPR theory states that molecules will attain whatever shape keeps the valence electrons of the central atom as far apart from one another as possible. We know that CH_4 has a tetrahedral shape because that shape allows carbon's valence electrons to be as far apart from each other as possible.

If you have a hard time picturing all of this in your mind, don't worry too much. Trying to picture 3–dimensional shapes can be very difficult for some people, while it is perfectly natural for others. As we go through more molecular shapes, we will give them all names. In addition, we will mention the angles between the chemical bonds. All you need to be able to do is learn those 2 things. It would be nice if you could also picture it in your mind, but we realize that this is a little too hard for some people. Hopefully, you will get the hang of this as we go along.

The tetrahedral shape is not the only shape a molecule might possess. For example, consider an ammonia molecule: NH_3. The Lewis structure for this molecule, which you have already drawn, is pictured below:

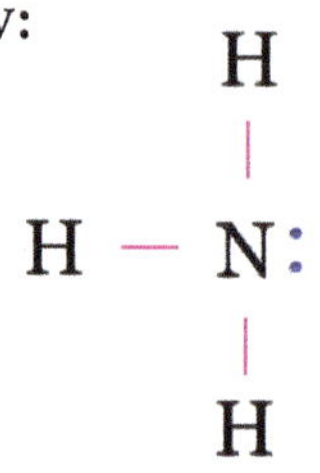

In trying to determine what kind of 3–dimensional shape a molecule attains, we have to look at the electron *groups* that surround the central atom. In this case, the central atom is nitrogen. How many groups of electrons surround the N atom? There are a total of 4. Three of them are bonds, represented by the dashes in the Lewis diagram. The fourth is a pair of nonbonding electrons. Even though these electrons do not form a bond, we still have to consider them because they are still electrons and will repel any other electron pair that comes close to them. The 4 electron groups will try to get as far away from each other as possible. As we saw above, when 4 groups of electrons try to get as far away from each other as possible, a tetrahedron is formed. This tetrahedron is a bit misshapen, as shown in Figure 5.2.

Figure 5.2
How NH_3 Acquires a Pyramidal Shape

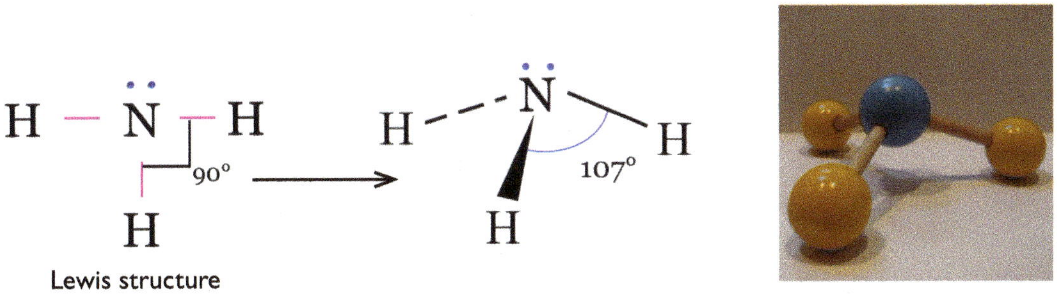

Lewis structure rotated to make 3-D drawing easier to see

3-dimensional molecule

3-dimensional model of NH_3's pyramidal shape

In the figure, we rotated the original Lewis structure so that the nonbonding electron pair is on top of the nitrogen atom. This is done simply to make the next drawing a little easier to understand. In that drawing, we tried to illustrate the 3–dimensional nature of the molecule. Once again, the dashed line indicates that the bond is behind the plane of the paper, while the heavy triangular line indicates that the bond lies in front of the plane of the paper. The nonbonding electron pair and the bond indicated by the solid line lie in the plane of the paper.

You may be tempted to call this a tetrahedron because it looks very similar to the situation with CH_4. However, there is a big difference. The top of the tetrahedron is not a bond in this case. It is a pair of nonbonding electrons. The top leg of the tetrahedron is missing, deforming the tetrahedron. The resulting shape looks a lot like a pyramid, with the H atoms as the pyramid's base and the N atom as its apex. We say that NH_3 has a pyramidal (pih ram' uh dul) shape. This pyramidal shape is easiest to understand by looking at the photograph in Figure 5.2, where the blue sphere represents the nitrogen atom, the yellow spheres represent hydrogen atoms, and the wooden sticks represent covalent bonds.

In a tetrahedron, the legs are 109° apart. In this shape, the legs are slightly closer together because the nonbonding electron pair tends to repel the bonds just a little more than the bonds repel one another. As a result, the bonds stay a little farther away from the nonbonding electron pair and a little closer to one another. The bond angle in this case is about 107°.

As you might imagine, we aren't done looking at the different shapes that a molecule can have. Next, we turn our attention to water:

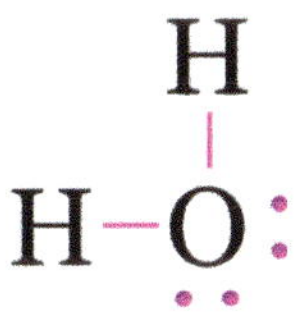

According to VSEPR theory, we must look at the electrons surrounding the central atom and determine what shape will allow the electron groups to stay as far apart from each other as possible. Once again, there are 4 groups of electrons around the central oxygen atom. Two of the groups are chemical bonds, while the other 2 are nonbonding electron pairs. Since a tetrahedron allows the 4 electron groups to stay as far apart from each other as possible, the basic shape of this molecule is also a tetrahedron. Once again, since 2 of the groups are nonbonding electron pairs, the tetrahedron is deformed.

FIGURE 5.3
How H_2O Gets Its Bent Shape

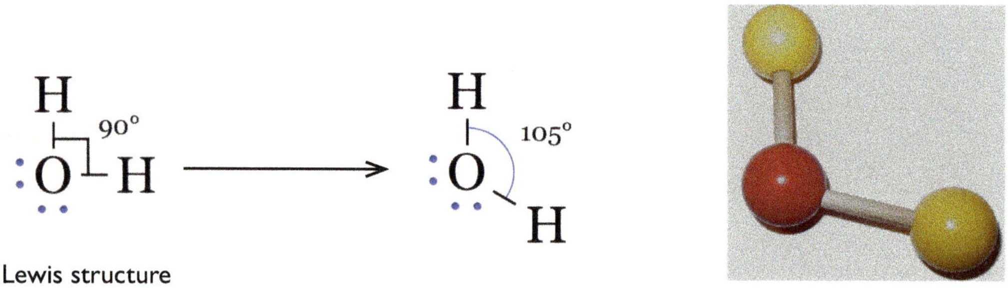

Lewis structure rotated to make the next drawing easier to see

3-dimensional molecule

model of H_2O's bent shape

It is easier to understand water's molecular shape if we first rotate the Lewis structure. When that is done, we put each group of electrons on a leg of the tetrahedron. In the second drawing above, each H atom is on a leg that is in the plane of the paper. One of the nonbonding electron pairs is on the tetrahedron leg that juts out in front of the paper, while the other is on the leg that sticks out behind the paper. Since these 2 nonbonding electron pairs are in the molecule, the tetrahedron looks like it is missing those 2 legs. What is left, then, is a shape that looks *bent*. As mentioned with NH_3, nonbonding electron pairs tend to repel more than bonding electron pairs. As a result, the 2 bonds in the water molecule are closer together than they would be in a tetrahedron or a pyramidal shape, so the bond angle is about 105°.

The 3 shapes we have considered so far are all built on the framework of a tetrahedron. The pyramidal shape is just a tetrahedron with 1 leg removed, while the bent shape is a tetrahedron with 2 legs removed. This tetrahedron framework exists because in the 3 molecules we have considered so far, the central atom has 4 groups of electrons around it. However, 4 groups of electrons around the central atom is not the only possible Lewis structure. As a result, we have more shapes to consider.

Let's look at a more complicated Lewis structure. In the previous module, you constructed the Lewis structure for CH_2O:

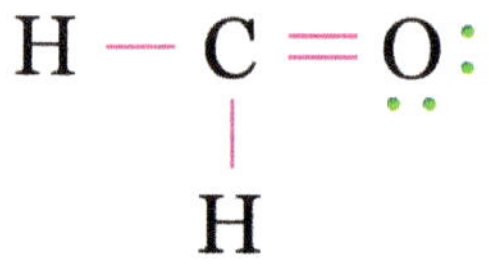

Notice that in this molecule, there are only 3 groups of electrons around the central carbon atom. There are 2 single bonds linking the H atoms to the C atom. That's 2 groups of electrons. There is also 1 double bond linking the O atom to the C atom. Although a double bond represents 4 electrons, it is still only 1 *group* of electrons. After all, a double bond is really only 1 bond; it is simply twice as strong as a single bond. In this molecule, there are only 3 groups of electrons surrounding the central carbon atom: 2 single bonds and 1 double bond. What happens in a situation like this?

FIGURE 5.4
The Trigonal Shape of CH_2O

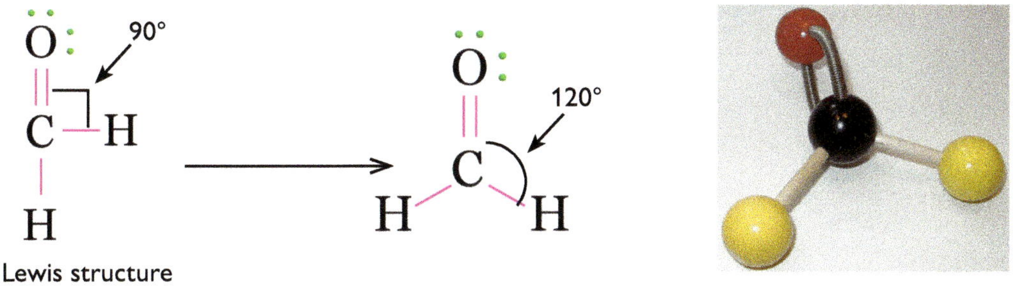

Lewis structure rotated to make the next drawing easier to see

3-dimensional molecule

model of CH_2O's trigonal shape

Remember, the guiding principle behind the VSEPR theory is that the groups of electrons try to get as far apart from one another as possible. The farthest that 3 groups of electrons can get from each other is 120°. The shape of this molecule is triangular, with the hydrogen and oxygen atoms each occupying a vertex of the triangle. We'll call this a trigonal (trig' uh nul) shape. Since all of the atoms lie in the same plane, it is sometimes called a trigonal planar shape. The picture shows a model of this geometry, with the black sphere representing carbon, the red sphere representing oxygen, and the yellow spheres representing hydrogen atoms. Notice that the model uses 2 springs to represent the double bond between oxygen and carbon. You will see that again in Figure 5.5.

Believe it or not, we are going to consider only 1 more molecular shape: linear. There are several more shapes that molecules can take, but the ones we are covering here are the most popular. If you take another year of chemistry, you will learn additional molecular shapes then. There are 2 ways for a molecule to become linear. The first should be pretty obvious. If a molecule consists of only 2 atoms, the molecule must be linear. After all, with only 2 points, no other shape than a line is possible. A molecule like F_2 must be linear:

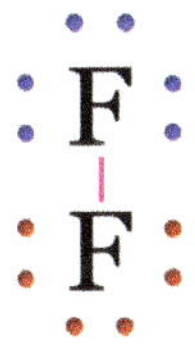

However, there is another way that a molecule might attain a linear shape. Figure 5.5 illustrates how this can happen.

FIGURE 5.5
Carbon Dioxide Is Also a Linear Molecule

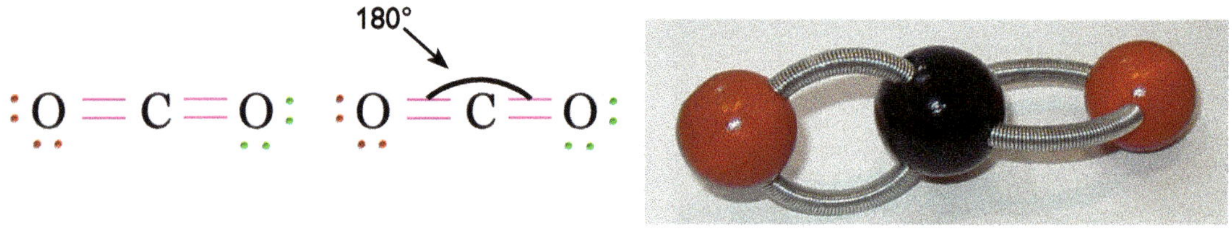

Lewis structure | 3-dimensional structure | model of CO_2's linear shape

In this molecule, there are only 2 groups of electrons around the central atom. Even though it looks like there are 4 bonds around the carbon, there are not. As mentioned before, a double bond is really only 1 bond; it is simply twice as strong as a single bond. Even though there are 4 *dashes* around the central carbon, there are really only 2 *bonds*. That means there are 2 groups of electrons. Two groups will get as far away from each other as possible when they form a straight line. This is a linear shape, with a bond angle of 180°. The 3–dimensional structure looks the same as the 2–dimensional structure because of the linear shape.

As mentioned earlier, if you are having trouble visualizing all of this, don't worry. We will give you a set of rules that will help you determine these molecular shapes. You need not be able to visualize this to solve the problems. You will be better off, however, if you are able to visualize the molecules. It will help you as you go to the last section of this module. In addition, much of the chemistry of life is based on molecular shape. If you wish to pursue chemistry as it applies to living creatures (organic chemistry, biochemistry, or pharmaceutical chemistry), you will need to be able to see the shapes of molecules in your head.

TABLE 5.2
Examples of Shapes of Molecules

Molecule	Lewis Structure	Bonded Pairs	Nonbonded Electron Pairs	Molecular Shape	
CO_2	:O=C=O:	2	0	linear	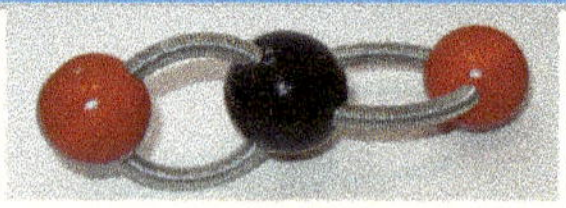
CH_2O	H–C=O: with H below C	3	0	trigonal	
CH_4	H–C–H with H above and below C	4	0	tetrahedral	

Molecule	Lewis Structure	Bonded Pairs	Nonbond-ed Electron Pairs	Molecular Shape	
NH_3	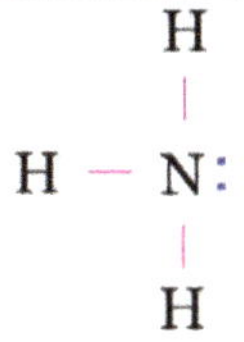	3	1	pyramidal	
H_2O	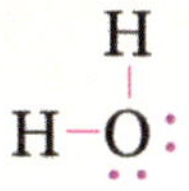	2	2	bent	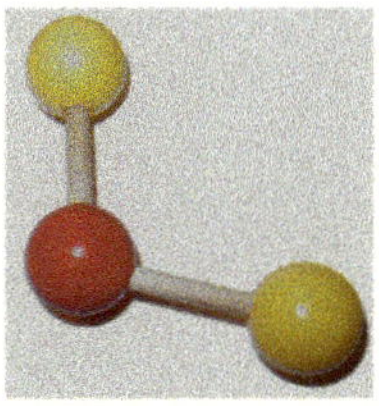

The key to VSEPR theory is to know how many groups of electrons are around the central atom. The first step, then, is to draw a Lewis structure. Once you have the correct structure, then you can count the number of groups around the central atom.

If there are only 2 groups around the central atom, then the molecule must be a linear shape with a 180° bond angle. If there are 3 groups of electrons around the central atom, then the molecule must be trigonal planar with a bond angle of 120°.

If there are 4 groups of electrons around the central atom, then there are 3 options. If all of the groups are used in bonds, then the molecule will have a tetrahedral shape with a 109° bond angle. If there are 3 bonds and 1 lone pair, then the molecule will have a pyramidal shape with a 107° bond angle. And finally, if there are 2 bonds and 2 lone pairs of electrons, then the molecule will have a bent shape with a 105° bond angle.

Let's go through Example 5.2 to determine the shapes of molecules so that you will see how we apply the rules you just read.

EXAMPLE 5.2

What are the shape and bond angle of a $PHCl_2$ molecule?

To answer this question, we must first determine the molecule's Lewis structure. Hopefully, you are familiar enough with Lewis structures to be able to get that far on your own:

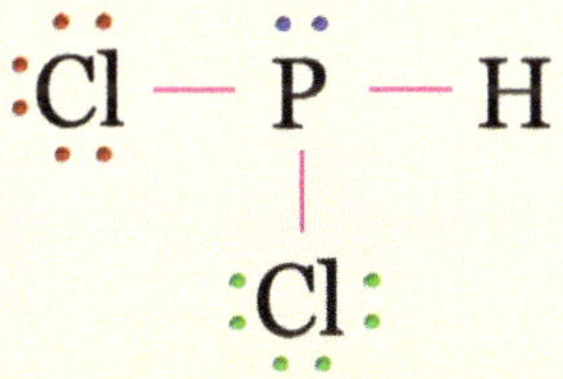

Now that we see the Lewis structure, we can tell that there are 4 groups of electrons around the central phosphorus atom. This means that the basic shape of the molecule is a tetrahedron. However, since there is 1 pair of nonbonding electrons on the phosphorus atom, this takes 1

leg away from the tetrahedron. The shape of the molecule is **pyramidal with a bond angle of 107°**. The molecule, then, would look something like this with the regular line planar to the paper, the dotted line going away, and the triangular line coming forward:

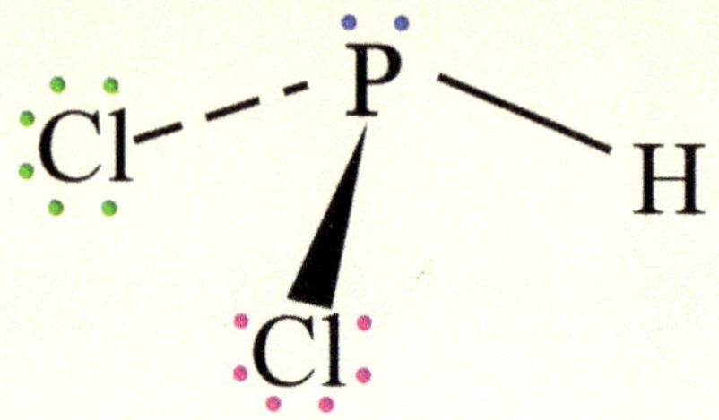

Before we leave this example, notice that the nonbonding electron pairs that surround the chlorine atoms played no role in determining the shape of the molecule because the shape of a molecule is based on its central atom. Only the electrons around the central atom come into play here. If you go on in chemistry, you will eventually start dealing with molecules that have so many atoms in them that it is difficult to determine the central atom. At that point, determining molecular shape can be very difficult. For this course, we will continue to concentrate on relatively simple molecules when talking about molecular shapes.

What are the shape and bond angle of a SiO_2 molecule?

First, we must determine the Lewis structure:

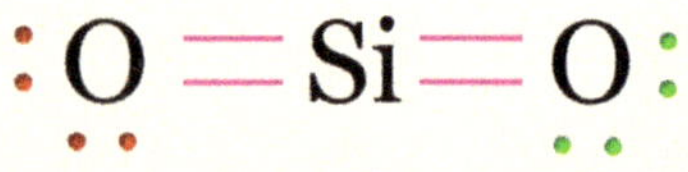

This is a simple molecule to deal with. It has only 2 groups of electrons (2 double bonds) around the central atom. This makes it a **linear molecule with a bond angle of 180°**.

What are the shape and bond angle for an OCl_2 molecule?

The Lewis structure is:

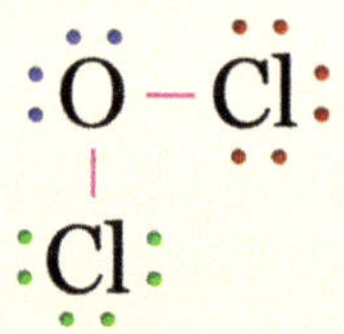

The central atom has 4 groups of electrons around it, making the basic shape a tetrahedron. Two of the groups of electrons are nonbonding electron pairs. This means that 2 legs of the tetrahedron are gone, resulting in a **bent shape with a bond angle of 105°**. The molecule would look something like this:

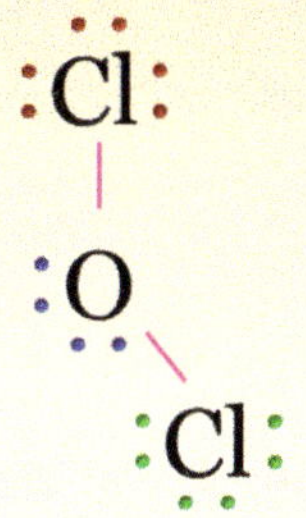

What are the shape and bond angle for a BF_3 molecule? (Note: Boron is an exception to the octet rule. It wants only 6 valence electrons, not 8.)

The Lewis structure is pretty easy. The boron atom has 3 valence electrons, and each fluorine atom needs an extra valance electron. The Lewis structure is:

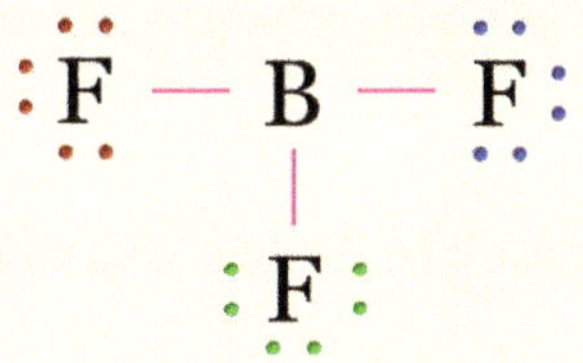

The central atom has 3 groups of electrons around it, making a **trigonal shape with a bond angle of 120°**. The molecule would look something like this:

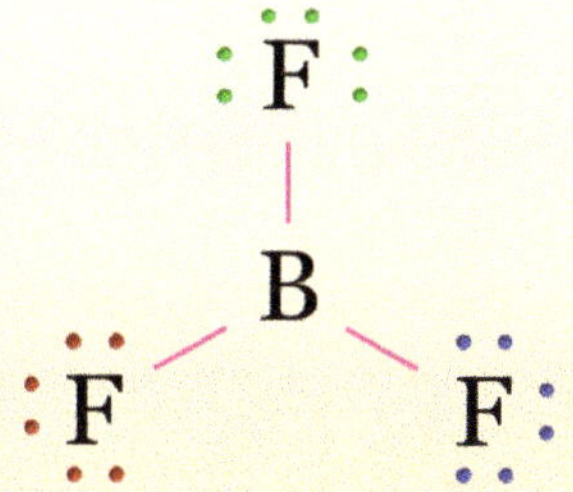

Make sure you understand the concept of molecular shape by answering "On Your Own" questions 5.4–5.8.

ON YOUR OWN

5.4 Determine the shape and bond angle of an oxygen molecule.

5.5 What are the shape and bond angle of $CHCl_3$?

5.6 What shape and bond angle are possessed by an arsenic trichloride molecule?

5.7 Determine the shape and bond angle of a sulfur dibromide molecule.

5.8 What are the shape and bond angle of a $SiCl_2O$ molecule?

NONPOLAR COVALENT AND POLAR COVALENT BONDS

In Module 4, we told you that covalent molecules could be further divided into 2 important subclasses. Now that you understand Lewis structures and molecular shapes, it is time to learn these 2 subclasses and their distinctions. To learn this, we must first go back to Lewis structures and think a bit about what they mean. For example, consider the Lewis structure for F_2:

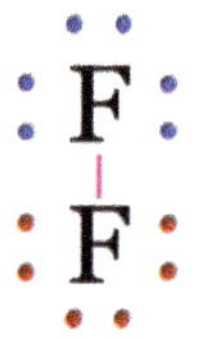

The dash between the fluorine atoms indicates a covalent bond, which is, in reality, a shared pair of electrons. But how do atoms share electrons?

As we learned in the last module, the fluorine atoms share electrons because they would each like to have 8 valence electrons. If there were metal atoms around, they would try to take the metal's electrons away so that they could each have their own 8 valence electrons. However, for whatever reason, no metal atoms could be found. As a result, the atoms decided to share electrons to have 8 valence electrons for at least some fraction of time.

They each have 8 valence electrons for a fraction of time because the shared electron pair is whirling around both atoms. For those moments in which the electrons are whirling around the first fluorine atom, that atom has 8 valence electrons. When that happens, the second atom has only 6 valence electrons. This is a problem for the second atom, so it tugs on the electrons, forcing them to whirl around it. Of course, at that point, the first fluorine atom wants the electron pair back, so it tugs on them, trying to get them to come back.

A tug–of–war is established over this shared electron pair. Since both of the atoms in this molecule are the same (they are both fluorine atoms), neither of them can win the tug–of–war. They each pull on the electron pair with the same strength. As a result, one fluorine atom has them exactly half of the time, while the other has them for the other half of the time. We could say then, that the molecular bond in this molecule is made up of a pair of electrons which are shared equally between the 2 atoms. When electrons in a chemical bond are shared *equally*, we call it a nonpolar covalent bond.

Nonpolar covalent bond—A covalent bond in which the electrons are shared equally between the atoms involved.

Contrast the situation in a fluorine molecule to that in a hydrogen chloride molecule (also referred to as hydrochloric acid):

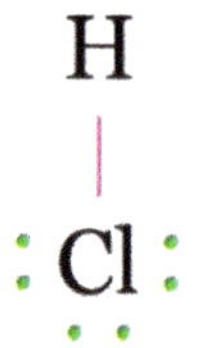

In this case, the electron pair that makes up the chemical bond is shared between 2 different atoms: a hydrogen atom and a chlorine atom. One of these atoms is, most likely, a little stronger at tugging on electrons than the other. Can you predict which atom that would be? In the last module, we learned about electronegativity, which determines the desire an atom has for extra electrons. Doesn't it make sense, then, that the atom with the higher electronegativity will also be able to tug on electrons with more force?

Since chlorine is to the right of hydrogen on the periodic table, we can assume that it has a higher electronegativity. You might think that since chlorine is also lower on the table than hydrogen, its electronegativity might be less than hydrogen's. If you look at the table, chlorine is much farther to the right of hydrogen than it is lower. The fact that electronegativity increases as you travel to the right on the table outweighs the fact that it decreases as you move down the table. Chlorine has a higher electronegativity than hydrogen and can tug on the electron pair more strongly than hydrogen can.

What is the result of this fact? If the chlorine atom is stronger, it can hold onto the electrons for a longer period of time than the hydrogen atom can. The electrons are no longer shared equally; they are shared unequally. When the electrons that make up a chemical bond are shared *unequally*, we call it a polar bond.

Polar bond—A covalent bond in which the electrons are shared unequally between the atoms involved.

So what's the big deal? Why be concerned about whether or not electrons are shared equally between atoms in a molecule? If you think about it, there is a rather important consequence if electrons are not shared equally in a molecule. For example, in the case of the HCl molecule, the electrons spend the majority of their time around the chlorine atom and only a small fraction of time around the hydrogen atom. As a result, the chlorine atom gets more than its fair share of electrons. In a way, it has an extra electron, at least for part of the time. What happens when an atom gets extra electrons? It becomes negatively charged. For some fraction of time, the chlorine atom is negatively charged. In the same way, the hydrogen atom doesn't get its fair share of electrons. As a result, it has a deficit of electrons, at least part of the time. When atoms have a deficit of electrons, they become positively charged. For some fraction of time, the hydrogen atom is positively charged.

Remember, these charges exist for only a fraction of time. As a result, the chlorine atom never develops a *full* negative charge, and the hydrogen atom never develops a *full* positive charge. We could picture this situation in the following way:

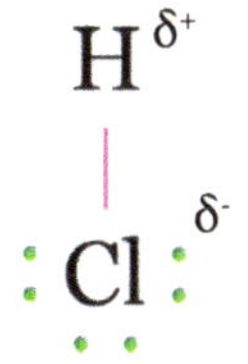

The symbol δ in the picture above is the lowercase Greek letter delta. Scientists and mathematicians often use this symbol to indicate a small fraction. The Lewis structure pictured above indicates that the hydrogen atom in the HCl molecule has a small fraction of a positive charge (a positive dipole or *partial* positive charge), whereas the chlorine atom has a small fraction of negative charge (a negative dipole or *partial* negative charge).

This is the situation in an HCl molecule. Even though the molecule has no true ions in it, there are still small electrical charges within the molecule itself. To see exactly what the consequences of this fact are, perform Experiment 5.1.

EXPERIMENT 5.1

PURPOSE: To compare polar covalent and nonpolar covalent compounds.

MATERIALS

- Glass of water
- Vegetable oil
- Styrofoam or paper cup
- Comb
- Pen
- Safety goggles

QUESTION: Can you observe the polar nature of molecules?

HYPOTHESIS: Write your hypothesis related to the polar nature of molecules.

PROCEDURE:

1. Use the pen to punch a small hole into the bottom of the cup. The smaller the hole, the better.
2. While holding it over the sink, pour some water into the cup from the glass. Water should start running out of the hole in the bottom of the cup. Make sure that the water is pouring out of the hole in a steady stream, not just dripping. If it is dripping, make your hole just a little bigger.
3. Once the water is pouring out of the cup in a steady stream, take the comb in your other hand and vigorously comb your hair. This is meant to accomplish the same thing it did in experiment 3.1: to make the comb develop an electrical charge. As mentioned in that experiment, if your hair is greasy, this may not work too well.
4. Once you have combed your hair for a few seconds, bring the comb (teeth first) near the stream of water. You should let the comb get very close to, but not actually touch, the water stream. What happens? The water stream should bend toward the comb. The more you combed your hair, the stronger the bend in the water should be.
5. Repeat this same experiment using vegetable oil instead of water. You may have to make the hole a little bigger this time because vegetable oil doesn't flow as easily as water does. Once you do the experiment, you should observe that the stream of vegetable oil doesn't bend like the water did. You may see the stream sputter a bit in reaction to the nearness of the comb, but the stream of oil does not bend.
6. Clean up and return everything to the proper place.

CONCLUSION: Write something about the polar nature of water and vegetable oil.

Why did the water stream bend while the oil stream didn't? The answer lies in the fact that water is a polar covalent compound, while vegetable oil is nonpolar covalent. A water molecule looks like this:

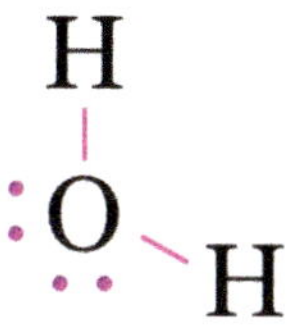

In this molecule, electrons are being shared between hydrogen atoms and an oxygen atom. Because oxygen is far to the right of hydrogen on the periodic table, it is more electronegative. As a result, it wins the tug–of–war for electrons and has more than its fair share of electrons. As a result, it develops a partial negative charge. The hydrogen atoms, then, have a slight electron deficit, giving them a partial positive charge. Once you combed your hair, the comb became negatively charged, attracting the partial positive charges on the hydrogen atoms. This caused the water molecules to move toward the comb, with their hydrogen atoms facing the comb and their oxygen atoms facing away from it:

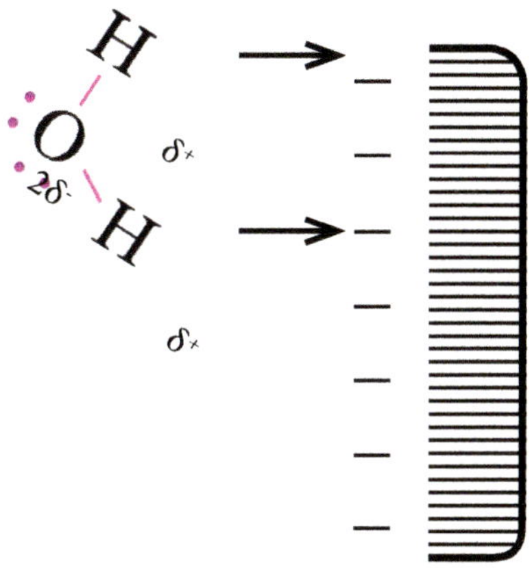

That's why the stream of water bent toward the comb. It was electrically attracted to it.

The vegetable oil, on the other hand, is made up of nonpolar covalent molecules. There are no net electrical charges in those molecules. They were not attracted to the comb, and therefore the stream of oil didn't bend toward it. If you happened to see the stream of oil sputter a little when the comb got near it, that's because there might have been a small amount of water in your vegetable oil. When the water was attracted to the comb, it started moving toward it. The rest of the stream (which was oil) was not attracted to the comb, so it still moved straight down. Since one part of the stream wanted to go one way and the other part wanted to go another, the stream sputtered. This might not have happened in your experiment, indicating that there was no water contaminating your vegetable oil.

Earlier, we defined polar and nonpolar bonds. Now let's define polar and nonpolar molecules.

Nonpolar molecule—A covalent molecule in which the electrons are evenly distributed throughout the whole molecule.

Polar molecule—A covalent molecule that has an imbalance in the distribution of electrons throughout the whole molecule.

To fully understand the difference between nonpolar covalent molecules and polar covalent molecules, you must be able to identify polar bonds as well as determine which direction the unequally shared electrons are pulled toward. To do this, chemists have developed a symbolic way of labeling polar bonds. When determining which bonds in a molecule are polar, we represent them with the following symbol:

The arrowhead in the symbol should point toward the atom with the largest electronegativity to indicate in what direction the unequally shared electrons are being pulled. The plus sign on the other end of the arrow, then, represents the electron deficit that results on the other atom.

Now that you know how to *label* a polar bond, how do you *determine* what bonds in a molecule are polar? Fortunately, that task is really easy. If a bond exists between 2 identical atoms (2 C atoms, for example), then it is not polar. If it exists between 2 different types of atoms (a C atom and an F atom, for example), then it is polar. Study Example 5.3 and then answer "On Your Own" question 5.9 to make sure you understand how to identify polar bonds and the direction in which the unequally shared electrons are pulled.

EXAMPLE 5.3

Identify and label the polar bonds in NH_3 (ammonia).

To identify the polar bonds and determine the charges that result, we must first look at the Lewis structure:

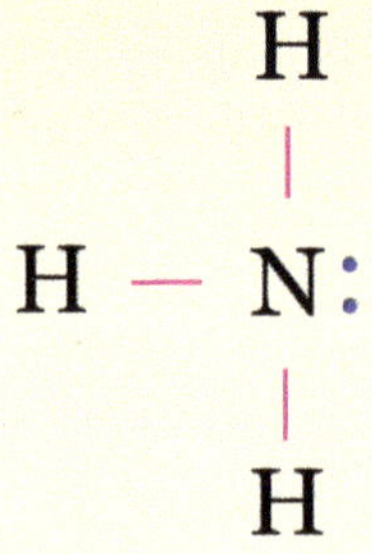

In this molecule, there are 3 bonds that will be polar since there are 3 bonds between 2 different types of atoms. To determine how to label the bonds, we must look at the electronegativity of the atoms involved. According to our rules, nitrogen must have a larger electronegativity because it lies far to the right of hydrogen on the periodic table and only a little lower than hydrogen. Since it gains a lot of electronegativity from being far to the right and loses only a little by being just a bit lower, the net result is that nitrogen has more electronegativity than hydrogen. The nitrogen atom gets more than its fair share of electrons from each of the 3 hydrogen atoms. This means that the arrows on the polar bond labels need to point toward the nitrogen atom:

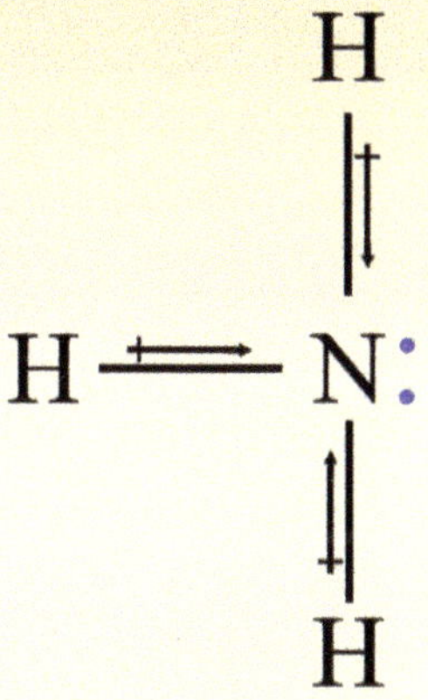

Identify and label the polar bonds in the C_2F_4 molecule.

First, we must look at the Lewis structure:

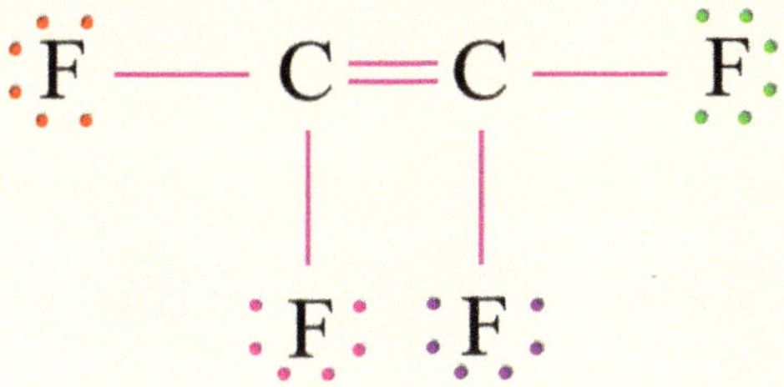

According to the way we defined polar bonds, the carbon–to–carbon double bond is obviously not polar. After all, the bond is between 2 identical atoms. Neither of them can win the tug–of–war for electrons. The 4 carbon–to–fluorine bonds are polar because they exist between different types of atoms. Since fluorine is more electronegative than carbon (as shown by its position on the periodic table), the arrows on our polar bond label must point toward the fluorine atoms:

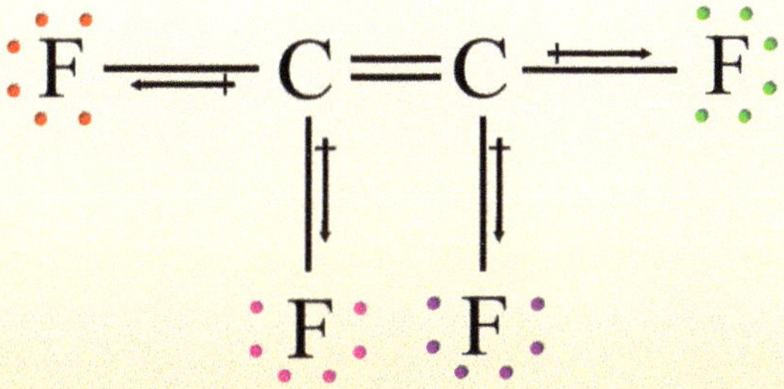

ON YOUR OWN

5.9 Identify and label the polar bonds in the following molecules:
a. CO_2 b. $CClF_3$

NONPOLAR COVALENT AND POLAR COVALENT MOLECULES

Remember, the important thing about polar covalent bonds and nonpolar covalent bonds is their effect on the molecule they are in. If a molecule is polar covalent, it contains electrical charges. On the other hand, a covalent molecule can have polar bonds but be nonpolar overall (with no net charge). Since this is a relatively important thing to know about a molecule, it is important to be able to determine whether a given molecule is polar covalent or nonpolar covalent.

So what makes a molecule polar covalent? It stands to reason that a molecule must contain polar bonds to be a polar covalent molecule. After all, if the bonds do not result in fractional charges, then there is no way that a molecule can become polar covalent. It might surprise you that polar bonds are not enough to ensure that a molecule is polar covalent. This is because, under certain circumstances, the polar bonds in a molecule can all work against each other and end up canceling each other out. When this happens, the molecule has an overall net of no charges in it, despite the fact that it has polar bonds, and the molecule itself is nonpolar covalent.

The best way to explain why this is the case is to give you an example of a molecule that has polar bonds but is nonpolar covalent. The perfect example of this is the CCl_4 molecule. The shape of this molecule is as follows:

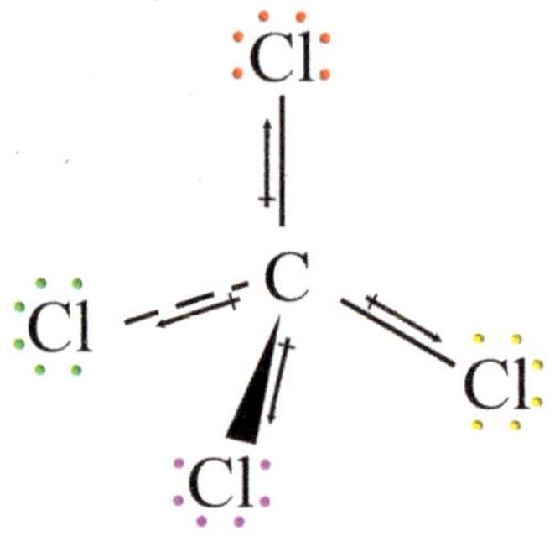

As shown, the chlorine atoms are more electronegative than the carbon. As a result, electrons are being pulled away from the carbon and toward the chlorine atoms. However, the chlorine atoms are all pulling from exactly opposite directions. As a result, the electrons are being pulled evenly in opposite directions.

What happens when something is pulled evenly in opposite directions? It ends up sitting still, of course. That's what happens to the electrons around the carbon in this molecule. Since they are all being pulled evenly in opposite directions, they end up staying put. Although this molecule has 4 polar covalent bonds, they end up canceling each other out, making the molecule overall nonpolar covalent. In other words, since the electrons around the carbon cannot move, there are no resulting charges in the molecule.

Notice that this is not what happens in the case of $AsCl_3$. This molecule has the following shape:

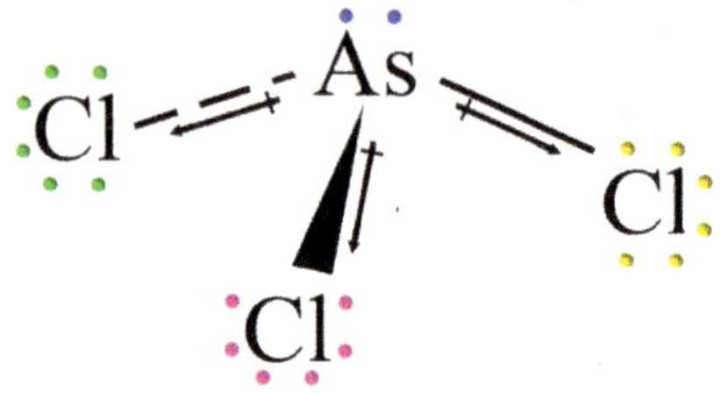

In this case, it should be obvious that the polar bonds do not cancel each other out. Since there is no bond on top of the As atom, there is nothing pulling electrons up. As a result, the 3 bonds exert a net pull downward. Since there is a net pull on the As atom's electrons, there will be charges in this molecule. We can therefore determine that the molecule is polar covalent.

For a molecule to be polar, 2 requirements must be met. First, the molecule must have polar bonds in it. Second, those polar bonds cannot cancel each other out. How can we determine whether or not polar bonds cancel each other out? We have to determine the *shape* of the molecule. If there is an equal pull on the central atom's electrons from opposite directions, then the polar bonds will end up canceling each other out and the molecule will be nonpolar covalent. On the other hand, if there is not an equal pull on the central atom's electrons from opposite directions, then the molecule will end up having a net fractional charge, making it polar covalent.

We must be careful in making this determination because there must be an *equal* pull on the central atom's electrons for the polar bonds to end up canceling each other out. This can be a little confusing unless you are paying close attention. For example, consider the molecule $CFCl_3$. The polar bonds look like this:

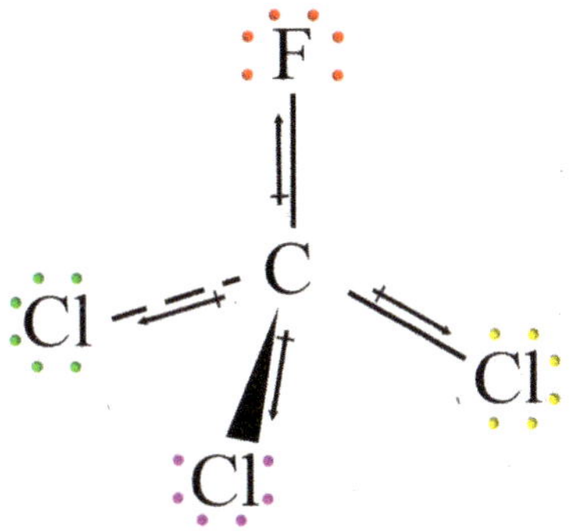

Looking at this molecule, you see that the electrons on the central carbon atom are, indeed, being pulled in opposite directions. But are the pulls equal? No, they are not. According to our rules of electronegativity, fluorine is more electronegative than chlorine because it is higher on the periodic table. The pull from the fluorine atom is harder than the pull from the chlorine atoms. While the carbon atom's electrons are being pulled in opposite directions, they are not being pulled equally, so the molecule is polar covalent.

What about the C_2F_4 molecule? We know from the example drawn above that it has polar bonds, but do they cancel out? If it is drawn this way, it appears that the overall molecule is polar—but this Lewis structure doesn't place the fluorine atoms in their correct position. The molecule is actually nonpolar overall or nonpolar covalent.

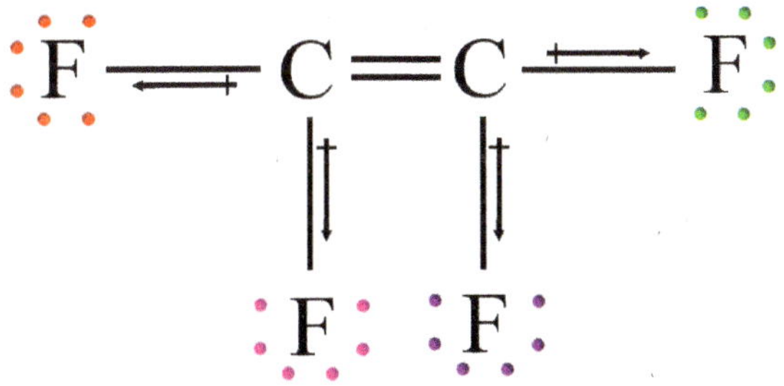

The correct Lewis diagram should be:

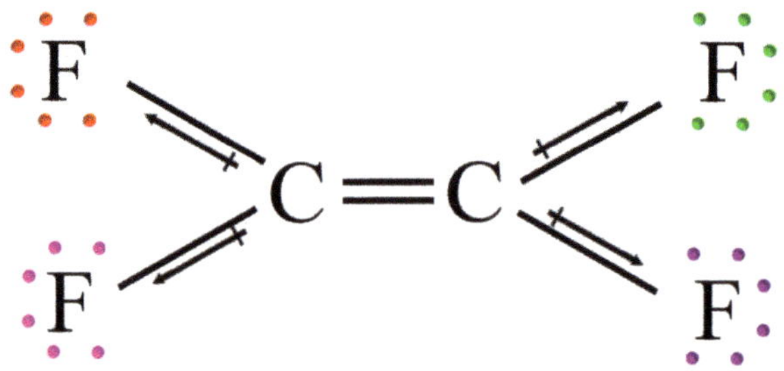

To reiterate, for a molecule to be polar covalent, the following conditions must be met:

- **The molecule must have polar bonds in it.**
- **Those polar bonds cannot be of equal strength and pull in opposite directions.**

Try to see how we apply these rules in Example 5.4, and then use them to answer "On Your Own" question 5.10.

EXAMPLE 5.4

Classify SiO_2 as polar covalent or nonpolar covalent.

To classify this molecule, we must first draw its Lewis structure:

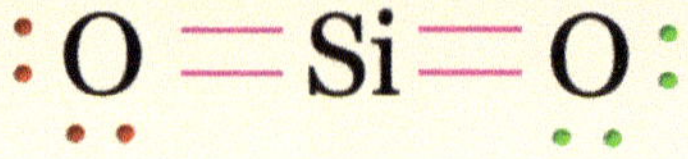

Since there are 2 groups of atoms around the silicon atom, this is a linear molecule. The shape we have drawn above is correct. Now that we know the shape, we can look at the bonds. According to their relative positions on the periodic table, oxygen is more electronegative than silicon. As a result, the silicon atom's electrons are being pulled toward the oxygen atoms:

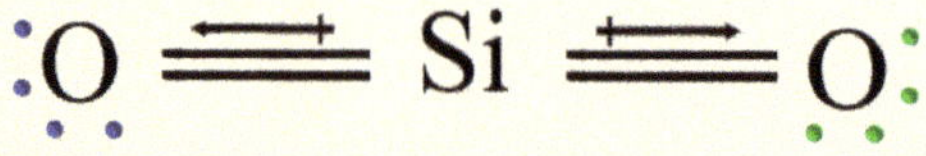

Notice that the electrons are being pulled in exactly opposite directions. In addition, they are all being pulled by oxygen atoms, so the pulls are equal. So these polar bonds cancel each other out, and the molecule is therefore **nonpolar covalent**.

Is OF_2 polar covalent or nonpolar covalent?

Once again, to solve this problem, we must look at the Lewis structure:

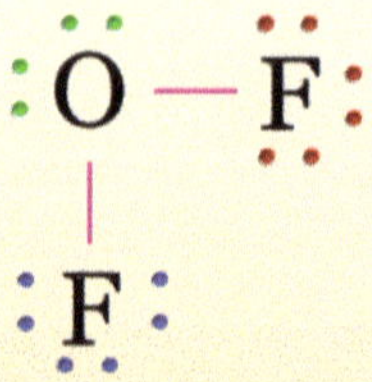

In this Lewis structure, there are 4 groups of electrons around the central atom. The base shape is a tetrahedron. However, this tetrahedron has 2 legs removed, making the shape of this molecule bent. Now that we know the shape, we can consider the bonds. Both bonds are polar, and since fluorine lies to the right of oxygen on the periodic table, the direction of the electron pull is as follows:

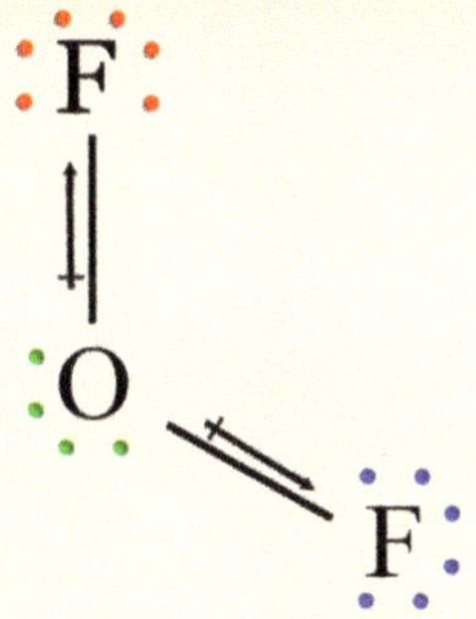

The electrons are not being pulled in exactly opposite directions, so there is a net pull on the oxygen atom's electrons, making the molecule **polar covalent**.

Is the HBr molecule polar covalent or nonpolar covalent?

The Lewis structure of HBr is:

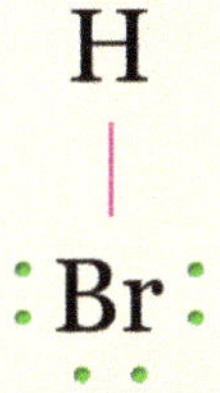

Since Br is far to the right of and only a little lower than H on the periodic table, the electron pull looks like this:

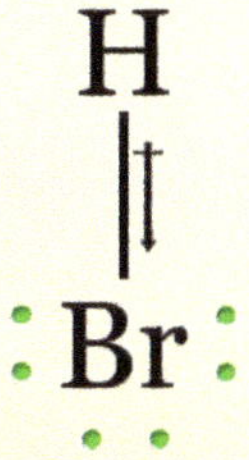

Since there is only one polar bond in this molecule, there is nothing to cancel it out. As a result, the molecule is **polar covalent**.

ON YOUR OWN

5.10 Classify the following molecules as polar covalent or nonpolar covalent:

a. Br_2 b. CO_2 c. H_2S
d. CF_2Cl_2 e. SiF_4 f. PH_3

THE PRACTICAL CONSEQUENCE OF WHETHER OR NOT A MOLECULE IS POLAR COVALENT

You might be saying to yourself, "So what? Now we can tell whether or not a molecule is nonpolar covalent or polar covalent. What's the big deal?" The difference between polar covalent compounds and nonpolar covalent compounds is of great practical significance. To understand why, do Experiment 5.2.

EXPERIMENT 5.2

PURPOSE: To compare the solubility of ionic compounds in polar covalent and nonpolar covalent compounds.

MATERIALS:

- 2 test tubes (Thin glasses will work, but they must be transparent.)
- Table salt
- Water
- Vegetable oil
- Safety goggles

QUESTION: What is the difference in solubility of ionic compounds in polar covalent and nonpolar covalent compounds?

HYPOTHESIS: Write something about what you expect about the solubility of polar and nonpolar covalent compounds in this experiment.

PROCEDURE:

1. Put enough salt into each test tube to cover the bottom of each. Put in enough so that it is easy to see that the salt is there, but do not put in too much.
2. Fill one of the tubes ¾ of the way with water and the other ¾ of the way with vegetable oil.
3. Cover the tubes with your thumbs and shake them vigorously for several minutes.
4. Allow them to stand for about a minute.
5. Hold the test tubes up to a light and look through them. In the one that you filled with water, the salt should have dissolved into the water. What about the one with vegetable oil, however? If you look closely, you should still see all of the salt either lying at the bottom of the tube or suspended in the oil. None of it dissolved.
6. Clean up and return everything to the proper place.

CONCLUSION: Write something about what you observed with respect to your hypothesis.

Why did the table salt dissolve in water but not in vegetable oil? Table salt, as we learned several modules ago, is an ionic compound. Therefore, we know that it is composed of ions, which possess electrical charge. Since water is a polar covalently bonded molecule, it also has charges in it, even if they are only partial charges. As we learned in Experiment 5.1, vegetable oil is nonpolar covalent. It has no electrical charges. Since it has no electrical charges, it cannot interact with (and therefore cannot dissolve) the table salt. This leads us to an important point about these 2 subclasses of covalent compounds:

Polar covalent compounds can dissolve other polar covalent compounds and ionic compounds because both types of compounds contain electrical charges.

Nonpolar covalent compounds, on the other hand, can dissolve only other nonpolar covalent compounds because no electrical charges are present.

These 2 chemical principles explain many household cleaning woes. For example, do you know why it is so hard to get grass and blood stains out of clothes? The reason is that grass and blood are both made of nonpolar covalent compounds. When you wash the clothes, you are washing them in water, which is polar covalent. As a result, the water cannot dissolve the grass or blood stains, so they tend to stay on the clothing. Adding soap to the water helps because soap contains long molecules that are ionic on one end and nonpolar covalent on the other. The reason that such a molecule can exist is a bit beyond the scope of this course, but at least you can see the advantage gained by adding such a molecule to your wash water. The ionic part of the soap molecule is attracted to the polar covalent water molecules, while the nonpolar covalent part of the soap molecule dissolves the grass or blood stain.

When you pretreat a stain with something like Shout, all you are doing is adding a lot of soap directly to the stain. The hope is that the nonpolar covalent part of the soap molecules will dissolve the entire stain because you are using so much of it right at the stain. Then, when the clothing is in the wash, the water will attract the ionic part of the soap molecule away from the fabric, pulling the stain away with it.

So you see that the classification of compounds as ionic, polar covalent, or nonpolar covalent can help us determine a lot about the behavior of certain compounds. Make sure you grasp this by answering "On Your Own" question 5.11.

ON YOUR OWN

5.11 Which of the following substances dissolve in water?

a. Br_2 b. CO_2 c. H_2S

d. CF_2Cl_2 e. SiF_4 f. PH_3

SUMMARY OF TABLES IN MODULE 5

Table 5.1: Important Polyatomic Ions

Table 5.2: Examples of Shapes of Molecules

ANSWERS TO THE "ON YOUR OWN" QUESTIONS

5.1 (a) $Ba(NO_3)_2$ is barium nitrate.
(b) $NaClO_3$ is sodium chlorate.
(c) $Ca(OH)_2$ is calcium hydroxide.
(d) LiCN is lithium cyanide.

5.2 (a) The name tells us that this substance is composed of the potassium ion and the sulfate ion. The periodic table tells us that the potassium ion has a charge of 1+, while we should have the sulfate ion memorized as having a 2- charge. Ignoring the signs and switching the numbers, we get:

$$K_2SO_4$$

No parentheses are needed here because there is only 1 sulfate ion.

(b) The name tells us that this substance is composed of the ammonium ion and the carbonate ion. Our memory tells us that the ammonium ion has a 1+ charge, and the carbonate ion has a 2- charge. Ignoring the signs and switching the numbers, we get:

$$(NH_4)_2CO_3$$

No parentheses are needed for the carbonate ion because there is only 1. However, since there are 2 ammonium ions, we must have parentheses around the NH_4.

(c) The name tells us that this substance is composed of the aluminum ion and the acetate ion. The periodic table tells us that the aluminum ion has a charge of 3+, while we should have the acetate ion memorized as having a 1- charge. Ignoring the signs and switching the numbers, we get:

$$Al(C_2H_3O_2)_3$$

5.3 We have to determine the chemical formulas:
Aluminum nitrite: Al is 3+ and NO_2 is 1-; therefore, the chemical formula is $Al(NO_2)_3$.
Sodium carbonate: Na is 1+ and CO_3 is 2-; therefore, the chemical formula is Na_2CO_3.
Aluminum carbonate: Al is 3+ and CO_3 is 2-; therefore, the chemical formula is $Al_2(CO_3)_3$.
Sodium nitrite: Na is 1+ and NO_2 is 1-; therefore, the chemical formula is $NaNO_2$.

5.4 The Lewis structure for an oxygen molecule is:

This, then, is a simple molecule. Because it is made up of only 2 atoms, it is linear with a bond angle of 180°.

5.5 The Lewis structure of $CHCl_3$ is:

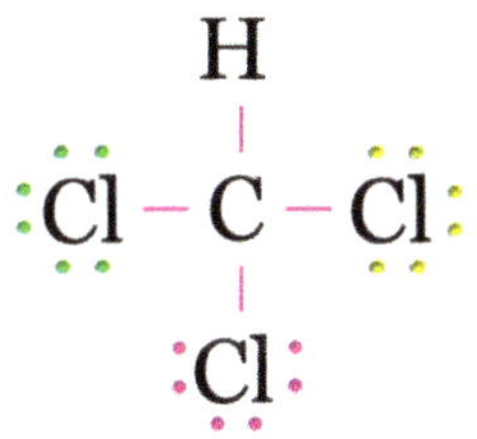

Looking at this structure, we see 4 groups of electrons around the central carbon atom. This means that the basic shape is that of a tetrahedron. Since there are no nonbonding electrons around the central atom, the final shape is tetrahedral with a bond angle of 109°. The molecule, then, would look something like this:

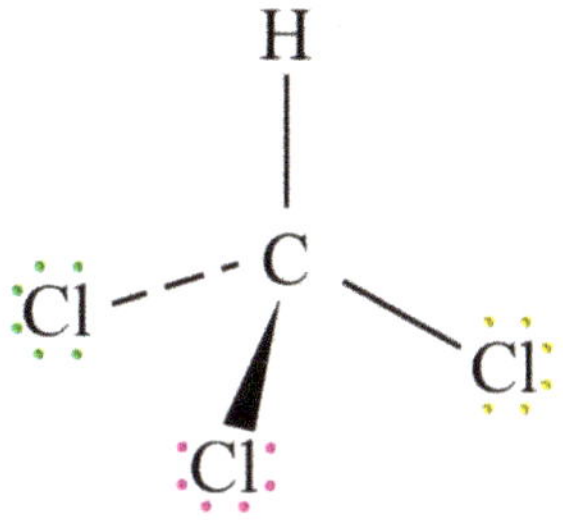

5.6 The name of this molecule indicates that its molecular formula is $AsCl_3$. The Lewis structure, then, is:

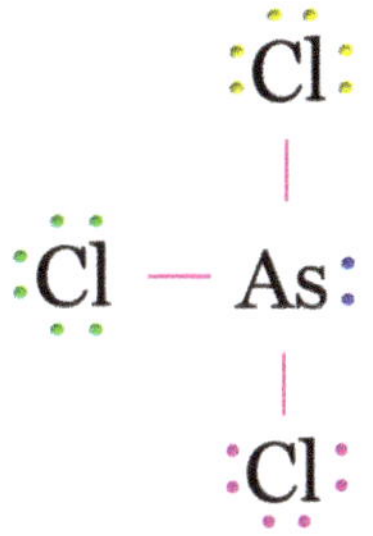

Based on this Lewis structure, we can see that there are 4 groups of electrons around the central atom, making the basic shape a tetrahedron. However, there is 1 pair of nonbonding electrons, taking one leg from the tetrahedron. This results in a pyramidal shape with a bond angle of 107°. The molecule would end up looking something like this:

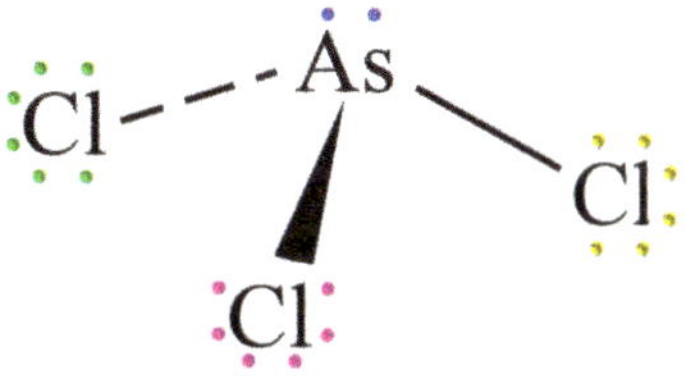

5.7 Sulfur dibromide has a formula of SBr_2, making its Lewis structure look like this:

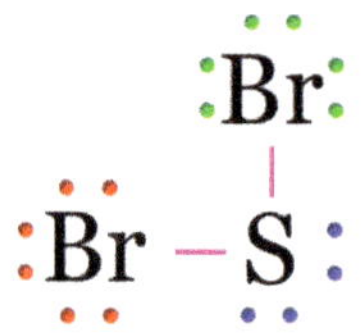

The Lewis structure indicates that the central sulfur atom is surrounded by 4 groups of electrons, making the basic shape a tetrahedron. Two of the groups of electrons are nonbonding electrons, taking 2 legs from the tetrahedron. The resulting shape is bent with a bond angle of 105°. The molecule, then, looks like this:

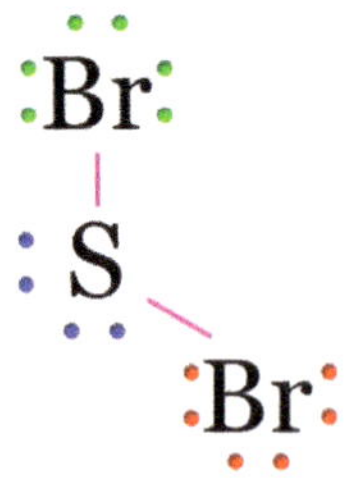

5.8 An $SiCl_2O$ molecule has the following Lewis structure:

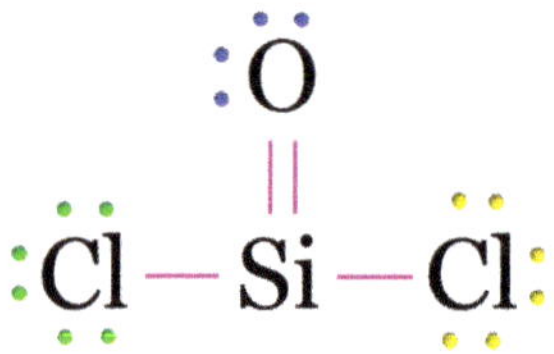

In this Lewis structure, there are only 3 groups of electrons around the central atom. Sure, 1 of those groups is a double bond, but a double bond is only one bond; it is just a lot stronger than a single bond. There are 3 bonds around the silicon atom, which means 3 groups of electrons. That makes this molecule trigonal with a bond angle of 120°. The molecule, then, looks like this:

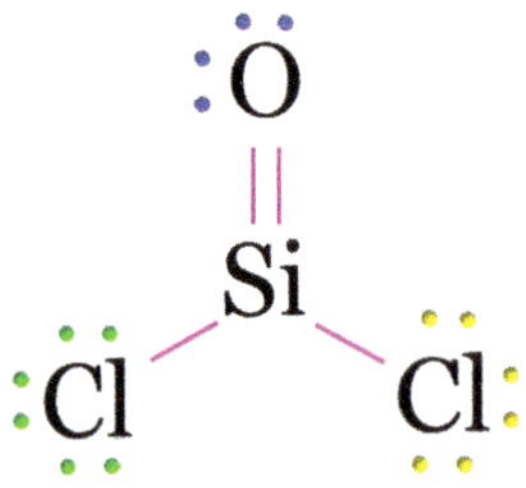

5.9 (a) To solve this problem, we must first look at the Lewis structure of the molecule:

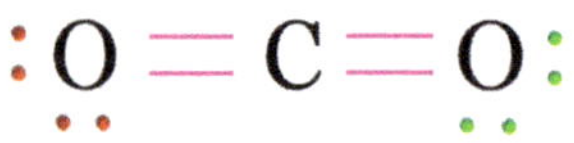

This molecule is linear since there are only 2 groups of electrons around the central atom. Those 2 double bonds both exist between atoms of different types. They are both polar. Since oxygen lies to the right of carbon, it is more electronegative, and therefore the arrows point toward the oxygen atoms:

This drawing, then, shows that the carbon atom's electrons are being pulled toward the oxygen atoms.

(b) The Lewis structure for $CClF_3$ is:

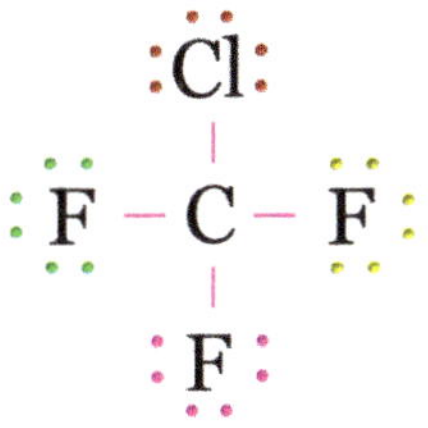

In this molecule, there are 4 groups of electrons around the central atom. The base geometry is a tetrahedron. Since all 4 groups are bonds, all legs of the tetrahedron exist, and therefore the molecule is tetrahedral. All 4 bonds are polar because each of them links different types of atoms. The fluorine atoms and chlorine atom are more electronegative than the carbon atom, so the arrows point toward the fluorine and chlorine:

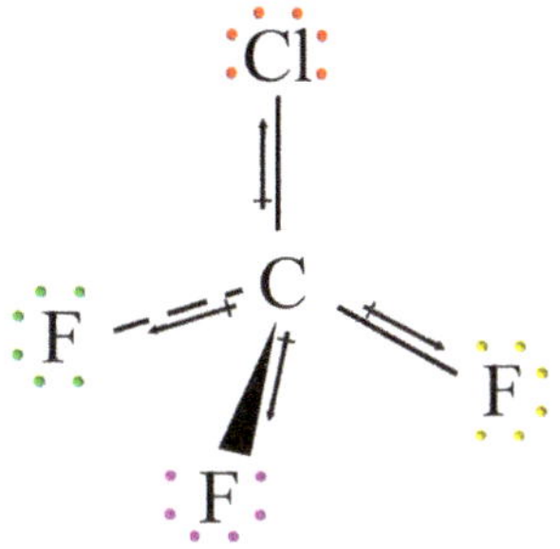

5.10 (a) We start each of these problems by looking at the Lewis structure:

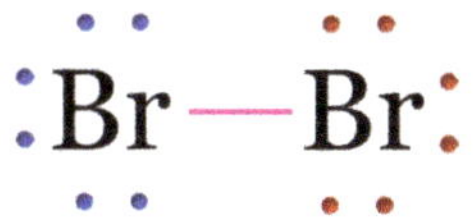

Since the only bond in this molecule is between identical atoms, there are no polar bonds in the molecule. As a result, the molecule cannot be polar covalent. Therefore, it is nonpolar covalent.

(b) We have already determined the electron pull of this molecule in an earlier problem:

There are 2 groups of electrons around the central atom (2 double bonds). This molecule is linear. Although there are polar bonds in this molecule, they pull the electrons in precisely opposite directions. As a result, the molecule cannot be polar covalent. It is therefore nonpolar covalent.

(c) The Lewis structure for this molecule is:

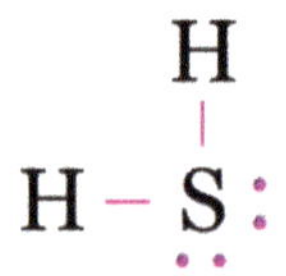

There are 4 groups of electrons surrounding the central atom (2 bonds, 2 nonbonding electron pairs). This means that the base geometry of the molecule is a tetrahedron. However, this tetrahedon has 2 legs removed, so the resulting shape is bent. Since the bonds exist between different types of atoms, they are polar. Sulfur is more electronegative than hydrogen, as shown by its position on the periodic table. The electron pull, then, looks like this:

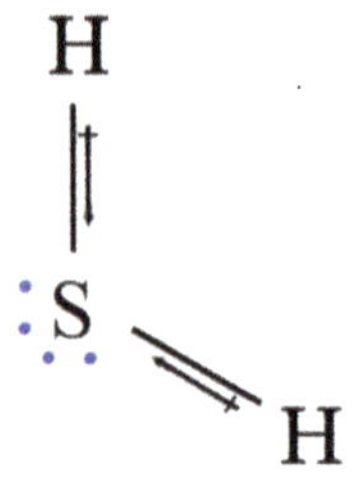

In this molecule, the polar bonds do not pull in exactly opposite directions, so they do not cancel each other out. As a result, this molecule is polar covalent.

(d) The Lewis structure for this molecule is:

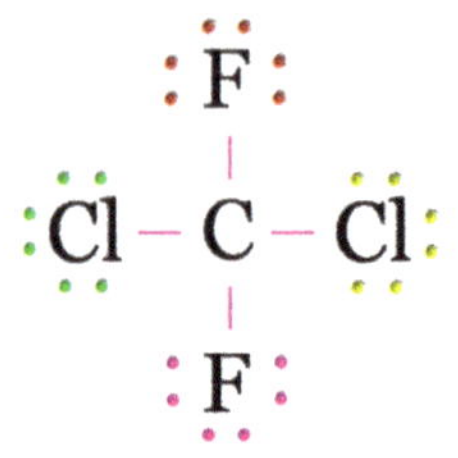

In this molecule, there are 4 groups of electrons around the central atom, making the base geometry tetrahedral. Since all 4 groups are bonds, all legs of the tetrahedron are present, so this is a tetrahedral molecule. All 4 bonds in this molecule are polar. In addition, both the Cl atoms and the F atoms are more electronegative than the C atom, so the electron pull looks like this:

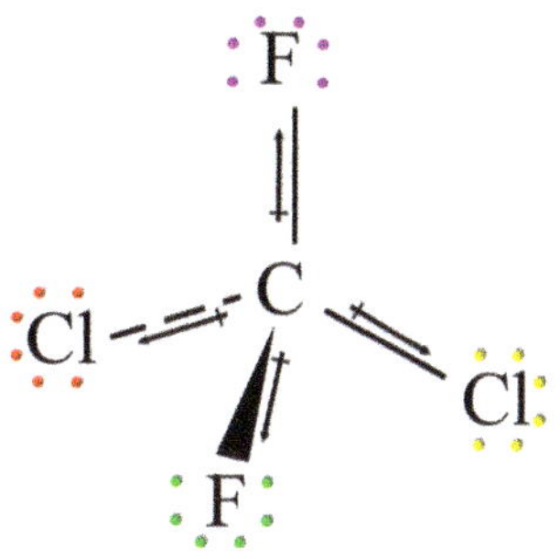

In this case, the carbon atom's electrons are being pulled in exactly opposite directions, but the pull is not equal. Since fluorine is more electronegative than chlorine, the pull from the fluorine atoms is a little bigger. As a result, there is a net pull on the carbon atom's electrons. This makes the molecule polar covalent.

We want to emphasize here how important it is to consider the *shape* of the molecule, not the Lewis structure, in determining the electron pull. In the Lewis structure drawn at the beginning of the problem, it looks like the electron pulls cancel out. After all, the F atoms are opposite each other, so they look like they may pull in opposite directions (one up and one down). In addition, the Cl atoms are opposite each other, so it looks like they may pull in opposite directions as well (one left and one right). However, the Lewis structure is not an accurate representation of the molecule's geometry. We have to figure out the geometry with VSEPR theory. Sure, we use the Lewis structure in VSEPR theory, so the Lewis structure is important. However, it is not the final word on the molecule's shape.

When we use VSEPR theory to determine that the molecule is tetrahedral, we see that the F atoms are not exactly opposite one another. One is in the plane of the paper, while the other one is pointed out of the plane of the paper toward us. The same is true of the Cl atoms. They are not exactly opposite because one is in the plane of the paper and the other is pointed out of the plane behind the paper. If you cannot visualize this, go back to Figure 5.1 and look at the model of CH_4. Imagine 2 of the yellow balls as fluorine atoms and the other 2 as chlorine atoms. There is no way you can get the fluorine atoms to cancel each other out, and there is no way you can get the chlorine atoms to cancel each other out. The only way that bonds in a tetrahedral molecule can cancel is if they are all identical.

(e) The Lewis structure for this molecule is:

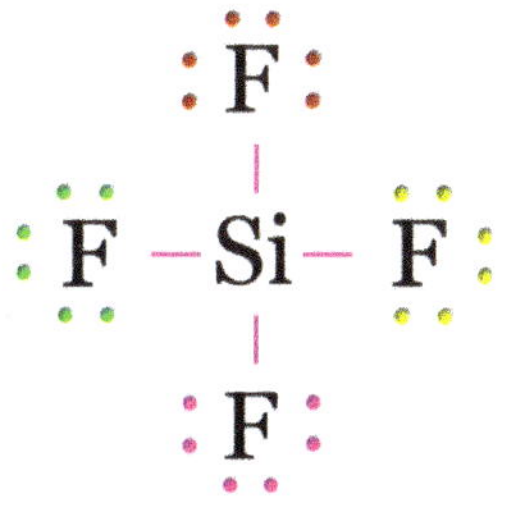

There are 4 groups of electrons around the central atom, and all of them are bonds. This is a tetrahedral molecule. All of the bonds in this molecule are polar, and since the fluorine is both to the right of and above silicon on the periodic table, fluorine is more electronegative than silicon. As a result, the electron pull looks like this:

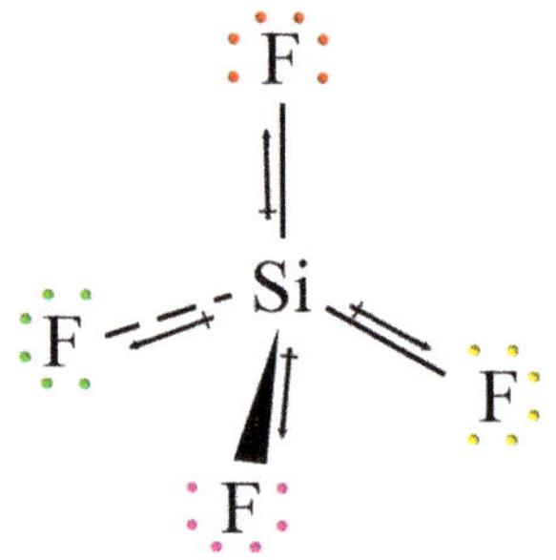

In this molecule, the silicon atom's electrons are all being pulled in opposite directions with equal strength. As a result, the polar bonds cancel each other out. This makes the molecule nonpolar covalent.

(f) The Lewis structure for PH_3 is:

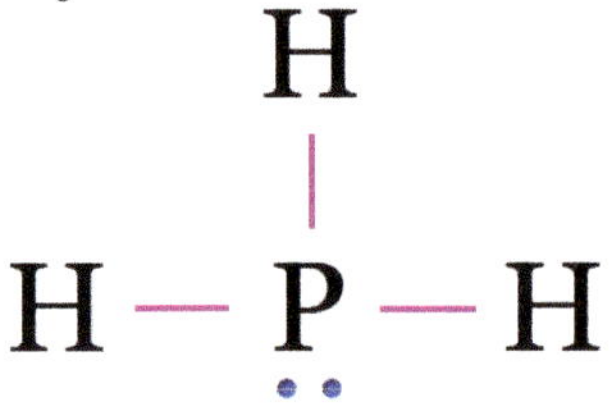

The bonds are all polar, and P is more electronegative than H. The electron pull looks like this:

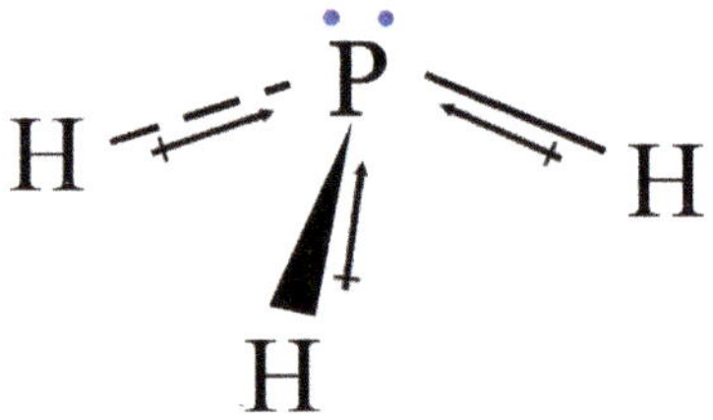

The bonds do not pull from opposite directions; therefore, this molecule is polar covalent.

5.11 We already classified each of these substances as either polar covalent or nonpolar covalent. The polar covalent substances will dissolve in water; the nonpolar covalent substances will not.

H_2S, CF_2Cl_2, and PH_3 will dissolve readily in water because they are polar covalent compounds. The rest nonpolar covalent and will NOT dissolve in water.

NOTE: All Lewis structure illustrations in this module are by Conner Hughes.

STUDY GUIDE FOR MODULE 5

REVIEW QUESTIONS

1. What is the difference between the ions you learned about in previous modules and the polyatomic ions you learned about in this module?

2. List the chemical formula (including charge) for each of the following ions:
   a. sulfate b. chlorite c. acetate d. sulfide e. dichromate f. phosphate

3. Give the names of the following ions:
   a. NO_2^- b. O^{2-} c. ClO_3^- d. CO_3^{2-} e. SO_3^{2-} f. CrO_4^{2-}

4. What does the acronym VSEPR stand for?

5. Explain VSEPR theory in your own words.

6. Why are the bond angles in a pyramidal shape smaller than the bond angles in a tetrahedral shape?

7. Explain the difference between polar and nonpolar covalent bonds.

8. If 2 different atoms are bonded together with a nonpolar bond, what can we say about the electronegativities of the 2 atoms?

9. "Oil and water don't mix" is a popular saying used to explain why very different people tend to not get along. Why don't oil and water mix?

10. Why does soap help wash away stains that water cannot wash away by itself?

PRACTICE PROBLEMS

1. Give the chemical formulas for the following compounds:
   a. potassium sulfite b. beryllium nitrate
   c. manganese (IV) carbonate d. aluminum chromate

2. Name the following compounds:
   a. $(NH_4)_2O$ b. $RbNO_2$ c. $Zn_3(PO_4)_2$ d. $CoPO_4$

3. Give the chemical formulas for each of the following compounds in a reaction: Calcium nitrate reacts with sodium carbonate to produce solid calcium carbonate (chalk) and sodium nitrate.

4. Determine the shape of a PCl_3 molecule. Give its bond angle and draw a picture of it.

5. Determine the shape of an H_2 molecule. Give its bond angle and draw a picture of it.

6. Determine the shape of a $SiCl_4$ molecule. Give its bond angle and draw a picture of it.

7. Determine the shape of an OF_2 molecule. Give its bond angle and draw a picture of it.

8. Determine the shape of a CS_2 molecule. Give its bond angle and draw a picture of it.

9. Classify each of the following molecules as ionic, polar covalent, or nonpolar covalent:
   a. $MgCl_2$ b. CF_3Cl c. CS_2 d. H_2 e. $ScCl_4$ f. PCl_3

10. Which of the substances in problem 9 would you expect to dissolve in water?

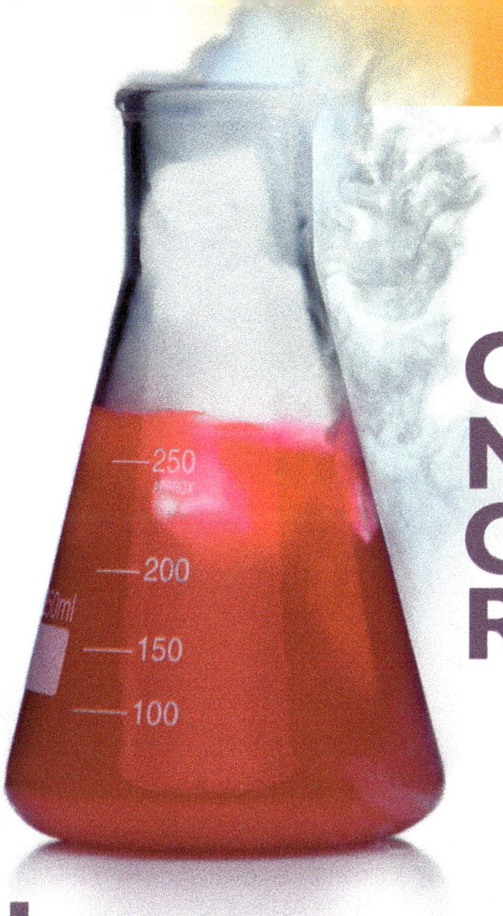

CHANGES IN MATTER AND CHEMICAL REACTIONS

In previous modules, we learned about the basic building blocks of matter: atoms and molecules. We learned that elements result when atoms of the same type are collected, and compounds result when molecules of the same type are collected. We also learned about what molecules look like, how they are shaped, and their tendencies (periodic properties). There are other ways atoms and molecules can combine to form other types of matter. In this module we will classify changes in matter and learn the criteria for this classification. In addition, we will start learning the basis of the most fundamental type of change matter can undergo: the chemical reaction.

first reactions

If you purchase a package of cookies at the grocery store, you will find an ingredient label. Your cookies might not look like eggs, but eggs have to be listed if they were used. Your cookies might not look like flour or sugar or any of the other ingredients listed, but the bakery has to account for each item that went into making the final product—your cookies. It's also that way in chemistry: you must always account for all the ingredients (called reactants) and the final product. Balancing both sides of a chemical equation is easy when you remember that what you started with in the beginning has to equal your final product.

CLASSIFYING CHANGES THAT OCCUR IN MATTER

In this module we will study how matter changes. Chemists classify the changes that matter undergoes into 2 large groups: physical change and chemical change.

Chemical change—A change that affects the type of molecules or atoms in a substance.

Physical change—A change in which the atoms or molecules in a substance stay the same.

The way to distinguish between these 2 types of changes is to think about the atoms or molecules involved. For example, if you rip a piece of paper in half, have you changed the molecules in the paper into different molecules? Both halves of the paper are still paper. All you did was split the sample of molecules in half. The molecules stayed the same; some were just separated from others. This is an example of physical change. Whenever you change a substance without altering its molecules or atoms, you have made a physical change.

FIGURE 6.1
Physical and Chemical Changes
Recycle photo by Lance Corporal Jessica Aranda (Public Domain). Fire photo by Ainhoa91 (CC BY-SA 3.0).

Tearing paper is an example of a physical change. Burning paper is an example of a chemical change.

On the other hand, suppose you burned that piece of paper. The molecules in the paper would slowly change into carbon dioxide, water vapor, and ash. This is an example of chemical change. If the type of atoms or molecules in a substance changes, the substance has undergone a chemical change.

Distinguishing between chemical and physical change is very important, so we are going to spend some time on it. Suppose we dissolve sugar in water. Is this a chemical change or a physical change? When the sugar dissolves in the water, it seems to disappear, but it hasn't changed its chemical makeup in any way. The sugar molecules are still sugar molecules; they are just evenly dispersed around the water now. Dissolving one substance in another is a physical change because the act of dissolving things does not affect the type of atoms or molecules involved.

think about this

Fire is used throughout Scripture as a metaphor. For example, Isaiah 64:1–2 says, "Oh that you would rend the heavens and come down, that the mountains might quake at your presence—as when fire kindles brushwood and the fire causes water to boil—to make your name known to your adversaries, and that the nations might tremble at your presence!" Think of how hot and consuming fire can be. Here we have a picture of the awesome power of God in a physical form to which we can relate.

On the other hand, remember what happened when you put lye into vinegar during Experiment 2.1? The lye did not *dissolve* in the vinegar; it *reacted* with the vinegar. You saw evidence of this because a color change occurred and heat was released. The color change was caused by the lye (NaOH) reacting with the active ingredient in vinegar ($C_2H_4O_2$) to make water (H_2O) and sodium acetate ($NaC_2H_3O_2$). The molecules involved actually changed. Before you put the lye in the vinegar, you had NaOH and $C_2H_4O_2$. After you mixed them and allowed them to react, you got H_2O and $NaC_2H_3O_2$. That was a chemical change.

Another way you can think about chemical and physical change is to consider whether a change can be easily reversed or not. Suppose you dissolved salt into water. Could you reverse that process? Could you recover the salt? Of course. To recover the salt, all you would have to do is boil off the water. In the same way, you could *easily* reverse the change that occurred when you ripped the paper in half. All you would need to do is put the 2 halves back together. Physical changes can usually be reversed relatively easily.

On the other hand, is there any way you could unburn paper? No. Once you have burned paper, there is no way to get the paper back. In the same way, once you reacted the lye with the vinegar in Experiment 2.1, you could not easily undo that change. If you boiled the liquid away, you would not find lye left over. Instead you would find sodium acetate. Chemical changes, then, are not easily reversed.

Although this way of distinguishing between chemical and physical change is convenient, it is not 100% correct. For example, chemical changes can be reversed, but it takes a chemical reaction to do so. In fact, in Module 15, you will see that in principle, most chemical reactions are reversible. However, for now, thinking about whether or not a change is easily reversible is a convenient way to distinguish between chemical and physical change. To become proficient at understanding the changes that occur in matter, you must be able to make this crucial distinction. To help you get more experience doing so, perform Experiment 6.1.

EXPERIMENT 6.1

PURPOSE: To distinguish between chemical and physical change.

MATERIALS:

- Water
- **Plastic** graduated cylinder (A rain gauge will work as well. A measuring cup may work, but if you have a frost-free freezer, the opening of whatever you use should be very small.)
- An egg in its shell
- Toilet bowl cleaner (The list of ingredients must include hydrochloric acid or hydrogen chloride. The Works was the brand used when this experiment was tested. You can use vinegar if you cannot find a proper toilet bowl cleaner, but the experiment will have to sit overnight for it to work. However, if you use vinegar, the egg will be much more interesting to observe!)
- Tall glass
- Sink
- Spoon
- Rubber cleaning gloves
- Safety goggles

QUESTION: Can chemical and physical changes be reversed?

HYPOTHESIS: Write down a hypothesis about whether chemical changes and physical changes can be reversed.

PROCEDURE:

1. The first part of this experiment will take quite a while to finish, so ideally you should start it now and finish the rest tomorrow. To start the first part of the experiment, simply fill your graduated cylinder with 30.0 mL of water (a quantitative measurement), and then put it in the freezer. (Note: If you have a frost-free freezer, cover the cylinder loosely with plastic wrap.) If you are using a measuring cup or a rain gauge, it doesn't really matter how much water you use, as long as the meniscus is sitting right on one of the measuring lines so that you know exactly what volume of water is in there.
2. Let the graduated cylinder remain in the freezer for at least several hours.
3. Place the tall glass in the sink. You need to do this because the experiment might cause suds to overflow the glass, and you will want the suds to fall right into the sink. ***Do not use aluminum foil around this experiment because aluminum will react with the toilet bowl cleaner and produce a nasty gas. Also, keep the glass in the sink throughout the entire experiment!***
4. Put on the rubber gloves and fill the glass about one-third full of toilet bowl cleaner. ***Be careful not to get any of the toilet bowl cleaner on your skin; it is caustic and will burn.*** If you do get it on your skin, just rinse it off with plenty of water.
5. Using the spoon, gently lower the egg into the toilet bowl cleaner and hold it completely under the liquid. Observe what is happening. A lot of bubbling should be going on. Since most toilet bowl cleaners contain detergent, some suds might also be forming. If you are using vinegar, you will not see the bubbling, and you will have to let it sit overnight rather than the few minutes discussed in the next step.
6. Allow this to go on for about 2 minutes, and then use the spoon to remove the egg from the glass. Rinse the egg with water and examine it. You should see that its shell is much thinner than it was before because the acid in the toilet bowl cleaner is reacting with the eggshell. When this happens, the calcium carbonate ($CaCO_3$) that makes up the eggshell is turned into calcium chloride, water, and carbon dioxide (which is making the bubbles you see). This is obviously a chemical change because the $CaCO_3$ is being changed into different molecules. This change is also not reversible. There is just no way to get the eggshell back once you've done this experiment.
7. If you do this experiment just right, you can strip away the eggshell and end up with just a thin, semitransparent membrane holding the egg together. This is much more likely if you use vinegar and wait overnight. If your egg still has a lot of shell on it, just put it back in the toilet bowl cleaner for a while and then pull it out again and rinse it off. Continue to do this until the shell is gone and a thin membrane is all that exists to hold the egg together. This can be a little tricky because the egg gets

pretty fragile and may break as you try to take it out of the toilet bowl cleaner.

8. After the graduated cylinder has been in the freezer at least 12 hours, pull it out and observe what has happened. As expected, the water has turned to ice, but read the volume. What has happened? The volume has increased. The water expanded as it froze. As we will learn a little later in this module, water is an exception to the general rule. Most substances decrease in volume when they freeze, but the volume of water increases upon freezing. This is a physical change. Think about the molecules involved. When water is liquid, its chemical formula is H_2O. After it freezes, it is still H_2O; it is just in solid form. Also, this change is reversible. Allow the graduated cylinder to stand out at room temperature until the ice melts. Then the volume will be back to 30.0 mL. This kind of change, called a phase change, will be discussed in great detail in a moment. For now, the important thing to realize is that phase changes are physical changes and do not affect the identity of the molecules or atoms that make up the substance being studied.
9. Clean up and return everything to the proper place.

CONCLUSION: What did you observe in this experiment? Write down some thoughts about what you qualitatively observed.

Now that you've had a bit of experience at identifying physical and chemical changes, try "On Your Own" question 6.1.

ON YOUR OWN

6.1 Classify the following changes as physical or chemical:

a. Steam condenses into water.

b. Gasoline burns in an automobile engine.

c. A barber cuts your hair.

d. Milk sours.

e. Sugar is dissolved in water.

f. Milk is poured on cereal.

g. An iron nail rusts.

PHASE CHANGES

As mentioned in Experiment 6.1, when a substance changes from solid to liquid (or vice versa), that change is referred to as a phase change. A change from liquid to gas (or vice versa) is also a phase change. These phase changes are quite interesting and worth discussing in some detail. The first thing to realize about phase changes is that they require only one thing: energy. If we want to change something from a solid to a liquid, all we need to do is add heat. Solid butter becomes liquid if we just heat it a little. Ice becomes water if it is allowed to warm up to room temperature. We call the transition from solid to liquid melting. The reverse process, liquid turning into solid, we call freezing. Instead of adding energy to a substance, if we take energy away from it, we can make the substance freeze. Likewise, going from liquid to gas, called boiling, requires energy, whereas going from gas to liquid, called condensing, requires that energy be removed. We could schematically depict this relationship as follows:

Solid photo by Darren Hester (CC BY-SA 2.5). Liquid photo © Jan Pietruszka. Gas photo © agoxa.

The 3 states of matter (solid, liquid, and gas) are called phases of matter, and that is why changes between them are called phase changes. Remember, these changes are physical changes. The molecules (or atoms) that make up the substance undergoing a phase change are not altered in any way. But if the molecules (or atoms) are not being changed, what is? Obviously, something must be changing for a phase change to take place. What can it be? Performing Experiment 6.2 might help you find out.

EXPERIMENT 6.2

PURPOSE: To condense steam

MATERIALS:

- Empty milk jug
- Water
- Small pot
- Stove
- Hot pads
- Safety goggles

QUESTION: Do gases take up more volume than liquids?

HYPOTHESIS: Write down your hypothesis about whether gases take up more volume than liquids.

PROCEDURE:

1. Put about 3 cups (a qualitative measurement) of water in the pot.
2. Heat the water on the stove. You will eventually hear the water boiling. Allow it to boil for quite some time, until steam is rising out of the pot.
3. Use a hot pad to remove the pot from the stove and pour the water into the milk jug.

4. Allow steam to rise out of the jug, and then put the lid on the milk jug and close it as tightly as possible.
5. Watch the jug for a while. You will start to hear it creak and groan, and then the jug will start collapsing in on itself. If this doesn't happen, it means that your jug has a leak in it somewhere, so you must start over with a new jug.
6. Clean up and return everything to the proper place.

CONCLUSION: What did you observe? Why did the jug collapse on itself? Write down some qualitative observations from this experiment.

Why did the jug collapse on itself? When you first poured the boiling water into the jug, water vapor began filling the jug, pushing any air that was in it out through the hole in the top. By the time you placed the lid on the jug, the jug was full of water in its gas phase. When water vapor cools, it condenses back into liquid water. When this happened in your experiment, the jug collapsed in on itself because water vapor takes up considerably more volume than liquid water. In its vapor form, the water was able to fill the jug. When it turned into liquid, the volume it required was significantly less. Since the lid was on the jug, no air could rush in and fill the void left behind by the condensing vapor, so the jug had to collapse in reaction to the fact that the condensing vapor was taking up significantly less volume than it had before.

We hope that this experiment convinced you of the main difference between phases of matter: the volume they occupy. With the exception of water and a handful of other substances (which will be discussed in a moment), most substances take up the least amount of volume in their solid phase, a little more volume in their liquid phase, and the largest amount of volume in their gas phase. Why? Remember, all substances are made up of individual atoms or molecules that come together and occupy the same general area. The reason these atoms or molecules come together in the first place is that they are attracted to one another. Sulfur atoms tend to stay with other sulfur atoms because they attract one another. In the same way, sugar molecules tend to hang around other sugar molecules because they are attracted to one another.

When things are attracted to one another, they tend to get close together. The closer atoms or molecules get to one another, the less volume is occupied by the substance that they comprise. In other words, when sugar molecules get close together, the sugar occupies little volume. When the sugar molecules get far apart, the substance occupies quite a bit of volume. Solids occupy the least volume, so when a substance is in its solid phase, the atoms or molecules that make up that substance are very close together. On the other hand, when a substance is in its gas phase, the atoms or molecules that make it up are very far away from one another.

For example, carbon dioxide (like all matter) can exist in any of the 3 phases we have discussed. Its most common phase is the gas phase. When we breathe out carbon dioxide, it is in its gas phase. In this phase, the carbon dioxide molecules are far away from one another. Under the right conditions, carbon dioxide can also become a liquid. When the carbon dioxide molecules get closer together, the gas starts to condense, and the carbon dioxide becomes a liquid. When this happens, the amount of space occupied by the

ON YOUR OWN

6.2 Natural gas companies store most of their natural gas in liquid form rather than in its gas phase, even though they must deliver it to their customers as a gas. Why do they do this?

same mass of carbon dioxide decreases significantly. Finally, carbon dioxide can also exist in the solid phase. We call solid carbon dioxide dry ice. Dry ice is formed when the molecules of carbon dioxide get as close as possible to one another. This further reduces the volume that the same mass of carbon dioxide occupies, making it a solid.

The obvious question is this: Why do the molecules in carbon dioxide get closer together or farther apart? If carbon dioxide molecules are attracted to one another, why don't they all just get as close as possible to one another, making carbon dioxide always a solid? To answer this question, we have to discuss the kinetic theory of matter. We will do that in the next section. First, answer "On Your Own" questions 6.2.

THE KINETIC THEORY OF MATTER

Did you know that right now, as you read this sentence, the molecules that make up the paper in this module and the molecules that make up the ink in these words are all moving very rapidly? You may not believe it, but it's true! The molecules or atoms that make up all of the matter we observe are always in constant motion. That is the basis of the kinetic theory of matter.

"Wait a minute," you might protest. "How in the world can the molecules which make up this paper be moving when the paper is sitting still?" That's a reasonable question. Atoms and molecules are so small that they cannot be seen, even with a very powerful microscope. As a result, any matter that we can actually see is made up of billions and billions and billions of atoms or molecules in random motion. For example, in the solid phase molecules (or atoms) vibrate back and forth a few billion times every second. Matter in the solid phase doesn't appear to be moving because for every molecule (or atom) that is vibrating one way, there is another molecule (or atom) vibrating another way. On average, solid matter cannot move anywhere because all its molecules or atoms are moving in different directions.

In the liquid phase, molecules and atoms do not just vibrate; they move around. If you were to look at a glass of water and track a single molecule, you would find that the molecule travels throughout the glass. It may start out on the bottom of the glass, but it will quickly move in one direction until

salt and light

Photo: Public Domain

George Stokes (1819–1903) was a British scientist who laid foundations for the study of how fluids move. He participated in the famous Gifford Lectures which sought to diffuse the knowledge of God. These lectures were given in several Scottish universities, and to be appointed lecturer was considered one of the greatest honors a Scottish scientist could obtain. Stokes said, "The evidence for the resurrection of Jesus Christ is never to be separated from a consideration of the character and teaching and works of Jesus Christ. The head and heart must go together" (Stokes 2010, 80).

it collides with the wall of the glass or with another water molecule. It will then switch directions and speed off in another direction until it collides with something else. The motion of a water molecule is a lot like a bumper car that continues moving in one direction until it hits something. It then changes direction and speeds off again until it hits something else. In the same way, a molecule of water will travel throughout the entire volume of the glass, continually running into things, changing directions, and speeding off again.

In the gas phase, things are just like the liquid phase, only the molecules travel faster and make fewer collisions with other molecules because they are farther apart from one another. The gas molecules simply speed around, collide into things, change directions, and speed off again. Whether we are talking about the solid phase, liquid phase, or gas phase, the molecules that make up matter are in some kind of random motion. Figure 6.2 illustrates this concept.

FIGURE 6.2
Random Atomic Motion in the Three Phases of Matter
Illustration by David Weiss

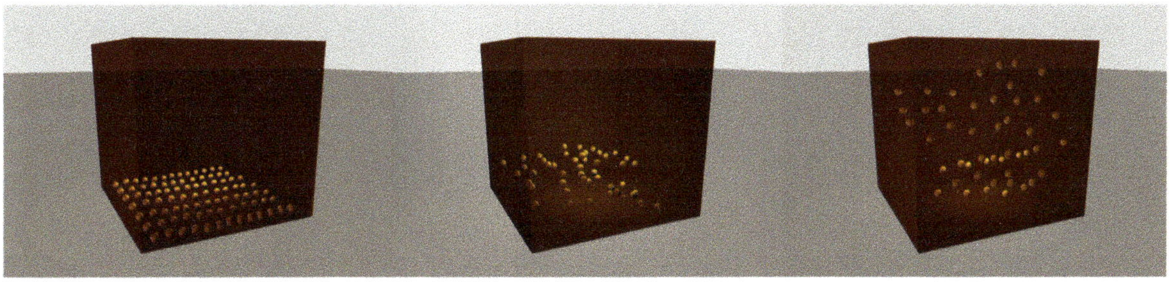

In the solid phase, atoms (or molecules) vibrate back and forth in random directions.

In the liquid phase, atoms (or molecules) move about, colliding with one another and the walls of their container.

In the gas phase, atoms (or molecules) move about faster than in the liquid phase. They are also much farther apart.

Since motion requires work, and work requires energy, the molecules in a substance must have energy to move around as they do. Where do they get this energy? They get it from the surroundings. The higher the temperature of the surroundings, the faster the molecules move. The faster the molecules move, the farther they get from the other molecules in the substance. When the temperature of a substance's surroundings increases, the molecules (or atoms) in that substance start to absorb energy. As they absorb energy, they start moving faster. As they move faster, they get farther away from their neighbors. As they get farther away from their neighbors, the substance eventually changes phase. This is why a substance changes from solid to liquid to gas as the temperature is increased. The temperature at which a substance changes from solid to liquid is called the melting point of the substance. The temperature at which a substance goes from its liquid phase to its gas phase is called the boiling point.

If you purchased the MicroChem kit mentioned in the Student Notes, you can get some experience measuring melting points by performing Experiment 2.

As a substance's surroundings decrease in temperature, energy travels out of the molecules (or atoms) in the substance and into the surroundings because heat always travels from a hot body to a cold body. If the surroundings decrease in temperature, the molecules (or atoms) in a substance will lose energy to their surroundings. Since they lose energy, their motion begins to decrease. As their motion decreases, the molecules (or atoms) get closer to their neighbors. This causes a substance to change from gas to liquid to solid as temperature decreases.

The kinetic theory of matter, then, can be summarized as follows:

Molecules and atoms are in constant motion.
The higher the temperature, the greater their speed.

Since we can never really see atoms or molecules, we cannot see that they move, either. However, making the assumption that molecules or atoms in a substance are always moving helps us explain a great deal about the behavior of matter. Perform Experiment 6.3 to get a visual understanding of the kinetic theory of matter.

EXPERIMENT 6.3

PURPOSE: To learn the relation between the speed and temperature of molecules.

MATERIALS:

- 2 glass canning jars or peanut butter jars, both the same size
- Food coloring (any color)
- Pan to boil water
- Stove
- Hot pad
- Safety goggles

QUESTION: Do molecules move faster at higher temperatures?

HYPOTHESIS: Write down your hypothesis about molecules moving faster at higher temperatures.

PROCEDURE:

1. Boil enough water so that the boiled water will fill one of the jars about halfway.
2. Once the water is boiling, take it off the stove (use the hot pad!) and pour it into one of the jars. Pour in enough water so that the jar is filled about halfway.
3. Fill the other jar about halfway with cold water from the tap.
4. Wait a few minutes so that the water in each jar is still.
5. Drop a single drop of food coloring into each jar. Observe what happens over the next several minutes.
6. Clean up and return everything to the proper place.

CONCLUSION: What did you see in the experiment? Write some qualitative observations.

What did you see in the experiment? If everything went well, you should have seen the drop of food coloring mix rapidly with the hot water, coloring the entire jar of water relatively quickly. In the jar of cold water, the food coloring should not have mixed nearly as well. This is an excellent illustration of the kinetic theory of matter. The molecules in the cold water were moving, but not nearly as rapidly as those in the hot water. When the food coloring was added to both jars, the water molecules began colliding with the molecules in the food coloring. This began spreading out the molecules in the food coloring, mixing them with the water. Since the water molecules in the hot water were moving faster than those in the cold water, the food coloring spread out much more rapidly in the hot water. The hot water changed color much more quickly than did the cold water because its molecules were moving more rapidly. Answer "On Your Own" question 6.3 to see if you understand this.

ON YOUR OWN

6.3 If a substance is heated 10 degrees, will its molecules be moving faster or slower? Will the molecules be closer together or farther apart?

DENSITY

Let's introduce one very important quantity that is measured quite frequently in the study of matter: density.

Density—An object's mass divided by the volume that the object occupies.

This definition for density can be mathematically represented with this equation:

$$\rho = \frac{m}{V} \qquad \text{Equation 6.1}$$

In the equation above, the ρ is the Greek letter rho and is typically the symbol used to represent density. The letter m stands for the mass of the object, and the letter V stands for the object's volume. From this equation, we can determine the units that describe density. Since mass is measured in grams and the most common volume unit in chemistry is mL, the equation says that the units for density will be the unit that results when grams are divided by mL, or $\frac{\text{grams}}{\text{mL}}$. In other words, we would call this unit grams per mL. Since mL and cm^3 are equivalent, we often see density expressed in grams per cm^3. These, of course, are not the only units for density. Any mass unit divided by any volume unit is a possible unit for density, but these 2 are the most common in chemistry.

Density is a very important quantity in chemistry. It gives you an idea of how tightly packed the matter is in an object. For example, suppose you had a ball made out of plastic and another ball of precisely the same size made out of lead. Which ball would be heavier? The lead ball would be much heavier than the plastic ball because there would be a lot more matter in the lead ball than in the plastic ball. This is because matter is packed very tightly together in lead, but matter is packed pretty loosely in plastic. For the same size ball, the lead ball is much heavier.

Density gives us a way to determine exactly how heavy the lead ball is compared to the plastic ball. Lead has a density of 11.4 grams per mL, whereas most plastic has a density of less than 1 gram per mL. Let's suppose we had a sample of plastic with a density of 0.54 grams per mL. Since the density of lead is 21 times (11.4 ÷ 0.54) greater than the density of the sample of plastic, we know that the matter inside lead is packed 21 times tighter than the matter inside the sample of plastic.

Every substance on earth has its own characteristic density. This is nice because it provides a way for us to identify a substance if we don't know exactly what it is. For example, years ago people used density to determine whether or not something that looked golden was really made out of pure gold. They would measure the substance's mass, measure its volume, and then divide the 2 numbers. If the result turned out to be 19.3 grams per mL (the density of pure gold), then they knew it was pure gold because no other substance on earth has that density. If the density was not 19.3 grams per mL, then they knew it was not pure gold.

Before showing you some examples of how to use density in problems, let's perform another experiment. Experiment 6.4 will teach you something very important about the physical meaning of density.

EXPERIMENT 6.4

PURPOSE: To compare the density of liquids.

MATERIALS:

- Water
- Vegetable oil
- Something that measures the volume of a liquid, preferably in mL or cm^3. A graduated cylinder would be ideal, but measuring cups will work as well.
- Maple syrup (Natural syrup does not work as well as something like Mrs. Butterworth's.)
- Large glass
- Mass scale, preferably one that reads in grams. (The scale should not go much over 500 grams, or it will be very difficult for you to read the mass of the objects in this experiment.)
- Safety goggles

QUESTION: How do we qualitatively and quantitatively determine the density of liquids?

HYPOTHESIS: Write a hypothesis about the densities of maple syrup, vegetable oil, and water.

PROCEDURE:

1. Measure the mass of the graduated cylinder or whatever you have that measures the volume of a liquid. Be sure to write it down with the correct precision. If you are using a standard mass scale from a grocery store, its scale is probably marked off in units of 10 grams, so you should be able to report the mass to a precision of 1 g.
2. Measure 50.0 mL (1/4 cup if you are using measuring cups) of syrup. Put the graduated cylinder (with the syrup in it) back on the scale and measure the total mass.

Subtract the mass of the graduated cylinder from this number (using our rules for significant figures) to get the mass of the syrup by itself. This method of measuring mass is called the **difference method**. Chemists often call it measuring the mass by difference.

3. Now that you have the mass of the syrup, and you know that its volume was 50.0 mL (because that's what you measured), divide the mass by the volume to get the density. Be sure to follow our significant figure rules when you do this! Finally, pour the syrup into the tall glass. Repeat this procedure for both the water and the vegetable oil.
4. Once you have measured the density of all 3 substances, look at the tall glass. You should see that the syrup is all at the bottom of the glass, the water forms a layer above that, and the vegetable oil is all in one layer on top!
5. Clean up and return everything to the proper place.

CONCLUSION: Write something about what you observed and whether it confirmed your hypothesis.

What explains the layering you saw in the glass? Look at the densities you calculated from your measurements. Which substance has the largest density? The syrup. Where was the syrup? It was on the bottom of the glass. In the same way, the vegetable oil has the smallest density and was at the top of the glass, while the water's density is between them, so it stays between the syrup and the vegetable oil. If we think about what density means, this should make sense. Density tells us how tightly packed the matter is inside a substance. Syrup's matter is very tightly packed, so it falls through the more loosely packed water and vegetable oil until it reaches the bottom of the glass. Water falls through the more loosely packed matter in vegetable oil but then is stopped by the more tightly packed matter in syrup. The vegetable oil could not fall through either water or syrup, so it stayed at the top of the glass.

Part of the question for the experiment was to "qualitatively and quantitatively determine the density of liquids." To qualitatively determine something is to observe it. In our experiment this was the visual observations you made of the layers of liquids. To quantitatively determine something is to measure it. In this case, you measured the density of 3 liquids. The qualitative observations showed us that denser liquids sink, and this should be confirmed by your calculations that showed the syrup to have the largest density.

Now that you have a better understanding of what density means, follow Example 6.1 and then answer "On Your Own" questions 6.4–6.6.

EXAMPLE 6.1

A gold miner has just found a nugget of pure gold. He measures its dimensions and then calculates its volume to be 0.125 L. Knowing that the density of gold is 19.3 grams per mL, calculate the mass of the miner's nugget.

The equation for density is:

$$\rho = \frac{m}{V}$$

However, the equation is not much good to us the way it is currently written. Since we would like to calculate the mass of the nugget, we need an equation that starts "m=." We can use algebra to rearrange the equation so that it reads:

$$m = \rho \times V$$

Now we can simply put in our numbers for volume and density, and we will get the mass, right? There's only one small problem. If we use the numbers as they are given in the problem, we would not get the correct answer. Why? If we did that, it would look like:

$$m = 19.3\ \frac{g}{mL} \times 0.125\ L$$

According to our rules for multiplying numbers with units, we multiply the numbers, and then we do the same with the units. Since we are calculating mass, we know that our answer should end up with the unit of grams. If we multiply these 2 units together, we don't get grams. To get grams, the mL in the denominator would have to cancel with the L in the volume measurement. That doesn't work. Instead, we need to change one of the units. It doesn't matter which one, but let's change the volume measurement into mL.

$$\frac{0.125\ \cancel{L}}{1} \times \frac{1\ mL}{.001\ \cancel{L}} = 125\ mL$$

Now that we have the proper units, we can put everything into the equation:

$$m = 19.3\ \frac{g}{\cancel{mL}} \times 125\ \cancel{mL} = 2{,}412.5\ g$$

We're still not quite done, however. According to our rules of significant figures, we can have only 3 significant figures in our final answer because there are 3 significant figures in each of our numbers. After rounding, the answer is **2,410 g**. In scientific notation, the answer is **2.41×10^3 g**. Either expression is correct, as they both have 3 significant figures.

ON YOUR OWN

6.4 Cubes of frozen rubbing alcohol are put into liquid rubbing alcohol. Will the cubes float or sink?

6.5 The density of silver is 10.5 grams per cm^3. A jeweler makes a silver bracelet out of 0.200 kg of silver. What is the bracelet's volume in mL?

6.6 A gold miner tries to sell some gold that he found in a nearby river. The person who is thinking about purchasing the gold measures the mass and volume of 1 nugget. The mass is 1.56 kg, and the volume is 0.081 liters. Is this nugget really made of gold? (Remember that the density of gold is 19.3 grams per mL.)

PHASE CHANGES IN WATER

Photo by Jeannie Fulbright.

Before we leave the subject of phase changes, we must spend a little time discussing water. As noted before, although almost all natural substances occupy less volume in the solid phase than in the liquid phase, solid water (ice) occupies more volume than liquid water. This is an exception to the phase change rule we discussed, and it is very good that this exception exists.

Because water occupies more volume in its solid phase than in its liquid phase, the density of ice is *less* than the density of liquid water. We know that less dense substances float on top of more dense substances. This means that ice floats on top of liquid water since its density is lower. This is why ice cubes float in a drink. This is also why ice forms on the *top* of a freezing lake. A layer of ice on top of a lake insulates the water below it, keeping it from freezing further. As a result, a lake never freezes completely. There is always plenty of water underneath the ice, allowing the fish and other organisms to stay alive.

If ice did not float on top of water, a lake would freeze from the bottom up. There would be nothing to insulate the unfrozen water, so most lakes would freeze solid each winter, killing most (if not all) of the life in the lake. Water–living organisms provide food for many different creatures, and water–living algae are responsible for a great deal of the photosynthesis that keeps the oxygen levels in the atmosphere constant. It is hard to imagine how life could exist on earth without the organisms that live in lakes.

In addition to the fact that water's exception to the rule makes it possible for fresh–water organisms to survive, many biochemists believe that it also makes it possible for life as we know it to exist. The same physical effect that causes water to be an exception to the phase change rule also keeps water a liquid at normal temperatures. This effect is called hydrogen bonding. Amazingly enough, all of the substances that are chemically similar to water (H_2S, H_2Se, H_2Te) are gases at normal temperatures. Water, on the other hand, is a liquid at those temperatures because it can participate in hydrogen bonding, which we will talk more about in Module 7. Most biochemists agree that if water were not a liquid at normal temperatures, life as we know it could not even exist!

CHEMICAL REACTIONS AND CHEMICAL EQUATIONS

Now that we have taken an in–depth look at physical change, it is time to turn our attention to chemical change. Remember, chemical change occurs when the molecules within a substance change into different molecules. For the rest of this module, we will examine how this change (called a chemical reaction) occurs and how to write these changes down in an abbreviated form called a chemical equation.

We will begin our examination of chemical change by looking at a process that is very important to today's economy: the burning of coal. When coal, which is mostly carbon, is burned, the energy released can be used to generate electricity. Currently, a little more than half the electricity produced in the United States is produced through coal burning, so it is a pretty important chemical change and deserves close examination. Before we can do this, we must review a concept that we discussed in Module 4—diatomic molecules.

Remember that elements H, N, O, F, Cl, Br, I, and At all exist as homonuclear diatomic molecules. This means that oxygen exists not as O but as O_2. Remember this as we continue our discussion of burning coal. When coal is burned, it combines with oxygen to make carbon dioxide. The energy released when this chemical change occurs is 30 kJ for every 1 gram of carbon that is burned. This energy output is sufficient to make electricity in an efficient manner.

This chemical change, called a chemical reaction, could be written as follows:

Carbon plus oxygen produces carbon dioxide.

Although this method of writing the chemical reaction tells us what happened, it is rather bulky. We would like to abbreviate it, making it smaller and easier to write. We will do so by writing the chemical formulas of the substances involved rather than their names. We will also replace the word *produces* with an arrow, $\rightarrow$. The abbreviated chemical reaction is:

$$C + O_2 \rightarrow CO_2$$

This gives us the same information as the previous representation, but in a much more condensed fashion. Notice that since oxygen is a homonuclear diatomic element, its chemical formula is O_2, not O. Carbon is not a homonuclear diatomic, so its chemical formula is simply C.

The abbreviations for chemical reactions, such as the one written above, are called chemical equations. Chemical equations form the basis for most of the work done in chemistry, so it is critically important that you understand what they are and how they are used. Of course, chemical equations get much more complicated than the one above, so we will need to spend a great deal of time learning about them.

First, you need to understand the framework of a chemical equation. The atoms and molecules that appear on the left side of the arrow represent the substances that exist before the chemical change takes place. We call these substances reactants because they are the substances that will react together. The atoms and molecules on the right side of the equation are called products because they represent the substances which are produced by the chemical reaction. A chemical equation is always written with reactants on the left side of the arrow and products on the right side.

$$\text{Reactants} \rightarrow \text{Products}$$

To give you an idea of what really goes on during a chemical reaction, Figure 6.3 attempts to illustrate, on the atomic level, the steps that take place when carbon reacts with oxygen to make carbon dioxide.

FIGURE 6.3
Burning Carbon at the Atomic Level
Illustrations by David Weiss

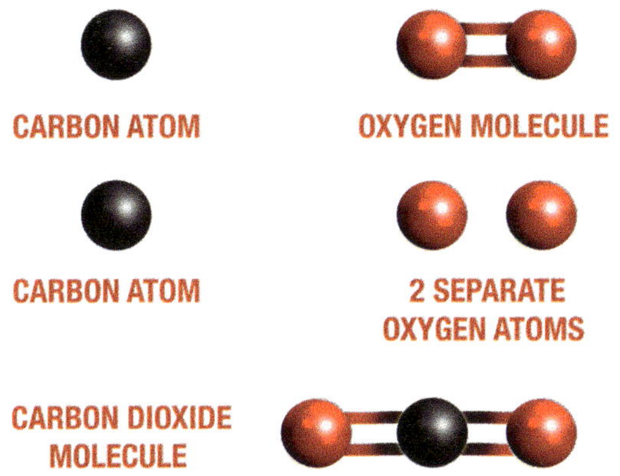

1. **The chemical reaction starts with just the reactants: a carbon atom and a homonuclear diatomic oxygen molecule.**
2. **The oxygen molecule splits apart to make 2 independent oxygen atoms.**
3. **The carbon then joins with both oxygen atoms, leaving just the product: a carbon dioxide molecule.**

This, then, is how a chemical reaction proceeds. The molecules must disassemble into their constituent atoms, and then the atoms recombine to form the products.

Now we can make the carbon/oxygen equation a little more meaningful by adding some information about the phase of each substance in the equation. For example, when we burn coal, the carbon is in its solid state, the oxygen is in its gaseous state, and the carbon dioxide produced is also in its gaseous state. We can add these facts in the equation by putting symbols in parentheses next to each substance:

$$C\ (s) + O_2\ (g) \longrightarrow CO_2\ (g)$$
$$\text{Reactants} \longrightarrow \text{Products}$$

Whenever you see a small letter enclosed with parentheses in a chemical equation, you know that it is a symbol telling you the phase of the substance. Table 6.1 lists the phase abbreviations used in chemical equations. Chemists do not always use phase symbols in their chemical reactions. Sometimes they are there; sometimes they are not. It is always preferable to include them because they provide important information. Don't be confused if you do not see them, though. You need to memorize the abbreviations for the phases in table 6.1.

TABLE 6.1
Phase Symbols

Phase	Symbol
Solid	(s)
Liquid	(l)
Gas	(g)
Aqueous (dissolved in water)	(aq)

If this were as complicated as things got, chemistry would be a breeze. However, things get a little tougher from here. Let's look at another important chemical reaction: the burning of methane (CH_4). Methane is the principal component of natural gas. If you have a gas stove, gas water heater, or gas furnace, this chemical reaction occurs quite frequently in your home. Once again, the combustion (burning) of methane releases quite a bit of energy. This energy can be used to cook, to heat water, or to heat your home. There are even some electrical plants that run on methane. For example, more than 45% of the electricity in Texas is generated by methane–burning power plants. The combustion of methane is another important chemical reaction.

The combustion of methane is a little more complicated than the combustion of coal because it makes 2 products: carbon dioxide and water. We could write this reaction as:

Methane + oxygen produces carbon dioxide + water.

Using chemical formulas and phase symbols to turn this into a chemical equation, we come up with:

$$CH_4\ (g) + O_2\ (g) \longrightarrow CO_2\ (g) + H_2O\ (g)$$

There are several things to note about the above equation. First, you might wonder about how we came up with the phase symbols. How did we know that CH_4 is a gas, for example? This knowledge just comes from experience. At this stage, we would not expect you to know what phase the reactants and products are in during a chemical reaction. The exceptions to this are oxygen and carbon dioxide. We have talked about these 2 substances so much that you should know by now that they are gases. If we expect you to put phase symbols with any other substances, we will tell you what phase they are in.

Second, you might notice that we listed water as a gas. You might think that we should have listed water as a liquid. After all, at normal temperatures, water is a liquid. This particular reaction releases a lot of energy, making its surroundings very hot. When the surroundings get hot, the water boils. Because this reaction is hot, it puts water into its gas phase, and that's why we listed it as a gas. In chemical reactions, sometimes water will be a liquid, sometimes it will be a gas, and sometimes it will be a solid. Once again, only experience can help you make that distinction.

Finally, this equation, as it is written now, is *wrong*. To see why it is wrong, study Figure 6.4.

FIGURE 6.4
The Combustion of Methane: The Wrong Approach
Illustrations by Striar Yunis

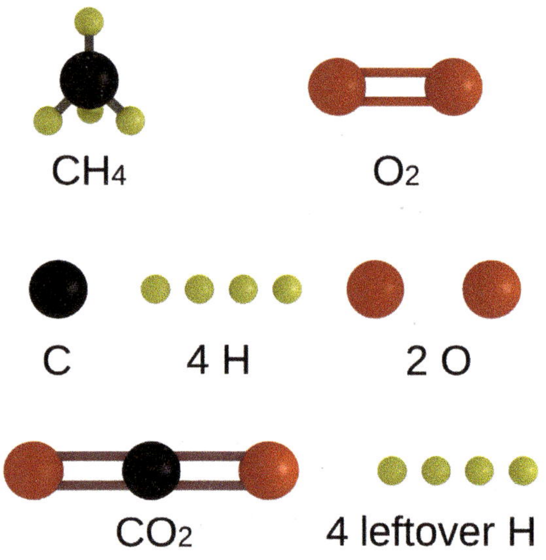

1. **We start with our reactants: a methane molecule and an oxygen molecule.**
2. **The oxygen molecule splits apart into 2 oxygen atoms, and the methane molecule splits apart into 1 carbon atom and 4 hydrogen atoms.**
3. **Now the atoms will try to recombine and form the products, CO_2 and H_2O.**

Do you see the problem in the figure? There are not enough oxygen atoms to get the job done. The chemical reaction makes both carbon dioxide and water. We were able to make the carbon dioxide molecule, but then we ran into a problem. To make H_2O, we need both hydrogen atoms and oxygen atoms. We have enough hydrogen atoms, but we don't have any oxygen atoms left after the CO_2 was formed. To solve this problem, we have to use another oxygen molecule. Figure 6.5 goes through this procedure again, but it starts with 2 oxygen molecules. As you will see, this works out much better!

FIGURE 6.5
The Combustion of Methane: The Correct Approach
Illustrations by Striar Yunis

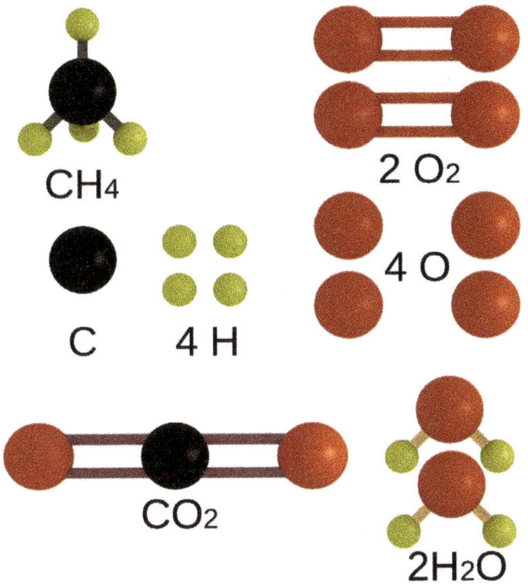

1. **We start with our reactants: a methane molecule and 2 oxygen molecules.**
2. **The oxygen molecules each split apart into 2 oxygen atoms, making a total of 4 oxygen atoms, and the methane molecule splits apart into 1 carbon atom and 4 hydrogen atoms.**
3. **Now the atoms will recombine and form the products, CO_2 and H_2O. This time, there is no problem because each of the extra 2 oxygen atoms can combine with 2 hydrogen atoms to make 2 water molecules.**

In the combustion of methane, then, 2 oxygen molecules must combine with 1 methane molecule. When this happens, 1 carbon dioxide molecule and 2 water molecules will be produced. We can add this information to the original equation so that it will be correct:

$$CH_4\ (g) + O_2\ (g) + O_2\ (g) \longrightarrow CO_2\ (g) + H_2O\ (g) + H_2O\ (g)$$

Now, at least, the chemical equation is correct. It tells us that our reactants are 1 methane molecule and 2 oxygen molecules, and our products are 1 carbon dioxide molecule and 2 water molecules. It is critically important for you to understand why the original equation is not correct and the above equation is correct. The way the original equation was written, we could not form the products that were listed, and we had extra atoms left over at the end. Neither of these things is possible in nature. A chemical reaction must start with its reactants, end up with its products, and have no extra atoms left over. The chemical equation used to describe a chemical reaction must reflect this fact.

We can clean up this chemical equation a little more by combining like terms. Remember from algebra that $x + x = 2x$. It's the same in chemistry. When we have several identical terms on the same side of a chemical equation, we can group them together.

Therefore:

$$O_2\ (g) + O_2\ (g) = 2O_2\ (g) \text{ and } H_2O\ (g) + H_2O\ (g) = 2H_2O\ (g).$$

Remember, just as in algebra, the terms must be identical before you can combine them. The chemical equation that describes the combustion of methane is:

$$CH_4\ (g) + 2O_2\ (g) \longrightarrow CO_2\ (g) + 2H_2O\ (g)$$

The above equation is called a balanced chemical equation. If a chemical equation is not balanced, then it is not correct. An unbalanced chemical equation will result in not all products being made or in leftover atoms (or both). Both of those results are impossible in nature, so unbalanced chemical equations are just not correct. Also, an unbalanced chemical equation violates the law of mass conservation. Since atoms cannot be created or destroyed, the total mass before a chemical reaction takes place depends on the number of each type of atom in the reactants. When the reaction is over, the products must include the same number of each type of atom, or the mass will not be the same. Since balancing chemical equations is so important, the rest of this module will be devoted to teaching you how to balance chemical equations. You cannot proceed to the next module until you learn this important technique, so please concentrate and practice.

Before we begin practicing how to balance chemical equations, let's make sure we understand everything that the balanced equation tells us. First, we know this equation tells us the reactants and products in the reaction. It also tells us what phase each substance is in during the reaction. Equally important, the chemical equation tells us the recipe we need to get this reaction to go. The big numbers to the left of each chemical formula tell us how many molecules of each substance are necessary for the reaction to work. As usual, if there is no number to the left of a chemical formula, we must assume it is 1. This balanced equation tells us that 1 CH_4 molecule will react with 2 O_2 molecules. These reactants will produce 1 CO_2 molecule and 2 H_2O molecules.

Remember not to confuse the big numbers on the left of the chemical formulas with the numbers in the subscripts. The numbers in the subscripts are part of the molecule's chemical formula. They tell us *what* molecules we are dealing with. The big numbers to the left of the chemical formulas (the coefficients), on the other hand, tell us *how many* of those molecules we are dealing with. This is an important distinction that you must understand to be able to balance chemical equations.

DETERMINING WHETHER OR NOT A CHEMICAL EQUATION IS BALANCED

To balance chemical equations, we must first learn to keep track of all atoms in an equation. An equation will always be balanced if the number of each type of atom on the reactants side of the equation is the same as the number of that type of atom on the products side of the chemical equation. For example, if a chemical equation has 5 H atoms on its reactants side, it must have 5 H atoms on its products side. There is a trick to counting the atoms, though, so we will go back to the balanced equation to teach you this trick.

Remember that the number to the left of each molecule tells us how many molecules are in the equation. The numbers in subscripts tell us how many of each atom are in the molecule. To count the number of atoms on each side of the equation, we must multiply the number to the left of the molecule by the subscript number that follows each atom.

This will tell us the total number of each atom on each side of the equation. Let's see how that works in this balanced equation. First, we will rewrite the equation so that we can refer to it easily:

$$CH_4\ (g) + 2O_2\ (g) \longrightarrow CO_2\ (g) + 2H_2O\ (g)$$

Now let's count the number of carbon atoms on the reactants side of this equation. To do this, multiply the number to the left of CH_4 by the subscript directly after C. Since there is no number to the left of CH_4, we assume it is 1. Also, since there is no subscript directly after C, we assume it is 1 as well. There is 1 x 1 = 1 carbon atom on the reactants side of the equation. To count the carbon atoms on the products side of the equation, we look at the only molecule that contains carbon on the products side: CO_2. Since there is no number to the left of CO_2 and no subscript after C, they are both 1. There is 1 x 1 = 1 carbon atom on the products side of the equation. Since the number of carbon atoms is the same on each side of the equation, the equation is balanced with respect to carbon.

Reactants Side	Products Side
#C: 1 x 1 = 1	#C: 1 x 1 = 1

Now we must count the H atoms to see if the equation is balanced with respect to hydrogen. On the reactants side, the only molecule that has H in it is CH_4. There is no number to the left of CH_4, so it must be 1. There is a subscript of 4 after H, so the number of H atoms on the reactants side of the equation is 1 x 4 = 4 hydrogen atoms. On the products side, the only molecule with H in it is H_2O. There is a 2 to the left of H_2O, and there is a subscript of 2 after H. Therefore, there are 2 x 2 = 4 hydrogen atoms on the products side of the equation. Since there are the same number (4) of hydrogen atoms on each side of the equation, the equation is balanced with respect to hydrogen.

Reactants Side	Products Side
#C: 1 x 1 = 1	#C: 1 x 1 = 1
#H: 4 x 1 = 4	#H: 2 x 2 = 4

Finally, we must determine whether or not the equation is balanced with respect to oxygen. On the reactants side, only the O_2 molecule has O atoms in it. There is a 2 to the left of O_2 and a subscript of 2 after O, so there are 2 x 2 = 4 oxygen atoms on the reactants side of the equation. What about the products side? This will be a little tougher because there are 2 molecules, CO_2 and H_2O, each of which contains O atoms. To count the total number of O atoms, we figure out how many come from each molecule, and then add those numbers together. This will give us the total number of O atoms on the products side of the equation. There is no number to the left of CO_2, so it must be 1. There is a subscript of 2 after O, so there are 1 x 2 = 2 oxygen atoms coming from the CO_2 molecule. There is a 2 to the left of H_2O and no subscript number after O, so it must be 1. Therefore, there are 2 x 1 = 2 oxygen atoms coming from the H_2O molecule. The total number of oxygen atoms, then, is the number of oxygen atoms that come from CO_2 plus the number of oxygen atoms that come from H_2O, or 2 + 2 = 4. Since there are the same number of oxygen atoms (4) on both sides of the equation, it is balanced with respect to oxygen.

Reactants Side	Products Side
#C: 1 x 1 = 1	#C: 1 x 1 = 1
#H: 4 x 1 = 4	#H: 2 x 2 = 4
#O: 2 x 2 = 4	#O: 2 x 1 + 1 x 2 = 4

Since the equation is balanced with respect to all of the atoms, it is completely balanced. Although this might seem a little tough at first, it gets easier with practice. Try to follow Example 6.2 and then answer "On Your Own" questionz 6.7. Please do not go on to the last part of the module until you have mastered this technique.

EXAMPLE 6.2

Determine whether or not the following equation is balanced:

$$2HCl\ (aq) + CaBr_2\ (s) \longrightarrow 2HBr\ (aq) + 2CaCl_2\ (s)$$

To determine whether or not this equation is balanced, we must count the number of atoms on each side of the equation:

Reactants Side	Products Side	
#H: 2 x 1 = 2	#H: 2 x 1 = 2	Since the number of H atoms is the same, the equation is balanced with respect to H.
#Cl: 2 x 1 = 2	#Cl: 2 x 2 = 4	There are more C atoms on the products side than the reactants side, so this equation is not balanced. We could stop here, but we will complete the process for the sake of illustration.

Reactants Side	Products Side	
#Ca: 1 x 1 = 1	#Ca: 2 x 1 = 2	The equation is also not balanced with respect to Ca.
#Br: 1 x 2 = 2	#Br: 2 x 1 = 2	The equation is balanced with respect to Br.

If an equation has even 1 type of atom that is not balanced, then the whole equation is not balanced. Therefore, **this equation is not balanced**.

Determine whether or not the following equation is balanced:

$$6CO_2 + 6H_2O \longrightarrow C_6H_{12}O_6 + 6O_2$$

Once again, we must count the atoms to see if they balance:

Reactants Side	Products Side	
#C: 6 x 1 = 6	#C: 1 x 6 = 6	The equation is balanced with respect to carbon.

#O: 6 x 2 + 6 x 1 = 18	#O: 1 x 6 + 6 x 2 = 18	Remember, since O appears in 2 molecules, we add the number of O atoms that come from each molecule. This makes the equation balanced with respect to O.
#H: 6 x 2 = 12	#H:1 x 12 = 12	The equation is balanced with respect to H.

Since the equation is balanced with respect to all its atoms, it is a **balanced chemical equation**.

ON YOUR OWN

6.7 Determine whether or not the following equation is balanced:
$2C_2H_6\text{ (g)} + 7O_2\text{ (g)} \longrightarrow 4CO_2\text{ (g)} + 6H_2O\text{ (g)}$

BALANCING CHEMICAL EQUATIONS

Now that you can count the atoms in a chemical equation, you are finally ready to balance chemical equations. When presented with a chemical equation that is not balanced, you must try to balance it by altering the number of molecules involved. You can alter the number of molecules involved by changing the numbers that appear to the left of each molecule. Those numbers can be changed in any way, as long as they are not made negative. *You cannot change any of the numbers that appear in the subscripts.*

If you think about what these numbers mean, this rule should make sense. When balancing chemical equations, all we are trying to do is get the number of molecules involved in the reaction to work out. The *number* of molecules is indicated by the numbers that appear to the left of the molecules. The numbers in the subscripts tell *which* molecules are involved. If you change any of those numbers, you are changing the substances in the chemical reaction, and that is not allowed. So remember, you can alter only the numbers that appear to the left of the molecules, not the subscript numbers. See if you understand these rules by following Example 6.3.

EXAMPLE 6.3

Balance the following equation:

$$N_2\text{ (g)} + H_2\text{ (g)} \longrightarrow NH_3\text{ (g)}$$

To balance this equation, we must get the same number of each type of atom on both sides of the equation. To see where we are, we will first count the number of atoms on each side:

Reactants Side	Products Side
#N: 1 x 2 = 2	#N: 1 x 1 = 1
#H: 1 x 2 = 2	#H: 1 x 3 = 3

Clearly, this equation is not balanced. How can we balance it? We must change the numbers in the equation so that each individual type of atom balances. Remember, we can't change the numbers in the subscripts; we can change only the numbers to the left of the molecules. Let's concentrate on the nitrogen atoms first. There are 2 N atoms on the reactants side and only 1 on the products side. To fix this, we must double the number of N atoms on the products side. The only way we are allowed to do this is to change the number to the left of the molecule. Since there is no number to the left of NH_3, it must be 1. To double the number of N atoms, we must double that number, so we change the equation to look like this:

$$N_2\ (g) + H_2\ (g) \longrightarrow 2NH_3\ (g)$$

Once again, we did not change the subscripted numbers at all. We just put a 2 in front of NH_3. We have emphasized this by making the 2 pink. This simply means there are 2 molecules of NH_3. Now when we count atoms again, we will see that the nitrogen atoms balance:

Reactants Side	Products Side
#N: 1 x 2 = 2	#N: 2 x 1 = 2
#H: 1 x 2 = 2	#H: 2 x 3 = 6

Now the nitrogen atoms are balanced. Of course, the hydrogen atoms are not balanced, so we must balance them next. Since there are 3 times as many hydrogen atoms on the products side, we must triple the number of hydrogen atoms on the reactants side. We can do this by tripling the number to the left of H_2:

$$N_2\ (g) + 3H_2\ (g) \longrightarrow 2NH_3\ (g)$$

Now look what happens when we count the atoms:

Reactants Side	Products Side
#N: 1 x 2 = 2	#N: 2 x 1 = 2
#H: 3 x 2 = 6	#H: 2 x 3 = 6

We now have a balanced chemical equation. The answer, then, is:

$$\mathbf{N_2\ (g) + 3H_2\ (g) \longrightarrow 2NH_3\ (g)}$$

Write a balanced chemical equation for the following reaction:

Propane (C_3H_8) reacts with oxygen to make carbon dioxide and water.

Don't let the fact that the reaction is written with words bother you. Just convert everything into equation form. We are given the chemical formula of propane, and we know that oxygen is a homonuclear diatomic, so its chemical formula is O_2. We also know how to go from a proper chemical name like carbon dioxide to the chemical formula CO_2. Finally, we know that the chemical formula for water is H_2O. We are ready to write the unbalanced chemical equation.

$$C_3H_8 + O_2 \longrightarrow CO_2 + H_2O$$

Since phases were not given, we cannot write the phase symbols in this equation. Now we must balance it. We start by counting the atoms:

Reactants Side	Products Side
#C: 1 x 3= 3	#C: 1 x 1 = 1
#H: 1 x 8 = 8	#H: 1 x 2 = 2
#O: 1 x 2 = 2	#O: 1 x 2 + 1 x 1 = 3

Since this equation is a little more complicated, we must use some strategy. If we were to start by balancing the O atoms, our job would be difficult because there are O atoms in 2 molecules on the products side. We would not know what numbers to put where. The general rule of thumb to follow when balancing equations is to start with atoms that appear in only 1 molecule on each side of the equation. These atoms will be easier to balance. So let's balance the C atoms first. Carbon appears only once on the reactants side (in C_3H_8) and only once on the products side (in CO_2). That makes it an excellent atom with which to start. There are 3 C atoms on the reactants side and only 1 on the products side, so we must triple the number of C atoms on the products side:

$$C_3H_8 + O_2 \longrightarrow 3CO_2 + H_2O$$

Now we count the atoms again:

Reactants Side	Products Side
#C: 1 x 3= 3	#C: 3 x 1 = 3
#H: 1 x 8 = 8	#H: 1 x 2 = 2
#O: 1 x 2 = 2	#O: 3 x 2 + 1 x 1 = 7

Now we can balance the H atoms because they also appear only once on each side of the equation. There are 8 on the reactants side of the equation and 2 on the products side, so we must quadruple the number of hydrogen atoms on the products side:

$$C_3H_8 + O_2 \longrightarrow 3CO_2 + 4H_2O$$

Reactants Side	Products Side
#C: 1 x 3 = 3	#C: 3 x 1 = 3
#H: 1 x 8 = 8	#H: 4 x 2 = 8
#O: 1 x 2 = 2	#O: 3 x 2 + 4 x 1 = 10

The only unbalanced atom left is O. Since oxygen appears in several molecules, we need to think about where we want to start changing the equation. If we changed the numbers in front of C_3H_8, CO_2, or H_2O, it would throw off the balancing that we already did. The only molecule we really want to work with is the O_2 molecule. That's fine, since we can balance the entire equation by multiplying O_2 by 5:

$$C_3H_8 + 5O_2 \longrightarrow 3CO_2 + 4H_2O$$

Now, as you can see, the equation is balanced:

Reactants Side	Products Side
#C: 1 x 3 = 3	#C: 3 x 1 = 3
#H: 1 x 8 = 8	#H: 4 x 2 = 8
#O: 5 x 2 = 10	#O: 3 x 2 + 4 x 1 = 10

So the balanced chemical equation is:

$$C_3H_8 + 5O_2 \longrightarrow 3CO_2 + 4H_2O$$

Try your hand at balancing chemical equations with "On Your Own" questions 6.8–6.10.

ON YOUR OWN

6.8 Hydrogen and oxygen can react together violently to make water. Write a balanced chemical equation for this reaction.

6.9 Solid calcium carbonate ($CaCO_3$) is made when solid calcium reacts with gaseous oxygen and solid carbon. What is the balanced chemical equation for this reaction?

6.10 Balance the following equation:
C_5H_{12} (l) + O_2 (g) $\longrightarrow$ CO_2 (g) + H_2O (g)

This represents your first introduction to the wonderful world of balancing chemical equations and a start on density. Like most of what we've discussed so far, this can get harder, as we will see in later modules. It is critically important that you understand things up to this point. If you have problems balancing the equations in the "On Your Own" questions and the practice problems, you can look at the extra practice problems in appendix B. Also, you can go to the library and check out another chemistry book. See how that book explains balancing chemical equations and try your hand at some of its problems and examples. You absolutely must be able to balance chemical equations before you can continue to Module 7!

SUMMARY OF KEY EQUATIONS AND TABLES IN MODULE 6

Equation 6.1: $\rho = \frac{m}{V}$ Density

Table 6.1: Phase Symbols

ANSWERS TO THE "ON YOUR OWN" QUESTIONS

6.1 (a) Steam is simply water vapor. When steam condenses back into liquid water, there is no change in the molecules. Also, this change can easily be reversed by simply boiling the water again. Either way you look at it, this is a physical change.

(b) When gasoline burns, it combines with oxygen to make carbon dioxide and water, just like when wood burns. Also, it is impossible to get the gasoline back once it's burned, so this change is irreversible. Either way you look at it, this is a chemical change.

(c) When a barber cuts your hair, he does not change the chemical makeup of the hair. He simply changes its length. This could also be reversed by gluing the hair back in place. It might look a little funny, but it still could be reversed. Either way you look at it, this is a physical change.

(d) Milk has its distinctive taste because of the molecules that make it up. The change in taste that occurs in sour milk is because the molecules have changed. Also, you cannot unsour milk. Either way you look at it, this is a chemical change.

(e) When sugar is dissolved in water, the water tastes slightly sweet. That is because the sugar is still there; it is just evenly distributed around the water. Also, this change can be reversed by boiling off the water. Either way you look at it, this is a physical change.

(f) When milk is poured on cereal, the only change is that the cereal gets soggy. No molecules have changed. Also, you can reverse this change by picking the cereal out of the milk and letting it dry. Either way you look at it, this is a physical change.

(g) When iron (Fe) rusts, the red color appears because the iron has reacted with the oxygen in the air to make Fe_2O_3, which we call rust. Also, there is no way to unrust a nail. You can scrape the rust away, but it will not turn back into iron. It will simply fall on the ground as a fine, red powder. Either way you look at it, this is a chemical change.

6.2 Since liquid natural gas occupies significantly less volume than natural gas in its gas phase, the gas company can store significantly larger amounts of natural gas when it is in its liquid form. The money natural gas companies save with the increased storage capacity makes up for the cost of cooling natural gas down to its liquid phase so that it can be stored.

6.3 When substances are heated, the atoms or molecules which make them up gain energy and move faster. As they move faster, they get farther apart.

6.4 Since all substances except water occupy less volume in the solid phase than in the liquid phase, frozen cubes of alcohol would be *more dense* than liquid alcohol. As a result, the frozen cubes of alcohol would sink in liquid alcohol.

6.5 Since we are trying to calculate volume, we must first use algebra to rearrange the equation so that it begins "V=." The first algebra step would be to get V out of the denominator. We cannot solve for a variable that is in the denominator of a fraction. The final algebra step will be to divide both sides by ρ. The rearranged equations look like:

$$\rho = \frac{m}{V}$$

$$V \times \rho = m$$

$$V = \frac{m}{\rho}$$

We must make sure that the units match. We need an answer with a volume unit, so the mass units must cancel. Right now, density is in grams per cm^3, while mass is in kg. So we first have to change kg into grams:

$$\frac{0.200\ \cancel{kg}}{1} \times \frac{1{,}000\ g}{1\ \cancel{kg}} = 2.00 \times 10^2\ g$$

Now we can divide 2.00×10^2 grams by 10.5 grams per cm^3:

$$V = \frac{2.00 \times 10^2\ \cancel{g}}{10.5\ \cancel{g}/cm^3} = 19.0\ cm^3 = 19.0\ mL$$

The answer was rounded to 19.0 because there are 3 significant figures in the mass measurement, and the volume unit was changed to mL since cm^3 and mL are equal.

6.6 Since we know gold's density only in the units of grams per mL, our mass has to be in grams and our volume has to be in mL:

$$\frac{1.56\ \cancel{kg}}{1} \times \frac{1{,}000\ g}{1\ \cancel{kg}} = 1{,}560\ g$$

$$\frac{0.081\ \cancel{L}}{1} \times \frac{1\ mL}{0.001\ \cancel{L}} = 81\ mL$$

Now we can divide these numbers to get the density:

$$\rho = \frac{1{,}560\ g}{81\ mL} = 19.259\ \frac{g}{mL}$$

The answer can have only 2 significant figures since 81 has only 2 significant figures. Since the density rounds to 19 g/mL, to the best of our precision we can say this is gold. If we had measured the volume with more precision, then our calculation could have been more precise.

6.7 To determine whether or not the equation is balanced, we must count the atoms on both sides of the equation:

Reactants Side	Products Side	
#C: 2 x 2 = 4	#C: 4 x 1 = 4	Balanced with respect to carbon
#H: 2 x 6 = 12	#H: 6 x 2 = 12	Balanced with respect to hydrogen
#O: 7 x 2 = 14	#O: 4 x 2 + 6 x 1 = 14	Balanced with respect to oxygen

This is a balanced chemical equation.

6.8 This reaction is written in words, but we can easily convert it to equation form. Remember, both hydrogen and oxygen are homonuclear diatomics.

$$H_2 + O_2 \longrightarrow H_2O$$

Since no phases were given, we cannot put phase symbols in the equation. Now we count the atoms:

Reactants Side	Products Side
#H: 1 x 2 = 2	#H: 1 x 2 = 2
#O: 1 x 2 = 2	#O: 1 x 1 = 1

The H atoms are already balanced, but the O atoms are not. The only way to balance the O atoms is to double them on the products side of the equation:

$$H_2 + O_2 \longrightarrow 2H_2O$$

Reactants Side	Products Side
#H: 1 x 2 = 2	#H: 2 x 2 = 4
#O: 1 x 2 = 2	#O: 2 x 1 = 2

This, of course, throws the H atoms off balance, but that can be fixed by doubling the number of H atoms on the reactants side:

$$2H_2 + O_2 \longrightarrow 2H_2O$$

Now, as you can see, this equation is balanced:

Reactants Side	Products Side
#H: 2 x 2 = 4	#H: 2 x 2 = 4
#O: 1 x 2 = 2	#O: 2 x 1 = 2

The balanced chemical equation is:

$$2H_2 + O_2 \longrightarrow 2H_2O$$

6.9 Remember that oxygen is a homonuclear diatomic. The words translate into the following chemical equation:

$$Ca\ (s) + O_2\ (g) + C\ (s) \longrightarrow CaCO_3\ (s)$$

This looks easy at first. Count the atoms:

Reactants Side	Products Side
#Ca: 1 x 1 = 1	#Ca: 1 x 1 = 1
#O: 1 x 2 = 2	#O: 1 x 3 = 3
#C: 1 x 1 = 1	#C: 1 x 1 = 1

Everything is balanced except the O atoms. The problem is how to balance them. There are 2 on the reactants side and 3 on the products side. The way to balance them is to find the least common multiple of each: 6. We need to get 6 O atoms on both sides of the equation. We do this by cross multiplying the products and reactants sides—tripling the O atoms on the reactants side and doubling them on the products side:

$$Ca\ (s) + 3O_2\ (g) + C\ (s) \longrightarrow 2CaCO_3\ (s)$$

Now we count atoms again:

Reactants Side	Products Side
#Ca: 1 x 1 = 1	#Ca: **2** x 1 = 2
#O: **3** x 2 = 6	#O: **2** x 3 = 6
#C: 1 x 1 = 1	#C: **2** x 1 = 2

Now the O atoms are balanced, but the C and Ca atoms are not. This is easy to fix since we need to just double the number of each on the reactants side:

$$2Ca\ (s) + 3O_2\ (g) + 2C\ (s) \longrightarrow 2CaCO_3\ (s)$$

Now everything is balanced:

Reactants Side	Products Side
#Ca: 2 x 1 = 2	#Ca: 2 x 1 = 2

#O: 3 x 2 = 6	#O: 2 x 3 = 6
#C: 2 x 1 = 2	#C: 2 x 1 = 2

The balanced chemical equation is:

$$2Ca\ (s) + 3O_2\ (g) + 2C\ (s) \rightarrow 2CaCO_3\ (s)$$

6.10 First we count atoms:

$$C_5H_{12}\ (l) + O_2\ (g) \rightarrow CO_2\ (g) + H_2O\ (g)$$

Reactants Side	Products Side
#C: 1 x 5 = 5	#C: 1 x 1 = 1
#H: 1 x 12 = 12	#H: 1 x 2 = 2
#O: 1 x 2 = 2	#O: 1 x 2 + 1 x 1 = 3

Once again, try to start with atoms that appear in only 1 molecule on each side of the equation. In this case, both C and H appear only once on each side, so we will start with them. To balance the C atoms, we need to multiply the number of C atoms on the products side by 5:

$$C_5H_{12}\ (l) + O_2\ (g) \rightarrow 5CO_2\ (g) + H_2O\ (g)$$

Now the C atoms are balanced:

Reactants Side	Products Side
#C: 1 x 5 = 5	#C: 5 x 1 = 5
#H: 1 x 12 = 12	#H: 1 x 2 = 2
#O: 1 x 2 = 2	#O: 5 x 2 + 1 x 1 = 11

To balance the H atoms, we need to multiply the H atoms on the products side by 6:

$$C_5H_{12}\ (l) + O_2\ (g) \rightarrow 5CO_2\ (g) + 6H_2O\ (g)$$

Now the H atoms balance:

Reactants Side	Products Side
#C: 1 x 5 = 5	#C: 5 x 1 = 5
#H: 1 x 12 = 12	#H: 6 x 2 = 12
#O: 1 x 2 = 2	#O: 5 x 2 + 6 x 1 = 16

To balance the O atoms, we just need to multiply the O atoms on the reactants side by 8:

$$C_5H_{12}\ (l) + 8O_2\ (g) \rightarrow 5CO_2\ (g) + 6H_2O\ (g)$$

Now the equation is completely balanced:

Reactants Side	Products Side
#C: 1 x 5 = 5	#C: 5 x 1 = 5
#H: 1 x 12 = 12	#H: 6 x 2 = 12
#O: 8 x 2 = 16	#O: 5 x 2 + 6 x 1 = 16

The balanced chemical equation is:

$$C_5H_{12}\ (l) + 8O_2\ (g) \rightarrow 5CO_2\ (g) + 6H_2O\ (g)$$

STUDY GUIDE FOR MODULE 6

REVIEW QUESTIONS

1. What is the difference between a chemical change and a physical change?
2. Explain the difference between dissolving and reacting. For example, baking soda dissolves in water, but baking soda reacts with vinegar.
3. Why are you warned about using aluminum foil in Experiment 6.1?
4. One type of phase change we did not discuss is sublimation. Sublimation occurs when a solid changes directly into a gas, without passing through a liquid phase. Does sublimation occur when something is heated or when something is cooled?
5. If a liquid goes through a phase change and all you know is that the molecules slowed down and moved closer together, what phase did the liquid turn into?
6. What makes water an exception to the phase change rule?
7. Explain what is wrong with this statement: While substances like CO_2 and H_2O can exist in 3 phases, it is impossible for metals to become gases.
8. Lead has a density of 11.4 grams per mL, whereas gold has a density of 19.3 grams per cc. If we were to make 2 identical statues, one out of gold and the other out of lead, which would be heavier?
9. A student measures the mass of an object as 2.32 grams and its volume as 34.56 mL. She then calculates the density to be 0.067129629. There are 2 things wrong with her value for density. What are they?
10. What makes a chemical equation balanced?
11. What law requires all chemical reactions to have balanced equations? Note: This law was discussed in a previous module.
12. Explain how a chemist could use the difference method in measuring the mass of a liquid.

PRACTICE PROBLEMS

1. Classify the following as physical or chemical changes:
   a. A precious vase is smashed to bits.
   b. Eggs, flour, sugar, milk, and vanilla are baked, making a cake.
   c. Dry ice turns into gas.
   d. Charcoal is burned in a grill.
   e. Kool–Aid is dissolved in water.

2. Lead has a density of 11.4 grams per mL. If you make a statue out of 3.45 L of lead, what is the statue's mass?

3. Gold has a density of 19.3 grams per cc. If a gold nugget has a mass of 45.6 kg, what is its volume?

4. Is the following equation balanced?

$$CaF_2\ (aq) + 2NH_4Cl\ (aq) \rightarrow CaCl_2\ (aq) + NH_4F\ (aq)$$

5. Balance the following equation:

$$HCl\ (aq) + Zn\ (s) \rightarrow ZnCl_2\ (aq) + H_2\ (g)$$

6. Balance the following equation:

$$NH_4ClO_3 + Al_2(SO_4)_3 \rightarrow Al(ClO_3)_3 + (NH_4)_2SO_4$$

7. In an automobile, nitrogen from the air reacts with oxygen in the engine to make nitrogen monoxide, a toxic pollutant. Write the balanced chemical equation for this reaction.

8. Liquid heptene, C_7H_{10}, can react with gaseous hydrogen to make liquid heptane, C_7H_{16}. Write the balanced chemical equation for this reaction.

9. Balance the following equation:

$$C_7H_{16}\ (l) + O_2\ (g) \rightarrow CO_2\ (g) + H_2O\ (g)$$

10. Balance the following equation:

$$CO_2\ (g) + H_2O\ (l) \rightarrow C_{12}H_{24}O_{12}\ (s) + O_2\ (g)$$

DESCRIBING CHEMICAL REACTIONS

In the last module, you learned the basics of chemical reactions. You learned how to recognize the difference between chemical and physical change, and how to write chemical change in terms of a chemical equation. You were then given a brief introduction to the wonderful world of balancing chemical equations.

In this module, we will continue our discussion of chemical equations by showing you how to predict certain types of chemical reactions. You will learn how to figure out the products of certain chemical reactions when you are given only the reactants, as well as how to determine the reactants when you are given only the products. In addition, you will learn how to count the atoms and molecules in a sample of matter. Counting the atoms and molecules in a substance is the first step in fully understanding the mathematics of a chemical equation.

first reactions

If your mom mixes chocolate into a bowl on the kitchen counter, you can probably predict that she is making something like brownies, not meatloaf. Likewise, if you walked into the kitchen at the end and only tasted the brownie, you could probably safely guess that there was no beef in it! With a little practice, it's just as easy in chemistry to figure out the products of certain chemical reactions when you are given only the reactants (ingredients), as well as to determine the reactants when you are given only the products.

THREE BASIC TYPES OF CHEMICAL REACTIONS

As we go through this course, you will learn many, many different types of chemical reactions. To get you started, we will first concentrate on 3 of the most basic types of chemical reactions: decomposition, formation, and combustion reactions. These reactions are relatively simple to understand, and the equations that describe them are reasonably easy to predict and balance.

DECOMPOSITION REACTIONS

Probably the easiest chemical reactions to predict are decomposition reactions. Consider the reaction pictured in Figure 7.1.

FIGURE 7.1
The Decomposition of Water
Photo by Dr. Jay L. Wile.

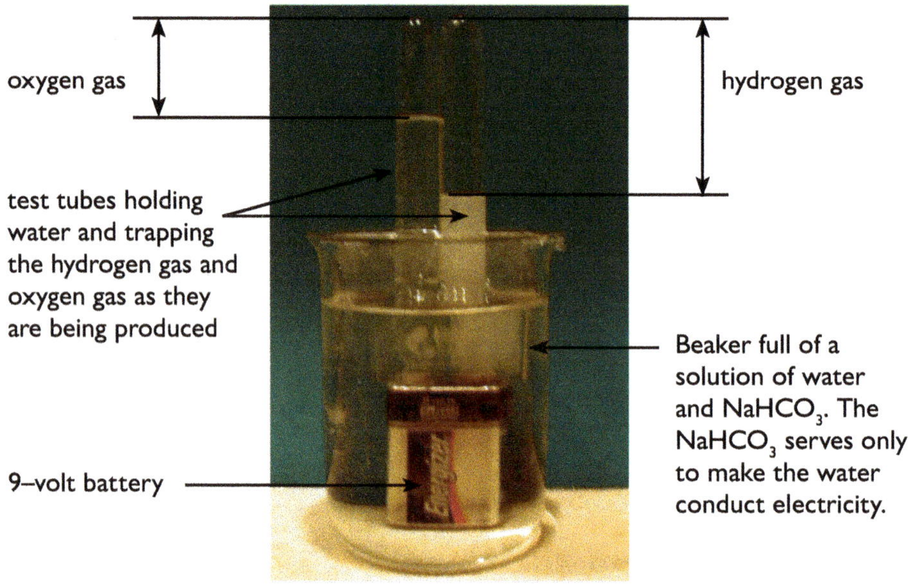

In the reaction shown, a battery is immersed in a solution of water and baking soda ($NaHCO_3$). Test tubes full of water are inverted over the posts of the battery. The test tube on the left is over the positive post of the battery, while the test tube on the right is over the negative post. Bubbles of oxygen gas form on the positive post of the battery, while bubbles of hydrogen gas form on the negative post of the battery. The bubbles then rise up into the test tubes, where they are trapped. The amount of space between the top of the inverted test tube and the water in the test tube is an indication of how much gas is collected. Notice that twice as much hydrogen has been produced as compared to oxygen. This is because, as the chemical formula tells us, there are 2 hydrogen atoms for every 1 oxygen atom in water.

The chemical reaction taking place in Figure 7.1 changes water into hydrogen and oxygen according to the following equation:

$$2H_2O\ (l) \longrightarrow 2H_2\ (g) + O_2\ (g)$$

If you are wondering why H_2 and O_2 are products rather than H and O, you have to remember that hydrogen and oxygen are homonuclear diatomic elements. Whenever a molecule has H or O (or any of the other homonuclear diatomics listed in Module 4), the decomposition reaction will produce homonuclear diatomic molecules. This reaction is an example of a decomposition reaction.

Decomposition reaction—A reaction that changes a compound into its constituent elements.

Notice how the water equation above fits the definition. A compound (H_2O) is changed into hydrogen (H_2) and oxygen (O_2), which are its constituent elements.

The nice thing about decomposition reactions is that their equations are relatively easy to predict and balance, given the compound that is undergoing the reaction. All you need to do is look at the chemical formula of the compound, split it up into its individual elements (being careful to recognize which ones are homonuclear diatomics), and then balance the equation. Example 7.1 shows how this is done. Be sure to follow this example closely because it presents a new facet of balancing chemical equations.

EXAMPLE 7.1

Write the chemical equation that describes the decomposition of ethanol (C_2H_6O).

To solve this problem, we analyze the problem and write a word equation for the reaction. We need to focus on the molecule that is undergoing decomposition. According to its chemical formula, it is made up of C, H, and O atoms. Therefore, its elements are carbon, hydrogen, and oxygen. This gives us the following word equation:

Ethanol decomposes to carbon, hydrogen, and oxygen.

The second step is to write the chemical equation, using chemical formulas to replace the names of the reactants and products. Hydrogen and oxygen are homonuclear diatomics; therefore, H_2 and O_2 will be produced. Carbon is not a homonuclear diatomic, so the decomposition will produce C. This gives us the unbalanced equation:

$$C_2H_6O \rightarrow C + H_2 + O_2$$

Reactants Side	Products Side
C: 2	C: 1
H: 6	H: 2
O: 1	O: 2

The third step is to balance the equation. First, we see 2 C atoms on the reactant side of the equation but only 1 on the product side. We therefore must double the C atoms on the product side:

$$C_2H_6O \rightarrow 2C + H_2 + O_2$$

Reactants Side	Products Side
C: 2	C: 2
H: 6	H: 2
O: 1	O: 2

Now the C atoms are balanced, but the H atoms are not. There are 6 on the reactant side and only 2 on the product side, so we need to triple the number of H atoms on the product side:

$$C_2H_6O \rightarrow 2C + 3H_2 + O_2$$

Reactants Side	Products Side
C: 2	C: 2
H: 6	H: 3 x 2 = 6
O: 1	O: 2

Now we need to address the problem with oxygen. This is tricky because there is 1 O atom on the reactant side and 2 O atoms on the product side. The way we would usually solve this problem is to double the number of O atoms on the reactant side. The problem is that to do that, we would have to multiply the C_2H_6O molecule by 2. That would throw the C and H atoms back out of balance. So . . . what do we do?

To take care of this problem, we have to use a bit of a trick by recognizing that there is another way to balance the O atoms. Instead of doubling them on the reactant side of the equation, we could halve them on the product side of the equation. In other words, we need only one–half of an O_2 molecule on the product side.

$$C_2H_6O \rightarrow 2C + 3H_2 + \tfrac{1}{2}O_2$$

Reactants Side	Products Side
C: 2	C: 2
H: 6	H: 6
O: 1	O: $\frac{1}{2}$ x 2 = 1

This balances the equation because now there is 1 oxygen atom on the reactant side and $\frac{1}{2}$ x 2 = 1 oxygen atom on the product side. The problem with this solution is that a fraction in the chemical equation doesn't make sense. After all, there can't be a fraction of a molecule. There can be only 0 molecules or 1 molecule, not one–half of a molecule.

To get rid of the fraction, we must do something we learned in algebra. We can do anything we want to one side of an equation as long as we do exactly the same thing to the other side of the equation. We could multiply the product side of the equation by 2 as long as we also multiply the reactant side of the equation by 2. What would this accomplish? Watch:

$$2 \times [C_2H_6O] \rightarrow 2 \times [2C + 3H_2 + \tfrac{1}{2}O_2]$$

Now we distribute the 2 by multiplying it with every term on both sides of the equation:

$$2C_2H_6O \rightarrow 4C + 6H_2 + 1O_2$$

When the 2 multiplies the one–half, we get 1, eliminating the fraction. Therefore, our balanced equation for the decomposition of ethanol is:

$$\mathbf{2C_2H_6O \rightarrow 4C + 6H_2 + O_2}$$

The final step is to make a final check on all atoms to make sure they are balanced.

Now the equation is balanced without fractions:

Reactants Side	Products Side
C: 2 x 2 = 4	C: 4 x 1 = 4
H: 2 x 6 = 12	H: 6 x 2 = 12
O: 2 x 1 = 2	O: 1 x 2 = 2

This little trick brings up an important fact about chemical equations. Chemical equations are just like mathematical equations. Anything we can do to a mathematical equation we can also do to a chemical equation. We can add, subtract, multiply, or divide both sides of a chemical equation by any number. This trick is such an important aspect of balancing equations that we will present one more problem to make sure you understand it.

Another important point about this example problem is that there is more than one way to balance an equation. Do you see another way of balancing this equation? If so, great! Normally, there will always be another path to take in balancing equations. Sometimes that different path may be easier than what we show in the solution, and sometimes it may be harder. But as long as you get to the correct balanced equation, the path does not matter. As you get more experience in balancing equations, you will start to recognize patterns that can help you balance the equations. Let's look at a common pattern that occurs with homonuclear diatomics in this equation. If you can recognize this pattern in the future, then it will help you balance equations more easily. Here is the unbalanced equation again:

$$C_2H_6O \rightarrow C + H_2 + O_2$$

Reactants Side	Products Side
C: 2	C: 1
H: 6	H: 2
O: 1	O: 2

The first thing we want to point out is that if we put any coefficient in front of the reactant, it will change every atom's count. No matter what number we put in front, the amount of C, H, and O will change on the left side. At this time we should do a quick survey to see which atom would be best to start with in balancing. If we start with carbon, then we simply need to change the C atom on the product side. If we start with hydrogen, then we will change the H atom on the product side. But if we start with oxygen, we will need to put a coefficient in front of the reactant. This will render any changes that we made to C and H useless. So we propose that we balance O first. Normally, we recommend saving O for last as it is usually the hardest atom to balance. But in this equation, O changes all of the other atoms. Doesn't it make sense to start with O?

In balancing O, we need to place a 2 in front of the molecule on the reactant side:

$$2C_2H_6O \rightarrow C + H_2 + O_2$$

Reactants Side	Products Side
C: 2 x 2 = 4	C: 1
H: 2 x 6 = 12	H: 2
O: 2 x 1 = 2	O: 2

MOD 7

Now, a 4 in front of the C atom on the product side will balance C:

$$2C_2H_6O \longrightarrow 4C + H_2 + O_2$$

Reactants Side	Products Side
C: 4	C: 4 x 1 = 4
H: 12	H: 2
O: 2	O: 2

Finally, a 6 in front of H_2 on the product side will finish balancing the equation:

$$\mathbf{2C_2H_6O \longrightarrow 4C + 6H_2 + O_2}$$

Reactants Side	Products Side
C: 4	C: 4
H: 12	H: 6 x 2 = 12
O: 2	O: 2

In the end, it does not matter how an equation is balanced as long as we get to the correct balanced equation. Pay attention when balancing equations because you will start recognizing patterns that will make it easier for you.

Write the balanced chemical equation for the decomposition of SiH_3F.

Since we have been given the formula, we can often skip the first step and move to the second step of writing the chemical equation. The elements in this molecule are Si, H, and F. H and F are homonuclear diatomics; thus, the decomposition reaction is:

$$SiH_3F \longrightarrow Si + H_2 + F_2$$

Reactants Side	Products Side
Si: 1	Si: 1
H: 3	H: 2
F: 1	F: 2

The third step is to balance this equation. We don't need to do anything to the silicon atoms since there is 1 Si atom on each side of the equation. However, there are 3 H atoms on the reactants side and only 2 on the products side. On the product side, what coefficient times 2 will equal 3? We can solve this as follows:

$$SiH_3F \longrightarrow Si + \tfrac{3}{2}H_2 + F_2$$

Reactants Side	Products Side
Si: 1	Si: 1
H: 3	H: $\frac{3}{2}$ x 2 = 3
F: 1	F: 2

We chose to balance the H atoms this way so that we wouldn't have to multiply SiH_3F by anything because that would destroy the balance between the Si atoms. Now we can balance the F atoms in a similar fashion:

$$SiH_3F \longrightarrow Si + \frac{3}{2}H_2 + \frac{1}{2}F_2$$

Reactants Side	Products Side
Si:1	Si: 1
H:3	H: 3
F: 1	F: $\frac{1}{2} \times 2 = 1$

The only problem remaining is that the fractions must be eliminated. To do this, we simply have to multiply both sides of the equation by some number that will get rid of the fractions. Since both fraction denominators are 2, we can eliminate both fractions by multiplying both sides of the equation by 2:

$$2 \times [SiH_3F] \longrightarrow 2 \times [Si + \frac{3}{2}H_2 + \frac{1}{2}F_2]$$

$$\mathbf{2SiH_3F \longrightarrow 2Si + 3H_2 + F_2}$$

The final step is to make a final check on all atoms to make sure they are balanced.

Reactants Side	Products Side
Si: 2 x 1 = 2	Si: 2 x 1 = 2
H: 2x 3 = 6	H: 3 x 2 = 6
F: 2 x 1 = 2	F: 1 x 2 = 2

Notice the list of atoms for the reactants side and products side. This is a trick you can use to help you balance equations. It helps you make sure you have the same number of atoms of each type on both sides of the equation.

Now see if you can do this in "On Your Own" questions 7.1–7.2.

Before we leave the topic of decomposition reactions, we must add a word of warning. The definition given for decomposition reactions is very narrow. Most chemists say that any reaction that has a single compound as the reactant and several elements or compounds as products is a decomposition reaction. For example, consider the reaction represented by this equation:

$$CaCO_3\ (s) \longrightarrow CaO\ (s) + CO_2\ (g)$$

Chemists consider this a decomposition reaction since it starts with a single compound

ON YOUR OWN

7.1 Write the chemical equation for the decomposition of nitrogen trichloride.

7.2 Write the chemical equation for the decomposition of $C_{12}H_{23}O_{11}$.

and breaks it into 2 different compounds. We will also consider reactions like the one shown above to be decomposition reactions. However, when you are asked to write the equation for a decomposition reaction, always write the equation for the reaction that produces only elements, like those given in the examples and the "On Your Own" problems.

If you purchased the MicroChem kit mentioned in the Student Notes, you can perform Experiment 7 to get some experience with decomposition reactions.

FORMATION REACTIONS

Formation reactions are really just the opposite of decomposition reactions.

Formation reaction—A reaction that starts with 2 or more elements and produces 1 compound.

A formation reaction has elements as its reactants and a compound as its product. An example of a formation reaction is the reverse of the water decomposition equation. Suppose we let hydrogen gas and oxygen gas react. They would form water according to this equation:

$$2H_2\ (g) + O_2\ (g) \longrightarrow 2H_2O\ (g)$$

Please understand that this is just the opposite of what was done in Figure 7.1. In that reaction, we started with water and added energy to it so that it would absorb the energy and decompose into hydrogen and oxygen. In this reaction, we are starting with hydrogen and oxygen and reacting them. The resulting reaction, shown in Figure 7.2, produces water and releases quite a bit of energy.

FIGURE 7.2
The Formation of Water
Photos: Flanders Video Productions

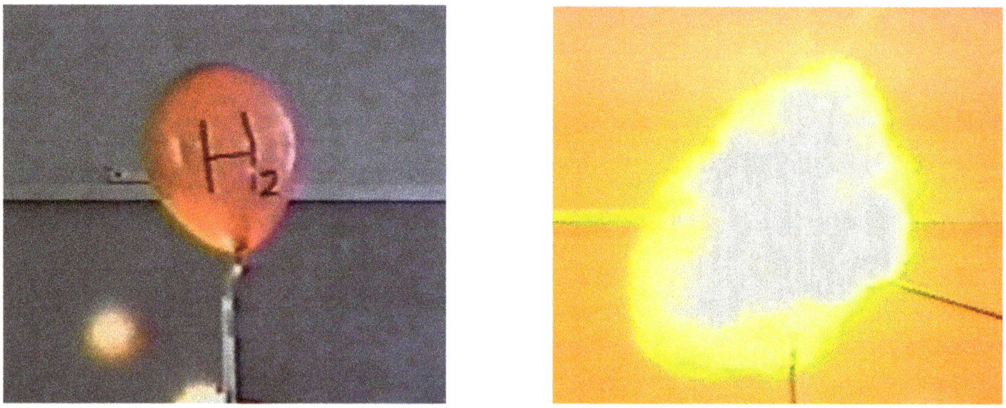

The balloon is full of hydrogen gas. There is oxygen in the air surrounding the balloon. A flame is brought near the balloon.

When the flame pops the balloon, the oxygen and hydrogen are allowed to mix. The heat from the flame starts the reaction, which produces water and energy in the form of an explosion.

Just like the definition given for a decomposition reaction, the definition given for a formation reaction is a bit narrow. Chemists consider any reaction that starts with several reactants and forms a single product to be a formation reaction. Once again, when you are asked to write chemical equations for formation reactions, write them for a reaction that has only elements as its reactants, as illustrated in the Example 7.2.

EXAMPLE 7.2

Write the chemical equation for the formation of $MgCO_3$.

The first step is to write a word equation if needed. Since we have the formula for this compound, we can move on to writing the chemical equation. This compound is made up of Mg, C, and O (a homonuclear diatomic). Thus, the reaction is:

$$Mg + C + O_2 \longrightarrow MgCO_3$$

Now we just have to balance the equation. The Mg and C atoms are already balanced. To balance the O atoms without disturbing Mg and C, we will have to use a fraction again:

$$Mg + C + \tfrac{3}{2}O_2 \longrightarrow MgCO_3$$

Now we can eliminate the fraction by multiplying both sides by 2:

$$2 \times [Mg + C + \tfrac{3}{2}O_2] \longrightarrow 2 \times [MgCO_3]$$

$$\mathbf{2Mg + 2C + 3O_2 \longrightarrow 2MgCO_3}$$

The last step is to make a final check on all atoms to make sure they are balanced.

The equation is balanced without fractions:

Reactants Side	Products Side
Mg: 2 x 1 = 2	Mg: 2 x 1 = 2
C: 2x1 = 2	C: 2 x 1 = 2
O: 3 x 2 = 6	O: 2 x 3 = 6

Make sure you understand formation reactions by answering "On Your Own" question 7.3.

ON YOUR OWN

7.3 Write a balanced chemical equation for the formation of $KClO_3$.

COMBUSTION REACTIONS

The final type of chemical reaction we want to discuss is called a *combustion* reaction. Have you heard of combustion before? If so, what do you think of when you hear the word? Most students tend to say explosions or burning something up when asked that question. The Merriam–Webster dictionary defines combustion as the act of burning. While this definition may work in our use of everyday English, it will not work in chemistry. Burning involves the release of heat and light in the form of fire or flame.

There are some combustion reactions in chemistry that don't release heat or light. So what is *combustion* then?

In chemistry, we define combustion as a chemical reaction where a substance reacts with oxygen gas. When you get to college chemistry, you will learn other types of combustion that don't involve oxygen, but for this course we will stick with reactions involving oxygen gas. Generally as long as oxygen gas is a reactant, the reaction can be classified as a combustion reaction.

In this module we will discuss 3 types of combustion reactions: the combustion of metals, complete combustion, and incomplete combustion. Each type will be given its own section in this module, so there will be plenty of information for you to learn about them.

COMBUSTION OF METALS

Did you know that metals can burn? If we were to heat the metal magnesium to a high enough temperature in the presence of oxygen, the following reaction would occur:

$$2Mg\ (s) + O_2\ (g) \longrightarrow 2MgO\ (s)$$

This is a combustion reaction releasing a lot of heat and very bright white light. Firework makers will burn magnesium when they want white light in their fireworks that explode high in the air. Many years ago, sparklers were made from magnesium, and they produced a much brighter light than the sparklers of today. But in addition to a brighter light, the sparklers released a lot of energy. In fact, they released so much energy that they would stay hot long after they burned out. If a child were to touch an exhausted sparkler, he could get burned very badly. The government outlawed the use of magnesium in hand–held fireworks such as sparklers for this safety concern.

Magnesium is not the only metal that can burn. There are many others, but they all tend to combust differently in some manner. For example, when sodium burns, it produces a bright yellow light. Also, when potassium burns the reaction produces a bright pale purple flame.

Some metals may react with oxygen but not necessarily produce a large amount of heat or a flame. It will still be a combustion reaction, but it happens so slowly that we may not even recognize that a chemical reaction is taking place. In Module 14, we will discuss the speed with which reactions take place. An example of a metal reacting with oxygen gas slowly would be iron. What happens to iron when it is left exposed to the air? It rusts, doesn't it? One of the products of the rusting process is an iron (III) oxide, Fe_2O_3. This molecule is formed by the following combustion reaction:

$$4Fe\ (s) + 3O_2\ (g) \longrightarrow 2Fe_2O_3$$

So combustion reactions don't have to release heat and light in the form of flame. But they have oxygen gas as a reactant.

COMPLETE COMBUSTION REACTIONS

Combustion is often used as another word for burning, so a combustion reaction will govern the process of burning. In general, to burn something, it must be chemically reacted with oxygen. About 90% of combustion reactions start with a compound that contains carbon and hydrogen, and they produce carbon dioxide and water. That kind of chemical reaction is called a complete combustion reaction. Why do we add the word *complete*? You will see momentarily.

Complete combustion reaction—A reaction in which O_2 is added to a compound containing carbon and hydrogen, producing CO_2 and H_2O.

An example of a complete combustion reaction is:

$$CH_2O\ (g) + O_2\ (g) \longrightarrow CO_2\ (g) + H_2O\ (g)$$

This is the standard form of a complete combustion reaction: O_2 is reacted with a compound that contains carbon and hydrogen (CH_2O in this case) to make CO_2 and H_2O. When something burns, a reaction similar to this one usually takes place. Follow Example 7.3 to make sure you understand how this works, then answer "On Your Own" question 7.4.

EXAMPLE 7.3

A butane pocket lighter works by evaporating and then burning the liquid butane (C_4H_{10}) stored in it. Write the balanced chemical equation for this complete combustion reaction.

According to our definition of complete combustion, oxygen is added to butane so that it produces carbon dioxide and water. First we write our word equation:

Butane reacts with oxygen to produce carbon dioxide and water.

Now we write the chemical equation for this reaction:

$$C_4H_{10} + O_2 \longrightarrow CO_2 + H_2O$$

We could go ahead and add phase symbols at this point. The problem told us that the butane was evaporated before it was burned, so the butane is a gas. We know that oxygen and carbon dioxide are gases, and we also know that the temperature associated with combustion is large enough to evaporate the water produced. Therefore:

$$C_4H_{10}\ (g) + O_2\ (g) \longrightarrow CO_2\ (g) + H_2O\ (g)$$

Now the only thing left to do is balance the equation. We need to quadruple the C atoms on the products side to balance them, and we must increase the number of H atoms on the products side by a factor of 5:

$$C_4H_{10}\ (g) + O_2\ (g) \longrightarrow 4CO_2\ (g) + 5H_2O\ (g)$$

In counting the O atoms, we see that there are 1 x 2 = 2 on the reactants side and 4 x 2 + 5 x 1 = 13 on the products side. To balance the O atoms without disturbing the C and H balance, we must use a fraction again:

$$C_4H_{10}\ (g) + \tfrac{13}{2}O_2\ (g) \longrightarrow 4CO_2\ (g) + 5H_2O\ (g)$$

This gives $\frac{13}{2}$x 2 = 13 O atoms on the reactants side and 4 x 2 + 5 x 1 = 13 O atoms on the products side. Now that the equation is balanced, all we have to do is get rid of the fraction:

$$2 \times [C_4H_{10}\ (g) + \tfrac{13}{2}\ O_2\ (g)] \longrightarrow 2 \times [4CO_2\ (g) + 5H_2O\ (g)]$$

$$2C_4H_{10}\ (g) + 13O_2\ (g) \longrightarrow 8CO_2\ (g) + 10H_2O\ (g)$$

The final step is to make a final check on all atoms to make sure they are balanced.

The equation is balanced without fractions:

Reactants Side	Products Side
C: 2 x 4 = 8	C: 2 x 4 = 8
H: 2 x 10 = 20	H: 10 x 2 = 20
O: 13 x 2 = 26	O: 8 x 2 + 10 x 1 = 26

ON YOUR OWN

7.4 Write a balanced chemical equation for the complete combustion of propionaldehyde (C_3H_6O).

INCOMPLETE COMBUSTION REACTIONS

Before we leave the subject of combustion, we need to learn one more thing. We know that to burn things, we need plenty of oxygen. If we start a candle burning and then cover it with a glass, the flame will eventually go out because it used up all of the oxygen in its surroundings. If there is plenty of oxygen, combustion takes place. If there is no oxygen, no combustion takes place. What happens somewhere in between? Let's suppose we try to burn something in an area where there is *some* oxygen, but not *plenty* of oxygen. What happens then?

When a substance containing carbon and hydrogen burns in an environment where there is limited oxygen, one of 2 chemical reactions known as incomplete combustion occurs. In these reactions, carbon dioxide is not produced. Instead, either carbon monoxide or just plain carbon is produced. For example, let's assume we try to burn natural gas (CH_4) in an area where there is limited oxygen. One of 2 reactions occurs:

$$2CH_4\ (g) + 3O_2\ (g) \longrightarrow 2CO\ (g) + 4H_2O\ (g)$$

or

$$CH_4\ (g) + O_2\ (g) \longrightarrow C\ (s) + 2H_2O\ (g)$$

The amount of oxygen in the area determines which reaction occurs. If there is still a reasonable amount of oxygen around, the first reaction will occur, forming carbon monoxide and water. If there is a serious shortage of oxygen in the area, the second reaction will occur, forming carbon (which we usually call soot) and water.

While carbon dioxide is generally not harmful to humans, carbon monoxide is poisonous to humans. Whenever you try to burn something in a limited supply of oxygen, you run the risk of producing a poisonous gas! Believe it or not, this happens all of the time in everyday life. The best example is an automobile.

For an automobile to run, energy must be generated. In the vast majority of automobiles today, this is done by burning gasoline. To burn gasoline, the automobile must have a steady supply of oxygen. To get this oxygen, the engine constantly pulls in air from the outside. The oxygen in the air is then used to burn the gasoline. The problem with this process is that an engine uses an *enormous* amount of energy, so it must burn an *enormous* amount of gasoline. The gasoline usually burns so quickly that it depletes the oxygen in the air faster than the engine can pull in more oxygen to replace it. When an automobile runs, it often burns gasoline in a limited supply of oxygen, causing incomplete combustion.

This incomplete combustion produces the poisonous gas carbon monoxide, as shown in the carbon monoxide product equation above. This is why you must always avoid being behind a car when it is running. Some of the combustion that occurs in any automobile engine is incomplete combustion. This produces carbon monoxide, which is ejected through the tailpipe of the car. Standing behind a car and breathing in the exhaust puts carbon monoxide, a poison, into your body. This is also why you should *never, ever* run a car engine in an enclosed area, such as a closed garage. The car exhaust contains carbon monoxide, which will build up in an enclosed area to the point that breathing the air could kill you.

The amount of carbon monoxide produced by automobiles got to be so large in the early 1970s that the U.S. government became concerned that too many people were breathing poisonous air. In 1975, the government required all new automobiles to be fitted with catalytic (kat' uh lih' tik) converters. A catalytic converter helps reduce the amount of carbon monoxide in automobile exhaust by converting the carbon monoxide into carbon dioxide. We will see exactly how this works in a later module. For now, you just need to know that a catalytic converter helps convert the carbon monoxide produced in an automobile into carbon dioxide, a nonpoisonous gas. Unfortunately, even the best catalytic converter is only about 90% efficient, so automobiles still produce carbon monoxide, making them a big source of pollution.

Carbon monoxide is not just produced in automobiles, however. Often, when it is very cold, natural gas furnaces have to burn their gas faster than the oxygen supply can be replenished. Under those conditions, carbon monoxide can also be produced. Of course, a large fraction of that carbon monoxide leaves the house through the chimney, but some of it can find its way inside.

Fireplaces are also a significant source of carbon monoxide. Often there is very poor airflow around a fireplace. When the airflow is weak, the fireplace will deplete its oxygen supply to the point where incomplete combustion occurs. In fact, many fireplaces deplete the oxygen supply so much that the second carbon product reaction occurs. The oxygen supply gets so limited that not even carbon monoxide can be made. In the end, only carbon is formed, leaving soot in the fireplace.

The more soot you see in your fireplace, the more concerned you should be about the airflow around it. If a large amount of soot is being produced, then the airflow around the fireplace must be very poor. Most likely, this means that a large amount of the combustion that takes place in the fireplace is incomplete combustion. This means that a significant amount of carbon monoxide is being made as well. Once again, a good fraction of that carbon monoxide leaves through the chimney, but some of it can get into your house.

Every year, a handful of people in the United States die because their furnaces or fireplaces produce too much carbon monoxide. Since carbon monoxide is odorless and tasteless, it is impossible to tell whether or not you are breathing it. Although this is a serious problem, there is no reason to get paranoid about it. If your furnace and fireplace are both well–ventilated, then there should be plenty of oxygen getting to them.

If you are concerned about the ventilation, there is a simple way to find out whether carbon monoxide production is a problem in your house. You can go to almost any hardware or department store and pick up a battery–powered carbon monoxide detector. There are several brands to choose from, and they all emit an alarm when the carbon monoxide levels get dangerously high. One word of caution: If you put your carbon monoxide detector in a garage, it will likely go off every time you start your car. This is not a problem because as soon as you pull the car out of the garage, the carbon monoxide level will decrease.

Incomplete combustion is a real problem today. It makes cars a serious source of pollution, and it can make a furnace or a fireplace dangerous. It is important to know the basics of incomplete combustion and its effects on humans. That is why we spent time on it in this module.

ATOMIC MASS

Now that you are firmly grounded in the concepts behind chemical equations, it is time to learn how chemists count atoms and molecules. Once you learn this new technique, another powerful aspect of chemical equations will be revealed. Let's not get too far ahead of ourselves, however. We first must learn how to count atoms and molecules.

Since we cannot see atoms or molecules, it might sound odd that they can be counted. Nevertheless, they can. Even though we cannot point to each atom or molecule and assign it a number, we can figure out how many atoms or molecules are in a sample of matter. We do this by measuring the mass of the sample and determining the types of atoms or molecules in that sample. Those 2 pieces of information can tell us how many atoms or molecules exist in the sample.

To see how this works, we must first look at the periodic table of elements and learn some more features of this important chart. Notice that each atom is represented by a box. Each box contains the atom's symbol and 2 numbers. The symbol tells us what atom the box represents, but what do the 2 numbers tell us?

The number that appears *above* the atomic symbol is called the atomic number. The atomic number tells how many protons that atom has in the nucleus. If you start at the top left of the chart and work your way across, you will see that the atomic number in each box is one higher than the atomic number in the box before it. This means that the periodic table is arranged in order of increasing number of protons. It is interesting that there are no gaps in the numbers. This means that we know the identity of every possible number of protons with a few exceptions toward the bottom of the table.

The number that appears *below* the atomic symbol is called the atomic mass, and it tells us how heavy the atom is. The bigger the atomic mass, the heavier the atom. If you look at the atomic masses listed on the chart, you will notice that they increase as you read the chart from left to right.

Of course, the first question that should come to your mind is this: What are the units?

After all, the chart says that a carbon atom has a mass of 12.0. What does that 12.0 mean? Is it 12.0 g? 12.0 kg? 12.0 mg? 12.0 slugs? The units used on the periodic table are special. The masses listed there are in atomic mass units. We usually abbreviate this as amu.

What is an amu? It is a unit of mass that was invented to help chemists deal with the mass of atoms. As you might expect, an atomic mass unit is very, very small. After all, atoms are so small that we cannot see them. They obviously cannot have too much mass since they are so small. Look at the atomic masses on the periodic table again. Aren't those numbers fairly easy to use? The mass in grams would be much more difficult to use. For example, the mass in grams of a carbon atom would be approximately 1.99×10^{-23} g. Which measurement would you like to use: 12.0 amu or 1.99×10^{-23} g? That really is not a tough choice, is it? The relationship between atomic mass units and grams is as follows:

salt and light

Dr. William Prout (1785–1850) recognized that the atomic weight of atoms could be converted into a scale made of whole numbers. These whole numbers would be integer multiples of the smallest fundamental particle, which he called a protyle. This idea became known as Prout's hypothesis. Ernest Rutherford named the positive part of the atom proton after Prout's basic particle.

Prout was commissioned to write the eighth volume of *The Bridgewater Treatises*, which were devoted to describing the work of God in creation. Prout's treatise was called *Chemistry, Meteorology, and the Function of Digestion, considered with reference to Natural Theology*. He took this opportunity to describe several examples, both living and nonliving, to argue for the existence of God, whom he called the Superior Chemist.

Painting of William Prout by Henry Wyndham Phillips.

$$1.00 \text{ amu} = 1.66 \times 10^{-24} \text{ g} \qquad \text{Equation 7.1}$$

This is a very important relationship, but it is not one that you need to memorize because we will give you this relationship (equation 7.1) on any assignment or test in which you need to use it. One thing you must remember about this relationship is that it is not exact. When you use this relationship in conversion problems, you must consider its 3 significant figures.

Even though you needn't memorize this relationship, it is important that you understand what it means. A gram is about the mass of a housefly. An atomic mass unit is 1.66×10^{-24} of that. This is an incredibly small mass! You might have thought that atoms and molecules were small, but did you have any idea that they were 1 trillion trillionths the mass of a single housefly? Since atoms and molecules have such small masses, the calculations we will perform on them will often use numbers expressed in scientific notation, as is the case with the number of grams in the above relationship. We dealt with scientific notation in Module 1, and you need to be familiar with it before you can proceed in this module. If you don't feel comfortable with scientific notation please review Module 1 or find another book and learn it. Now that we know the relationship between atomic mass units and grams, let's get an idea of exactly how much mass certain atoms have. First, though, look through Example 7.4, and then answer "On Your Own" question 7.5.

EXAMPLE 7.4

What is the mass of 1 nitrogen atom in grams? (1.00 amu = 1.66×10^{-24} g)

To figure out the mass of an atom, we must first look at the periodic chart. According to the chart, a nitrogen atom has a mass of 14.0 amu. To determine the mass in grams, all we have to do is a conversion. We know from equation 7.1 that 1.00 amu = 1.66×10^{-24} g. Thus, according to what we learned in Module 1:

$$\frac{14.0\ \cancel{\text{amu}}}{1} \times \frac{1.66 \times 10^{-24}\text{g}}{1.00\ \cancel{\text{amu}}} = 2.32 \times 10^{-23}\text{ g}$$

So the mass of one nitrogen atom is 14.0 amu, or **2.32×10^{-23} g**. If you can't figure out how to do the arithmetic to get the right answer, refer to a math book or your calculator's instruction book. We will be using scientific notation in equations like this quite a bit, so you need to know how it's done.

ON YOUR OWN

7.5 What is the mass of one lithium atom in kg? (1.00 amu = 1.66×10^{-24} g)

MOLECULAR MASS

What about molecules? If we can find the mass of an atom by simply looking at the periodic chart, how do we get the mass of a molecule? If you think about it, the answer becomes pretty obvious. A molecule is made when atoms join together. According to the law of mass conservation, the mass of the atoms before they join together should be the same as the mass of the atoms after they join together. If you just add all the masses of all of the atoms in a molecule, you'll end up with the mass of that molecule, which is called the molecular mass. See if you can follow this logic in Example 7.5, then answer "On Your Own" question 7.6.

EXAMPLE 7.5

What is the molecular mass of $CHCl_3$ in amu?

According to the chemical formula, a $CHCl_3$ molecule has 1 carbon atom (which has a mass of 12.0 amu according to the chart), 1 hydrogen atom (which has a mass of 1.01 amu), and 3 chlorine atoms (each of which has a mass of 35.5 amu). To figure out the mass of the molecule, we simply add the masses of 1 carbon atom, 1 hydrogen atom, and 3 chlorine atoms:

$$\text{Mass of } CHCl_3 = 1 \times 12.0\text{ amu} + 1 \times 1.01\text{ amu} + 3 \times 35.5\text{ amu}$$

$$\text{Mass of } CHCl_3 = \mathbf{119.5\ amu}$$

Notice that all we really had to do was find the mass of each atom in the molecule, multiply by the number of that atom in the chemical formula, and then add it all up.

Let's spend a moment on significant figures. In reality, we are just adding numbers in this problem. It is true that we are multiplying 35.5 by 3, but that 3 is exact. We cannot have some fraction of an atom. There are exactly 3 chlorine atoms in a $CHCl_3$ molecule. So, we are really just adding 35.5 three times (3 x 35.5 = 35.5 + 35.5 + 35.5). In this problem, then, we must follow the rule of addition and subtraction when determining the significant figures. We keep the lowest *precision* when we add. Although 1.01 has its last significant figure in the hundredths place, both 12.0 and 35.5 have their last significant figure in the tenths place. Therefore, we must report the answer to the tenths place. Therefore, a $CHCl_3$ molecule has a mass of **119.5 amu**. Notice that this has more significant figures than any of the atomic masses used in the equation, but that does not matter. In addition and subtraction, we consider only precision.

What is the mass of a $Si_6F_{12}O_6$ molecule in grams? (1.00 amu = 1.66×10^{-24} g)

According to the chart, Si has a mass of 28.1 amu, F has a mass of 19.0 amu, and O has a mass of 16.0 amu. The mass of the molecule in amu is:

$$\text{Mass of } Si_6F_{12}O_6 = 6 \times 28.1 \text{ amu} + 12 \times 19.0 \text{ amu} + 6 \times 16.0 \text{ amu} = 492.6 \text{ amu}$$

Once again, we report our answer to the tenths place because the masses that we are adding together all have their last significant figure in the tenths place. Converting to grams:

$$\frac{492.6 \cancel{\text{ amu}}}{1} \times \frac{1.66 \times 10^{-24} \text{ g}}{1.00 \cancel{\text{ amu}}} = 8.18 \times 10^{-22} \text{ g}$$

At this point, we must count significant figures because we are multiplying. The conversion relationship has 3 significant figures, so the answer can have only 3 significant figures. Thus, a $Si_6F_{12}O_6$ molecule has a mass of 492.6 amu, or **8.18×10^{-22} g**.

THE MOLE CONCEPT

It is now time to learn how chemists can count the number of atoms or molecules in a sample of matter. To do this, chemists use a concept known as the mole. The mole that we use in chemistry is not a furry little creature that lives underground. Instead, chemists use the term *mole* to refer to a specific number of atoms. If we have 6.02×10^{23} atoms, we have a mole of atoms. If we have 6.02×10^{23} molecules, we have a mole of molecules.

ON YOUR OWN

7.6 What is the mass of a sodium sulfate molecule in g? (1.00 amu = 1.66×10^{-24} g)

This kind of terminology might sound strange at first, but we use it often in our day–to–day language. For example, if you go to the doughnut shop to get some doughnuts, you usually buy them in groups that we call a dozen. When we say "one dozen doughnuts," we really mean 12 doughnuts. Likewise, if you ask the person at the doughnut shop for 3 dozen doughnuts, he will pick out 36 doughnuts. Thus, *dozen* is just a word that means 12.

In the same way, if someone has a party and tells you that 12 couples attended the party, you would know that they meant there were 24 people who came to the party. The word *couple* means 2. A set of 12 couples, then, is 24 people. The word *mole* simply means

6.02×10^{23}. All of these words—*dozen*, *couple*, and *mole*—refer to a group of things. The only difference is what number of things they refer to. *Couple* and *dozen* refer to manageable numbers (2 and 12). Why does *mole* refer to such a big number (6.02×10^{23})?

Remember why we use the word *mole*. We use it to count atoms and molecules in a sample of matter. Since atoms and molecules are so small, there obviously must be a huge number of them in any given sample of matter. For example, a 12–ounce glass of water contains 1.14×10^{25} molecules of water. This is a *huge* number. The word we use to count molecules or atoms must refer to a huge number as well. One mole, therefore, represents 602,000,000,000,000,000,000,000 (6.02×10^{23}) atoms or molecules.

Therefore, if we wanted to tell you how many molecules of water were in a 12–ounce glass, we could do it in 2 ways. We could tell you that there are 1.14×10^{25} molecules, as we did above, or we could tell you that there are 18.9 moles of water molecules in that glass $\left(\frac{1.14 \times 10^{25}}{6.02 \times 10^{23}}\right)$. Again, just like the situation we had with amu and grams, 18.9 moles is a much easier number to use than 1.14×10^{25} atoms. This is exactly the same as telling how many doughnuts you bought at the store. We could say you bought 36 doughnuts, or you bought 3 dozen doughnuts $\left(\frac{36}{12}\right)$. In each case, we can say the number of objects directly, or we can use a word which then can be used to figure out how many objects there were. In chemistry, using the word *mole* is a much easier way to report how many atoms or molecules exist in a sample, so that's what we will use most often.

This rather odd number (6.02×10^{23}) is called Avogadro's (ah vuh gah' drohs) number because the chemist whose work led to its discovery was an Italian scientist named Amedeo Avogadro. Avogadro's number represents how many atoms are in a sample of an element when that sample has the same mass in grams as the element's mass in amu. In other words, the mass of a carbon atom is 12.0 amu. A 12.0–*gram* sample of carbon contains 6.02×10^{23} atoms. Similarly, the atomic mass of sulfur is 32.1 amu, and a 32.1–gram sample of sulfur contains 6.02×10^{23} atoms of sulfur.

Regardless of the atom you choose, if you gather a sample with a mass in grams equal to the mass of the atom in amu, you have 6.02×10^{23} atoms in that sample. Thus (look at the periodic chart to see where we are getting these numbers), 4.0 grams of helium, 52.0 grams of Cr, 55.8 grams of Fe, and 197.0 grams of Au all have exactly 6.02×10^{23} atoms in each sample. Or, to put it another way, each of those samples contains 1 mole of atoms.

$$1 \text{ mole atoms} = 6.02 \times 10^{23} \text{ atoms} \qquad \text{Equation 7.2}$$

This same relationship works for molecules as well. A water molecule has a molecular formula of H_2O. This means that its molecular mass is $2 \times 1.01 + 16.0 = 18.0$ amu. If we have an 18.0–gram sample of water, there are 6.02×10^{23} molecules (1 mole) of water in it. Similarly, the molecular mass of table salt (NaCl) is 58.5 amu. Thus, a 58.5–gram sample of table salt contains 6.02×10^{23} molecules or 1 mole of NaCl molecules.

$$1 \text{ mole molecules} = 6.02 \times 10^{23} \text{ molecules} \qquad \text{Equation 7.3}$$

This leads us to a very, very important fact: You must pay attention to the context of the question in order to choose which unit to put on the atomic mass.

The atomic mass in amu will give the mass of 1 atom.
The atomic mass in grams will give the mass of 1 mole of that atom.

In the same way:

A molecule's mass in amu will give the mass of 1 molecule.
The molecule's mass in grams will give the mass of 1 mole of that molecule.

These are the 2 most important concepts presented in this module. Commit them to memory and be sure that you understand what they mean.

FIGURE 7.3
An Illustration of the Mole Concept

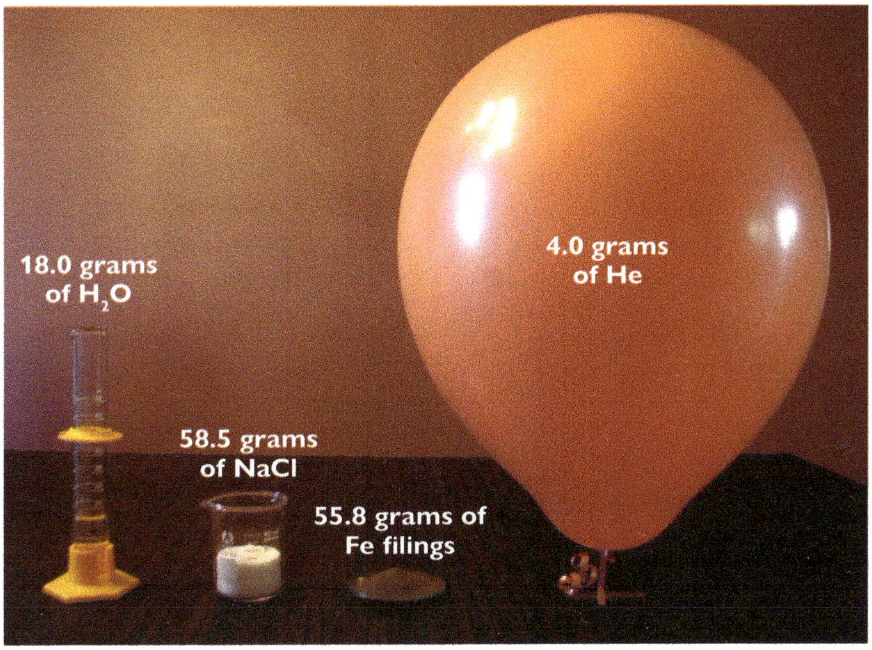

Each of these samples has exactly the same number of atoms or molecules in it. The balloon contains 4.0 g of helium. That means there is 1 mole of helium atoms in the balloon. There are 55.8 grams of iron filings, meaning that there is 1 mole of iron atoms. Since there are 58.5 grams of table salt, there is 1 mole of table salt. There are 18.0 grams of water, which means there is 1 mole of water. Another way to look at this is that there are 6.02×10^{23} atoms of helium, 6.02×10^{23} atoms of iron, 6.02×10^{23} molecules of NaCl, and 6.02×10^{23} molecules of water.

If you are still having trouble grasping the mole concept, think about it this way: An atom of boron has a mass of 10.8 amu. This also means that it takes 10.8 grams of that element to get a mole of boron atoms. In other words, 10.8 grams boron = 1 mole boron. Thus, once we know the mass of an atom, we know a conversion relationship that relates grams to moles. For another atom, this conversion relationship exists, but the number of grams is different. If we need to know how many moles of copper (Cu) atoms are in a sample of copper, we must use the conversion relationship between grams of copper and moles of copper. Since the chart tells us that the mass of a copper atom is 63.5 amu, then we know that 63.5 grams Cu = 1 mole Cu. Try to follow this logic in Example 7.6.

EXAMPLE 7.6

How many moles of fluorine atoms exist in a 50.0 g sample of fluorine?

To solve this problem, we first need to find out a relationship between grams and moles. We do this by looking at the periodic table, which tells us that a fluorine atom has a mass of 19.0 amu. As we just learned, this also means:

19.0 grams fluorine = 1 mole fluorine (conversion factor)

To convert between grams and moles, we simply have to use that relationship:

$$\frac{50.0\ \cancel{\text{g fluorine}}}{1} \times \frac{1\ \text{mole fluorine}}{19.0\ \cancel{\text{g fluorine}}} = 2.63\ \text{moles fluorine}$$

So a 50.0 g sample of fluorine contains **2.63 moles** of fluorine atoms. Notice that we kept 3 significant figures in the answer. This is because the 1 in "1 mole fluorine" is an integer and is therefore exact, giving it an infinite number of significant figures.

People are encouraged to limit their table salt (NaCl) intake to 5.000 grams per day. How many moles of NaCl is that?

To get the conversion relationship between grams and moles in this case, we must first determine the mass of an NaCl molecule. According to the periodic chart:

NaCl mass = 1 x 23.0 amu + 1 x 35.5 amu = 58.5 amu

According to what we just learned, this also means:

58.5 g NaCl = 1 mole NaCl (conversion factor)

Now we can do our conversion:

$$\frac{5.000\ \cancel{\text{g NaCl}}}{1} \times \frac{1\ \text{mole NaCl}}{58.5\ \cancel{\text{g NaCl}}} = 0.0855\ \text{moles NaCl}$$

So people are to limit their intake of NaCl to **0.0855 moles**. Notice that in this case, we could keep only 3 significant figures because of the 58.5 g NaCl. Thus, even though the 1 mole part of the conversion relationship is considered exact, the other half of the relationship is not, and therefore we must count *its* significant figures. We can take this problem further and determine the number of molecules.

How many molecules of NaCl are in a 5.000 gram sample?

As we analyze this problem, we already know that 5.000 grams of NaCl is 0.0855 moles of NaCl. We also know the relationship between moles and molecules from equation 7.3 that 1 mole = 6.02×10^{23} molecules.

Now we set up our problem and can do our conversion:

$$\frac{0.08550\ \cancel{\text{mole NaCl}}}{1} \times \frac{6.02 \times 10^{23}\text{molecules NaCl}}{1\ \cancel{\text{mole NaCl}}} = 5.15 \times 10^{22} \text{ molecules NaCl}$$

Don't get lost in the math and forget what we are doing here. These may seem like pointless conversions, but they are far from it! We are converting something that is very easy to measure (mass) into something that tells us how many moles, atoms, or molecules are in our sample. This is a very important and useful thing to be able to do. Every time we can report the number of moles of atoms (or molecules) in a sample, we are really reporting the number of atoms (or molecules) in that sample. Why is this so useful? Perform Experiment 7.1 to find out.

EXPERIMENT 7.1

PURPOSE: To measure the width of a molecule.

MATERIALS:

- Safety goggles
- Eyedropper
- Stirring rod (or spoon)
- Water
- Large glass (at least 16 ounces)
- Graduated cylinder (Measuring cups and measuring spoons can be used, but they will be less precise.)
- Dishwashing liquid (It must be the kind used for washing dishes by hand)
- Large bowl (It should have a diameter larger than 10 inches but smaller than 12 inches. If you don't have a bowl in that size range, then use one larger than 12 inches; it will just make it a little harder to use the ruler.)
- Pepper
- Ruler

QUESTION: Can the width of a molecule be measured?

Disclaimer: Dishwashing liquid formulations have changed over time making it difficult to find one which contains sodium stearate. This experiment can still be completed with the assumption the dish soap contains a theoretical sodium stearate concentration. It will work regardless, with the difference that in the end, the molecule's width estimate will not be accurate for sodium stearate.

HYPOTHESIS: Write down what you think the width of a molecule is.

PROCEDURE:

1. We've already learned that atoms and molecules are small, but exactly how small are they? We're going to use the mole concept we just learned to get a rough idea. To do this, you first need to fill your large bowl almost full of water. Leave only a centimeter or so from the top of the water to the rim of the bowl. Put the bowl of water where it will be easy to work with but won't be disturbed while you perform the next few steps.
2. Quantitatively measure 5.0 mL (1 teaspoon) of the dishwashing liquid into your graduated cylinder.
3. Leaving the dishwashing liquid in the graduated cylinder, fill the cylinder to the 50.0 mL (¼ cup) mark with water. Do it slowly to reduce the formation of bubbles.
4. Pour the entire contents of the cylinder into the large glass.
5. Now that the cylinder is empty, fill it up to the 50.0 mL (¼ cup) mark with water and pour it into the same glass.
6. Repeat this step 6 more times, so that a total of 395.0 mL (2 cups) of water and 5 mL (1 teaspoon) of dishwashing liquid are in the tall glass.
7. Stir the contents of the glass slowly but thoroughly. By performing the preceding steps, you have **diluted** 5.0 mL of dishwashing liquid to 400.0 mL.
8. Next, you need to calibrate your eyedropper. To do this, first clean and dry your graduated cylinder.
9. Use your eyedropper to transfer 10.0 mL of the diluted dishwashing liquid *(drop by drop)* into the graduated cylinder. Count how many drops it takes to fill the cylinder to the 10.0 mL mark.
10. Divide the number 10.0 by the number of drops you counted. The result represents how many mL are in each drop of your eyedropper. Your answer should be somewhere between 0.060 mL and 0.020 mL. You have used your graduated cylinder to calibrate your eyedropper so that you can figure out how many mL of liquid are in each drop of the eyedropper.
11. Take the pepper over to the bowl of water. Since it has just been sitting for a while, the water in the bowl should be very calm.
12. Lightly sprinkle the pepper onto the surface of the water in the bowl so that there are a few dots of pepper on the surface of the water in each part of the bowl. It is very important that you don't sprinkle *too much* pepper on the water's surface, however. You want only a slight dusting of pepper on the surface of the water.
13. Use your eyedropper to put *1 single drop* of diluted dishwashing liquid into the very center of the bowl. Do this by holding the dropper a few centimeters above the surface of the water at the center of the bowl and carefully allowing 1 drop to fall into the bowl. Observe what happens. The pepper in the center of the bowl will be pushed outward in a roughly circular form. Thus, you will see a circle of clear water form, outlined by the pepper that has been pushed away.
14. Quickly use your ruler to measure the diameter of this circle. You can do this by laying the ruler across the top of the bowl and looking over it at the circle. Try to do this quickly because the circle will tend to collapse, and you want to measure its largest diameter.

15. Why did the circle form in the pepper? The pepper was floating on top of the water, evenly distributed around the surface of the water. When the diluted dishwashing liquid (a mixture that contains several compounds) was dropped into the bowl, most of the compounds in the mixture started to sink into the water and mix with it. One component of the dishwashing liquid (sodium stearate) did not sink. Its chemical nature caused it to float on top of the water. While it floated on the water, it began to spread out in a circle, pushing the pepper out of its way. This is why you saw a clear circle form with the pepper at the edge. As long as there wasn't too much pepper on top of the water, the sodium stearate would continue to spread out until it formed a single circular layer of molecules sitting on top of the water.
16. You can now use the mole concept and the size of the circle to determine the width of the molecules. After all, you can use the size of the circle to determine its area. Then you can use the mole concept to determine how many molecules were in the circle. That way, you can determine the area that each molecule must have occupied. You can then use that area to determine the width of the molecule.
17. First, you need to know how many molecules were in the circle. You have all the information we need to calculate that. First, take the volume of 1 drop that you calculated when you calibrated your eyedropper. Multiply that volume by the density of your diluted dishwashing liquid, which is 1.00 grams/mL. This gives you the number of grams of solution you dropped into the water.
18. Of course, only a tiny fraction of the drop that you put in the water was sodium stearate, the chemical responsible for making the circle. As long as you did your dilution right, about 0.0125% of the solution was sodium stearate. Multiply the number of grams you just calculated by 0.000125. This will tell you how many grams of sodium stearate you added to the bowl. Your answer should be somewhere between 2.5×10^{-6} grams and 7.5×10^{-6} grams.
19. The chemical formula for sodium stearate is $NaC_{18}H_{35}O_2$. Use this information to calculate the molecular mass of sodium stearate, just like we did in the first part of Example 7.5.
20. Use the molecular mass of sodium stearate to convert the grams of sodium stearate into moles of sodium stearate, just like we did in Example 7.6. Your answer should be somewhere between 8.0×10^{-9} moles and 2.9×10^{-8} moles.
21. To determine how many molecules were in the circle, multiply the number of moles you just calculated by the conversion factor in equation 7.3. After all, each mole contains 6.02×10^{23} molecules. If you multiply the number of moles by 6.02×10^{23}, you get the number of molecules in the sample. Your answer should be somewhere between 4.8×10^{15} and 2.0×10^{16} molecules within your circle. If you don't exactly understand every step you took to get this number, don't worry. You just need to understand the concept: By knowing the mass and using the mole concept, you can calculate the number of molecules that were in the circle we observed.
22. Now you need to know the area of the circle. That's easy. Find the diameter of the circle (in centimeters) and divide it by 2. This gives you the radius of the circle. Now put that value in the equation for the area of a circle, Area = πr^2. You do this by squaring the radius you calculated and then multiplying by 3.14. Now you have the area of the circle in cm^2.

23. Divide that area by the number of molecules you calculated above. The result gives you the area occupied by each individual molecule.
24. Now that you have the area occupied by each individual molecule, you can estimate the molecules' width. To do this, assume that the molecules are roughly squares. This is not really true, but it at least gives you an easy way to determine the width. The length of the sides of a square is equal to the square root of the area. Take the square root of the area that you calculated in the previous step, and you have an estimation of the width of a molecule.
25. Look at your answer. It should be on the order of a few times 10^{-8} cm or a few times 10^{-7} cm. This is a very small number. The average human hair has a width of about 10^{-2} cm. The width of a molecule is 100,000 to 1,000,000 times smaller than the width of a single human hair!
26. Clean up and return everything to the proper place.

CONCLUSION: Write something about what you learned in this experiment. What you should gain from this experiment is that the mole concept is a very powerful tool, allowing us to count atoms and determine things about them that cannot be determined by direct observation.

ON YOUR OWN

7.7 How many moles of scandium atoms are in a 50.0–gram sample of scandium?

7.8 How many moles of glucose ($C_6H_{12}O_6$) are in a 2.50 kg sample?

Now that you have seen some examples that deal with the mole concept and have also performed an experiment that uses the mole concept, make sure you have a firm grasp of it by answering "On Your Own" questions 7.7–7.8.

It is important to note that chemists don't always convert from grams to moles. Often it is necessary to go the other way. Sometimes chemists have already figured out how many moles are in a sample, and they need to calculate the mass of the sample. This type of conversion uses the same concepts you have just learned, but it is useful for you to see it worked out in Example 7.7.

EXAMPLE 7.7

A chemist determines that a sample of pure ammonia (NH_3) contains 12.331 moles of NH_3. What is the mass of the sample?

Notice that in this case, we already have the number of moles, but we are looking for the number of grams. To solve this, we have to get the conversion relationship between grams and moles. The molecular mass of NH_3 = 1 x 14.0 amu + 3 x 1.01 amu = 17.0 amu. This means:

$$17.0 \text{ g } NH_3 = 1 \text{ mole } NH_3$$

We can then use this relationship in our standard factor–label conversion:

$$\frac{12.331\ \cancel{\text{moles NH}_3}}{1} \times \frac{17.0\text{ g NH}_3}{1\ \cancel{\text{mole NH}_3}} = 2.10 \times 10^2\text{g NH}_3$$

Thus, a sample that contains 12.331 moles of NH_3 molecules has a mass of **2.10×10^2 g.** Notice that this problem is really no different from the problems we did previously. We just used our conversion relationship in a different way. Notice also that since we know the molecular mass to 3 significant figures, we can report the answer to only 3 significant figures. Scientific notation is the only way we can report this answer properly because 210 has only 2 significant figures.

So you see that whether a chemist is converting from grams to moles or from moles to grams, the first step is to determine the mass of the molecule or atom involved. Then the chemist must use that mass as a conversion relationship. Make sure you understand this process by answering "On Your Own" questions 7.9–7.10.

ON YOUR OWN

7.9 What is the mass of a nitrogen trioxide sample if it contains 0.46 moles?

7.10 What is the mass of a sample that contains 5.5×10^{23} atoms of sulfur, given 1 mole = 6.02×10^{23} atoms?

Before you leave this section, make sure you understand the importance of what you have just learned. When you convert from grams to moles, you are taking a measurement that is very easy to obtain (mass) and converting it into something that tells you how many molecules or atoms are present in a sample of matter (moles). This is an enormously useful thing to be able to do. These conversion problems are not useless exercises that are designed to make sure you understand the factor–label method! Instead, these problems are designed to make sure you understand one of the most powerful tools in all of chemistry. We will use this tool over and over and over again, so you must become very familiar with it.

The mole is such an important concept in chemistry that some people have tried to make an official holiday celebrating this event. It is called National Mole Day, and it starts at 6:02 a.m. on October 23 (10/23). If you think about Avogadro's number, then you should see why National Mole Day is defined that way.

USING THE MOLE CONCEPT IN CHEMICAL EQUATIONS

Now that you have a reasonable grasp of the mole concept, you are ready to see why it is considered such a powerful tool. The real importance of the mole concept is revealed when we look at how it relates to chemical equations. Remember that a chemical equation is really a recipe. By looking at the coefficients to the left of the substances in the equation, we can learn how much of each substance we need to make a reaction occur. For example, the following methane combustion reaction:

$$CH_4\ (g) + 2O_2\ (g) \longrightarrow CO_2\ (g) + 2H_2O\ (g)$$

tells us that when 1 molecule of CH_4 is reacted with 2 molecules of O_2, the products will be 1 molecule of CO_2 and 2 molecules of H_2O. The coefficients in the chemical equation tell us the quantity of substance that we either use or make.

Now suppose we ran the reaction represented by the above equation twice. We could say that 2 molecules of CH_4 reacted with 4 molecules of O_2, producing 2 molecules of CO_2 and 4 molecules of H_2O. Since the reaction ran twice, we can simply double the list of ingredient and products. There is another way to say this, though. We could also say that 1 couple of CH_4 molecules reacted with 2 couples of O_2 molecules, producing 1 couple of CO_2 molecules and 2 couples of H_2O molecules. When stated in this way, we are simply using the fact that the word *couple* means 2.

Let's suppose that instead of running this reaction twice, we run it a total of 12 times. Then we could say that 12 molecules of CH_4 reacted with 24 molecules of O_2, producing 12 molecules of CO_2 and 24 molecules of H_2O. Alternatively, we could say that 1 dozen CH_4 molecules reacted with 2 dozen O_2 molecules, producing 1 dozen CO_2 molecules and 2 dozen H_2O molecules. This way, we are using the word *dozen* to refer to groups of 12 molecules.

Finally, suppose we ran this reaction 6.02×10^{23} times. If we did this, we could say that 1 mole of CH_4 molecules reacted with 2 moles of O_2 molecules, producing 1 mole of CO_2 molecules and 2 moles of H_2O molecules. This is why the mole concept is so important.

In chemical equations, the coefficients that appear to the left of each substance give the ratio of the substances used in the chemical equation.

Think about this fact for a minute: As long as you can balance chemical equations, you can determine the recipe necessary for making that chemical reaction occur! That's a very powerful technique to understand.

Please understand that the coefficients in a chemical equation *do not* give us the recipe in grams. We cannot use the methane combustion reaction equation to say that 1 gram of CH_4 reacts with 2 grams of O_2, producing 1 gram of CO_2 and 2 grams of H_2O. That is just not the case. The coefficients in the chemical equation tell us how many *molecules* react and are produced. Since the word *mole* just represents a group of molecules, they also tell us how many moles react and are produced. A chemical equation is a wonderfully useful thing, but to use it properly, we can use only moles, not grams. If we stick with the concept of moles, we can do quite a lot. In fact, once we have a balanced chemical equation, we can convert from one substance in the equation to any other substance in that equation at will! Let's see how to do that in Example 7.8.

EXAMPLE 7.8

The decomposition of hydrogen peroxide into water and oxygen is described by the following chemical equation:

$$2H_2O_2\ (l) \longrightarrow 2H_2O\ (l) + O_2\ (g)$$

If a chemist begins with 2.4 moles of H_2O_2, how many moles of O_2 will be produced?

The chemical equation tells us that 2 moles of H_2O_2 will decompose into 1 mole of O_2. We can think of this fact as a conversion relationship. For this particular chemical reaction,

$$2 \text{ moles } H_2O_2 = 1 \text{ mole } O_2$$

We have to stress that this relationship is good only for *this particular reaction*. Another reaction that has hydrogen peroxide and oxygen in it will, most likely, have a completely different relationship between these 2 substances.

To find out how many moles of oxygen are formed, all we have to do is convert moles of hydrogen peroxide into moles of oxygen. The equation above represents a conversion relationship like any other conversion relationship. We can therefore use the factor–label method to convert between moles of H_2O_2 and moles of O_2:

$$\frac{2.4 \cancel{\text{ moles } H_2O_2}}{1} \times \frac{1 \text{ mole } O_2}{2 \cancel{\text{ moles } H_2O_2}} = 1.2 \text{ moles } O_2$$

Thus, when 2.4 moles of hydrogen peroxide decompose, they produce **1.2 moles of oxygen gas**. The relationship that we develop from the chemical equation is considered to be exact. After all, it comes from the coefficients in the chemical equation, and those are integers. Therefore, both numbers in the conversion relationship have an infinite number of significant figures. As a result, the only number with error in it is the original amount of H_2O_2. Its number of significant figures determines the number of significant figures we can report in the answer.

It may seem odd that we can convert from one substance to another in a chemical equation. Nevertheless, we can. After all, what does a chemical reaction do? It *converts* one or more substances into different substances. Thus, we can use the factor–label method in this type of conversion, just like we can in unit conversions.

You might also find it odd that we did not have to use any information about the water produced. Water is produced in the reaction, but the amount of water produced does not really affect the amount of oxygen produced, and vice versa. Both of them are dependent on the amount of hydrogen peroxide that decomposes, but neither is dependent on the other. This is an important aspect of chemical equations. In general, we needn't know everything about the substances in a chemical equation to arrive at an answer to a problem. Usually, we can relate any one substance in a chemical equation to any other substance in that equation. There are some exceptions to this general rule (of course), and we will discuss those exceptions in detail in the next module.

Let's do one more problem just to make sure you really understand how to use the coefficients in a chemical equation.

If a chemist has 0.234 moles of C_2H_6O, how many moles of carbon will she get by decomposing the sample?

We are not given the chemical equation in this problem, but we can figure it out. The problem says that the chemist is decomposing C_2H_6O. The chemical reaction, then, is:

$$C_2H_6O \rightarrow C + H_2 + O_2$$

Remember, both hydrogen and oxygen are homonuclear diatomics, but carbon is not. The chemical formula for carbon is C, but the chemical formula for hydrogen is H_2, and the chemical formula for oxygen is O_2. Balancing the equation gives us:

$$2C_2H_6O \rightarrow 4C + 6H_2 + O_2$$

This chemical equation tells us that 2 moles of C_2H_6O will decompose into 4 moles of C. Thus:

$$2 \text{ moles } C_2H_6O = 4 \text{ moles C}$$

Now that we have our conversion relationship for this particular reaction, we can simply convert:

$$\frac{0.234 \cancel{\text{ moles } C_2H_6O}}{1} \times \frac{4 \text{ moles C}}{2 \cancel{\text{ moles } C_2H_6O}} = 0.468 \text{ moles C}$$

Our chemical equation, then, tells us that when 0.234 moles of C_2H_6O decompose, they produce **0.468 moles of C**. Once again, the numbers in the conversion relationship are exact since they come from the chemical equation. Therefore, the number of significant figures in the original measurement determines the number of significant figures in the answer.

Hopefully, you are beginning to appreciate the power behind the mole concept. It not only allows us to count the number of atoms or molecules in a sample of matter, but it also allows us to predict the quantities of substances that are produced in a chemical reaction. See if you can perform the latter of these 2 techniques in "On Your Own" question 7.11.

ON YOUR OWN

7.11 A chemist obtains a 100.0 moles of ammonia (NH_3). If he allows this sample to decompose, how many moles of hydrogen gas will be produced?

The section we have just completed is an introduction into the most important aspect of chemistry: relating the quantities of chemical substances in a chemical reaction. This subject, commonly called stoichiometry (stoy' kee ahm' uh tree), will be the main subject of the next module. There isn't a chemist in the world who doesn't use stoichiometry all of the time, so you must learn this subject inside and out.

If you have a good handle on how to convert from grams to moles, how to convert from moles to grams, and how to convert from moles of one substance to moles of another substance by using a chemical equation, then you are well on your way to understanding stoichiometry. If, on the other hand, you are unsure of one of these 3 techniques, you need to review what you have read, look at the extra problems in appendix B, or seek out alternative resources. You cannot really be successful in the next module (not to mention the rest of the course) unless you are comfortable with these 3 techniques.

SUMMARY OF KEY EQUATIONS AND TABLES IN MODULE 7

Equation 7.1: 1.00 amu = 1.66×10^{-24} g
Relationship of atomic mass units and grams

Equation 7.2: 1 mole atoms = 6.02×10^{23} atoms
Avogadro's number

Equation 7.3: 1 mole molecules = 6.02×10^{23} molecules
Relationship of moles to molecules

Atomic mass in amu = mass of 1 atom

Atomic mass in grams = mass of 1 mole of that molecule

Molecular mass in amu = mass of 1 molecule

Molecular mass in grams = mass of 1 mole of that molecule

ANSWERS TO THE "ON YOUR OWN" QUESTIONS

7.1 Nitrogen trichloride is NCl_3, which contains 2 elements: N and Cl. Both nitrogen and chlorine are homonuclear diatomics.

$$NCl_3 \longrightarrow N_2 + Cl_2$$

Now we just have to balance the equation. To balance the N, we need a 2 in front of NCl_3.

$$2\ NCl_3 \longrightarrow N_2 + Cl_2$$

To balance the Cl, we need a 3 in front of Cl_2 on the product side:

$$2\ NCl_3 \longrightarrow N_2 + 3\ Cl_2$$

$$2\ NCl_3 \longrightarrow N_2 + 3\ Cl_2$$

7.2 This molecule contains C, H, and O. Hydrogen and oxygen are both homonuclear diatomics, but carbon is not.

$$C_{12}H_{23}O_{11} \longrightarrow C + H_2 + O_2$$

To balance, we start by making 12 C atoms on the products side:

$$C_{12}H_{23}O_{11} \longrightarrow 12C + H_2 + O_2$$

To balance the H without disturbing the C, we need to get 23 H atoms on the products side:

$$C_{12}H_{23}O_{11} \longrightarrow 2C + \tfrac{23}{2}H_2 + O_2$$

To balance the O without disturbing the C or the H, we need to get 11 O atoms on the products side:

$$C_{12}H_{23}O_{11} \longrightarrow 12C + \tfrac{23}{2}H_2 + \tfrac{11}{2}O_2$$

Now we just need to get rid of the fractions:

$$2 \times [C_{12}H_{23}O_{11}] \longrightarrow 2 \times [12C + \tfrac{23}{2}H_2 + \tfrac{11}{2}O_2]$$

$$2C_{12}H_{23}O_{11} \longrightarrow 24C + 23H_2 + 11O_2$$

7.3 The compound is composed of K, Cl, and O. Cl and O are homonuclear diatomics, but K is not.

$$K + Cl_2 + O_2 \longrightarrow KClO_3$$

The K atoms are currently balanced. To balance the Cl atoms, we need a 2 in front of $KClO_3$:

$$K + Cl_2 + O_2 \longrightarrow 2\ KClO_3$$

Balancing O will require a 3 in front of O_2:

$$K + Cl_2 + 3O_2 \longrightarrow 2\ KClO_3$$

K is not balanced. We need 2 K atoms on the reactant side, and this will balance the equation.

$$2K + Cl_2 + 3O_2 \longrightarrow 2\ KClO_3$$

7.4 A complete combustion reaction involves adding oxygen while producing carbon dioxide and water:

$$C_3H_6O + O_2 \longrightarrow CO_2 + H_2O$$

Now all we have to do is balance the equation. Balancing the C and H is easy:

$$C_3H_6O + O_2 \longrightarrow 3CO_2 + 3H_2O$$

To balance the O, we should try not to disturb the balance of the C and H. There is 1 O atom in C_3H_6O, which means we need 8 more O atoms on the reactant side. We can get this by putting a 4 in front of O_2 :

$$C_3H_6O + 4O_2 \longrightarrow 3CO_2 + 3H_2O$$

7.5 First, we look at the table and see that lithium atoms have a mass of 6.94 amu. All we need to do now is convert. This conversion is a little more difficult than the one in Example 7.4 because we are asked to make a conversion to kg. The problem is that we have a relationship between amu and g, not amu and kg. We need to make 2 conversions. The first will convert amu to g, and the second will convert g to kg. We will do it with successive factors as explained in Module 1:

$$\frac{6.94\ \cancel{amu}}{1} \times \frac{1.66 \times 10^{-24}\ \cancel{g}}{1.00\ \cancel{amu}} \times \frac{1kg}{1{,}000\ \cancel{g}} = 1.15 \times 10^{-26}\ kg$$

So the mass of a lithium atom is 6.94 amu, or 1.15×10^{-26} kg.

7.6 The formula for sodium sulfate is Na_2SO_4. This molecule has 2 Na atoms (23.0 amu), 1 S atom (32.1 amu), and 4 O atoms (16.0 amu). The molecular mass, then, is:

$$\text{Mass of } Na_2SO_4 = 2 \times 23.0\ amu + 32.1\ amu + 4 \times 16.0\ amu = 142.1\ amu$$

Converting to grams:

$$\frac{142.1\ \cancel{amu}}{1} \times \frac{1.66 \times 10^{-24} g}{1.00\ \cancel{amu}} = 2.36 \times 10^{-22}\ g$$

The mass is 2.36×10^{-22} g. Note that the conversion relationship between grams and amu limits the number of significant figures we can have. As discussed in the module, the conversion relationship has only 3 significant figures, so we must count those when using it.

7.7 The periodic table tells us that 1 scandium atom has a mass of 45.0 amu. Therefore:

$$45.0 \text{ grams Sc} = 1 \text{ mole Sc}$$

This is the conversion relationship we need for converting grams of scandium into moles of scandium:

$$\frac{50.0 \cancel{\text{g Sc}}}{1} \times \frac{1 \text{ mole Sc}}{45.0 \cancel{\text{g Sc}}} = 1.11 \text{ moles Sc}$$

There are 1.11 moles in the sample.

7.8 To do these conversions, we need to know the mass of the molecule:

$$\text{Mass of } C_6H_{12}O_6 = 6 \times 12.0 \text{ amu} + 12 \times 1.01 \text{ amu} + 6 \times 16.0 \text{ amu} = 180.1 \text{ amu}$$

Remember, we are adding here. When adding (or subtracting), we do not count significant figures. Instead, we report our answer to the same decimal place as the least precise number in the problem. Most of the masses have their last significant figure in the tenths place. The mass of hydrogen has its last significant figure in the hundredths place. The answer must be reported to the tenths place. This means:

$$180.1 \text{ g } C_6H_{12}O_6 = 1 \text{ mole } C_6H_{12}O_6$$

This is the conversion relationship we need for converting grams of glucose into moles of glucose. Before we can use it, though, we need to convert kg of glucose into g of glucose. By now, you should be able to do that on your own. The conversion yields 2.50×10^3 g glucose. Now we can do the grams–to–moles conversion:

$$\frac{2.50 \times 10^3 \cancel{\text{g } C_6H_{12}O_6}}{1} \times \frac{1 \text{ mole } C_6H_{12}O_6}{180.1 \cancel{\text{g } C_6H_{12}O_6}} = 13.9 \text{ moles } C_6H_{12}O_6$$

There are 13.9 moles in the sample.

7.9 This is just another conversion problem, but this time we are converting moles into grams. We still need to determine the formula of nitrogen trioxide, which is NO_3. Now we need to find the molecular mass of NO_3 :

$$\text{Mass of } NO_3 = 1 \times 14.0 \text{ amu} + 3 \times 16.0 \text{ amu} = 62.0 \text{ amu}$$

This means:

$$62.0 \text{ grams } NO_3 = 1 \text{ mole } NO_3$$

Now we can do our conversion:

$$\frac{0.46 \cancel{\text{moles } NO_3}}{1} \times \frac{62.0 \text{ g } NO_3}{1 \cancel{\text{moles } NO_3}} = 29 \text{ g } NO_3$$

The mass is 29 g.

7.10 This is a bit more complicated conversion problem but still just a conversion problem. This time we are converting molecules into grams. We have the moles to molecules conversion (1 mole = 6.02×10^{23} atoms) but need to determine the molecular mass of sulfur.

The atomic mass of S is listed as 32.1 amu, which means:

$$32.1 \text{ grams of S} = 1 \text{ mole S}$$

Now we can set up our equation and do our conversion:

$$\frac{5.5 \times 10^{23} \cancel{\text{atoms S}}}{1} \times \frac{1 \cancel{\text{mole S}}}{6.02 \times 10^{23} \cancel{\text{atoms S}}} \times \frac{32.1 \text{ g S}}{1 \cancel{\text{mole S}}} = 29 \text{ g}$$

5.5×10^{23} atoms of S will have a mass of 29 grams.

7.11 To solve this problem, we must first find the balanced chemical equation for this reaction. According to the problem, the reaction is the decomposition of NH_3:

$$NH_3 \rightarrow N_2 + H_2$$

Now we have to balance it:

$$2NH_3 \rightarrow N_2 + 3H_2$$

Remember what this tells you. This tells you the *recipe* for the reaction *in moles*. So, 2 moles of NH_3 decompose into 1 mole of N_2 and 3 moles of H_2. The relationship between NH_3 and H_2, then, is given by:

$$2 \text{ moles } NH_3 = 3 \text{ moles } H_2$$

This is a conversion relationship that allows us to convert from moles of one substance into moles of another:

$$\frac{100.0 \cancel{\text{moles } NH_3}}{1} \times \frac{3 \text{ moles } H_2}{2 \cancel{\text{moles } NH_3}} = 150.0 \text{ moles } H_2$$

The answer is 150.0 moles.

STUDY GUIDE FOR MODULE 7

REVIEW QUESTIONS

1. Can an element undergo a decomposition reaction?

2. Classify the following reactions as decomposition, formation, complete combustion, or none of these:
   a. $C_6H_{12}O_6$ (s) + $6O_2$ (g) ⟶ $6CO_2$ (g) + $6H_2O$ (g)
   b. $2H_2$ (g) + O_2 (g) ⟶ $2H_2O$ (l)
   c. NaCl (s) + $AgNO_3$ (aq) ⟶ AgCl (s) + $NaNO_3$ (aq)
   d. 2HBr (l) + $Ca(OH)_2$ (s) ⟶ $2H_2O$ (l) + $CaBr_2$ (aq)
   e. $2NaHPO_4$ (s) ⟶ 2Na (s) + H_2 (g) + 2P (s) + $4O_2$ (g)

3. Which of the following is *not* a formation reaction?
   a. H_2CO_3 + 2KOH ⟶ $2H_2O$ + K_2CO_3
   b. CO_2 (g) + H_2O (g) ⟶ H_2CO_3 (l)
   c. 2S (s) + Cl_2 (g) + $5F_2$ (g) ⟶ $2SF_5Cl$

4. Which of the following is *not* a combustion equation?
   a. $2C_3H_8O$ + $9O_2$ ⟶ $6CO_2$ + $8H_2O$
   b. H_2CO_2 (s) ⟶ CO_2 (g) + H_2O (g)
   c. CH_2 (g) + $2O_2$ (g) ⟶ CO_2 (g) + $2H_2O$ (g)
   d. 2 Ca (s) + O_2 (g) ⟶ 2 CaO (s)

5. What is the difference between complete combustion and incomplete combustion?

6. What does a catalytic converter do for an automobile?

7. Which has more mass: 100 hydrogen atoms, 4 sulfur atoms, or 1 lanthanum (La) atom?

8. What do the following things have in common?
   32.1 g of sulfur, 40.1 g of calcium, and 60.1 g of SiO_2

9. What is Avogadro's number and what does it represent?

10. In the following equation, what are two ways the numbers in front of each molecule can be interpreted? Use relationships and terms learned in this module to explain your answer.
    2 C_3H_8O + 9 O_2 ⟶ 6 CO_2 + 8 H_2O

PRACTICE PROBLEMS

1.00 amu = 1.66×10^{-24} g
1 mole = 6.02×10^{23} atoms or molecules

1. Write a balanced chemical equation for the decomposition of $RbNO_3$.

2. Write a balanced chemical equation for the formation of $NaHSO_4$.

3. An automobile uses the energy gained from burning evaporated gasoline to run. Write a balanced chemical equation for the complete combustion of C_8H_{18}, an important component of gasoline. Include phase symbols in your equation.

4. What is the mass of an argon atom in grams?

5. What is the mass of an aluminum dichromate molecule in kg?

6. How many moles of Au atoms are in 2.56 kg?

7. How many moles of $NaHCO_3$ are in a 125–gram sample of the compound?

8. What is the mass of a copper (II) chloride sample if it contains 0.344 moles of the compound?

9. How many atoms are in 457 grams of tungsten (W)?

10. In the decomposition of 1.2 moles of dinitrogen pentaoxide, how many moles of oxygen would be formed?

STOICHIOMETRY

first reactions

Although the math in this section may get to you, remember the simple explanation that if we gave you three mittens to wear, you still only have two hands to put them on. There will be one mitten unused. Stoichiometry is simply this principal in chemistry. You will need to evaluate how much of each material you are given and which pieces limit the production of additional product. Always remember your equations need to account for even the material left over.

In a previous module, you learned how to count atoms and molecules using the mole concept. You were then given a brief introduction to the relationship between the mole concept and the mathematics of chemical equations. You found out that, using the mole concept, you can calculate the amount of product formed in a chemical reaction if you are given the number of moles of reactant material.

In this module, we will expand on the mathematical relationships within chemical equations. We will see that, using the mole concept, we can relate the quantity of any substance in a chemical equation to the quantity of any other substance in that same equation. We will also see that, given the proper conversion steps, we can develop relationships between the masses of those substances as well. As you learned in a previous module, relating the quantities of different substances in a chemical equation is called stoichiometry, and it is the single most important concept in the entire course. Please read this module carefully and thoroughly because we will constantly go back to the concepts presented here. You must be firmly grounded in this material to be able to proceed in the course.

MOLE RELATIONSHIPS IN CHEMICAL EQUATIONS

In a previous module, you did a few problems in which you related the number of moles of a reactant to the number of moles of products formed in a chemical reaction. That was your first step in the process of understanding the mole relationships between substances in a chemical equation. Now you are ready for the next step. As noted in the previous module, chemical equations are really the same as mathematical equations. Anything we can do to one, we can do to the other. We will use this fact to help us take the next step.

In the stoichiometry problems you performed previously, you were always given a decomposition reaction and the amount of reactant, and you were always asked to determine the amount of product that was formed. This may have left you with the impression that stoichiometry must always relate the amount of reactant to the amount of product. This simply isn't true. Like mathematical equations, chemical equations can be used to relate the left side of the equation to the right side of the equation, *and vice versa.*

For example, consider the mathematical relationship below:

$$y + 3 = x$$

If you were told that y = 4, then you could determine the value for x, right? You would simply replace y with 4 in the equation, and you would end up determining that x = 7. In this case, you related the terms on the left side of the equation to the term on the right side of the equation.

Suppose that you were given the value for x instead of y. Suppose you were told that x = 7. Could you solve for y? Of course you could. You would simply replace x with 7 in the equation, and you would use your algebra skills to determine that y = 4. In this case, you related the term on the right side of the equation to those on the left side of the equation.

Chemical equations behave in exactly the same way. We do not always have to give you the amount of reactants and have you calculate the amount of products formed. Instead, we could tell you how much product was formed, and you could use that information to determine the quantities of the reactants involved. See if you can follow this reasoning through Example 8.1.

salt and light

Are you beginning to appreciate the study of chemistry? Do you ever wonder where these chemical reactions came from? Well, they came from the research of past chemists. A chemist might spend years running different chemical reactions, or he might find a way to run a known chemical reaction more efficiently. One great American chemist who spent many years in a research laboratory running chemical reactions was **Charles Stine (1882–1954)**, the director of research for DuPont Chemical Company. He developed a chemical reaction that produced a more stable explosive of dynamite. Also, he and Wallace Carothers developed DuPont's laboratory for research that produced Carothers' most famous product, nylon. Stine was a dedicated Christian and wrote his views on the Gospel and Christianity in a book titled *A Chemist and His Bible*. He wrote, "The world about us, far more intricate than any watch, filled with checks and balances of a hundred varieties, marvelous beyond even the imagination of the most skilled scientific investigator, this beautiful and intricate creation, bears the signature of its Creator, graven in its works" (quoted in Morris 1988, 84).

Photo courtesy of DuPont.

EXAMPLE 8.1

A chemist forms ammonium sulfate, $(NH_4)_2SO_4$, an excellent fertilizer, by combining ammonia (NH_3) and sulfuric acid (H_2SO_4). If she ends up making 12.2 moles of ammonium sulfate, how many moles of ammonia were used?

To be able to do any stoichiometry problem, we must first figure out the chemical equation. According to the problem, ammonium sulfate is the product, while sulfuric acid and ammonia are the reactants:

$$NH_3 + H_2SO_4 \rightarrow (NH_4)_2SO_4$$

Before we can balance the equation, let's review the chemical formula for ammonium sulfate (the product). Notice that it has parentheses in it. Do you remember what that means? You already learned the *physical* meaning, and now we will learn the *practical* meaning. The 2 outside the parentheses applies to *every atom within the parentheses.* We must multiply that 2 with the numbers inside the parentheses. That's the only way we can find out how many N atoms and H atoms are in the molecule. When we multiply the 2 with the numbers inside the parentheses, we get $N_2H_8SO_4$. In this molecule, there are really 2 N atoms and 8 H atoms. You will need to get used to this because you will see it again. Whenever you see parentheses in a molecule, you must distribute the subscript after the parentheses to every atom within the parentheses. That's the only way you will know how many of each atom is in the molecule.

Now that we have that out of the way, we can go back to work. We first need to balance the equation. Remember, there are 2 N atoms and 8 H atoms in the product. So the balanced chemical equation is:

$$2NH_3 + H_2SO_4 \rightarrow (NH_4)_2SO_4$$

The problem gives us the number of moles of ammonium sulfate and asks us how many moles of ammonia were used in the reaction. According to the chemical equation, the relationship between ammonia and ammonium sulfate is:

$$2 \text{ moles } NH_3 = 1 \text{ mole } (NH_4)_2SO_4$$

Using this relationship, we can convert from moles of ammonium sulfate to moles of ammonia:

$$\frac{12.2 \cancel{\text{ moles } (NH_4)_2SO_4}}{1} \times \frac{2 \text{ moles } NH_3}{1 \cancel{\text{ mole } (NH_4)_2SO_4}} = 24.4 \text{ moles } NH_3$$

The chemist must have started with **24.4 moles of NH_3**.

The first thing you should notice about this example is that there is nothing new here, except for the parentheses in the chemical formula of ammonium sulfate. We have used exactly the same kind of reasoning that we used in the previous module—we just went from products to reactants instead of reactants to products. The relationships between substances in a chemical equation are independent of which side of the equation we begin with.

The problem asked how much NH_3 was used to make the ammonium sulfate, but we could also figure out how much sulfuric acid was used. After all, the chemical equation relates moles of sulfuric acid to moles of ammonium sulfate:

$$1 \text{ mole } H_2SO_4 = 1 \text{ mole } (NH_4)_2SO_4$$

We can use that relationship to convert from moles of ammonium sulfate to moles of sulfuric acid:

$$\frac{12.2 \cancel{\text{ moles } (NH_4)_2SO_4}}{1} \times \frac{1 \text{ mole } H_2SO_4}{1 \cancel{\text{ mole } (NH_4)_2SO_4}} = 12.2 \text{ moles } H_2SO_4$$

We learn from the chemical equation that the chemist had to start with 24.4 moles of ammonia and 12.2 moles of sulfuric acid to make 12.2 moles of ammonium sulfate.

So you see that the quantity of just one substance in the equation has the potential to give us the quantities of all other substances in the chemical equation. This gives you some idea of how powerful the tool of stoichiometry can be. Check your stoichiometry skills with "On Your Own" question 8.1.

ON YOUR OWN

8.1 A chemist performs the following reaction to make sodium nitrate:

$2NaNO_2(s) + O_2(g) \rightarrow 2NaNO_3(s)$

How many moles of oxygen gas must the chemist use to make 45.0 moles of $NaNO_3$? How many moles of $NaNO_2$ are necessary?

LIMITING REACTANTS AND EXCESS COMPONENTS

If all of this seems rather easy, rest assured that this subject gets a lot harder. For example, so far we have worked only with decomposition or formation reactions. Those kinds of reactions are a little easier to deal with because for both of these reaction types, 1 side of the equation has only 1 substance. In decomposition reactions, there is only 1 reactant, whereas in formation reactions, there is only 1 product. Stoichiometry is a bit more difficult when multiple substances appear on both sides of the equation. Perform Experiment 8.1 to see why.

EXPERIMENT 8.1

PURPOSE: To determine which reactant is the limiting reactant.

MATERIALS:

- Baking soda
- Vinegar
- String or tape measure
- Graduated cylinder (or measuring cups and spoons)
- Ruler
- Round balloon
- Plastic 2-liter bottle (or other large bottle)
- Mass scale
- Funnel or butter knife (see step 2)
- Safety goggles

QUESTION: Which reactant runs out first?

HYPOTHESIS: Write down your hypothesis related to limiting reactants.

PROCEDURE:

1. Quantitatively measure 250.0 mL (1¼ cups) of vinegar into the clean 2-liter plastic bottle.
2. Fill the balloon with 10 grams of baking soda. You can do this with a funnel, or if you have a hard time doing it that way, use a butter knife. Use the butter knife to pick up some baking soda, and then carefully push the knife into the balloon. Once the knife is in the balloon, tip them both so that the baking soda on the knife falls into the balloon. It is best to do this over the sink because you will spill a significant amount of baking soda. As you fill the balloon, keep measuring its mass until you have 10.0 grams of baking soda.
3. Once you have 10.0 grams of baking soda in the balloon, attach the balloon to the bottle as shown below, making sure that the lip of the balloon covers the bottle opening completely so that gas cannot escape.

4. Now hold the balloon upright so that the baking soda falls into the vinegar. You will see bubbles form, and the balloon will inflate. If you hear gas escaping from the balloon, then you must redo the experiment, being more careful to fasten the lip of the balloon tightly to the opening of the bottle. For this experiment to work, no gas can escape from the setup!
5. Carefully pick up the bottle and swirl the vinegar around so that it reacts with all of the baking soda. Also, gently shake the balloon from side to side to make sure that all of the baking soda has fallen into the vinegar. After a few minutes, the bubbles should die down.
6. Once the bubbles have died down, wrap the string once around the biggest part of the balloon. Use your fingers to mark exactly how much of the string was needed to go around the balloon once, then measure the length of that part of the string. This will give you a rough value for the circumference of the balloon. If you have a tape measure, use it to measure the circumference directly.
7. Repeat this experiment twice. The first time you repeat it, use 30.0 grams of baking soda rather than 10.0 grams, and the second time you repeat it, use 60.0 grams of baking soda. Each time you should use 250.0 mL of vinegar. *Do not reuse the old vinegar!* Clean the bottle out each time you have finished and use 250.0 mL of fresh vinegar each time.
8. Once you have all 3 circumference measurements, clean up and return everything to the proper place.

CONCLUSION: Write something about what this experiment illustrates.

In each trial you ran, the chemical reaction that occurred was:

$$\underset{\text{acetic acid in vinegar}}{C_2H_4O_2\,(l)} + \underset{\text{baking soda}}{NaHCO_3\,(s)} \rightarrow H_2O\,(l) + CO_2\,(g) + \underset{\text{sodium acetate}}{NaC_2H_3O_2\,(aq)}$$

The carbon dioxide that was formed in the reaction is the gas that filled the balloon. According to the chemical equation, each mole of baking soda that you add to the acetic acid should produce another mole of gas. This should result in the balloon inflating more. You might expect that in each trial you ran, the balloon would get bigger. If you look at your measurements for the circumference of the balloon, that's not what you should see. The circumference of the balloon in the second trial should have been significantly greater than the circumference of the balloon in the first trial. However, in the third trial, although you added twice as much baking soda as you did in the second trial, the circumference of the balloon should not be significantly larger than it was in the second trial. Why?

In the first trial, there were more moles of acetic acid in the bottle than the moles of baking soda that you added. When the reaction was over, leftover acetic acid was waiting to react with more baking soda. However, since there was no more baking soda, the reaction stopped, and the balloon stopped filling up. In the second trial, you added slightly more moles of baking soda compared to the moles of acetic acid. The acetic acid ran out first that time. When the reaction was over, then, there was no leftover acetic acid, but there was a little bit of leftover baking soda. When you went to the third trial, adding all of that extra baking soda did not increase the amount of gas produced at all. After all, the acetic acid ran out after reacting with the first 30.0 grams of baking soda. All of that extra baking soda had nothing to react with, so it produced nothing extra.

This demonstrates the concept of limiting reactants. When there is more than one reactant in a chemical equation, one of those reactants will run out first, stopping the entire reaction. We call that reactant the *limiting* reactant because once it runs out, the reaction cannot continue. Since the reaction cannot continue, no more products will form.

The limiting reactant limits the amount of product that can be formed in the chemical reaction.

Looking back over this experiment, you can see that in the first trial, baking soda was the limiting reactant. In that trial, the baking soda ran out first, stopping the formation of any more product. When more baking soda was added in the second trial, the balloon got larger because there was still acetic acid to react with, and therefore more product was made. In the final trial, the acetic acid was the limiting reactant, stopping the production of carbon dioxide and leaving plenty of leftover baking soda. So even though there was a lot of baking soda in the reaction, the balloon could not fill with more product because the reaction ran out of acetic acid.

As we begin to do more complex stoichiometry problems, it will be necessary to tell which reactant is the limiting reactant. Once you know that, you can use its quantity to predict the quantity of products formed. In more complex chemical equations, the relationships between quantities of substances depend only on the quantity of the *limiting reactant*. As a result, you can essentially ignore the quantities of the other reactants because the limiting reactant determines how much product can be formed in the reaction.

FULLY ANALYZING CHEMICAL EQUATIONS

Now that you understand the concept of limiting reactants, we can discuss how to fully analyze a chemical equation. Consider the following equation:

$$Mg\ (s) + 2HCl\ (aq) \rightarrow MgCl_2\ (aq) + H_2\ (g)$$

In this equation, there are 2 reactants and 2 products. These reactants and products all relate to one another. For example, if we told you the number of moles of *any one substance* in the equation, you could use it to calculate the moles of *any other substance* in the equation. This may sound a little strange because you are used to algebra, where a single equation can provide you with the value of only a single unknown. In chemistry, a single equation can be used to determine many unknowns.

For example, suppose we told you that the above reaction was started with 1 mole of Mg. According to the equation, 1 mole of Mg reacts with 2 moles of HCl. One thing you can determine from the reaction is that for all of the Mg to be used up, 2 moles of HCl must be added. After all, if less than 2 moles of HCl were added, HCl would be the limiting reactant and there would be leftover Mg. Alternatively, if more than 2 moles of HCl were added, all of the Mg would be used up and there would be leftover HCl. That would make Mg the limiting reactant.

One thing that you can determine from a chemical equation is the amount of other reactant(s) necessary to completely use up a given quantity of the given reactant. When reactants are added in this way, so that all reactants are used up, we say that they reacted completely. This means that all reactants were added in just the right quantities necessary to be completely used up. We would say that 2 moles of HCl are needed in this reaction for 1 mole of Mg to react completely.

Of course, we can also use the quantity of Mg to determine how much product is formed. As long as we know that the Mg reacts completely, we know that it is the limiting reactant. We can then use its number of moles to determine the number of moles of products that form. According to the reaction equation, 1 mole of Mg will make 1 mole of $MgCl_2$. We know that the reaction we have been discussing will make 1 mole of $MgCl_2$. In addition, we can also say that since the reaction equation tells us that 1 mole of Mg also produces 1 mole of H_2, we know that 1 mole of H_2 is produced in the reaction as well. Let's see how this pans out in Example 8.2.

EXAMPLE 8.2

An antacid tablet, with an active ingredient of $CaCO_3$, neutralizes excess HCl in your stomach according to the following reaction:

$$CaCO_3\ (s) + 2HCl\ (aq) \rightarrow CaCl_2\ (aq) + H_2O\ (l) + CO_2\ (g)$$

If 0.0075 moles of $CaCO_3$ are in the tablet, how many moles of HCl can it neutralize? When it neutralizes that amount of stomach acid, how much $CaCl_2$, H_2O, and CO_2 are produced?

In this problem, we are given the quantity of one of the reactants and are asked how much of the other reactant it can react with. This is just like asking how many moles of HCl are necessary to use up all of the $CaCO_3$. We first look at the chemical equation and develop a relationship between HCl and $CaCO_3$:

$$1 \text{ mole } CaCO_3 = 2 \text{ moles HCl}$$

Now we can simply use this relationship to convert from moles of $CaCO_3$ to moles of HCl:

$$\frac{0.0075 \cancel{\text{ moles } CaCO_3}}{1} \times \frac{2 \text{ moles HCl}}{1 \cancel{\text{ mole } CaCO_3}} = 0.015 \text{ moles HCl}$$

This tells us that an antacid tablet neutralizes **0.015 moles of HCl**. We can learn a lot more from this equation, though. Since we have determined that $CaCO_3$ will be completely used up, we know that it is the limiting reactant. The quantity of limiting reactant is also related to the quantity of all products. According to the equation:

$$1 \text{ mole } CaCO_3 = 1 \text{ mole } CaCl_2$$

$$1 \text{ mole } CaCO_3 = 1 \text{ mole } CO_2$$

$$1 \text{ mole } CaCO_3 = 1 \text{ mole } H_2O$$

We can use all of these relationships to convert from the given number of moles of $CaCO_3$ into moles of each product

$$\frac{0.0075 \cancel{\text{ moles } CaCO_3}}{1} \times \frac{1 \text{ moles } CaCl_2}{1 \cancel{\text{ mole } CaCO_3}} = \mathbf{0.0075 \text{ moles } CaCl_2}$$

$$\frac{0.0075 \cancel{\text{ moles } CaCO_3}}{1} \times \frac{1 \text{ moles } CO_2}{1 \cancel{\text{ mole } CaCO_3}} = \mathbf{0.0075 \text{ moles } CO_2}$$

$$\frac{0.0075 \cancel{\text{ moles } CaCO_3}}{1} \times \frac{1 \text{ moles } H_2O}{1 \cancel{\text{ mole } CaCO_3}} = \mathbf{0.0075 \text{ moles } H_2O}$$

With just the quantity of one reactant given, then, we can calculate the needed amounts of all of the other substances in the chemical equation. Try this yourself with "On Your Own" question 8.2.

RELATING PRODUCTS TO REACTANTS IN CHEMICAL EQUATIONS

There are, of course, many other ways to analyze a chemical equation. One way chemists often use stoichiometry is to determine the quantity of a reactant necessary

ON YOUR OWN

8.2 The first step in a long process that converts iron ore into pure iron involves converting iron ore into iron oxide. This step is accomplished through the following chemical reaction:

$$\underset{\text{iron ore}}{4FeS_2\ (s)} + 11O_2\ (g) \rightarrow \underset{\text{iron oxide}}{2Fe_2O_3\ (s)} + 8SO_2\ (g)$$

If an iron manufacturer starts with 2,512.0 moles of iron ore, how many moles of oxygen will be needed to react completely with the iron ore? How many moles of iron oxide will be made?

to make a certain amount of product. For example, suppose you were told to go into a laboratory and make 2 moles of $FeCl_3$ (a chemical used in electronics manufacturing) using the following reaction:

$$2FeTiO_3\ (s) + 4HCl\ (l) + Cl_2\ (g) \rightarrow 2FeCl_3\ (aq) + 2TiO_2\ (s) + 2H_2O\ (l)$$

How many moles of each reactant would you need to start with? If you knew you had to end up with 2 moles of $FeCl_3$, then you should be able to use the chemical equation to determine how much starting material you would need.

After all, the equation, as it is written, tells us that 2 moles of $FeTiO_3$ react with 4 moles of HCl and 1 mole of Cl_2 to make 2 moles of $FeCl_3$, 2 moles of TiO_2, and 2 moles of water. You know the recipe for making 2 moles of $FeCl_3$. Therefore, you can conclude that the minimum amount of each reactant needed is 2 moles of $FeTiO_3$, 4 moles of HCl, and 1 mole of Cl_2. If only 1 mole of HCl were used, then the HCl would run out long before the other reactants, limiting the amount of product you would make. If, on the other hand, 5 moles of HCl were used, it would not affect the amount of product made because the other 2 reactants would run out, leaving excess HCl.

Another way we can use chemical equations is to relate the amount of product we desire to the minimum quantities of reactants we would need. Let's see how this applies to a more complicated problem in Example 8.3.

EXAMPLE 8.3

A chemist wants to make ether ($C_4H_{10}O$) for a medical anesthetic. He uses the following reaction:

$$2C_2H_6O\ (l) + H_2SO_4 (l) \rightarrow C_4H_{10}O\ (g) + H_4SO_5\ (aq)$$

If the chemist needs 134.9 moles of ether, what is the minimum number of moles of C_2H_6O and H_2SO_4 that he needs?

In this problem, we are simply trying to relate the quantity of a product to the minimum quantities of each reactant needed. As usual, all we have to do is develop relationships between the quantity we know and the quantities we want to know. The chemical equation tells us:

$$2 \text{ moles } C_2H_6O = 1 \text{ mole } C_4H_{10}O$$

We can use this relationship to convert between moles of ether desired and moles of C_2H_6O required:

$$\frac{134.9 \cancel{\text{moles } C_4H_{10}O}}{1} \times \frac{2 \text{ moles } C_2H_6O}{1 \cancel{\text{mole } C_4H_{10}O}} = 269.8 \text{ moles } C_2H_6O$$

To determine how much H_2SO_4 is needed, we use the relationship between ether and H_2SO_4 as given in the equation:

$$1 \text{ mole } H_2SO_4 = 1 \text{ mole } C_4H_{10}O$$

Therefore:

$$\frac{134.9\ \cancel{\text{moles } C_4H_{10}O}}{1} \times \frac{1\ \text{moles } H_2SO_4}{1\ \cancel{\text{mole } C_4H_{10}O}} = 134.9\ \text{moles } H_2SO_4$$

To make 134.9 moles of ether, the chemist must have a minimum of **269.8 moles of C_2H_6O and 134.9 moles of H_2SO_4**. Note that we could also use the chemical equation to determine how many moles of H_4SO_5 would be produced. The problem did not ask us to determine this, so we won't.

See if you can do a similar problem in "On Your Own" question 8.3.

ON YOUR OWN

8.3 Citric acid ($C_6H_8O_7$), a component of fruit drinks, jams, and jellies, is produced using the following reaction:

$$C_{12}H_{22}O_{11} + H_2O + 3O_2 \rightarrow 2C_6H_8O_7 + 4H_2O$$

If a fruit drink manufacturer decides she needs 2.8×10^5 moles of citric acid, what is the minimum amount of reactants she will need?

USING CHEMICAL EQUATIONS WHEN THE LIMITING REACTANT IS IDENTIFIED

One other way that chemists use stoichiometry is to determine the amount of products made when the limiting reactant is identified. For example, if a chemist wants to make silver chloride (a substance used in photography), he might use the following reaction:

$$Ag_2CO_3\ (s) + 2HCl\ (aq) \rightarrow 2AgCl\ (s) + H_2O\ (l) + CO_2\ (g)$$

If the chemist were to buy the 2 reactants in the equation (HCl and Ag_2CO_3), he would notice a huge difference in price. While the HCl is relatively cheap, the Ag_2CO_3 is incredibly expensive. To make sure that he doesn't waste any money, he might want to use a lot of extra HCl to be confident that all of the Ag_2CO_3 reacted fully. After all, he wouldn't be too concerned if there were some leftover HCl to throw away since it is relatively cheap. However, if he didn't use enough HCl and as a result had leftover Ag_2CO_3 to throw away, that would be a big waste of money.

When a chemist adds lots of extra reactant to make sure that the other reactant is the limiting reactant, we say that he is adding excess reactant. In our example above, the chemist would make sure that Ag_2CO_3 is the limiting reactant by adding excess HCl. Since excess HCl is used, all of the Ag_2CO_3 is certain to react, making it the limiting reactant. Once we know that Ag_2CO_3 is the limiting reactant, we can use its number of moles to calculate the number of moles of each product.

In the above reaction, if the chemist bought 1 mole of Ag_2CO_3, he would want to buy more than 2 moles of HCl. Since the equation says that 2 moles of HCl are required to react with 1 mole of Ag_2CO_3, using more than 2 moles (2.5 moles, for example) would ensure that there was excess HCl, making Ag_2CO_3 the limiting reactant. Since the equation says that 1 mole of Ag_2CO_3 makes 2 moles of AgCl, the chemist will make 2 moles of AgCl when 1 mole of Ag_2CO_3 is reacted with excess HCl. If you are a little confused here, follow Example 8.4 closely. It should help clear things up.

EXAMPLE 8.4

Some ores contain a small amount of precious metals that are in compounds with many other elements. For example, one common ore found on earth is $KAgC_2N_2$. Since silver (Ag) is a precious metal, it is useful to be able to extract silver from this ore. The following reaction accomplishes this task:

$$2KAgC_2N_2\text{ (aq)} + Zn\text{ (s)} \rightarrow 2Ag\text{ (s)} + ZnC_2N_2\text{ (aq)} + 2KCN\text{ (aq)}$$

If a chemist adds 5.61 moles of $KAgC_2N_2$ to excess Zn, how much pure silver (Ag) can be made?

Since excess Zn is added, $KAgC_2N_2$ is the limiting reactant. Since we already learned that the limiting reactant determines the amount of all products formed, all we need to do is to develop a relationship between $KAgC_2N_2$ and Ag and then use that relationship to convert from one to the other.

According to the chemical equation:

$$2\text{ moles } KAgC_2N_2 = 2\text{ moles Ag}$$

Using this relationship to convert:

$$\frac{5.61 \cancel{\text{ moles } KAgC_2N_2}}{1} \times \frac{2\text{ moles Ag}}{2 \cancel{\text{ moles } KAgC_2N_2}} = 5.61\text{ moles Ag}$$

This tells us that **5.61 moles of Ag** would be made. Realize that we could do the same type of conversion to determine how much KCN and ZnC_2N_2 were produced because the number of moles of the limiting reactant can be converted into the number of moles of all products.

ON YOUR OWN

8.4 Calcium chloride 6–hydrate ($CaCl_2H_{12}O_6$) is a substance used to melt snow on streets and sidewalks. It is made from calcium carbonate ($CaCO_3$) using the following reaction:

$$CaCO_3\text{ (s)} + 2HCl\text{ (aq)} + 5H_2O\text{ (l)} \rightarrow CaCl_2H_{12}O_6\text{ (s)} + CO_2\text{ (g)}$$

If 7.20 moles of HCl are added to excess $CaCO_3$ and excess water, how many moles of calcium chloride 6–hydrate will be made?

Now try "On Your Own" question 8.4.

We hope you are beginning to see that there is really very little difference among all of the problems we have discussed so far. In each case, we are using the relationship given by the chemical equation to convert the number of moles of one substance into the number of moles of another substance. The only difference among these problems is what we are trying to learn by doing this. Sometimes we are trying to determine how much reactant is needed to react fully with another reactant, sometimes we are trying to see how much product would be formed, and sometimes we are trying to determine how much starting material we would need to end up with a certain amount of product. In each case, we use the same mathematical tools to determine the answer.

This is the way stoichiometry works. The numbers that appear to the left of the substances in a chemical equation are used to develop a conversion relationship between the substance with a quantity we know and the substance with a quantity we do not know. If done properly, the number of moles of any one substance in the equation can be converted into the number of moles of any other substance in the equation.

Since the numbers that appear to the left of each substance in the equation are so important in stoichiometry, we give them a name. We call these numbers stoichiometric (stoy' kee uh meh' trik) coefficients, and they provide a way of converting from the number of moles of one substance in the equation to the number of moles of any other substance in the equation.

If you purchased the MicroChem kit mentioned in the Student Notes, you can perform Experiment 4 to get some more experience with stoichiometry and limiting reactants.

VOLUME RELATIONSHIPS FOR GASES IN CHEMICAL EQUATIONS

The stoichiometric coefficients in a chemical equation do not relate just the number of moles of each substance in a chemical equation. Quite a while ago, a French chemist named Joseph Louis Gay–Lussac determined that these coefficients also relate the *volumes of gases* in a chemical equation. For example, consider the formation of ammonia (NH_3) according to the following equation:

$$3H_2\ (g) + N_2\ (g) \rightarrow 2NH_3\ (g)$$

Three liters of hydrogen gas react with 1 liter of nitrogen gas to form 2 liters of ammonia gas. Since these numbers correspond exactly to the stoichiometric coefficients of each substance in the equation, Gay–Lussac hypothesized that the stoichiometric coefficients relate the volumes (remember, liter is a volume unit) of the gases in a chemical equation as well as the number of moles.

Gay–Lussac performed many experiments on many different reactions to confirm his hypothesis. In the end, it became generally accepted as a scientific principle and is often called Gay–Lussac's law:

Gay–Lussac's law—The stoichiometric coefficients in a chemical equation relate the volumes of gases in the equation as well as the number of moles of substances in the equation.

We will use Gay–Lussac's law in solving problems as well, but remember this important fact:

Gay–Lussac's law can be used in stoichiometry only if *the substances being compared in the problem are gases.*

For example, consider the following reaction:

$$C_2H_2\ (g) + 2H_2\ (g) \rightarrow C_2H_6\ (g)$$

In this reaction, we can say that 1 mole of C_2H_2 reacts with 2 moles of H_2 to make 1 mole of C_2H_6. In addition, we could use Gay–Lussac's law to say that 1 liter of C_2H_2 gas reacts with 2 liters of hydrogen gas to make 1 liter of C_2H_6 gas. We can use Gay–Lussac's law in this case because every substance that we mentioned is a gas.

On the other hand, consider the following equation:

$$CaO\ (s) + 2HCl\ (g) \rightarrow CaCl_2\ (s) + H_2O\ (g)$$

In this reaction, we would say that 1 mole of CaO reacts with 2 moles of HCl to make 1 mole of $CaCl_2$ and 1 mole of H_2O. The most we can say using Gay–Lussac's law is that 2 liters of HCl gas will make 1 liter of H_2O gas because Gay–Lussac's law can be used only on gases. If we know that excess CaO was used, we could convert the number of liters of HCl to the number of liters of H_2O. We could say nothing about the number of liters of $CaCl_2$ made because $CaCl_2$ is not a gas.

When using Gay–Lussac's law, we must make certain that each substance we use in our conversion is a gas. If not, then we cannot use Gay–Lussac's law and we cannot relate the volumes of substances in the chemical equation. See if you can follow this reasoning in Example 8.5.

EXAMPLE 8.5

An automobile engine burns gaseous octane (C_8H_{18}). If an automobile completely combusts 12 liters of gaseous octane in excess oxygen, how many liters of carbon dioxide will be produced?

First, we must write a balanced chemical equation for the complete combustion of octane. According to our definition of complete combustion, the reaction would be:

$$C_8H_{18}\ (g) + O_2\ (g) \rightarrow CO_2\ (g) + H_2O\ (g)$$

Balancing the equation gives us:

$$2C_8H_{18}\ (g) + 25O_2\ (g) \rightarrow 16CO_2\ (g) + 18H_2O\ (g)$$

Since the substances we are interested in (octane and carbon dioxide) are both gases, we can use Gay–Lussac's law to relate their volumes:

$$2 \text{ liters of } C_8H_{18} = 16 \text{ liters } CO_2$$

Now we can use this relationship to convert:

$$\frac{12\ \cancel{\text{liters } C_8H_{18}}}{1} \times \frac{16 \text{ liters } CO_2}{2\ \cancel{\text{liters } C_8H_{18}}} = 96 \text{ liters } CO_2$$

So we find out that burning 12 liters of octane in excess oxygen will make **96 liters of carbon dioxide**.

We could just as easily have used moles in this problem. If the problem had started by giving us the number of moles of octane burned, then we could have calculated the number of moles of carbon dioxide produced. What Gay–Lussac's law allows us to do is *also* answer the problem if liters (or any other volume unit) are used instead of moles, so long as all the substances we are interested in are gases.

There is one more thing you must realize. Even though Gay–Lussac's law is a useful tool, there are many things it cannot do. One thing it cannot do is convert from moles to liters. For example, if you see a problem that gives the quantity of a substance in moles, Gay–Lussac's law is not capable of converting it to liters. Although there is a way to accomplish this task, we will not discuss it for a while. Gay–Lussac's law can be used only to convert liters of one gas in a chemical equation to liters of another gas in that chemical equation. Make sure you understand this by answering "On Your Own" question 8.5.

ON YOUR OWN

8.5 A chemist decides to produce carbon dioxide gas using the following reaction:

Na_2CO_3 (s) + 2HCl (g) →
2NaCl (s) + CO_2 (g) + H_2O (g)

If the chemist wants to make 8.20 liters of CO_2 gas, how many liters of HCl gas must be used?

MASS RELATIONSHIPS IN CHEMICAL EQUATIONS

What we have learned so far is that the stoichiometric coefficients in chemical equations relate the number of moles of each substance in that chemical equation. If the substances are gases, the stoichiometric coefficients can also relate their volumes. This makes stoichiometry a powerful tool in chemistry.

Unfortunately, we rarely measure the quantity of substances in moles. Usually, when asked to determine how much of a certain substance we have, the easiest thing to do is to measure the substance's mass. In other words, we usually report the quantity of a substance in grams, not moles. To make stoichiometry really useful, we have to determine a way to calculate relationships between the masses of substances in a chemical equation.

For example, suppose we were asked to prepare 50.0 grams of CuI with the following chemical reaction:

$$2CuCl_2 + 4KI \rightarrow 2CuI + 4KCl + I_2$$

How would we go about determining the amount of $CuCl_2$ and KI that we need to accomplish our goal? We couldn't do it exactly the way we have done previously because we can use stoichiometry only to relate the number of *moles* of product to the number of *moles* of reactant. In this case, we were not given the number of moles of product. We were given the mass of the product. So what should we do?

In the previous module, we learned how to convert from grams to moles. If we did that, then we would have the number of moles of product we need, and that could then be used to determine the number of moles of reactants we need. When we had accomplished that, we would probably still not be done. We cannot purchase chemicals by the mole.

Instead, chemicals are purchased by the gram. If we wanted to order the necessary quantities of reactants, we would have to determine how many grams of each are required. Once again, in the previous module we learned how to convert from moles to grams. Once we determined how many moles of each reactant we needed, we could then convert to grams. See how this works in Example 8.6.

EXAMPLE 8.6

A chemist needs to produce 100.0 grams of $GaCl_3$ using the following reaction:

$$2Ga + 6HCl \rightarrow 2GaCl_3 + 3H_2$$

How many grams of Ga and HCl are necessary to accomplish this?

The problem asks us to relate the quantity of a product to the required quantities of the reactants. If the quantity of $GaCl_3$ were given in moles, we could use the techniques we have already learned to start solving the problem, much like we solved Example 8.3. Unfortunately, the amount of $GaCl_3$ was not given in moles; it was given in grams. To fix this problem, the first thing we need to do is convert from grams of $GaCl_3$ to moles of $GaCl_3$, as we learned in the previous module:

$$\text{mass of } GaCl_3 = 69.7 \text{ amu} + 3 \times 35.5 \text{ amu} = 176.2 \text{ amu}$$

Remember, when adding and subtracting, we do not count significant figures; we look at decimal place. Both of these masses have their last significant figure in the tenths place, so the answer should have its last significant figure in the tenths place. As you should recall from the previous module, the mass also tells us:

$$1 \text{ mole } GaCl_3 = 176.2 \text{ grams } GaCl_3$$

We can use this relationship to convert from grams of $GaCl_3$ to moles of $GaCl_3$:

$$\frac{100.0 \cancel{\text{grams } GaCl_3}}{1} \times \frac{1 \text{ mole } GaCl_3}{176.2 \cancel{\text{grams } GaCl_3}} = 0.5675 \text{ moles } GaCl_3$$

Now that we have the number of moles of $GaCl_3$, we can use the chemical equation to determine how many moles of each reactant we need. According to the chemical equation:

$$2 \text{ moles Ga} = 2 \text{ moles } GaCl_3$$

$$6 \text{ moles HCl} = 2 \text{ moles } GaCl_3$$

Using these relationships to convert:

$$\frac{0.5675 \cancel{\text{moles } GaCl_3}}{1} \times \frac{2 \text{ moles Ga}}{2 \cancel{\text{moles } GaCl_3}} = 0.5675 \text{ moles Ga}$$

$$\frac{0.5675 \cancel{\text{moles } GaCl_3}}{1} \times \frac{6 \text{ moles HCl}}{2 \cancel{\text{moles } GaCl_3}} = 1.703 \text{ moles HCl}$$

As we learned before, this tells us the minimum quantity of reactants needed to make 100.0 grams of $GaCl_3$. We aren't done yet because the problem asked us to calculate the number of grams of each reactant. Since we have the number of moles already, we can use the skills we learned in Module 7 to convert this answer to grams.

Let's start with Ga:

$$\text{mass of Ga} = 69.7 \text{ amu}$$

This means:

$$1 \text{ mole Ga} = 69.7 \text{ grams Ga}$$

Therefore:

$$\frac{0.5675 \cancel{\text{moles Ga}}}{1} \times \frac{69.7 \text{ grams Ga}}{1 \cancel{\text{mole Ga}}} = 39.6 \text{ grams Ga}$$

Now we can move on to HCl:

$$\text{mass of HCl} = 1.01 \text{ amu} + 35.5 \text{ amu} = 36.5 \text{ amu}$$

Again, this means:

$$1 \text{ mole HCl} = 36.5 \text{ grams HCl}$$

Therefore:

$$\frac{1.703 \cancel{\text{moles HCl}}}{1} \times \frac{36.5 \text{ grams HCl}}{1 \cancel{\text{mole HCl}}} = 62.2 \text{ grams HCl}$$

So we learn from this complicated process that to make 100.0 grams of $GaCl_3$, **we must start with a minimum of 39.6 grams of Ga and 62.2 grams of HCl.**

The previous example may have seemed a little long and complicated at first, but as you examine the steps we took, it should begin to make sense. After all, we know that stoichiometry allows us to relate the number of moles of one substance in a chemical equation to the number of moles of another substance in the chemical equation. To use stoichiometry, *we must have our quantities in moles*. Therefore, if a problem gives us the amount of a substance in grams, we must first convert to moles before we can use stoichiometry.

Once we have the substance's quantity in moles, we can use the chemical equation and its stoichiometric coefficients to convert the number of moles of that substance into the number of moles of any other substance in the chemical equation. If the problem asks us to put our answer into grams, then we must convert from the moles we just calculated back into grams.

In case you are having a little trouble picturing all of this in your mind, let's illustrate the procedure. Consider Figure 8.1.

FIGURE 8.1

Mass Relationships in Chemical Equations

Balance illustration by Megan Whitaker / Mole illustration from MasterClips collection

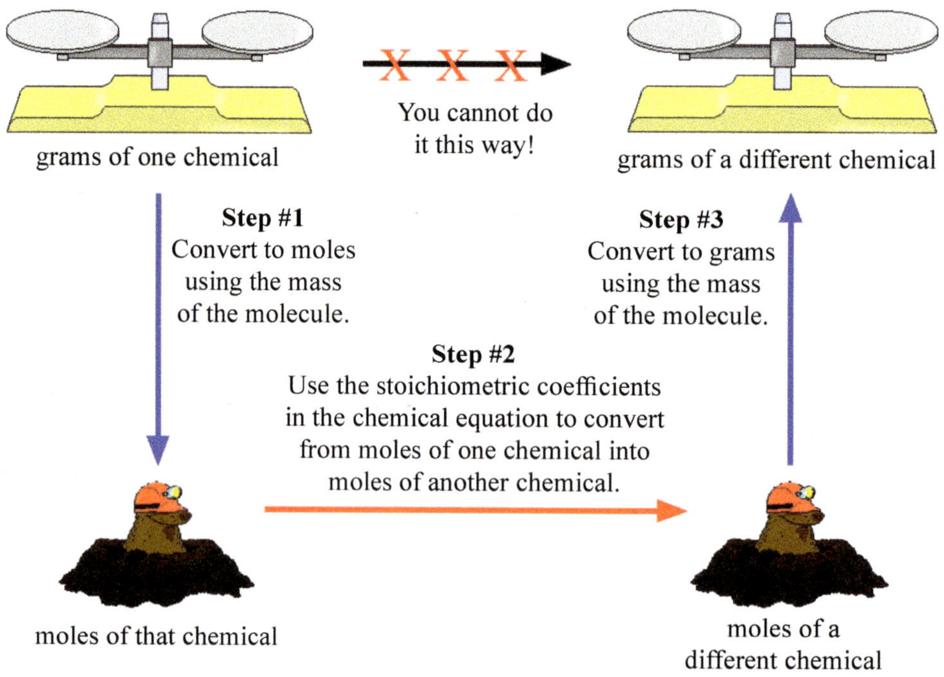

In this figure, the balance represents the mass of a chemical, while the cartoon mole represents (what else?) the number of moles of a chemical. Since we can measure the mass of a sample pretty easily, we would like to relate the mass of one chemical in a chemical equation to the mass of another chemical in that same chemical equation. However, we cannot just convert from the mass of one chemical into the mass of another. There is nothing that will give us a conversion relationship between the masses of the 2 different chemicals. However, we *do* have something that will give us the conversion relationship in *moles*. The chemical equation does that. So, to determine the mass relationship between one chemical and another in a chemical equation, we must *convert to moles*.

We do this by starting with the chemical whose mass is given. Using the chemical formula, we can calculate the mass of a single molecule, and that tells us how many grams it takes to make a mole. We can use that information to convert from grams into moles. Once we have the amount of the chemical in *moles*, we can use the chemical equation. The stoichiometric coeffecients in the chemical equation give us a conversion relationship between moles of one chemical in the equation and moles of any other chemical in the equation. We can use that information to convert from moles of the given chemical to moles of any other chemical in the chemical equation.

Of course, knowing the number of moles of the other chemical is nice, but in the laboratory, it is hard to measure the number of moles in a sample. We usually need to convert moles back into grams. We can do that, of course, by using the chemical formula to calculate the mass of a single molecule. That tells us the number of grams it takes to make a mole, which will allow us to convert from moles back into grams.

To relate the masses of chemicals in a chemical equation, we must go via an indirect route. Since the chemical equation gives us the conversion relationship *in moles*, we must first convert from grams to moles. Then we can use the chemical equation to convert from one chemical into another. At that point, we have the moles of the new chemical, not the mass. Therefore, we must then convert from moles back into grams. we can relate the masses of chemicals in a chemical equation only in this roundabout way. We want to really make sure that you have this idea down pat, so read through Example 8.7.

EXAMPLE 8.7

The human body produces the energy it requires to live by burning simple sugars such as $C_6H_{12}O_6$ according to the following chemical equation:

$$C_6H_{12}O_6 + 6O_2 \rightarrow 6CO_2 + 6H_2O$$

This reaction produces a lot of heat, which is used to power all of the body's motion. When the sugar is burned, the CO_2 produced is exhaled, while the H_2O produced is mostly eliminated through urine. If the average adult human burns about 1.2 kg of simple sugars each day in excess oxygen, how many grams of water does the average human eliminate every day as a result of this process?

This problem asks us to relate the quantity of $C_6H_{12}O_6$ burned in the body to the quantity of H_2O produced. It also tells us that the O_2 is in excess, indicating that $C_6H_{12}O_6$ is the limiting reactant. All we need to do is convert from the quantity of the limiting reactant to the quantity of product. Unfortunately, we can do this only if our quantities are in moles. So the first thing we must do is convert the quantity of $C_6H_{12}O_6$ from kilograms to moles. To do this, we must first convert 1.2 kilograms into 1.2×10^3 grams. Then, we can convert to moles:

$$\text{mass of } C_6H_{12}O_6 = 6 \times 12.0 \text{ amu} + 12 \times 1.01 \text{ amu} + 6 \times 16.0 \text{ amu} = 180.1 \text{ amu}$$

This means:

$$1 \text{ mole } C_6H_{12}O_6 = 180.1 \text{ grams } C_6H_{12}O_6$$

Which allows us to convert to moles :

$$\frac{1.2 \times 10^3 \cancel{\text{grams } C_6H_{12}O_6}}{1} \times \frac{1 \text{ mole } C_6H_{12}O_6}{180.1 \cancel{\text{grams } C_6H_{12}O_6}} = 6.7 \text{ moles } C_6H_{12}O_6$$

Now that we have moles of $C_6H_{12}O_6$, we can use stoichiometry to determine how much water will be produced:

$$1 \text{ mole } C_6H_{12}O_6 = 6 \text{ moles } H_2O$$

$$\frac{6.7 \cancel{\text{moles } C_6H_{12}O_6}}{1} \times \frac{6 \text{ moles } H_2O}{1 \cancel{\text{mole } C_6H_{12}O_6}} = 4.0 \times 10^1 \text{ moles } H_2O$$

So we now know that burning 1.2 kg of $C_6H_{12}O_6$ makes 4.0×10^1 moles of H_2O. Note that since the number of moles must have 2 significant figures, the answer must be expressed in scientific notation. This is not the final answer because the problem asks how many grams of water are produced, not how many moles. So now we have to convert from moles to grams:

$$\text{mass of } H_2O = 2 \times 1.01 \text{ amu} + 16.0 \text{ amu} = 18.0 \text{ amu}$$

$$1 \text{ mole } H_2O = 18.0 \text{ grams } H_2O$$

$$\frac{4 \times 10^1 \cancel{\text{moles } H_2O}}{1} \times \frac{18.0 \text{ grams } H_2O}{1 \cancel{\text{mole } H_2O}} = 7.2 \times 10^2 \text{ grams } H_2O$$

In the end, we find out that the average human eliminates **7.2×10^2 grams of water** each day as a result of burning 1.2 kg of simple sugars.

Toxic acid spills, which occur regularly in industrial settings, are usually cleaned up with a substance known as barium hydroxide, $Ba(OH)_2$. For example, if hydrochloric acid (HCl, a toxic acid) is spilled, barium hydroxide neutralizes it with the following reaction:

$$2HCl\ (l) + Ba(OH)_2\ (s) \rightarrow BaCl_2\ (aq) + 2H_2O\ (l)$$

The products of this reaction are not nearly as toxic as the reactants, so as long as enough barium hydroxide is used so that all of the HCl reacts, a spill that is potentially very toxic gets neutralized. If 540.0 grams of HCl are spilled, what is the minimum number of grams of barium hydroxide necessary to completely neutralize the spill?

This problem simply gives us the mass of one reactant and asks what mass of the other reactant is necessary for a complete reaction. Once again, we can use stoichiometry to figure this out, but only if the quantity of reactant is given in moles. Our first job, then, is to convert from grams of HCl to moles of HCl:

$$\text{mass of HCl} = 1.01 \text{ amu} + 35.5 \text{ amu} = 36.5 \text{ amu}$$

$$1 \text{ mole HCl} = 36.5 \text{ grams HCl}$$

$$\frac{540.0 \cancel{\text{grams HCl}}}{1} \times \frac{1 \text{ mole HCl}}{36.5 \cancel{\text{grams HCl}}} = 14.8 \text{ moles HCl}$$

Now that we have the quantity of HCl in moles, we can do stoichiometry:

$$2 \text{ moles HCl} = 1 \text{ mole } Ba(OH)_2$$

$$\frac{14.8 \cancel{\text{moles HCl}}}{1} \times \frac{1 \text{ mole } Ba(OH)_2}{2 \cancel{\text{moles HCl}}} = 7.40 \text{ moles } Ba(OH)_2$$

This tells us how much barium hydroxide is needed. However, this problem specifically asks how many grams of barium hydroxide are needed, so now we have to convert from moles to grams. Remember, since there are parentheses in this chemical formula, we must multiply the 2 after the parentheses with every atom inside the parentheses. This molecule contains 1 barium atom, but 2 oxygen atoms and 2 hydrogen atoms.

$$\text{mass of Ba(OH)}_2 = 137.3\text{ amu} + 2 \times 16.0\text{ amu} + 2 \times 1.01\text{ amu} = 171.3\text{ amu}$$

$$1\text{ mole Ba(OH)}_2 = 171.3\text{ grams Ba(OH)}_2$$

$$\frac{7.40\ \cancel{\text{moles Ba(OH)}_2}}{1} \times \frac{171.3\text{ grams Ba(OH)}_2}{1\ \cancel{\text{mole Ba(OH)}_2}} = 1.27 \times 10^3\text{ grams Ba(OH)}_2$$

This problem shows us that to clean up 540.0 grams of HCl, at least **1.27×10^3 grams of barium hydroxide** are necessary.

Problems such as these are the toughest stoichiometry problems that you will be asked to perform in this course. If you understand the examples, all you need to do is cement your stoichiometry skills by answering "On Your Own" questions 8.6–8.8. If you are not comfortable with these types of problems, you need to explore stoichiometry further with another resource. Check the course website to see if one of the links related to this module explains stoichiometry in a way that is easier for you to understand. If you need more practice solving problems like this, you can find more problems in appendix B.

It is extremely important for you to understand stoichiometry so that you can solve problems like these. Since all of chemistry is built upon developing relationships between substances in chemical equations, you could say that all of chemistry is based on stoichiometry! Every college chemistry course stresses this subject for that very reason, so it is important for you to learn it now.

ON YOUR OWN

8.6 The insecticide DDT ($C_{14}H_9Cl_5$), which is now banned because of its deleterious environmental effects, was manufactured with this chemical reaction:

$$2C_6H_5Cl + C_2HCl_3O \rightarrow C_{14}H_9Cl_5 + H_2O$$

If 500.0 grams of C_6H_5Cl are reacted with excess C_2HCl_3O, how many grams of DDT will be produced?

8.7 One type of rocket engine uses the following reaction to develop its thrust:

$$7H_2O_2\ (l) + N_2H_4\ (s) \rightarrow 2HNO_3\ (aq) + 8H_2O\ (l)$$

If a rocket engine has 1,120 kg of N_2H_4 in it, what is the minimum mass of H_2O_2 necessary to allow the N_2H_4 to react completely?

8.8 Welding torches burn a chemical called acetylene (C_2H_2), which is made with the following reaction:

$$CaC_2 + 2H_2O \rightarrow Ca(OH)_2 + C_2H_2$$

How many grams of water are necessary to make at least 250.0 grams of acetylene?

USING STOICHIOMETRY TO DETERMINE CHEMICAL FORMULAS

In Module 2 you were introduced to the use of chemical formulas, and since that time you have become very familiar with them. You can use them to name compounds, figure out the mass of a compound, count the molecules in a compound, and perform stoichiometric calculations. Have you ever wondered how we got these chemical formulas?

After all, when we say that the chemical formula for water is H_2O, we are saying that

2 hydrogen atoms and 1 oxygen atom are linked together to make a water molecule. How do we know that? As we said in a previous module, it's really not possible to see atoms, so there is no way that we could look at a water molecule and point out the 2 hydrogen atoms and the 1 oxygen atom. So how do we know that they are there?

Remember in Module 7 when we measured the width of a molecule without ever seeing it? We used the mole concept to count the number of molecules in a circle, and we used the size of the circle to determine the width of the molecules. In other words, we measured the width of a molecule *indirectly*. That same basic principle has guided the way in which chemists have determined chemical formulas over the years.

When a compound (molecule) is decomposed, all of the elements (atoms) in that compound are separated. Once they are separated, their mass can be measured. Using the mole concept, the mass allows us to calculate how many atoms of each element existed in the compound. Once we have counted all of the atoms in the compound, we have essentially determined its chemical formula. In a nutshell, that's how we can determine a molecule's chemical formula without ever seeing the atoms that make it up. If this all sounds a little confusing, don't worry about it. The whole concept will get a lot easier when you see Example 8.8.

EXAMPLE 8.8

A chemist decomposes an unknown compound. The products of this decomposition are 12.00 grams of carbon and 4.04 grams of hydrogen. What is the chemical formula of the unknown compound?

In looking at the information given, the first thing we should notice is that the unknown compound contains only carbon and hydrogen. If it contained other types of atoms, there would have been more products in the decomposition. This particular compound includes only C and H, so its chemical formula must be C_xH_y, where x and y are numbers that we must determine. The decomposition reaction, then, could be written:

$$C_xH_y \rightarrow C + H_2$$

Remember, we don't know x and y, but we do know that elemental carbon has a chemical formula of C and that hydrogen (a homonuclear diatomic) has a formula of H_2. Regardless of what values exist for x and y, the products must be C and H_2.

For this to be a proper equation for a decomposition reaction, it must be balanced. The problem is that since we do not have x or y, balancing will be a little difficult. That's where the experimental data comes in. The problem says that 12.00 grams of carbon were produced. The stoichiometric coefficients in a chemical formula simply represent the number of moles, so let's convert the grams of carbon into moles of carbon:

mass of C = 12.0 amu

1 mole C = 12.0 g C

$$\frac{12.00 \cancel{\text{g C}}}{1} \times \frac{1 \text{ mole C}}{12.0 \cancel{\text{g C}}} = 1.00 \text{ moles C}$$

Since the stoichiometric coefficients simply represent the number of moles of the substance they multiply, and we also know that 1 mole of carbon was produced in this decomposition, we could say that the stoichiometric coefficient for C is 1.

In the same way, we could calculate the number of moles of hydrogen produced:

$$\text{mass of } H_2 = 2 \times 1.01 \text{ amu} = 2.02 \text{ amu}$$

$$1 \text{ mole } H_2 = 2.02 \text{ g } H_2$$

$$\frac{4.04 \cancel{\text{g } H_2}}{1} \times \frac{1 \text{ mole } H_2}{2.02 \cancel{\text{g } H_2}} = 2.00 \text{ moles } H_2$$

Once again, since the stoichiometric coefficients really just represent the number of moles, and since we know that there were 2.00 moles of H_2 produced, the stoichiometric coefficient for H_2 is 2. Therefore, our chemical equation becomes:

$$C_xH_y \rightarrow C + 2H_2$$

Now the problem of figuring out x and y is a snap! For the equation to balance, there must be 1 C on both sides of the equation. Therefore, x = 1. For the equation to balance, there must be 4 H atoms on both sides of the equation, making y = 4. Therefore, the equation is:

$$CH_4 \rightarrow C + 2H_2$$

The chemical formula of the unknown compound is **CH_4**.

Let's review what just happened so that we understand why this method works. First, a decomposition reaction is carried out that separates a compound into its constituent elements. This tells us what atoms go in the chemical formula and allows us to write an unbalanced chemical equation for the decomposition. The masses of each individual element then allow us to calculate the moles of each element, and that tells us the stoichiometric coefficients that go next to each element on the products side of the equation. Once we have that, then we can figure out the chemical formula necessary to make the equation balance.

EMPIRICAL AND MOLECULAR FORMULAS

Before you start thinking that this is all a little too easy, there are 2 wrinkles that get thrown into this. First, there is an inherent assumption we make in this type of experiment which you probably didn't notice. The assumption was that the stoichiometric coefficient next to the C_xH_y on the reactants side of the equation is a 1. After all, we said the only way we could balance the equation was if x = 1 and y = 4, but that is simply not true. There are other possibilities. Consider the following equations:

$$\tfrac{1}{2}C_2H_8 \rightarrow C + 2H_2$$

$$\tfrac{1}{3}C_3H_{12} \rightarrow C + 2H_2$$

$$\frac{1}{4}C_4H_{16} \rightarrow C + 2H_2$$

Each of these equations is balanced, so any of the chemical formulas listed on the reactants side of the equation represents a *possible* chemical formula for our unknown compound. Of course we would have to get rid of the fractions to make these *proper* chemical equations, but nevertheless, they do balance. So the example we just did doesn't really measure the exact chemical formula of the compound. It just determines that the compound must have a formula like CH_4, C_2H_8, C_3H_{12}, C_4H_{16}, and so on.

Notice that these chemical formulas are all multiples of each other. Each one of them has 1 C atom for every 4 H atoms. Therefore, our experiment did help us learn something about the chemical formula. It told us that the chemical formula of this compound must have 1 C atom for every 4 H atoms. This rules out a host of other possibilities and leads us to only a few possible formulas, which is more than we knew in the beginning.

This leads us to developing a distinction between chemical formulas. When we know exactly how many atoms of each type exist in a molecule, we will say that we know the molecular formula of the compound. On the other hand, when we only know the ratio of atoms in a compound (1 C atom for every 4 H atoms), we call that an empirical (im peer' uh kul) formula. Just to make sure you understand the distinction, we will write out formal definitions.

Molecular formula—A chemical formula that provides the number of each type of atom in a molecule.

Empirical formula—A chemical formula that provides a simple, whole–number ratio for the atoms in a molecule.

For example, if we tell you that the molecular formula for benzene is C_6H_6, you know that there are 6 C atoms and 6 H atoms in each molecule of benzene. However, instead of telling you its molecular formula, we could just tell you that benzene's empirical formula is CH (1 C for every 1 H). This would tell you that a benzene molecule has 1 C atom and 1 H atom *or* 2 C atoms and 2 H atoms *or* 3 C atoms and 3 H atoms *or* 4 C atoms and 4 H atoms *or* 5 C atoms and 5 H atoms *or* 6 C atoms and 6 H atoms, and so on. Study Example 8.9 and answer "On Your Own" question 8.9 to make sure you understand the difference between empirical and molecular formulas.

EXAMPLE 8.9

The molecular formulas of 3 compounds are listed below. Determine the empirical formula for each compound:

a. C_4H_{12} b. $Na_2S_2O_8$ c. $K_2Cr_2O_7$

The empirical formula of a compound represents the simplest, whole–number ratio of the atoms in the molecule. To get the simplest, whole–number ratio, we must divide the molecular formulas by any common factor that exists between the subscripts:

(a) The numbers 4 and 12 have a common factor of 4; we divide both subscripts by 4 and get **CH_3**. This represents the simplest, whole–number ratio of C atoms and H atoms in the molecule.

(b) The numbers 2, 2, and 8 have a common factor of 2, so we must divide all subscripts by 2 and get **$NaSO_4$**.

(c) The numbers 2, 2, and 7 have no common factors, so **$K_2Cr_2O_7$ is already an empirical formula**. This particular example shows that empirical formulas can be molecular formulas.

ON YOUR OWN

8.9 Which of the following molecular formulas are also empirical formulas? If the formula is not an empirical formula, write the empirical formula for that molecule.

a. Al_6O_9 b. PbO_2 c. $MgCO_3$
d. $C_8H_{18}O_2$ e. $Si_3H_9O_2$

Experiments such as the one in Example 8.8 give only empirical formulas, not molecular formulas. Even though empirical formulas do not contain as much detailed information about a molecule as do molecular formulas, they are better than nothing. It is possible to convert from a molecule's empirical formula to its molecular formula, if we happen to know the molecule's mass. Think about it: To go from a molecular formula to an empirical formula, what do we have to do? We have to find the common factor and divide by it. For example, in Example 8.9a, we noted that the common factor in the molecular formula was 4. When we divided by 4, we turned the molecular formula into an empirical formula.

If we want to turn an empirical formula into a molecular formula, what will we have to do? We will have to *multiply* by the common factor. How do we know what that common factor is? We can determine it from the molecule's mass. Example 8.10 shows how to do that.

EXAMPLE 8.10

A compound's empirical formula is determined by experiment to be CH_2O. In a separate experiment, the molecule's mass was determined to be 180.1 amu. What is the compound's molecular formula?

Remember, to go from an empirical formula to a molecular formula, we must multiply by the common factor. So this problem really boils down to figuring out what the common factor is. We figure that out by looking at the mass. If the molecular formula really were CH_2O, the mass would be:

$$\text{mass of } CH_2O = 1 \times 12.0 \text{ amu} + 2 \times 1.01 \text{ amu} + 1 \times 16.0 \text{ amu} = 30.0 \text{ amu}$$

This isn't anywhere close to the correct mass of 180.1 amu. We need to multiply the numbers in the empirical formula by some factor. How do we find that factor? Think about it. We know that we need a mass of 180.1 amu. The empirical formula has a mass of 30.0 amu. To determine the common factor, we simply divide the mass that we need by the mass of the empirical formula:

$$\text{common factor} = \frac{\text{mass of molecule}}{\text{mass of empirical formula}} \times \frac{180.1 \cancel{\text{amu}}}{30.0 \cancel{\text{amu}}} = 6.00$$

This tells us that if we multiply each number in the empirical formula by 6, we will get a molecule whose empirical formula is CH_2O and whose mass is 180.1 amu:

$$\text{molecular formula} = C_{1x6}H_{2x6}O_{1x6} = C_6H_{12}O_6$$

The molecular formula, then, is $\mathbf{C_6H_{12}O_6}$.

If you are a little shaky on this, don't worry. We will go over another problem just like it in a moment. However, before we do that, let's define a term that you will often see in problems like this one. Sometimes, chemists talk about a compound's molar mass instead of its molecular mass. It is important to know the definition of that term.

Molar mass—The mass of one mole of a given compound.

Think about what this definition means. Suppose we have a mole of $C_6H_{12}O_6$. What would its mass be? The molecule has a mass of 180.1 amu. That also means it takes 180.1 grams to make a mole. The molar mass, then, is 180.1 grams.

In other words, the molar mass is just the conversion relationship that goes from grams to moles. If we know the molecular mass of a compound, we know its molar mass. For example, the molecular mass of CH_4 is 16.0 amu. That means it takes 16.0 grams to make 1 mole. The molar mass, therefore, is 16.0 grams. In the same way, if we know the molar mass of a compound, we know its molecular mass. For example, the molar mass of KCl is 74.6 grams. That means its molecular mass is 74.6 amu.

Why do chemists use the term *molar mass*? We use it because molar mass is typically what we measure in the lab. It is hard to measure the mass of a single molecule because a single molecule has very little mass (on the order of 10^{-24} grams). The molecular mass is hard to measure. It *can* be measured; it just requires some pretty sophisticated equipment. However, it is pretty easy to measure the mass of a mole of molecules because a mole of molecules has a mass that is somewhere between several grams and a few hundred grams. That can be measured on a typical lab scale. We typically *measure* the molar mass of a compound and then *infer* from that its molecular mass.

See if you can put this concept together with Example 8.11.

EXAMPLE 8.11

A compound's empirical formula is PO_3. If its molar mass is 158 g, what is its molecular formula?

The problem tells us that the mass of 1 mole of this compound is 158 grams. This tells us that the molecular mass of the compound must be 158 amu. To get the molecular formula from the empirical formula, we must determine the common factor. To do that, we divide the molecular mass by the mass of the empirical formula. We must first determine the mass of the empirical formula:

$$\text{mass of } PO_3 = 1 \times 31.0 + 3 \times 16.0 = 79.0 \text{ amu}$$

To get the common factor, we divide this into the molecular mass:

$$\text{common factor} = \frac{\text{mass of molecule}}{\text{mass of empirical formula}} \times \frac{158 \cancel{\text{ amu}}}{79 \cancel{\text{ amu}}} = 2.0$$

Therefore, to get the molecular formula, we must multiply each number in the formula by 2:

$$\text{molecular formula} = P_{1\text{x}2}O_{3\text{x}2} = \mathbf{P_2O_6}$$

Answer "On Your Own" question 8.10 to make sure you understand how to convert empirical formulas into molecular formulas.

ON YOUR OWN

8.10 In one experiment, the empirical formula of a compound is determined to be $NaPO_4$. In a separate experiment, its molar mass is determined to be 354 grams. What is its molecular formula?

MORE COMPLICATED EXPERIMENTS FOR DETERMINING CHEMICAL FORMULAS

Remember, there were 2 wrinkles in the problem of determining chemical formulas. The first was the distinction between molecular and empirical formulas. The second is the fact that most experiments that determine empirical formulas are much harder to analyze than the experiment presented in Example 8.8, which had very easy numbers to work with. In Example 8.12, you will see that the numbers don't always work out so nicely.

EXAMPLE 8.12

A sample of brown-colored gas is decomposed into 2.34 g of nitrogen and 5.34 g of oxygen. What is the empirical formula of the compound?

Since only nitrogen and oxygen are products in the decomposition, the formula of the compound must be N_xO_y. Nitrogen and oxygen are both homonuclear diatomics, so the decomposition reaction looks like this:

$$N_xO_y \rightarrow N_2 + O_2$$

To determine the stoichiometric coefficients on the products side, we calculate the number of moles of the elements produced. Let's start with nitrogen:

$$\text{mass of } N_2 = 2 \times 14.0 \text{ amu} = 28.0 \text{ amu}$$

$$1 \text{ mole } N_2 = 28.0 \text{ g } N_2$$

$$\frac{2.34 \cancel{\text{g } N_2}}{1} \times \frac{1 \text{ mole } N_2}{28.0 \cancel{\text{g } N_2}} = 0.0836 \text{ moles } N_2$$

Now let's figure out the number of moles of oxygen produced:

$$\text{mass of } O_2 = 2 \times 16.0 \text{ amu} = 32.0 \text{ amu}$$

$$1 \text{ mole } O_2 = 32.0 \text{ g } O_2$$

$$\frac{5.34 \cancel{\text{g } O_2}}{1} \times \frac{1 \text{ mole } O_2}{32.0 \cancel{\text{g } O_2}} = 0.167 \text{ moles } O_2$$

Since the number of moles of each product is its stoichiometric coefficient, our equation looks like this:

$$N_xO_y \rightarrow 0.0836\ N_2 + 0.167\ O_2$$

This is where the wrinkle occurs. Unlike in Example 8.8, these numbers did not work out nicely. This time, the stoichiometric coefficients are not integers. To be able to determine an empirical formula, we need integer numbers of atoms. We get that by dividing both of the coefficients by the smallest number present:

$$N_xO_y \rightarrow \frac{0.0836}{0.0836}\ N_2 + \frac{0.167}{0.0836}\ O_2$$

Do you see how this works? When we divide 0.0836 by 0.0836, we get 1! When we divide 0.167 by 0.0836, we get 2. Dividing by the smallest number has turned these decimals into integers:

$$N_xO_y \rightarrow N_2 + 2O_2$$

The only way to balance the equation is if x = 2 and y = 4. The formula, then, is N_2O_4. This is not the answer, however. Remember, these kinds of problems can determine only the *empirical* formula. They cannot determine the molecular formula. Even though the answer is a *possible* molecular formula, we cannot be sure that it is the *correct* molecular formula. Since these experiments can determine only empirical formulas, we must convert this into an empirical formula by dividing by the common factor, which is 2. This means the empirical formula is **NO_2**.

This example shows that when we solve problems like this one, if the stoichiometric coefficients turn out not to be integers, we can simply divide by the smallest of them, and that will give us a reasonable chemical equation. Balancing the equation then gives us the chemical formula. If the chemical formula is not an empirical formula, we can then convert it to one. See if you can do this with "On Your Own" question 8.11.

We hope you now have an appreciation for how useful stoichiometry is. This powerful tool allows us to analyze chemical equations and determine empirical formulas. We will see in future modules that there are many, many other uses of stoichiometry, and that's why we say you must understand the concepts presented in this module to be successful in chemistry.

ON YOUR OWN

8.11 A compound is decomposed into 28.0 g sulfur and 27.8 g oxygen. What is its empirical formula?

ANSWERS TO THE "ON YOUR OWN" QUESTIONS

8.1 The chemical equation tells us:

$$1 \text{ mole } O_2 = 2 \text{ moles } NaNO_3$$

We can use it as a conversion relationship to convert moles of $NaNO_3$ into moles of O_2:

$$\frac{45.0 \cancel{\text{moles } NaNO_3}}{1} \times \frac{1 \text{ mole } O_2}{2 \cancel{\text{moles } NaNO_3}} = 22.5 \text{ moles } O_2$$

Note that the numbers that come from the chemical equation are integers, which means they are exact. Therefore, 45.0 is the only number with a finite amount of significant figures. Since there are 3 significant figures in 45.0, we must have 3 in the answer. To make 45.0 moles of sodium nitrate, then, the chemist must start with 22.5 moles of oxygen.

We can also use the chemical equation to develop a relationship between moles of $NaNO_2$ and moles of $NaNO_3$:

$$2 \text{ moles } NaNO_2 = 2 \text{ moles } NaNO_3$$

Using this in a conversion:

$$\frac{45.0 \cancel{\text{moles } NaNO_3}}{1} \times \frac{2 \text{ moles } NaNO_2}{2 \cancel{\text{moles } NaNO_3}} = 45.0 \text{ moles } NaNO_2$$

Once again, the 45.0 determines the significant figures. To make 45.0 moles of sodium nitrate, a chemist must have 22.5 moles of oxygen and 45.0 moles of $NaNO_2$.

8.2 The first thing we are asked to do is calculate how many moles of oxygen are necessary to use up all of the iron ore. According to the equation:

$$4 \text{ moles } FeS_2 = 11 \text{ moles } O_2$$

Converting from iron ore to oxygen:

$$\frac{2{,}512.0 \cancel{\text{moles } FeS_2}}{1} \times \frac{11 \text{ moles } O_2}{4 \cancel{\text{moles } FeS_2}} = 6{,}908.0 \text{ moles } O_2$$

In addition, the problem asks us to calculate how much iron oxide will be produced. Once again, since all substances in a chemical equation relate to all other substances in that equation, we can simply convert from moles of iron ore to moles of iron oxide. The equation tells us that 4 moles FeS_2 = 2 moles Fe_2O_3:

$$\frac{2{,}512.0 \cancel{\text{moles } FeS_2}}{1} \times \frac{2 \text{ moles } Fe_2O_3}{4 \cancel{\text{moles } FeS_2}} = 1{,}256.0 \text{ moles } Fe_2O_3$$

In the end, then, the reaction will require 6,908.0 moles of oxygen and will produce 1,256.0 moles of iron oxide. We could also have calculated how much sulfur dioxide was produced, but the problem did not ask us to, so we won't.

8.3 The problem asks us to calculate the minimum amount of each reactant needed to produce 2.8 x 10⁵ moles of citric acid. All we have to do is determine the relationship between citric acid and each of the reactants and then use those relationships to convert:

To determine the amount of $C_{12}H_{22}O_{11}$ needed:

$$1 \text{ mole } C_{12}H_{22}O_{11} = 2 \text{ moles } C_6H_8O_7$$

$$\frac{2.8 \times 10^5 \cancel{\text{moles } C_6H_8O_7}}{1} \times \frac{1 \text{ mole } C_{12}H_{22}O_{11}}{2 \cancel{\text{moles } C_6H_8O_7}} = 1.4 \times 10^5 \text{ moles } C_{12}H_{22}O_{11}$$

To determine the amount of water needed:

$$1 \text{ mole } H_2O = 2 \text{ moles } C_6H_8O_7$$

$$\frac{2.8 \times 10^5 \cancel{\text{moles } C_6H_8O_7}}{1} \times \frac{1 \text{ mole } H_2O}{2 \cancel{\text{moles } C_6H_8O_7}} = 1.4 \times 10^5 \text{ moles } H_2O$$

To determine the amount of oxygen needed:

$$3 \text{ moles } O_2 = 2 \text{ moles } C_6H_8O_7$$

$$\frac{2.8 \times 10^5 \cancel{\text{moles } C_6H_8O_7}}{1} \times \frac{3 \text{ moles } O_2}{2 \cancel{\text{moles } C_6H_8O_7}} = 4.2 \times 10^5 \text{ moles } O_2$$

To make the citric acid, then, the manufacturer must use 1.4×10^5 moles of $C_{12}H_{22}O_{11}$, 1.4×10^5 moles of water, and 4.2×10^5 moles of oxygen.

8.4 Since the problem tells us that both $CaCO_3$ and water are added in excess, HCl is the limiting reactant. The amount of HCl determines the amount of product. To solve the problem, then, we need to convert from the amount of HCl to the amount of product using the relationship given in the chemical equation.

The chemical equation tells us:

$$2 \text{ moles HCl} = 1 \text{ mole } CaCl_2H_{12}O_6$$

Using this relationship to convert:

$$\frac{7.20 \cancel{\text{moles HCl}}}{1} \times \frac{1 \text{ mole } CaCl_2H_{12}O_6}{2 \cancel{\text{moles HCl}}} = 3.60 \text{ moles } CaCl_2H_{12}O_6$$

So 3.60 moles will be produced. Once again, we *could* also use the information above to determine how much CO_2 is produced because the number of moles of limiting reactant can be used to determine the number of moles of all products.

MOD 8

8.5 The problem asks us to relate the volumes of 2 gases in a chemical equation. Since both are gases, we can use Gay–Lussac's law. The stoichiometric coefficients in the chemical equation relate the volumes of the gases:

$$2 \text{ liters of HCl} = 1 \text{ liter of } CO_2$$

Using this relationship to convert :

$$\frac{8.20 \text{ liters } CO_2}{1} \times \frac{2 \text{ liters HCl}}{1 \text{ liter } CO_2} = 16.4 \text{ liters HCl}$$

So long as Na_2CO_3 is added in excess, you need 16.4 liters of HCl to make 8.20 liters of CO_2.

8.6 In reading this problem, the first thing you should notice is that you were told one reactant is in excess. This means that the other reactant (C_6H_5Cl) is the limiting reactant. Once you recognize that, you can relate its number of moles to the number of moles of any other substance in the chemical equation. Of course, you were not given the number of moles; you were given the number of grams. There is no conversion relationship between grams of C_6H_5Cl and grams of $C_{14}H_9Cl_5$. So the first thing you must do is convert from grams to moles. Only then can you use the chemical equation.

$$\text{Mass of } C_6H_5Cl = 6 \times 12.0 \text{ amu} + 5 \times 1.01 \text{ amu} + 1 \times 35.5 \text{ amu} = 112.6 \text{ amu}$$

$$1 \text{ mole of } C_6H_5Cl = 112.6 \text{ g } C_6H_5Cl$$

$$\frac{500.0 \text{ g } C_6H_5Cl}{1} \times \frac{1 \text{ mole } C_6H_5Cl}{112.6 \text{ g } C_6H_5Cl} = 4.440 \text{ moles } C_6H_5Cl$$

Now that you have the number of moles of limiting reactant, you can relate that to any substance in the equation using stoichiometry:

$$2 \text{ moles } C_6H_5Cl = 1 \text{ mole } C_{14}H_9Cl_5$$

$$\frac{4.440 \text{ moles } C_6H_5Cl}{1} \times \frac{1 \text{ mole } C_{14}H_9Cl_5}{2 \text{ moles } C_6H_5Cl} = 2.220 \text{ moles } C_{14}H_9Cl_5$$

This gives you the amount of product, but this problem asked us to calculate the grams of product, not the moles. So now you have to take the number of moles and convert to grams:

$$\text{Mass of } C_{14}H_9Cl_5 = 14 \times 12.0 \text{ amu} + 9 \times 1.01 \text{ amu} + 5 \times 35.5 \text{ amu} = 354.6 \text{ amu}$$

$$1 \text{ mole } C_{14}H_9Cl_5 = 354.6 \text{ g } C_{14}H_9Cl_5$$

$$\frac{2.220 \text{ moles } C_{14}H_9Cl_5}{1} \times \frac{354.6 \text{ g } C_{14}H_9Cl_5}{1 \text{ mole } C_{14}H_9Cl_5} = 787.2 \text{ g } C_{14}H_9Cl_5$$

There will be 787.2 g of product.

8.7 This problem asks you to relate the quantity of one reactant to the quantity of the other reactant that will allow for everything to be used up. This is easy to do with stoichiometry, but first you must convert the number of grams of reactant to moles of reactant because stoichiometry can only be done with moles:

$$\frac{1{,}120\ \cancel{kg}}{1} \times \frac{1{,}000\ g}{1\ \cancel{kg}} = 1.12 \times 10^6\ g$$

$$\text{Mass of } N_2H_4 = 2 \times 14.0\ \text{amu} + 4 \times 1.01\ \text{amu} = 32.0\ \text{amu}$$

$$1\ \text{mole of } N_2H_4 = 32.0\ g\ N_2H_4$$

$$\frac{1.12 \times 10^6\ \cancel{g\ N_2H_4}}{1} \times \frac{1\ \text{mole } N_2H_4}{32.0\ \cancel{g\ N_2H_4}} = 3.50 \times 10^4\ \text{moles } N_2H_4$$

Note that the number of moles must be expressed in scientific notation because there must be 3 significant figures. That can only be done in scientific notation because the 0 needs a decimal point to make it significant. Now that you have the number of moles of reactant, you can relate that to the minimum amount of the other reactant in the equation using stoichiometry:

$$1\ \text{mole } N_2H_4 = 7\ \text{moles } H_2O_2$$

$$\frac{3.50 \times 10^4\ \cancel{\text{moles } N_2H_4}}{1} \times \frac{7\ \text{moles } H_2O_2}{1\ \cancel{\text{mole } N_2H_4}} = 2.45 \times 10^5\ \text{moles } H_2O_2$$

This gives you the amount of reactant, but now you have to take the number of moles and convert to grams:

$$\text{Mass of } H_2O_2 = 2 \times 1.01\ \text{amu} + 2 \times 16.0\ \text{amu} = 34.0\ \text{amu}$$

$$1\ \text{mole } H_2O_2 = 34.0\ g\ H_2O_2$$

$$\frac{2.45 \times 10^5\ \cancel{\text{moles } H_2O_2}}{1} \times \frac{34.0\ g\ H_2O_2}{1\ \cancel{\text{mole } H_2O_2}} = 8.33 \times 10^6\ g\ H_2O_2$$

The answer is 8.33×10^6 g.

8.8 In this problem, you are asked to calculate the minimum amount of reactant necessary to make a certain amount of product. You must assume that the reactant you have to calculate will be the limiting reactant. You can do this with stoichiometry, once you convert from grams of product to moles of product:

$$\text{Mass of } C_2H_2 = 2 \times 12.0\ \text{amu} + 2 \times 1.01\ \text{amu} = 26.0\ \text{amu}$$

$$1\ \text{mole of } C_2H_2 = 26.0\ g\ C_2H_2$$

$$\frac{250.0\ \cancel{g\ C_2H_2}}{1} \times \frac{1\ \text{mole } C_2H_2}{26.0\ \cancel{g\ C_2H_2}} = 9.62\ \text{moles } C_2H_2$$

Now that you have the number of moles of product, you can relate that to any substance in the equation using stoichiometry:

$$1\ \text{mole } C_2H_2 = 2\ \text{moles } H_2O$$

$$\frac{9.62 \cancel{\text{ moles } C_2H_2}}{1} \times \frac{2 \text{ moles } H_2O}{1 \cancel{\text{ mole } C_2H_2}} = 19.2 \text{ moles } H_2O$$

This gives you the amount of water, but now you have to take the number of moles and convert to grams:

$$\text{Mass of } H_2O = 2 \times 1.01 \text{ amu} + 1 \times 16.0 \text{ amu} = 18.0 \text{ amu}$$

$$1 \text{ mole } H_2O = 18.0 \text{ g } H_2O$$

$$\frac{19.2 \cancel{\text{ moles } H_2O}}{1} \times \frac{18.0 \text{ g } H_2O}{1 \cancel{\text{ mole } H_2O}} = 346 \text{ g } H_2O$$

The minimum mass of water necessary is 346 g.

8.9 For a chemical formula to be empirical, it must represent a simple, whole–number ratio of atoms. Therefore, if a chemical formula is an empirical formula, the subscripts cannot have a common factor.

(a) This is *not* an empirical formula because 6 and 9 have a common factor of 3. To get the empirical formula, you would divide by that common factor and get Al_2O_3.

(b) This *is* an empirical formula because 1 and 2 have no common factors.

(c) This *is* an empirical formula because 1, 1, and 3 have no common factors.

(d) This is *not* an empirical formula because 8, 18 and 2 have a common factor of 2. To get the empirical formula, you would divide by that common factor and get C_4H_9O.

(e) This *is* an empirical formula because 3, 9, and 2 have no common factors.

8.10 The molar mass is 354 grams. That means it takes 354 grams to make a mole, so the molecular mass must be 354 amu. To get from the empirical formula to the molecular formula, you must determine the common factor by which to multiply. To do that, you need to divide the molecular mass by the mass of the empirical formula. You need to start by determining the mass of the empirical formula:

$$\text{Mass of } NaPO_4 = 1 \times 23.0 \text{ amu} + 1 \times 31.0 \text{ amu} + 4 \times 16.0 \text{ amu} = 118.0 \text{ amu}$$

Now you can determine the common factor:

$$\text{common factor} = \frac{\text{mass of molecule}}{\text{mass of empirical formula}} \times \frac{354 \cancel{\text{ amu}}}{118.0 \cancel{\text{ amu}}} = 3.0$$

To get from the empirical formula to the molecular formula, then, you must multiply by 3:

$$Na_{1x3}P_{1x3}O_{4x3} = Na_3P_3O_{12}$$

In Module 5 we learned that PO_4 is the polyatomic ion, phosphate. If you recognize it, you will know that we can correctly write the answer as $Na_3(PO_4)_3$

8.11 Since only sulfur and oxygen were products in the decomposition, the chemical formula is S_xO_y. Oxygen is a homonuclear diatomic, so the decomposition reaction looks like this:

$$S_xO_y \rightarrow S + O_2$$

To get the stoichiometric coefficients, you need to determine the number of moles produced in the decomposition reaction. Start with sulfur:

$$\text{Mass of S} = 32.1 \text{ amu}$$

$$1 \text{ mole S} = 32.1 \text{ g S}$$

$$\frac{28.0 \cancel{\text{ g S}}}{1} \times \frac{1 \text{ mole S}}{32.1 \cancel{\text{ g S}}} = 0.872 \text{ moles S}$$

Now you can determine the number of moles of oxygen produced:

$$\text{Mass of } O_2 = 2 \text{ x } 16.0 \text{ amu} = 32.0 \text{ amu}$$

$$1 \text{ mole } O_2 = 32.0 \text{ g } O_2$$

$$\frac{27.8 \cancel{\text{ g } O_2}}{1} \times \frac{1 \text{ mole } O_2}{32.0 \cancel{\text{ g } O_2}} = 0.869 \text{ moles } O_2$$

So the equation looks like this:

$$S_xO_y \rightarrow 0.872 \text{ S} + 0.869 \text{ } O_2$$

The problem with this equation is that the coefficients are not integers. You need to turn them into integers to be able to determine the empirical formula of the reactant. You do that by dividing by the smallest number:

$$S_xO_y \rightarrow \frac{0.872}{0.869} \text{ S} + \frac{0.869}{0.869} = O_2$$

When you divide 0.872 by 0.869, you get 1.003452.... Since there are only 3 significant figures, the answer is 1.00. Since 0.869 divided by 0.869 is also 1.00, the chemical equation now looks like this:

$$S_xO_y \rightarrow S + O_2$$

The formula that balances the equation, then, is SO_2. This is already an empirical formula since 1 and 2 have no common factors.

STUDY GUIDE FOR MODULE 8

REVIEW QUESTIONS

1. Define stoichiometry and explain why it is such a useful tool for chemists.

2. Explain what a limiting reactant is and why it is important in stoichiometry.

3. A chemist experiments with the following reaction:

   $$14HCl + K_2Cr_2O_7 \rightarrow 2KCl + 2CrCl_3 + 3Cl_2 + 7H_2O$$

   If he adds 15 moles of HCl to 1 mole of $K_2Cr_2O_7$, what is the limiting reactant?

4. State Gay–Lussac's law and explain how it is used.

5. A chemist wants to perform the following reaction:

   $$CaCO_3\ (s) + 2HCl\ (g) \rightarrow CaCl_2\ (aq) + CO_2\ (g) + H_2O\ (l)$$

   For which substances can she use Gay–Lussac's law to relate them to one another?

6. If a stoichiometry problem gives you the quantity of a substance in grams, what is the first thing you must do to solve the problem? Why?

7. What is the difference between molecular formulas and empirical formulas?

8. Which of the following are empirical formulas?

   a. $C_{12}H_{24}O_{10}$ b. $K_2S_2O_3$ c. CaN_2O_6

9. A compound has a molecular formula of $C_{14}H_{21}O_7$. What is its empirical formula?

10. A compound has a molecular formula of H_3PO_4. What is its empirical formula?

PRACTICE PROBLEMS

1. When 21.3 moles of C_3H_8 burn in excess oxygen, how many moles of CO_2 will be formed? Assume this is complete combustion.

2. Silver tarnishes when exposed to oxygen and dihydrogen monosulfide. The chemical reaction is as follows:

$$\underset{\text{silver}}{4Ag\ (s)} + 2H_2S\ (g) + O_2 \rightarrow \underset{\text{tarnish}}{2Ag_2S\ (s)} + 2H_2O\ (l)$$

   Every household has some H_2S in the air (it smells like rotting eggs), but there is usually only a small quantity. H_2S is almost always the limiting reactant in this reaction. If a silver spoon has 0.0030 moles of tarnish on it, how many moles of H_2S was it exposed to?

3. Freon, a very useful refrigerant, is produced in the following reaction:

$$3CCl_4\ (g) + 2SbF_3\ (s) \rightarrow \underset{\text{Freon}}{3CCl_2F_2\ (g)} + 2SbCl_3\ (s)$$

   If a chemist wants to make 3.0×10^6 moles of Freon using excess carbon tetrachloride, how many moles of antimony trifluoride will he need?

4. In the reaction from problem 3, suppose the chemist wanted to make 120.0 liters of Freon using excess antimony trifluoride. How many liters of carbon tetrachloride would the chemist need?

5. To make a phosphorus fertilizer, agricultural companies use the following reaction:

$$Ca_3P_2O_8 + 2H_2SO_4 + 4H_2O \rightarrow \underset{\text{fertilizer}}{CaH_4P_2O_8} + 2CaH_4SO_6$$

   If 9.80×10^3 grams of H_2SO_4 are reacted with excess $Ca_3P_2O_8$ and H_2O, how many grams of fertilizer can be made?

6. Phosgene, $COCl_2$, was once used as a chemical weapon in war. When humans breathe in the gas, it reacts with the water in their lungs in the following way:

$$COCl_2 + H_2O \rightarrow CO_2 + 2HCl$$

   The product HCl burns the lining in the lungs, making them unable to function. By breathing in phosgene, a person will eventually suffocate. Assuming that 5.00 grams of HCl in the lungs is deadly, how many grams of phosgene would a person need to breathe to die?

7. $Mg(OH)_2$ is the active ingredient in many laxatives. It can be made with the following reaction:

$$Mg\ (s) + 2H_2O\ (l) \rightarrow Mg(OH)_2\ (s) + H_2\ (g)$$

If a laxative manufacturer has 2.8×10^5 grams of Mg, how many grams of water must be used to have all of the Mg react to form $Mg(OH)_2$?

8. A compound has an empirical formula of $CHBr_2$. If its molar mass is 345.6 grams, what is its molecular formula?

9. An unknown compound was decomposed into 316 g carbon, 26.3 g hydrogen, and 208 g oxygen. What is its empirical formula?

10. A dry–cleaning compound was decomposed into 145 g carbon and 855 g chlorine. In a separate experiment, its molar mass was determined to be 166.0 grams. What is the molecular formula of the compound?

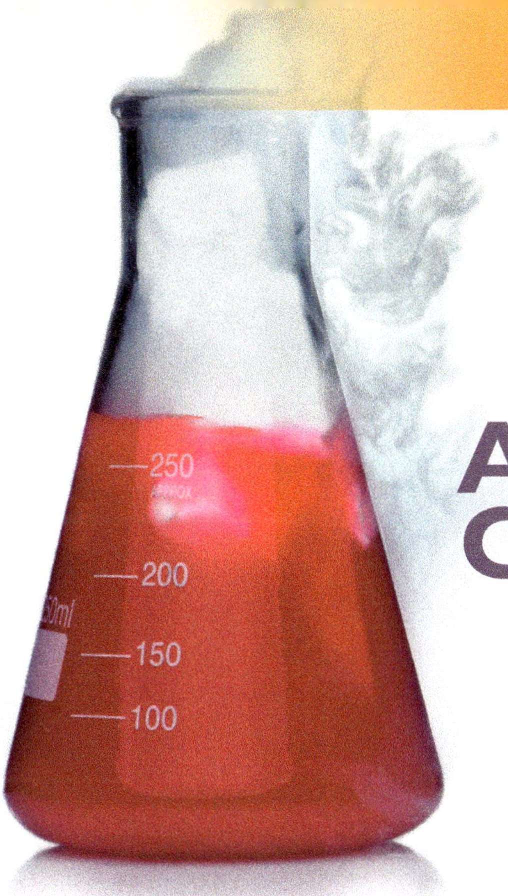

ACID-BASE CHEMISTRY

The vast majority of chemistry revolves around chemical reactions. To study reactions, we must remember the skills we learned in the last 2 modules because we will be using the mole concept, stoichiometry, unit conversions, ionic compound naming, and chemical equation balancing skills in this module. If you have trouble understanding some of this module, you might need to go back and review some of those concepts.

first reactions

As we understand more of the interactions of chemistry, we begin to classify the properties we expect they will have. Ions of atoms behave in certain ways, acids behave in certain ways, and so forth. We can see the work of a sovereign, unchanging God in the way the world is so regular that by studying it through science, like chemistry, we can predict reaction outcomes.

You have already learned about 3 different classes of chemical reactions: formation, decomposition, and combustion. As you might expect, there are several more categories of chemical reactions. As we go on through this course, we will touch on some of them. This module concentrates on one of the most important and fundamental classes of chemical reactions: the acid-base reaction.

ACIDS AND BASES

To understand acid-base reactions, we must first understand what an acid is and what a base is. When you read the word *acid*, an image probably pops up in your mind. You might think of an acid as a liquid chemical that can eat up anything it touches. Although some acids can eat up certain materials, it is not true of the vast majority of acids. Many of the things we eat and drink contain acids. Lemon juice in lemonade, for example, is an acid. So an acid is not necessarily a corrosive, dangerous chemical. It is often a beneficial and even tasty substance.

Chemists typically use 2 methods to define acids. Acids have the same general properties, so one way to define an acid is by its characteristics. There is also a strict chemical definition for acids. In this module, you need to learn both types of definitions. We will start with the characteristics of an acid, and later on we will discuss the chemical definition.

In general, acids have the following properties:

1. Acids taste sour.

2. Acids are covalent compounds that conduct electricity when added to water.

3. Acids turn blue litmus paper red.

First, even though we include the taste of an acid in this list, you should never taste a substance to determine what it is. Many man-made substances, as well as many naturally occurring substances, are lethal poisons for human beings. In general, you should never taste something that you cannot clearly identify as safe to eat. Some acids will cause severe damage to your body if you try to taste them. *Never* taste something in a chemistry experiment unless you are expressly instructed to do so by your teacher!

The second characteristic of acids is a bit narrow. When you take college chemistry, you may learn about some ionic compounds that are classified as acids. However, for this course we will concentrate only on the covalent acids. Having covalent compounds conduct electricity might seem to contradict something we learned in Module 2. In that module, we said that *ionic* compounds conduct electricity when dissolved in water, while *covalent* compounds *do not* conduct electricity when dissolved in water. In general, that's true. However, acids are an exception to that general rule. Even though acids are covalent compounds (composed of atoms that share electrons rather than form ions), they can conduct electricity when added to water. Later in this module we will learn *how* this happens.

You won't really be able to understand the third characteristic of acids until you know what litmus (lit' mus) is. Litmus is part of a broad class of compounds known as indicators.

Indicator—A substance that turns one color in the presence of acids and another color in the presence of bases.

Litmus has the property of turning red whenever it is in the presence of an acid. We could say that litmus is an excellent indicator of the presence of an acid. You will be doing your own litmus tests on household chemicals shortly.

Acids are rarely discussed in chemistry without also discussing bases. When you see the chemical definitions of these two classes of compounds, you will see why. In general, bases have the following characteristics:

1. Bases taste bitter.

think about this

Lye is a base commonly used in cleaning products, and forms of it can be very powerful. Yet even the most powerful detergents cannot wash away the sin that has corrupted mankind. Jeremiah 2:22 declares, "Though you wash yourself with lye and use much soap, the stain of your guilt is still before me, declares the Lord GOD." Only through Jesus Christ's perfect life, His death on the cross, and His resurrection, could sins be washed away for those who have faith in Him. 1 John 1:7 says, "But if we walk in the light, as he is in the light, we have fellowship with one another, and the blood of Jesus his Son cleanses us from all sin."

2. Bases are slippery to the touch when dissolved in water.

3. Bases turn red litmus paper blue.

Of course, the same warnings about tasting acids also apply to bases. Some bases are just as caustic and dangerous as the most dangerous acids. Remember, you should *never* taste any chemical in an experiment unless you are expressly instructed to do so.

The characteristics of acids and bases are important enough that you should remember them. To help you do this, as well as to help you learn about the nature of certain household chemicals, perform Experiment 9.1.

EXPERIMENT 9.1

PURPOSE: To examine common household examples of acids and bases.

MATERIALS:

- Red litmus paper
- Blue litmus paper
- Apple
- Orange juice or soda pop
- Toilet bowl cleaner (Both The Works and Lime-Away have been tested, but any toilet bowl cleaner designed to combat lime should work.)
- Bar soap (Make sure it isn't labeled "pH balanced.")
- All-purpose cleaner (Windex and 409 have been tested, but any spray cleaner not specifically designed for toilets should work.)
- Powdered drain cleaner (like Red Devil Lye) or scouring powder (like Comet)
- 4 test tubes (Small cups will work.)
- Watch glass (A small saucer will work.)
- Stirring rod
- Rubber gloves (Wear gloves whenever you use powdered drain cleaners and toilet bowl cleaners because these chemicals are caustic.)
- Safety goggles

QUESTION: Can you identify common household acids and bases?

HYPOTHESIS: Write a hypothesis related to whether the items in the supply list are acids or bases.

PROCEDURE:

1. Take a bite out of the apple and eat it. Would you say that the taste was more bitter or sour? Based on that, would you say that the apple has acid or base in it?
2. Confirm your guess by pressing some blue litmus paper against the part of the apple exposed by your bite. Do the same with some red litmus paper. You should see that the red litmus paper does not change colors, but the blue litmus paper changes to red.

Apples contain acid.

3. Rinse your hands to get off any apple juice that might be on them.
4. Lay out your watch glass and spray some all-purpose cleaner on it.
5. Rub 2 fingers against the wet watch glass to feel the consistency of the all-purpose cleaner. Slippery, isn't it? What kind of substance does that mean it is?
6. Confirm your guess with red and blue litmus paper by placing the papers directly on the watch glass. You should see that the blue litmus paper does not change colors, but the red litmus paper changes to blue. This means that the all-purpose cleaner contains base.
7. Rinse your hands to get off the all-purpose cleaner.
8. Shave some soap off the bar of soap and put the shavings into a test tube.
9. Fill the test tube about ¾ full with the shavings, then add just enough water to fill the tube almost to the top.
10. Cover the tube with your thumb and shake it vigorously for about 2 minutes. Based on your experience with soap, would you think it is an acid or base?
11. Confirm your guess with red and blue litmus paper by dipping the paper into the test tube. The red litmus should turn blue, and the blue litmus should not change colors. Soap contains base. You should have guessed this based on the fact that soap is slippery when exposed to water.
12. Rinse your hands and fill another test tube with the orange juice or soda pop. Based on your previous experience with this drink (think about the taste), do you expect it to be a base or an acid?
13. Confirm your guess by dipping red and blue litmus paper into the test tube. The blue litmus should turn red and the red litmus should not change colors, indicating that the drink contains acid.
14. Rinse off your hands. .
15. Put the gloves on. Fill a test tube with the toilet bowl cleaner. ***Be careful when you do this because toilet bowl cleaner can be very caustic. Do not get any on your skin!***
16. Test this substance by dipping red and blue litmus paper into the test tube. The blue litmus should turn red, and the red litmus should not change colors. The toilet bowl cleaner contains acid.
17. With your gloves still on, add a few crystals of powdered drain cleaner to a test tube. If you are using scouring powder, fill a test tube ¾ full with it.
18. Add enough water to fill the test tube almost full.
19. Stir the test tube contents with a stirring rod.
20. Test the solution you just made with red and blue litmus paper. The red paper should turn blue, and the blue paper should not change colors. This means that powdered drain cleaners contain base.
21. Clean up and return everything to the proper place.

CONCLUSION: Write something about what you observed and whether it confirmed your hypothesis.

Although you have identified the fact that some of these household chemicals contain acid and some contain base, you still don't know *why* they contain acid or base. Don't get too far ahead of yourself. As you begin to understand the chemistry of acids and bases, you will begin to understand why they are so useful for household items.

THE CHEMICAL DEFINITIONS OF ACIDS AND BASES

Now that we know something about the *properties* of acids and bases, it is time to learn how chemists *define* these substances. According to chemists, an acid is defined as follows:

Acid—A molecule that donates protons or (H^+) ions.

A base, on the other hand, has this definition:

Base—A molecule that accepts protons or (H^+) ions.

Now do you see why acids and bases are always discussed together? If an acid is a molecule that donates H^+ ions, it won't be able to do anything unless there is a base around to accept the H^+ ions. Acids cannot do their job without the help of bases, and bases cannot do their job unless they have an acid to work with.

Before we show you how to determine whether or not a substance is an acid or a base, we want you to think for a moment about what is being donated and accepted in an acid-base reaction. Acids donate H^+ ions, and bases accept them. What is an H^+ ion? The hydrogen *atom* consists of a single proton and a single electron. The hydrogen *ion* (H^+) is a hydrogen atom that has *lost an electron*. What is left when a hydrogen atom (which has one proton and one electron) loses an electron? Only a proton is left. An H^+ ion is just a single proton. Since acids donate H^+ ions, they really donate protons. In the same way, since bases accept H^+ ions, they really accept protons. As a result, some chemists call acids proton donors and bases proton acceptors. We will use the terms *H^+ donor* and *H^+ acceptor*, but you need to know that those other equivalent terms exist.

salt and light

Portrait: Public Domain

Johannes Baptista van Helmont (1579-1644) was a Flemish scientist who performed research in chemistry, physiology, and medicine. He is responsible for giving science the term *gas*, from the Greek word for chaos. He improved the processes used to isolate sulfuric, hydrochloric, and nitric acids. In his physiological studies, he found that alkali salts neutralize acids. Walter Pagel, author of *The Religious and Philosophical Aspects of van Helmont's Science and Medicine*, wrote that van Helmont's Christianity compelled him to study science.

If you're not quite sure what it means to be a molecule that donates H^+ ions or accepts H^+ ions, a few examples should clear things up a bit. Consider hydrochloric acid (HCl). When this acid is mixed with water, the following reaction occurs:

$$HCl\ (aq) + H_2O\ (l) \rightarrow H_3O^+\ (aq) + Cl^-\ (aq)$$

What happened in this reaction? It might help to look at it from a Lewis structure point of view. When we mix HCl and water, it looks like this:

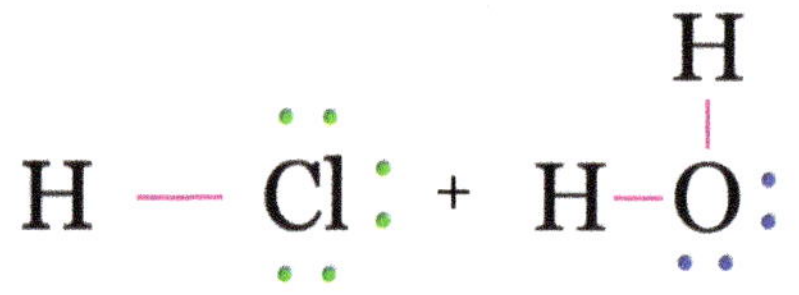

Remember, the bonds pictured here are shared electrons:

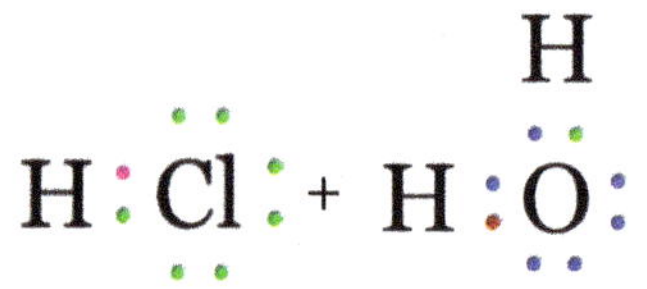

Think for a moment about the electron pair that is being shared by the hydrogen atom and the chlorine atom in the HCl molecule. The electronegativity of the chlorine is significantly larger than that of the hydrogen. As a result, these shared electrons spend most of their time around the Cl atom. Suppose, then, that the H atom wanted a fairer share of electrons. Is there any way it could attain this goal?

Yes, there is. You see, the oxygen atom has 2 extra pairs of electrons. The hydrogen in the HCl molecule can leave the electron pair it is currently sharing and give it entirely to the chlorine atom:

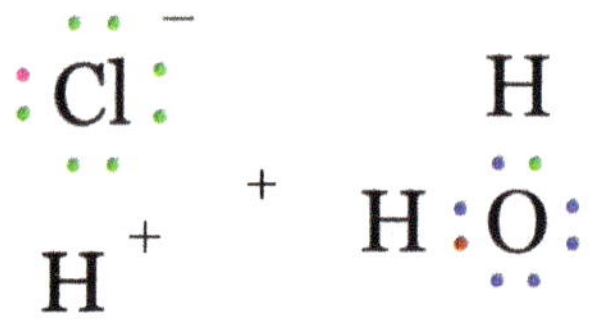

Since the hydrogen is leaving the electron pair behind, it now has no valence electrons. It usually has 1 valence electron, so it is 1 electron shy of its normal state. This makes it a 1+ ion. In the same way, the chlorine atom now has sole possession of the electron pair, so it now has 8 valence electrons. Compared to its normal number of valence electrons (7), this is 1 too many, making it a 1- ion.

Under normal circumstances, of course, the hydrogen atom would not do this. It will do it only because it can get a better share of electrons if it attaches to the oxygen atom on the water molecule:

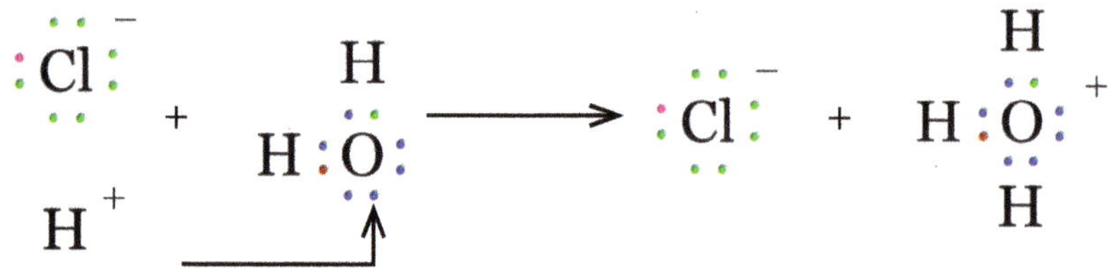

In the end, then, the hydrogen atom that was formerly in the HCl molecule left its shared electron pair with the chlorine, making an H^+ ion. This ion then attached itself to a nonbonding electron pair on the oxygen of the water molecule. Of course, since the hydrogen ion possessed a 1+ charge, that charge is now imposed on the molecule of which it became a part. This new polyatomic ion, H_3O^+, is called the hydronium (hy droh' nee uhm) ion, and it plays a crucial role in most acid-base chemistry. This is another polyatomic ion you must memorize.

Looking back on our original chemical equation, we should now be able to determine which compound was the acid in the reaction and which was the base:

$$\underset{\text{acid}}{HCl\ (aq)} + \underset{\text{base}}{H_2O\ (l)} \rightarrow H_3O^+\ (aq) + Cl^-\ (aq)$$

Since the H^+ *left* the HCl molecule, we could say that the H^+ was *donated* by the HCl. Things that donate H^+ are called acids. HCl acted as an acid in this reaction. In the same way, the H^+ ended up *joining* the H_2O molecule, making the H_2O molecule an H^+ *acceptor*. This means that the H_2O was acting as a base.

In a chemical equation, then, it is relatively simple to determine which reactant is the acid and which is the base. All we have to do is look at where the H^+ ion went. The molecule that lost 1 or more H^+ ions is the acid, and the molecule that gained 1 or more H^+ ions is the base. Before we see some examples of how this is done, we want to introduce 2 other concepts. First, consider the following chemical equation:

$$H_2O\ (l) + NH_3\ (aq) \rightarrow NH_4^+\ (aq) + OH^-\ (aq)$$

Which reactant is the acid in this case and which is the base? The NH_3 in the reaction above was turned into NH_4^+. The only way that could happen would be for the NH_3 to gain (or accept) an H^+. NH_3 is the base. In the same way, H_2O became OH^- by losing (or donating) an H^+, making it the acid. But wasn't H_2O a base in the reaction represented by the first acid-base reaction equation shown? How is it possible for H_2O to be an acid in one reaction and a base in another?

There are certain compounds that chemists call amphiprotic (am' fuh proh' tik). These compounds have the interesting ability to act as a base in some reactions and an acid in others.

Amphiprotic compounds—Compounds that can act as either an acid or a base, depending on the situation.

Water is the most common amphiprotic substance, but there are others that you will see from time to time in this course. Don't let it confuse you if we discuss a compound in one context where it is an acid and then talk later about it behaving like a base. If that happens, you know you have run into an amphiprotic compound. Please note that some chemists use the term amphoteric (am' foh tehr' ik) instead of amphiprotic. The two terms mean essentially the same thing, so you might see one in some textbooks and the other in other textbooks.

THE BEHAVIOR OF IONIC COMPOUNDS IN AQUEOUS SOLUTIONS

The second concept that must be introduced before you can begin analyzing acid-base reactions is something that we touched on way back in Module 2. Remember we said that a compound that conducts electricity when dissolved in water is an ionic compound? To understand acid-base reactions, we must learn *why* ionic compounds conduct electricity when dissolved in water.

For a substance to conduct electricity, it must contain electrical charges that are free to move. Since pure water is a polar covalent compound, it does have electrical charges, but they are not free to move. Remember from Module 5 that a water molecule is polar and has a fractional positive charge on its hydrogen atoms and a fractional negative charge on its oxygen atom. These charges are not free to move because the hydrogen atoms and oxygen atom are all bonded together. Although water does have electrical charges, they are not free to move. They are stuck together.

When an ionic compound is dissolved in water, the situation changes. Ionic compounds dissolve in water by splitting up into their individual ions. This is an important fact worth remembering.

When ionic compounds dissolve, they split into their constituent ions.

When dissolved in water, an NaCl molecule will split up into an Na^+ ion and a Cl^- ion. Because they separate, these ions are free to move and thus can conduct electricity. That's why ionic compounds conduct electricity when dissolved in water.

We can now see why acids are covalent compounds that conduct electricity when dissolved in water. Look back at the reaction described by the first acid-base reaction equation. Notice how the acid reacted with water to make 2 ions: H_3O^+ and Cl^-. Once those 2 ions are formed, they are able to move freely because they are separate ions. When acids are mixed with water, the result is a solution that has ions that are free to move, and the solution therefore conducts electricity.

IDENTIFYING ACIDS AND BASES IN CHEMICAL REACTIONS

Now that we have fully discussed the ins and outs of ions and their relationship to acids and bases, let's see how all of this pans out in Example 9.1.

EXAMPLE 9.1

Determine which reactant is the acid and which is the base in the following chemical equation:

$$H_2SO_4\ (aq) + CaCO_3\ (aq) \rightarrow Ca^{2+}\ (aq) + HCO_3^-\ (aq) + HSO_4^-\ (aq)$$

This looks like a very complicated chemical equation, but if we notice that $CaCO_3$ is an ionic compound, we know it cannot be the acid because acids are covalent compounds. Eliminating $CaCO_3$ means H_2SO_4 has to be the acid.

Here is another approach to determine the acid and base in a reaction. Since $CaCO_3$ is an ionic compound dissolved in water, we know that it splits up into its constituent ions. You should be able to name this compound. It is called calcium carbonate. This means it is composed of the calcium ion and the carbonate ion. You should have the carbonate ion memorized as CO_3^{2-}. The calcium ion is a single-atom ion. Since Ca is in group 2A of the periodic chart, the ion is Ca^{2+}. The equation looks like this:

$$H_2SO_4\ (aq) + Ca^{2+}\ (aq) + CO_3^{2-}\ (aq) \rightarrow Ca^{2+}\ (aq) + HCO_3^-\ (aq) + HSO_4^-\ (aq)$$

Even though there are more components in this equation, it is easier to analyze now. We see that the H_2SO_4 lost an H^+ ion to become HSO_4^-. This means that **H_2SO_4 is the acid** since it donated an H^+. We also see that nothing happened to the Ca^{2+} because it is the same on both sides of the equation. We call ions like this **spectator ions** because they don't do anything in the reaction. They just seem to sit and watch the other compounds react. Finally, we see that CO_3^{2-} gained an H^+ to become HCO_3^-. Since CO_3^{2-} gained an H^+, we could call it the base. However, our original equation did not have CO_3^{2-} written in it; it had $CaCO_3$. Since $CaCO_3$ is the source of CO_3^{2-}, we can say that **$CaCO_3$ is the base.**

$$\underset{\text{acid}}{H_2SO_4\,(aq)} + Ca^{2+}\,(aq) + \underset{\text{base}}{CO_3^{2-}\,(aq)} \rightarrow Ca^{2+}\,(aq) + HCO_3^-\,(aq) + HSO_4^-\,(aq)$$

This is how we figure out which reactant is the acid and which is the base in an acid-base reaction. First, we examine the chemical equation and look for any ionic compounds. If they exist, we split them up into their ions. We then see what molecule lost an H^+ ion and what molecule gained an H^- ion. The molecule that gained the ion is the base, and the one that lost it is the acid. If an ion gained or lost an H^+ ion, the ionic compound from which it came is identified as the acid or base. Follow this method in 2 more examples.

Identify the acid and base in the following reaction:

$$HCl\,(aq) + NaOH\,(aq) \rightarrow H_2O\,(l) + NaCl\,(aq)$$

To solve this problem, we first take all ionic compounds and split them up into their ions. HCl is covalent, but NaOH is ionic. You should recognize the OH part of the compound. That's the hydroxide ion, OH^-. This is sodium hydroxide. That means it is composed of the sodium ion (Na^+ because Na is in group 1A) and the hydroxide ion (OH^-, which you should have memorized from Module 5). NaOH splits up into an Na^+ ion and an OH^- ion when it dissolves in water. In the same way, H_2O is covalent, but NaCl (sodium chloride) is ionic. There are no polyatomic ions in NaCl, so it just splits into a sodium ion (Na^+) and a chloride ion (Cl^-):

$$HCl\,(aq) + Na^+\,(aq) + OH^-\,(aq) \rightarrow H_2O\,(l) + Na^+\,(aq) + Cl^-\,(aq)$$

Now that we have all ionic compounds split into their ions, we can analyze the equation to see where the H^+ ion went. We can see that the Na^+ ion didn't change from the reactants side to the products side. Na^+ is a spectator ion and has no real role in the reaction. On the other hand, the HCl molecule turned into Cl^-. This means it lost an H^+ ion, and therefore **HCl is the acid**. In addition, the OH^- changed into H_2O. The only way that can happen is for it to gain an H^+. We could call the OH^- ion the base. Unfortunately, OH^- is not present in the original equation. However, we know that the OH^- ion came from NaOH, so **NaOH is the base**.

$$\underset{\text{acid}}{HCl\,(aq)} + Na^+\,(aq) + \underset{\text{base}}{OH^-\,(aq)} \rightarrow H_2O\,(l) + Na^+\,(aq) + Cl^-\,(aq)$$

Another approach to answering this question is to recognize that NaOH is ionic and therefore could not be the acid. Since this reaction was identified as an acid-base reaction, HCl would have to be the acid. Be careful when using this approach because the chemical reaction must be clearly identified as an acid-base reaction before we can start eliminating ionic compounds from being acids.

Determine the acid and base in the following reaction:

$$2HNO_3\ (aq) + Mg(OH)_2\ (aq) \rightarrow Mg(NO_3)_2\ (aq) + 2H_2O\ (l)$$

To analyze this equation, we must split up any dissolved ionic compounds into their components. The ionic compounds in this equation are $Mg(OH)_2$ and $Mg(NO_3)_2$. $Mg(OH)_2$ is magnesium hydroxide; thus, it is composed of the magnesium ion (Mg^{2+} because Mg is in group 2A) and the hydroxide ion (OH^-, which you should have memorized). We must be a little careful here, however. How many magnesium ions are in this molecule? Only one. But how many hydroxide ions are in the molecule? There are 2, as shown by the subscript of 2 after the parentheses that contain the polyatomic ion. Since there are 2 hydroxide ions in the molecule, we must be sure to put that information into our new chemical equation:

$$2HNO_3\ (aq) + Mg^{2+}\ (aq) + 2OH^-\ (aq) \rightarrow Mg(NO_3)_2\ (aq) + 2H_2O\ (l)$$

Since the magnesium hydroxide molecule has 1 magnesium ion and 2 hydroxide ions, we used a coefficient of 2 next to the hydroxide ion to show that. Of course, we are not done dealing with the ionic compounds because we have not split $Mg(NO_3)_2$ into its constituent ions yet.

You should recognize the NO_3 in $Mg(NO_3)_2$. It represents the nitrate ion, NO_3^-, which you should have memorized from Module 5. This is the magnesium nitrate ion, which is composed of 1 Mg^{2+} ion and 2 NO_3^- ions. When we split up the ions, then, we must put a coefficient of 2 next to the nitrate ion:

$$2HNO_3\ (aq) + Mg^{2+}\ (aq) + 2OH^-\ (aq) \rightarrow Mg^{2+}\ (aq) + 2NO_3^-\ (aq) + 2H_2O\ (l)$$

Now we can finally analyze the reaction. The Mg^{2+} ion did not change, so it is a spectator. The HNO_3 turned into NO_3^-. This means it lost an H^+ ion, and therefore **HNO_3 is the acid.** The OH^- ion ended up becoming H_2O. The only way this can happen is for it to gain an H^+ ion. Since the OH^- ion came from the $Mg(OH)_2$ in the original equation, **$Mg(OH)_2$ is the base.**

$$\underset{\text{acid}}{2HNO_3}\ (aq) + Mg^{2+}\ (aq) + \underset{\text{base}}{2OH^-}\ (aq) \rightarrow Mg^{2+}\ (aq) + 2NO_3^-\ (aq) + 2H_2O\ (l)$$

Again, this equation is clearly identified as an acid–base reaction, so the ionic compound, $Mg(OH)_2$, cannot be the acid. By default, it is the base, and HNO_3 is the acid.

Even though this may seem like a long, drawn–out process, it is important to be able to analyze acid–base equations in this way because it builds on so many of the concepts we have learned. First, you have to be able to name ionic compounds, including those that contain polyatomic ions. That means you still must have all of the polyatomic ions from table 5.1 memorized. In addition, you have one more polyatomic ion (the hydronium ion) to memorize. Second, you have to be able to interpret chemical formulas of ionic compounds in terms of charge. Third, you have to be able to recognize what subscripts and parentheses in a chemical formula mean. This is all review. If you are having trouble with these skills, review the modules that explained them. The only new skill you need here is the ability to determine where the H^+ ion went. Try to see if you can do these kinds of problems with "On Your Own" questions 9.1–9.3.

ON YOUR OWN

9.1 Determine the acid and base in the following chemical reaction:

$HBr\ (aq) + NH_3\ (aq) \rightarrow NH_4^+\ (aq) + Br^-\ (aq)$

9.2 Determine the acid and base in the following chemical reaction by dividing up the ionic compounds into their ions:

$3C_2H_4O_2\ (aq) + Al(OH)_3\ (aq) \rightarrow Al(C_2H_3O_2)_3\ (aq) + 3H_2O\ (l)$

9.3 Determine the acid and base in the following chemical reaction:

$H_3PO_4\ (aq) + 3KOH\ (aq) \rightarrow 3H_2O\ (l) + K_3PO_4\ (aq)$

RECOGNIZING ACIDS AND BASES FROM THEIR CHEMICAL FORMULAS

Hopefully, you are beginning to see a pattern. With the exception of one ($C_2H_4O_2$), all of the acids we have used so far have an H atom as the very first atom in their chemical formula. In addition, with the exception of one (NH_3), all of the bases we have used so far contain the hydroxide ion. This is a general rule. With a few notable exceptions, most acids have an H atom as the first atom in their chemical formula. Therefore, while we might expect HF to be an acid (because its hydrogen atom is the first in the chemical formula), we would not expect CH_4 to be an acid (because its hydrogen atoms are not first in the chemical formula). In the same way, most bases have the hydroxide ion in them. We would expect RbOH to be a base (because it contains the hydroxide ion), whereas we would predict that CH_3OH is not a base (because, although there is an OH in it, there is not a hydroxide ion since the compound is not ionic and therefore cannot contain ions).

If there is such a simple way to identify acids and bases, why did we show you such a complicated way? First, you must be able to handle the exceptions to these general rules. Second, the next section asks you to *predict* the products of a reaction between any given acid and any given base. The steps you have gone through will help you immensely in this undertaking. Finally, the only way you truly can appreciate the chemical definitions of acids and bases is to see how they pan out in chemical reactions as you did in the section above. Now that you know the easier way to predict whether a compound is an acid or a base, you are free to use it whenever it applies.

Before we leave this section, you need to know one more thing about acids. They have their own names that do not follow the rules for naming compounds that you learned in Module 2. For example, according to the standard means of naming compounds, the molecule HCl would be named hydrogen monochloride. We would name it that way because it is a covalent compound, and covalent compounds use the prefix system to get their names. However, since it is an acid, chemists usually call it hydrochloric acid.

There are several acids like this. Table 9.1 summarizes several of the more popular acids and their names. If we ever refer to an acid by its name in a problem, we will be sure to put its chemical formula in parentheses, and we have taught you other ways to identify acids and bases, so please do not spend any time trying to memorize them. However, as time goes on, you will probably get to know some of these acids by their names because you will see them again and again. If you do have the time and ability to memorize the names and formulas of these acids, then it will make the questions where you are asked to identify the acid and base much easier!

TABLE 9.1
Common Acids

Formula	Name	Formula	Name
HCl	hydrochloric (hy' droh klor' ik) acid	H_2SO_4	sulfuric (sul fyur' ik) acid
HBr	hydrobromic (hy' droh brohm' ik) acid	H_3PO_4	phosphoric (fos for' ik) acid
HF	hydrofluoric (hy' droh flor' ik) acid	H_2CO_3	carbonic (kar bon' ik) acid
HNO_3	nitric (nye' trik) acid	$C_2H_4O_2$	acetic (uh see' tik) acid

Just to give you an idea of where you might find these acids, a diluted form of hydrochloric acid is sold at hardware stores for heavy–duty cleaning. It is often used to get tough lime stains out of bathtubs, toilets, and swimming pools. (You'll see why later on.) In the hardware store, it is called muriatic acid, and it is about 30% HCl and 70% water. Also, very dilute hydrochloric acid is often used in toilet bowl cleaners. Carbonic acid is a compound in soda pop that results from the carbonation process. It is mostly responsible for soda's acidic taste. Acetic acid is the main component of vinegar, giving vinegar its nasty smell. Finally, sulfuric acid is the biggest chemical export for the United States because it is synthesized in the U.S. and sold to both domestic and foreign industries.

You will be happy to know that although chemists do not follow the standard naming rules for acids, they usually do when it comes to bases. The base $Sr(OH)_2$ is simply called strontium hydroxide, which is the proper name for this ionic compound. The only time this isn't the case is when a base is not an ionic compound. For example, we already learned that NH_3 is a base although it doesn't have a hydroxide ion. This covalent compound would have the proper name *nitrogen trihydride*, but chemists usually call it by its nickname, *ammonia*, because it's been known for a very long time by that name.

PREDICTING THE REACTIONS THAT OCCUR BETWEEN ACIDS AND BASES

Now that you've had some experience working with acid–base reactions, it is time for you to start predicting the results of mixing acids with bases. This shouldn't be too hard if you understand the previous section. After all, you now know how to recognize most acids and bases. In addition, you know what acids and bases do. There is really only one more thing that you need to learn before you can begin predicting the results of such reactions on your own.

"On Your Own" question 9.3 touched on this subject. In it, you saw that phosphoric acid (H_3PO_4) donated 3 H^+ ions. Many acids have the ability to donate more than one H^+ ion. Acids that have this ability are generally called polyprotic (pah' lee proh' tik) acids.

Polyprotic acid—An acid that can donate more than one H+ ion.

If you remember the earlier discussion about what an H^+ ion is, you will understand the term *polyprotic*. After all, an H^+ ion is simply a proton. *Polyprotic* refers to the ability to donate more than one (poly) proton (protic).

Chemists often get even more specific in referring to these acids. Phosphoric acid (H_3PO_4) is often called a triprotic (try' proh tik) acid because it can donate 3 protons (H^+ ions). On the other hand, sulfuric acid (H_2SO_4) is often referred to as a diprotic (dye' proh tik) acid because it can donate 2 protons (H^+ ions). For polyprotic acids, the number of hydrogen atoms that begin the chemical formula tells how many H^+ ions the acid will donate. H_2CO_3 (carbonic acid) is a diprotic acid.

From table 9.1, acetic acid, $C_2H_4O_2$, is the only acid where the number of hydrogen atoms in the formula does not match how many H^+ ions are donated. The reason for this is that once $C_2H_4O_2$ loses 1 H^+ ion, $C_2H_3O_2^{1-}$ will be left over. This is the acetate ion, and it is very stable dissolved in water and therefore does not need or want to lose any more H^+ ions. The rest of the acids from table 9.1 will do the same thing; they will lose enough H^+ ions to get to an ion that will be stable when dissolved in water.

Now that we have this new information under our belts, let's see how to predict the products of chemical reactions involving acids and bases. For example, suppose we had hydrofluoric acid (HF) and potassium hydroxide (KOH). Clearly, HF is an acid since it has a hydrogen atom as the first atom in its chemical formula, and potassium hydroxide is a base because it contains a hydroxide ion. We know this is an acid–base reaction.

The first thing we do, then, is split up the ionic compound. This will allow us to see exactly what we're dealing with:

$$HF + K^+ + OH^- \rightarrow ?$$

The acid (HF) is supposed to donate an H^+ ion. Where do you suppose that ion would like to go? Since it is positive, it will be attracted to the negatively charged hydroxide ion. The H^+ ion will be donated to the OH^- ion. When that happens, the 1+ charge on the hydrogen ion will cancel the 1- charge on the hydroxide ion, and the result will be a neutral molecule with 2 hydrogen atoms and 1 oxygen atom: H_2O. The H^+ ion came from a neutral HF molecule, however. If a positive charge is taken from a neutral substance, the substance becomes negatively charged. When the HF molecule donates its H^+ ion, it becomes an F^- ion. In the end, then, something like this happened:

$$HF + K^+ + OH^- \rightarrow F^- + K^+ + H_2O$$

H^+

That's the actual reaction. However, chemists often don't care to write out all of those ions, even though they exist in solution. Instead, we often regroup the ions into the ionic compounds that formed them. The more commonly used version of this equation would be:

$$HF + KOH \rightarrow KF + H_2O$$

This is how we can predict the reactions that occur between acids and bases.

This method for determining the reactions that occur between acids and bases can become very cumbersome when we are dealing with polyprotic acids or bases that contain more than one hydroxide ion [$Mg(OH)_2$, for example]. We will use the results of the above discussion to develop a shortcut approach to determining the products of an acid–base reaction.

Notice the 2 products formed in the above discussion. The acid and base reacted to form an ionic compound and water. When acids react with bases that are composed of hydroxide ions (which represent the vast majority of bases in nature), they will always produce water. This should make sense to you. If an acid always donates H^+ ions, and if those ions have the opportunity to react with OH^- ions, clearly the product will be water. There will most likely be another product as well.

If a base contains a hydroxide ion, it must also have a positive ion. This positive ion will be left behind once the hydroxide ion reacts with the hydrogen ion. In addition, a negative ion will be left behind once the H^+ ion leaves the acid. The positive ion left over from the base and the negative ion left over from the acid can form a new ionic compound. The products of acid–base reactions are (almost always) water and an ionic compound. Chemists use the general term *salt* to refer to ionic compounds. Although most people use the term salt to refer to table salt (the ionic compound NaCl), chemists use the term much more broadly as an all-inclusive term that refers to most ionic compounds. Chemists often say that the following general guideline can be used to determine the products of an acid–base reaction:

$$\text{ACID} + \underset{\text{(hydroxide)}}{\text{BASE}} \rightarrow \text{SALT} + \text{WATER}$$

Looking at the acid–base reaction we went through above, you can see that it fits this general pattern:

$$\begin{array}{ccccccc} \text{HF} & + & \text{KOH} & \rightarrow & \text{KF} & + & \text{H}_2\text{O} \\ \text{acid} & + & \text{base} & \rightarrow & \text{salt} & + & \text{water} \end{array}$$

In general, most acid–base reactions will fall into this general pattern. To predict the products of an acid–base reaction, all we have to realize is that one of them will be water. The other will be an ionic compound made up of the positive ion from the base and the negative ion left over from the acid. This type of acid–base reaction is often called a neutralization reaction because the acid and base neutralize each other, producing water and a salt. Let's see how this pans out with Example 9.2.

EXAMPLE 9.2

Give the balanced chemical equation that represents the reaction between HCl and RbOH.

We know that when dealing with bases that contain the hydroxide ion (as RbOH does), the general reaction is:

$$\text{ACID} + \underset{\text{(hydroxide)}}{\text{BASE}} \rightarrow \text{SALT} + \text{WATER}$$

The only difficulty is determining what the salt is. According to our rules, we take the positive ion from the base (Rb^+) and the negative ion left over when the H^+ leaves the acid (Cl^-) and form an ionic compound. Since the charges are the same on these 2 ions, we can simply ignore them in the chemical formula. The ionic compound that forms between these 2 ions is RbCl. The reaction, then, is:

$$\mathbf{HCl + RbOH \rightarrow RbCl + H_2O}$$

Notice the skills we have to use to solve these problems. First we have to recognize the acid and the base. Then we have to be able to break the base into its ions. Then we need to determine what ion is left behind when the acid donates its H^+ ion. Finally, we have to be able to determine the chemical formula of the resulting ionic compound. Remember, you learned how to do that last step in Module 4, and you practiced it again in Module 5 with polyatomic ions. If you have forgotten how to do that, you might want to go back and review those modules.

Write the chemical equation that describes the reaction between sulfuric acid (H_2SO_4) and NaOH.

Once again, we know that one of the products in this reaction is water. What is the other? The positive ion from the base is Na^+. The acid in question is diprotic, so it wants to donate *both* of its H^+ ions. Once H_2SO_4 donates both of its H^+ ions, the SO_4^{2-} ion (which you should recognize as the sulfate ion) remains. The other product is the ionic compound that forms between Na^+ and SO_4^{2-}. To get the formula of that ionic compound, we switch the charges and drop the signs, giving us Na_2SO_4. The equation, then, is:

$$H_2SO_4 + NaOH \rightarrow Na_2SO_4 + H_2O$$

If you look at this equation carefully, you will see that it is not balanced. Therefore, to answer this question properly, we must balance it.

$$\mathbf{H_2SO_4 + 2NaOH \rightarrow Na_2SO_4 + 2H_2O}$$

This is the reaction that occurs between sulfuric acid and sodium hydroxide.

The acid–hydroxide base reactions can be balanced quickly by comparing the acid to the base. H_2SO_4 is a diprotic acid, but NaOH only has 1 OH^- ion to complete a water molecule. We can see that there will have to be 2 OH^- ions to accept the 2 H^+ ions donated by the acid. Don't forget that 2 H^+ ions and 2 OH^- ions will form 2 water molecules. If this method of balancing does not make sense to you, then go ahead and balance the equations in the manner you were taught back in Module 6.

What is the reaction that occurs between HNO_3 and $Al(OH)_3$?

Once again, we know that one of the products in this reaction will be water. In addition, the positive ion from the base is Al^{3+}, while the negative ion left over from the acid will be NO_3^- (which you should recognize as the nitrate ion). Switching the charges and dropping the signs gives us $Al(NO_3)_3$ as the chemical formula of the salt:

$$HNO_3 + Al(OH)_3 \rightarrow Al(NO_3)_3 + H_2O$$

Now all we have to do is balance the equation. Monoprotic acid plus 3 OH^- ions means we need a 3:1 ratio between the reactants and 3 water molecules will be produced.

$$\mathbf{3HNO_3 + Al(OH)_3 \rightarrow Al(NO_3)_3 + 3H_2O}$$

> **ON YOUR OWN**
>
> 9.4 Give the chemical equation that represents the reaction between each of the following pairs of reactants:
>
> a. $HClO_3$ and KOH
> b. HBr and $Ca(OH)_2$
> c. H_3PO_4 and $Mg(OH)_2$

Now answer "On Your Own" question 9.4 to be sure you understand this concept.

Now that you know how to predict the reactions between acids and bases, you can begin to see why acids and bases are so useful as household chemicals. For example, you found out in Experiment 9.1 that the all–purpose cleaner you tested contained a base. The vast majority of messes that need to be cleaned up are made of food and other substances that contain acids. When you spray a mess with all–purpose cleaner, the acids that make up the mess are turned into salt and water, which can then be picked up with a rag or paper towel.

In Experiment 9.1, you also found out that the toilet bowl cleaner was made of acid. If all–purpose cleaners are bases, why are toilet bowl cleaners acid? The major problem in toilet bowls is lime stains. Lime (CaO) becomes a base when mixed with water. So to clean up a stain involving base, what do you need? An acid, of course.

THE REACTIONS BETWEEN ACIDS AND COVALENT BASES

The method you just learned is adequate for predicting the reactions that occur between acids and hydroxide bases, as long as the bases contain a hydroxide ion. But what if an acid wants to react with a covalent base? What can we do to predict the subsequent reaction? Predicting these kinds of reactions is a bit easier than what you just learned.

For example, suppose we mix sulfuric acid with water. We know that water is an amphiprotic substance. Therefore, it can act as an acid or a base. In the presence of an acid, water will act as a base. The problem is that water does not contain the hydroxide ion. In fact, it has no ions because water is a covalent compound. How, then, do we determine the reaction that occurs? All we need to do is fall back on the definitions of acids and bases.

Acids want to donate an H^+ ion, while bases want to accept an H^+ ion. It is easy to see why the OH^- ion is a base because by accepting the H^+ ion, it will form H_2O, a very stable and neutral molecule. But how can a covalent base accept an H^+ ion since the covalent base molecule is already neutral? To answer this question, let's go back to Lewis structures.

H^+ is an ion on the search for 2 electrons to reach its octet. The covalent base needs to have 2 electrons to share with the H^+ ion. Take a look at the Lewis structure of a common covalent base, ammonia, NH_3.

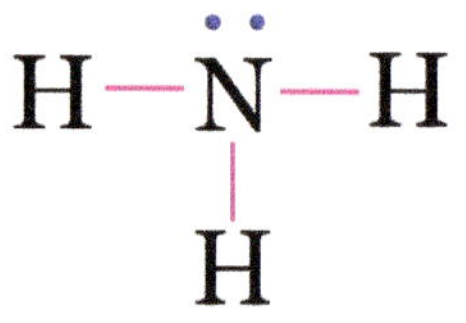

2–Dimensional

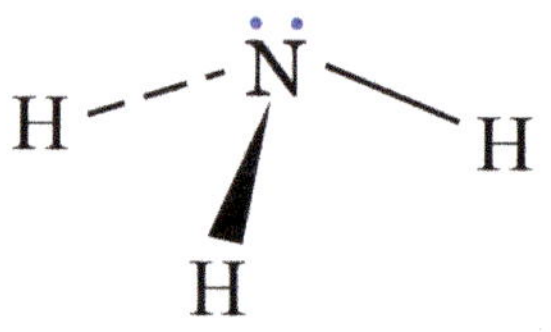

3–Dimensional

Can you see where the H^+ is going to go? The lone pair of electrons on the central atom are all exposed for the H^+ ion to get in there and share those 2 electrons. This will be a common feature for all covalent bases, so we can make a general rule for covalent bases. There are exceptions to this rule (What chemistry rule doesn't have exceptions, right?), but the exceptions will not be covered in this course.

Covalent bases accept only 1 H^+ ion. Diprotic and triprotic acids require more than 1 covalent base to accept their H^+ ions.

Since H_2SO_4 is an acid, it wants to donate H^+ ions, and since H_2O will act as a base in the presence of acids, it wants to accept an H^+ ion. H_2SO_4 will donate an H^+ ion to H_2O and form H_3O^+. The problem is that H_2SO_4 wants to donate 2 H^+ ions, but water can accept only 1 H^+ ion. The sulfuric acid will react with 2 H_2O molecules to form 2 H_3O^+ ions. After donating 2 H^+ ions, the H_2SO_4 becomes SO_4^{2-}. Thus, the overall reaction is:

$$H_2SO_4 + 2H_2O \rightarrow 2H_3O^+ + SO_4^{2-}$$

Notice the difference between this kind of acid–base reaction and the ones we did in the previous section. In the previous section, a covalent acid reacted with an ionic base to form an ionic compound (a salt) and a covalent compound (water). In this kind of reaction, a covalent acid reacted with a covalent base to form 2 ions. Make sure you understand this concept by studying Example 9.3 and answering "On Your Own" questions 9.5–9.6.

EXAMPLE 9.3

What reaction occurs between H_3PO_4 and ammonia (NH_3), which is a base?

In this case, we are reacting a triprotic acid with a covalent base. We just transfer an H^+ ion to the base. However, since the acid wants to donate 3 H^+ ions but the base can accept only 1, we have to do this 3 times. Once H_3PO_4 donates its 3 H^+ ions, it becomes PO_4^{3-}. The reaction, then, is:

$$H_3PO_4 + 3NH_3 \rightarrow 3NH_4^+ + PO_4^{3-}$$

What reaction occurs between $H_2Cr_2O_7$ and water?

Water is an amphiprotic compound, so it will act as a base in the presence of an acid. This particular acid is diprotic, so it will donate 2 H^+ ions to 2 water molecules and make 2 H_3O^+ ions. Once it does this, it will become $Cr_2O_7^{2-}$, so the reaction is:

$$H_2Cr_2O_7 + 2H_2O \rightarrow 2H_3O^+ + Cr_2O_7^{2-}$$

Before we leave this section, step back and look at what you have accomplished. Given only the reactants in an acid–base reaction, you can write down the chemical equation and therefore determine the products that will be formed. You already learned how to do this for 3 other types of reactions: combustion, formation, and decomposition. At this point in your

ON YOUR OWN

9.5 What is the chemical equation that describes the reaction between H_2SO_4 and ammonia (NH_3)?

9.6 What reaction occurs between HI and water?

chemistry education, you can already predict the products of 4 different types of chemical reactions, given just the reactants. That's quite a feat! It took human beings about 2,500 years to get to the point where they could do such a thing. After less than a year of study, you now have that ability.

MOLARITY

Now that we know what reactions occur between acids and bases, we need to understand another very important concept that relates not only to acids and bases but all chemistry that occurs in combination with water.

The Works is a specific toilet bowl cleaner that is used to get rid of lime stains in a toilet bowl. However, hardware stores sell a product called muriatic acid which claims (accurately) to get rid of lime stains that normal toilet bowl cleaners cannot get rid of. Each of these products has only one active ingredient: hydrochloric acid (HCl). If they both have the same active ingredient, why is one much more powerful than the other?

The answer is quite simple. Muriatic acid is about 30% HCl and about 70% water. The Works, on the other hand, is only about 3% HCl and about 97% water. Even though they both have HCl, muriatic acid is a stronger cleaner because the HCl in it is *more concentrated.* The concentration of a chemical in a product is just as important as the type of chemical.

In chemistry, we can express concentration in several different ways, depending on what we need to know. In the example above, we used percentage as an expression of concentration. The higher the percentage of HCl, the more concentrated the HCl. Usually concentration is discussed in terms of amount divided by volume. For example, the unit of grams/mL (grams per milliliter) is a concentration unit because it tells how much (grams) of a substance is in a certain volume (mL). The most popular way of expressing concentration for chemists is with the unit moles/liter (moles per liter), which is called molarity (moh lehr' uh tee).

Molarity (M)—A concentration unit that tells how many moles of a substance are in a liter of solution. It is determined by dividing the number of moles of a substance by the number of liters of solution.

$$M = \frac{\#\text{ moles}}{\#\text{ liters}} \qquad \text{Equation 9.1}$$

It should make sense to you why molarity is the most popular concentration unit for chemists. After all, the most convenient way for a chemist to express the quantity of a substance is in moles since that way it can be directly used in stoichiometry problems.

How do we use molarity? Let's go back to our discussion of muriatic acid. If you go to the hardware store to buy muriatic acid, you will usually find it in 4–liter bottles. A 4–liter bottle of muriatic acid contains 1,400 grams of HCl dissolved in enough water to fill up a 4–liter bottle. Since we know that the molecular mass of HCl is 36.5 amu, we know that it takes 36.5 grams of HCl to make 1 mole. We can figure out how many moles of HCl are in that 4–liter bottle:

$$\frac{1{,}400\ \cancel{\text{g HCl}}}{1} \times \frac{1\text{ mole HCl}}{36.5\ \cancel{\text{g HCl}}} = 38\text{ moles HCl}$$

Now that we know how many moles of HCl are in the bottle of muriatic acid, we can calculate the concentration of HCl in terms of molarity by dividing that number of moles by the volume of the solution (4.00 liters):

$$\text{Concentration of HCl} = \frac{\text{\# moles}}{\text{\# liters}} = \frac{38\ \text{moles}}{4.00\ \text{liters}} = 9.5\ \text{M}$$

You might ask what good this does. What does a concentration of 9.5 M mean? The concentration of an acid (or base) indicates how strong it is. For example, the concentration of HCl in The Works toilet bowl cleaner is about 0.8 M. That means that muriatic acid is over 10 times more acidic than The Works. If you are trying to get out lime stains, muriatic acid is more than 10 times better at getting the job done than The Works. Of course, the flip side to this is that muriatic acid is also much more caustic and can therefore be more dangerous to work with. Spilling some of The Works on your hands is not so bad, but spilling muriatic acid on your hands will burn your skin quite painfully.

We will see another use of the concept of concentration in a moment, but first you need to be sure that you truly understand how to calculate molarity with Example 9.4, followed by "On Your Own" question 9.7.

EXAMPLE 9.4

What is the concentration (in units of molarity) of NaOH if there are 2.6 moles of the substance in a solution with a volume of 534 mL?

To get molarity, we must divide the number of moles of NaOH by the number of liters of solution. We already have the number of moles of NaOH, but we do not have the number of liters of solution; instead, we have the volume in mL. First we must convert from mL to liters:

$$\frac{534\ \cancel{\text{mL}}}{1} \times \frac{0.001\ \text{L}}{1\ \cancel{\text{mL}}} = 0.534\ \text{L}$$

Now that we have both moles of NaOH and liters of solution, we can just use equation 9.1 to get concentration in units of molarity:

$$\text{M} = \frac{\text{\# moles}}{\text{\# liters}} = \frac{2.6\ \text{moles NaOH}}{0.534\ \text{L}} = 4.9\ \frac{\text{moles NaOH}}{\text{L}} = 4.9\ \text{M}$$

Thus, our NaOH solution has a concentration of **4.9 M**.

If a student dissolves 124.3 grams of HNO_3 in enough water to make 250.0 mL of solution, what is the resulting molarity of HNO_3?

To get to molarity, we must have moles of acid and liters of solution. Right now we have neither, so we need to convert grams into moles and mL into liters. First, we'll convert grams of HNO_3 to moles:

Molecular mass of HNO_3 = (1 x 1.01 amu) + (1 x 14.0 amu) + (3 x 16.0 amu) = 63.0 amu

1 mole HNO_3 = 63.0 grams HNO_3

$$\frac{124.3\ \cancel{g\ HNO_3}}{1} \times \frac{1\text{ mole } HNO_3}{63.0\ \cancel{g\ HNO_3}} = 1.97\text{ moles } HNO_3$$

Now we'll convert mL to liters:

$$\frac{250.0\ \cancel{mL}}{1} \times \frac{0.001\text{ L}}{1\ \cancel{mL}} = 0.2500\text{ L}$$

Now that we have moles and liters, we can finally calculate molarity using equation 9.1:

$$M = \frac{\#\text{ moles}}{\#\text{ liters}} = \frac{1.97\text{ moles } HNO_3}{0.2500\text{ liters}} = 7.88\ \frac{\text{moles } HNO_3}{\text{liter}} = 7.88\text{ M}$$

Our HNO_3 has a concentration of **7.88 M**.

ON YOUR OWN

9.7 Determine the concentration of acid or base for each of the following solutions:

a. 0.21 moles HCl in 0.346 L of solution

b. 40.2 grams of H_2SO_4 dissolved in enough water to make 6.8 liters of solution

c. 5.55 grams of KOH dissolved in enough water to make 125.0 mL of solution

THE DILUTION EQUATION

So far, you have learned to calculate the concentration of an acid or base when you are given the number of grams (or moles) of the acid or base and the total volume of the solution that contains it. This is a very useful way to calculate concentration, but it is not adequate by itself. Most of the time, chemists do not make up acid or base solutions by weighing a certain number of grams and then dissolving the acid or base into a solution. A much more common way for chemists to make acid and base solutions is by dilution.

Dilution—Adding water to a solution to decrease the concentration.

For example, when chemists buy HCl, they usually buy concentrated HCl, which is a liquid that comes in bottles. The HCl in these bottles has a concentration of 12.0 M. If a chemist wants to make a 2.0 M solution of HCl, it is easier to just dilute the 12.0 M until its concentration decreases to 2.0 M, rather than measure the correct number of grams of HCl.

To be able do this, the chemist dilutes a small volume of the highly concentrated solution with water. For this to be successful, the chemist must know *how much* of the original solution to use and *how much* water to add. Chemists can calculate this information with a very simple equation, called the dilution equation:

$$M_1V_1 = M_2V_2 \qquad \text{Equation 9.2}$$

In this equation, M_1 refers to the original concentration of the solution, and V_1 stands for the volume of the original solution that is going to be diluted. M_2 refers to the desired concentration, while V_2 refers to the final volume of the new, diluted solution. Let's see how this equation is used in Example 9.5, and then try "On Your Own" question 9.8.

EXAMPLE 9.5

A chemist wants to dilute 12.0 M HCl to make 500.0 mL of a 3.5 M HCl solution. How will he do this?

This is a dilution, so we obviously must use the molar dilution equation. The original concentration of the HCl is 12.0 M, so that is M_1. The chemist wants the diluted solution to have a concentration of 3.5 M, so that is M_2. He needs 500.0 mL of this 3.5 M solution, so that is V_2. The only thing we don't know in the molar dilution equation is V_1, so we can solve for it:

$$M_1V_1 = M_2V_2$$

$$12.0\text{ M} \times V_1 = 3.5\text{ M} \times 500.0\text{ mL}$$

$$V_1 = \frac{3.5\cancel{\text{M}} \times 500.00\text{ mL}}{12.0\cancel{\text{M}}} = 1.5 \times 10^2\text{ mL}$$

Notice that since we are dividing and multiplying by molarity, it cancels, leaving us with the unit mL. This is good since we are solving for volume. So what does this answer mean? It means that the chemist needs to take **1.5×10^2 mL of the original solution and dilute it with water until it reaches 500.0 mL of volume**. This will give him 500.0 mL of a 3.5 M solution.

A chemist takes 2.5 liters of 5.6 M NaOH and dilutes it with enough water to make 10.0 liters. What is the new concentration of the NaOH?

In this case, we are not told what the chemist wants; we are told what he does: He dilutes a solution of NaOH. We are told the original concentration (M_1 = 5.6 M) and volume (V_1 = 2.5 liters), as well as the final volume (V_2 = 10.0 liters). To determine the final concentration, we just use the molar dilution equation:

$$M_1V_1 = M_2V_2$$

$$5.6\text{ M} \times 2.5\text{ L} = M_2 \times 10.0\text{ L}$$

$$M = \frac{5.6\text{ M} \times 2.5\cancel{\text{L}}}{10.0\cancel{\text{L}}} = \mathbf{1.4\text{ M}}$$

Once again, the units work out. Since we are both multiplying and dividing by liters, they cancel out, leaving molarity, which is the unit of concentration.

ON YOUR OWN

9.8 A chemist needs to make 500.0 mL of a 2.5 M solution of sulfuric acid. Looking on the shelves, she finds a bottle of 10.0 M sulfuric acid. How can the desired solution be made?

THE IMPORTANCE OF CONCENTRATION IN CHEMISTRY

Before we move on to applying the concept of concentration to stoichiometry, let's make one point very clear. We have used the subject of acids and bases to introduce the concept of concentration, but don't think for a moment that this concept is limited to acids and bases. Concentration is a very important topic that will come up time and time again throughout this course because the concentration of a chemical often determines its practical effect.

The topic in which concentration probably plays its most crucial role is that of chemical toxicity. For example, do you remember learning about ozone in Module 4? You were told that ozone is a toxic poison for human beings and most other forms of life. Would it surprise you to know, then, that you are breathing ozone *right now, as you read this page*? If ozone is an extremely toxic poison and you are breathing ozone right now, why isn't it affecting you? The reason it doesn't affect you is that its concentration is too low for its poisonous effects to occur. Although ozone is a poison, you can still breathe it, as long as its concentration is very low. This keeps it from being able to perform effectively enough, making it safe for you to inhale.

In fact, you are breathing in several different poisons right now. No matter where you are, the air that you are breathing contains nitrogen monoxide, nitrogen dioxide, ozone, sulfur dioxide, sulfur trioxide, and carbon monoxide. These are all toxic poisons, but their concentration is low in most areas of the country. Even though the poisons exist, they are not dangerous because their concentration is not large enough to allow them to perform effectively. If you live in places like Los Angeles, some air pollutants (especially nitrogen monoxide) can reach high enough concentrations that they can begin to harm your body. That's when you should begin getting concerned.

You might be surprised to learn that the pendulum can also swing the other way. At one time or another you have probably learned that your body needs certain vitamins to continue to function properly. This is quite true. Did you also know that some vitamins (especially vitamins A, D, E, and K) can become *poisonous* to your body if their concentrations become too high? Some people have become very sick and have even died because they have taken *too many* of these vitamins! This is how important the subject of concentration is. At low concentrations, vitamins are beneficial (indeed, necessary) for your body. However, if their concentrations build up, they begin working against the body and can kill you, or at least make you very sick.

There is no reason to get paranoid about this. Even if you take vitamin supplements, you are probably not getting anywhere near toxic levels. It is important to make sure that you are not taking *more* than the recommended dosage of your vitamin supplement. This is where people start to get into trouble. They think that if they double or triple the recommended dosage, they are doubling or tripling the benefits. That may be true of some vitamins (vitamin C, for example), but not most vitamins (especially vitamins A, D, E and K). There has been a lot of study about vitamins, and that's where the recommended dosages for your vitamin supplements come from. As long as you follow those directions, you should be fine.

USING CONCENTRATION IN STOICHIOMETRY

Another place where concentration plays a critical role is in stoichiometry. This is because the concentration of an acid or a base gives us a way of determining the number of moles of acid or base that we are adding to a chemical reaction. Suppose, for example, we wanted to perform the following chemical reaction:

$$2HCl + Ba(OH)_2 \rightarrow BaCl_2 + 2H_2O$$

This will produce barium chloride. If we have 100.0 mL of muriatic acid and an excess of barium hydroxide, we should be able to use stoichiometry to determine how much barium chloride would be produced in the reaction. The fact that we added 100.0 mL of acid tells how much acid we added, but it doesn't help in terms of stoichiometry. For stoichiometry to work, we need to know how many *moles* of HCl we added.

Since we know that the concentration of HCl in muriatic acid is 9.5 M (we calculated it earlier), we can use that information to convert the 100.0 mL into moles. After all, the molarity tells the number of moles of HCl per liter of muriatic acid. We already have the number of mL of muriatic acid we used. If we convert that into liters, we can then multiply the concentration by the volume to get the number of moles of acid we added:

$$\frac{100.0\ \cancel{\text{mL}}}{1} \times \frac{0.001\ \text{L}}{1\ \cancel{\text{mL}}} = 0.1000\ \text{L}$$

$$\frac{9.5\ \text{moles HCl}}{1\ \cancel{\text{L}}} \times \frac{0.1000\ \cancel{\text{L}}}{1} = 0.95\ \text{moles HCl}$$

Now that we have the number of moles of HCl added, we can figure out how many grams of $BaCl_2$ were produced by using stoichiometry:

$$\frac{0.95\ \text{moles}\ \cancel{\text{HCl}}}{1} \times \frac{1\ \text{mole}\ BaCl_2}{2\ \text{moles}\ \cancel{\text{HCl}}} = 0.48\ \text{moles}\ BaCl_2$$

$$\frac{0.48\ \cancel{\text{moles}\ BaCl_2}}{1} \times \frac{208.3\ \text{g}\ BaCl_2}{1\ \cancel{\text{mole}\ BaCl_2}} = 1.0 \times 10^2\ \text{g}\ BaCl_2$$

When 100.0 mL of muriatic acid reacts with an excess of barium hydroxide, 1.0×10^2 g of barium chloride is produced.

It is important that you see the relationship between molarity and moles:

When *number of moles* is *divided* by volume (in liters), we get *molarity*.

When *molarity* is *multiplied* by volume (in liters), we get *number of moles*.

Getting this relationship straight in your mind will help you to use concentration, especially in the stoichiometry problems in Example 9.6 and "On Your Own" question 9.9.

MOD 9

EXAMPLE 9.6

One way to make sodium sulfate is through the reaction of sulfuric acid (H_2SO_4) with sodium hydroxide (NaOH). How many grams of sodium sulfate can be produced if 500.0 mL of 1.6 M sodium hydroxide is reacted with an excess of sulfuric acid?

To solve any stoichiometry problem, we must first figure out the balanced chemical equation. According to the problem, our reactants are H_2SO_4 and NaOH. When they react, a salt and water will be produced. The salt will be composed of the positive ion from the base (Na^+) and the negative ion left over when the acid gets rid of all of its H^+ ions. In this case, that will be the SO_4^{2-} ion. Thus, the reaction is:

$$H_2SO_4 + NaOH \rightarrow Na_2SO_4 + H_2O$$

Of course, this equation is not balanced, so that is the first thing to do:

$$H_2SO_4 + 2NaOH \rightarrow Na_2SO_4 + 2H_2O$$

Now that we have a balanced equation, we can start the stoichiometry. We are told that the sulfuric acid is in excess, so we know that NaOH is the limiting reactant. Therefore, we need to know the number of moles of NaOH to be able to predict how much sodium sulfate is formed. To do this, we will multiply concentration by volume:

$$\frac{1.6 \text{ moles NaOH}}{1 \cancel{\text{ L NaOH}}} \times \frac{0.5000 \cancel{\text{ L NaOH}}}{1} = 0.80 \text{ moles NaOH}$$

We can now use this information to calculate the number of moles of sodium sulfate produced:

$$\frac{0.80 \cancel{\text{ moles NaOH}}}{1} \times \frac{1 \text{ mole } Na_2SO_4}{2 \cancel{\text{ moles NaOH}}} = 0.40 \text{ moles } Na_2SO_4$$

Now that we have the number of moles of sodium sulfate produced, we can convert back to grams:

$$\frac{0.40 \cancel{\text{ moles } Na_2SO_4}}{1} \times \frac{142.1 \text{ g } Na_2SO_4}{1 \cancel{\text{ mole } Na_2SO_4}} = 57 \text{ g } Na_2SO_4$$

ON YOUR OWN

9.9 Barium hydroxide is often used to clean up toxic acid spills in industrial settings. If a 10.0 liter container of 12.0 M nitric acid (HNO_3) is broken and the acid spills all over the floor, how many grams of barium hydroxide will be needed to clean up the spill?

Remember, the vast majority of this should be review. You should already know how to do the stoichiometry part of this problem. The only new things here are determining the equation (which you learned in an earlier section) and using concentration to figure out the number of moles of limiting reactant.

ACID–BASE TITRATIONS

One specific application of stoichiometry to acid–base reactions is the process known as titration.

Titration—The process of slowly reacting a base of unknown concentration with an acid of known concentration (or vice versa) until just enough acid has been added to react with all of the base. This process determines the concentration of the unknown base (or acid).

In other words, if we have an acid of unknown concentration, we can figure out the concentration by titrating it against a base of known concentration. This is done by slowly adding the base to the acid until we have added just enough base to react completely with all of the acid present. Once we know that this has happened, we can use stoichiometry to determine the concentration of the acid. To be able to do this, there must be a way to determine when we have added just enough base to react completely with all of the acid present. This is accomplished through the use of indicators.

At the beginning of this module, we introduced the concept of indicators, and you used litmus as an acid–base indicator. There are several different types of indicators that turn one color in the presence of a base and a different color in the presence of an acid. (You will make your own in Experiment 9.2.) Let's suppose, for the sake of argument, that we were using litmus as an indicator. Litmus is red in the presence of an acid, so if we added litmus to an acid of unknown concentration, it would turn red.

Let's suppose you slowly started adding base to your acid. At first, the litmus would stay red because as you added base, it would immediately react with the acid. The litmus, then, would still see acid in the solution and would stay red. However, what would happen after all of the acid had reacted with the base? No acid would be left. Therefore, the *very next drop* of base that we added would turn the litmus from red to blue. This would indicate that all of the acid was gone, and we would finish our titration. When the indicator changed color, we would call this the endpoint of the titration.

If this sounds a bit confusing to you at this point, don't worry. Just perform Experiment 9.2, and you will understand much better after you are finished.

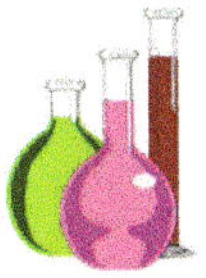

If you purchased the MicroChem kit mentioned in the Student Notes, you can perform Experiment 12 in place of or in addition to this experiment.

EXPERIMENT 9.2

PURPOSE: To determine the concentration of ammonia.

MATERIALS:

- Eyedropper
- Mass scale
- Distilled water (available at any grocery store)
- Sheet of unlined white paper
- Stirring rod (or a small spoon)
- A few leaves of red cabbage (It must be red cabbage, not regular cabbage.)
- 2 beakers (If you don't have beakers, use a short, fat, transparent glass and a small pot to boil water in.)
- Flame or stove for heating
- Graduated cylinder (Measuring cups and spoons will work, but the experiment will be much harder.)
- Clear ammonia solution (This is sold with the cleaning supplies in most supermarkets. It must be clear. A colored solution will mess up the endpoint.)
- Clear vinegar (Colored vinegar will mess up the endpoint.)
- Safety goggles

QUESTION: What is the concentration of a household ammonia solution?

HYPOTHESIS: Write a hypothesis related to this experiment. Consult the definition of titration if you need help making a hypothesis.

PROCEDURE:

1. Rinse your beakers, graduated cylinders, and medicine dropper thoroughly with distilled water. Any contamination in them will mess up the endpoint of the experiment.
2. Put the red cabbage leaves in the small beaker and fill it with about 70 mL (a qualitative measurement) of distilled water. If you are using a boiling pot, you will probably need to add 3 times as much water because the water will evaporate more quickly in a pot than in the small beaker.
3. Boil the cabbage leaves and water in the beaker for about 3 minutes.
4. Allow the beaker to cool, and then carefully remove the leaves from the beaker. The water should now be a pinkish-red color. If it is blue-green, that indicates that you contaminated your water a bit. This is your indicator. In the presence of acid (or neutral water), it is a pinkish-red color. In the presence of base, it is a blue-green color.
5. Use the graduated cylinder to add 10.0 mL (a quantitative measurement) of the clear ammonia to the large beaker. (If you are using measuring spoons, this is 2 teaspoons.)
6. Add 90.0 mL (about ½ cup) of distilled water to the beaker. Stir the solution.
7. Add a little less than half of the indicator to the beaker. The indicator should immediately turn greenish-blue. This indicates the presence of a base, ammonia (NH_3). You will now do a titration to determine the concentration of the ammonia in the solution that you bought.
8. Rinse the graduated cylinder thoroughly with distilled water.
9. Measure the mass of the graduated cylinder.
10. Fill the graduated cylinder with 50.0 mL (a quantitative measurement) of vinegar and measure the new mass of the cylinder plus the vinegar.
11. Determine the mass of the vinegar by difference.
12. Use the mass and the volume (50.0 mL) to calculate the density of vinegar. (If you are using measuring cups, you can do the same thing with ¼ cup. You must also calibrate your medicine dropper the way you did in experiment 7.1, using ¼ cup as a good approximation of 50.0 mL).
13. Place the beaker full of ammonia solution on the white paper so that it is easy to see the color of the solution.
14. Fill your medicine dropper from the graduated cylinder full of vinegar.
15. Squirt the vinegar into the ammonia solution and stir. Watch for any color change. (If you are using measuring cups, you cannot do it this way. Read the following instructions, but perform only the instructions in step 26.) Continue to do this until the color of the solution turns pinkish. It will not turn as vividly pinkish-red as the original color of the indicator, but it should turn noticeably pink. This tells you that you passed the endpoint of the titration.
16. Read the number of mL of vinegar left in the graduated cylinder and, by difference, determine how much vinegar you added in the titration.
17. What you just did is referred to as a rough titration. You ended up adding more acid

than necessary, but you did it fairly quickly. Now you need to start over and do a careful titration.

18. Pour the titration solution down the drain, and thoroughly rinse the beaker with distilled water.
19. Fill the beaker with 10.0 mL of ammonia and 90.0 mL of distilled water.
20. Stir the solution, and then add most of what is left of the indicator solution.
21. Put the beaker back on the white paper and fill the graduated cylinder with precisely 50.0 mL of vinegar.
22. Now you can begin the careful titration. Use your medicine dropper to quickly add vinegar to the beaker until you get to within 5 mL of the amount you added during the rough titration.
23. Now you need to slow down. Add the vinegar drop by drop and stir thoroughly between drops. Keep careful watch on the color. The color should change from green to blue, then to a very light blue, and then to a pinkish violet. The pinkish violet color tells you that you have reached the endpoint. The best thing to look for is the very first hint of pink color that continues to exist after you have stirred the solution. As soon as you see that, you know you have reached the endpoint.
24. Squirt the remainder of the medicine dropper's contents back into the graduated cylinder.
25. Read the graduated cylinder and, by difference, calculate the number of mL you added to reach the true endpoint of the titration.
26. If you are not using measuring cups, skip this step. If you are using measuring cups, you must do a careful titration the first time. Add the vinegar drop by drop, making sure to keep track of the number of drops that you added. Watch for the endpoint as detailed in the previous steps. When you reach it, the number of drops times your dropper calibration will tell you the number of mL you needed to add to get to the endpoint.
27. Clean up and return everything to the proper place.

CONCLUSION: Write something about what you observed and whether it confirmed your hypothesis.

Before we tell you how to analyze your experimental results, we want to review exactly what you did so that you do not get lost in the details of the experiment. You took an ammonia solution (a base) with an unknown concentration and titrated it with a vinegar solution (an acid), whose concentration is known. By using an indicator and adding the vinegar drop by drop, you should have been able to determine exactly when you had added enough acid (vinegar) to neutralize all of the base (ammonia). That's what you do in a titration. Now that you have done that, you can do the calculation that will tell you the concentration of the ammonia.

First, realize that the reaction going on in this experiment is:

$$C_2H_4O_2 + NH_3 \rightarrow NH_4^+ + C_2H_3O_2^-$$

This reaction tells us that for every 1 mole of acid you added, 1 mole of base reacted. When you reached the endpoint, you also knew that you had added just enough acid to react with all of the base present. You can use this information to calculate the concentration of the ammonia in the reaction.

Start with the number of moles of acid you added. This is easy. Multiply the density of the vinegar (which you measured in the first part of the experiment) by the number of mL you added to reach the endpoint. This will tell you how many grams of vinegar you added. The acid makes up only 5.00% of the vinegar, so multiply your answer by 0.0500. This will tell you how many grams of acid you added. Use the chemical formula of the acid in the equation above to convert the number of grams of acid added into the number of moles of acid added.

Now that you know how many moles of acid you added, you can figure out how many moles of base were originally present. In the equation, we see that every 1 mole of acid requires 1 mole of base. The fact that you reached the endpoint of the titration shows that all of the acid added reacted with all of the base present. Therefore, the number of moles of base present is equal to the number of moles of acid added.

The concentration of a base is defined as the number of moles of base (which you just figured out) divided by the volume of the solution (10.0 mL or 0.0100 liters). You divide by 10.0 mL because that's the volume of the ammonia solution; the 90.0 mL of water you added was just to make the color easier to see. You want to know the concentration of the ammonia solution that you bought. The volume of that solution was 10.0 mL. Divide the number of moles by the number of liters, and you have the concentration of the ammonia solution. Check your calculations against the sample calculations in the solutions and tests guide to make sure you did everything correctly.

Titrations like the one you just performed are a very common tool in laboratory chemistry. Most laboratories use slightly more sophisticated equipment than you did. Instead of a graduated cylinder and an eyedropper, for example, most titrations are carried out with a buret (byoor eht'). This piece of laboratory glassware is a long, glass tube with a volume scale on it and a dropper at the bottom. It is more precise than a graduated cylinder because it is able to measure volume to 0.01 mL. A buret is quite expensive, however.

Notice that there is *absolutely nothing new* to learn to understand titrations. Titrations are simply stoichiometry experiments that involve acids and bases. The special thing about these stoichiometry problems is that there are not really any limiting reactants. By being able to identify the endpoint of the titration, you have identified the point at which exactly enough of one reactant is added to react completely with the other reactant. You can use either of the reactants to perform any stoichiometry you wish to perform. Using the titration you just did as a guide, finish this module by answering "On Your Own" question 9.10.

ON YOUR OWN

9.10 125 mL of nitric acid (HNO_3) with unknown concentration is titrated against magnesium hydroxide with a concentration of 2.3 M. If 35.4 mL of base are required to reach the endpoint, what was the concentration of the acid?

SUMMARY OF KEY EQUATIONS AND TABLES IN MODULE 9

Equation 9.1:	$M = \frac{\text{\# moles}}{\text{\# liters}}$	Molarity
Equation 9.2:	$M_1V_1 = M_2V_2$	Molar dilution
Table 9.1:	Common Acids	

ANSWERS TO THE "ON YOUR OWN" QUESTIONS

9.1 We usually split up the ionic compounds into their component ions. However, since there are no metals in the entire reaction, there are also no ionic compounds. Looking at the equation, then, it is rather easy to see what has happened. The HBr has lost its H^+ ion to become a Br^- ion, while the NH_3 has gained an H^+ ion to become an NH_4^+ ion. This means that HBr is the acid and NH_3 is the base.

9.2 To solve this problem, we split up the ionic compounds into their constituent ions. The aluminum hydroxide [$Al(OH)_3$] on the reactants side is composed of 1 aluminum (Al^{3+}) ion and 3 hydroxide (OH^-) ions. The aluminum acetate on the products side is composed of 1 aluminum (Al^{3+}) ion and 3 acetate ($C_2H_3O_2^-$) ions. Notice that to be able to split both of these ionic compounds into their constituent ions, we needed to recognize the polyatomic hydroxide and acetate ions. This will be true throughout this module. If you have forgotten the polyatomic ions in table 5.1, go back and rememorize them. After we split up the ionic compounds, the equation looks like this:

$$3C_2H_4O_2\ (aq) + Al^{3+}\ (aq) + 3OH^-\ (aq) \rightarrow Al^{3+}\ (aq) + 3C_2H_3O_2^-\ (aq) + 3H_2O\ (l)$$

Now we can see what happened. The $C_2H_4O_2$ lost an H^+ ion to become $C_2H_3O^{2-}$, while the OH^- ion gained an H^+ ion to become H_2O. The OH^- ion came from the $Al(OH)_3$; thus, $C_2H_4O_2$ is the acid, and $Al(OH)_3$ is the base.

9.3 To solve this problem, we can split up the ionic compounds into their constituent ions. The potassium hydroxide (KOH) on the reactants side is composed of 1 potassium (K^+) ion and 1 hydroxide (OH^-) ion. The potassium phosphate on the products side is composed of 3 potassium (K^+) ions and 1 phosphate (PO_4^{3-}) ion. So the equation looks like this:

$$H_3PO_4\ (aq) + 3K^+\ (aq) + 3OH^-\ (aq) \rightarrow 3H_2O\ (l) + 3K^+\ (aq) + PO_4^{3-}\ (aq)$$

Notice how we got $3K^+$ and $3OH^-$ on the reactants side of the equation. The original equation tells us that we had 3 KOH molecules. When we split up the KOH into its 2 ions, we must carry that 3 to each of the ions.

Now we can see what happened. The H_3PO_4 molecule lost 3 H+ ions to become $PO4^{3-}$, while each of the 3 OH^- ions gained an H^+ ion to become 3 H_2O molecules. The OH^- ion came from the KOH molecule; thus, H_3PO_4 is the acid, and KOH is the base. Don't let it bother you that more than 1 H^+ ion was traded between the acid and base. This can (and often does) happen.

An alternative approach to solving this problem is to recognize KOH as ionic. Acids are covalent, so KOH cannot be the acid. This leaves H_3PO_4 as the acid and KOH as the base.

9.4 (a) One of the products of this reaction will be water. The other will be the ionic compound that forms between the positive ion of the base (K^+) and the negative ion left behind from the acid. When an H^+ ion leaves $HClO_3$, a ClO_3^- ion remains. Since the charges of these ions are of equal magnitude, we ignore the numbers in the chemical formula, and therefore the salt formed will be $KClO_3$.

$$HClO_3 + KOH \rightarrow KClO_3 + H_2O$$

This chemical equation is already balanced, so we are finished.

(b) This reaction will produce water and the ionic compound that forms between a Ca^{2+} ion and a Br^- ion. To get that chemical formula, we switch the charges and drop the signs to get $CaBr_2$.

$$HBr + Ca(OH)_2 \rightarrow CaBr_2 + H_2O$$

This equation is not balanced, of course, so now we have to balance it.

$$2HBr + Ca(OH)_2 \rightarrow CaBr_2 + 2H_2O$$

(c) Once again, this reaction will produce water as well as the ionic compound that forms between the positive ion from the base and the negative ion formed when the 3 H^+ ions leave this triprotic acid. The positive ion from the base is Mg^{2+}. When 3 H^+ ions leave the H_3PO_4 acid, they leave behind a PO_4^{3-} ion. The ionic compound that forms between Mg^{2+} and PO_4^{3-} is $Mg_3(PO_4)_2$. The reaction, therefore, is:

$$H_3PO_4 + Mg(OH)_2 \rightarrow Mg_3(PO_4)_2 + H_2O$$

This equation is not balanced, so we must balance it now:

$$2H_3PO_4 + 3Mg(OH)_2 \rightarrow Mg_3(PO_4)_2 + 6H_2O$$

MOD 9

9.5 Sulfuric acid is a diprotic acid that wants to donate 2 H^+ ions. Those ions will be donated to 2 NH_3 molecules, leaving behind an SO_4^{2-} ion:

$$H_2SO_4 + 2NH_3 \rightarrow 2NH_4^+ + SO_4^{2-}$$

9.6 Hydroiodic acid will force the amphiprotic water to act as a base. The HI will donate an H^+ ion to the water, forming an H_3O^+ ion. When it does that, HI will become an I^- ion:

$$HI + H_2O \rightarrow H_3O^+ + I^-$$

9.7 (a) To get concentration, we must have moles divided by liters. We already have moles and liters. We must first convert to liters so we can calculate concentration:

$$M = \frac{\text{\# moles}}{\text{\# liters}} = \frac{0.21 \text{ moles HCl}}{0.346 \text{ L}} = 0.61 \frac{\text{moles HCl}}{\text{L}} = 0.61 \text{ M}$$

The concentration is 0.61 M.

(b) To get to molarity, we need moles and liters. We already have liters, so all we have to do is convert grams to moles:

Molecular mass of H_2SO_4 - (2 x 1.01 amu) + (1 x 32.1 amu) + (4 x 16.0 amu) = 98.1 amu

$$1 \text{ mole } H_2SO_4 = 98.1 \text{ g } H_2SO_4$$

$$\frac{40.2 \cancel{\text{g } H_2SO_4}}{1} \times \frac{1 \text{ mole } H_2SO_4}{98.1 \cancel{\text{g } H_2SO_4}} = 0.410 \text{ moles } H_2SO_4$$

Now we can calculate concentration:

$$M = \frac{\# \text{ moles}}{\# \text{ liters}} = \frac{0.410 \text{ moles } H_2SO_4}{6.8 \text{ L}} = 0.060 \frac{\text{moles } H_2SO_4}{\text{L}} = 0.060 \text{ M}$$

The concentration is 0.060 M.

(c) To get concentration, we must have moles and liters. The problem gives us grams and mL, so we must make 2 conversions:

Molecular mass of KOH = (1 x 39.1 amu) + (1 x 16.0 amu) + (1 x 1.01 amu) = 56.1 amu

1 mole KOH = 56.1 grams KOH

$$\frac{5.55 \cancel{\text{g KOH}}}{1} \times \frac{1 \text{ mole KOH}}{56.1 \cancel{\text{g KOH}}} = 0.0989 \text{ moles KOH}$$

$$\frac{125.0 \cancel{\text{mL}}}{1} \times \frac{0.001 \text{ L}}{1 \cancel{\text{mL}}} = 0.1250 \text{ L}$$

Now we can calculate molarity:

$$M = \frac{\# \text{ moles}}{\# \text{ liters}} = \frac{0.0989 \text{ moles KOH}}{0.1250 \text{ L}} = 0.791 \frac{\text{moles KOH}}{\text{L}} = 0.791 \text{ M}$$

The concentration is 0.791 M.

9.8 We can tell that this is a dilution problem because the chemist has a solution of high concentration that she wants to make into a solution of low concentration. That's what dilution does. The original concentration of the sulfuric acid is 10.0 M, so that is M_1. The chemist wants the diluted solution to have a concentration of 2.5 M, so that is M_2. She also knows that she needs to have 500.0 mL of this 2.5 M solution. Since the 500.0 mL is the desired volume of the final solution, that is V_2. The only thing we don't know in the dilution equation is V_1, so we can solve for it:

$$M_1V_1 = M_2V_2$$

$$10.0 \text{ M} \times V_1 = 2.5 \text{ M} \times 500.0 \text{ mL}$$

$$V_1 = \frac{2.5 \cancel{\text{M}} \times 500.0 \text{ mL}}{10.0 \cancel{\text{M}}} = 130 \text{ mL}$$

The chemist needs to take 130 mL of the original solution and dilute it with water until it reaches 500.0 mL of volume. This will produce 500.0 mL of a 2.5 M solution.

9.9 To solve any stoichiometry problem, we must first figure out the balanced chemical equation. According to the problem, our reactants are HNO_3 and barium hydroxide. What is the chemical formula of barium hydroxide? Remember from Module 5 that the polyatomic hydroxide ion is OH^-. Barium is in group 2A of the periodic chart, so it develops a 2^+ charge in ionic compounds. To determine the chemical formula of barium hydroxide, then, we must switch the charges and drop the signs. Thus, the chemical formula is $Ba(OH)_2$.

When $Ba(OH)_2$ and HNO_3 react, a salt and water will be produced. The salt will be composed of the positive ion from the base (Ba^{2+}) and the negative ion left over when the acid gets rid of all of its H^+ ions. In this case, that will be the NO_3^- ion. Once again, to get the chemical formula of the compound that forms between these 2 ions, we switch the charges and ignore the signs to get $Ba(NO_3)_2$. Thus, the reaction is:

$$HNO_3 + Ba(OH)_2 \rightarrow Ba(NO_3)_2 + H_2O$$

Of course, this equation is not balanced, so that is the first thing we must do:

$$2HNO_3 + Ba(OH)_2 \rightarrow Ba(NO_3)_2 + 2H_2O$$

Now that we have a balanced equation, we can start the stoichiometry. We need to clean up the nitric acid, so we know that it is the limiting reactant. Therefore, we need to know the number of moles of nitric acid to be able to predict how much barium hydroxide is needed. To do this, we will multiply concentration by volume:

$$\frac{12.0 \text{ moles } HNO_3}{1.00 \text{ L}} \times \frac{10.0 \text{ L}}{1} = 1.20 \times 10^2 \text{ moles } HNO_3$$

Notice that we must use scientific notation when reporting this answer because we must report it to 3 significant figures. Scientific notation is the only way we can do that. We can now use this information to calculate the number of moles of barium hydroxide needed:

$$\frac{1.20 \times 10^2 \cancel{\text{ moles } HNO_3}}{1} \times \frac{1 \text{ mole } Ba(OH)_2}{2 \cancel{\text{ moles } HNO_3}} = 60.0 \text{ moles } Ba(OH)_2$$

Now that we have the number of moles of barium hydroxide needed, we can convert back to grams:

$$\frac{60.0 \cancel{\text{ moles } Ba(OH)_2}}{1} \times \frac{171.3 \text{ g } Ba(OH)_2}{1 \cancel{\text{ mole } Ba(OH)_2}} = 1.03 \times 10^4 \text{ g } Ba(OH)_2$$

Thus, 1.03×10^4 g of barium hydroxide are needed. This answer could be reported in decimal notation as well because 10,300 g also has 3 significant figures.

9.10 Remember, titrations are just stoichiometry problems, so first we have to come up with a balanced chemical equation. To do that, we must first figure out the chemical formula of the magnesium hydroxide. Magnesium is in group 2A of the periodic chart, so it takes a charge of 2^+ in ionic compounds. You should have memorized the hydroxide ion as OH^-. To get the chemical formula, we switch the charges and ignore the signs to get $Mg(OH)_2$. The equation, therefore, looks like this:

$$HNO_3 + Mg(OH)_2 \rightarrow Mg(NO_3)_2 + H_2O$$

We must now balance the equation:

$$2HNO_3 + Mg(OH)_2 \rightarrow Mg(NO_3)_2 + 2H_2O$$

Now we can do the stoichiometry. Since the endpoint was reached, we know that exactly enough base was added to react with all of the acid. First, then, we calculate how many moles of base were added:

$$\frac{2.3 \text{ moles } Mg(OH)_2}{1 \cancel{L}} \times \frac{0.0354 \cancel{L}}{1} = 0.081 \text{ moles } Mg(OH)_2$$

We can now use the chemical equation to determine how many moles of acid were present:

$$\frac{0.081 \cancel{\text{moles } Mg(OH)_2}}{1} \times \frac{2 \text{ moles } HNO_3}{1 \cancel{\text{mole } Mg(OH)_2}} = 0.16 \text{ moles } HNO_3$$

Now that we have the number of moles of acid present, we simply divide by the volume of acid to get concentration:

$$M = \frac{\text{\# moles}}{\text{\# liters}} = \frac{0.16 \text{ moles } HNO_3}{0.125 \text{ L}} = 1.3 \text{ M}$$

The concentration is 1.3 M.

NOTE: All Lewis structure illustrations in this module are by Connor Hughes.

STUDY GUIDE FOR MODULE 9

REVIEW QUESTIONS

1. State whether the substance described is most likely an acid or a base:
   a. A substance that tastes bitter
   b. A substance that feels slippery
   c. A substance that turns blue litmus paper red

2. Identify the acid in the following reaction:

$$PH_3 + CH_4O \rightarrow CH_3O^- + PH_4^+$$

3. Some books define a base as a proton acceptor. Why is this just as correct a definition as the one presented in this book?

4. Which of the following is (are) polyprotic acid(s)?
   a. HNO_3 b. H_2CO_3 c. $Ca(OH)_2$ d. H_3PO_4

5. Which of the following units could be used as concentration units?
   a. $\frac{mL}{meter}$ b. $\frac{grams}{m^3}$ c. $\frac{moles}{sec}$ d. $\frac{moles}{mL}$

6. A chemist wants to make a solution of 3.4 M HCl. He can find 2 solutions of HCl on the shelf. One has a concentration of 6.0 M, while the other has a concentration of 2.0 M. Which solution can the chemist use to make the desired acid?

7. When the compound $Al_2(CO_3)_3$ dissolves in water, what ions does it form and how many of each ion are present?

8. Consider the following reactions:

$$HCO_3^- + NaOH \rightarrow Na^+ + CO_3^{2-} + H_2O$$

and

$$H_2O + HCO_3^- \rightarrow H_2CO_3 + OH^-$$

   Which substance in the 2 equations is acting as an amphiprotic substance?

9. In acid–base chemistry, what 2 uses does an indicator serve?

10. In a titration, what is the significance of the endpoint?

MOD 9

PRACTICE PROBLEMS

1. A chemist mixes sulfuric acid (H_2SO_4) with aluminum hydroxide. Give the balanced chemical equation that represents the resulting reaction.

2. Give the balanced chemical equation for the reaction that occurs between nitric acid (HNO_3) and calcium hydroxide.

3. Given that ammonia (NH_3) is a base, determine the reaction that occurs between H_2CO_3 and ammonia and give the balanced chemical equation.

4. When HBr is mixed with water, what chemical reaction occurs? Give the balanced chemical equation.

5. Give the concentration (in M) of each of the following solutions:
   a. 3.51 moles of HNO_3 are dissolved in enough water to make 1.2 liters of solution.
   b. 234.1 grams of KOH are dissolved in enough water to make 345 mL of solution.
   c. 4.1 grams of H_3PO_4 are dissolved in enough water to make 45 mL of solution.

6. A chemist has 12.0 M HBr and would like to use it to make 500.0 mL of 3.5 M HBr. Describe how she should do this.

7. A chemist makes a stock solution of NaOH by dissolving 950.0 grams in enough water to make 2.00 liters of solution. If, later on, he wants to use this stock solution to make 100.0 mL of 0.10 M NaOH, what would he need to do?

8. A chemist reacts 50.0 mL of 4.5 M carbonic acid (H_2CO_3) with excess potassium hydroxide. How many grams of potassium carbonate are produced?

9. A chemist needs to know the concentration of some NaOH that is in the laboratory. To find this out, she titrates a 50.0 mL sample of the solution with 3.5 M HCl. If it takes 34.3 mL of the HCl to reach the titration endpoint, what is the concentration of the NaOH solution?

10. A solution of H_2SO_4 is titrated against 1.2 M KOH. A total of 345.1 mL of KOH must be added to a 500.0 mL sample of H_2SO_4 to reach the endpoint. What is the concentration of the H_2SO_4?

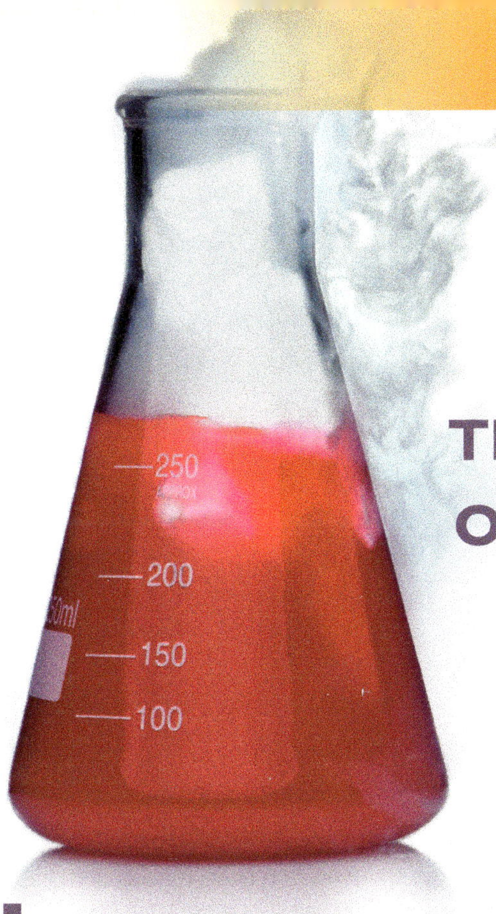

THE CHEMISTRY OF SOLUTIONS

first reactions

Probably the biggest questions in chemistry are "Why?" and "How?" Why does salt dissolve in water and sand does not? The beach would look very different if sand reacted in the same manner as salt. How does milk mix in with water, yet oil floats on top? These are great questions. You might feel like a toddler with all of your new whys, but this time you are equipped to answer them!

In the previous module, we spent some time learning about acids and bases. We learned that acids and bases are often dissolved in water to make an acidic or basic solution. Although the word *solution* was used several times, the concept was not fully explained. It is now time to do that. Quite a lot of interesting chemistry is involved in the simple task of making a solution. That's what we will study in this module.

One of the first things you have to know about solutions is some terminology. A chemist usually makes a solution by mixing things together. If the mixture is homogeneous, we generally say that one of the substances dissolved in the other substance. The substance that we are dissolving is called the **solute** (sahl' yoot), and the substance in which we are dissolving the solute is called the **solvent** (sahl' vent). When solute and solvent are mixed together, they are called a **solution**. When we dissolve table salt in water, the table salt is the solute, the water is the solvent, and the resulting salt water is the solution.

It is important to note that even though we are most familiar with dissolving solids (like table salt) into liquids (like water), solutes do not have to be solids. For example, one of the steps in the manufacture of soda pop involves dissolving gaseous carbon dioxide into water. This is what gives soda pop its fizz. In this case, the solute (carbon dioxide) is a gas, not a solid.

As a side note, this same procedure also gives soda pop its acidic nature, which you discovered in the previous module. When carbon dioxide dissolves in water, a fraction of the molecules will undergo the following reaction:

$$CO_2\ (g) + H_2O\ (l) \rightarrow H_2CO_3\ (aq)$$

You should recognize the product of this reaction as carbonic acid. This is what turned the blue litmus paper red in experiment 9.1.

Did you know that solvents do not necessarily have to be liquids? We are very familiar with water dissolving substances like sugar and table salt. Fingernail polish remover is a liquid solvent that dissolves fingernail polish so it can be removed quickly and easily. However, can other phases be solvents? Yes, gases can be solvents! For example, methane gas is transported through pipes to homes to be burned as a fuel. However, methane is colorless and odorless, so it would be very hard to detect gas leaks that can make a very dangerous and explosive situation. So natural gas companies add a small amount of a chemical with a very strong odor so we can detect leaks by smell. In this mixture, methane is the solvent and the odorant is the solute.

HOW SOLUTES DISSOLVE IN SOLVENTS

Now that you know the terminology, you can begin to learn the science. The first thing you might ask yourself is why solutes dissolve in solvents. After all, when you buy table salt ($NaCl$) at the grocery store, it is a solid. When you mix it with water, why does it go into the aqueous state? Why doesn't it just stay solid? The answers to these important questions involve the polar covalent nature of the water and the ionic nature of the $NaCl$.

First, remember that $NaCl$ is an ionic compound composed of Na^+ ions and Cl^- ions. Unlike covalent molecules, the ionic $NaCl$ molecule has no chemical bond because the atoms that make it up do not share electrons. Instead, the Na atom has donated its electrons to the Cl atom. As a result, the Na ion is positive and the Cl ion is negative. *The electronic attraction between these opposite charges is the only thing that holds the Na^+ and Cl^- ions together.* Therefore, a sample of solid ionic $NaCl$ would look something like Figure 10.1.

FIGURE 10.1
A Solid Sample of NaCl
3D Illustration by Michael Dailey

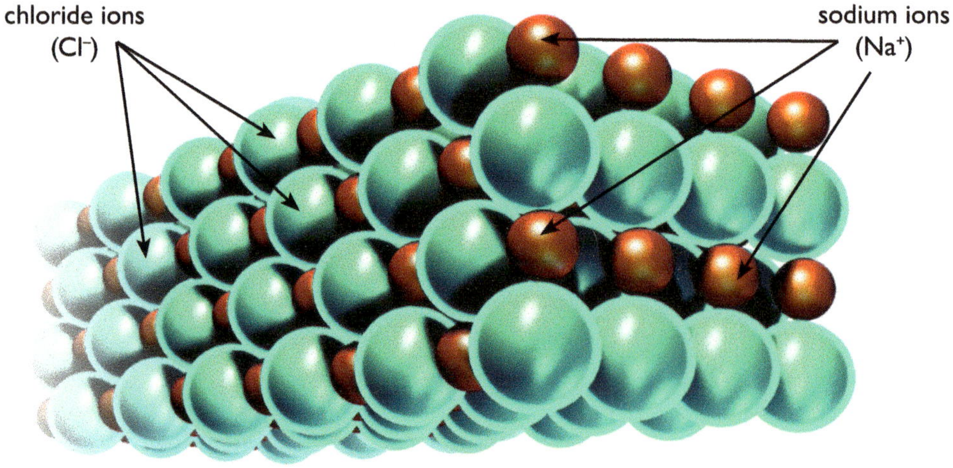

Sodium chloride ($NaCl$) is a solid because the Na^+ and Cl^- ions are very close to one another, due to their mutual electronic attraction. As we learned in Module 6, the only way to get $NaCl$ out of its solid phase is to pull these ions away from each other. This is where the water comes in.

Remember, because water is a polar covalent compound, it has fractional positive charges on its hydrogen atoms and a fractional negative charge on its oxygen atom similar to what is illustrated in Figure 10.2.

FIGURE 10.2
The Polar Nature of a Water Molecule
3D Illustration by Michael Dailey

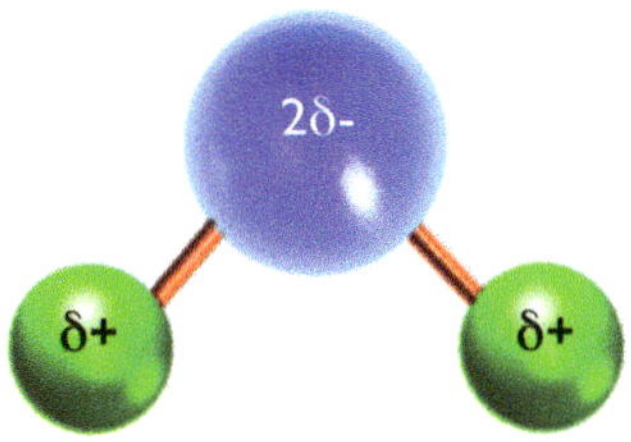

When water and NaCl are mixed, the fractional positive charges on the water's hydrogen atoms are attracted to the negative charges on the chloride ions. In the same way, the fractional negative charge on the water's oxygen atom is attracted to the positive charges on the sodium ions.

To get these charges closer to one another, the water molecules muscle in between the ions in NaCl, so the ions are not as strongly attracted to each other as they were before. This allows the water molecules to surround the ions and pull them away from each other. Figure 10.3 illustrates this

FIGURE 10.3
Water Dissolving Sodium Chloride
3D Illustration by Michael Dailey

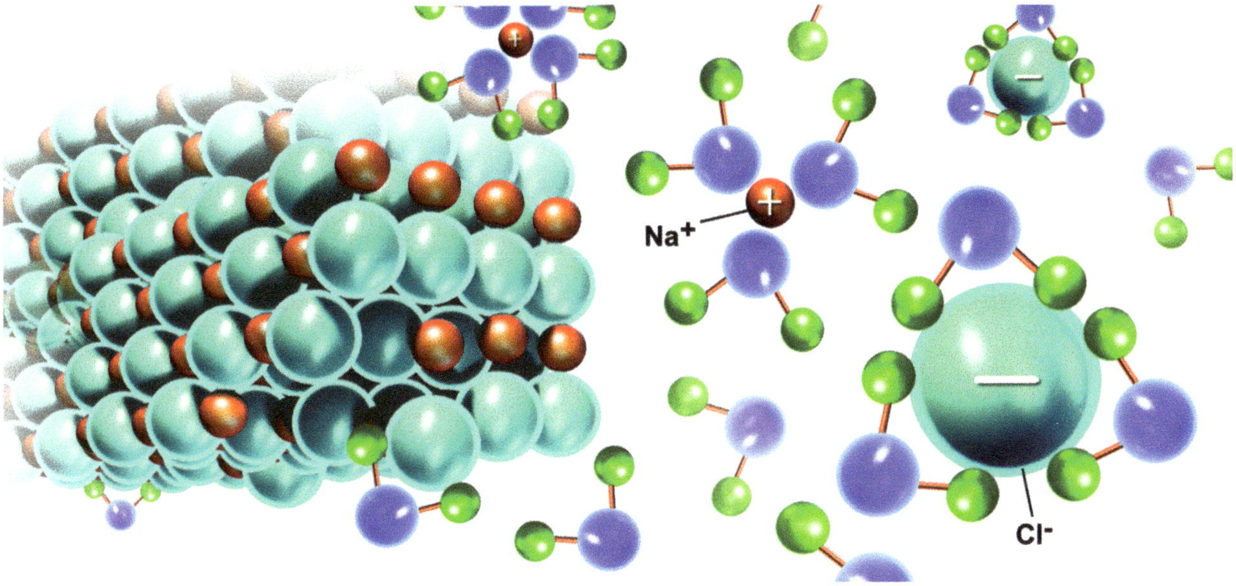

Once the ions get far apart from each other, they no longer form a solid; instead, they are dissolved, and we say that they are in the aqueous state.

Notice *how* the water molecules surround the ions on the right side of the figure. The sodium ions (red spheres) are positive. The water molecules surround the sodium ions so that their oxygen atoms point toward the sodium ions. This, of course, should make sense. After all, the oxygen atom in a water molecule carries a partial negative charge. The partial negative charge on the oxygen atom is attracted to the positive charge on the sodium ion, and the water molecules orient themselves so that their oxygen atoms are as close as possible to the sodium ions. In the same way, the water molecules surround the chloride ions (blue-green spheres) so that their hydrogen atoms are as close as possible to the chloride ions. This happens because the partial positive charges on the hydrogen atoms are attracted to the negative charge on the chloride ion.

This is how ionic solids dissolve in water. Because of the attraction between the water molecules and the ions in the ionic solid, the water molecules get between the ions, pushing them away from each other. Soon the ions are so far away from each other that they can move freely. This makes the ionic solid change its phase to aqueous.

Does this mean that all ionic solids dissolve in water? No, it does not. Remember, for the solid to dissolve in water, the water molecules must first get between the ions. In some ionic solids, the attraction between the ions is so great that the water molecules simply cannot force their way in between the ions. As a result, we say that the compound is insoluble in water. An example of this is aluminum hydroxide, $Al(OH)_3$. Even though this compound will not dissolve in water, it is used in manufacturing antacids because of its basic properties.

If you purchased the MicroChem kit mentioned in the Student Notes, you can perform Experiment 5 to get some experience making water-insoluble ionic compounds.

Now that you see how ionic solids dissolve in water, you need to see how covalent solids do it. The first thing to remember is that purely covalent solids *do not* dissolve well in water because they have no electrical charges in them and thus cannot interact easily with water. Polar covalent solids, on the other hand, can dissolve in water because they possess fractional charges. However, these charges cannot separate because they reside on atoms that are bonded to each other. As a result, even though polar covalent molecules can dissolve in water, they cannot split up into smaller parts. This means that the molecules dissolve as individuals.

The way that this happens is much the same as what happens with ionic solids. Let's look at how solid PH_3 dissolves in water. As we learned in Module 5, PH_3 is a polar covalent solid. Therefore, the atoms within the molecule possess fractional charges, as shown in Figure 10.4.

FIGURE 10.4
The Polar Nature of a PH_3 Molecule
3D Illustration by Michael Dailey

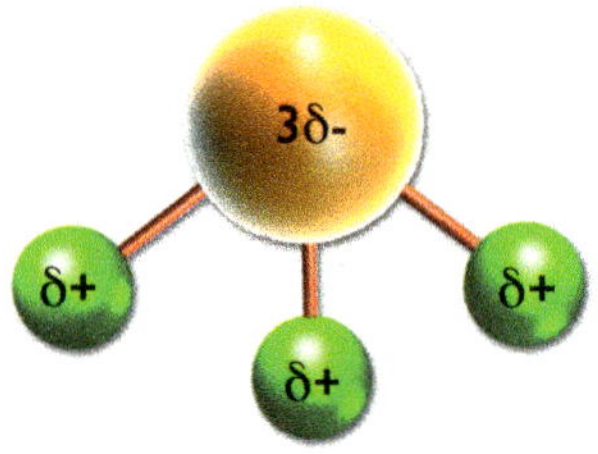

These fractional charges attract the opposite fractional charges in the water molecules. In the same way as before, the water molecules muscle their way in between the PH_3 molecules. They surround the individual PH_3 molecules, orienting their negatively charged oxygen atoms toward the PH_3's positively charged hydrogen atoms. In addition, the water's positively charged hydrogen atoms are oriented toward the PH_3's negatively charged phosphorous atom. Once they are surrounded by the water molecules, the PH_3 molecules are pulled away from each other. When this happens, the compound changes from the solid phase into the aqueous phase, as shown in Figure 10.5.

FIGURE 10.5
PH_3 after It Has Been Dissolved in Water
3D Illustration by Michael Dailey

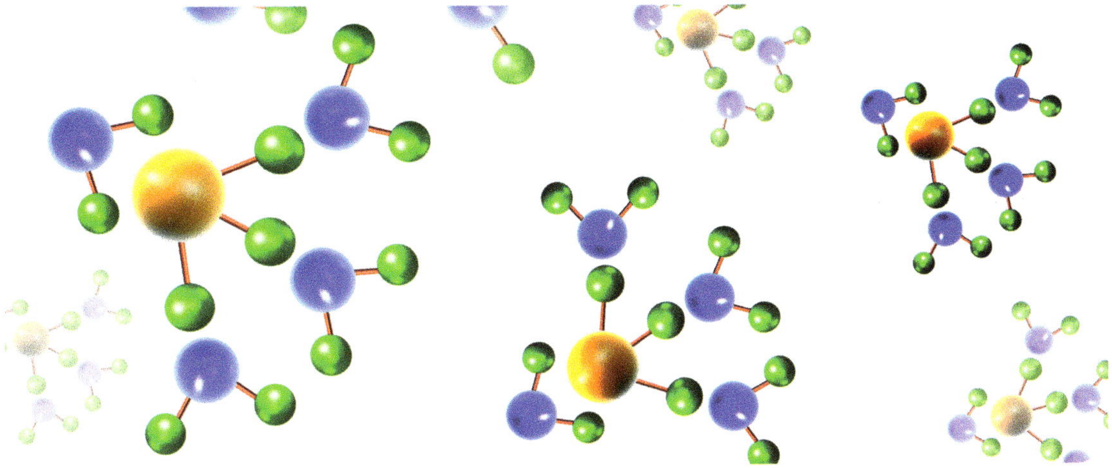

Notice *how* the water molecules surround the phosphorus trihydride molecules in Figure 10.5. The water molecules orient themselves so that their oxygen atoms point toward the hydrogen atoms in the PH_3 molecules because the partial negative charge on the oxygen atom is attracted to the partial positive charges on PH_3's hydrogen atoms. In the same way, the hydrogen atoms on the water molecules are pointed toward the phosphorus atoms because the partial positive charges on the hydrogen atoms are attracted to the partial negative charge on the phosphorus atom.

There are similarities and differences between the way in which ionic solids dissolve in water and the way in which polar covalent solids dissolve in water. In both cases, water molecules are attracted to the charges in the solute. Also, both types of solute change from the solid phase to the aqueous phase because the water molecules come between the solute molecules, surround them, and then pull them apart. The main difference between these 2 types of compounds is that while ionic compounds break apart into smaller pieces (their individual ions) when dissolved, polar covalent compounds do not. Polar covalent compounds dissolve by separating each individual molecule from its neighbors.

This same reasoning can be used to explain how any ionic or polar covalent solid dissolves in any polar solvent. What about non-polar covalent compounds, however? How do they dissolve in other non-polar covalent compounds? If the solvent molecules are attracted to the solute molecules, then the same thing that we have just discussed can also happen. The solvent molecules will muscle in between the solute molecules, pulling them far apart from each other. However, the precise mechanisms for how this happens are a bit too advanced for this course.

Of course, everything we have discussed so far has been focused on solid solutes dissolving into liquid solvents. As mentioned already, solutes can also be gases or liquids. For example, when dishwashing liquid is mixed with water, the dishwashing liquid is a solute that is dissolved in water (the solvent). The way in which liquid solutes dissolve in liquid solvents is similar to the way in which solid solutes dissolve in liquid solvents.

As is the case with solid solutes, the solvent molecules must be attracted to liquid solute molecules for the solute to dissolve. This allows the solvent molecules to muscle in between the solute molecules. However, in liquid solutes, the solute molecules are not very close to each other to begin with. It isn't very hard for the solvent molecules to get between them. Remember, in solids, the molecules (or ions) are close together. As a result, solvent molecules must work harder to get between the solid solute molecules (or ions). Also, unlike what happens when solids dissolve, the molecules of a liquid solute are not drawn very far away from each other when the solvent gets between them. After all, the solute is a liquid already, so the molecules are already pretty far apart from each other and do not have to be separated very much. They have to be drawn apart only far enough to allow the solvent molecules to get between them.

Based on this information, you would expect that liquid solutes dissolve in liquid solvents more readily than solid solutes do. You probably knew this fact already, but at least now you know why this is the case. Remember that not all liquids can dissolve in other liquids. As mentioned before, the solvent molecules must be attracted to the solute molecules before the solute can be dissolved. As a result, polar covalent liquid solutes will dissolve only in polar covalent liquid solvents, and non-polar covalent liquid solutes will dissolve only in non-polar covalent liquid solvents. This is why oil and water don't mix. Oil is non-polar covalent, while water is polar covalent.

What about gaseous solutes? How do they dissolve in liquid solvents? If you think about it, there is a fundamental difference between solid or liquid solutes and gaseous solutes. In the case of solid solutes, the solute molecules (or ions) must be pulled away from each other to convert them to the liquid state. In the case of liquid solutes, the molecules must also be pulled away from each other, but not by very much. In the case of gaseous solutes, the solute molecules are not pulled apart. In fact, to dissolve into a liquid solvent, the gaseous solute molecules must *be brought closer together*.

Once again, this fact should make sense if you think about it. The main difference between the phases of matter is the distance between the molecules. Molecules in the solid phase tend to be closer together; molecules in the liquid phase tend to be farther apart; and molecules in the gaseous phase tend to be very far apart. To dissolve a gas into a liquid, the gaseous molecules must be brought closer together. You can think of this as precisely the opposite of dissolving a solid solute. When dissolving a gaseous solute, the solvent molecules do not muscle in between the solute molecules. Instead, they attract the solute molecules to them, like a magnet attracts small bits of iron. As a result, the gaseous molecules come closer together.

Table 10.1 summarizes the similarities and differences in the processes that dissolve solid, liquid, and gaseous solutes. It is important to understand and remember these facts.

TABLE 10.1
Summary of the Methods by Which Solutes Dissolve in Liquid Solvents

Solute Phase	Requirements for Dissolving
Solid	The solvent molecules must be attracted to the solute molecules so strongly that the solvent molecules can get between the solute molecules (or ions) and pull them far apart from each other.
Liquid	The solvent molecules need to be attracted to the solute molecules only a small amount because the solvent does not need to separate the solute molecules very much. The solvent merely needs to get between the solute molecules.
Gas	The solvent molecules must be attracted to the solute molecules enough to pull the solute molecules closer to one another.

SOLUBILITY

Now that we see *how* solutes dissolve in solvents, the next question you might ask is how much solute can dissolve in a solvent. The answer to that is a bit involved. First, let's get the vocabulary down. Solubility refers to how much solute dissolves in a solvent

Solubility— The maximum amount of solute that can dissolve in a given amount of solvent.

The solubility of a solute is generally reported in terms of grams of solute that will dissolve in 100.0 g of solvent (grams per hundred). In other words, if a solute has a solubility of 10.0 grams per hundred, it means that as much as 10.0 grams of solute can be dissolved for every 100.0 grams of solvent.

If you think about it, the solubility of each solute must be unique to that solute. For example, it is quite easy to dissolve table salt (NaCl) in water. At room temperature, NaCl has a solubility of 37 grams per hundred. We can dissolve 37 grams of NaCl in 100 grams of water. Under the same conditions, baking soda ($NaHCO_3$) has a solubility of just 12 grams per hundred. Baking soda does not dissolve in water nearly as well as table salt does. In addition, the solubility of each of these solutes changes if we change the solvent. The solubility of both table salt and baking soda decreases if we try to dissolve them in alcohol rather than water.

These facts should make sense. After all, the solubility of each solute depends fundamentally on the fact that the solvent molecules are attracted to the solute molecules. Therefore, each solute molecule has its own unique solubility in each solvent because the amount of attraction between solute and solvent changes depending on the identity of each. We come, then, to our first rule about solubility:

The solubility of any solute depends on both the identity of the solute and the identity of the solvent.

We will develop 2 more rules about solubility, and it will be very important for you to keep them in mind as you go through the rest of this module.

First, let's learn some more terminology. As mentioned, at room temperature, we can dissolve 37 grams of table salt into 100 grams of water. However, if we added any more than 37 grams of table salt, the excess would not dissolve. If we add 50.0 grams of table salt to 100.0 grams of water, 37 grams of it would dissolve, and the other 13 grams would just fall to the bottom of the container, unable to dissolve. If we were then to filter away all of that excess salt, we would have a solution that contains as much salt as can possibly be dissolved in that amount of water. We would call this a saturated solution.

Saturated solution—A solution in which the maximum amount of solute has been dissolved.

You will make and use a saturated solution in the next experiment.

FIGURE 10.6
Saturation of a Solution
Illustration: Striar Yunis.

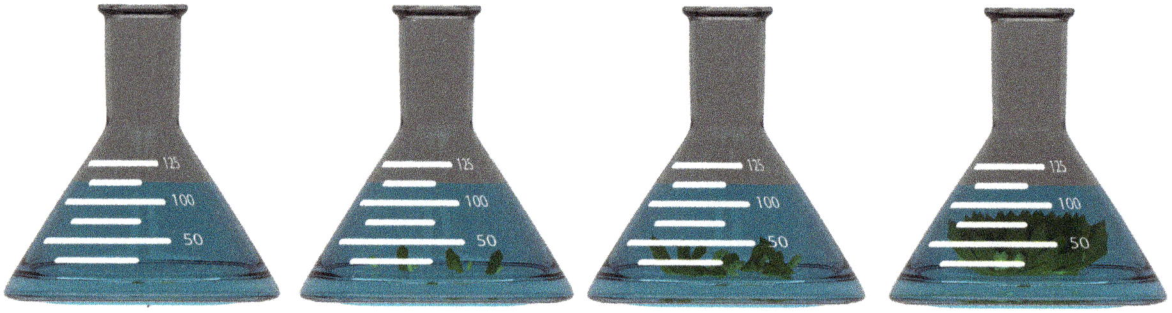

When a solvent cools, solubility decreases and excess solute comes out of the solution. Pictures 2–4 show a saturated solution.

Now let's go back to developing our 2 additional rules about solubility. The solubility of solutes depends not only on the identity of the solute and solvent, but also on certain conditions under which the solution is made. Perform Experiment 10.1 to see how the solubility of a solid solute depends on the temperature at which the solution is made.

EXPERIMENT 10.1

PURPOSE: To determine the effect of temperature on the solubility of solid solutes.

MATERIALS:

- 250 mL beaker (A short, fat glass or canning jar might work, but ***be careful.*** Glasses tend to crack when subjected to the temperature extremes of this experiment. If you don't have a beaker, you may want to just read this experiment, unless you aren't afraid of losing a glass or 2.)
- 100 mL beaker (Another short, fat glass or canning jar might work.)
- Graduated cylinder (A ¼ measuring cup will work.)
- Flame heater (A stove will work.)
- Stirring rod (A spoon will work.)

- Thermometer
- Mass scale
- Salt
- Filter paper (You can cut circles out of the bottom of a coffee filter to make your own filter paper.)
- Funnel
- Oven mitt
- Safety goggles

QUESTION: How does temperature effect solubility of solutes?

HYPOTHESIS: Write a hypothesis related to this experiment.

PROCEDURE:

1. Use your mass scale to measure 50.0 grams of table salt.
2. Put the salt into the 250 mL beaker.
3. Add 50.0 mL (¼ cup) of water to the salt and stir it vigorously with your stirring rod.
4. Allow the solution to stand while you fold the filter paper and put it in the funnel the way you learned in experiment 2.3..
5. Heat the solution with your flame heater. As it is heating, stir the solution carefully with your thermometer. Wear the oven mitt on the hand that is holding the thermometer!
6. When the temperature of the solution reaches 95°C, take the thermometer out of the solution and use your hand that has the oven mitt on it to pick up the beaker.
7. With your other hand, hold the funnel with the filter paper over the 100 mL beaker.
8. Carefully pour the solution onto the filter paper in the funnel. Make sure the level of the solution in the funnel never rises above the top of the filter paper. Continue to filter the solution in this way until you have about 20 mL of clear solution in the beaker below the funnel.
9. Look at your solution in the 100 mL beaker. It should be clear because all undissolved salt should have been filtered out by the filter paper. This is a saturated solution of table salt.
10. Place that solution in your freezer for about 20 minutes. After that, remove the solution from the freezer and examine it again. Now there should be crystals of salt lying on the bottom of the beaker.
11. Clean up and return everything to the proper place.

CONCLUSION: Write something about what happened in this experiment and whether it confirmed your hypothesis.

What happened in the experiment? After filtering the solution, you had a saturated solution of table salt, but the solution was at a temperature near 95°C. When you placed the solution in the freezer, it cooled to a temperature of less than 0°C. Obviously, the table salt is less soluble in cold water than in hot water because some of the table salt had to leave the solution and fall to the bottom of the beaker. The process by which a solid solute leaves a solution and turns back into its solid phase is called precipitation.

From the experiment, then, we see that the solubility of a solute also depends on temperature. For solid solutes, the solubility usually increases with increasing temperature, as we saw in the experiment. This makes sense because for a solid solute to dissolve, the solvent molecules must be able to move between the solute molecules (or ions) and drive them away from each other. We will learn more about heat and energy in a later module, but for now you need to understand that heating a compound involves adding energy to it. We can conclude, then, that the molecules in hot solvents have more energy than the molecules in cold solvents. With more energy, the solvent molecules can be more successful at getting between the solid molecules (or ions), making the solute dissolve more readily.

What about liquids and gases? How does the temperature of the solvent affect their solubility? The temperature of the solvent has little effect on the solubility of a liquid solute. After all, the solvent really doesn't have to work very hard to dissolve a liquid solute, so its energy is of little consequence. In the case of gaseous solutes, the solubility is dramatically affected by the solvent temperature, as is illustrated in Experiment 10.2.

EXPERIMENT 10.2

PURPOSE: To determine the effect of temperature on the solubility of a gas.

MATERIALS:

- 2 beakers (Two saucepans will work.)
- A test tube (A tall, thin glass will work, but it must fit easily in the saucepans.)
- Flame heater (A stove will work.)
- Ice
- Cold soda pop (Pepsi, Coke, Sprite, etc. It must be carbonated.)

QUESTION: How does temperature effect solubility of a gas?

HYPOTHESIS: Write a hypothesis related to this experiment.

PROCEDURE:

1. Fill one beaker ¾ full with hot water.
2. Put it on the flame until the water begins to boil.
3. While you are waiting for the water to boil, fill the other beaker ½ full with ice, then add enough water so that the beaker is ¾ full with the ice water mixture.
4. Once the water starts boiling, extinguish the flame so that the water is hot, but not boiling.
5. Fill the test tube ¾ full with the soda pop. Set the test tube in the beaker that contains the hot water. Observe what happens. In a few moments, the soda pop should begin bubbling.
6. Watch the soda pop bubble for a moment, then pull it out of the hot water.
7. Place it in the beaker that contains ice water. Observe what happens. In a few

moments, the soda pop should stop bubbling again.

8. After a moment, return the test tube to the hot water. The soda pop should start bubbling again.
9. Repeat this process a couple of times, noticing that each time the soda pop bubbles in the hot water but stops bubbling when it reaches the ice water.
10. Clean up and return everything to the proper place.

CONCLUSION: Write something about what you observed and whether it confirmed your hypothesis. Why did the soda pop behave in this way? Was the soda pop boiling?

The boiling point of soda pop is higher than that of water. If the water in the beaker was not boiling, there is no way that the soda pop was boiling. Also, your experience with boiling water should tell you that it takes quite a while to get cold water to boil. In this experiment, the soda pop began bubbling just moments after it hit the hot water.

If the soda pop wasn't boiling, what happened in this experiment? As mentioned before, soda pop has carbon dioxide dissolved in it. Carbon dioxide is a gas, so soda pop is a solution made by dissolving carbon dioxide (a gaseous solute) into the pop (a liquid solvent). When you put the test tube into the beaker that contained the hot water, you were, in effect, increasing the temperature of the solvent. The bubbling that you saw was the gaseous carbon dioxide leaving the solution. When the pop got warmer, the carbon dioxide became less soluble and had to leave the solution. On the other hand, when you put the test tube into the beaker that contained the cold water, you lowered the temperature of the solvent. This increased the solubility of the carbon dioxide, so no more gas had to escape the solution, and the bubbling ceased. Unlike solid solutes, then, gaseous solutes are less soluble in hot solvents than in cold solvents.

Experiments 10.1 and 10.2 lead us to our next rules of solubility shown in table 10.2.

TABLE 10.2
The Effect of Temperature on Solubility

Solute Phase	Rules for the Effect of Temperature
Solid	The solubility of solid solutes usually increases with increasing temperature.
Liquid	The solubility of liquid solutes is not affected by temperature.
Gas	The solubility of gases decreases with increasing temperature.

Make sure you understand both the rules and the reasons we discussed that explain the rules. Notice also that the first rule says the solubility of solid solutes usually increases with increasing temperature. Some solids dissolve better in cooler solvents. However, we will not dwell on those exceptions in this course.

The last solubility rule deals with the effect of pressure on solute solubility. Think about what happens when you open a fresh can or bottle of soda pop. You usually hear a loud pop or fizz. If you look at the soda after it is opened, it bubbles. What does this tell you? The fact that the soda pop fizzes when the lid is opened indicates that it is stored at a high pressure. When the lid is opened, the air is pushed out by that pressure, causing the fizzing sound.

ON YOUR OWN

10.1 A student is asked to dissolve 50.0 grams of potassium chromate (an orange-yellow solid) in 250 mL of water. He is having a hard time getting all of the solid to dissolve, but he is not allowed to add any more water. What should he do to try to make sure all 50.0 grams dissolve?

10.2 A chemistry lab is storing several bottles of a solution made by dissolving methane gas in octane, a nonpolar liquid. The chemist notices that every once in a while, the lids from the bottles pop off. She reasons that the lids are popping off because the gas that is dissolved in the liquid is escaping from solution. What can she do to stop the lids from popping off?

10.3 A student is doing an experiment that involves dissolving a gas in water. He notices that even though the temperature in the lab is exactly the same every day of the year, there are some days in which more gas can dissolve in water as compared to other days. What, most likely, is changing from day to day to produce this effect?

The bubbles that occur after the lid is opened indicate that carbon dioxide is leaving the solution. The solubility of the carbon dioxide decreases as soon as the lid is opened. This shows that the solubility of gaseous solutes decreases when the pressure at which the solution is stored is lowered. This is another rule of solubility:

Increasing pressure increases the solubility of gases.

Pressure does not affect the solubility of liquids or solids.

The best way to understand this last rule is to remember how gases dissolve. For gases to dissolve into liquids, the gas molecules must get closer together. If the pressure surrounding the solution is high, then the molecules are forced a bit closer together, helping them dissolve. See if you can apply the 3 rules you just learned by answering "On Your Own" questions 10.1–10.3.

ENERGY CHANGES THAT OCCUR WHEN MAKING A SOLUTION

Now that you have learned about the physical changes that occur when solutes dissolve in solvents, it is time to investigate how the energy of the solute and solvent changes as well. Begin this investigation by performing Experiment 10.3.

EXPERIMENT 10.3

PURPOSE: To investigate a solute that releases heat when dissolved.

MATERIALS:

- Beaker (A short, fat glass will do.)
- Lye (This is commonly sold with the drain cleaners in hardware stores and

supermarkets. Make sure that your bottle is labeled 100% lye. If you cannot find lye, you can order it online. It may be listed under its chemical name, sodium hydroxide. ***Always use this chemical in a well-ventilated area.***

- Rubber gloves
- Water
- Sink
- Tablespoon
- Safety goggles

QUESTION: Can a solute release heat when dissolved?

HYPOTHESIS: Write a hypothesis related to this experiment.

PROCEDURE:

1. Put on the gloves and safety goggles.
2. Put the beaker in the sink directly under the faucet.
3. Slowly turn on the water and add enough water to your beaker so that the water level in it is about 1.5 centimeters high.
4. Measure 3 tablespoons of lye and slowly add it to the beaker.
5. Stir the solution with the tablespoon until all or most of the lye dissolves.
6. Take one glove off and ***carefully*** touch the outside of the beaker, near the bottom. Most likely, the beaker will feel warm. It might be quite hot, so be careful.
7. Continue to stir the solution with the other hand, and periodically touch the outside of the beaker near the bottom to see how hot it is getting. Once again, ***be careful*** when you touch the beaker with your bare hand because it can get very hot!
8. Eventually, the solution might get so warm that you can no longer comfortably touch the beaker. At that point, put the glove back on.
9. Turn on the cold water, allowing it to slowly fill and overflow the beaker. Keep the water running for several minutes, allowing a large amount of the lye to be flushed down the drain.
10. Tip the beaker over and rinse it out.
11. Clean up and return everything to the proper place.

CONCLUSION: Write something about what you observed and whether it confirmed your hypothesis.

What happened? The beaker became warm because lye releases heat when it dissolves. This heat is transferred to the water, causing the water's temperature to increase. Eventually, the water might have become hot enough to boil.

The process you observed in the previous experiment is called an exothermic (ek' soh ther' mik) process. The prefix *exo-* means out, and *therm* refers to heat; so exothermic refers to a process that lets out heat. The opposite of this kind of process is one that absorbs heat and is called endothermic (en' doh ther' mik):

Exothermic process—A process that releases heat.

Endothermic process—A process that absorbs heat.

These 2 terms will appear several times throughout the rest of this course, so you need to learn them.

Now that we know the terminology, we can discuss what happened in the experiment. Strong bases such as NaOH (the principal component of lye) dissolve exothermically. This means that they release heat, which raises the temperature of the surroundings. That's why the beaker got warm. Exothermic processes tend to heat up their surroundings. Endothermic processes, on the other hand, absorb heat from their surroundings. They take energy away from the surroundings, which reduces the temperature. Endothermic processes, then, cool down their surroundings.

This relates to our discussion of solubility because most solids dissolve in an endothermic fashion. In other words, most solids must absorb energy to dissolve properly. When we dissolve a solid in water, the temperature of the solution decreases. Often, the decrease is so small that we simply do not notice it. For example, sodium chloride (table salt) dissolves in water endothermically. Therefore, the temperature of the solution decreases once the table salt dissolves. However, the change is so small that we do not really notice the effect. However, if we were to dissolve ammonium nitrate in water, we would definitely notice the temperature change because ammonium nitrate dissolves very endothermically in water.

The fact that some solids (such as ammonium nitrate) dissolve endothermically has a very practical application, illustrated in Figure 10.7.

FIGURE 10.7
Instant Cold and Hot Packs

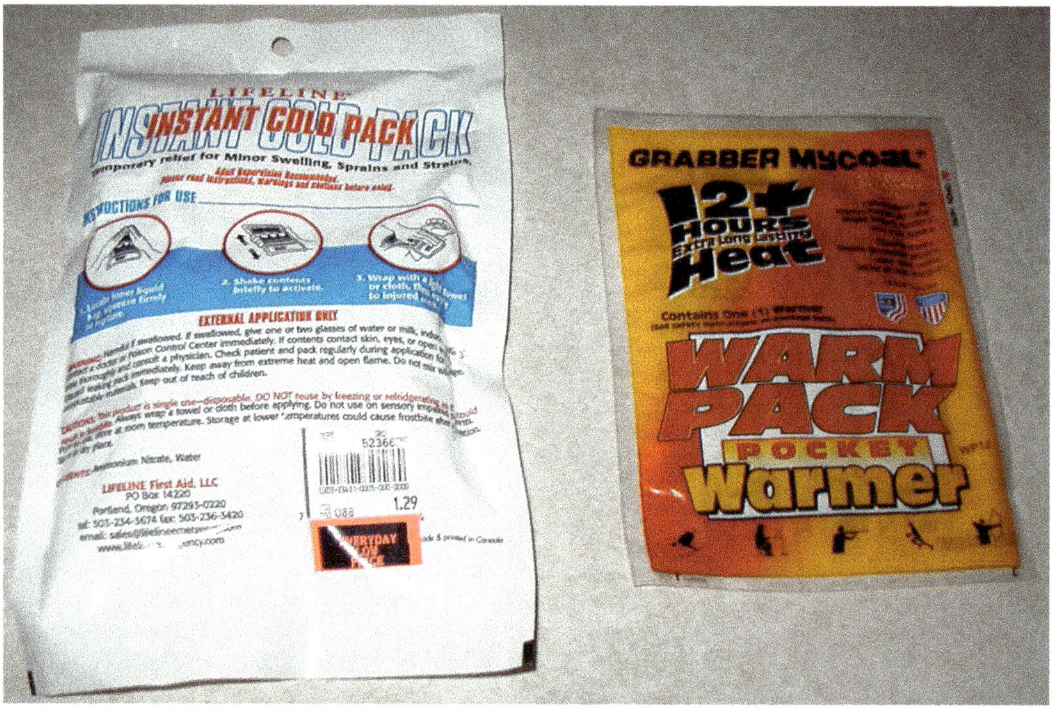

If you have ever watched a soccer game, you might have noticed that when a player twists or sprains an ankle, the team doctor might pull out a plastic bag such as the one pictured on the left side of the figure, squeeze it, shake it, and put it on the player's ankle. This very handy bag, available at drugstores and sporting goods stores, is a temporary ice pack.

The bag contains water and crystals of ammonium nitrate (NH_4NO_3). The water is contained in a fragile inner bag. When the outer bag is squeezed hard, the inner bag breaks, mixing the water and ammonium nitrate. As the bag is shaken, the ammonium nitrate begins to dissolve in the water. Since ammonium nitrate dissolves endothermically in water, the bag gets cold. It gets so cold that it is a good substitute for ice, without all of the mess.

A similar product (shown on the right side of Figure 10.7) provides a temporary source of heat. It is available at most stores that sell hunting and fishing equipment. This product contains iron, water, salt, and a few other chemicals. When the outer plastic bag is opened, the contents are exposed to air, and the iron begins to react with the oxygen in the air. This chemical reaction is exothermic, providing a great deal of heat. The packet pictured in the figure can stay warm up to 12 hours and is often used by hunters and fishermen to keep their hands warm on cold days.

APPLYING STOICHIOMETRY TO SOLUTIONS

In the last module, we learned about molarity and its applications to acid-base chemistry. In that module, we saw that molarity was an important concept in any chemistry that involves solutions. We want to emphasize this by showing you a typical example of how we would apply stoichiometry to problems involving solutions in Example 10.1.

EXAMPLE 10.1

A chemist wishes to make calcium carbonate (the principal component in Tums antacid as well as chalk) by using the following reaction:

$$Na_2CO_3\ (aq) + Ca(OH)_2\ (aq) \rightarrow CaCO_3\ (s) + 2NaOH\ (aq)$$

If the chemist mixes 50.0 mL of 1.23 M Na_2CO_3 with an excess of calcium hydroxide, how many grams of calcium carbonate will be produced?

This is just a stoichiometry problem. We can tell this because we are being asked to determine the amount of one substance when we are given the amount of another substance. The only way to do that is by stoichiometry. To do stoichiometry, we must first get the amount in moles. The amount is given in concentration and volume. To turn that into moles, we must convert mL into liters so that our volume units are consistent with our concentration unit (remember, M means moles/liter):

$$\frac{50.0\ \cancel{mL}}{1} \times \frac{0.001\ L}{1\ \cancel{mL}} = 0.0500\ L$$

Now we can convert from volume and concentration into moles:

$$\frac{1.23 \text{ moles } Na_2CO_3}{1 \cancel{L}} \times 0.0500 \cancel{L} = 0.0615 \text{ moles } Na_2CO_3$$

Now that we have moles, we can do stoichiometry:

$$\frac{0.0615 \cancel{\text{moles } Na_2CO_3}}{1} \times \frac{1 \text{ mole } CaCO_3}{1 \cancel{\text{mole } Na_2CO_3}} = 0.0615 \text{ moles } CaCO_3$$

This is not quite the answer we need. We were asked to figure out how many grams of calcium carbonate were produced, so we have to convert from moles back to grams:

$$\frac{0.0615 \cancel{\text{moles } CaCO_3}}{1} \times \frac{100.1 \text{ grams } CaCO_3}{1 \cancel{\text{mole } CaCO_3}} = 6.16 \text{ grams } CaCO_3$$

This means that **6.16 g of calcium carbonate** will be produced.

How many liters of a 0.245 M solution of NaCl will be necessary to make 100.0 grams of $AlCl_3$ according to the following reaction?

$$3NaCl\ (aq) + Al(NO_3)_3\ (aq) \rightarrow AlCl_3\ (s) + 3NaNO_3\ (aq)$$

Once again, we can tell that this is a stoichiometry problem because we are asked to convert from the amount of one substance to the amount of another. To do stoichiometry, we must convert to moles. Since we have the number of grams of $AlCl_3$, we will use that as our starting point:

$$\frac{100.0 \cancel{\text{g } AlCl_3}}{1} \times \frac{1 \text{ mole } AlCl_3}{133.5 \cancel{\text{g } AlCl_3}} = 0.7491 \text{ moles } AlCl_3$$

Now that we have moles, we can use stoichiometry to determine how much NaCl is needed:

$$\frac{0.7491 \cancel{\text{moles } AlCl_3}}{1} \times \frac{3 \text{ moles NaCl}}{1 \cancel{\text{mole } AlCl_3}} = 2.247 \text{ moles NaCl}$$

This tells us how much NaCl we need, but it doesn't answer the question. The question asks how many liters of a 0.245 M solution are needed. To determine this, we must remember that M means moles per liter. Therefore, the concentration of a solution is a conversion relationship that allows us to relate the number of moles to the number of liters. We can therefore do the following conversion:

$$\frac{2.247 \cancel{\text{moles NaCl}}}{1} \times \frac{1 \text{ Liter NaCl}}{0.245 \cancel{\text{moles NaCl}}} = 9.17 \text{ L}$$

9.17 liters of the NaCl solution will be needed.

Now make sure you can do stoichiometry with solutions by answering "On Your Own" questions 10.4–10.5.

MOLALITY

When we first introduced the concept of concentration, we told you that there are many units we can use to express it. The units we choose depend on the situation we are in at the time. In a moment, you will learn 2 final concepts about solutions, and both of those concepts require a new unit of concentration. That unit is called molality (moh lal' ih tee).

Molality (m)—The number of moles of solute per kilogram of solvent.

$$m = \frac{\text{\# moles solute}}{\text{\# kg solvent}} \qquad \text{Equation 10.1}$$

Do not get confused between molarity and molality. The words look very similar, but their meanings are quite different.

Remember first what molarity means. It means the number of moles of solute per *liter of solution*. If we take 145 mL of water and add 25 g of NaCl to it, the NaCl added to the water will increase the total volume of the solution. Let's suppose that adding 25 g of NaCl to 145 mL of water increases the volume to 150 mL. To get molarity, we would divide the number of moles of NaCl added by the *total* number of liters of the solution: 0.15 liters. To find molarity, we divide the number of moles of solute by the *total volume of solution*.

ON YOUR OWN

10.4 The following reaction is carried out in the laboratory:

$5H_2O_2(aq) + 2KMnO_4(aq) + 3H_2SO_4(aq) \rightarrow 5O_2(g) + 2MnSO_4(aq) + K_2SO_4(aq) + 8H_2O(l)$

If 200.0 mL of 0.750 M H_2O_2 is reacted with an excess of the other reactants, how many grams of oxygen will be produced?

10.5 One way to purify a mixture of 2 metals is to add to the mixture something that reacts with only one of the metals and turns it into an aqueous compound. The mixture can then be filtered, isolating the other metal. A mixture of magnesium and aluminum was treated in such a way to isolate the magnesium. Sodium hydroxide, which reacts only with aluminum, was added:

$2Al\ (s) + 2NaOH(aq) + 6H_2O(l) \rightarrow 2NaAl(OH)_4(aq) + 3H_2(g)$

The aluminum was converted to an aqueous salt, and the magnesium was not part of the reaction and stayed as a solid. The solution was then filtered, isolating the pure magnesium. If a mixture of these 2 metals contained 1.50 kg of Al, how many liters of a 8.10 M NaOH solution would be needed to make sure that all aluminum was taken out of the mixture?

Compare that to molality. In determining the molality of a solution, we instead take the number of moles of solute and divide by the number of *kilograms of solvent*. First, we are not concerned about volume when calculating molality. Instead, we are concerned about mass. Second, we do not divide by the total mass of the solution. Instead, we divide only by the mass of the *solvent*. Molality gives us a ratio between moles of solute and mass of solvent, whereas molarity compares the moles of solute to the volume of the entire solution. This may sound like a subtle distinction, but it is very important.

In the next section, you will learn why molality is a useful concentration unit. For now, it is important to make sure that you understand how to calculate the molality of a solution. Study Example 10.2 and answer "On Your Own" question 10.6 to make sure you have a firm grasp on the concept of molality.

EXAMPLE 10.2

A chemist mixes 3.4 moles of NaCl with 4.5 kg of water. What is the molality of the resulting solution?

According to equation 10.1, we can calculate the molality of a solution by dividing the number of moles of solute by the number of kilograms of solvent. We have both of those numbers, so we just apply equation 10.1:

$$m = \frac{\#\text{ moles solute}}{\#\text{ kg solvent}} = \frac{3.4\text{ moles NaCl}}{4.5\text{ kg water}} = 0.76\ \frac{\text{moles NaCl}}{\text{kg water}} = 0.76\text{ m}$$

We would say, then, that the solution has a molality of **0.76 m.**

If a solution is made by adding 45.0 g of $ZnCl_2$ to 150.0 g of water, what is the molality?

To use equation 10.1, we need to have the number of moles of solute and the number of kg of solvent. In this problem, we are given neither. We have to convert to the right units. First, we will convert grams of solute to moles of solute:

$$\frac{45.0\ \cancel{\text{g ZnCl}_2}}{1} \times \frac{1\text{ mole ZnCl}_2}{136.4\ \cancel{\text{g ZnCl}_2}} = 0.330\text{ moles ZnCl}_2$$

Next we need to convert grams of solvent into kg of solvent:

$$\frac{150.0\ \cancel{\text{g}}}{1} \times \frac{1\text{ kg}}{1{,}000\ \cancel{\text{g}}} = 0.1500\text{ kg}$$

Now that we have moles of solute and kg of solvent, we simply need to apply equation 10.1:

$$m = \frac{\#\text{ moles solute}}{\#\text{ kg solvent}} = \frac{0.330\text{ moles ZnCl}_2}{0.1500\text{ kg water}} = 2.20\ \frac{\text{moles ZnCl}_2}{\text{kg water}} = \mathbf{2.20\text{ m}}$$

ON YOUR OWN

10.6 Calculate the molality of the following solutions:

a. 4.6 moles of NaOH dissolved in 2.4 kg of water

b. 125.0 g of KNO_3 dissolved in 1.22 kg of water

c. 65.2 g of lithium bromide dissolved in 750.0 g of water

Remember, at this point your ability to calculate molality doesn't mean very much because you have no clear indication of how molality can be used. That's okay. In the next 2 sections, you will see that molality is a very useful concentration unit. At that point, you will see that the examples and "On Your Own" questions you just finished were not a series of useless exercises.

FREEZING-POINT DEPRESSION

Have you ever noticed that during snowy and icy conditions, street crews tend to spread salt on the roads? Perhaps you or your parents have done the same thing after you have shoveled snow off your sidewalk or driveway. Why do people do this? The answer, you think, is quite obvious. We spread salt on the roads, driveways, and sidewalks to melt the ice and snow that make them slippery. This is very true, but have you ever stopped to think about *why* salt melts the ice and snow? That's what you're going to find out by performing Experiment 10.4.

EXPERIMENT 10.4

PURPOSE: To measure the freezing-point depression.

MATERIALS:

- 2 beakers (Two short, fat glasses will do.)
- Graduated cylinder (A ¼ measuring cup will do.)
- Table salt
- A ¼ measuring teaspoon
- Thermometer
- Safety goggles

QUESTION: Does a salt increase or decrease temperature in a solvent?

HYPOTHESIS: Write a hypothesis related to this experiment.

PROCEDURE:

1. Fill each of your beakers with 50 mL (¼ cup) of tap water (a qualitative measurement).
2. In one of the beakers, dissolve ¼ teaspoon of salt. Mark that beaker in some way so that you know it is the one that contains the saltwater solution.
3. Place both beakers in the freezer. Check on the beakers every few minutes to see when they begin to freeze.
4. When the salt water is half frozen, remove both beakers. If you didn't keep a careful eye on the beakers and they are both frozen solid, don't worry. Just set them out until they are about half melted. At this point, you should have a mixture of ice and liquid in both beakers.
5. Use your thermometer to carefully stir the mixture of ice and salt water for about 3 minutes and then read the temperature of the thermometer. It should be less than 0°C.
6. Rinse the thermometer and stir the mixture of ice and regular tap water for the same amount of time. Read the temperature. It should be closer to 0°C.
7. Clean up and return everything to the proper place.

CONCLUSION: Write something about what you observed and whether it confirmed your hypothesis.

Remember, water freezes at 0.00°C. If thoroughly mixed, an ice-water solution will always be at 0.00°C, the freezing point of water. Since the temperature of the ice and saltwater mixture was lower than 0.00°C, you can conclude that the *freezing* point of salt water is lower than 0.00°C. The freezing point of a saltwater solution is lower than the freezing point of water. This turns out to be a general rule. When a solute is dissolved in a solvent, the solution always has a *lower* freezing point than the pure solvent. In this experiment, your solvent was water. Pure water has a freezing point of 0.00°C. However, when you added salt to the water in the second beaker, you were adding a solute to the solvent. The resulting solution had a lower freezing point. This phenomenon is known as freezing-point depression and is very useful in the chemistry of solutions.

Hopefully, you see why salt is spread on icy and snowy roads, driveways, and sidewalks. The salt becomes a solute, mixing with the water. This lowers the freezing point of the water on the roads, so the weather has to get *much* colder to freeze the water. This makes icy conditions much less likely.

FIGURE 10.8
Salt Helps Snow and Ice Melt
Photo: Petr Magera (CC BY 2.0)

Although we told you that solutes tend to reduce the freezing points of solvents, we still haven't told you *why* this happens. That explanation requires some arguments based on energy. First, we must remember why a liquid freezes. In its liquid state, the molecules of a substance have a significant amount of energy. They are free to move around, so they are constantly traveling around their container. To get into the solid state, the molecules have to slow down and stop moving around so that they only vibrate back and forth. The only way that this can happen is to take energy away from the molecules. That's what we are doing when we cool down a substance.

As a substance cools, its molecules lose energy. As they lose energy, they start to slow down. Eventually, they lose enough energy so that they almost stop moving around altogether. At that point, the liquid begins to turn into a solid. When a solute is dissolved in a solvent, the solute molecules (or ions) also begin moving around in the solution. Since the solvent molecules are attracted to the solute molecules, a chase ensues. The mutual attraction between the solvent and solute adds more energy to the chase. To get all of the molecules to slow down and stop moving around, *more energy must be removed*. The only way that this can happen is by lowering the temperature further.

In other words, the mutual attraction between solute and solvent tends to add energy to the molecules in the solution. Since freezing involves pulling energy out of the molecules in a liquid, the freezing temperature of a solution must be less than that of the solvent. This is because more energy must be pulled out of the molecules in a solution to freeze the solution as compared to the energy that must be pulled out of a pure solvent to freeze the solvent.

Hopefully, you now not only have learned the concept of freezing-point depression, but you also have a reasonable understanding of *why* it happens. This is still not quite enough to master the topic. We can use mathematics to predict the amount by which the freezing point of a solvent is lowered when a solute is added. This is where the concept of molality plays a role.

When a solute is dissolved in a solvent, the resulting change in the freezing temperature is easily determined using equation 10.2:

$$\Delta T = -i \times K_f \times m \qquad \text{Equation 10.2}$$

In this equation, the Δ symbol is the capital Greek letter delta. In science, the delta symbol means change in. So ΔT means change in temperature. The symbol i refers to the number of molecules (or ions) that the solute splits into when it dissolves, while the m refers to the molality of the solution. The symbol K_f is called the freezing-point depression constant; it is a physical constant that depends solely on the chemical nature of the solvent. Like many physical constants, each solvent has its own unique value of K_f. For example, water has a K_f of 1.86°C/m, while acetic acid (the active ingredient in vinegar) has a K_f of 3.90°C/m.

Why the strange units on K_f? Look at equation 10.2. What units should ΔT end up in? Since ΔT is a temperature measurement, its units should be °C. The symbol i has no units because it is just a number, and the symbol m refers to molality. The K_f must cancel the molality unit and replace it with a temperature unit. That's why K_f has such a strange unit attached to it.

If you're not sure exactly what everything in equation 10.2 means, don't worry. Example 10.3 should clear everything up.

EXAMPLE 10.3

When 15.0 grams of NaCl are added to 100.0 grams of water, what is the change in the water's freezing point? (K_f of water = 1.86°C/m)

In a freezing-point depression problem, we must use equation 10.2. However, to use that equation, we must know K_f, i, and m. Right now, we know only K_f. However, we have been given enough information to determine both i and m. We just got through calculating molalities in the previous section, so that shouldn't be too hard:

$$\frac{15.0\ \cancel{\text{g NaCl}}}{1} \times \frac{1 \text{ mole NaCl}}{58.5\ \cancel{\text{g NaCl}}} = 0.256 \text{ moles NaCl}$$

$$\frac{100\ \text{g}}{1} \times \frac{1\ \text{kg}}{1{,}000\ \text{g}} = 0.1000\ \text{kg}$$

$$m = \frac{0.256\ \text{moles NaCl}}{0.1000\ \text{kg water}} = 2.56\ \text{m}$$

To figure out i, we just have to think about how NaCl dissolves. The symbol i is defined as the number of molecules or ions that the solute splits into when it dissolves. NaCl, being an ionic compound, splits up into its ions when dissolved.

$$NaCl(s) \rightarrow Na^{+}(aq) + Cl^{-}(aq)$$

Each molecule of NaCl has 1 sodium ion and 1 chloride ion, so i = 2.

Now that we have all of the components of equation 10.2, we can use it:

$$\Delta T = -i \times K_f \times m = -2 \times 1.86\ \frac{^{\circ}\text{C}}{\text{m}} \times 2.56\ \text{m} = -9.52^{\circ}\text{C}$$

First, let's make sure that we understand what our answer means. The negative sign means that the freezing point lowered, which means that the freezing point of the NaCl solution is **9.52°C lower than the normal freezing point of water**.

Notice that we kept 3 significant figures in the answer, although the symbol i has only 1 significant figure. This is because the value for i is exact. There are exactly 2 (not 1.999 or 2.001) ions when NaCl dissolves. As a result, this number has no effect on significant figures. When you are solving these kinds of problems, ignore the value of i when calculating significant figures.

What is the freezing point of a solution that is made by mixing 2.5 g $Al(NO_3)_3$ in 150.0 g of water? (K_f of water = 1.86°C/m)

In a freezing-point depression problem, we must use equation 10.2. To use that equation, we must know K_f, i, and m. Right now, we know only K_f. However, we have been given enough information to determine both i and m. Let's start with m:

$$\frac{2.5\ \text{g Al(NO}_3)_3}{1} \times \frac{1\ \text{mole Al(NO}_3)_3}{213.0\ \text{g Al(NO}_3)_3} = 0.012\ \text{moles Al(NO}_3)_3$$

$$\frac{150.0\ \text{g}}{1} \times \frac{1\ \text{kg}}{1{,}000\ \text{g}} = 0.1500\ \text{kg}$$

$$m = \frac{0.012\ \text{moles Al(NO}_3)_3}{0.1500\ \text{kg water}} = 0.080\ \text{m}$$

To figure out i, we just have to think about how $Al(NO_3)_3$ dissolves. The symbol i is defined as the number of molecules or ions that the solute splits into when it dissolves. Being an ionic compound, $Al(NO_3)_3$ splits up into its ions when dissolved. According to its chemical formula, each molecule of aluminum nitrate has 1 aluminum ion and 3 nitrate ions.

$$Al(NO_3)_3\ (s) \rightarrow Al^{3+}\ (aq) + 3\ NO_3^-\ (aq)$$

The total number of ions is 4, so i = 4.

Now that we have all of the components of equation 10.2, we can use it:

$$\Delta T = -i \times K_f \times m = -4 \times 1.86\ \frac{°C}{\cancel{m}} \times 0.080\ \cancel{m} = -0.60°C$$

Realize that this is not the answer to the problem. The problem asks what the freezing point of the solution is. What we found was the change in the freezing point. You are supposed to have memorized that water freezes at 0.00°C. Our answer indicates that the freezing point of this solution is 0.60°C lower than that. So the answer to our problem is that the freezing point of the solution is **-0.60°C**.

What is the freezing point of a solution that is made by dissolving 100.0 grams of table sugar ($C_{12}H_{22}O_{11}$) into 950 grams of water? (K_f of water = 1.86$\frac{°C}{m}$)

We start solving this problem by calculating m:

$$\frac{100.0\ \cancel{g\ C_{12}H_{22}O_{11}}}{1} \times \frac{1\ \text{mole}\ C_{12}H_{22}O_{11}}{342.2\ \cancel{g\ C_{12}H_{22}O_{11}}} = 0.2922\ \text{moles}\ C_{12}H_{22}O_{11}$$

$$\frac{950\ \cancel{g}}{1} \times \frac{1\ \text{kg}}{1{,}000\ \cancel{g}} = 0.95\ \text{kg}$$

$$m = \frac{0.2922\ \text{moles}\ C_{12}H_{22}O_{11}}{0.95\ \text{kg water}} = 0.31\ m$$

To figure out i, we just have to think about how table sugar dissolves. According to its formula, there are no metals present, so table sugar is not ionic. It does not split up into ions. It dissolves 1 molecule at a time; therefore, i = 1.

Now that we have all of the components of equation 10.2, we can use it:

$$\Delta T = -i \times K_f \times m = -1 \times 1.86\ \frac{°C}{\cancel{m}} \times 0.31\ \cancel{m} = -0.58°C$$

The solution has a freezing point that is 0.58°C lower than the freezing point of the solvent. Since the solvent is water, which normally has a freezing point of 0.00°C, this solution has a freezing point of **-0.58°C**.

ON YOUR OWN

10.7 To prevent the water in a car's cooling system from freezing, people add antifreeze to the water in the radiator. This antifreeze is typically ethylene glycol ($C_2H_6O_2$), a slightly polar covalent molecule. What is the freezing point of a radiator solution made from 4.00 kg water and 2.00 kg ethylene glycol? (K_f = 1.86°C/m)

10.8 If a solution is made by dissolving 3.00 grams of magnesium chlorite into 30.0 grams of acetic acid ($C_2H_4O_2$), what is the freezing point? (K_f of acetic acid = 3.90°C/m; freezing point of pure acetic acid = 16.6°C)

Make sure you understand how to do these problems by answering "On Your Own" questions 10.7–10.8.

Before we leave this section, we need to discuss terminology. We have used the term *freezing point*, whereas other books may substitute the term *melting point*. Even though you might think that these terms are different, they are, in fact, the same. After all, any substance will freeze at the same temperature at which it melts. Whether we call it a freezing point or a melting point really just depends on direction. If a substance is turning from liquid to solid, we usually call it a freezing point. If, on the other hand, the substance is turning from solid to liquid, we call it a melting point. The terms are equivalent, so some books use them interchangeably. Do not get confused if you see that terminology.

BOILING-POINT ELEVATION

A long time ago, as a child, I remember watching my grandmother cook my favorite meal, chicken and noodles with mashed potatoes. Before she added the chopped potatoes to the water, she added salt to the nearly boiling water. I asked her why she added the salt, and she said that it makes the water boil faster. This is a myth that has been around for a very long time. This leads us to another property of solutions.

Boiling-point elevation is another property of solutions that is closely tied to freezing-point depression. When a solute is dissolved in a solvent, the boiling point of the resulting solution is *higher* than that of the pure solvent. Once again, this can be understood by looking at the way in which solutes dissolve.

Solutes dissolve in solvents because there is an attraction between them. When we are trying to boil a liquid, what are we trying to do? We are trying to pull the molecules far away from each other so that the substance will change from a liquid to a gas. If a solute is present, the molecules of the solvent are attracted to the molecules of the solute. This makes the solvent molecules harder to pull apart. Since they are harder to pull apart, more energy must be added to the solution to make it boil. In the end, then, the boiling temperature increases.

Boiling-point elevation is governed by an equation that is almost identical to that of the one for freezing-point depression:

$$\Delta T = i \times K_b \times m \qquad \text{Equation 10.3}$$

In this equation, ΔT, i, and m are the same as they were in equation 10.2. The only differences between equation 10.2 and equation 10.3 are that there is no negative sign in equation 10.3 and that we use the boiling-point elevation constant (K_b) instead of the freezing-point depression constant (K_f).

Both of these differences should make sense. First, the boiling point is elevated, not depressed, making ΔT positive, not negative. Second, boiling is a completely different process from freezing. We should therefore expect substances to have a K_b that is quite different from K_f. For water, K_b = 0.512°C/m. Even though the equations for freezing-point depression and boiling-point elevation are a bit different, their application is the same. Study Example 10.4 and then answer "On Your Own" questions 10.9–10.10 to be sure you understand this.

EXAMPLE 10.4

In cooking pasta, a chef adds 20 grams of table salt (NaCl) to 1.45 kg of water and brings the water to a boil. What is the temperature of this boiling solution? (K_b of water = 0.512°C/m)

The addition of a solute (NaCl) to the water will elevate the boiling point. To see how the boiling point is elevated, we must use equation 10.3. To do this, we must calculate i and m:

$$\frac{20 \cancel{\text{ g NaCl}}}{1} \times \frac{1 \text{ mole NaCl}}{58.5 \cancel{\text{ g NaCl}}} = 0.3 \text{ moles NaCl}$$

$$\frac{0.3 \text{ moles NaCl}}{1.45 \text{ kg water}} = 0.2 \text{ m}$$

Once again, i just refers to how the substance splits up when it dissolves. NaCl, being an ionic compound, splits up into its 1 sodium ion and 1 chloride ion. So i = 2:

$$\Delta T = i \times K_b \times m = 2 \times 0.512 \frac{°C}{\cancel{m}} \times 0.2 \cancel{\text{ m}} = 0.2°C$$

We must think about what this tells us. The ΔT that we just solved for tells us at how many °C above the normal boiling point the solution boils. You are supposed to have memorized that water boils at 100.0°C, so the boiling point of this solution is **100.2°C**.

Now that you have a basic idea of how things dissolve and the effects of solutes on solvents, you are ready to proceed with a few different subjects in the next couple of modules. Like most of the subjects we have discussed, we have only scratched the surface of solubility in this module. If you pursue more study in the field of chemistry, you will find that there is much more depth to this intriguing subject than what was discussed here.

ON YOUR OWN

10.9 Acetic acid has a K_b of 2.93°C/m and a normal boiling point of 118.1°C. What would be the boiling point of a solution made by dissolving 100.0 g of $CaCO_3$ in 400.0 g of acetic acid?

10.10 If you wanted to boil a solution made by dissolving 265.0 grams of phosphorus trihydride into 2.0 kg of water, to what temperature would you have to raise it? (K_b of water = 0.512°C/m)

SUMMARY OF KEY EQUATIONS AND TABLES IN MODULE 10

Equation 10.1: $m = \frac{\text{\# moles solute}}{\text{\# kg solvent}}$ Molality

Equation 10.2: $\Delta T = -i \times K_f \times m$ Freezing-point depression

Equation 10.3: $\Delta T = i \times K_b \times m$ Boiling-point elevation

Table 10.1: Summary of the Methods by Which Solutes Dissolve in Liquid Solvents

Table 10.2: The Effect of Temperature on Solubility

ANSWERS TO THE "ON YOUR OWN" QUESTIONS

10.1 Since the solubility of most solids increases as the temperature of the solution increases, the student should try heating up the solvent a bit. This should increase the solid's solubility. It may, of course, decrease it because we do not know whether or not this particular solute is an exception to the general rule. However, you should assume it will follow the rule unless otherwise indicated.

10.2 If the gas is escaping the solution, the solubility of the gas is not quite large enough under the conditions of the storeroom. The chemist needs to increase the solubility of the gas. The way to do this is to reduce the temperature of the area that stores the bottles. This could be done by putting the bottles in a refrigerator or just turning off the heat to the storeroom. Note that she should not just tape the lids down to keep them from popping off. If the gas were really escaping the solution, the pressure would continue to build even after the lids were taped down. Eventually, the pressure buildup could be great enough to explode the bottles!

10.3 Since temperature is not the issue, we must consider pressure. Increased pressure increases the solubility of gases in liquids. Since the atmospheric pressure varies every day, the pressure in the room probably varies with it. This means that the pressure in the room is probably varying. On days when the pressure is high, more gas can dissolve in the water. On days when the pressure is low, less gas will dissolve. If the student really needs a constant amount of gas dissolved in liquid, he may have to control the pressure in the lab as well as the temperature.

10.4 This is just a stoichiometry problem. We can tell this because we are being asked to determine the amount of one substance when we are given the amount of another substance. The only way to do that is by stoichiometry. To do stoichiometry, we must first get our amount in moles. Right now, the amount is given in concentration and volume. We must turn that into moles. To do that, though, we must convert mL into liters, so that our volume unit is consistent with our concentration unit (remember, M means moles/L).

$$\frac{200.0 \cancel{\text{mL}}}{1} \times \frac{0.001 \text{ L}}{1 \cancel{\text{mL}}} = 0.2000 \text{ L}$$

Now we can convert from volume and concentration into moles:

$$\frac{0.750 \text{ moles } H_2O_2}{1 \cancel{\text{L}}} \times 0.2000 \cancel{\text{L}} = 0.150 \text{ moles } H_2O_2$$

Now that we have moles, we can do stoichiometry:

$$\frac{0.150 \cancel{\text{moles } H_2O_2}}{1} \times \frac{5 \text{ moles } O_2}{5 \cancel{\text{moles } H_2O_2}} = 0.150 \text{ moles } O_2$$

This is not quite the answer we need. We were asked to figure out how many grams of oxygen were produced, so we have to convert from moles back to grams:

$$\frac{0.150 \text{ moles } O_2}{1} \times \frac{32.0 \text{ g } O_2}{1 \text{ mole } O_2} = 4.80 \text{ grams } O_2$$

Therefore, 4.80 g of oxygen will be produced.

10.5 Once again, we can tell that this is a stoichiometry problem because we are asked to convert from the amount of one substance to the amount of another. To do stoichiometry, we must convert to moles. Since we have the number of kg of Al, we will use that as our starting point. First we must convert kg to g:

$$\frac{1.50 \text{ kg}}{1} \times \frac{1000 \text{ g}}{1 \text{ kg}} = 1.50 \times 10^3 \text{ g Al}$$

$$\frac{1.50 \times 10^3 \text{ g Al}}{1} \times \frac{1 \text{ mole Al}}{27.0 \text{ g Al}} = 55.6 \text{ moles Al}$$

Now that we have moles, we can use stoichiometry to determine how much NaOH is needed:

$$\frac{55.6 \text{ moles Al}}{1} \times \frac{2 \text{ moles NaOH}}{2 \text{ moles Al}} = 55.6 \text{ moles NaOH}$$

This tells us how much NaOH we need, but it doesn't answer the question. The question asks how many liters of a 8.10 M solution are needed. To determine this, we must remember that M means moles per liter. The molarity of a solution is a conversion relationship that allows us to relate the number of moles to the number of liters. We can therefore do the following conversion:

$$\frac{55.6 \text{ moles NaOH}}{1} \times \frac{1 \text{ L NaOH}}{8.10 \text{ moles NaOH}} = 6.86 \text{ L of solution}$$

This means that 6.86 liters of the NaOH solution would be needed.

10.6 (a) Since the problem already gives us the number of moles of solute and the number of kg of solvent, this problem is just a straight application of equation 10.1:

$$\text{molality} = \frac{\text{\# moles solute}}{\text{\# kg solvent}} = \frac{4.6 \text{ moles NaOH}}{2.4 \text{ kg water}} = 1.9 \frac{\text{moles NaOH}}{\text{kg water}} = 1.9 \text{ m}$$

(b) This problem is a bit more difficult because, although we are given the number of kg of solvent, we are given the number of grams of solute. So we first must convert grams to moles:

$$\frac{125.0 \text{ g } KNO_3}{1} \times \frac{1 \text{ mole } KNO_3}{101.1 \text{ g } KNO_3} = 1.236 \text{ moles } KNO_3$$

Now that we have moles of solute and kg of solvent, we can use equation 10.1:

$$\text{molality} = \frac{\text{\# moles solute}}{\text{\# kg solvent}} = \frac{1.236 \text{ moles } KNO_3}{1.22 \text{ kg water}} = 1.01 \frac{\text{moles } KNO_3}{\text{kg water}} = 1.01 \text{ m}$$

The molality is 1.01 m.

(c) In this problem, we are given neither moles of solute nor kg of solvent, so we must convert both. But first we need the formula of lithium bromide. Li is a 1^+ ion and Br is a 1^- ion. Therefore, the formula will be LiBr.

$$\frac{65.2 \cancel{\text{ g LiBr}}}{1} \times \frac{1 \text{ mole LiBr}}{86.8 \cancel{\text{ g LiBr}}} = 0.751 \text{ moles LiBr}$$

$$\frac{750.0 \cancel{\text{ g}} \text{ water}}{1} \times \frac{1 \text{ kg}}{1000 \cancel{\text{ g}}} = 0.7500 \text{ kg water}$$

Now that we have moles of solute and kg of solvent, we can use equation 10.1:

$$\text{molality} = \frac{\text{\# moles solute}}{\text{\# kg solvent}} = \frac{0.751 \text{ moles LiBr}}{0.7500 \text{ kg water}} = 1.00 \frac{\text{moles LiBr}}{\text{kg water}} = 1.00 \text{ m}$$

The molality is 1.00 m.

10.7 In a freezing-point depression problem, we must use equation 10.2. To use that equation, we must know K_f, i, and m. Right now, we only know K_f. However, we have been given enough information to calculate both i and m. First let's calculate m:

$$\frac{2.00 \times 10^3 \cancel{\text{ g } C_2H_6O_2}}{1} \times \frac{1 \text{ mole } C_2H_6O_2}{62.1 \cancel{\text{ g } C_2H_6O_2}} = 32.2 \text{ moles } C_2H_6O_2$$

$$m = \frac{32.2 \text{ moles } C_2H_6O_2}{4.00 \text{ kg water}} = 8.05 \text{ m}$$

To figure out i, we just have to realize that according to its formula, ethylene glycol is not ionic. Therefore, i = 1.

Now that we have all of the components of equation 10.2, we can use it:

$$\Delta T = -i \times K_f \times m = -1 \times 1.86 \frac{°C}{\cancel{m}} \times 8.05 \cancel{\text{ m}} = -15.0°C$$

So the freezing point is 15.0°C lower than that of normal water, or -15.0°C.

10.8 The formula for magnesium chlorite is $Mg(ClO_2)_2$. In a freezing-point depression problem, we must use equation 10.2. To use that equation, we must know K_f, i, and m. Right now, we only know K_f. However, we have been given enough information to calculate both i and m. To calculate m:

$$\frac{3.00\ \cancel{g\ Mg(ClO_2)_2}}{1} \times \frac{1\ \text{mole}\ Mg(ClO_2)_2}{159.3\ \cancel{g\ Mg(ClO_2)_2}} = 0.0188\ \text{moles}\ Mg(ClO_2)_2$$

$$\frac{30\ \cancel{g}\ \text{acetic acid}}{1} \times \frac{1\ kg}{1000\ \cancel{g}} = 0.0300\ kg\ \text{acetic acid}$$

$$m = \frac{0.0188\ \text{moles}\ Mg(ClO_2)_2}{0.0300\ kg\ \text{acetic acid}} = 0.627\ m$$

To figure out i, we just have to think about how $Mg(ClO_2)_2$ dissolves. Being an ionic compound, it splits up into its ions when dissolved. Each molecule of magnesium chlorite has 1 magnesium ion and 2 chlorite ions, so i = 3.

Now that we have all of the components of equation 10.2, we can use it:

$$\Delta T = -i \times K_f \times m = -3 \times 3.90\ \frac{°C}{\cancel{m}} \times 0.627\ \cancel{m} = -7.34°C$$

Realize that *this is not the answer to the problem.* The problem asks what the freezing point of the solution is. What we found was the *change in the freezing point.* We are told that pure acetic acid has a freezing point of 16.6°C. Our answer indicates that the freezing point of this solution is 7.34°C lower than that. The answer to our problem is that the freezing point of the solution is 16.6 - 7.34 = 9.3°C.

10.9 To calculate boiling points, we must use equation 10.3. To do that, we must know i and m. To calculate m:

$$\frac{100.0\ \cancel{g\ CaCO_3}}{1} \times \frac{1\ \text{mole}\ CaCO_3}{100.1\ \cancel{g\ CaCO_3}} = 0.9990\ \text{moles}\ CaCO_3$$

$$\frac{400.0\ \cancel{g}\ \text{acetic acid}}{1} \times \frac{1\ kg}{1000\ \cancel{g}} = 0.4000\ kg\ \text{acetic acid}$$

$$m = \frac{0.9990\ \text{moles}\ CaCO_3}{0.4000\ kg\ \text{acetic acid}} = 2.498\ m$$

Since calcium carbonate is an ionic compound, it dissolves by splitting up into its calcium ion and its carbonate ion. So i = 2.

$$\Delta T = i \times K_b \times m = 2 \times 2.93\ \frac{°C}{\cancel{m}} \times 2.498\ \cancel{m} = 14.6°C$$

This means that the boiling point of the solution is 14.6°C *higher* than that of pure acetic acid. The boiling point of pure acetic acid was given as 118.1°C, so the boiling point of this solution is 118.1 + 14.6 = 132.7°C.

10.10 The addition of a solute (phosphorus trihydride) to the water will elevate the boiling point. To see how the boiling point is elevated, we must use equation 10.3. To do this, we must calculate i, m, and a formula for phosphorus trihydride. The formula is PH_3.

$$\frac{265.0\ \cancel{g\ PH_3}}{1} \times \frac{1\ \text{mole}\ PH_3}{34.0\ \cancel{g\ PH_3}} = 7.79\ \text{moles}\ PH_3$$

$$\frac{7.79\ \text{moles}\ PH_3}{2.0\ \text{kg water}} = 3.9\ m$$

Once again, i just refers to how the substance splits up when it dissolves. Since PH_3 is covalent, it splits up into 1 molecule at a time when it dissolves. So i = 1.

$$\Delta T = i \times K_b \times m = 1 \times 0.512\ \frac{°C}{\cancel{m}} \times 3.9\ \cancel{m} = 2.0°C$$

We must think about what this tells us. The ΔT that we just solved for tells us how many °C *above the normal boiling point* that the solution boils. You are supposed to have memorized that water boils at 100.0°C, so the boiling point of this solution is 102.0°C.

STUDY GUIDE FOR MODULE 10

REVIEW QUESTIONS

1. When you dissolve sodium fluoride in water, which compound is the solute and which is the solvent?

2. Contrast the way ionic compounds dissolve in water with the way that polar covalent compounds dissolve in water.

3. What is wrong with this statement: It does not matter what the solvent is in determining the solubility.

4. What is a saturated solution?

5. Within the topic of solutions, what is precipitation?

6. What kind of solute (solid, liquid, or gas) dissolves best under high-temperature conditions?

7. What kind of solute (solid, liquid, or gas) dissolves best under high-pressure conditions?

8. What kind of solute (solid, liquid, or gas) has a solubility that is least affected by the conditions under which the solution is made?

9. If a solid solute is not dissolving well in water, what can you do to increase its solubility?

10. If a gas solute is not dissolving well in water, what can you do to increase its solubility?

11. If a chemist makes a solution in a beaker using water and a solute that dissolves endothermically in water, would you expect the beaker to feel hot or cold?

12. What is the difference between molality and molarity?

13. If you wanted to protect water from freezing, which compound would accomplish this best: $NaNO_3$, $Mg(NO_3)_2$, or $Al(NO_3)_3$?

PRACTICE PROBLEMS

1. Hot tubs generally contain water at a temperature of 38°C. To keep the water clean, several solid solutes are dissolved in the water. Many hot tub manufacturers warn that if the water gets too cold, it will get cloudy and fill with a fine solid. Why?

2. An open glass of soda pop sitting on the counter gets flatter (less fizzy) than the same open glass of soda pop in the refrigerator. Why?

3. The following reaction is performed in a lab:

$$3Na_2SO_4\ (aq) + 2Al(NO_3)_3\ (aq) \rightarrow Al_2(SO_4)_3\ (s) + 6NaNO_3\ (aq)$$

If 690.0 mL of 2.50 M aluminum nitrate is added to an excess of sodium sulfate, how many grams of aluminum sulfate will be produced?

4. Aqueous sodium bromide and silver nitrate will react to form solid silver bromide (a compound used in photography) and sodium nitrate:

$$NaBr\ (aq) + AgNO_3\ (aq) \rightarrow AgBr\ (s) + NaNO_3\ (aq)$$

If a chemist needs to produce 150.0 grams of silver bromide, how many liters of 3.5 M silver nitrate must be added to an excess of sodium bromide?

5. One popular test for lead (Pb^{2+}) contamination in drinking water is the chloride test. This test involves adding NaCl to water. Any Pb^{2+} in the water will immediately react with the Cl^-, forming a solid precipitate:

$$Pb^{2+}\ (aq) + 2NaCl\ (aq) \rightarrow PbCl_2\ (s) + 2Na^+\ (aq)$$

Measuring the mass of this precipitate can tell you what concentration of Pb^{2+} is in the water. If an excess of NaCl is added to 3.2 liters of water and 50.0 grams of $PbCl_2$ are produced, what was the molarity of Pb^{2+} in the water?

6. If a solution is made by mixing 75.0 moles of potassium hydroxide with 4.50 kilograms of water, what is the solution's molality?

7. If a solution is made by mixing 70.0 g of $CaBr_2$ with 1,314 g of water, what is the solution's molality?

8. If you want to lower water's freezing point 18°C by adding NaCl, what must the molality of the salt solution be? (K_f for water is 1.86°C/m)

9. What is the freezing point of a solution made by mixing 42.0 grams of NH_3 with 420.0 grams of water? (K_f for water is 1.86°C/m)

10. What is the boiling point of a solution made by mixing 75.0 g $ZnCl_2$ with 375.0 grams of water? (K_b for water is 0.512°C/m)

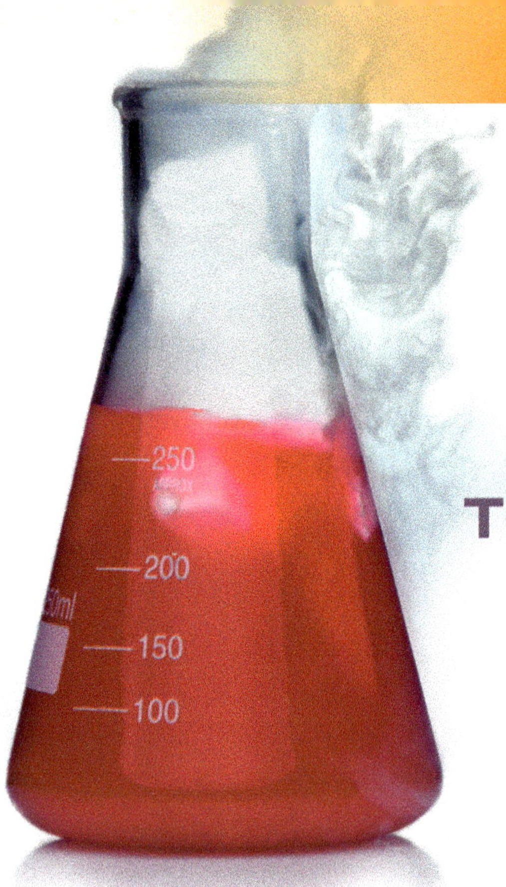

THE GAS PHASE

Since the last module concentrated on chemistry as it happens in the liquid phase, it is only natural that this module should emphasize the gas phase. Gases are an important part of God's creation, and we see them every day in the sky. We learned in Module 2 that the mixture of gases that makes up the earth's atmosphere is perfectly designed to support life. In some of the experiments you have performed, gases were produced as part of the chemical reaction. In addition, we have learned that Gay-Lussac's law allows us to relate the volumes of gases that appear within a chemical equation. So we have already touched on the subject of gases quite a bit.

In this module, we will learn more about how the physical characteristics of gases change in response to changes in the environment. In addition, we will learn a way to relate the number of moles of a gas to such things as the volume that the gas occupies, the pressure it exerts, and the temperature of its surroundings. This particular technique will become very useful in helping us perform stoichiometric calculations on chemical equations that contain gases.

first reactions

How does the water evaporate out of wet clothes hung on a clothesline? Why don't the clothes evaporate with the water? Scientists have spent many years learning about the physical properties of gases. Meteorologists use this knowledge to predict the weather. Aerospace engineers use the knowledge for determining combustion. Even soda pop makers use the knowledge to carbonate your drinks.

THE DEFINITION OF PRESSURE

We learned in Module 1 that to fully understand matter, scientists measure different properties of that matter. For solids and liquids, scientists measure their mass and volume. For gases, they measure not only mass and volume but also the pressure of the gas. So what is pressure? Let's start with a definition.

Pressure—The force per unit area exerted on an object.

Mathematically, we would write:

$$P = \frac{F}{A} \qquad \text{Equation 11.1}$$

where P is the pressure, F is the force exerted on an object, and A is the area over which this force is applied.

Although equation 11.1 is very important, you do not have to memorize it because chemists don't really use it. When you take physics, this equation will become much more important for you to know. For now, the only reason we present this equation is to illustrate the standard unit for pressure. The unit for force is called the Newton, and your experience in geometry should indicate that the unit for area is m^2. Pressure, then, has the units of Newtons/m^2. This unit is named the Pascal (abbreviated Pa) to honor Blaise Pascal (Pas kal'), a French mathematician, scientist, and philosopher. Pascal was a brilliant man who revolutionized the study of probability as well as the sciences of chemistry and physics. In addition, he was a celebrated philosopher who provided several convincing arguments for the validity of the Christian faith.

The important thing that you have to realize now is that a gas exerts pressure on any object that comes into contact with it. How does a gas exert pressure? In module 6, we learned about the kinetic theory of matter. This theory states that the molecules (or atoms) that make up a substance are in constant motion. In solids, the molecules (or atoms) vibrate back and forth. In the liquid phase, molecules (or atoms) move around their container, staying relatively close to one another. In the gas phase, the molecules (or atoms) move around their container quickly, staying far apart from each other.

Since gas molecules move around quickly, they tend to collide with anything that gets into their way. If an object comes into contact with a gas, the molecules (or atoms) that make up the gas will begin striking the object, exerting a force. The average force per unit area that these collisions exert on the object is called the gas pressure.

Since our atmosphere is composed of a mixture of gases, the atmosphere must exert pressure on any object with which it comes into contact. Every person, animal, and object on earth is subject to a pressure exerted on it by the atmosphere. This atmospheric pressure varies a bit, depending on the weather conditions, but on average, it is equal to 101.3 kiloPascals (kPa). Since atmospheric pressure is reasonably constant, we use this value as an alternative unit for measuring pressure. We say that a gas exerts 1.000 atmosphere (abbreviated atm) if it exerts 101.3 kPa of pressure. Mathematically, we would say:

$$1.000 \text{ atm} = 101.3 \text{ kPa} \qquad \text{Equation 11.2}$$

Chemists use both of these units to measure pressure, so you will have to be comfortable with both of them.

Just to make things a bit more confusing, chemists often use 2 other units for pressure: millimeters of mercury (abbreviated mm Hg) and torr. If you take physics, you will learn more about these units. For now, all you need to know is that they are alternative units for pressure and they are both equal. Their relationship to atmospheres is given in table 11.1.

TABLE 11.1
Units of Pressure

Atmosphere (atm)	kiloPascals (kPa)	Millimeters of mercury (mm Hg)	Torr (torr)
1.000 atm	101.3 kPa	760.0 mm Hg	760.0 torr

Chemists use 4 units for pressure: Pa, atm, torr, and mm Hg. You need to know that all of these are pressure units, and you also need to know how they relate to atmospheres. In other words, you need to know the relationships presented in table 11.1.

But how is pressure measured? If we could measure the force with which the gas pushes and measure the area over which that force is applied, then we could use equation 11.1 to calculate the pressure. Are there any instruments that can be used to measure the pressure of a gas instead of calculating it? Yes, a barometer measures the pressure of the atmosphere. A manometer measures the pressure of a gas in a container. Pressure gauges are used by tradesmen like plumbers to measure the pressure of gases inside pipes.

FIGURE 11.1
A Manometer
Photo by Zaphod (CC-BY-SA 3.0)

BOYLE'S LAW

Now that you know the definition of pressure and the various units used to measure it, you can begin to see how the pressure of a gas is affected by the conditions under which it is stored. In the mid-1600s, an Irish scientist named Robert Boyle did several experiments to determine how the pressure of a sample of gas relates to its volume. He determined that as long as the temperature of the gas stayed the same throughout the experiment, the product of a gas's volume and its pressure is always the same. In mathematical terms:

$$PV = \text{constant} \qquad \text{Equation 11.3}$$

This relationship is called Boyle's law. It tells us that since the product of pressure and volume must stay constant, when the volume of a gas is increased, the pressure will decrease. Conversely, if the volume of a gas is decreased, its pressure will increase. Of course, it works the other way as well. If the pressure of a gas is increased, its volume will decrease in response. If the pressure is decreased, the volume of the gas will increase. This is illustrated in Figure 11.2.

FIGURE 11.2

Boyle's Law

Illustration by David Weiss

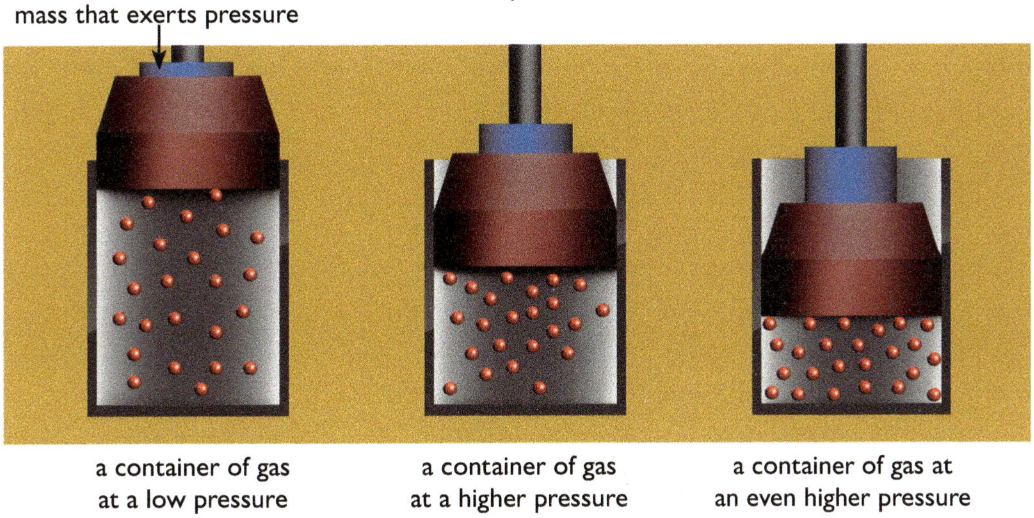

In this figure, a gas is contained in a piston with a top that is free to move up and down. In the drawing on the far left, a small mass is sitting on top of the piston. The weight of the mass pushes down on the top of the piston, exerting a pressure on the gas. In the middle drawing, the only thing that has changed is the size of the mass. The amount of gas, temperature, size of the piston, etc., are all the same as they were in the drawing on the left. However, since the mass in the middle drawing is larger, it exerts a larger pressure on the gas. As a result, the volume of the gas decreases. In the drawing on the right, the mass has been increased even more. This results in an even larger pressure, which further reduces the volume of the gas. That's Boyle's law in action.

If we think about the kinetic theory of matter and the definition of pressure, this law should make sense. After all, if we increase the volume that a gas occupies, the molecules (or atoms) of the gas can spread out. They will make fewer collisions with any object that comes into contact with the gas. Based on our definition of pressure, fewer collisions will result in lower pressure. On the other hand, if we decrease the volume that a gas can occupy, the molecules (or atoms) of the gas cannot spread out. Since they have less space to move around in, they will collide with objects in the gas a lot more often. Based on our definition of pressure, more collisions result in higher pressure.

Boyle did not have the benefit of understanding his law based on the kinetic theory of matter because the kinetic theory of matter wasn't around in the mid-1600s. Instead, he wrote his law based solely on experiments. He studied several gases and found that if he doubled the pressure on a container of gas, the volume of the gas within the container decreased by a factor of 2. After hundreds of such experiments on several different gases, he assumed that his law was correct, even though he didn't have a good understanding as to why.

FIGURE 11.3
Volume Affects Pressure
Photo by Triddle (Public Domain)

There are many sizes of gas tanks available. With other factors being constant, as the volume of gas increases in a tank, the pressure must decrease.

There is an alternative way of expressing Boyle's law. Since equation 11.3 tells us that the product of a gas's volume and its pressure is always the same, we could also say that the product of a gas's volume and its pressure under one set of conditions must be equal to the product of the gas's volume and its pressure under any other set of conditions. Mathematically, we write:

$$P_1V_1 = P_2V_2 \qquad \text{Equation 11.4}$$

where P_1 and V_1 are the pressure and volume of a gas under one set of conditions, and P_2 and V_2 are the pressure and volume of the same gas under another set of conditions. It is important for you to understand that equations 11.3 and 11.4 both say the same thing. They are both ways of mathematically representing Boyle's law; they just make the statement in different ways.

Although Boyle's law is important to know, you do not need to memorize equations 11.3 and 11.4. Boyle's law can be combined with the law you are going to learn next (Charles's law). When the 2 laws are combined, the mathematical equation that results is the important one to remember. So you need to understand Boyle's law only from a conceptual point of view. We will put the mathematics with the concepts a little later on.

If you purchased the MicroChem kit mentioned in the Student Notes, you can perform Experiment 8 in that kit to get more experience with Boyle's law.

CHARLES'S LAW

In the late 1700s, a French scientist named Jacques Charles began building on the work of Robert Boyle by investigating the effect of temperature on a gas's volume. He found that when a gas's temperature was raised under conditions of constant pressure, the volume occupied by the gas increased as well. This is shown in Figure 11.4.

salt and light

Painting of Robert Boyle: Public Domain

Robert Boyle (1627–1691) , was a philosopher, chemist, physicist, inventor, and Christian noted for his writings on theology. He is considered by many to be the first modern chemist and the founder of modern chemistry because he turned the unscientific field of alchemy into a scientific endeavor. He is best known for Boyle's law, which describes the relationships between pressure and volume of gas. He was the first to define the term *element* properly, he invented a primitive kind of litmus paper, he demonstrated that water expands when it freezes, and he provided strong evidence for the existence of atoms.

Boyle was a devout Christian. While many scientists have difficulty reconciling faith and science, he believed that all truth comes from God, so understanding scientific truths will lead the scientist to God. He turned down the opportunity to become the head of Eton College because it would have required him to become a minister, and he thought his writings on religious subjects would have greater impact coming from a layman rather than a paid minister. He wrote many books, all of which show his Christianity. He spent large sums of money promoting the spread Christianity in the East, including paying the expenses to translate the Bible or portions of it into various languages because he felt strongly that the Bible should be available in the common language of the people.

Boyle's Christian faith provided the motivation for changing alchemy into a scientific discipline. Boyle considered science to be a means of glorifying God, and he was eager to uncover the secrets of God's creation. Rather than trying to turn common metals such as lead into precious metals such as gold (which was alchemy's aim), he wanted to learn as much as he could about how God had designed the world. Boyle's attitude towards science and faith is best summed up in a quote from his final address to the Royal Society, an organization of scientists in England. He told the assembled scientists, "Remember to give glory to the One who authored nature" (Graves 1996, 63). Sadly, many scientists today ignore Boyle's admonition, and science has suffered as a result. Boyle wanted to make sure that the important aspects of his work continued after his death, so his will established a lecture series designed to convince people of the reality of the Christian faith.

FIGURE 11.4
Charles's Law
Illustration by David Weiss

In the figure, a container of gas at a certain pressure is placed in ice water so that its temperature is low. The container is then put over a flame to increase its temperature. During the entire time, the pressure is constant because the size of the mass on the top of the piston is the same. Notice that when the temperature is warmer, the gas occupies a greater volume. If you perform this kind of experiment with different gases and different temperatures, you can produce a graph that looks like the one in Figure 11.5.

FIGURE 11.5
The Effect of Temperature on the Volume of a Gas at Constant Pressure

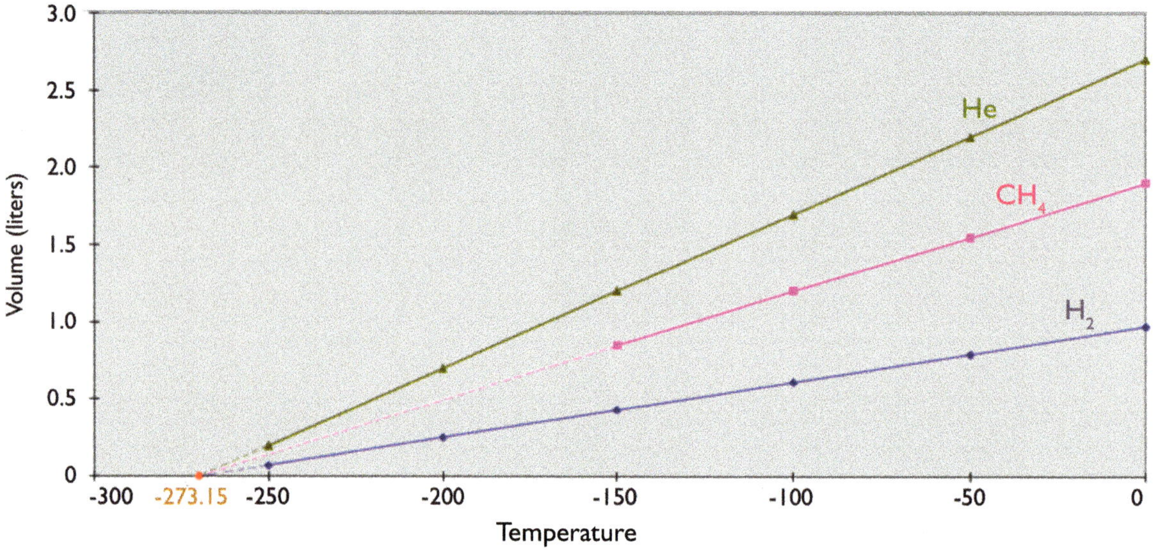

If you study the figure, you should see 2 important aspects of the data. First, the volume of a gas seems to depend linearly on its temperature. If the temperature of a gas is doubled, its volume doubles. Conversely, if the temperature is halved, the volume of the gas decreases by a factor of 2 as well. This fact is called Charles's law:

Charles's law—At constant pressure, the temperature and volume of a gas are linearly proportional.

This law can be mathematically expressed as follows:

$$\frac{V}{T} = \text{constant} \qquad \text{Equation 11.5}$$

In this equation, V is the volume of the gas, and T is its temperature. Here's how this equation expresses Charles's law: If the ratio of volume and temperature must remain constant, then when temperature is increased, volume must increase proportionally to keep the quotient the same.

In terms of the kinetic theory, you should see why Charles's law makes sense. If the temperature of a gas increases, the gas molecules (or atoms) have more energy and want to move around a lot more quickly. However, Charles specifically constructed his experiments so that the pressure exerted by the gas could not change. The only way a gas can move around more quickly without making more collisions with other objects (and hence raising the pressure) is to spread out. If the gas molecules spread out and occupy a larger volume, they can move around more quickly without colliding with other objects more frequently. This will keep the gas pressure from rising.

Charles's law can also be expressed in terms of an equation similar to equation 11.4. Since volume divided by temperature must always remain the same, the volume divided by the temperature under one set of conditions should be the same as the volume divided by the temperature under another set of conditions:

$$\frac{V_1}{T_1} = \frac{V_2}{T_2} \qquad \text{Equation 11.6}$$

Remember that equations 11.5 and 11.6 really say the same thing; they just express it in different ways. Also, as was the case with Boyle's law, you do not need to memorize these 2 equations. Instead, the combined gas law that you will learn in the next section will be the important equation to remember.

Before we move on to the combined gas law, we must look once more at Figure 11.5. After all, there were 2 important aspects to the data in that figure, but so far we have discussed only one. Notice that each gas has a different line on the graph. The line that goes through helium's data is higher on the graph than the one that goes through methane's (CH_4) data. In addition, the line that goes through methane's data is higher on the graph than the one that goes through hydrogen's data. If we follow the lines backward far enough (see the dashed part of each line), they all converge to the same temperature when the volume of the gas reaches 0. That temperature is -273.15°C.

No matter what gas you put on that graph, the line that goes through that gas's data points will always hit -273.15°C when the volume of the gas is 0. If you think about that fact for a minute, you will come to the remarkable conclusion that no matter how hard we try, we can never cool something to a lower temperature than -273.15°C. After all, a gas (which is made up of matter) can never reach a volume of 0. If that were the case, the gas would no longer be matter because, by definition, all matter takes up some space. Since we can never get to 0 volume, we can never get to the place where all of the lines in Figure 11.5 converge.

This might sound a little odd. After all, to get a substance's temperature a bit lower, all you need to do is cool it a little more, right? Well, that's right up to a certain point. However, once the temperature of a substance gets near -273.15°C, it gets harder and harder to cool. In the end, you can never quite get there. For centuries, scientists have tried to cool things to a temperature of -273.15°C or lower, but they have never succeeded.

You should recognize the number 273.15. In the very first module, you learned that to convert temperature from Celsius to Kelvin, you had to add 273.15 to the Celsius temperature. You also learned that the Kelvin temperature is called an absolute temperature scale because we can never reach 0.00 Kelvin. Now you know why that is the case. Since we can never reach -273.15°C, and since the Kelvin temperature scale is determined by adding 273.15 to all Celsius temperatures, we can never reach (or go below) 0.00 Kelvin. As of 2013, the lowest temperature ever reached was 0.00000005 K. (See the Apologia book extras website mentioned in the Student Notes for more information.)

If you purchased the MicroChem kit mentioned in the Student Notes, you can perform experiment 9 in that kit to get more experience with Charles's law and absolute zero.

Before we leave this section, let's digress a bit. Notice that in Figure 11.5, the portions of the lines closest to -273.15°C are dashed to indicate that no data have been collected for the temperatures on that portion of the line. In other words, for helium and hydrogen, there are no data points for temperatures lower than -250°C. For methane (CH_4), there are no data points for temperatures lower than -150°C. At those low temperatures, it is nearly impossible to get an accurate measurement of volume for the gases in the graph. As a result, the experiments that give us the data presented in Figure 11.5 could not be performed at temperatures significantly lower than those listed.

If there are no data points in that region of the graph, why do we draw in a dashed line? Well, the 3 sets of data follow a very steady trend from 0°C all the way down to -250°C. Since the data seem to follow the lines so well for such a big temperature range, *there is no reason to expect that they will stop following that trend.* As a result, even though there are no data points below -250°C, it is probably safe to assume that the data follow the same trend, and we continue to draw the line without any data points to guide it. This process of assumption is called extrapolation (ik strap' uh lay' shun).

Extrapolation—Following an established trend in the data even though there is no data available for that region.

Whether you realize it or not, you use the process of extrapolation all of the time.

science and creation

Some scientists take extrapolation to an extreme and get into all sorts of trouble. This is how the theory of evolution got started. Charles Darwin, an English naturalist who lived in the mid-to-late-1800s, noticed that living organisms had the ability to change and adapt to their environment. While studying the Galapagos Islands, he noticed that finches (a type of bird) lived on each island, but the finches were all different. On islands where the birds could find food by sticking their beaks into crevices in the trees and logs, the finches had long, narrow beaks. On islands where woods were scarce, the finches had short, fat, hard beaks that allowed them to burrow for food. He imagined that at one time, both of these types of finches were the same; that when the finches began living on separate islands, their species began to adapt to the different food sources; and that after many, many generations, they developed different kinds of beaks that were appropriate for the different food supply on each island.

Darwin was absolutely right on this point. Today, scientists have shown conclusively that species do have the ability to adapt and change in response to their environment. For example, in 1977, there was a major drought on Daphne, one of the Galapagos Islands. Researchers had been measuring the beak sizes of finches on that island for some time, and they continued to measure beak sizes long after the drought. They found that the very next generation of finches on the island had beaks that were (on average) about 5% larger than the generation of finches that existed prior to the drought. Since the drought caused a shortage of seeds on the island, the finches with larger beaks were better able to crack open the few, tough seeds left on the island. The size of the finches' beaks varied in response to the drought. In 1983, strong rains on the same island resulted in an abundance of seeds. Sure enough, scientists who were measuring beak sizes noticed that (on average) the next generation of finches had smaller beak sizes. Once again, the finches adapted to a change in their environment. Since seeds were plentiful, a large beak provided no specific advantage

Science and Creation Continued

for survival. This means the finch beaks began to decrease in size again. These 2 instances really showed that Darwin's idea was right. The population of finches could indeed adapt to changes in their surroundings from generation to generation.

Even though Darwin was right when he said that species have the ability to change and adapt to their environment, he was dead wrong when he tried to extrapolate from his data. He said that since species have the ability to change, they should be able to change into a different species. In other words, if a population of finches can, through several generations, slowly develop different beaks, why can't they also develop different wings, heads, bodies, and feet so that they change into eagles? Why can't finches, after many years of such change, develop into a completely new species? Darwin thought that this could really happen. This idea became the foundation for the theory of evolution; and it is, unfortunately, still believed by many scientists today. The problem is that Darwin made a big mistake in extrapolation. He took small changes that he observed in animals and extrapolated them into huge changes. Darwin noticed that the finches on the Galapagos Islands had changed their beaks, their feather color, and (to some extent) their body sizes to adapt to the environment of each different island. Those kinds of changes are very small compared to the kinds of changes necessary to turn a finch into a completely different species of bird! So Darwin took a small amount of data and tried to make a huge extrapolation with it. This is precisely the opposite of what a careful scientist does. A careful scientist will make only small extrapolations based on large sets of data. Darwin was not careful and, as a result, has caused all sorts of problems for both the scientific community and the population as a whole.

Extrapolation is a good and necessary part of science, but you must be careful when you do it. In general, an extrapolation is good as long as the amount of extrapolation is small compared to the available data. For example, in Figure 11.5, we have data that spans a temperature range of 0°C to -250°C. That's a range of 250 degrees. In the end, we had to extrapolate an additional 23.15 degrees to get all of the lines to converge. Compared to 250 degrees, an additional 23.15 degrees is pretty small, so we can say that our extrapolation is probably okay.

THE COMBINED GAS LAW

Now that you have seen both Boyle's law and Charles's law, it is time to combine them. Remember, the experiments that Boyle performed involved changing the volume of a gas and measuring its pressure or vice versa. During all of those experiments, the temperature was held constant. Likewise, when Charles performed his experiments, he changed the temperature of a gas and measured its volume or vice versa. During those experiments, the pressure exerted by the gas was held constant.

Even though the experiments performed by Boyle and Charles were quite different, we can combine their results. The product of a gas's pressure and volume, divided by the gas's temperature, remains the same despite the conditions under which the gas is stored. Mathematically, we write:

$$\frac{PV}{T} = \text{constant} \qquad \text{Equation 11.7}$$

As we did with both Boyle's law and Charles's law, we can restate this mathematical equation as follows:

$$\frac{P_1V_1}{T_1} = \frac{P_2V_2}{T_2} \qquad \text{Equation 11.8}$$

In this equation, P_1, V_1, and T_1 represent the pressure, volume, and temperature of a gas under one set of conditions, while P_2, V_2, and T_2 represent the pressure, volume, and temperature of the gas under any other set of conditions.

What happens to equation 11.8 if pressure is held constant during a change in volume and temperature? When the same value is on both sides of an equation, that value cancels from the equation. If P cancels from that equation, then we would be left with Charles's law.

$$\frac{V_1}{T_1} = \frac{V_2}{T_2} \qquad \text{Equation 11.6}$$

The same thing happens if temperature is held constant during a change in pressure and volume. Temperature will cancel from both sides, leaving Boyle's law.

$$P_1V_1 = P_2V_2 \qquad \text{Equation 11.4}$$

Looking back on equations 11.4 and 11.6, it should be rather apparent how equation 11.8 combines both Boyle's law and Charles's law. As noted previously, this particular equation is important because it allows us to predict how the characteristics of a gas change when the conditions to which it is exposed change. Since this equation is so powerful, it is imperative that you memorize it and become proficient in its use.

As you will soon see, equation 11.8 is very simple to use, but there is one very important fact that you must remember:

When using any equation in this module, you must always use the Kelvin temperature scale.

This is a very important fact. Notice that equations 11.5 through 11.8 all contain a temperature term in the denominator of a fraction. What do the rules of mathematics tell us about variables that appear in the denominator of a fraction? Those variables can never equal 0 because the fraction would become undefined at that point. As we have discussed, the Kelvin temperature scale is defined so that we can never reach a value of 0.00 Kelvin. As a result, this is the *only* temperature scale that can be used in the equations presented in this module.

Even though it is very important to make sure that we keep the temperatures in this equation in units of Kelvin, there are no constraints on the units for pressure and volume. As long as both pressures (P_1 and P_2) have the same units, it doesn't matter whether those units are atm, torr, Pascal, or mm Hg. In the same way, as long as both volumes (V_1 and V_2) have the same units, it doesn't matter whether those units are L, mL, cm^3, or m^3. Let's see how this works in Example 11.1.

EXAMPLE 11.1

A weather balloon is filled with helium before it is launched. On the ground, the gas has a temperature of 25°C and a pressure of 1.0 atm. The volume of the balloon is 4.5 m^3. As the balloon rises, the pressure and temperature change to 0.30 atm and 15°C. What is the new volume of the balloon?

This problem gives us the pressure, temperature, and volume of a gas under one set of conditions. The conditions are then changed, and the new pressure and temperature are given. Using equation 11.8, we can use this information to determine the new volume:

$$\frac{P_1V_1}{T_1} = \frac{P_2V_2}{T_2}$$

In this problem, P_1 = 1.0 atm, V_1 = 4.5 m^3, and T_1 = 25°C, whereas P_2 = 0.30 atm and T_2 = 15°C. Since we are solving for the new volume, V_2 is the unknown variable in the equation. Before we can use equation 11.8, we must convert the temperatures into Kelvin:

$$T_1 = 25 + 273.15 = 298 \text{ K}$$

$$T_2 = 15 + 273.15 = 288 \text{ K}$$

Remember, since we are adding in these equations, we do not count significant figures. Instead, we look at decimal places. Since the gas temperatures have their last significant figure in the ones place, the answer must be reported to the ones place as well.

Now that we have the proper units for temperature, we can use algebra to rearrange equation 11.8 and solve for V_2:

$$V_2 = \frac{P_1V_1T_2}{T_1P_2}$$

Sticking in the values for the known variables:

$$V_2 = \frac{1.0 \cancel{\text{atm}} \times 4.5 \text{ m}^3 \times 288 \cancel{\text{K}}}{298 \cancel{\text{K}} \times 0.30 \cancel{\text{atm}}} = 14 \text{ m}^3$$

Notice that as long as the pressures have the same unit, they cancel. We could have used any pressure units, as long as they were the same. In addition, whatever units V_1 has will be the units of V_2. Therefore, the volume of the balloon at a high altitude will be **14 m^3**.

A gas is stored in a 3.00 liter container at a pressure of 456 torr. If the temperature stays constant, what will the pressure of the gas be if the volume is decreased to 0.75 liters?

In this problem, we are given only V_1 (3.00 liters), P_1 (456 torr), and V_2 (0.75 liters). You might think, therefore, that we do not have enough information to solve equation 11.8. However, if we look at the equation, we should see something:

$$\frac{P_1V_1}{T_1} = \frac{P_2V_2}{T_2}$$

The problem tells us that the temperature remains constant, which means that $T_1 = T_2$. So temperatures will cancel out, leaving Boyle's law:

$$\frac{P_1V_1}{\cancel{T_1}} = \frac{P_2V_2}{\cancel{T_2}}$$

$$P_1V_1 = P_2V_2$$

Don't forget that if a problem tells us that one of the variables stays constant, that variable will cancel on both sides of the equation.

Now that we have a useful equation, we can solve for P_2, plug in our numbers, and calculate our answer:

$$P_2 = \frac{P_1V_1}{V_2} = \frac{456 \text{ torr} \times 3.00\cancel{\text{L}}}{0.75\cancel{\text{L}}} = 1.8 \times 10^3 \text{ torr}$$

So when the volume is decreased under conditions of constant temperature, the pressure increases to **1.8×10^3 torr**.

ON YOUR OWN

11.1 A gas is stored in a metal cylinder with a volume that cannot change. If the gas is packed into the cylinder at a temperature of 5.00°C and under a pressure of 3,115 kPa, what will the pressure inside the cylinder be when it is stored at 30.0°C?

11.2 At room temperature (25.0°C) and normal atmospheric pressure (1.00 atm), the average person's lungs have a total capacity of about 5.5 liters. If a man takes a deep breath (filling his lungs to capacity) and then dives deep into a lake where the temperature is 17.0°C and the pressure is 895 torr, what will be the new volume of the air in his lungs?

11.3 If you fill a balloon with air to a volume of 3.21 liters at a temperature of 22.0°C, what volume will the balloon have if it is stuck in the freezer (-5.0°C)? Assume that the pressure stays constant.

Check your understanding of these equations by answering "On Your Own" questions 11.1–11.3

IDEAL GASES

Now that you have some experience working with the combined gas law, it is important to learn *when you can* use this law and *when you can't*. This very useful law cannot be applied to all gases under all conditions. For this law to apply, the gas in question must behave in an ideal manner. Such a gas is called an ideal gas and has the following properties:

1. The molecules (or atoms) that make up the gas are very small compared to the total volume available to the gas.

2. The molecules (or atoms) that make up the gas must be so far apart from one another that there is no attraction or repulsion between them.

3. The collisions that occur between the gas molecules (or atoms) must be elastic. In other words, when the molecules (or atoms) collide, no energy can be lost in the collision. Also, no energy can be lost when the molecules (or atoms) of the gas collide with the walls of the container.

To use the combined gas law (or any equations presented in this module), you will have to know that the gas you are working with is ideal. How will you know that? It's rather simple. Gases tend to behave ideally at high temperatures and low pressures. If you think about it, that should make sense.

After all, for a gas to be ideal, its molecules (or atoms) must be very small compared to the total volume of the gas, and they must be far apart from each other. So the larger total volume that the gas occupies, the more ideal its behavior will be. According to Boyle's law, when do gases occupy a large volume? At low pressures. In the same way, Charles's law tells us that gases tend to occupy large volumes at high temperatures. This means gases tend to behave ideally at high temperatures and low pressures.

You might be asking yourself what we mean by high temperatures and low pressures. Is 10°C a high temperature? What about 0.9 atmospheres? Is that a low pressure? To help us determine whether or not a given pressure or temperature is high or low, we define something called standard temperature and pressure, which we abbreviate as STP.

Standard temperature and pressure (STP)—A temperature of 273 K and a pressure of 1.00 atm.

This is a very important definition because we use it as a reference to determine whether or not a gas will behave in an ideal manner. If the pressure is significantly larger than standard pressure, we say that the pressure is too high to guarantee that a gas will behave in an ideal manner. In the same way, if the temperature is significantly lower than standard temperature, we say that the temperature is too low to ensure ideal behavior. In other words:

A gas with a temperature that is close to (or larger than) 273 K and a pressure that is near (or lower than) 1.00 atm will behave in an ideal fashion. Under those conditions, we can use all of the equations in this module.

It is important that you be able to recognize whether or not a gas will behave ideally based on its pressure and temperature.

DALTON'S LAW OF PARTIAL PRESSURES

Now that we know when a gas behaves ideally, we can learn a few more laws that relate to ideal gases. The next one we will learn about is Dalton's law of partial pressures.

Dalton's law of partial pressures—When 2 or more ideal gases are mixed together, the total pressure of the mixture is equal to the sum of the pressures of each individual gas.

We could restate Dalton's law mathematically as follows:

$$P_T = P_1 + P_2 + P_3 + \ldots \qquad \text{Equation 11.9}$$

where P_T is the total pressure of the gas, and P_1, P_2, etc. are the pressures due to each individual gas. These individual pressures are often referred to as the partial pressures of each gas.

Dalton's law might look very simple, but it actually tells us a great deal. For example, suppose we have 2 steel containers that each contain a gas at a pressure of 5 atms. Let's say that container 1 holds oxygen gas at a pressure of 5 atms, while container 2 holds nitrogen gas at the same pressure. Dalton's law tells us that if we were to transfer all of the nitrogen from container 2 into container 1, the resulting pressure in container 1 would be 10 atms. Sounds simple, doesn't it?

Consider another situation. Suppose that container 1 and container 2 each hold 5 atms of oxygen gas. What is the resulting pressure when we transfer all of the gas from container 2 into container 1? It is still 10 atms. Whether container 2 holds oxygen, nitrogen, or any other ideal gas at a pressure of 5 atms, the resulting pressure when the contents of the second container are transferred to the first container will remain 10 atms. This tells us something important:

The pressure of an ideal gas does not depend on the identity of the gas.
It depends on the quantity of that gas.

We will use this fact in a moment to restate Dalton's law in a completely different way. For now, we need to learn an important application of Dalton's law: dealing with vapor pressure.

VAPOR PRESSURE

If you were to fill a glass with water and set it out on a table, what would eventually happen? All of the water would eventually evaporate and the glass would end up empty. Think for a moment about why this occurs. Since the water evaporates, we know that it turns from its liquid phase into its gas phase. How does it do that? After all, the water isn't *boiling*. If the water isn't boiling, why is it changing from liquid to gas?

Remember, according to the kinetic theory of matter, all the molecules in a glass of liquid water are moving around. Some of those molecules move around pretty quickly, while others move rather slowly. As the molecules move around, some of them will encounter the surface of the water. If they are moving fast enough, they can hit the surface of the water with enough energy to escape from the liquid. When they do that, they turn from the liquid phase into the gas phase. So evaporation occurs because molecules of a liquid are sometimes able to escape from the surface of that liquid, provided they have sufficient energy to do so.

Since this is happening continuously, water molecules are continuously hitting the surface of the water and escaping. At any one time, there is always some vapor that sits on the top of a sample of liquid. That vapor is composed of the molecules that have escaped the liquid by crashing through the surface and turning into a gas. Every gas exerts pressure, so the vapor that sits above a sample of liquid exerts a certain amount of pressure.

Vapor pressure—The pressure exerted by the vapor that sits on top of any liquid.

In other words, any container of liquid has gas floating on top of it. The pressure that this gas exerts on its surroundings is called vapor pressure.

As you might expect, the vapor pressure of any liquid depends on the chemical nature of that liquid. For example, gasoline has a very large vapor pressure because it evaporates rather easily. On the other hand, water has a reasonably low vapor pressure because it does not evaporate as easily. What you may not know is that the vapor pressure of any given liquid also depends on temperature. For example, table 11.2 lists the vapor pressure of water at several different temperatures.

TABLE 11.2
The Vapor Pressure of Water

Temperature (°C)	Vapor Pressure (torr)	Temperature (°C)	Vapor Pressure (torr)
20	17.5	29	30.0
21	18.7	30	31.8
22	19.8	40	55.3
23	21.1	50	92.5
24	22.4	60	149.4
25	23.8	70	233.7
26	25.2	80	355.1
27	26.7	90	525.8
28	28.3	100	760.0

Notice the general trend in this table. As the temperature increases, so does the vapor pressure. Based on the kinetic theory of matter, this makes sense. As the temperature increases, the molecules in a sample of water increase in energy. Since each molecule has more energy, any molecule that reaches the surface of the liquid has an increased chance of escaping the liquid and turning into a gas. As a result, there is more gas floating on top of the liquid, increasing the vapor pressure. This is a general rule:

The vapor pressure of any liquid increases with increasing temperature.

To determine a liquid's vapor pressure, we must know both the identity of the liquid and its temperature.

Look at the vapor pressure of water at 100°C. It is equal to 760.0 torr. Does that number mean anything to you? According to Table 11.1, 760.0 torr is the same as 1.000 atm, which is normal atmospheric pressure. A temperature of 100°C should also mean something to you. It is the boiling point of water. It isn't a coincidence that the vapor pressure of water is 1.000 atm at its boiling point. That is the true definition of boiling point.

Boiling point—The temperature at which the vapor pressure of a liquid is equal to normal atmospheric pressure.

Have you ever heard that water boils at a lower temperature at higher altitudes? This definition tells why. At high altitudes, atmospheric pressure is measurably lower than 760 torr, and the temperature to which water must be heated for its vapor pressure to equal atmospheric pressure is lower. As a result, the boiling point of water is lower.

Now that you understand vapor pressure, we can finally relate it to Dalton's law. When chemists make a gas, that gas is often collected over a liquid. For example, in experiment 8.1, you mixed baking soda and vinegar. The reaction produced gaseous carbon dioxide, which you collected in a balloon. Think for a minute about what the balloon contained. Did it contain pure carbon dioxide? No, it didn't. You see, the carbon dioxide you collected was in the presence of liquid water. Since liquid water was present, its vapor was also present, so the gas inside the balloon was a mixture of carbon dioxide *and* water vapor.

How much of the mixture was carbon dioxide and how much was water vapor? That's where Dalton's law can help. After all, Dalton's law tells us that the total pressure of the gas mixture is just the sum of the partial pressure of carbon dioxide and the partial pressure of water. If we know the total pressure, we can look up the partial pressure of the water vapor by simply looking at table 11.2. Then we can use algebra to determine the partial pressure of carbon dioxide. This tells us how much carbon dioxide was in the balloon.

If you don't quite understand, don't worry. Example 11.2 will allow you to see exactly how all of this works. Before we go through that example, we need to introduce a bit of terminology. Chemists often produce gas in chemical reactions that contain water. The gas bubbles out of the water, and then the chemist collects it in some container, like a balloon. Since this happens so often, chemists have a phrase for it. When a chemist makes gas in the presence of water, we call it either collecting gas over water or collecting gas in the presence of water. So whenever you see one of those phrases in a problem, you must realize that the gas collected is a mixture of the gas produced and water vapor. To find out how much of the gas was produced, you must use Dalton's law to account for the vapor pressure of water. Example 11.2 shows how that is done.

EXAMPLE 11.2

A chemist produces hydrogen gas in an experiment and collects that gas in the presence of water. If the total pressure of gas collected was 1.115 atm and the temperature was 26°C, what was the pressure of hydrogen gas collected?

This problem tells us that a chemist collects hydrogen gas in the presence of water. This means that the gas collected is actually a mixture of hydrogen gas and water vapor. Dalton's law tells us that the total pressure of that mixture is simply the sum of the individual pressures of water vapor and hydrogen:

$$P_T = P_{\text{hydrogen}} + P_{\text{water vapor}}$$

The problem tells us what the total pressure is, and since we are also given the temperature, we can use table 11.2 to determine the vapor pressure of the water. According to the table, water at a temperature of 26°C has a vapor pressure of 25.2 torr. Our pressure units do not match. The total pressure is in atm, while the pressure of the water vapor is in torr. We will fix that by converting torr to atm:

$$\frac{25.2\ \cancel{\text{torr}}}{1} \times \frac{1\ \text{atm}}{760.0\ \cancel{\text{torr}}} = 0.0332\ \text{atm}$$

Now that we have consistent units, we can use Dalton's law:

$$1.115\ \text{atm} = P_{\text{hydrogen}} + 0.0332\ \text{atm}$$

Using algebra, we can then solve for P_{hydrogen}:

$$P_{\text{hydrogen}} = 1.115\ \text{atm} - 0.0332\ \text{atm} = 1.082\ \text{atm}$$

The chemical reaction produced **1.082 atm** of hydrogen gas.

Make sure you understand this by answering "On Your Own" question 11.4.

ON YOUR OWN

11.4 A chemist collects carbon dioxide over water. The pressure of gas in the collection container was 861 torr. If the temperature was 22.0°C, what pressure of carbon dioxide was collected? (Refer to table 11.2.)

AN ALTERNATIVE STATEMENT OF DALTON'S LAW

The way we have used Dalton's law in the previous examples and "On Your Own" questions is not the only way you can use this important law. There is an alternative way of expressing it. Before we can learn this, we must learn to use yet another concentration unit called the mole fraction. Mole fraction is always used in the context of a mixture. Mathematically, the mole fraction of any component in a mixture (abbreviated as X) can be defined in the following manner:

$$X = \frac{\text{\# of moles of compound}}{\text{total \# of moles in the mixture}} \qquad \text{Equation 11.10}$$

Before we learn what mole fraction tells us, let's make sure we understand its definition by studying Example 11.3 and answering "On Your Own" question 11.5.

EXAMPLE 11.3

A mixture of gases contains 3.0 moles of oxygen gas, 5.0 moles of nitrogen gas, and 1.0 mole of neon gas. What is the mole fraction of each component in the mixture?

According to equation 11.10, we calculate the mole fraction of any component in a mixture by taking the number of moles of that component and dividing by the total number of moles in the mixture. According to the problem:

Total number of moles = 3.0 moles + 5.0 moles + 1.0 mole = 9.0 moles

Now that we have the total number of moles in the mixture, we can simply use equation 11.10:

$$X_{O_2} = \frac{3.0 \text{ moles}}{9.0 \text{ moles}} = 0.33$$

$$X_{N_2} = \frac{5.0 \text{ moles}}{9.0 \text{ moles}} = 0.56$$

$$X_{Ne} = \frac{1.0 \text{ mole}}{9.0 \text{ moles}} = 0.11$$

Notice a couple of things about mole fraction. First, since the mole unit cancels in the equation, mole fraction has no units. Remember, we call this a **dimensionless quantity**. Second, notice that the 3 mole fractions add up to 1.00. That should always be the case. If you add the mole fractions of each component in a mixture, the sum must equal 1. Because of significant figures, they might add to 0.99 or 1.01, but in general, they should add to 1.

ON YOUR OWN

11.5 A tank of compressed air contains 768.0 grams of oxygen, 2,496.0 grams of nitrogen, and 45.6 grams of argon. What is the mole fraction of each component in this mixture?

Now that you have learned to use the equation for mole fraction, it is time to learn what it means. The mole fraction tells us what fraction of the total number of molecules (or atoms) in the mixture is represented by the component of interest. For example, suppose we have a gas mixture that contains carbon dioxide and water vapor. If the mole fraction of carbon dioxide is 0.10, it means that one-tenth (0.10) of the molecules in the mixture are carbon dioxide molecules. So for every 100 molecules in the mixture, 10 will be carbon dioxide molecules and the rest will be water molecules.

We need to understand mole fraction because we can use this concept to restate Dalton's law. Remember, Dalton's law tells us that the pressure of a gas does not depend on the identity of the gas. It depends only on the quantity. If we have a mixture of gases, the partial pressure of each gas should depend only on the amount of that gas in the mixture. Mathematically, we would say:

$$P_1 = X_1 P_T \qquad \text{Equation 11.11}$$

This equation tells us that the partial pressure of any gas in a mixture depends on the mole fraction of that gas and the total pressure of the gas mixture.

Think about what this means: If we have a mixture of nitrogen and oxygen and we know the total pressure that the mixture exerts on its container, we can calculate the partial pressure of each gas by multiplying the gas's mole fraction by the total pressure. If this seems a little confusing right now, don't worry. Example 11.4 should flesh it all out.

EXAMPLE 11.4

A gas mixture contains 40.0 grams of carbon monoxide and 10.0 grams of nitrogen dioxide. If the total pressure of this gas mixture is 1.2 atms, what is the partial pressure of each individual gas?

Equation 11.11 tells us that if we know the total pressure of the gas mixture and the mole fraction of each component, we can calculate the partial pressure of each component. We were given the total pressure of the gas mixture, and we were also given enough information to determine the mole fraction of each component of the mixture. To determine the mole fractions, we must first convert from grams to moles:

$$\frac{40.0\ \cancel{\text{g CO}}}{1} \times \frac{1\ \text{mole CO}}{28.0\ \cancel{\text{g CO}}} = 1.43\ \text{moles CO}$$

$$\frac{10.0\ \cancel{\text{g NO}_2}}{1} \times \frac{1\ \text{mole NO}_2}{46.0\ \cancel{\text{g NO}_2}} = 0.217\ \text{moles NO}_2$$

We can now use this information to calculate the mole fractions of each gas:

$$\text{Total moles} = 1.43\ \text{moles} + 0.217\ \text{moles} = 1.65\ \text{moles}$$

$$X_{CO} = \frac{1.43\ \cancel{\text{moles}}}{1.65\ \cancel{\text{moles}}} = 0.867$$

$$X_{NO_2} = \frac{0.217\ \cancel{\text{moles}}}{1.65\ \cancel{\text{moles}}} = 0.132$$

Notice that the mole fractions do not quite add up to 1. Rather, they add up to 0.999. This is just due to significant figures, so don't worry about it. Now that we have the mole fractions, we can use equation 11.11 to get the partial pressures of each gas:

$$P_{CO} = 0.867 \times 1.2\ \text{atms} = 1.0\ \text{atm}$$

$$P_{NO_2} = 0.132 \times 1.2\ \text{atms} = 0.16\ \text{atm}$$

This mixture, then, is made up of **1.0 atm of CO and 0.16 atm of NO_2**.

A gas mixture contains 456 torr of oxygen, 112 torr of CO_2, and 501 torr of nitrogen. What is the mole fraction of each component in the mixture?

At first, you might think that this problem gives the wrong information. After all, to calculate mole fraction, don't we need either moles or grams? Not when we have equation 11.11. Since this equation relates pressure and mole fraction, we can calculate mole fraction as long as we are given the pressure.

According to the equation, to calculate the mole fraction of any component in the mixture, all we have to know is the total pressure of the mixture and the partial pressure of the component. Since we were given the partial pressure of each component, we can use equation 11.9 to calculate the total pressure:

$$P_T = P_{oxygen} + P_{carbon\ dioxide} + P_{nitrogen}$$

$$P_T = 456\ \text{torr} + 112\ \text{torr} + 501\ \text{torr} = 1{,}069\ \text{torr}$$

Now that we have both the total pressure of the mixture and the partial pressure of each component, we can use equation 11.11 to calculate the mole fraction of each component:

$$P_1 = X_1 \cdot P_T$$

$$X_1 = \frac{P_1}{P_T}$$

Plugging in all 3 sets of numbers:

$$X_{oxygen} = \frac{456\ \cancel{\text{torr}}}{1{,}069\ \cancel{\text{torr}}} = 0.427$$

$$X_{carbon\ dioxide} = \frac{112\ \cancel{\text{torr}}}{1{,}069\ \cancel{\text{torr}}} = 0.105$$

$$X_{nitrogen} = \frac{501\ \cancel{\text{torr}}}{1{,}069\ \cancel{\text{torr}}} = 0.469$$

ON YOUR OWN

11.6 A sample of artificially made air contains 150.0 grams of nitrogen and 40.0 grams of oxygen. If the pressure of the air sample is 760.0 torr, what is the partial pressure of each gas?

11.7 A chemist produces hydrogen gas in a chemical reaction and collects it over water at 21.0°C. If the pressure of the gas inside the collection vessel is 815 torr, what is the mole fraction of the hydrogen gas in the vessel?

We hope that you are beginning to see how powerful Dalton's law can be. We can use it to determine the pressures of individual components in a mixture or, more importantly, we can use the partial pressure of a gas to determine its concentration in terms of mole fraction. Make sure you understand this by answering "On Your Own" questions 11.6–11.7.

THE IDEAL GAS LAW

The last thing we need to learn about gases is probably the most important. As we have seen in other modules, the most important thing we can learn about a substance is the number of moles of that substance. Usually, we can get this information by knowing the identity of the substance and its mass. It is very difficult to measure the mass of a gas, and it would be nice to be able to relate the number of moles of a gas to things that are easy to measure. The ideal gas law does this. The ideal gas law is:

$$PV = nRT \qquad \text{Equation 11.12}$$

In this equation, P is the pressure of the gas, V is its volume, T is its temperature, and n is the number of moles of that gas. R refers to a special physical constant called the ideal gas constant. It is a relatively important quantity in the study of chemistry.

The ideal gas constant can have several values, depending on what units are used in equation 11.12. For our purposes, we will learn only one value:

$$R = 0.0821 \frac{L \cdot atm}{mole \cdot K}$$

As with all centered blue facts, you are expected to memorize this. Before you do that, make sure you know what it means. The ideal gas constant never changes. However, the value of the constant is different if we use different units.

For example, suppose we use m^3 as our volume unit. If we multiply length times width times height, we get volume. If the length, width, and height are measured in meters, the volume unit will be m^3. Also, suppose we measure pressure in Pascals. We will use a different value for the ideal gas constant because we are using different units. For those units, the value of the ideal gas constant is 8.315 $(Pa \cdot m^3)/(mole \cdot K)$, which can also be expressed as 8.315 $J/(mole \cdot K)$. In this course, you will usually work in liters and atmospheres. That's why you were asked to memorize the value of R in those units.

Did you know that the ideal gas constant, R, can be determined by experiment? Perform Experiment 11.1 to determine how close to R = 0.0821 $(L \cdot atm)/(mole \cdot K)$ you can get.

EXPERIMENT 11.1

PURPOSE: To use an ideal gas to determine the value of R.

MATERIALS:

- Kitchen sink or large bucket
- Towels or napkins
- 250 mL beaker or a measuring cup with mL markings
- Butane lighter (A cigarette lighter with a clear case will work best.)
- Mass scale
- Thermometer
- Safety goggles
- Weather report that contains the atmospheric (sometimes called barometric) pressure for the day

QUESTION: Can the value of R be determined by experiment?

HYPOTHESIS: Write a hypothesis about how close to the theoretical value of R you can get.

PROCEDURE:

1. Fill your kitchen sink or large bucket with warm water. Allow the water to sit for half an hour so that it will attain room temperature.
2. Obtain the current atmospheric pressure where you live. You can get this information at almost any weather website like www.weather.com. If you can't find a weather report or access the Internet, assume that the atmospheric pressure is 1.00 atm. Convert the measurement to atm, if needed.

3. Submerge the butane lighter underwater and allow it to sit there for 1 minute. This will allow any air bubbles in the lighter to escape. The air bubbles could get trapped with the gas later in this experiment, causing your volume measurement to be incorrect.
4. Remove the butane lighter and dry it off with a towel or napkin. Get as much water off the butane lighter as possible.
5. Measure the mass of the butane lighter. Record the mass in your notebook.
6. After the water in the sink has sat for at least half an hour, measure the temperature of the water with your thermometer, and record your result.
7. Submerge the 250 mL beaker or a measuring cup with mL markings into the sink or bucket and allow it to fill completely with water with no air bubbles. You may have to tip the beaker up slightly to allow the air to escape.
8. Hold the beaker upside down in the water. If some of the beaker sticks up above the water level, that is okay.
9. Submerge the butane lighter under the water and put the mouth of the lighter directly under the opening of the beaker.
10. Push the red button on the lighter to allow the gas to escape into the beaker. If some of the gas escapes the beaker and moves into the atmosphere, then you will need to pull the lighter out, dry it off, and measure the mass again. Replace the mass from step 5 with this measurement.
11. As the beaker is filling with gas, the water level will drop. Stop filling the beaker when the water level is around 200 mL. If your mass scale cannot measure more than 0.5 g differences, then double the volume of gas collected to 400 mL.
12. Pull the lighter out of the water and set it aside.
13. With the beaker opening still submerged, move the beaker up or down in the water until the water level inside the beaker is even with the water level outside of the beaker. This will ensure that the pressure of the gas inside the container is equal to atmospheric pressure.
14. Read the volume of the gas when the water levels are even and record your measurement.
15. Dry off the lighter with a towel and measure its new mass. Record this mass in your notebook.
16. In your notebook, use PV = nRT to solve for R.
17. To find the moles of butane, n, you need to find the mass of butane expelled from the lighter. Take the mass of the lighter from step 5 and subtract the mass measured in step 15. The result will be the mass in grams of butane in the beaker. You will need to find the molar mass of butane, C_4H_{10}, to convert the grams of butane to moles of butane.
18. You now have all of the measurements needed to solve for R. Record the result you obtain for R in your notebook and compare it to the known value of R.
19. Clean up and return everything to the proper place.

CONCLUSION: Write something about what happened in this experiment and whether it confirmed your hypothesis.

Looking at the ideal gas equation, you can see that it relates 4 pieces of information about a gas. It relates a gas's volume, pressure, temperature, and number of moles. If we are given 3 of those 4 pieces of information, we can use equation 11.12 to determine the fourth. Notice how this works in Example 11.5

EXAMPLE 11.5

A gas is in a 12.2-liter container at a pressure of 1.05 atm and a temperature of 25°C. How many moles of gas are present?

This problem gives us P, V, and T and wants us to calculate n. To do this, we start with equation 11.12 and rearrange it using algebra to solve for n:

$$PV = nRT$$

$$n = \frac{PV}{RT}$$

We can't just go plugging in our numbers right away. First we have to examine our units. Notice that volume is in liters and pressure is in atm. That tells us we will use R = 0.0821 L · atm/(mole · K). However, since R has temperature units of K (and since we already know that all temperatures in this module must be converted to K), we need to turn 25°C into 298 K. Now we can plug in our numbers:

$$n = \frac{1.05\ \cancel{\text{atm}} \cdot 12.2\ \cancel{\text{L}}}{0.0821\ \frac{(\cancel{\text{L}} \cdot \cancel{\text{atm}})}{\text{mole} \cdot \cancel{\text{K}}} \cdot 298\ \cancel{\text{K}}} = 0.524 \text{ moles}$$

This container contains **0.524 moles of gas.**

So you see that using the ideal gas law is rather simple, as long as you have your units correct. Make sure you understand how this works by answering "On Your Own" question 11.8.

ON YOUR OWN

11.8 How many liters does 1.00 mole of an ideal gas occupy at STP?

One final note: Mathematically if the number of moles remains constant ($n_1 = n_2$), the ideal gas law equation becomes the combined gas law because the ideal gas law constant and moles cancel out:

$$PV = nRT$$

$$nR = \frac{PV}{T}$$

$$\frac{P_1V_1}{T_1} = \frac{P_2V_2}{T_2}$$

So from the ideal gas law, we can derive the combined gas law equation 11.8.

USING THE IDEAL GAS LAW IN STOICHIOMETRY

By now you have probably guessed that we can take virtually any subject in chemistry and relate it back to stoichiometry. We can do it with the ideal gas equation too! Remember, stoichiometry is done in moles. If we can figure out a way to determine the number of moles of gas we have, we can do stoichiometry on that gas. Perform Experiment 11.2 to see how this is done.

EXPERIMENT 11.2

PURPOSE: To use the ideal gas equation to determine the amount of acid in vinegar.

MATERIALS:

- Mass scale
- Plastic 2-liter bottle
- Round balloon with an 8-inch diameter
- Vinegar
- Baking soda
- Seamstress's tape measure (A piece of string and a ruler will work as well.)
- Thermometer
- Safety goggles
- Weather report that contains the atmospheric (sometimes called barometric) pressure for the day (You can get this information at almost any weather website like www.weather.com. If you can't find a weather report or access the Internet, assume that the atmospheric pressure is 1.00 atm.)

QUESTION: Can you determine the amount of gas using the ideal gas equation?

HYPOTHESIS: Write a hypothesis related to this experiment.

PROCEDURE:

1. Clean out the 2-liter bottle and fill it with 200.0 grams of vinegar (a quantitative measurement). The best way to do this is to measure the mass of your bottle with the scale and then slowly add vinegar while the bottle is still on the scale. You have added 200.0 grams of vinegar when the mass reaches 200.0 more grams than the original mass of the bottle.
2. Fill the balloon with at least 30 grams of baking soda. The baking soda is the excess reactant here, so more than 30 grams is fine.
3. Once you have filled the bottle and the balloon, attach the balloon to the bottle as you did in experiment 8.1. Once again, be sure that you have an airtight seal between the bottle and the balloon.
4. Hold the balloon up so that the baking soda mixes with the vinegar.
5. Gently shake the balloon so that all of the baking soda falls into the vinegar.
6. Once the bubbling subsides, gently swirl the contents of the bottle to make sure that the baking soda has fully mixed with the vinegar.
7. Measure the radius of the balloon. To accomplish this, you need to measure the distance around the widest part of the balloon. You can use a seamstress's tape measure to do this. If you don't have such a tape measure, wrap a piece of string once around the

widest part of the balloon. Carefully cut the string so that it is just the length necessary to wrap around the balloon, then use a ruler to measure the length of the string.

8. If your ruler or tape measure is marked in inches, convert to centimeters by multiplying your measurement by 2.54. Once you have the measurement in centimeters, divide it by 10. This will give you the circumference of the balloon in decimeters.
9. Divide the circumference of the balloon in decimeters by 2π. Since the circumference of a sphere is $2\pi r$, dividing the circumference of the balloon by 2π gives the balloon's radius.
10. Take the radius you just calculated and plug it into the volume equation for a sphere:

$$V = \frac{4}{3}\pi r^3$$

11. Your answer is the volume of the balloon in dm^3, which (believe it or not) is the same as liters. So by measuring the circumference of the balloon, you were able to calculate the volume in liters. This represents the volume of the gas produced in the reaction.
12. As we learned in experiment 8.1, the reaction that occurs when baking soda and vinegar are mixed is:

$$\underset{\text{acetic acid in vinegar}}{C_2H_4O_2\ (l)} + \underset{\text{baking soda}}{NaHCO_3\ (s)} \rightarrow H_2O\ (l) + CO_2\ (g) + \underset{\text{sodium acetate}}{NaC_2H_3O_2\ (aq)}$$

13. The calculation that you just performed gives you the volume of gas produced in this reaction. Because the gas was collected over water, it will be a mixture of water vapor and carbon dioxide.
14. Look at the weather report to determine the atmospheric pressure for today. If you can't find it, 1.00 atm is a good assumption. It is likely that the atmospheric pressure will be listed with a unit of inches. That represents inches of mercury. If you divide the pressure in inches by 29.92, you will get the pressure in atmospheres.
15. The atmospheric pressure corresponds to the pressure of the gas inside your balloon because the atmosphere is pushing against the balloon. The problem is that there is not only CO_2 in the balloon, but there is also water vapor. To determine the pressure of the CO_2 in the balloon, you must subtract the vapor pressure of water. To do this, measure the temperature of the room and determine the vapor pressure of water for that temperature by looking at table 11.2. Convert that pressure to atm by dividing by 760.0 and subtracting it from the atmospheric pressure. The answer represents the pressure of the CO_2 formed in the experiment.
16. Convert your measured temperature to K. This will give you the temperature of the CO_2 gas formed in the experiment.
17. Now look at what you have. You know the volume of CO_2 gas (in liters), the pressure of the gas (in atm), and the temperature of the gas (in K). You can use that information and the ideal gas law (with $R = 0.0821 \frac{L \cdot atm}{mole \cdot K}$) to determine the number of moles of gas produced. Once you have that, you can use it in stoichiometry to determine how much of the limiting reactant (acetic acid) you had.
18. According to the equation, for every 1 mole of CO_2 produced, 1 mole of acetic acid should have reacted. This means the number of moles you just calculated is equal to the number of moles of acetic acid you started with.
19. The molecular mass of acetic acid is 60.0 amu, so multiply the number of moles you just calculated by 60.0. This gives you the number of grams of acetic acid you started with.
20. Clean up and return everything to the proper place.

CONCLUSION: Write something about what happened in this experiment and whether it confirmed your hypothesis.

Don't get so lost in the math that you fail to realize what you did. By measuring the pressure, volume, and temperature of a gas, you were able to calculate the number of moles. That allowed you to perform stoichiometric calculations to learn about the other substances in the reaction you just studied. If you aren't quite sure exactly why you did all of the math, don't worry about that, either. Real-life experiments are always more difficult than the problems that you are expected to do.

Now that you've had a little practical experience with using the ideal gas law in stoichiometry, study Example 11.6, which has the same level of difficulty as the problems that we expect you to be able to do. Then answer "On Your Own" questions 11.9–11.10.

EXAMPLE 11.6

One way that hydrogen gas can be produced commercially is to react a strong acid with a metal:

$$2HCl\ (aq) + Zn\ (s) \rightarrow H_2\ (g) + ZnCl_2\ (aq)$$

If 512 grams of HCl are reacted with excess Zn at T = 24°C and P = 0.95 atm, what volume of hydrogen gas will be produced?

In this stoichiometry problem, we are given the amount of limiting reactant and asked to calculate how much product will be made. This is nothing new; it just uses the things you have already learned in a different way. We first convert the amount of limiting reactant to moles:

$$\frac{512\ \cancel{\text{g HCl}}}{1} \times \frac{1\text{ mole HCl}}{36.5\ \cancel{\text{g HCl}}} = 14.0\text{ moles HCl}$$

We can then use stoichiometry to determine the number of moles of H_2 produced:

$$\frac{14.0\ \cancel{\text{moles HCl}}}{1} \times \frac{1\text{ mole }H_2}{2\ \cancel{\text{moles HCl}}} = 7.00\text{ moles }H_2$$

Now we get to the point where we apply the things we learned in this module. Instead of asking how many moles or how many grams of H_2, the problem asks how many liters. For that, we need to use the ideal gas law. Looking at our units, we will need to convert temperature into Kelvin so that the units for R will work out. Once we do that, we can then use the ideal gas law:

$$PV = nRT$$

$$V = \frac{nRT}{P} = \frac{7.00\ \cancel{\text{moles}} \times 0.0821\ \frac{\text{L}\cdot\cancel{\text{atm}}}{\cancel{\text{mole}}\cdot\cancel{\text{K}}} \times 297\ \cancel{\text{K}}}{0.95\ \cancel{\text{atm}}} = 180\text{ L}$$

The reaction, then, will produce **180 liters** of hydrogen.

To make magnesium hydroxide, a major component in some stomach antacids, hot water vapor is passed over solid magnesium:

$$2H_2O\ (g) + Mg(s) \rightarrow Mg(OH)_2\ (s) + H_2\ (g)$$

If 4.50 liters of water vapor at 200.0°C and 4.94 atm are passed over excess magnesium, how many grams of magnesium hydroxide will be produced?

In this problem, we are once again given the amount of limiting reactant and asked to determine how much product is made. The problem here is that the amount of limiting reactant is not given in grams or moles; instead, it is given in P, V, and T. So we must use the ideal gas law to determine the number of moles of limiting reactant. Looking at the units, we must convert the temperature to Kelvin, and then we can use the ideal gas law to solve for the number of moles:

$$n = \frac{PV}{RT} = \frac{4.94\ \cancel{\text{atm}} \cdot 4.50\ \cancel{\text{L}}}{0.0821\ \frac{\cancel{\text{L}} \cdot \cancel{\text{atm}}}{\text{mole} \cdot \cancel{\text{K}}} \times 473.2\ \cancel{\text{K}}} = 0.572\ \text{moles}$$

Now that we have the moles of limiting reactant, this becomes a stoichiometry problem:

$$\frac{0.572\ \cancel{\text{moles } H_2O}}{1} \times \frac{1\ \text{mole } Mg(OH)_2}{2\ \cancel{\text{moles } H_2O}} = 0.286\ \text{moles } Mg(OH)_2$$

$$\frac{0.286\ \cancel{\text{moles } Mg(OH)_2}}{1} \times \frac{58.3\ \text{g } Mg(OH)_2}{1\ \cancel{\text{mole } Mg(OH)_2}} = 16.7\ \text{g } Mg(OH)_2$$

16.7 g of magnesium hydroxide will be produced.

Now that you are done studying ideal gases, you might wonder how chemists deal with nonideal gases. After all, we said earlier that gases behave ideally only when the pressure is near or below standard pressure and the temperature is near or above standard temperature. What happens at a pressure and temperature where gases do not perform ideally? Nonideal gases are very difficult to deal with, so you will not have to worry about them in this course. However, since STP is very near room temperature and pressure, it is safe to assume that all gases under conditions of room temperature and pressure are ideal.

ON YOUR OWN

11.9 In an automobile engine, gasoline (an important component of which is octane, C_8H_{18}) is burned to make gaseous carbon dioxide and water. A gallon of octane has a mass of about 3.2×10^3 grams. If a gallon of octane is burned in excess oxygen at P = 1.00 atm and T = 180.0°C, how many liters of CO_2 are produced? The unbalanced chemical equation is:

$$C_8H_{18}\ (l) + O_2\ (g) \rightarrow CO_2\ (g) + H_2O\ (g)$$

11.10 Methanol (commonly called wood alcohol) can be produced as follows:

$$CO\ (g) + 2H_2\ (g) \rightarrow CH_4O\ (l)$$

If 250.0 liters of hydrogen gas are reacted with excess CO at STP, how many grams of CH_4O can be produced?

SUMMARY OF KEY EQUATIONS AND TABLES IN MODULE 11

Equation 11.1:	$P = \frac{F}{A}$	Pressure, force, and area relationship
Equation 11.3:	$PV = \text{constant}$	Boyle's law
Equation 11.4:	$P_1V_1 = P_2V_2$	Boyle's law
Equation 11.5:	$\frac{V}{T} = \text{constant}$	Charles's law
Equation 11.6:	$\frac{V_1}{T_1} = \frac{V_2}{T_2}$	Charles's law
Equation 11.7:	$\frac{PV}{T} = \text{constant}$	Combined gas law
Equation 11.8:	$\frac{P_1V_1}{T_1} = \frac{P_2V_2}{T_2}$	Combined gas law
Equation 11.9:	$P_T = P_1 + P_2 + P_3 + \ldots$	Dalton's law of partial pressures
Equation 11.10:	$X = \frac{\text{\# of moles of compound}}{\text{total \# of moles in the mixture}}$	Mole fraction
Equation 11.11:	$P_1 = X_1P_T$	Dalton's law using mole fractions
Equation 11.12:	$PV = nRT$	Ideal gas law
	$R = 0.0821\frac{L \cdot atm}{mole \cdot K}$	Ideal gas law constant

Table 11.1: Units of Pressure

Table 11.2: The Vapor Pressure of Water

ANSWERS TO THE "ON YOUR OWN" QUESTIONS

11.1 This problem asks us to predict how a gas will change when we change some of the conditions under which it is stored. This means that we need to use the combined gas law (equation 11.10).

$$\frac{P_1V_1}{T_1} = \frac{P_2V_2}{T_2}$$

According to this problem, P_1 = 3,115 kPa, T_1 = 5.00°C, and T_2 = 30.0°C. Also, the problem states that the cylinder's volume cannot change; the volumes cancel out:

$$\frac{P_1\cancel{V_1}}{T_1} = \frac{P_2\cancel{V_2}}{T_2}$$

We can now rearrange the equation to solve for the new pressure:

$$\frac{P_1T_2}{T_1} = P_2$$

Before we put in the numbers and determine the new pressure, we first have to convert the temperatures to Kelvin because all temperatures in this equation must be in Kelvin:

$$T_1 = 5.00 + 273.15 = 278.15\text{ K}$$

$$T_2 = 30.0 + 273.15 = 303.2\text{ K}$$

Now we can put in the numbers and determine the new pressure:

$$\frac{3115\text{ kPa} \times 303.2\cancel{\text{K}}}{278.15\cancel{\text{K}}} = 3396\text{ kPa}$$

The new pressure, then, is 3396 kPa.

11.2 This is obviously another combined gas law problem, with P_1 = 1.00 atm, V_1 = 5.5 liters, T_1 = 25°C, P_2 = 895 torr, and T_2 = 17°C. The problem asks us to determine the new volume, so we have to rearrange equation 11.8 to solve for V_2:

$$V_2 = \frac{P_1V_1T_2}{T_1P_2}$$

Before we can plug in the numbers, we need to convert the temperatures to Kelvin:

$$T_1 = 25.0 + 273.15 = 298.2\text{ K}$$

$$T_2 = 17.0 + 273.15 = 290.2\text{ K}$$

Additionally, we need to make the pressure units the same. We can do this by converting torr into atm or vice versa. In this case, we convert atm to torr:

$$P_1 = \frac{1.00\ \text{atm}}{1} \times \frac{760.0\ \text{torr}}{1.000\ \text{atm}} = 7.60 \times 10^2\ \text{torr}$$

Now we can plug in the numbers:

$$V_2 = \frac{7.60 \times 10^2\ \text{torr} \times 5.5\ \text{liters} \times 290.2\ \text{K}}{298.2\ \text{K} \times 895\ \text{torr}}$$

$$V_2 = 4.55\ \text{liters}$$

So when the person dives deep, the volume of his lungs decreases to 4.5 liters.

11.3 In this combined gas law problem, we are asked to calculate the volume of a balloon under conditions of constant pressure. We need to rearrange the equation to solve for V_2, realizing that the pressures cancel because they are the same:

$$\frac{P_1V_1}{T_1} = \frac{P_2V_2}{T_2}$$

$$V_2 = \frac{V_1T_2}{T_1}$$

To be able to calculate the new volume, we first need to get the temperatures into Kelvin:

$$T_1 = 22.0 + 273.15 = 295.2\ \text{K}$$

$$T_2 = -5.0 + 273.15 = 268.2\ \text{K}$$

Now we can plug in our numbers:

$$\frac{3.21\ \text{L} \times 268.2\ \text{K}}{295.2\ \text{K}} = 2.92\ \text{L}$$

Putting a 3.21 L balloon in a freezer shrinks the balloon to a volume of 2.92 liters.

11.4 This problem requires us to recognize that when a gas is collected over water, the gas is contaminated with water vapor. So the 861 torr of gas is the total pressure of carbon dioxide *plus* water vapor. To determine the partial pressure of water vapor, we need only look at table 11.2. At 22.0°C, the vapor pressure of water is 19.8 torr. Dalton's law now becomes:

$$P_T = P_{\text{carbon dioxide}} + P_{\text{water vapor}}$$

$$861\ \text{torr} = P_{\text{carbon dioxide}} + 19.8\ \text{torr}$$

$$P_{\text{carbon dioxide}} = 861\ \text{torr} - 19.8\ \text{torr} = 841\ \text{torr}$$

Only 841 torr of carbon dioxide were collected.

11.5 Mole fraction is defined as the number of *moles* of component divided by the total number of *moles*. Right now, the problem gives us *grams*, not moles. We must first convert from grams to moles:

$$\frac{768.0 \cancel{\text{g } O_2}}{1} \times \frac{1 \text{ mole } O_2}{32.0 \cancel{\text{g } O_2}} = 24.0 \text{ moles } O_2$$

$$\frac{2496.0 \cancel{\text{g } N_2}}{1} \times \frac{1 \text{ mole } N_2}{28.0 \cancel{\text{g } N_2}} = 89.1 \text{ moles } N_2$$

$$\frac{45.6 \cancel{\text{g Ar}}}{1} \times \frac{1 \text{ mole Ar}}{39.9 \cancel{\text{g Ar}}} = 1.14 \text{ moles Ar}$$

Realize that we had to use N_2 and O_2 for nitrogen and oxygen because they are both homonuclear diatomics. Now that we have the number of moles of each component, we can calculate the total number of moles in the mixture:

Total number of moles = 24.0 moles + 89.1 moles + 1.14 moles = 114.2 moles

Plugging that into equation 11.10:

$$X_{O_2} = \frac{24.0 \cancel{\text{moles}}}{114.2 \cancel{\text{moles}}} = 0.210$$

$$X_{N_2} = \frac{89.1 \cancel{\text{moles}}}{114.2 \cancel{\text{moles}}} = 0.780$$

$$X_{Ar} = \frac{1.14 \cancel{\text{moles}}}{114.2 \cancel{\text{moles}}} = 0.00998$$

The mole fractions do not add up to exactly 1 because of significant figures.

11.6 This problem gives us the total pressure and the amount of each gas, and asks us to determine the partial pressures of those gases. This is clearly a Dalton's law problem using equation 11.11. We must therefore use the amounts of each gas to determine the mole fraction of each component:

$$\frac{150.0 \cancel{\text{g } N_2}}{1} \times \frac{1 \text{ mole } N_2}{28.0 \cancel{\text{g } N_2}} = 5.36 \text{ moles } N_2$$

$$\frac{40.0 \cancel{\text{g } O_2}}{1} \times \frac{1 \text{ mole } O_2}{32.0 \cancel{\text{g } O_2}} = 1.25 \text{ moles } O_2$$

Total moles = 5.36 moles + 1.25 moles = 6.61 moles

$$X_{N_2} = \frac{5.36 \cancel{\text{moles}}}{6.61 \cancel{\text{moles}}} = 0.811$$

$$X_{O_2} = \frac{1.25 \cancel{\text{ moles}}}{6.61 \cancel{\text{ moles}}} = 0.189$$

Your calculations of the mole fractions may not add up to exactly 1 because of significant figures. Now we can finally use equation 11.11:

$$P_1 = X_1 \times P_T$$

$$P_{N_2} = 0.811 \times 760.0 \text{ torr} = 616 \text{ torr}$$

$$P_{O_2} = 0.189 \times 760.0 \text{ torr} = 144 \text{ torr}$$

11.7 This problem combines what we learned about vapor pressure and what we learned about using Dalton's law to calculate mole fraction. We are told that the total pressure of the gas once it is collected is 815 torr. However, since it was collected over water, we should recognize that it is contaminated with water vapor. So we really have a mixture of water vapor and hydrogen gas. According to table 11.2, water vapor has a pressure of 18.7 torr at 21.0°C. We can use this information and equation 11.9 to calculate the partial pressure of hydrogen in this mixture:

$$P_T = P_{\text{hydrogen}} + P_{\text{water vapor}}$$

$$815 \text{ torr} = P_{\text{hydrogen}} + 18.7 \text{ torr}$$

$$P_{\text{hydrogen}} = 815 \text{ torr} - 18.7 \text{ torr} = 796 \text{ torr}$$

Now that we have the total pressure and the partial pressure of hydrogen, we can figure out the mole fraction of hydrogen:

$$P_{\text{hydrogen}} = X_{\text{hydrogen}} \times P_T$$

$$X_{\text{hydrogen}} = \frac{P_{\text{hydrogen}}}{P_T} = \frac{796 \cancel{\text{ torr}}}{815 \cancel{\text{ torr}}} = 0.977$$

11.8 In this problem, we are given pressure and temperature (STP means 1.00 atm and 273 K) and the number of moles. We are then asked to calculate V. We can do this by rearranging the ideal gas law:

$$PV = nRT$$

$$V = \frac{nRT}{P}$$

Once again, we have to examine our units. In this case, they work because temperature is in Kelvin:

$$V = \frac{1.00 \cancel{\text{ moles}} \times 0.0821 \frac{\text{L} \cdot \cancel{\text{atm}}}{\cancel{\text{mole} \cdot \text{K}}} \times 273 \cancel{\text{ K}}}{1.000 \cancel{\text{ atm}}} = 22.4 \text{ L}$$

At STP, then, 1 mole of an ideal gas occupies 22.4 liters.

11.9 In this stoichiometry problem, we are given the amount of limiting reactant and asked to calculate how much product will be made. First, we must balance the equation:

$$2C_8H_{18}(l) + 25O_2(g) \rightarrow 16CO_2(g) + 18H_2O(g)$$

Now we can convert the amount of limiting reactant to moles:

$$\frac{3.2 \times 10^3 \text{ g octane}}{1} \times \frac{1 \text{ mole octane}}{114.2 \text{ g octane}} = 28 \text{ moles octane}$$

We can then use stoichiometry to determine the number of moles of CO_2 produced:

$$\frac{28 \text{ moles octane}}{1} \times \frac{16 \text{ moles } CO_2}{2 \text{ moles octane}} = 2.2 \times 10^2 \text{ moles } CO_2$$

Now we need to use the ideal gas law. Looking at our units, we will need to convert temperature into Kelvin before we use the ideal gas law:

$$PV = nRT$$

$$V = \frac{nRT}{P} = \frac{2.2 \times 10^2 \text{ moles} \cdot 0.0821 \frac{L \cdot atm}{mole \cdot K} \cdot 453.2 \text{ K}}{1.00 \text{ atm}} = 8.2 \times 10^3 \text{ L}$$

11.10 In this problem, we are once again given the amount of limiting reactant and asked to determine how much product is made. The problem here is that the amount of limiting reactant is not given in grams or moles; instead, it is given in P, V, and T. This means we must use the ideal gas law to determine the number of moles of limiting reactant. Since STP means P = 1.00 atm and T = 273 K, the ideal gas law becomes:

$$n = \frac{PV}{RT} = \frac{1.00 \text{ atm} \cdot 250.0 \text{ L}}{0.0821 \frac{L \cdot atm}{mole \cdot K} \cdot 273 \text{ K}} = 11.2 \text{ moles}$$

Now that we have the moles of limiting reactant, this becomes a stoichiometry problem:

$$\frac{11.2 \text{ moles } H_2}{1} \times \frac{1 \text{ mole } CH_4O}{2 \text{ moles } H_2} = 5.60 \text{ moles } CH_4O$$

$$\frac{5.60 \text{ moles } CH_4O}{1} \times \frac{32.0 \text{ g } CH_4O}{1 \text{ mole } CH_4O} = 179 \text{ g } CH_4O$$

179 g of CH_4O are produced in this reaction.

STUDY GUIDE FOR MODULE 11

REVIEW QUESTIONS

1. Define pressure and name 3 units used to measure it.

2. Write Boyle's law in your own words.

3. Which of the laws you studied in this module leads to the Kelvin temperature scale?

4. What general rule must a careful scientist follow when extrapolating data?

5. What are the properties of an ideal gas?

6. Under what conditions do gases behave ideally?

7. What are standard temperature and pressure?

8. Two steel containers of equal volume contain 2 different gases at the same temperature. The first container holds 1.3 moles of N_2 gas, and the second contains 1.3 moles of SO_3 gas. Compare the pressures of the 2 containers.

9. A liquid's temperature is lowered from 50°C to 10°C. Did the liquid's vapor pressure increase, decrease, or stay the same during that time?

10. The mole fraction of nitrogen gas in compressed air is 0.78. If you have 1,000 molecules of compressed air, how many will be nitrogen molecules?

PRACTICE PROBLEMS

1. A balloon is filled with hydrogen at a temperature of 22.0°C and a pressure of 812 mm Hg. If the balloon's original volume was 1.25 liters, what will its new volume be at a higher altitude, where the pressure is only 625 mm Hg? Assume the temperature stays the same.
2. A bacterial culture isolated from sewage produced 23.0 mL of methane gas at 23.0°C and 795 torr. What would the volume of the methane be at STP?
3. A steel container with a volume that cannot change is filled with 95.0 atms of compressed air at 20.0°C. It sits out in the sun all day, and its temperature rises to 55.0°C. What is the new pressure?
4. In an experiment similar to experiment 8.1, a chemist collects carbon dioxide over water at 24°C. If the gas in the collection vessel has a pressure of 790 torr, what is the pressure of carbon dioxide in the vessel? (Use table 11.2.)
5. As part of an environmental experiment, the mixture of gases released from a smokestack is collected in a balloon. If the mixture is analyzed and found to contain 25.0 g of SO_2 gas, 13.1 g of NO gas, and 3.5 g of SO_3 gas, what is the mole fraction of each component in the mixture?
6. If the total pressure of the gas mixture in the problem above was 1.2 atm, what was the partial pressure of each gas?
7. A steel container is filled with 10.00 atms of nitrogen, 4.00 atms of oxygen, and 1.00 atms of argon. What is the total pressure in the vessel and the mole fraction of each gas?
8. What is the volume of 32.2 grams of chlorine gas at 1.00 atm and 25.0°C?
9. Silver tarnishes in the presence of oxygen and hydrogen sulfide according to the following equation:

$$4Ag(s) + 2H_2S(g) + O_2(g) \rightarrow \underset{\text{tarnish}}{2Ag_2S(s)} + 2H_2O(l)$$

   Usually, H_2S is the limiting reactant. If a silver spoon is placed in a container that holds 62.1 mL of H_2S at 1.15 atm and 23.0°C, how many grams of silver on the spoon will tarnish?
10. One common rocket engine uses the following reaction:

$$7H_2O_2\ (aq) + N_2H_4\ (g) \rightarrow 2HNO_3\ (g) + 8H_2O\ (g)$$

   If the rocket engine contains 2.89 kilograms of H_2O_2 and an excess of N_2H_4, what volume of water vapor will be produced when the rocket fires up to a temperature of 500.0°C and a pressure of 1.5 atm?

ENERGY, HEAT, AND TEMPERATURE

You now have an excellent foundation in chemical reactions from the last few modules. It's time to talk about energy, heat, and temperature. Have you ever sat down and watched a fire? The flames dancing along the wood, the crackling sounds, and the soft light make fire a pleasant source of extra heat. But have you ever wondered what exactly is going on while the flames consume the wood? What happens to the wood during a fire, and what is the nature of the ash that the wood leaves behind? Where does the heat come from? What makes the flames?

first reactions

Probably one of the most interesting things you'll contemplate in this course is that energy cannot be created or destroyed. It can only change form. As you read this module, spend some time noticing all of the energy changing form around you. You might be astounded at how much is changing once you are aware.

The answers to these questions lie in a study of the energy involved in chemical reactions. During the course of a fire, the matter that makes up the wood undergoes a chemical change. Some of the matter in the wood changes into water vapor, some of it changes into a gas (carbon dioxide), and most of the rest of it changes into what we call ash.

The details of this change are very interesting and will be discussed in several upcoming modules. For now, we want to concentrate on understanding the nature of the flames. How did the flames appear? Flames are a form of energy. Whenever matter undergoes a chemical change, that change is accompanied by either a release or an absorption of energy. As we go through this module, we will learn more about the nature of this energy, how to measure it, and how it relates to the study of chemistry.

ENERGY AND HEAT

Before we can get started, we have to make sure we understand what the term *energy* really means. Although we have used this term in previous modules and you have probably used this word several times in normal conversation, you probably could not define it. Here is the definition we will work with for this course:

Energy—The ability to do work.

This definition probably doesn't mean a lot to us until we also define work:

Work—The force applied to an object times the distance that the object travels parallel to that force.

Let's think about these 2 definitions for a moment. First, let's concentrate on work. Our definition of work tells us that for work to occur, there must be motion. After all, the object must travel a distance parallel to the force that is applied, or the work will be the force times 0, which is 0. As strange as it might sound, when you push against a wall with all of your might, *you are doing no work*. Only if the wall moved would you be doing work. You might exhaust yourself trying, but unless you got the wall to move, you would not be doing any work because to do work, the object to which you applied a force must move some distance parallel to that force.

Now that we understand the definition of work, we can see that the definition of energy makes a lot more sense. Energy is, essentially, the ability to cause motion. We mentioned earlier that the flames in a fire are a form of energy. Based on our definition of energy, then, the flames in a fire must be able to cause motion. What motion can the flames cause?

For one thing, we can see smoke rising up out of the fire. The energy in the flames is causing that motion. Also, if we were to put a pot of water on the flames, it would eventually start to boil. The energy in the flames causes the motion of the bubbles that result from boiling. Finally, think about the fact that we can feel heat coming from a fire. The only way that the heat can get to us is by traveling; so the fire's very act of warming us involves motion. Now that we're on the subject of heat, we had better come up with a good definition of it:

Heat—Energy that is transferred as a consequence of temperature differences.

One thing we see from this definition is that heat is really energy. This should make sense if you think about it. Energy is, essentially, the ability to create motion. If we used the heat from a fire to boil water, the heat would create motion; therefore, heat must be energy. What we see from the definition is that heat is a special form of energy—energy that is on its way from a hot body to a cold body. Heat always travels from hot to cold. If there is no difference in temperature, there is no heat.

FIGURE 12.1
Fire Producing Heat and Light
Photo © Crestock

Let's go back to discussing the fire. We know that the fire is sending out heat because we can feel it. Where did that heat come from? Since heat is energy that is being transferred from a hot body to a cold body, there must be energy in the fire that is being transferred to the things surrounding the fire. Where did that energy first come from?

The energy we feel coming from the fire was there long before the fire started. The wood that made up the fire and the oxygen in the air surrounding the fire contained all of the energy that the fire transfers to us in the form of heat. So when a pile of wood is just sitting in a fireplace, it has an enormous amount of energy in it. That energy has the potential of heating an entire room for a long period of time. Why doesn't the wood by itself feel hot, then? If there is energy contained in a pile of wood, why don't we feel heat coming from it? Why do we first have to light the wood and make a fire before we feel heat? Because the energy is trapped, or stored, inside the wood and cannot be transferred to us until the wood is burned. When the wood burns, the energy that has been stored there is released and transferred to us in the form of heat. Only then will it feel hot!

This discussion leads us to one of the most important concepts in all of science: the first law of thermodynamics (ther' moh dye nam' iks).

THE FIRST LAW OF THERMODYNAMICS

Let's learn what is probably the most important law of science:

The first law of thermodynamics—Energy cannot be created or destroyed. It can only change form.

This law tells us that God made only so much energy for the entire universe. This energy is used to do all of the work that has to be done for the universe to exist. Since humans are not as smart or powerful as God, there is no way that we can make energy. We cannot destroy it either.

"Wait a minute," you might say. "Aren't we creating energy when we burn something? Before we start the fire, the wood is not hot, but afterward, it is. Isn't that creating energy?" No. Remember what we said earlier: The energy was already in the wood. It was stored there, unable to move to us and warm us. The act of starting the fire merely *released* the stored energy that was *already there*. Once the stored energy was released, it became heat and was able to warm us.

In terms of the first law of thermodynamics, then, when we start a fire, we are changing the *form* of the energy that is already in the wood and the surrounding air. There are basically 2 forms of energy:

Potential energy—Energy that is stored.

Kinetic energy—Energy that is in motion.

In our fire example, then, the wood had a lot of *potential* energy. The energy was there, but it was being stored. When the fire got started, that potential energy was changed, or converted, into *kinetic energy*. So in a fire, energy is not created; it is simply changed from potential to kinetic energy.

In chemistry, the difference between potential energy and kinetic energy is very straightforward. Heat is considered a type of kinetic energy. After all, heat is defined as energy that is being transferred as a result of temperature differences. For something to be transferred, it must be moved. Since kinetic energy is energy in motion, heat is kinetic energy. On the other hand, all substances have energy stored in them, so all substances have potential energy. These are not the only types of potential and kinetic energy in chemistry, however. In later modules, we will introduce other types of potential and kinetic energy. For now, these ideas will suffice.

UNITS FOR MEASURING HEAT AND ENERGY

Now that we know the definitions of heat and energy, the forms of energy, and a law of chemistry that governs energy, it is time to learn how we measure energy. To do that, we must first understand the units that govern energy measurement. In the metric system, energy is measured in a unit called the Joule. This unit is named after James Prescott Joule, a brilliant scientist who worked during the mid-1800s. Joule, a devout Christian, did pioneering research on energy, and we honor him by using his name as the metric unit for energy.

To give you an idea of how much energy is in one Joule, lifting a 102-gram object straight up (at a constant speed) for 1.0 meter takes 1.0 Joule of energy. Remember how much each of these metric units is worth. One gram is approximately the mass of a paperclip. So a 102-gram object has the same mass as about 102 paperclips (not all that much). One meter is just a little longer than a yardstick. So 1.0 Joule is the energy needed to lift about 102 paperclips up in the air (at a constant speed) just a bit higher than a yard.

How do we measure energy? That's a tricky question. When in potential form, energy is very difficult to measure. There are ways of doing it, but they go beyond the scope of this course. On the other hand, it is relatively easy to measure energy in its kinetic form. We will spend a considerable amount of time on the measurement of heat. To measure heat, we must first measure temperature. It is important to understand that temperature and heat are not the same, but they are related to each other. That's why we can use temperature to measure heat. We all know

salt and light

Photo: Public Domain

James Prescott Joule (1818–1889) was an English scientist who worked in the study of heat. Joule approached research with a belief that all of creation should be understood with simplicity, and his experiments were designed with this simplicity in mind. But don't let the term *simplicity* fool you; Joule was able to measure heat to a precision never seen before. He was the first to develop a conversion factor between heat and mechanical work, which led to the development of the first law of thermodynamics. This quotation from Joule summarizes his belief in God the Creator: "After the knowledge of, and obedience to, the will of God, the next aim must be to know something of His attributes of wisdom, power, and goodness as evidenced by his handiwork. It is evident that an acquaintance with natural laws means no less than an acquaintanceship with the mind of God therein expressed." (Quoted in Graves 1996, 133)

MOD 12

that to measure temperature we use a thermometer. But how does a thermometer work?

A thermometer makes use of the fact that substances tend to expand as they get warmer. Such a substance (usually mercury or alcohol) contained in a column of glass begins to expand as it gets warmer (as heat transfers into it). Since the only way it can expand is upward, it begins to move up the glass column. The distance that it moves up the column is directly related to the temperature.

In the end, then, we measure temperature *indirectly*. Instead of finding some way to measure the temperature directly, we find it easier to measure the expansion of a substance and relate it back to temperature. This is quite common in science. Most measurements we make are, in fact, measurements of something else that we then can use to relate back to the physical quantity in which we are interested. Perform Experiment 12.1 to calibrate a thermometer and learn about thermometers.

EXPERIMENT 12.1

PURPOSE: To calibrate your thermometer and confirm the boiling and freezing temperatures of water.

MATERIALS:
- A thermometer that you intend to use in future experiments (preferably marked in Celsius)
- A 250 mL beaker (A small glass container will do, but make sure it is safe for boiling water.)
- Water (preferably distilled water, which can be purchased at any supermarket)
- Ice (preferably crushed)
- Graph paper or computer with spreadsheet program
- An alcohol burner, or something like a hot plate or stove, to heat the water
- Safety goggles

QUESTIONS: At what temperature does water boil and freeze? Are all thermometers the same?

HYPOTHESIS: Write something about what you think will be the accuracy of your thermometer and about the outcome of this experiment.

PROCEDURE:
1. Fill the beaker three-quarters full with ice (preferably crushed).
2. Add enough water so that the ice begins to float. Use your thermometer to *carefully* stir the ice and water for 5 minutes. This will allow the ice to fully mix with the water.
3. After 5 minutes, read your thermometer (without removing it from the ice water) and write down the temperature reading. Remember that according to the rules we developed for reading a scale, you should be able to estimate between the dashes marked on the thermometer. Since most thermometers are marked off in increments

of 1 degree, you should be able to write down your measurement to the nearest tenth of a degree.

4. If your thermometer is Celsius, it should read 0.0 degrees. If it is Fahrenheit, it should read 32.0 degrees. It is very likely that your thermometer will not read the correct temperature. Most thermometers are not marked very accurately when they are made, so your thermometer may be as much as 2 degrees off. This is okay, as long as you write down the reading so that you can refer back to it often because you will use it to correct any inaccuracy in your thermometer. For example, let's say your thermometer reads 1.8 degrees Celsius. This tells you that your thermometer reads 1.8 degrees too hot. In the future, when you measure temperature, you will correct for this inaccuracy by subtracting 1.8 from your measurements. Whatever you have to add or subtract to your measurement so that it turns into the correct value (0.0 degrees Celsius or 32.0 degrees Fahrenheit), you will do the same thing to all of your future measurements. This is another example of *calibration*.
5. Now that you have a calibrated thermometer, it is time to demonstrate the fact that any mixture of ice and water is always at the same temperature once it is thoroughly mixed. Slowly begin to heat the container, while constantly stirring carefully with your thermometer. Keep your flame, stove, or hot plate on low. If you add heat too quickly, you will not be able to stir quickly enough to distribute the heat evenly throughout the ice water.
6. While heat is being added to the ice water, read the temperature every minute. Write down both the time of the reading and the temperature, being sure to use your calibration to correct your thermometer. For this experiment to work properly, you must keep stirring the ice water mixture at all times except for the instances in which you read the thermometer. In addition, *be sure that your thermometer never touches the edges or bottom of the glass container*. Keep the thermometer in the ice water, but do not allow it to come in contact with the container.
7. As you watch your experiment, you should see that even though the ice starts melting, the temperature of the ice water does not change. If your temperature is changing, then that is due to the impreciseness of the red liquid inside your thermometer. If you had a mercury thermometer, you would not see a rise in temperature (but this red liquid is not toxic like mercury is).
8. Continue to heat the container and make your temperature readings every minute until all of the ice has melted. Note the time at which all of the ice is gone.
9. After that, continue for at least 5 more minutes. If you are using a beaker (or a glass container that is safe for boiling), you can continue this experiment until the water begins boiling vigorously. If you do this, it may take a while. To speed things up, you could increase the heat source to add heat more quickly. Continue to make temperature measurements until you have taken 4 measurements that are close to within 2 degrees of each other.

★ **Graphing your results on graph paper** - If you are going to use a computer to graph your data, then skip ahead to step 15.

10. Take your temperature measurements and use graph paper to graph all of the results you obtained up to 5 minutes after all of the ice melted. Put the time (in minutes) on

the x-axis and the temperature on the y-axis. Your y-axis should have a range of -1.0°C to 5°C, and your x-axis should start at 0 minutes and end 5 minutes after all of the ice melted. If any of your temperature measurements within this time frame were greater than 5°C, then the maximum on your y-axis should be equal to the maximum temperature. Your graph should look something like this:

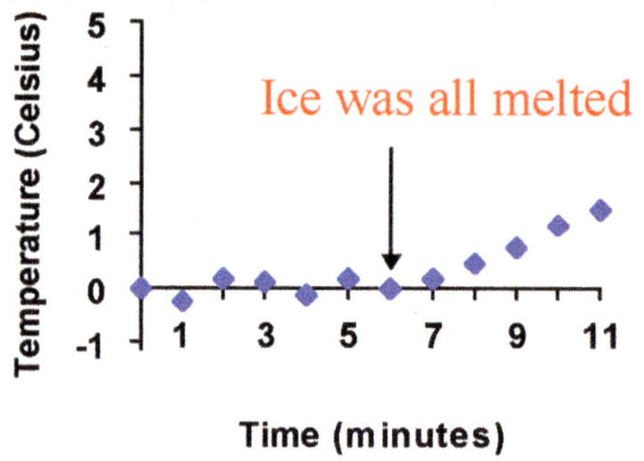

11. You should notice a couple of things in this graph. First of all, notice how the first several data points bounce up and down right around a temperature of 0.0°C. This is called **experimental scatter** and is simply a reflection of the fact that all measurements have some inaccuracy. Any single measurement may not be all that accurate, but the more measurements you have, the more accurate your final result will be. Since all of the data points scatter around 0.0°C, you can assume that the ice water stayed at 0.0°C during the entire time ice was present. Second, notice that after 7 minutes, the temperature of the water began a slow but steady increase. This represents the point at which all of the ice was melted. Only after the ice was melted did the temperature of the water increase. Therefore, no matter how much ice was present, the ice water mixture was always at a temperature of 0.0°C! Even though the ice water was being heated, it would not change its temperature until all of the ice had melted. The reason for this surprising result is pretty simple: It takes energy for ice to melt. It also takes energy to increase temperature. When a mixture of ice and water is heated, all of the energy it receives is used to melt the ice, and none of it will be used to raise the temperature. Once all of the ice is melted, then the incoming energy can be used to increase the temperature of the water.
12. If you continued the experiment through boiling, make another graph that includes that data as well. The graph should look something like this:

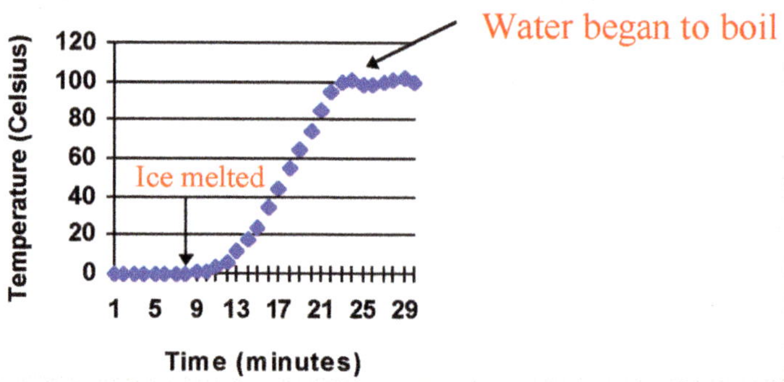

13. Notice how the data levels off at 100.0°C. Once again, this is because when the water begins boiling, all of the energy that is being transferred to the water is being

used to push the water molecules apart from each other, creating the motion of boiling as liquid water transitions to water vapor. As long as the water continues to boil, it will always stay at the same temperature. Graphs like the one pictured above are often called **heating curves**.

14. If your graph does not level off at precisely 100.0°C, don't worry about it. Remember from Module 11, the boiling temperature of water depends a bit on the atmospheric pressure. If the atmospheric pressure is a little higher than normal on the day you perform this experiment, then the boiling temperature will be slightly higher than 100.0°C. If the atmospheric pressure is a little lower than normal, then your boiling temperature will be slightly less than 100.0°C. One other thing to consider is that the atmospheric pressure depends on your elevation. If you live somewhere with a high elevation like Denver, Colorado, your graph will probably level off well under 100.0°C.

★ **Graphing your results with a computer** - You may skip these steps if you graphed your data on a piece of graph paper.

15. You will need a spreadsheet program like Microsoft Excel or Apple Numbers. If you have never used this program before, then you should watch tutorial videos on how to use the spreadsheet program. Your program's Help feature should have tutorials for new users. Or you could search the web for tutorials on spreadsheets. As of the time this book was printed, there were several free tutorial videos that teach how to use spreadsheet programs.
16. In the first column of your spreadsheet, enter Time in the top cell followed by the minutes that you took measurements.
17. Enter your temperature measurements in the second column. Your data table should look like this if you are using Microsoft Excel.

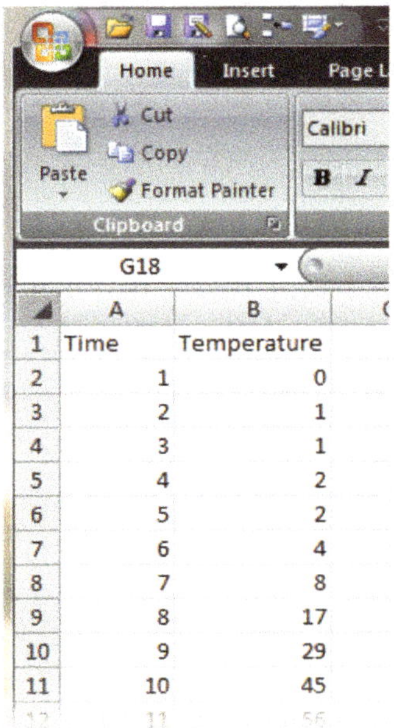

	A	B
1	Time	Temperature
2	1	0
3	2	1
4	3	1
5	4	2
6	5	2
7	6	4
8	7	8
9	8	17
10	9	29
11	10	45

18. Highlight all times and temperatures for 5 minutes after your ice melted. Use your spreadsheet program's graphing function to make an X-Y plot (sometimes called a scatter plot).

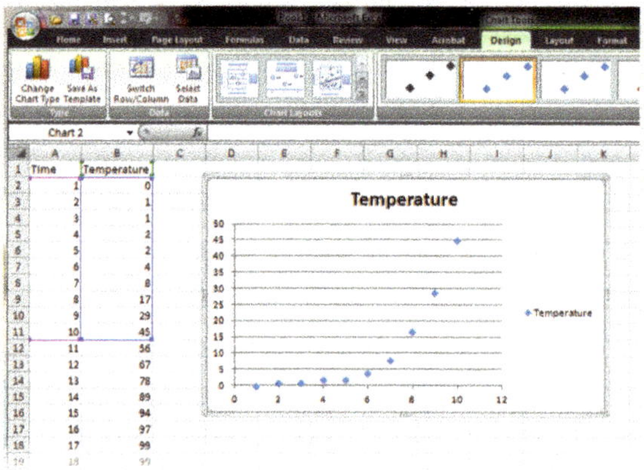

19. You should notice a couple of things in this graph. First, notice how the first several data points bounce up and down right around a temperature of 0.0°C. This is called experimental scatter and is simply a reflection of the fact that all measurements have some inaccuracy. Any single measurement may not be all that accurate, but the more measurements you have, the more accurate your final result will be. Since all of the data points scatter around 0.0 degrees, you can assume that the ice water stayed at 0.0°C during the entire time ice was present. Second, notice that after 5 minutes, the temperature of the water began a slow but steady increase. This represents the point at which all of the ice was melted. Only after the ice was melted did the temperature of the water increase. So no matter how much ice was present, the ice water mixture was always at a temperature of 0.0°C. Even though the ice water was being heated, it would not change its temperature until all of the ice had melted. The reason for this surprising result is pretty simple: It takes energy for ice to melt. It also takes energy to increase temperature. When a mixture of ice and water is heated, all of the energy it receives is used to melt the ice, and none of it will be used to raise the temperature. Once all of the ice is melted, the incoming energy can be used to increase the temperature of the water.
20. Make a second graph including all of your data. Your graph should look like this.

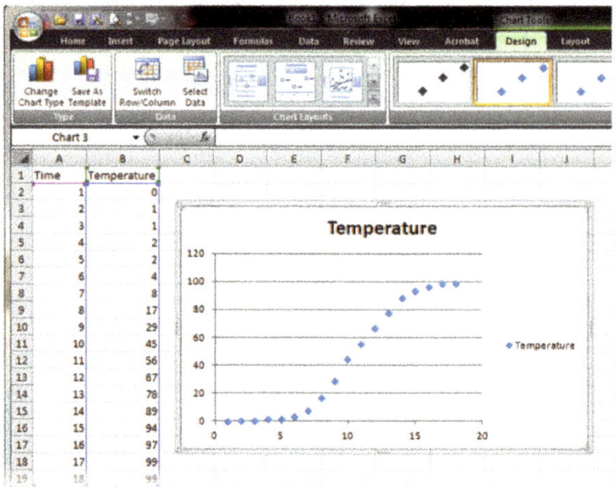

21. Notice how the data levels off at 100.0°C. Once again, this is because when the water begins boiling, all of the energy that is being transferred to the water is being used to push the water molecules apart from each other, creating the motion of boiling as liquid water transitions to water vapor. As long as the water continued

to boil, it would always stay at the same temperature. Graphs like the one pictured above are often called **heating curves**.

22. If your graph does not level off at precisely 100.0°C, don't worry about it. Remember from Module 11, the boiling temperature of water depends a bit on the atmospheric pressure. If the atmospheric pressure is a little higher than normal on the day you perform this experiment, the boiling temperature will be slightly higher than 100.0°C. If the atmospheric pressure is a little lower than normal, then your boiling temperature will be slightly less than 100.0°C. One other thing to consider is that the atmospheric pressure depends on your elevation. If you live somewhere with a high elevation like Denver, Colorado, your graph will probably level off well under 100.0°C

CONCLUSION: How accurate is your thermometer? How close to freezing and boiling point of water is it? Write something about its accuracy (how close to 0.0°C and 100.0°C) and precision (how many decimal spaces it can be read to).

salt and light

Photo: Public Domain

Lord Kelvin, William Thomson, First Baron Kelvin (1824–1907) was a mathematical physicist and engineer. His work was extremely important in the formulation of the first and second laws of thermodynamics and in the emerging discipline of modern physics. He had extensive interests as an engineer and inventor, including work on the electric telegraph and the mariner's compass, which had previously been limited in reliability. He is widely known for realizing there is a lower limit to temperature, called absolute zero, and the unit we know as Kelvin or K which was named in his honor. Lord Kelvin won the Copley Medal and Royal Medal for his work and was knighted by Queen Victoria. He was the first United Kingdom scientist to be elevated to the House of Lords. A strong Christian, Lord Kelvin gave a famous address to the Christian Evidence Society, a United Kingdom Christian apologetics organization founded in 1870 to fight the atheism in Victorian society. Lord Kelvin was a professor of natural philosophy at the University of Glasgow and started his classes with prayer. He had a strong faith in God and the Bible and is noted for saying, "Do not be afraid of being free thinkers. If you think strongly enough, you will be forced by science to the belief in God." (Quoted in Kelvin Library 2008)

THE CALORIE UNIT

We discussed temperature units in Modules 1, 10 and 11, and now we know how temperature is measured, so we can begin our discussion of measuring heat. The first thing we need to know is that chemists sometimes use a unit other than the Joule for measuring heat. This unit is the calorie and is defined as follows:

calorie (cal)—The amount of heat necessary to warm 1 gram of water 1 degree Celsius.

You might wonder why chemists sometimes use the calorie unit to measure heat. After all, since heat is energy, it could be measured in Joules. That's true, but chemists didn't always know that heat was just another form of energy. They would measure heat with the calorie unit and energy with a different unit. When James Joule demonstrated that heat was just another form of energy, it was a major breakthrough in our understanding of energy. It was such a profound accomplishment that we honor him by naming the metric energy unit after him. However, since old habits die hard, chemists sometimes use the calorie to measure heat, even though the standard metric unit is the Joule.

Since Joules and calories really measure the same thing, there must be a relationship between them:

1 calorie = 4.184 Joules

Since it takes a little more than 4 Joules to make a calorie, you can see that the calorie is a bigger energy unit than the Joule. This relationship can be used to convert between calories and Joules using the factor-label method that we employed in Module 1. Note that the 4.184 is not exact, which means its significant figures must be taken into account when using it in a conversion.

Most people have heard of calories before. If you look at the nutritional information on any food-product label, you will see a listing of how many Calories are contained in one serving of the product. Those who are watching their weight are usually quite conscious of this number. Do the Calories that you see on a food label have anything to do with the calories we are discussing here? Yes.

FIGURE 12.2
A Food Nutrition Label
Image: U.S. Food and Drug Administration

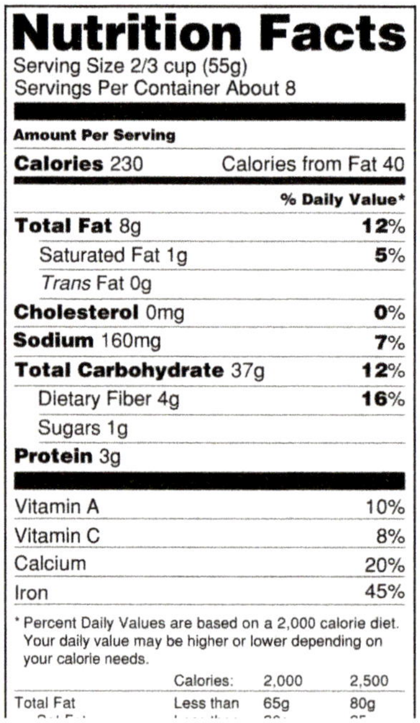

Nutrition Facts
Serving Size 2/3 cup (55g)
Servings Per Container About 8

Amount Per Serving

Calories 230	Calories from Fat 40
	% Daily Value*
Total Fat 8g	**12%**
Saturated Fat 1g	**5%**
Trans Fat 0g	
Cholesterol 0mg	**0%**
Sodium 160mg	**7%**
Total Carbohydrate 37g	**12%**
Dietary Fiber 4g	**16%**
Sugars 1g	
Protein 3g	
Vitamin A	10%
Vitamin C	8%
Calcium	20%
Iron	45%

* Percent Daily Values are based on a 2,000 calorie diet. Your daily value may be higher or lower depending on your calorie needs.

	Calories:	2,000	2,500
Total Fat	Less than	65g	80g

When you eat food, one of the most important things that the food can give you is energy. Most of the things that you do (walking, running, breathing, even sitting down and reading a book) involve work. Since work requires energy, your body needs a steady supply of it. Your body gets the energy it needs from the food you eat. For example, the Calories listed on the label of a soup can measure how much energy a serving of that soup provides to your body. Your body can then use that energy for any work it needs to do.

For some reason, the food Calorie (notice we use a capital C here) is not the same as a chemistry calorie (written with a lowercase c). Food generally contains an enormous amount of energy; therefore, the calorie as we defined it earlier is simply too small a unit to use when measuring the energy content of food. Instead, 1 food Calorie is equal to 1,000 chemistry calories.

1 food calorie (Cal) = 1,000 chemistry calories (cal)

Based on our discussion of the metric system, we should call the food Calorie a kilocalorie. Unfortunately, food scientists do not like to use that term, so they drop the *kilo-* prefix and just use a capital C instead. It is a little confusing, but we will just have to live with it.

Think for a moment about what all of this means: We defined a calorie as the amount of energy required to heat 1 gram of water 1 degree Celsius. We now know that a food Calorie is 1,000 of those calories and can heat up 1,000 grams of water 1.0 degree Celsius. Also consider that in an normal day, the average human body uses 2,500 food Calories to do all its work. That's enough energy to heat 2.5 *million* grams of water 1.0 degree Celsius. That's a lot of energy! The amazing thing is that God has designed our bodies to use all of that energy without overheating!

MEASURING HEAT

Although chemists have traditionally measured heat in calories, most modern chemistry textbooks use the standard metric unit, the Joule. Since you will probably have to use Joules in your college chemistry course, it is a good idea for you to start using them now. Most of the problems in this course will use Joules rather than calories. However, we will have you use calories occasionally, just to keep them in your mind.

As mentioned at the beginning of the module, when heat is given to an object, it can cause that object's temperature to increase. If we measure the temperature of an object and notice that it begins to rise, we can assume that the object is absorbing some heat. Alternatively, if we notice that the temperature is decreasing, that tells us the object is releasing (losing) heat. The temperature increase or decrease we measure should be, in some way, proportional to the amount of heat the object absorbs or releases.

The temperature change in an object is related to the amount of heat it absorbs or releases by the following equation:

$$q = m \cdot c \cdot \Delta T \qquad \text{Equation 12.1}$$

In this equation, q stands for the amount of heat absorbed or released, m represents the mass of the object, and c is the specific heat of the object, which will be discussed momentarily. The last symbol in the equation, ΔT (pronounced Delta T), stands for the change in temperature and is defined as the difference between the final temperature and the initial temperature:

$$\Delta T = T_{final} - T_{initial} \qquad \text{Equation 12.2}$$

As we saw in Module 10, the Δ symbol means change in. So whenever you see a Δ in front of a variable, it means you are looking for a change in that variable.

The variable c that we used in equation 12.1 above is a very important quantity in the study of energy. It is called specific heat capacity or, usually, specific heat. The physical meaning of specific heat is a very important concept to learn.

Specific heat—The amount of heat necessary to raise the temperature of 1 gram of a substance by 1 degree Celsius.

In other words, specific heat tells how easy it is to heat a substance. For example, if we put a metal plate on the burner of a stove and we put a wooden plate on another burner of the stove, which one do you think would get hot faster? As long as the wood did not catch on fire, the metal plate would get much hotter much faster because the metal plate has a lower specific heat than the wood. Since metals have low specific heats, a small amount of heat changes the temperature significantly. On the other hand, wood has a relatively large specific heat, so it takes more heat to produce the same temperature change in wood.

Like density, each substance has a unique specific heat. Table 12.1 lists some common substances and their specific heats. You needn't memorize any of these numbers; they are simply provided for illustration. When doing problems, you may need to refer back to this table.

TABLE 12.1
Specific Heats of Common Substances

Substance	Specific Heat (J / g°C)
Aluminum	0.9000
Copper	0.3851
Glass	0.8372
Gold	0.1260
Iron	0.4521
Silver	0.2330
Water	4.184
Zinc	0.387

From table 12.1, compare the specific heats of gold and copper. If we had 1 gram of both gold and copper and added the same amount of heat to each, then the gold would have a higher temperature. The specific heat of gold says that if we add 0.1260 J of energy to 1 gram, then the temperature will be 1°C. What if we added that same amount, 0.1260 J, to copper? The temperature would not have risen 1 degree yet! What would happen if we added an additional 0.1260 J to the gold and copper? The gold would increase another degree but the copper would not have increased 1°C yet. So with 0.2560 J added, gold has increased by 2°C, while copper has not increased its temperature by 1°C yet because 0.3851 J would be needed to do that.

To determine whether or not you understand the concept of specific heat, answer "On Your Own" questions 12.1–12.2.

ON YOUR OWN

12.1 Which substance takes less energy to heat up: gold or zinc? (Refer to table 12.1.)

12.2 A 5.0 gram piece of copper and a 5.0 gram piece of iron are placed side by side on a burner. After 3.0 minutes , the iron reaches a temperature of 100.0° C. Will the copper temperature be higher or lower than 100.0°C? (Refer to table 12.1.)

Now that we know what specific heat is, we need to learn its units. Looking back at equation 12.1, you should recognize the units that we will use for all quantities in that equation except the specific heat. The units for heat (q) will either be calories or Joules; the units for mass (m) will be something like grams, kilograms, milligrams, etc.; and the units for temperature will be °F, °C, or K. What will the units for specific heat be? That can be determined from the equation.

Let's suppose we wanted to measure heat with the units of Joules. All of the units on the right side of equation 12.1 would have to work out to a final unit of Joules. If we measured temperature in °C and mass in grams, those units would have to somehow cancel and leave the unit of Joules in their place so that heat (q) would have the proper unit. How does this happen? We select the correct units for specific heat to obtain the correct final answer in Joules.

The only way for °C to cancel out is for that same unit to be in the denominator of the unit for specific heat. That way, when specific heat is multiplied by the change in temperature, °C would cancel. In the same way, for grams to cancel, they would also have to appear in the denominator of the unit for specific heat. On the other hand, since the final answer must be in Joules, then Joules must appear in the numerator of the unit for specific heat. The unit for specific heat in this case must be J/(g · °C).

Of course, if we decided to measure heat in calories instead of Joules, that would change the unit for specific heat. The units would then become cal/(g · °C). If we wanted to measure the mass in kg, the specific heat unit would be cal/(kg · °C). In general, then, the unit for specific heat will have an energy unit (calories or Joules) in the numerator, a mass unit (grams, kg, mg, etc.) in the denominator, and a temperature unit (°C, °F, or K) in the denominator. See if you can follow this type of reasoning in Example 12.1.

EXAMPLE 12.1

To cook an omelet, a 5.0 kg iron skillet must be heated from room temperature (25.0°C) to 150.0°C. How many Joules of heat must be used to accomplish this?

To solve this, we use equation 12.1 and table 12.1. The table gives us the specific heat of iron in (J/g · °C). Therefore, we must use temperature in °C, mass in grams, and we will get our answer in Joules. So first we have to convert 5.0 kg into grams:

$$\frac{5.0\ \cancel{\text{kg}}}{1} \times \frac{1{,}000\text{ g}}{1\ \cancel{\text{kg}}} = 5.0 \times 10^3\text{ g}$$

Since there are 2 significant figures in our mass measurement, we can have only 2 significant figures after the conversion, so our mass is 5.0×10^3 grams. Note that scientific notation is the only way we can express the mass properly since the 0 must be significant. If we had used 5,000 grams, there would be only 1 significant figure. Now that we have everything in the proper units, we can use equation 12.1:

$$q = 5.0 \times 10^3\text{g} \times 0.4521 \frac{\text{J}}{(\text{g}\cdot{}^\circ\text{C})} \times (150.0^\circ\text{C} - 25.0^\circ\text{C})$$

$$q = 5.0 \times 10^3 \cancel{g} \times 0.4521 \frac{J}{(\cancel{g \cdot °C})} \times 125.0\cancel{°C}$$

$$q = 280{,}000\ J$$

Notice that since we used all of the same units that we found in our specific heat, everything except Joules cancels out.

Pay attention to the significant figures here. Notice that to calculate ΔT, we had to subtract 25.0°C from 150.0°C. When we did that, we had to follow the rules of addition and subtraction. Both temperatures went out to the tenths place, so the answer can go out to the tenths place as well. This means the value for ΔT has 4 significant figures. In the end, we had to multiply that number by 5.0×10^3 g, which has only 2 significant figures, so the final answer can have only 2 significant figures. We could have expressed the answer in scientific notation (2.8×10^5 J) as well. Either **280,000** or **2.8×10^5 J** would be correct.

If a 305.4 gram mass of copper that is initially at 75°F is given 1,234.0 Joules of energy, what will its final temperature be?

To solve this problem, we must solve for ΔT. Since we already know the initial temperature, solving for ΔT will allow us to calculate the final temperature. The first thing we have to do is use algebra to rearrange the equation:

$$\Delta T = \frac{q}{m \cdot c}$$

Once we do that, we now have to think about units. The only unit we have specific heat in is J/(g· °C). This tells us, we need to convert our temperature into °C:

$$°C = \frac{5}{9}(°F - 32)$$

$$°C = \frac{5}{9} \times (75 - 32) = 24°C$$

Since our temperature in °F had only 2 significant figures, our final answer can have only 2, so it was rounded to 24. Now that we have all of our numbers in the correct units, we can use the first equation to calculate ΔT:

$$\Delta T = \frac{1{,}234.0\cancel{J}}{305.4\cancel{g} \times 0.3851 \frac{\cancel{J}}{\cancel{g} \cdot °C}}$$

$$\Delta T = 10.49°C$$

Since all of the numbers in the equation have at least 4 significant figures, ΔT can also have 4 significant figures. Now that we have ΔT, we can use its definition to solve for the final temperature:

$$\Delta T = T_{final} - T_{initial}$$

$$T_{final} = \Delta T + T_{initial}$$

$$T_{final} = 10.49°C + 24°C = 34°C$$

Notice that in this case, we are adding, so we look at decimal place. The value for ΔT goes out to the hundredths place, but the value for $T_{initial}$ goes out to the ones place. So the answer can go out only to the ones place. The chunk of copper, then, reaches a final temperature of **34°C.**

See if you really understand this by answering "On Your Own" questions 12.3–12.4.

ON YOUR OWN

12.3 A 432 g aluminum rod is heated to an unknown initial temperature. If, while cooling down to a temperature of 24.0°C, the rod releases 612 Joules of energy, what was its initial temperature? (Hint: When an object loses energy, its q is negative. We will discuss this fact in detail shortly.)

12.4 If a 16.0 g object requires 85 J of heat to raise its temperature by 22°C, what is the specific heat of the object?

CALORIMETRY

Even though we mentioned that you needn't memorize any of the specific heats listed in table 12.1, there is one specific heat that you do need to memorize: the specific heat of water. This is an easy one to memorize because it relates back to the definition of a calorie. Remember that we defined a calorie as the amount of heat necessary to raise 1 gram of water 1.0 degree Celsius. If that is the definition of a calorie, what does that tell us about the specific heat of water?

The specific heat of water is $1.000 \frac{cal}{g \cdot C°}$ **or** $4.184 \frac{J}{g \cdot C°}$

These 2 numbers are very important to chemists, so you need to memorize them.

So how is specific heat measured? By looking at equation 12.1, we see that 3 measurements are needed to calculate specific heat. If we measure the mass of an object, its temperature change, and how much energy it has experienced, then we can calculate heat with those measurements. How can we measure mass then? We use a balance to measure grams! Okay, that was easy, but how can we measure ΔT? We need 2 different temperatures. We measure temperature with a thermometer. This works well for liquids and gases, but what about solids like gold or copper? How can we measure the temperature of solids? Thermometers don't work well for this.

What are we going to do? We will measure the temperature of the solid's environment. If we allow the metal to sit in one place for a while, then heat will transfer between the metal and its environment until they are the same temperature. To determine the specific heat, then, we will need 2 different environments at 2 different temperatures. Then ΔT will be the difference between the 2 temperatures.

But now we have reached the hard part of determining the specific heat of an object. We need to know how much energy that object is holding. In equation 12.1, we need to know q. There is no such instrument that can measure heat, so how are we going to measure q? We are going to measure how much heat the object's environment absorbs because this amount of heat had to come from the object! The experimental process that chemists use to do this is called calorimetry (kal uh rim' uh tree) and is illustrated by Figure 12.3.

FIGURE 12.3
Calorimetry
Illustration by David Weiss

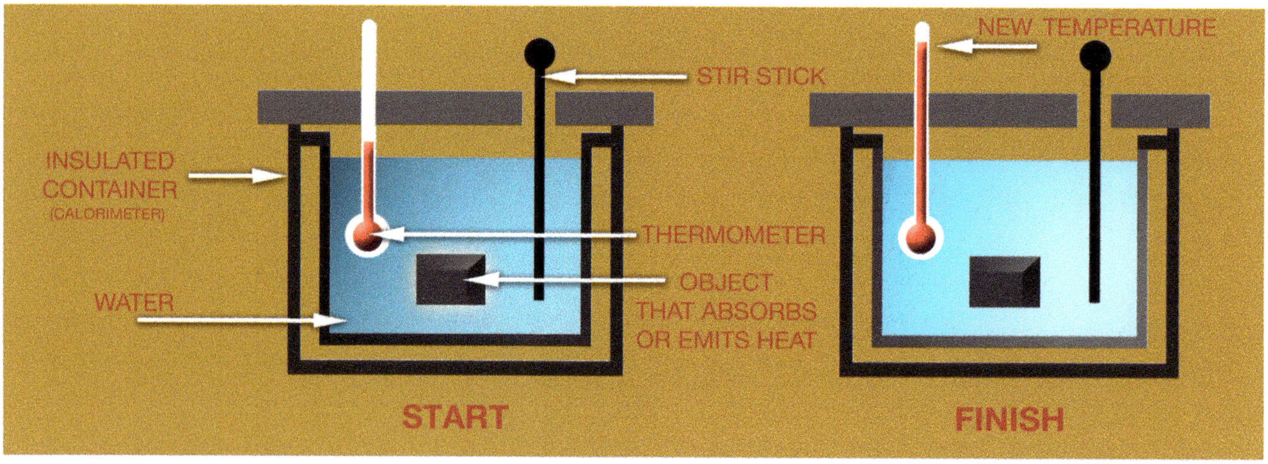

In calorimetry, chemists use an apparatus known as a calorimeter (kal uh rim' uh ter) A calorimeter is an insulated container that holds a known mass of water. The temperature of the water is constantly measured by a thermometer, and there is usually a rod that stirs the water to keep the heat distributed evenly. When a hot object is placed in the calorimeter, it begins to emit heat. That heat is absorbed by the water and the calorimeter, causing the temperature measured by the thermometer to increase. Because heat is being transferred from the object to the water and the calorimeter, the water and calorimeter will begin to warm up and the object will begin to cool down. At some point, they will reach the same temperature, and heat will no longer be transferred from the object to the water and calorimeter. At this point, the temperature measured by the thermometer will stop increasing. This tells the chemist that the experiment is finished.

The difference in temperature between the end of the experiment and the beginning of the experiment can then be used as the ΔT in equation 12.1. Since we know the specific heat of water, the specific heat of the calorimeter, and the mass of each, we can use equation 12.1 to calculate the total amount of heat that the object emitted. If you think about it, we can write a simple equation to relate the heat gained or lost by the object, the water, and the calorimeter.

$$-q_{object} = q_{water} + q_{calorimeter} \qquad \text{Equation 12.3}$$

You can think of this equation as the calorimetry equation. It tells us that any heat gained by the calorimeter and water must be lost by the object placed in the calorimeter. This makes sense in light of the first law of thermodynamics. Energy cannot be created or destroyed. If the water and calorimeter gain energy, that energy must come from somewhere. It comes from the hot object.

Why is there a negative sign in front of q_{object}? Think about it. The object is *releasing* heat. That means it is *losing* energy. If it is losing energy, its q will be negative. However, the water and the calorimeter are *gaining heat*. That means they will each have a *positive* q. To make the 2 sides equal, then, we must put a negative sign in front of q_{object}. You will see how this works in a moment.

In the end, then, if we want to measure the amount of heat released by a substance, we can surround that substance with water and a calorimeter, and measure the change in temperature of the water and the calorimeter. Equations 12.1, 12.2, and 12.3 will then tell us exactly how much heat was released. We can also measure the specific heat of a substance with calorimetry. Example 12.2 should help you see how this is done.

EXAMPLE 12.2

A chemical reaction is carried out in a 5.0 gram calorimeter. The calorimeter has a specific heat of 1.95 J/(g · °C) and is filled with 150.0 grams of water. If the temperature of the water was 24.0°C before the reaction took place, and it rose to 32.0°C by the end of the experiment, how much heat was released by the reaction?

Equation 12.3 tells us that:

$$-q_{object} = q_{water} + q_{calorimeter}$$

Equation 12.1 gives us a way to calculate q_{water} and $q_{calorimeter}$. After all, equation 12.1 tells us that to calculate the heat gained or lost by a substance, we need to know the mass, specific heat, and change in temperature of the substance. We know all of those things for both the calorimeter and the water. First, let's calculate ΔT with equation 12.2. Since the water and calorimeter start out in contact with each other and remain in contact with each other the entire time, they are at the same temperature throughout the course of the experiment. So they start out at the same temperature (24.0°C), and they end up at the same temperature (32.0°C). This means they have the same ΔT:

$$\Delta T = T_{final} - T_{initial} = 32.0°C - 24.0°C = 8.0°C$$

Notice that since we are subtracting here, we must look at decimal place to determine what digits to keep. Since both temperatures go out to the tenths place, our answer can go out to the tenths place. Now that we know ΔT for both the water and the calorimeter, we can calculate the heat they gained. Let's start with the calorimeter. We know its mass and specific heat, and we just determined its ΔT:

$$q_{calorimeter} = m \cdot c \cdot \Delta T$$

$$q_{calorimeter} = 5.0\,\cancel{g} \times 1.95\,\frac{J}{\cancel{g} \cdot \cancel{°C}} \times 8.0\,\cancel{°C}$$

$$q_{calorimeter} = 78\ J$$

We can now calculate the heat gained by the water using the mass of the water, the specific heat of the water (which you should memorize), and the same ΔT:

$$q_{water} = m \cdot c \cdot \Delta T$$

$$q_{water} = 150.0\,\cancel{g} \times 4.184\,\frac{J}{\cancel{g} \cdot \cancel{°C}} \times 8.0\,\cancel{°C}$$

$$q_{water} = 5.0 \times 10^3\ J$$

Now that we know the q for both the water and the calorimeter, we can use equation 12.3:

$$-q_{object} = q_{water} + q_{calorimeter} = 5.0 \times 10^3 \text{ J} + 78 \text{ J} = 5{,}078 \text{ J}$$

$$q_{object} = \mathbf{-5{,}100 \text{ J}}$$

Notice the significant figures here. The q for the water has its last significant figure in the hundreds place. Don't let the scientific notation fool you. The 0 is to the right of the decimal, but that is still the hundreds place, which can be seen if we take it out of scientific notation. The q for the calorimeter has its last significant figure in the ones place. Based on the rules of addition and subtraction, then, the final answer can have its last significant figure in the hundreds place.

What does the negative sign mean? It means that the object *lost* energy. That makes sense because the water and calorimeter warmed up, which means they gained energy. Since they gained energy, the object must have lost energy, which is why it has a negative q.

Although calorimetry is often used to determine the heat released by a chemical reaction (as was done in Example 12.2), it can also be used to determine the specific heat of an object. In fact, the specific heats listed in table 12.1 were measured using calorimetry. To see how this is done, study Example 12.3.

EXAMPLE 12.3

A 28.5 gram block of an unknown metal is heated to a temperature of 151.7°C. It is then dropped into the same calorimeter that was used in the previous example (m = 5.0 g; c = 1.95 J/g °C; mass of water = 150.0 grams). If the initial temperature of the water was 22.0°C and its final temperature was 24.2°C, what is the specific heat of the metal and what is its identity?

Once again, we can use equation 12.3, but we must first calculate all of the q's that we can. We can first calculate the ΔT of the water and calorimeter:

$$\Delta T = T_{final} - T_{initial} = 24.2°\text{C} - 22.0°\text{C} = 2.2°\text{C}$$

We can now calculate q for both the water and calorimeter because we have all the information we need for equation 12.1. Let's start with the calorimeter:

$$q_{calorimeter} = m \cdot c \cdot \Delta T$$

$$q_{calorimeter} = 5.0 \cancel{\text{g}} \times 1.95 \frac{\text{J}}{\cancel{\text{g} \cdot °\text{C}}} \times 2.2 \cancel{°\text{C}}$$

$$q_{calorimeter} = 21 \text{ J}$$

Now we can move on to the water:

$$q_{water} = m \cdot c \cdot \Delta T$$

$$q_{water} = 150.0 \cancel{\text{g}} \times 4.184 \frac{\text{J}}{\cancel{\text{g} \cdot °\text{C}}} \times 2.2 \cancel{°\text{C}}$$

$$q_{water} = 1{,}400\text{ J}$$

Now that we have calculated the q's of the water and calorimeter, we can use equation 12.3 to determine the q of the metal. Since the metal is the object in the calorimeter, we will replace the word object with the word metal:

$$-q_{metal} = q_{water} + q_{calorimeter} = 1{,}400\text{ J} + 21\text{ J} = 1{,}421\text{ J}$$

$$q_{metal} = -1{,}400\text{ J}$$

Before we go on, look at the significant figures here. The q_{water} has its last significant figure in the hundreds place, and the $q_{calorimeter}$ has its last significant figure in the ones place. By the rule of addition then, we can report the answer only to the hundreds place. This means the q of the calorimeter did not affect the answer. It gained so little heat that its effect was too small to matter in terms of the significant figures we are allowed to keep. We will come back to this point after we finish the example.

Okay, then, what do we have now? We have q_{metal}. How does that help? We now have all of the variables needed to use equation 12.1 to calculate its specific heat. Since the metal was in contact with the water and the calorimeter after the experiment ended, the metal's final temperature was the same as the water's final temperature. Its initial temperature was quite different from that of the water and calorimeter. The ΔT of the metal will be different from the ΔT of the water and calorimeter:

$$\Delta T_{metal} = T_{final} - T_{initial} = 24.2^\circ\text{C} - 151.7^\circ\text{C} = -127.5^\circ\text{C}$$

Notice that this ΔT is negative because the object cooled down. Now we can rearrange equation 12.1 using algebra and solve for the specific heat of the metal:

$$c = \frac{q}{m \cdot \Delta T}$$

$$c = \frac{-1{,}400\text{ J}}{28.5\text{ g} \times (-127.5^\circ\text{C})}$$

$$c = \mathbf{0.39\ \frac{J}{g \cdot {}^\circ C}}$$

Looking at table 12.1, we can see that the metal must have been **copper** because the specific heat that we determined is the same (to 2 significant figures) as the specific heat of copper.

That was a long problem, so let's review how we were able to solve it. When we read the problem, we noticed it was a calorimetry problem, so we knew that equations 12.1–12.3 would be involved. Based on the data, we recognized that we could calculate the heat absorbed by the water and the calorimeter. Once we calculated those heats, we plugged them into equation 12.3 so that we could solve for the heat lost by the metal. Once we did that, we simply used equation 12.1 again with all of the metal's data in it. Since the only thing we did not know in that equation was the metal's specific heat, we could solve for it.

Notice one more thing about Example 12.3: the heat absorbed by the calorimeter was so small (21 Joules), that it did not even show up when we added it to the heat absorbed by the water (1,400 J). This happens often in calorimetry experiments. As a result, we can often disregard the heat absorbed by the calorimeter when doing calorimetry experiments. If that is the case, then, we can treat $q_{calorimeter}$ in equation 12.3 as being equal to 0. This simplifies the analysis of calorimetry experiments quite a bit. So if we tell you to ignore the calorimeter in a calorimetry problem, just set $q_{calorimeter}$ equal to 0. However, if we give you information about the calorimeter (its mass or specific heat, for example), then you cannot ignore the calorimeter, and you must solve the problem the way we did in Example 12.3.

To help you better understand the process of calorimetry, do Experiment 12.2.

EXPERIMENT 12.2

PURPOSE: To measure the specific heat of a metal.

MATERIALS:

- A calibrated thermometer
- A scale that reads mass, preferably in grams
- 2 Styrofoam cups
- A chunk of metal that has mass of at least 30 grams (a lead sinker or a very large steel nut, for example)
- Boiling water (either in a pot or a beaker)
- Kitchen tongs
- Safety goggles

QUESTION: Can the specific heat be measured of a metal?

HYPOTHESIS: Write a hypothesis related to this experiment.

PROCEDURE:

1. Heat water in a pan or beaker until it is boiling vigorously.
2. While you are waiting for the water to boil, measure the mass of the metal with your scale. Remember to report your answer to one more decimal place than what is marked off by the scale.
3. Once the water on the stove is boiling, drop the metal into the pot and let it sit there for about 5 minutes. This will heat the metal to the temperature of the boiling water (100.0°C).
4. While the metal is heating, nest one Styrofoam cup inside the other. This will be your calorimeter. Since Styrofoam is a good insulator, 2 Styrofoam cups make an excellent calorimeter.
5. Measure the mass of this calorimeter with your scale. After you do this, fill the calorimeter about three-quarters full of room-temperature water. Measure the mass again, and determine the mass of the water by difference.
6. Place your thermometer in the water for 3 minutes, then read the temperature. Use the calibration determined in Experiment 12.1 to correct for any errors in your

thermometer. Once the metal has been in the boiling water for 5 minutes, quickly pull it out of the boiling water with the kitchen tongs and transfer it to the calorimeter.

7. Stir the water carefully with your thermometer and periodically read the temperature without lifting the thermometer out of the water. Continue this process until the temperature stops increasing. Write down the final temperature.
8. Now you can use equation 12.1 to determine how much heat the metal transferred to the water. In this case, ignore the heat absorbed by the calorimeter. Use the change in temperature in °C, the mass of the water in grams, and the specific heat of water as 4.184 J/(g · °C). This will then put your answer in Joules. Be sure to use the correct number of significant figures in determining the heat absorbed by the water.
9. Now you can use equation 12.3. Since we are ignoring the calorimeter in the experiment, $q_{calorimeter} = 0$. You calculated q_{water} in the previous step, so you can determine q_{metal}. It will come out negative because the metal *lost* energy.
10. In step 11 you will need to use equation 12.1 to calculate the specific heat of the metal. Before you can do that, you need to know the mass and ΔT of the metal. You measured its mass. What is its ΔT? Since the metal was in boiling water, its initial temperature was 100.0°C. Since it was in contact with the water and the calorimeter at the end of the experiment, its final temperature was the same as the final temperature of the water. With those 2 numbers, you can calculate ΔT. This number will also turn out negative, canceling the negative sign on the q.
11. Now that you have q, m, and ΔT for the metal, you can use equation 12.1 to calculate c, just as we did in Example 12.3. Be sure to keep track of significant figures!
12. Clean up and return everything to the proper place.

CONCLUSION: Write something about what you observed and whether it confirmed your hypothesis. What should your specific heat be? Metals have small specific heats, so it should probably be less than 1 J/(g · °C).

Now that you've seen 2 examples of calorimetry problems and you've done an experiment on calorimetry, you should be able to answer "On Your Own" questions 12.5–12.7.

ON YOUR OWN

12.5 A calorimeter holds 188.0 grams of water at an initial temperature of 23.5°C. A 55.0 g piece of metal at 100.0°C is dropped into the calorimeter, and the final temperature of the water is 28.0°C. What is the specific heat of the metal? Ignore the calorimeter in this problem.

12.6 A calorimeter (m = 8.5 g; c = 2.31 J/g°C) holds 120.0 grams of water at an initial temperature of 23.0°C. A 45.0 g piece of metal at 110.0°C is dropped into the calorimeter, and the final temperature of the water is 29.2°C. What is the specific heat of the metal?

12.7 A calorimetry experiment is performed with a calorimeter that has a specific heat of 37.7 J/g°C. The calorimeter has a mass of 1.4 g and is filled with 125 g of an unknown liquid that is initially at a temperature of 21.7°C. If a 50.0 g chunk of copper at 106.0°C is dropped in the calorimeter and the liquid's temperature increases by 2.8°C, what is the specific heat of the unknown liquid?

This concludes your module on heat energy. However, don't think we've left this topic for good. What you have learned in this module is just an introduction to the fascinating topic of energy. We will revisit this subject again in the next module and delve deeper into the wonders of how energy is used in God's creation. For now, be sure you understand everything you've read by answering the review questions and the practice problems at the end of this module.

SUMMARY OF KEY EQUATIONS AND TABLES IN MODULE 12

Equation 12.1: $q = m \cdot c \cdot \Delta T$

Calculating heat change in terms of specific heat

Equation 12.2: $\Delta T = T_{final} - T_{initial}$ Calculating temperature change

Equation 12.3: $-q_{object} = q_{water} + q_{calorimeter}$ Heat change of the object

Table 12.1: Specific Heats of Common Substances

ANSWERS TO THE "ON YOUR OWN" QUESTIONS

12.1 Gold has a lower specific heat than zinc. This means that gold is easier to heat up because it takes less energy to raise its temperature by 1°C.

12.2 Since both pieces are the same mass, and on the same burner for the same amount of time, they are each getting the same amount of heat. Therefore, the piece made out of the substance which is more difficult to heat will be the cooler one. Since iron has a larger specific heat, it is more difficult to heat and will be cooler than the copper. That means the copper temperature will be higher than 100.0°C.

12.3 As we look at this problem, we can see that we are given q, m, and (from table 12.1) c. If we use equation 12.1, we can calculate ΔT. Once we know ΔT, we can use it and the final temperature (24.0°C) to get the unknown initial temperature. So our first step is to rearrange equation 12.1 to solve for ΔT:

$$\Delta T = \frac{q}{m \cdot c}$$

As mentioned in the hint, since the aluminum rod releases energy, its q is negative. The mass is given, and since we know that the rod is aluminum, we can use table 12.1 to get the specific heat:

$$\Delta T = \frac{-612 \, \cancel{J}}{432 \, \cancel{g} \times 0.9000 \, \frac{\cancel{J}}{\cancel{g} \cdot °C}}$$

$$\Delta T = -1.57°C$$

Now that we have ΔT, we can get the initial temperature by rearranging equation 12.2:

$$T_{initial} = T_{final} - \Delta T = 24.0°C - (-1.57°C) = 25.6°C$$

Since our final temperature had the lowest precision, our answer must also have the same precision. So the initial temperature was 25.6°C.

12.4 In this problem we are given q, m, and ΔT and asked to solve for c. So this is simply a rearrangement of equation 12.1:

$$c = \frac{q}{m \cdot \Delta T}$$

$$c = \frac{85 \text{ J}}{16.0 \text{ g} \times 22°C}$$

$$c = 0.24 \frac{\text{J}}{\text{g} \cdot °C}$$

12.5 We are asked to find the specific heat of the metal. What is needed to solve for c in equation 12.2? We need q, m, and ΔT for the metal. We are given m, and we can calculate ΔT, but q for the metal is not given. What do we do now? Since we can ignore the calorimeter, the heat released by the metal will equal the heat absorbed by the water. So we need to calculate q_{water}:

$$q_{water} = m \cdot c \cdot \Delta T$$

$$q_{water} = 188.0\text{ g} \times 4.184 \frac{\text{J}}{\text{g} \cdot {}^\circ\text{C}} \times (28.0^\circ\text{C} - 23.5^\circ\text{C})$$

$$q_{water} = 188.0\cancel{\text{g}} \times 4.184 \frac{\text{J}}{\cancel{\text{g} \cdot {}^\circ\text{C}}} \times 4.5\cancel{^\circ\text{C}}$$

$$q_{water} = 3{,}500\text{ J}$$

Notice that we have to determine the significant figures of ΔT by looking at precision because we are subtracting at that point. Since both temperatures go out to the tenths place, ΔT must go out to the tenths place as well. When we then multiply m, c, and ΔT, the 2 significant figures in ΔT limit the answer to 2 significant figures. This is a good time to mention that when you get to college, your professor may have a different view of how to handle significant figures. For example, some college texts would say not to round until the very end, so 3 significant figures would be their answer. However, for this course we will follow the pattern laid out in this example.

Since we can assume $q_{calorimeter} = 0$, we can now determine q_{metal}:

$$-q_{metal} = q_{water} + q_{calorimeter} = 3{,}500\text{ J} + 0$$

$$q_{metal} = -3{,}500\text{ J}$$

The value for qmetal is negative because the metal lost energy. We can calculate the ΔT of the metal. Since the metal started out at 100.0°C and ended up at the same final temperature as the water, ΔT is:

$$\Delta T_{metal} = T_{final} + T_{initial} = 28.0^\circ\text{C} - 100.0^\circ\text{C} = -72.0^\circ\text{C}$$

Once again because we are subtracting, the significant figures here are determined by precision. We now have all the information we need to calculate the specific heat of the metal:

$$c = \frac{q}{m \cdot \Delta T}$$

$$c = \frac{-3{,}500\text{ J}}{55.0\text{g} \times (-72.0^\circ\text{C})}$$

$$c = 0.88 \frac{\text{J}}{\text{g} \cdot {}^\circ\text{C}}$$

12.6 Again, we are asked for c_{metal}, but the q_{metal} is not given. The heat released by the metal will be absorbed by the water and calorimeter. We are given the details of the calorimeter in this problem, and we are not told that we can ignore it. To solve this problem, we will need to calculate the heat gained by the water and the heat gained by the calorimeter. Let's start with the calorimeter:

$$q_{calorimeter} = m \cdot c \cdot \Delta T$$

$$q_{calorimeter} = 8.5 \text{ g} \times 2.31 \frac{\text{J}}{\text{g} \cdot {}^\circ\text{C}} \times (29.2^\circ\text{C} - 23.0^\circ\text{C})$$

$$q_{calorimeter} = 8.5 \cancel{\text{g}} \times 2.31 \frac{\text{J}}{\cancel{\text{g} \cdot {}^\circ\text{C}}} \times 6.2 \cancel{{}^\circ\text{C}}$$

$$q_{calorimeter} = 120 \text{ J}$$

Now we can move on to the water:

$$q_{water} = m \cdot c \cdot \Delta T$$

$$q_{water} = 120 \cancel{\text{g}} \times 4.184 \frac{\text{J}}{\cancel{\text{g} \cdot {}^\circ\text{C}}} \times 6.2 \cancel{{}^\circ\text{C}}$$

$$q_{water} = 3{,}100 \text{ J}$$

Now that we have calculated the q's of the water and calorimeter, we can use equation 12.3 to determine the q of the metal.

$$-q_{metal} = q_{water} + q_{calorimeter} = 3{,}100 \text{ J} + 120 \text{ J} = 3{,}220 \text{ J}$$

$$-q_{metal} = 3{,}200 \text{ J}$$

Since the q of the water has its last significant figure in the hundreds place, the answer must be reported to the hundreds place. That's why 3,220 was rounded to 3,200. Now that we have q_{metal}, we can determine ΔT for the metal and then use equation 12.1 to determine the specific heat:

$$\Delta T_{metal} = T_{final} - T_{initial} = 29.2^\circ\text{C} - 110.0^\circ\text{C} = -80.8^\circ\text{C}$$

Notice that this ΔT is negative because the object cooled down. Now we can rearrange equation 12.1 using algebra and solve for the specific heat of the metal:

$$c = \frac{q}{m \cdot \Delta T}$$

$$c = \frac{-3{,}200 \text{ J}}{45.0 \text{ g} \times (-80.8^\circ\text{C})}$$

$$c = 0.88 \frac{\text{J}}{\text{g} \cdot {}^\circ\text{C}}$$

12.7 When we look at this problem, we see that we're trying to discover the specific heat of the liquid in the calorimeter, not the specific heat of the metal, as we did earlier. To find the liquid's specific heat, we need to find out q_{liquid}. How can we do that? We have enough information to calculate q_{metal} and $q_{calorimeter}$, so we can start there.

To calculate q_{metal}, we need to know the ΔT for the metal. To get that, we need to know the initial and final temperatures. The initial temperature is given, but what is the final temperature? The water started out at 21.7°C and increased a total of 2.8°C as a result of the experiment. The final temperature of the water, which is also the final temperature of the metal, is 24.5°C. Instead, the specific heat of the metal can be found in table 12.1 ($c = 0.3851 \frac{J}{g°C}$).

$$q_{metal} = m \cdot c \cdot \Delta T$$

$$q_{metal} = 50.0g \times 0.3851 \frac{J}{g \cdot °C} \times (24.5°C - 106.0°C)$$

$$q_{metal} = 50.0\cancel{g} \times 0.3851 \frac{J}{\cancel{g} \cdot \cancel{°C}} \times (-81.5\cancel{°C})$$

$$q_{metal} = -1{,}570 \text{ J}$$

Notice that the heat is negative. You should expect that because the metal lost energy. We can now determine the heat gained by the calorimeter. Notice that ΔT for the water and calorimeter is given since the problem told us how much the temperature increased. Since it increased by 2.8°C, ΔT = 2.8°C:

$$q_{calorimeter} = m \cdot c \cdot \Delta T$$

$$q_{calorimeter} = 1.4\cancel{g} \times 37.7 \frac{J}{\cancel{g} \cdot \cancel{°C}} \times 2.8°\cancel{C}$$

$$q_{calorimeter} = 150 \text{ J}$$

Now we can use equation 12.3 to determine q_{liquid}. We should use q_{liquid} instead of q_{water} in the calorimetry equation.

$$-q_{metal} = q_{liquid} + q_{calorimeter}$$

$$q_{liquid} = -q_{metal} - q_{calorimeter} = -(-1{,}570 \text{ J}) - 150 \text{ J} = 1{,}420 \text{ J}$$

Since both measurements have their last significant figure in the tens place, the answer must be reported to the tens place. So we can keep our answer as 1,420 J.

We were given the mass of the liquid (125 g) and the ΔT of the liquid (2.8°C), so we can use the q_{liquid} that we just calculated to determine the specific heat of the liquid:

$$c = \frac{q}{m \cdot \Delta T}$$

$$c = \frac{1{,}420\ \text{J}}{125\text{g} \times 2.8°\text{C}}$$

$$c = 4.1\ \frac{\text{J}}{\text{g} \cdot °\text{C}}$$

STUDY GUIDE FOR MODULE 12

REVIEW QUESTIONS

1. A man pushes a stalled car for 2 minutes. During the first 1.5 minutes, the car does not move, but for the last 0.5 minutes the man pushes even harder and succeeds in getting the car to move. Describe if and when work is being done.

2. Classify each of the following as having either potential or kinetic energy, or both:
   a. A gold bar
   b. Sunlight
   c. An apple
   d. A hurricane

3. A chemist watches the temperature of a vat of liquid and notices that it increases over time. What can the chemist conclude from this observation?

4. If a hot object is totally insulated from the outside world (totally unable to interact with anything else), will its temperature ever be able to change?

5. What is experimental scatter?

6. Heat is added to 2 identical objects that are initially at the same temperature. If the first absorbs 50.0 cal and the second absorbs 50.0 J, which gets hotter?

7. Some diet books say that an excellent way to lose weight is to drink ice-cold water. Why?

8. Examine the heating curve for the following unknown substance:

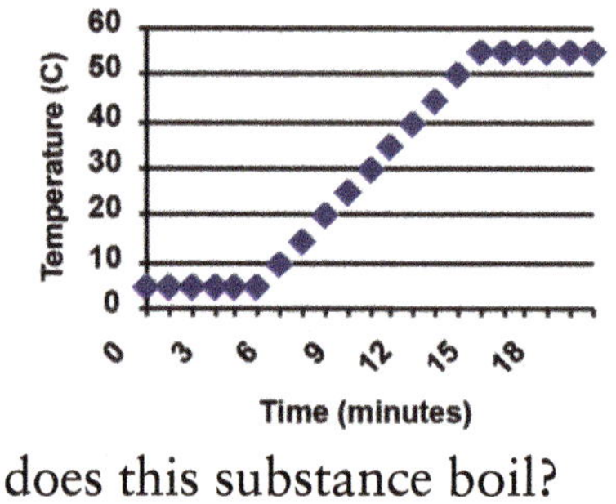

   At what temperature does this substance boil?

9. The specific heat of aluminum is roughly 2 times that of iron. Equal masses of aluminum and iron start out at room temperature. Heat is added to each mass at the same rate for the same amount of time. The iron's temperature increased by 900°C. What was the temperature increase of the aluminum?

10. Why should a calorimeter always be made from an insulating material?

PRACTICE PROBLEMS

1. The average person uses 2,500.0 Cal of energy per day. How many Joules is that? Remember that a food calorie (Cal) is 1,000 chemistry calories (cal).

2. How many Joules does it take to heat 15.1 kg of glass from 15°C to 45°C? (Refer to table 12.1.)

3. 124.1 g of an unknown substance absorbs 50.0 kJ of heat and increases its temperature by 36.3°C. What is its specific heat? (Remember that k is the abbreviation for kilo, so kJ stands for kiloJoules.)

4. A 245 g piece of copper at room temperature (25°C) loses 456.7 Joules of heat. What will its final temperature be?

5. Review question 7 mentioned that drinking ice-cold water is a way of burning excess Calories. Calculate how many Calories are burned when a 12-ounce (3.40×10^2 g) glass of water at 0.0°C is warmed up to body temperature (37.0°C).

6. How many grams of silver are needed to absorb 1.00 kJ of energy changing the temperature from 33.0°C to 41.8°C?

7. A calorimeter is filled with 150.0 g of water at 24.1°C. A 50.0 g sample of a metal at 100.0°C is dropped in this calorimeter and causes a temperature increase of 5.4°C. What is the specific heat of the metal? Ignore the calorimeter in this problem.

8. Now let's take the calorimeter into consideration for problem 7. The calorimeter has a mass of 2.0 g and a specific heat of 3.5 J/(g · °C). Recalculate the specific heat of the metal.

9. A 150.0 g sample of copper [0.3851 J/(g · °C)] was heated to 200.0°C. It is dropped into a 5.0 g calorimeter [c = 50.00 J/(g · °C)] with 175.0 g of liquid. The liquid was initially at 25.0°C, and the temperature increased to 31.2°C by the end. What is the specific heat of the liquid?

10. A 345.1 g sample of copper at 100.0°C is dropped into a 4.5 g calorimeter made of an unknown substance. If the calorimeter has 150.0 grams of water in it and the temperature changed from 24.2°C to 25.1°C, what is the specific heat of the calorimeter? (Refer to table 12.1.)

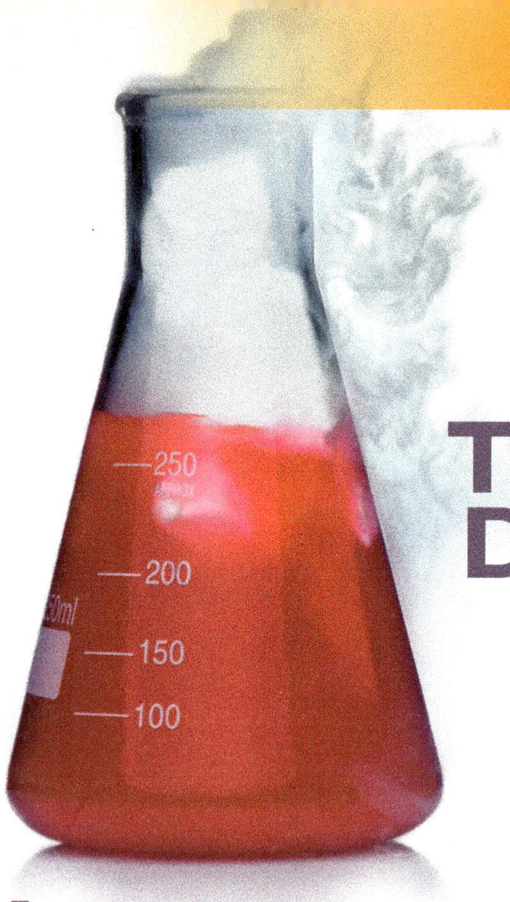

THERMO-DYNAMICS

In Module 10 we learned about endothermic and exothermic processes. Specifically, when solutes dissolve exothermically, they release energy, causing the solution to heat up. Similarly, when solutes dissolve endothermically, they absorb energy from their surroundings, causing the solution to cool down. Remember that when a solute dissolves in a solvent, no *chemical* change is occurring. In Module 6 we learned that when things dissolve, they undergo a *physical* change. In this module, we will examine exothermic and endothermic *chemical* reactions.

first reactions

Doing things requires energy. Stand up. That took energy. Sit down. That too took energy. Think about the energy of reactions. Sometimes a reaction produces heat, and a beaker will feel hot. Sometimes a reaction absorbs heat, and a beaker will feel cold. You've learned, however, that energy is neither created nor destroyed. It only changes form. It's a fascinating fact about creation. "Is there anything of which one can say, 'Look! This is something new?' It was here already, long ago; it was here before our time" (Ecclesiastes 1:10, NIV).

As you might expect, chemical changes can also be accompanied by either a release or absorption of energy. In fact, almost all chemical changes either release or absorb energy, so the definitions of *exothermic* and *endothermic* can also be applied to chemical reactions. In this module, we will learn how this is done.

ENTHALPY

Up to this point, our discussions of chemical reactions have focused on the substances that participate in those reactions. It is now time to turn our attention to the energy change that accompanies reactions. As mentioned above, chemical reactions can be exothermic or endothermic. If a reaction is exothermic, energy is released in the process. This means that we can think of energy as a *product* in the chemical reaction. After all, if energy is released during the reaction, it is as if the reaction produced energy. In the same way, if a chemical reaction is endothermic, we can think of energy as a reactant in the chemical equation because heat is required to make the reaction go.

Remember, the first law of thermodynamics tells us that energy cannot be created or destroyed. Energy is never really produced by a chemical reaction, and neither is it really consumed. Nevertheless, for the purpose of our discussion, we will think of energy and chemical reactions in the following way:

When a reaction is exothermic, energy will be thought of as a product in the reaction. If the reaction is endothermic, energy will be considered a reactant.

Once again, make sure you understand that this is just a convenient way of thinking about the energy changes that accompany chemical reactions. It is not really correct because reactants in a chemical reaction are consumed and products are produced. The first law of thermodynamics tells us that energy cannot be consumed or produced; it can only change forms.

If chemical reactions don't really produce or consume energy, where does the energy come from? When we perform an exothermic chemical reaction, the entire reaction will get hot. Since energy is required to raise the temperature of any object, we know that energy had come from somewhere. If the reaction didn't really produce the energy, how did it get there? The answer to this requires us to remember something that we learned in the last module.

Remember when we first discussed energy? We learned that there are 2 types of energy: potential and kinetic. Potential energy is defined as energy that is stored, while kinetic energy is defined as energy in motion. In chemistry, we can think of potential energy as the energy that is stored in the chemical bonds of a substance. Kinetic energy, on the other hand, can be thought of as the heat associated with absorbing or releasing during a chemical reaction. The substances that are a part of the chemical reaction contain potential energy.

If we think about things in this way, we can finally see where the energy associated with a chemical reaction comes from. If a chemical reaction is exothermic, some of the potential energy that is stored in the reactants is converted into kinetic energy. This kinetic energy then heats the things surrounding the reaction, making the reaction hot. On the other hand, if a chemical reaction is endothermic, some of the kinetic energy in the surroundings is absorbed and converted into the potential energy stored in the products of the reaction. Since this situation causes the things surrounding the chemical reaction to lose kinetic energy, they feel cold.

This is the real explanation for where the energy in a chemical reaction comes from. Potential energy in the chemical bonds of the substances involved can be converted into kinetic energy or vice versa. However, we will still use the convention that exothermic reactions have energy as a product and endothermic reactions have energy as a reactant. Even though it is not physically accurate, this convention is too convenient to ignore.

If this discussion has all been a bit too theoretical for you, let's bring it down to practical terms with a couple of examples. If we react HCl (which you should recognize as an acid) with NaOH (which you should recognize as a base), the following chemical reaction will occur:

$$HCl + NaOH \rightarrow H_2O + NaCl$$

Based on our discussion of acid-base chemistry in Module 9, you should have been able to predict the products of this reaction, given only the reactants. If we were to perform this reaction, we would feel the reaction vessel get very hot because this reaction is exothermic. Since we now know this is an exothermic reaction, we could rewrite the above equation as follows:

$$HCl + NaOH \rightarrow H_2O + NaCl + \text{energy}$$

This is what we mean when we say that we will consider energy to be a product in an exothermic reaction.

In the same way, if we mix NH_4SCN and $Ba(OH)_2(H_2O)_8$ together, the following reaction will occur:

$$2NH_4SCN + Ba(OH)_2(H_2O)_8 \rightarrow 2NH_4OH + Ba(SCN)_2 + 8H_2O$$

When we perform this reaction, we will feel the reaction vessel get cold, indicating that this is an endothermic reaction. Since we know it is endothermic, we could rewrite the equation in the following way:

$$2NH_4SCN + Ba(OH)_2(H_2O)_8 + \text{energy} \rightarrow 2NH_4OH + Ba(SCN)_2 + 8H_2O$$

This is what we mean when we say that we will consider energy as a reactant in endothermic reactions.

Now that we understand how to think about the energy changes that accompany chemical reactions, it is time to learn some more terminology. Every substance in nature contains potential energy. The energy stored in a substance is called the enthalpy (en' thuhl pee) of that substance. For some odd reason, chemists use H to symbolize enthalpy. In a chemical reaction, substances are both consumed and produced. As a result, enthalpy changes in a chemical reaction. When this happens, energy is either released or absorbed. We call the energy associated with a chemical reaction (the same energy that appears in the previous equations) the change in enthalpy.

Change in enthalpy (ΔH)—The energy change that accompanies a chemical reaction.

Since enthalpy is abbreviated with an H, the change in enthalpy is symbolized by ΔH.

Because chemists often get very sloppy with our terminology, even though ΔH represents the change in enthalpy, we often refer to it as just enthalpy. So we might call the ΔH of a particular reaction the enthalpy of the reaction, when it really should be referred to as the change in enthalpy of that reaction. You will therefore have to be comfortable with ΔH being referred to in both ways. Calling it enthalpy is wrong because enthalpy refers to the energy associated with a *particular* substance, not a group of substances that undergo change. Nevertheless, this sloppy terminology exists in chemistry, so you must get used to it.

Remember, ΔH represents the energy change associated with a chemical reaction. As a result, it will have an energy unit attached to it. Remember from the last module, that the 2 major energy units are Joules and calories. These units are related to each other as follows:

$$1 \text{ calorie} = 4.184 \text{ Joules} \qquad \text{Equation 13.1}$$

Since most chemical reactions involve quite a bit of energy, the units you will probably see attached to ΔH are either kiloJoules (kJ) or kilocalories (kcal).

However, kJ and kcals are not the only units you will see attached to ΔH. As you might expect, the energy released or absorbed by a chemical reaction depends on the *amount* of reactants that are consumed. After all, if we burn 1.00 kg of wood, we will produce more heat than if we burn 1.00 g of wood. As a result, the value of ΔH depends on the amount of reactants undergoing the chemical change. To account for this, chemists sometimes define the ΔH of a reaction on a mole-by-mole basis. You will sometimes see ΔH with the units kJ/mole or kcals/mole. These units indicate that the energy change represented by ΔH occurs when 1 mole of reactant gets consumed in the reaction. Therefore, a ΔH of 50 kJ/mole simply means that after 1 mole of reactant undergoes the chemical change, there is an energy change of 50 kJ.

You not only have to recognize the units that are attached to ΔH, but you also need to understand that ΔH can be either positive or negative. By convention, we say that:

ΔH is positive for endothermic reactions, and
ΔH is negative for exothermic reactions.

Table 13.1 gives another way to remember this relationship.

TABLE 13.1
ΔH Relationships in Reactions

Reaction is	ΔH is	Heat is	Potential energy stored in the reaction's products is
Endothermic	Positive (+)	Absorbed	Greater than the reactants
Exothermic	Negative (-)	Released	Less than the reactants

If we are told that a reaction has a ΔH of -123 kJ/mole, we know that the reaction is exothermic (ΔH is negative), and we can think of the reaction as producing 123 kJ of energy for every mole of reactant consumed. On the other hand, a reaction with a ΔH of 56 kJ/mole is endothermic and will have 56 kJ as a reactant. Let's see how we use all of this terminology in Example 13.1.

EXAMPLE 13.1

When CH_4 undergoes complete combustion, the change in enthalpy of the reaction is -803.1 kJ. Write a balanced chemical equation representing this process. Include energy in your equation.

This problem first requires us to remember what a complete combustion reaction is. Remember that complete combustion was defined as reacting a substance with oxygen to produce carbon dioxide and water. This means that the unbalanced chemical equation is:

$$CH_4 + O_2 \rightarrow CO_2 + H_2O$$

Balancing the equation gives us:

$$CH_4 + 2O_2 \longrightarrow CO_2 + 2H_2O$$

According to the example, ΔH is negative, implying that the reaction is exothermic and the energy is a product:

$$CH_4 + 2O_2 \longrightarrow CO_2 + 2H_2O + \textbf{803.1 kJ}$$

ON YOUR OWN

13.1 Write balanced chemical equations for each of the following processes, being sure to include energy as either a reactant or a product:

a. The formation of ammonium nitrate, which has a ΔH of -365.1 kJ.

b. The decomposition of manganese (IV) oxide, which has a ΔH of 519.7 kJ.

c. The reaction between H_2CO_3 and KOH, which releases 21 kcals.

Answer "On Your Own" question 13.1, which also provides you with some review because it requires you to predict certain chemical equations. If you don't remember how to predict formation, decomposition, and combustion equations, refer to Module 7. If you don't remember how to figure out the acid-base reaction, refer to Module 9.

DETERMINING ΔH FOR A CHEMICAL REACTION BY EXPERIMENT

Now that you understand what the enthalpy of a chemical reaction is and how it relates to the chemical equation, the next logical step is to figure out how we can measure it. After all, in both the example and the "On Your Own" question from the last section, we told you the ΔH for several different chemical reactions. How did we figure them out? Well, of course, we looked in a few books. But how did the authors of those books measure each ΔH?

There are 3 ways to determine the ΔH of a chemical reaction. The first method is experimental. In the last module, you did an experiment using calorimetry. This technique can be applied to chemical reactions as a method of determining ΔH. In a calorimetry experiment, we measure the temperature change of water inside a calorimeter. If we run a chemical reaction inside a calorimeter, then we can use that temperature change to calculate the energy absorbed or released inside the calorimeter. To refresh your memory on calorimetry and to learn how ΔH can be experimentally determined, perform Experiment 13.1.

EXPERIMENT 13.1

PURPOSE: To determine the ΔH of a chemical reaction.

MATERIALS:

- 2 Styrofoam coffee cups
- Thermometer

- Vinegar
- Mass scale
- Measuring tablespoon and ½ teaspoon
- Lye (This is commonly sold with the drain cleaners in hardware stores and supermarkets. Make sure that your bottle is labeled 100% lye. If you cannot find lye, you can order it online. It may be listed under its chemical name, sodium hydroxide.) **Always use this chemical in a well-ventilated area.**
- Safety goggles

QUESTION: Can you determine the ΔH of a chemical reaction?

HYPOTHESIS: Write a hypothesis for this experiment.

PROCEDURE:

1. To perform this experiment, you must be able to measure a relatively small mass of lye. The problem is that your mass scale is not very good at measuring small masses. What you need to do is figure out how many grams are in a teaspoon of lye. Then you can use measuring spoons as a quick way of measuring a small mass of lye. To do this, measure the mass of 10 tablespoons of lye.
2. Remembering that each tablespoon represents 3 teaspoons, realize that the mass you just measured is the mass of 30 teaspoons of lye. Now divide that mass by 30. The result is the mass of 1 teaspoon of lye. You should recognize this process as another example of calibration. You have just calibrated your measuring spoons to tell you the mass of lye.
3. In a moment, you will add ½ teaspoon of lye to some vinegar. Calculate what mass that is by taking the mass of 1 teaspoon of lye and dividing by 2. This will be the mass of lye that you will use in the experiment.
4. Now you can begin the real experiment. Nest the 2 coffee cups as you did in experiment 12.2 to make a calorimeter.
5. Pour 100.0 mL (½ cup) of vinegar into your calorimeter.
6. Place your thermometer in the vinegar and let it sit for 3 minutes. After that time, record the temperature of the vinegar. This will be the initial temperature in the experiment.
7. Add ½ teaspoon of lye to the vinegar, and begin stirring the mixture with your thermometer.
8. As time goes on, the following reaction will take place:

$$\underset{\text{lye}}{NaOH\ (s)} + \underset{\text{acetic acid in vinegar}}{C_2H_4O_2\ (aq)} \rightarrow H_2O\ (l) + NaC_2H_3O_2\ (aq)$$

9. This reaction is exothermic; therefore, you should see the temperature begin to rise. Read the temperature every 30 seconds. Once the temperature falls or stays constant for 2 consecutive readings, you can end the experiment.
10. Now you are ready to do your calculations. Remember, calorimetry experiments use the change in temperature to determine the heat transferred. The equation we use is:

$$q = mc\Delta T$$

In this equation, q is the amount of heat transferred, m is the mass of the sample being heated, c is the specific heat capacity of the thing being heated, and ΔT is the change in temperature. If you recall your calorimetry experiment from the last module, you will remember that we can sometimes ignore the calorimeter. You will do that here, so all you need to consider is the contents of the calorimeter.

11. The specific heat (c) of vinegar is 4.1 J/g °C.
12. The mass of the vinegar in the calorimeter will be its volume (100.0 mL) times its density (which equals 0.99 g/mL). In addition to this mass, you must add the mass of ½ teaspoon of lye because that was also in the calorimeter. In the end, then, the mass of the calorimeter contents is the mass of the vinegar plus the mass of the lye added.
13. Subtract the initial temperature in the experiment from the final temperature to get ΔT.
14. Now you can calculate the heat absorbed by the contents of the calorimeter. Multiply the specific heat given in step 11 by the mass you calculated in step 12, and then multiply it by the ΔT you calculated in step 13. This gives you the heat absorbed by the calorimeter's contents.
15. Since the heat absorbed by the calorimeter's contents came from the chemical reaction, you have just calculated the ΔH of the reaction.
16. You aren't quite finished, however. The ΔH you calculated was the amount of energy released when the mass of lye that *you used in the experiment reacted with vinegar.* It would be more useful to determine the ΔH per mole of NaOH that reacts. To do that, all you have to do is take the ΔH you just calculated and divide it by the number of moles of NaOH that you used.
17. Take the mass of lye and divide it by the molar mass of NaOH. That will give you the number of moles of NaOH that you used in the reaction.
18. Divide the ΔH that you calculated in step 14 by the number of moles you calculated in step 17. The result is the ΔH of the reaction in J/mole.
19. Clean up and return everything to the proper place.

CONCLUSION: Write something about the outcome and your hypothesis.

We wanted you to perform Experiment 13.1 to get you to appreciate how ΔH has been measured over the years. Hundreds of thousands of ΔH measurements are compiled in today's chemistry literature. This means that chemists have done hundreds of thousands of experiments just like the one you performed to provide this information. You should now appreciate how much work this is! Finally, performing the experiment should have shown you how useful calorimetry is for the pursuit of chemical knowledge. Now that we are done with all that, it is time for you to learn some easier methods for determining the ΔH of a chemical reaction.

DETERMINING THE ΔH OF A CHEMICAL REACTION USING BOND ENERGIES

When a chemical reaction occurs, the valence electrons of the atoms that make up the reactants must rearrange themselves into the configurations necessary to make the products This process requires a lot of chemical bond breaking and chemical bond making. For example, let's go back to the burning of methane:

$$CH_4 + 2O_2 \rightarrow CO_2 + 2H_2O$$

The Lewis structures of the molecules in this equation would look like this:

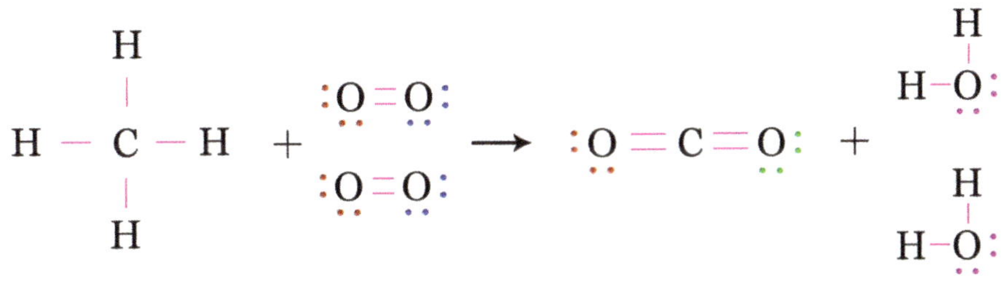

Looking at the equation in this form, we can see that for the reactants to turn into the products, several bonds have to be broken and several others have to be formed.

For example, look at the CH_4 molecule. It contains 1 carbon atom bonded to 4 hydrogen atoms by 4 separate single bonds. To form the products, those 4 bonds must be broken. That way, the hydrogen atoms will be free to bond with the oxygen atoms to make water, while the carbon atom will be free to bond to the other oxygen atoms to make carbon dioxide. For this to happen, of course, the double bonds that hold the 2 oxygen atoms together in each oxygen molecule must also be broken. Therefore, to make this chemical reaction work, all the bonds on the reactants side of the equation must be broken.

Once the bonds are broken, the job is only halfway done. Now all the bonds on the products side must be formed. This means that 2 double bonds between the carbon atom and the oxygen atoms in carbon dioxide must be formed. In addition, 4 single bonds between the hydrogen atoms and the other oxygen atoms must be formed to make 2 water molecules. After all this happens, the chemical reaction is finished.

Think for a moment about the energy required to make all this happen. First, it takes energy to break bonds. This should make sense to you. After all, to break anything, we have to expend some energy. As the bonds on the reactants side of the equation are being broken, energy is being absorbed by the reactants. What may not make as much sense to you (but is nevertheless true) is that when bonds are formed on the products side of the equation, energy is released. Each time a bond forms on the products side of the equation, the product being formed releases energy. If we look at the balance between the energy absorbed and the energy released, we will see why some chemical reactions are endothermic and others are exothermic.

For example, suppose a chemical reaction involves reactants with very strong bonds and products with very weak bonds. Since the reactant bonds are strong, it takes a lot of energy to break them. So the reactants tend to absorb a lot of energy. On the products side of the reaction, the bonds are weak. This means that when they are formed, the products release only a small amount of energy. In this situation, the reactants had to absorb lots of energy, and the products only released a little. On balance, the chemical reaction absorbed more energy than it released. This reaction would be considered endothermic.

On the other hand, consider a chemical reaction in which the reactants contain weak bonds and the products contain strong bonds. In this situation, the reactants would not need to absorb much energy because their bonds would be easy to break. On the other hand, when strong bonds are formed, a lot of energy is released. Therefore, this chemical reaction would release more energy than it absorbed. This makes the reaction, on balance, an exothermic reaction.

One thing that you should see from this discussion is that whether a reaction is exothermic or endothermic, energy is *both absorbed and released during the course of the reaction*. The difference between exothermic and endothermic reactions depends on how much energy is absorbed and how much energy is released. In exothermic reactions, more energy is released than absorbed. In endothermic reactions, more energy is absorbed than released.

Of course, for any of this discussion to be of practical use, there has to be some way of determining whether a set of chemical bonds is weak or strong. If we want to predict whether or not a chemical reaction is exothermic or endothermic, we need to know how strong all the bonds in the reaction are. How is that done? Believe it or not, chemists can measure the strength of a chemical bond, which is typically called the bond energy. The process by which bond energies are measured is a bit beyond the scope of this course, but the bond energies themselves are easily understood and applied to this course. Table 13.2 summarizes several chemical bond energies.

TABLE 13.2
Bond Energies

Chemical Bond	Bond Energy (kJ/mole)	Chemical Bond	Bond Energy (kJ/mole)
C-H	411	H-H	432
C-Cl	327	H-O	459
C-C	346	H-Cl	428
C-N	305	O=O	494
C-O	358	N≡N	942
C=C	602	N-H	386
C=O	799	Cl-Cl	240
C≡C	835	Br-Br	190

Notice the unit used to measure bond energy. It is the same as the unit used to measure ΔH: kJ/mole. Realize, of course, that the unit could also be kcals/mole or, for that matter, any energy unit divided by moles. What does this unit mean when applied to the measurement of bond strength? It tells how much energy is required to break 1 mole of those bonds or how much energy will be released when 1 mole of those bonds is formed.

For example, suppose we had 1 mole of Cl_2 molecules. We could write the Lewis structure of Cl_2 and find out that the 2 Cl atoms that make up this molecule are held together with 1 Cl-to-Cl single bond. According to our table, the energy of that bond is 240 kJ/mole. This means 2 things. First, it means that if we wanted to break apart 1 mole of Cl_2 molecules, we would have to expend 240 kJ of energy to break all their bonds. Likewise, this number also tells us that if we formed 1 mole of Cl_2 molecules by letting 2 moles of Cl atoms bond together, 240 kJ of energy would be released in the process. See how the bond energy tells us not only how much energy is required to break bonds, but also how much energy is released to form bonds?

Before we see how these bond energies can be applied to learn about the energetics of a chemical reaction, we want to point out that you needn't memorize anything on table 13.2. No chemist has these things memorized, but a good chemist knows exactly where to look them up. That's what we will expect of you. Throughout the rest of this module, several more tables will be presented. You will not be required to know any of these tables, but you will be required to look at them to solve problems. All the tables you will need to solve the practice problems and the test are included in appendix A near the end of the text. Feel free to use that appendix whenever you solve any problems, including the ones on the test.

Now that we have this table of bond energies, how do we use it to analyze the energetics of a chemical reaction? It's relatively simple. As mentioned before, to determine whether or not a chemical equation is exothermic or endothermic, all we have to do is figure out how much energy is absorbed to break the reactants' bonds and how much energy is released when the products' bonds form. The difference between the 2 will be the ΔH of the reaction. Mathematically:

ΔH = Energy required to break bonds - Energy released when bonds form   Equation 13.2

The table of bond energies gives us all the information we need to use equation 13.2 as long as we know what bonds are breaking and what bonds are forming. To do this, of course, we need to be able to look at a chemical equation in terms of its Lewis structures. That's the only way we will know what kinds of bonds are broken and what kinds of bonds are formed. This means to solve the problems in this module, you will need to be able to do Lewis structures. If you're a bit shaky on those, review Module 4.

Before we show you a couple of examples of this kind of problem, we need to point out one last thing. Look at equation 13.2 for a moment. Suppose the energy absorbed to break bonds is greater than the energy released when bonds are formed. In this case, the chemical reaction will be endothermic. What sign will ΔH have? It will be positive. On the other hand, in an exothermic reaction, the energy absorbed to break bonds is less than the energy released when bonds are formed. When this happens, equation 13.2 tells us that ΔH will be negative and is the same as our previous designations of the signs of ΔH. If we calculate the ΔH from equation 13.2 and it ends up positive, this tells us that the reaction is endothermic. A negative ΔH indicates an exothermic reaction. Let's see how this works in Example 13.2.

EXAMPLE 13.2

To make ammonia, nitrogen and hydrogen are reacted at high temperatures according to this equation:

$$N_2(g) + 3H_2(g) \longrightarrow 2NH_3(g)$$

What is the ΔH for this reaction? Is it endothermic or exothermic?

This problem requires the use of equation 13.2. To use this equation, we need to know what bonds are being broken and what bonds are being formed. So we need to rewrite the chemical equation in terms of Lewis structures:

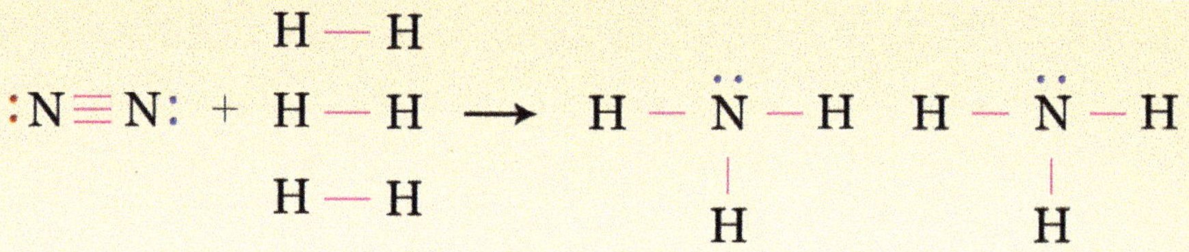

Make sure you understand each of these Lewis structures. You have to be able to draw Lewis structures to solve this kind of problem. Also, realize why we drew 3 Lewis structures of H_2 and 2 Lewis structures of NH_3. We did that because the chemical equation tells us that we have 3 H_2 molecules and 2 NH_3 molecules, so we had to draw that many Lewis structures to be able to see all the bonds involved in this chemical reaction.

Now that we can see all the bonds involved in the chemical equation, we can look in table 13.2 and find the energy of each bond. The first bond we see is a nitrogen-to-nitrogen triple bond. According to table 13.2, this bond has an energy of 942 kJ/mole. The other bonds involved in the reaction are an H-H bond (energy is 432 kJ/mole) and an N-H bond (energy is 386 kJ/mole). Now that we have the bonds involved in the chemical reaction and their energies, we are ready to use equation 13.2.

Remember, though, to calculate the total amount of energy required to break all the reactant bonds, we need to count *how many* of each bond we need to break. According to our Lewis structure equation, there is 1 N≡N bond, so the energy required to break that bond is (1 mole) x (942 kJ/mole). There are 3 H-H bonds. Since each individual H-H bond has an energy of 432 kJ/mole, the total amount of energy needed to break these bonds is (3 moles) x (432 kJ/mole). Finally, when we look at the products side, we see a total of 6 N-H bonds. Each of those bonds has an energy of 386 kJ/mole. So, the products release (6 moles) x (386 kJ/mole) of energy when they form. Equation 13.2 looks like this:

$$\Delta H = [(1\,\cancel{\text{mole}} \times 942\,\frac{\text{kJ}}{\cancel{\text{mole}}}) + (3\,\cancel{\text{moles}} \times 432\,\frac{\text{kJ}}{\cancel{\text{mole}}})] - (6\,\cancel{\text{moles}} \times 386\,\frac{\text{kJ}}{\cancel{\text{mole}}}) = -78\text{ kJ}$$

The change in enthalpy for this reaction is **-78 kJ**, and since ΔH is negative, the reaction is **exothermic**.

Notice a couple of things about the way we solved this problem. First, let's deal with the units. Remember, the stoichiometric coefficients in a chemical equation refer to the number of moles. In the equation, 1 mole of N_2 reacts with 3 moles of H_2 to make 2 moles of NH_3. In the multiplication, the moles cancel, leaving just kJ as the unit for ΔH. This is typically the way we report the change in enthalpy for a reaction. We simply list the energy produced when the reactants are combined in exactly the mole amounts given by the stoichiometric coefficients.

Now let's deal with the significant figures. Remember, the stoichiometric coefficients in a chemical equation are exact. What we are really doing in equation 13.2 is adding and subtracting. We are adding one 942 with three 432s and then subtracting six 386s. As a result, we must use the rule of addition and subtraction when determining the significant figures here. Since all of the bond energies have their last significant figure in the ones place, our answer must be reported to the ones place. That's why the answer is -78 kJ.

The complete combustion of methyl alcohol (CH_3OH) is used to power some experimental automobiles. What is the ΔH of that reaction?

In this problem, we are not even given a chemical equation, so our first job is to get an equation. According to the definition of complete combustion, the unbalanced chemical equation is:

$$CH_3OH + O_2 \rightarrow CO_2 + H_2O$$

Balancing the equation gives us:

$$2CH_3OH + 3O_2 \rightarrow 2CO_2 + 4H_2O$$

Next, we must draw Lewis structures for each molecule. Before we do that, look at the way methyl alcohol's chemical formula has been written to help you make the Lewis structure. We could have referred to methyl alcohol as CH_4O. Often chemists will write a chemical formula in such a way as to illustrate the Lewis structure. That's what we have done here. The 3 H atoms are grouped together next to the C atom to make it clear that those 3 H atoms are bonded to that C atom. The last H atom is written after the O atom to make it clear that it is bonded to the O atom. This should help you in writing the Lewis structure:

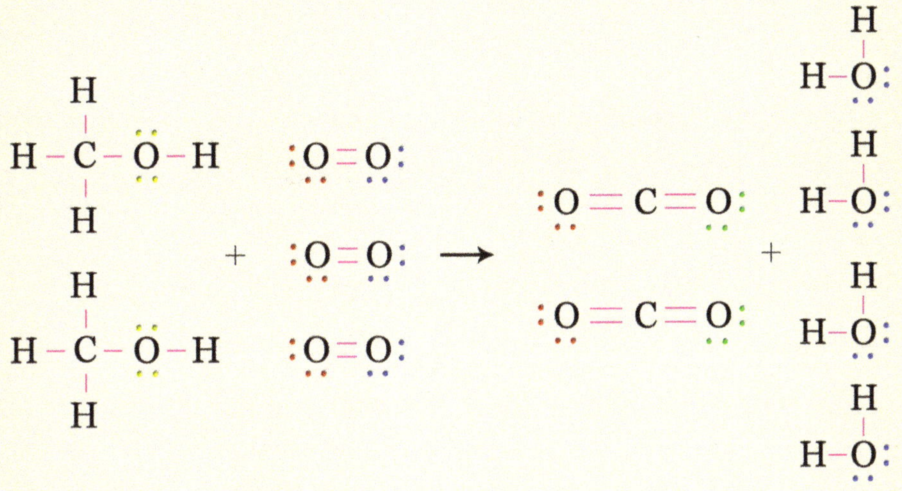

Now that we have the Lewis structures, we can count up all of the bonds, look up the bond energies in table 13.2, and apply equation 13.2.

On the reactants side of the equation, there are 6 C-H bonds (each worth 411 kJ/mole), 2 C-O bonds (each worth 358 kJ/mole), 2 O-H bonds (each worth 459 kJ/mole), and 3 O=O bonds (each worth 494 kJ/mole).

On the products side of the equation, there are 4 C=O bonds (which will release 799 kJ/mole each) and 8 O-H bonds (which will release 459 kJ/mole).

In the end, then, equation 13.2 looks like this:

$$\Delta H = [(6 \text{ moles}) \times (411 \frac{\text{kJ}}{\text{mole}}) + (2 \text{ moles}) \times (358 \frac{\text{kJ}}{\text{mole}}) + (2 \text{ moles}) \times (459 \frac{\text{kJ}}{\text{mole}}) +$$

$$(3 \text{ moles}) \times (494 \frac{\text{kJ}}{\text{mole}})] - [(4 \text{ moles}) \times (799 \frac{\text{kJ}}{\text{mole}}) - (8 \text{ moles}) \times (459 \frac{\text{kJ}}{\text{mole}})]$$

$$\Delta H = -1{,}286 \text{ kJ}$$

In the end, then, **ΔH = -1,286 kJ** for this reaction, which makes it exothermic. Of course, you should have already realized that this was an exothermic reaction because combustion is always exothermic.

Notice a couple of things in these 2 problems. First, we refer to the oxygen-to-hydrogen bonds as O-H bonds, but we use the bond energy in table 13.2 for the H-O bonds because the bond energy is independent of the order of the atoms. The H-O bond energy is the same as the O-H bond energy. Second, notice that we subtract each of the terms from the products side. It is important to remember to do that. Each term from the reactants side must be added to the equation, while each term from the products side must be subtracted.

Now that you have seen how we use table 13.2 and equation 13.2 to calculate ΔH, make sure you understand the process by answering "On Your Own" questions 13.2-13.3. Remember, one place you can really mess up is in making your Lewis structures. Be very careful that the Lewis structures you are using are correct. Otherwise, your answer will definitely be wrong!

ON YOUR OWN

13.2 Carbon tetrachloride is a widely used industrial solvent. What is the ΔH of the formation reaction? Is it exothermic or endothermic?

13.3 In an acetylene torch, C_2H_2 is burned to make the hot, blue flame that cuts or welds metal. Assuming that the combustion is complete, what is the ΔH of this reaction?

HESS'S LAW

Before we begin this section, it is necessary to warn you that the discussion which follows gets a little difficult. If you feel lost while you are reading, don't worry too much about it. Read the section through and pick up as much as you can. Before we get to the example problems, we will summarize what you need to know from this section, which is significantly less than what we are going to cover. In fact, what you need to know is rather simple, even though the discussion that leads up to it is not. However, for a complete discussion of the topic, it is necessary to provide you with all the following information, and your ability in chemistry will determine how much of the information you actually absorb.

So far, we have discussed one experimental method of determining the ΔH of a reaction and one mathematical or theoretical method. Remember that we said there were 3 ways to measure the ΔH of a reaction. The third method makes use of Hess's law, named after Germain Henri Hess, a famous Russian chemist whose experimental work led to its development:

Hess's law—Enthalpy is a state function.

A state function is any quantity that depends solely on the final destination, not on the way we get to that destination. For example, Busch Gardens is an amusement park in Williamsburg, Virginia. People from all over the mid-Atlantic United States come to this park for a fun-filled day of rides, shows, and other attractions. Two people can leave from the same city and take completely different routes to the park. One of them may take a route that is made up mostly of highways. Most likely, he will have to travel a little out of the way to stay on the highway, but since highway travel is faster, he might end up saving time. On the other hand, if someone wants to take the straightest route to Busch Gardens, he will have to travel the back roads. One person might travel the highway route, and the other might travel the back roads. In the end, they will both get to Busch Gardens and have the same amount of fun. In this case, fun could be described as a state function. Regardless of the route that each person took, they both got to the same location (Busch Gardens) and therefore had the same amount of fun once they arrived.

In this scenario, fun might be a state function, but many other things will not be state functions. For example, the number of miles traveled will not be a state function. The person who took the straightest route to Busch Gardens will have traveled fewer miles than the person who took the highway route. Time will also not be a state function. Most likely, one of the routes to Busch Gardens will take longer than the other. So some of the quantities associated with a trip to Busch Gardens depend on the path taken and therefore are not state functions, while other quantities are independent of the path taken and therefore are state functions.

Hess's law tells us that enthalpy is also a state function. In other words, the amount of energy contained in a substance is independent of how it is made. For example, each of the chemical equations below represents a way of making water:

$$2H_2 + O_2 \rightarrow 2H_2O$$

$$2H_2O_2 \rightarrow 2H_2O + O_2$$

Suppose we made 2 moles of water using hydrogen (H_2) and oxygen (O_2) in the first equation and measured the enthalpy of that water. If we used peroxide (H_2O_2) from the second equation instead and isolated just the water, the enthalpy of that water would be exactly the same as the enthalpy of the water we made using hydrogen and oxygen in the first equation. Both chemical reactions represent 2 different paths that we could use for making water. Regardless of which path we used, the enthalpy of the water we created would be exactly the same. That's what Hess's law tells us.

How does that help us calculate the ΔH of a chemical reaction? Well, remember what ΔH represents. It represents the *change in enthalpy* that occurs within a chemical reaction. If we somehow knew the enthalpy of each substance in a chemical equation, we could simply take the difference of the enthalpies of the products and the enthalpies of the reactants to get the change in enthalpy of the reaction, which would be ΔH. Hess's law tells us that the enthalpy of a given substance is *independent* of the way it was made. Therefore, if we have some standard way of measuring the enthalpy of substances, we could use that information to calculate the ΔH of any reaction that contained those substances, regardless of how they were made in preparation for that particular chemical reaction.

It is very hard to actually measure the enthalpy of an individual substance. The only thing that chemists can measure is the change in enthalpy that occurs in a chemical reaction. You might think that this whole discussion was in vain because we can't actually measure the enthalpy of an individual substance. However, there is a way that we can get around this problem. It's the enthalpy of formation (abbreviated as ΔH_f):

Enthalpy of formation (ΔH_f)—The ΔH of a formation reaction.

Remember that a formation reaction is defined as a reaction that forms a single substance from simpler components. Therefore, the enthalpy of formation of any given substance is just the ΔH of the formation reaction that creates that substance. This, of course, is something that can be measured experimentally. All we have to do is run the formation reaction in a calorimeter (like Experiment 13.1), and we can find the enthalpy of formation.

How does this help? We define the enthalpy of any element in its elemental form to be 0. If we use elements in their elemental form for the reactants of a formation reaction, then the total enthalpy of the reactants in a formation reaction will be 0. In addition, since there is only one product in a formation reaction (the substance of interest), the ΔH of the formation equation is, in essence, the enthalpy of that substance. Having the ΔH_f of a substance is equivalent to having the enthalpy of the substance. Therefore, if we had a table of the ΔH_f of several substances, we could calculate the ΔH of any reaction that uses those substances!

Before we give you that table, we have to introduce a point of terminology. The ΔH of a chemical reaction can change depending on the conditions under which the reaction occurs. When we report ΔH, it is important to specify the conditions under which the reaction has been run. Most chemistry is done at room temperature (25°C) and normal atmospheric pressure (1.00 atm), so we call these standard conditions. Note that this is *not* standard temperature and pressure (STP), which is defined as 0.00°C and 1.00 atm. Instead, a temperature of 25°C and a pressure of 1.00 atm are considered the *standard conditions*.

In this course, every ΔH you will calculate will refer to standard conditions. In table 13.3, you will find the ΔH_f for several molecules under standard conditions. To indicate that they are under standard conditions, we include a degree symbol, and we call them standard enthalpies of formation. A symbol of $\Delta H_f°$ indicates that the enthalpy of formation was measured at 25°C and 1.00 atm.

TABLE 13.3
Standard Enthalpies of Formation in kJ/mole

Substance	$\Delta H_f°$	Substance	$\Delta H_f°$	Substance	$\Delta H_f°$	Substance	$\Delta H_f°$
H_2O (g)	-242	NO_2 (g)	33.2	$Ca(OH)_2$ (s)	-987	C_2H_6O (l)	-278
H_2O (l)	-286	NH_3 (g)	-45.9	$CaCO_3$ (s)	-1207	C_2H_4 (g)	52.5
SO_3 (g)	-396	CO_2 (g)	-394	C_2H_6 (g)	-84.7	C_4H_{10} (g)	-126
NO (g)	90.3	CaO (s)	-635	CH_3OH (l)	-239	H_2SO_4 (l)	-814

Notice a couple of things about these standard enthalpies of formation. First, they can be either positive or negative. This should make sense to you. After all, they are just the ΔHs of formation reactions under standard conditions. Some formation reactions are exothermic, while others are endothermic. Also, notice that the enthalpy of formation depends on the state of the substance. For example, the enthalpy of formation of liquid water is different from the enthalpy of formation of gaseous water. Once again, this should make sense once you realize that it takes energy to turn water from a liquid to a gas. The heat released by a reaction that makes gaseous water should be less than the heat released by a reaction that makes liquid water because the first reaction will have to expend some of its extra energy to turn water from a liquid into a gas.

Now that we have a table that gives us the enthalpies of formation for several substances, we can define the standard change in enthalpy (ΔH°) of any reaction that contains those substances as:

$$\Delta H^\circ = \Sigma\, \Delta H_f^\circ \text{ (products)} - \Sigma\, \Delta H_f^\circ \text{ (reactants)} \qquad \text{Equation 13.3}$$

As you may know from algebra, the Σ symbol stands for summation. Equation 13.3 tells us that the ΔH° of any chemical reaction is simply the sum of the ΔH_f° for the products minus the sum of the ΔH_f° for the reactants. It is important to make sure that you use products minus reactants because that will keep the sign of ΔH° consistent with our previous definitions. That way, a negative ΔH° will still indicate an exothermic reaction and a positive ΔH° will still indicate an endothermic reaction. Equation 13.3 is often referred to as Hess's law because it is a direct consequence of that law.

Since we are using standard enthalpies of formation in equation 13.3, the change in enthalpy that we calculate using that equation will also refer to standard conditions. That's why there is a degree symbol after the ΔH in equation 13.3. When we use equation 13.3 and table 13.3, then, the change in enthalpy that we determine will be valid only under standard conditions. Since most of the chemistry we do in this course takes place at room temperature and normal atmospheric pressure, this is really not much of a limitation.

Now that you've made it through this rather deep discussion of Hess's law, it is time to summarize the things that you really need to understand. First, you need to know the definition of Hess's law and what a state function is. You also need to know the definition of enthalpy of formation and equation 13.3. Finally, you just have to remember:

The standard enthalpy of formation of an element in its elemental form is 0.

Chemicals like Ca (s), Na (s), K (s), and Fe (s) all have $\Delta H_f^\circ = 0$. In addition, the homonuclear diatomics like H_2 (g), N_2 (g), O_2 (g), and Cl_2 (g) have $\Delta H_f^\circ = 0$. Let's see how all of this pans out in Example 13.3.

EXAMPLE 13.3

What is the ΔH° of the following reaction?

$$2NO\text{ (g)} + O_2\text{ (g)} \longrightarrow 2NO_2\text{ (g)}$$

Is it an exothermic or an endothermic reaction?

Looking at table 13.3, we have the ΔH_f° for all of the substances in this equation except $O_2(g)$. However, we know that any element in its elemental form has $\Delta H_f^\circ = 0$. Since $O_2(g)$ is the elemental form of oxygen (because oxygen is a homonuclear diatomic and a gas at room temperature), we know that its $\Delta H_f^\circ = 0$. Now we know the ΔH_f° of all substances in the equation. Remember, when looking for the ΔH_f° of a substance, we must match both the substance and the phase. For example, if this chemical reaction contained NO (l), we could not solve the problem because our table does not list NO (l); it lists only NO (g). In addition, if the reaction used $O_2(l)$ instead of $O_2(g)$, we could not solve the problem because we are working under standard conditions. Under standard conditions, oxygen is a gas, not a liquid. If the equation used $O_2(l)$, we could not assume that its ΔH_f° is 0. However, since we do have all of the standard enthalpies of formation that we need, we can simply apply equation 13.3:

$$\Delta H^\circ = (2\ \cancel{\text{moles}} \times 33.2\ \frac{\text{kJ}}{\cancel{\text{mole}}}) - [(2\ \cancel{\text{moles}} \times 90.3\ \frac{\text{kJ}}{\cancel{\text{mole}}}) + (1\ \cancel{\text{mole}} \times 0\ \frac{\text{kJ}}{\cancel{\text{mole}}})]$$

$$\Delta H^\circ = -114.2\ \text{kJ}$$

Notice that we had to multiply each ΔH_f° by the stoichiometric coefficient of the substance in the chemical equation. Also notice that we added all of the terms from the products side and subtracted all of the terms from the reactants side. This is the proper way to apply equation 13.3 to a chemical reaction. This means that the ΔH° for this reaction is **-114.2 kJ**, and it is therefore **exothermic**.

Some gasoline companies put a small amount of liquid ethyl alcohol (C_2H_6O) into their gasoline. What is the ΔH° for the complete combustion of this chemical?

We could solve this problem using bond energies as explained in the previous section. However, to illustrate the use of Hess's law, we will solve it using the concepts discussed in this section. First, we write the balanced chemical equation:

$$C_2H_6O\ (l) + 3O_2\ (g) \longrightarrow 2CO_2\ (g) + 3H_2O\ (g)$$

Notice that we added phase symbols in this equation. We must have them to use table 13.3. In Module 7, we told you to assume that the water and carbon dioxide produced in combustion reactions were in the gas phase, so there is nothing new here.

Looking in table 13.3, we can find the ΔH_f° for every substance in its proper phase except O_2 (g). However, since it is an element in its elemental form, the ΔH_f° is 0. We can now apply equation 13.3:

$$\Delta H^\circ = [(2\ \cancel{\text{moles}} \times -394\ \frac{\text{kJ}}{\cancel{\text{mole}}}) + (3\ \cancel{\text{moles}} \times -242\ \frac{\text{kJ}}{\cancel{\text{mole}}})] -$$

$$[(1\ \cancel{\text{mole}} \times -278\ \frac{\text{kJ}}{\cancel{\text{mole}}}) + (3\ \cancel{\text{moles}} \times 0\ \frac{\text{kJ}}{\cancel{\text{mole}}})]$$

The ΔH° for this reaction is **-1,236 kJ**.

So you see that even though the discussion of Hess's law was very complicated, the application of it is very simple. If you understand the examples you have just read and can answer "On Your Own" questions 13.4–13.5, you have understood this section well enough to move on.

Before we leave this section, we need to answer a question that you should be asking yourself. Since you now have 2 mathematical or theoretical methods of determining ΔH, which should you use? In one of the example problems from this section and one of the "On Your Own" questions, we specifically told you to use Hess's law instead of bond energies. Using Hess's law is a bit more accurate since it also takes into account the phase of the substances involved. If you have the ΔH_f° of each substance in the equation with the proper phase, you should always use Hess's law. This is good because Hess's law is less work than determining ΔH from bond energies!

ON YOUR OWN

13.4 When you use a pocket cigarette lighter (hopefully not to light cigarettes!), you are burning gaseous butane (C_4H_{10}). Assuming it is complete combustion, what is the ΔH° for this reaction? Do not use bond energies to solve this problem; use Hess's law.

13.5 To make the shells that house and protect them, many shellfish take solid lime (CaO) from the ocean floor and react it with gaseous carbon dioxide that has been exhaled by marine organisms. The resulting calcium carbonate is the major component of their shells. What is the ΔH° for this reaction?

The only time you should use bond energies is if you don't have every ΔH_f° necessary to solve the problem or if the problem doesn't tell you the phases of the substances in the reaction. After all, if you don't know the phase, you can't use table 13.3! When you use bond energies to calculate ΔH, it is important to realize that your answer is only an approximation. The only way to accurately calculate the ΔH of a reaction is by taking the phase of each substance into account. You can do that only by using Hess's law.

You should use table 13.3 when you solve all the problems in this module, including the test problems. You will also find a copy of table 13.3 in appendix A at the end of the text. Make sure you use that appendix when you are working on problems from this module!

APPLYING ENTHALPY TO STOICHIOMETRY

Now that we know 2 different ways to determine the ΔH of a chemical reaction, we are ready to apply this newly learned concept. Where better to apply it than to the wonderful world of stoichiometry? As we have said many times before, stoichiometry is the backbone of chemistry; therefore, we try to relate nearly everything that we study (in one way or another) to this terribly important subject.

Remember, once we have determined both the value and the sign of ΔH, we can include it in the equation, as you did in "On Your Own" question 13.1. Once something becomes part of a chemical equation, it can be used in stoichiometry. Study Example 13.4 to see how this is done, and then answer "On Your Own" questions 13.6–13.7 to cement this new skill into your head.

EXAMPLE 13.4

We determined previously that the ΔH° for the combustion of ethyl alcohol is -1,236 kJ. How much energy can be produced by burning 150.0 grams of ethyl alcohol in excess oxygen?

From our last example problem, we determined that the balanced chemical equation for the combustion of ethyl alcohol is:

$$C_2H_6O\ (l) + 3O_2\ (g) \longrightarrow 2CO_2\ (g) + 3H_2O\ (g)$$

Once we determined the ΔH° of the reaction, we could have simply put it in the chemical equation. Since ΔH° turned out to be negative, energy can be considered a product in the reaction.

$$C_2H_6O\ (l) + 3O_2\ (g) \longrightarrow 2CO_2\ (g) + 3H_2O\ (g) + 1{,}236\ kJ$$

We can start doing stoichiometry as soon as we convert the amount of ethyl alcohol into moles:

$$\frac{150.0\ \cancel{g\ C_2H_6O}}{1} \times \frac{1\ mole\ C_2H_6O}{46.1\ \cancel{g\ C_2H_6O}} = 3.25\ moles\ C_2H_6O$$

According to our equation, 1 mole of C_2H_6O will give us 1,236 kJ of energy. Therefore, our stoichiometric relationship is:

$$1\ mole\ C_2H_6O = 1{,}236\ kJ$$

Using that relationship gives us:

$$\frac{3.25\ \cancel{moles\ C_2H_6O}}{1} \times \frac{1{,}236\ kJ}{1\ \cancel{mole\ C_2H_6O}} = \mathbf{4{,}020\ kJ}$$

One way of producing butane (C_4H_{10}) for cigarette lighters is to force ethane (C_2H_6) to react with itself:

$$2C_2H_6\ (g) \longrightarrow C_4H_{10}\ (g) + H_2\ (g)$$

If we have 1.23 kg of ethane and would like to turn it all into butane, how many kJ of energy would be necessary?

Since this question asks about the energy associated with a chemical reaction, we must determine ΔH° to solve the problem. This is easy to do since we have all of the information related to this problem available in table 13.3. According to Hess's law,

$$\Delta H° = (1\ \cancel{mole} \times -126\ \frac{kJ}{\cancel{mole}}) + (1\ \cancel{mole} \times 0\ \frac{kJ}{\cancel{mole}}) - (2\ \cancel{moles} \times -84.7\ \frac{kJ}{\cancel{mole}}) = 43\ kJ$$

Since ΔH° is positive, energy can be thought of as a reactant in this problem:

$$2C_2H_6\ (g) + 43\ kJ \longrightarrow C_4H_{10}\ (g) - H_2\ (g)$$

Now we can use the energy in stoichiometry. First, we must convert the amount of ethane into moles:

$$\frac{1.23 \times 10^3\ \cancel{g\ C_2H_6}}{1} \times \frac{1\ mole\ C_2H_6}{30.1\ \cancel{g\ C_2H_6}} = 40.9\ moles\ C_2H_6$$

Then we use the chemical equation:

$$2\ moles\ C_2H_6 = 43\ kJ$$

We can use that relationship, then, to convert from moles of ethane to kJ of energy:

$$\frac{40.9 \text{ moles } C_2H_6}{1} \times \frac{43 \text{ kJ}}{2 \text{ moles } C_2H_6} = 880 \text{ kJ}$$

ENERGY DIAGRAMS

Have you ever heard the saying "A picture is worth a thousand words"? We have developed a way to draw a picture of the energy associated with a chemical reaction. This picture, called an energy diagram, tells a chemist about a chemical reaction without using any numbers or words. For example, suppose we were studying the following chemical reaction:

$$C_2H_6O(l) + H_2(g) \longrightarrow C_2H_6(g) + H_2O(g)$$

We could use Hess's law to calculate the $\Delta H°$ of this reaction and communicate this result to other chemists. The sign of $\Delta H°$ would tell them whether the reaction was exothermic or endothermic, giving them an idea of the energetics of the reaction.

Instead of calculating $\Delta H°$, we could communicate the energetics of this reaction by simply drawing a picture that looks something like Figure 13.1.

ON YOUR OWN

13.6 Ethane gas (C_2H_6) has been considered as a fuel for automobiles because, like octane, it releases a lot of energy when it is burned. How many kJ of energy are released when 500.0 grams of ethane undergo complete combustion?

13.7 When mined, coal contains a small amount of sulfur. When that coal is burned in a power plant, the sulfur burns right along with it, producing SO_3 gas. When the SO_3 gas reacts with the water in the air, it produces sulfuric acid according to the following equation:

$$H_2O\ (g) + SO_3\ (g) \longrightarrow H_2SO_4\ (l)$$

The sulfuric acid produced in this manner falls to the earth as acid rain. How much energy is produced when 25.0 grams of H_2SO_4 are produced in this way?

FIGURE 13.1

An Energy Diagram for the Reaction $C_2H_6O\ (l) + H_2(g) \longrightarrow C_2H_6(g) + H_2O\ (g)$

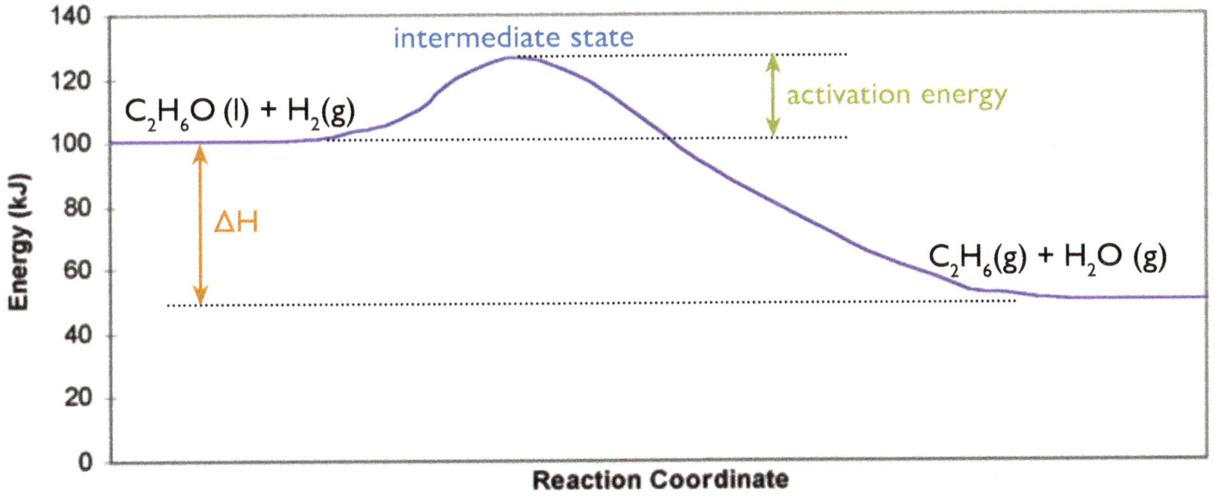

Notice that this picture is a graph. On the vertical (y) axis of the graph, the energy of the substances in the chemical reaction is plotted. On the horizontal (x) axis of the graph, we plot the reaction coordinate, which tells us how close the reaction is to completion. For example, when the reaction coordinate is 0, we say that the reaction hasn't started yet; therefore, only the reactants are present. That's why only the reactants are listed on the left side of the graph. As x gets larger, the reaction starts. The reactants come together and interact. At the top of the hump, which represents the midpoint of the reaction, we say that some intermediate state (sometimes called the activated complex) exists. By the time x reaches the far right side of the axis, the reaction has completed, and we have only products. That's why only the products are listed on the right side of the graph. When you look at the horizontal part of this graph, just remember that the far left side of the graph represents the reaction before it begins, while the far right side represents the end of the reaction. The points in between represent the reaction as it progresses.

Now let's look at the vertical (y) axis, which represents the energy of the substances in the chemical reaction. For this particular reaction, notice that the reactants (the left side of the graph) are much higher on the vertical axis than the products (the right side of the graph). This tells us that the reactants have more energy than the products. What does this tell us about the ΔH of the reaction? If the reactants have more energy than the products, then once the reaction completes, the excess energy from the reactants has to go somewhere. Where does it go? It gets released into the surroundings, usually in the form of heat, making this energy diagram that tells us that the reaction is exothermic.

The energy diagram tells us a bit more than that, however. The difference between the vertical (y) position of the products and the vertical (y) position of the reactants tells us how much energy was released into the surroundings. That's why we have labeled the vertical difference between the reactants and the products as ΔH. If we can read the graph well enough, we can calculate the value of ΔH from it.

This graph gives us more information than just the ΔH of the reaction. Between the left and right sides of the graph is a big hump. This hump represents the energy of the intermediate state of the reaction. Notice that the energy of the intermediate state of the reaction is higher than the energy of the reactants and the products. This is always the case. The intermediate state of the reaction has more energy than either the reactants or the products. This tells us that to reach the intermediate state of the reaction, some extra energy has to be put into the reaction. This energy, which is necessary to start the reaction, is called activation energy.

Activation energy—The energy necessary to start a chemical reaction.

All reactions, even exothermic ones, require a certain amount of activation energy to get the reaction started.

It is good that all chemical reactions require some amount of activation energy to start. After all, to burn paper, the only 2 things we need are paper and oxygen. The paper you are looking at right now in this book is in the presence of oxygen, but it is not bursting into flames because the chemical reaction between paper and oxygen requires some activation energy. If we were to provide that activation energy (by lighting a match and sticking it under the paper, for example), the reaction could begin and the paper would burst into flames. The activation energy is represented by the difference between the energy of the reactants and the energy of the intermediate state, as labeled in the graph.

The explanation for why all chemical reactions have an activation energy is really quite simple. If you think about it, for any chemical reaction to proceed, the reactants must get close enough to each other to start rearranging their electrons. After all, as we have said many times before, a chemical reaction is just a rearrangement of the valence electrons in the molecules that make up the reactants. The problem is that under normal conditions, molecules don't like to get close enough to do this because the electrons that must be rearranged repel each other. For the molecules in a chemical reaction to get close enough to rearrange their electrons, they must be given enough energy to overcome the repulsion that their electrons feel from each other. This energy is what we call activation energy. Each reaction has its own unique activation energy since each set of reactants has a different repulsive force between their electrons.

As you can see, energy diagrams can be very informative. Of course, the energy diagram you just studied is not the only type of energy diagram that exists. For example, consider the energy diagram in Figure 13.2.

FIGURE 13.2
An Energy Diagram for an Endothermic Reaction

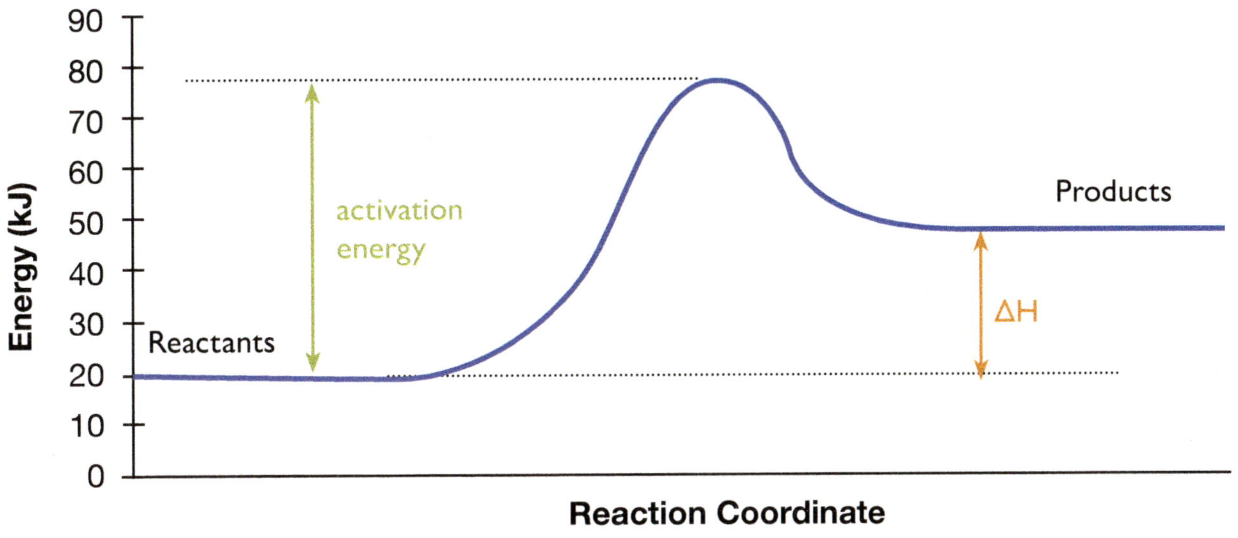

This diagram looks similar to the one presented earlier, but there is one important difference. In this case, the energy of the reactants is *lower* than the energy of the products. This tells us that the reactants have to absorb energy to reach the energy of the products. In other words, this is an endothermic reaction. Once again, the activation energy and ΔH can be read directly from the graph, as is indicated.

Study Example 13.5 and answer "On Your Own" question 13.8 to make sure you understand how to interpret these energy diagrams.

EXAMPLE 13.5

Consider the following energy diagrams:

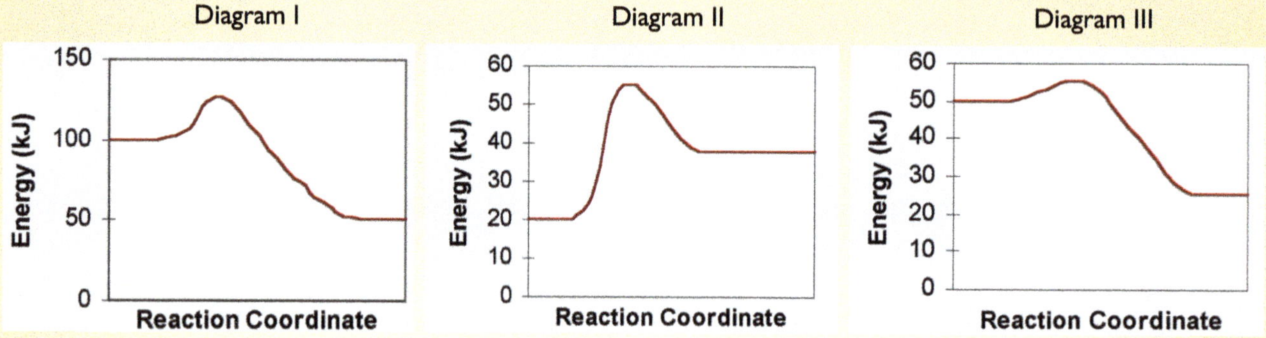

a. Which diagram represents the most exothermic reaction? What is its ΔH?

To solve this problem, we are looking for the diagrams that represent exothermic reactions. Any diagram in which the reactants (the left side of the diagram) have a higher vertical position than the products (the right side of the diagram) represents an exothermic reaction. Both I and III represent exothermic reactions. The most exothermic is the one that has the largest difference between the vertical position of the reactants and products. Therefore, **diagram I represents the most exothermic reaction**. To get the ΔH of the reaction, we must read the difference in energy between the products and the reactants. The products (on the right side of the graph) are at an energy (y-axis position) of 50 kJ, while the reactants (on the left side of the graph) are at a position of 100 kJ. To calculate ΔH, we take the energy of the products and subtract the energy of the reactants:

$$\Delta H = 50\text{ kJ} - 100\text{ kJ} = -50\text{ kJ}$$

Notice that by taking the energy of the products minus the energy of the reactants, we have the proper sign for ΔH. This makes the final answer, **ΔH = -50 kJ**.

b. Which diagram represents the reaction that is easiest to get started?

The reaction that is easiest to start has the lowest activation energy. So we are looking for the reaction that has the lowest difference in energy between the reactants and the intermediate state. This is clearly **diagram III**.

c. Which diagram represents an endothermic reaction? What is its ΔH?

An endothermic reaction is one in which the products have a higher energy than the reactants. This is the case in **diagram II**. Once again, to get ΔH, we take the energy of the products (which is about 40 kJ) minus the energy of the reactants (which is about 20 kJ). **ΔH = 20 kJ**.

THE SECOND LAW OF THERMODYNAMICS

In the last module, you learned about the first law of thermodynamics, which tells us that energy cannot be created or destroyed; it can only change form. Now it's time to learn the second law of thermodynamics. First, you need to learn about the fascinating subject of entropy (en' truh pee):

Entropy—A measure of the disorder that exists in any system.

ON YOUR OWN

13.8 Draw an energy diagram for the following reactions. (Since we do not give you any measurements, your axes should not be labeled with numbers.)

Reaction A: This reaction is hard to start but gets very, very hot once it does start.

Reaction B: This reaction is easy to start but gets only slightly warm during the reaction.

Reaction C: This reaction is hard to start but gets slightly cold during the reaction.

Reaction D: This reaction is easy to start and gets very, very cold during the reaction.

Since entropy is a measurement, it must have units. These units are always an energy unit divided by both moles and Kelvin. J/(mole·K), kcals/(mole·K) and kJ/(mole·K) are all acceptable units for measuring entropy, which is abbreviated with a capital S.

It might seem odd that chemists can measure disorder, but disorder is one of the most fundamental things we can learn about a system. For example, suppose a priceless vase is on a shelf somewhere and something causes the vase to fall onto the floor and shatter into a million pieces. Which situation do you think represents more disorder? Clearly, when the vase is shattered, it is more disordered than when it was sitting on the shelf in one piece. While the vase is sitting on the shelf, it is in a state of low entropy because it is not very disordered. However, when the vase shatters, it moves into a state of high entropy because now it is very disordered.

FIGURE 13.3
Entropy
Left photo by Karen (CC-BY 2.0). Right photo by James from Somerville, USA (CC-BY 2.0).

Marketplace at low entropy. Marketplace at high entropy.

Although you can imagine a vase falling from a shelf and shattering into a million pieces, can you ever imagine the reverse process happening? Can you imagine a situation where a shattered vase spontaneously rearranges itself into a whole vase again? Of course not! Even though you know that this will never happen, do you know why? This cannot happen because of the second law of thermodynamics:

The second law of thermodynamics—The entropy of the universe must always either increase or remain the same. It can never decrease.

Think for a moment about how this law would apply to our vase. If the shattered vase were to spontaneously rearrange itself into a whole vase again, it would be going from a high entropy state to a low entropy state. This would mean that the entropy of the universe would decrease because the entropy of the vase decreased. This would violate the second law. That's why broken things do not spontaneously fix themselves.

Now do you see why the concept of entropy is so important to a scientist? To a large extent, the second law of thermodynamics determines what types of events are possible in nature and what types of events are not possible. This law is just as important and just as fundamental to scientists as is the first law of thermodynamics. However, this law is much harder to understand than the first law of thermodynamics, so there is a lot of confusion about what the second law of thermodynamics prohibits and what it does not prohibit. At the end of this section, we will explain to you exactly how the second law of thermodynamics should be interpreted. For now, you need to have a firm grounding in the concept of entropy, so let's start with that.

As we said before, entropy is a measure of the disorder in a system. As a chemist, you should be very comfortable in determining whether a given state represents high entropy or low entropy. In the case of the vase, it was rather simple to determine which state was high entropy and which was low entropy. However, there are more subtle differences in entropy of which you must be aware.

First and foremost, the entropy of a substance changes when the substance changes phase. In the solid phase, the molecules (or atoms) that make up a substance are held in place and are only allowed to vibrate. Since the molecules cannot move around very much, they are not very disordered. As a result, the solid state of matter has the lowest amount of entropy. In the liquid phase, molecules are free to move around. Their movement is somewhat restricted because they must stay relatively close to their neighbors. As a result, the entropy of the liquid state is larger than that of the solid state. The gaseous state is the phase with the highest entropy because the molecules (or atoms) that make up a substance move freely and quickly. You need to remember the relative entropy of the phases of matter:

The solid state has the least amount of entropy associated with it. The entropy of the liquid state is higher, and the entropy of the gaseous state is the highest of all.

Although this fact is very important, there are other things to consider besides the phase in which a substance finds itself.

In general, whenever a substance's temperature increases, its entropy increases. As temperature increases, the motion of the molecules (or atoms) in a substance increases as well. The faster molecules (or atoms) begin moving, the harder it is to keep track of them. As a result, the entropy increases. This is another important fact:

The entropy of a system increases with increasing temperature.

If we have 2 identical samples of matter and raise one's temperature, the sample with the higher temperature is more disordered.

Finally, the more matter there is in a system, the more disordered it is. This should make sense based on your own experience. Consider your own room. If you had only a few things to store in your room, it would probably be very neat and tidy. However, the more things you have in your room, the harder it is to keep your room neat. The same problem exists in nature.

The entropy of a system increases as the matter it contains increases.

A sample of gas that contains 5 moles has more entropy than a sample of the same gas that contains only 3 moles.

Now that you know some general rules about entropy, it is time that you learned how a chemist applies these concepts. First, since it is very difficult to measure the entropy of a system, chemists are usually more interested in the change in entropy that a system undergoes. This should sound a little familiar to you. We have already learned that enthalpy is very hard to measure, whereas the change in enthalpy of a system is easier to measure. The same thing holds true for entropy, but chemists are usually more interested in ΔS than they are in S. In fact, we can even restate the second law of thermodynamics in terms of ΔS:

$$\Delta S_{universe} \geq 0 \qquad \text{Equation 13.4}$$

You should be able to see how equation 13.4 restates the second law of thermodynamics. If the entropy of the universe must always either increase or stay the same, then the change in entropy of the universe must be either 0 or positive.

What chemists concentrate on, of course, is trying to determine the ΔS of a chemical reaction. There are 2 ways that chemists typically do this. The first is a qualitative method that doesn't allow us to calculate a *number* for the ΔS of a chemical reaction. Rather, this method allows us to determine the *sign* of ΔS for a chemical reaction. The sign of ΔS is an important thing to know because it relates to the second law of thermodynamics, as illustrated by equation 13.4.

To determine the sign of ΔS for a chemical reaction, all we have to do is look at the number of molecules (or atoms) on each side of the equation as well as the phase of the substances involved. We can then use some of the facts we learned above. In general, the side of the reaction that contains the most molecules (or atoms) in the gas phase has the highest entropy. If no gases are present, then the side that has the most molecules (or atoms) in the liquid or aqueous phase has the highest entropy. Finally, if all substances in the chemical equation are solids, then the side with the largest number of molecules has the highest entropy. If none of these situations is true, then we simply cannot determine the sign of ΔS this way. If this all sounds a bit confusing, don't worry. Example 13.6 should clear up everything.

EXAMPLE 13.6

For each reaction below, determine the sign of ΔS:

a. $Zn(s) + HCl(aq) \longrightarrow ZnCl_2(aq) + H_2(g)$

The reactants side of the equation has 0 molecules in the gas phase, while the products side of the reaction has 1 molecule in the gas phase. So, the products side has a higher entropy. This means that when the reaction occurs, the substances go from a lower entropy state to a higher entropy state. This means the **ΔS is positive**.

b. $N_2(g) + 3H_2(g) \longrightarrow 2NH_3(g)$

The reactants side of this equation contains 4 molecules in the gas phase (1 nitrogen molecule and 3 hydrogen molecules). The products side contains only 2 molecules in the gas phase (2 ammonia molecules). Therefore, the reactants side of this equation has the highest entropy. This means that as the reaction occurs, the chemicals go from a higher entropy to a lower entropy. So our **ΔS is negative**. You might be tempted to say that since ΔS is negative for this reaction, the reaction can never proceed. After all, the second law of thermodynamics says that $\Delta S_{universe}$ must always be greater than or equal to 0. Using the second law of thermodynamics is a bit more complicated than that, however. We will soon see that a reaction can proceed even if its ΔS is negative. Some very important conditions must be met first, however.

c. $2NH_4NO_3\ (s) + Ba(OH)_2(H_2O)_8\ (s) \longrightarrow Ba(NO_3)_2\ (aq) + 2NH_4OH\ (aq) + 8H_2O\ (l)$

This equation has no molecules or atoms in the gas phase on either side of the chemical equation. We examine the number of molecules or atoms in the liquid or aqueous phase. Based on that, the products side clearly has the highest entropy. This means that **ΔS is positive**.

ON YOUR OWN

13.9 Determine the sign of ΔS for the following reactions:

a. $NaC_2H_3O_2(aq) + H_2O\ (l) + CO_2(g) \longrightarrow C_2H_4O_2\ (aq) + NaHCO_3\ (s)$

b. $2H_2O\ (g) + O_2\ (g) \longrightarrow 2H_2O_2\ (g)$

c. $2AgOH\ (s) + Mg(NO_3)_2\ (aq) \longrightarrow 2AgNO_3\ (aq) + Mg(OH)_2\ (aq)$

Make sure you can determine the sign of ΔS for a chemical reaction by answering "On Your Own" question 13.9.

If we need a little more information than just the sign of ΔS, there is a way to determine a number for the change in entropy of a chemical reaction. Even though it is very difficult to measure the entropy of an individual substance, it can be done. Table 13.4 tabulates the absolute entropies of the same substances contained in table 13.3. This, of course, makes determining ΔS very simple. You can use an equation very similar to equation 13.3 to calculate the change in entropy between the products and the reactants, which is the definition of ΔS:

$$\Delta S^\circ = \Sigma\ S^\circ\ (\text{products}) - \Sigma\ S^\circ\ (\text{reactants}) \qquad \text{Equation 13.5}$$

Determining the change in entropy of a chemical reaction is as simple as using Hess's law, provided we have the proper table. There is one difference between calculating ΔS and ΔH, however. When calculating ΔH, we follow the rule that the ΔH_f of an element is 0. When calculating ΔS, the S of an element is *not* 0. This is very important:

All substances, even elements, have an absolute entropy.

TABLE 13.4
Standard Absolute Entropies in J/(mole· K)

Substance	S°	Substance	S°	Substance	S°	Substance	S°
H_2O (g)	189	NO_2 (g)	240	$Ca(OH)_2$ (s)	76.1	C_2H_6O (l)	161
H_2O (l)	69.9	NH_3 (g)	193	$CaCO_3$ (s)	92.9	C_2H_4 (g)	219
SO_3 (g)	257	CO_2 (g)	214	C_2H_6 (g)	230	C_4H_{10} (g)	260
NO (g)	211	CaO (s)	38.2	CH_3OH (l)	127	H_2SO_4 (l)	157

Notice a few things about this table. The entropies listed here are all positive. This is because the table does not contain values of ΔS; it contains values of S. As a result, they all must be positive because *all* substances have some level of disorder which means they have some absolute entropy. Second, notice that compared to the values of ΔH tabulated in table 13.3, these numbers are quite small. Table 13.3 contained numbers with the unit of kJ/mole, whereas these numbers are in J/(mole·K). This means that these numbers are about 1,000 times smaller than the ones tabulated in table 13.3.

Since the use of table 13.4 and equation 13.5 is exactly the same as that of Hess's law, we will not give you an example of how to solve a problem using them. Instead, just use the same ideas you learned before to answer "On Your Own" question 13.10.

ON YOUR OWN

13.10 In "On Your Own" problem 13.5, you determined the ΔH° of the following reaction:

$CaO\ (s) + CO_2\ (g) \longrightarrow CaCO_3\ (s)$

Use what you just learned to calculate ΔS° for this same reaction.

THE PROPER APPLICATION OF THE SECOND LAW OF THERMODYNAMICS

As we said before, even though the second law of thermodynamics is very, very, important in science, it is also very misunderstood and misused. For example, you might think that the second law of thermodynamics makes evolution impossible. After all, the whole theory of evolution is predicated on the belief that simple (disordered) life forms can evolve into complex (ordered) life forms. The second law of thermodynamics states that disorder must always stay the same or increase. Evolution requires just the opposite, so the second law of thermodynamics says that evolution cannot happen, right?

Although this argument sounds quite nice, it is, nevertheless, wrong. There are spontaneous processes that occur every day that result in a decrease of disorder. For example, what happens to a bucket of water when it is left outside on a cold, winter day? It freezes. As we learned earlier, the solid state of matter has lower entropy than the liquid state. When water freezes, it goes from a more disordered state to a less disordered state. In other words, for this process, ΔS is negative. This, on the surface, seems to violate the second law, which states that $\Delta S_{universe}$ must always be greater than or equal to 0.

The process of freezing water only *seems* to contradict the second law of thermodynamics because we have considered only *part* of the process. You see, when water freezes, it affects the molecules that surround it. Think about it this way. When water freezes, its molecules slow down and don't move as much as they did in the liquid state. This means that they lose energy. Where does that energy go? It is released into the surroundings. As a result, when water freezes, its surroundings heat up a bit. This may sound odd, but it is nevertheless true. The very process of freezing requires a release of energy. When a system releases energy, its surroundings heat up. So when water freezes, the things surrounding the water heat up.

What happens to a substance's entropy when it heats up? The entropy increases, resulting in the entropy of the things surrounding the water increasing. This increase in entropy of the surroundings offsets the decrease in entropy of the water. As a result, the *total* ΔS of the universe (water + its surroundings) does not change. This is in agreement with the second law of thermodynamics because it says that ΔS can be positive, *or 0*.

When we look at a process in terms of the second law of thermodynamics, we *must* look at everything that is affected. This includes the system that we are interested in, *plus everything else that is affected by it*. Let's imagine one more example. Suppose you come home one afternoon and realize that your house is a mess, so you decide to clean it up. When you start to clean your house, you begin to transform it from a disordered (high entropy) state into a relatively ordered (low entropy) state. At first glance, then, cleaning the house seems to violate the second law of thermodynamics. What are we missing?

As you clean up the house, you begin to move around. This agitates the molecules that make up the surrounding air, causing them to move faster. This increases their entropy. Plus, as you begin to work, your body starts to burn calories to supply you with the energy you need. This heats up your body, increasing its entropy. If you were able to measure the amount of increased entropy in the air

science and creation

How does entropy apply to the discussion on evolution? If an advanced species (which is highly ordered) did evolve from a simple species (which is less ordered), then that organism would have, indeed, experienced a decrease in entropy. To make this consistent with the second law of thermodynamics, the organism's surroundings would have to become more disordered so that the decrease in entropy of the organism could be offset. This is the only way that evolution could be consistent with the second law of thermodynamics.

Suppose a single-celled creature were to evolve into a cell that would cooperate with other cells. Evolutionists say that this could happen if that cell's DNA were to undergo a mutation that would make the DNA a little more advanced. The problem is that the very instant that the DNA mutated and became more ordered, there would have to be a corresponding increase in entropy in the organism's surroundings so that the *total* entropy of the universe would still increase or at least stay the same. No scientist on earth can come up with a single idea about *how* that can happen. As far as anyone can tell, mutating DNA does not significantly increase the entropy of the surroundings.

So you see that the second law of thermodynamics doesn't really disprove evolution, but it does attach some pretty strong conditions to it. The second law says that *if* evolution were to occur, then *each step of that evolution* must be accompanied by an *immediate* increase in the entropy of the surroundings. If anyone ever comes up with a mechanism by which this might happen, evolution could at least be made *consistent* with the second law. Currently, no one can come up with even a vague notion about how this might happen. As a result, the current theories of evolution are not consistent with the second law. This does not mean that the second law *disproves* these theories, however. It just means that no one has come up with the proper mechanism. In our opinion, such a mechanism does not exist.

and your body, you would find out that it is larger than the decrease in entropy caused by the fact that you cleaned house. In the end, then, the *total* entropy of the universe increased, in agreement with the second law.

GIBBS FREE ENERGY

As we have seen, if a process is to occur, it must result in no change to the entropy of the universe or an increase in the entropy of the universe. To determine whether or not this happens, we must consider not only the entropy of the process, but also the entropy of any surroundings that this process affects. This might sound like a daunting task. However, as far as chemical reactions go, it is rather easy!

We can already determine the ΔS of the chemical reaction as we learned earlier. We are halfway there! The only other thing we have to do to determine whether or not a reaction will really happen is to determine the effect that it has on the entropy of its surroundings. This is relatively simple as well.

The main way that a chemical reaction can alter the entropy of its surroundings is through its enthalpy. If a chemical reaction has a negative ΔH, it releases energy into its surroundings. As a result, the entropy of the surroundings increases. Alternatively, if a chemical reaction has a positive ΔH, it absorbs energy from its surroundings, decreasing their entropy. So to determine whether or not a chemical reaction is consistent with the second law of thermodynamics, we need only look at its ΔS and ΔH. The value of ΔS will tell us about the change in entropy of the reaction itself, and the ΔH will allow us to learn about the change in entropy of the reaction's surroundings.

There is a very simple formula that does all of this for us. It relates ΔS and ΔH into something called the Gibbs free energy, which we label as ΔG:

$$\Delta G = \Delta H - T \times \Delta S \qquad \text{Equation 13.6}$$

In this equation, T is the temperature of the surroundings (in Kelvin). The Gibbs free energy can tell us whether or not a reaction is consistent with the second law. If it is, we say that the reaction is spontaneous because we know that it can proceed. If it is not consistent with the second law, we call it not spontaneous because the reaction cannot occur.

As long as the ΔG for any reaction is negative, the reaction is consistent with the second law. If the ΔG is positive, the reaction is not consistent with the second law. This important fact must be remembered:

If $\Delta G < 0$, then the reaction is consistent with the second law and is spontaneous.

If $\Delta G > 0$, then the reaction cannot proceed because it violates the second law.

This statement is a bit of an approximation. You see, there are ways that a reaction can affect its surroundings other than through ΔH. Since the equation for ΔG does not consider any other factors, it is not an exact way of determining the spontaneity of a reaction. However, it is such a good approximation that most chemists are content to use it, so we definitely will. If you take advanced chemistry, you will learn a more precise way of relating ΔG to chemical reactions.

Let's look at equation 13.6 and see what it tells us. If a reaction is to be spontaneous, then ΔG has to be negative. How can this happen? If ΔH is negative and ΔS is positive, the first term in equation 13.6 will be negative and the second term ($T \cdot \Delta S$) will be positive. If a negative term has a positive term subtracted from it, then the result will *always* be negative. Therefore, whenever a reaction is exothermic (has a negative ΔH) and also has a positive ΔS, it will *always* be a spontaneous reaction. On the other hand, if ΔH is positive and ΔS is negative, ΔG will *always* be positive. Therefore, reactions that are endothermic and have a negative ΔS will *never* be spontaneous.

This is one qualitative way of determining whether or not a reaction is spontaneous. If we know whether the reaction is exothermic or endothermic, we can look at the equation and determine the sign of its ΔS. If the reaction is endothermic and ΔS is negative, the reaction will never occur. If the reaction is exothermic and ΔS is positive, the reaction will always be spontaneous.

The 2 possibilities we just discussed are not the only possibilities for a chemical reaction. Suppose the ΔH of a chemical reaction is negative and the ΔS is also negative. In this case, the first term in equation 13.6 will be negative, but the $T \cdot \Delta S$ term will also be negative. This means that we will have one negative number subtracted from another negative number. Sometimes, the result will be positive (indicating a nonspontaneous reaction), and sometimes the result will be negative (indicating a spontaneous reaction). In those cases, we have to calculate a value of ΔG to determine whether or not the reaction is spontaneous.

Of course, calculating a value for ΔG isn't too hard since we already know how to calculate ΔH and ΔS. One way to calculate ΔG is to calculate ΔH and ΔS for a reaction and plug the result into equation 13.6. There is another way to calculate ΔG. Because ΔG is a state function,we can use Hess's law. Earlier in this module, our job was made easier by using standard enthalpies and standard absolute entropies. Similarly, chemists have also calculated standard Gibbs free energies of formation (ΔG_f°). Using ΔG_f° from the table below with Hess's law, we can calculate standard Gibbs free energies (ΔG°).

TABLE 13.5
Standard Gibbs Free Energies of Formation in kJ/mole

Substance	ΔG_f°	Substance	ΔG_f°	Substance	ΔG_f°	Substance	ΔG_f°
CH_4 (g)	-50.8	NO_2 (g)	240	$CaSO_4$ (s)	-1,320	C_2H_6O (l)	-175
C_6H_6 (l)	125	H_2O (l)	-237	$CaCO_3$ (s)	-1,129	$CaCl_2$ (s)	-750
CO (g)	-137	H_2O (g)	-229	H_2CO_3 (aq)	-623	HCl (aq)	-131
CO_2 (g)	-394	CaO (s)	38.2	CH_3OH (l)	-166	$Ca(OH)_2$ (s)	-897

The numbers in this table will allow us to use equation 13.7 as a means of determining ΔG:

$$\Delta G^\circ = \Sigma\, \Delta G_f^\circ \text{ (products)} - \Sigma\, \Delta G_f^\circ \text{ (reactants)} \qquad \text{Equation 13.7}$$

Remember, the ΔG_f° of any element in its elemental form is 0, as is its ΔH_f°. Since we have 2 different methods for calculating ΔG, does it matter which we use? Yes, it does. You see, the values given for ΔG_f° in table 13.5 are all for standard conditions: 298 K and 1 atm of pressure (pressure of Earth's atmosphere). So if we are working with a reaction at standard conditions, we can use equation 13.7 and table 13.5 to calculate ΔG^o. Anytime you are working on a reaction at a temperature other than 298 K, you need to use equation 13.6. Let's see how this works in example13.7.

EXAMPLE 13.7

In Example 13.3, we studied the following chemical reaction:

$$2NO\,(g) + O_2\,(g) \rightarrow 2NO_2\,(g)$$

and found that its $\Delta H°$ was -114.2 kJ. If the $\Delta S°$ is -147 $\frac{J}{K}$, what is the ΔG at 532°C (805 K)? Is it a spontaneous reaction at that temperature?

In this problem, we are working with a reaction at a temperature other than 298 K, so we must use equation 13.6. This isn't hard, though, because we already know $\Delta S°$ and $\Delta H°$. The only thing we have to notice is that their units are not, at this point, consistent. The value for $\Delta S°$ has Joules in it, while the value for $\Delta H°$ has kJ in it. We must therefore change one set of units. We will choose to convert $\Delta S°$:

$$\frac{-147\,\cancel{J}}{K} \times \frac{1\text{ kJ}}{1000\,\cancel{J}} = -0.147\,\frac{\text{kJ}}{K}$$

Now we can use equation 13.6:

$$\Delta G = \Delta H - T \times \Delta S = -114.2\text{ kJ} - (805\cancel{K} \times -0.147\frac{\text{kJ}}{\cancel{K}}) = 4\text{ kJ}$$

Notice that since ΔS has the Kelvin temperature unit in it, the temperature we use in the equation must also be in Kelvin. Since this ΔG is positive, we can say that this reaction is **not spontaneous**. The decrease in entropy of the reaction is not offset by the increase in entropy of the surroundings caused by the release of heat.

Is there any temperature at which the above reaction does become spontaneous?

Since ΔG must be negative for the reaction to be spontaneous, this question really asks for the temperature at which ΔG falls below 0. Since $\Delta G = \Delta H - T\times\Delta S$, that means we are looking for the following situation:

$$\Delta H - T \times \Delta S < 0$$

$$-114.2\,\frac{\text{kJ}}{\text{mole}} - \left(T \times -0.147\frac{\text{kJ}}{\text{mole}\cdot K}\right) < 0$$

We can solve inequalities just like regular algebraic equations, as long as we remember that if we multiply or divide both sides by a negative, we must switch the direction of the inequality sign:

$$T \times 0.147 \frac{\text{kJ}}{\text{mole}\cdot\text{K}} < 114.2 \frac{\text{kJ}}{\text{mole}}$$

$$T < \frac{114.2 \frac{\cancel{\text{kJ}}}{\cancel{\text{mole}}}}{0.147 \frac{\cancel{\text{kJ}}}{\cancel{\text{mole}}\cdot\text{K}}}$$

$$T < 777\ \text{K}$$

For any temperature less than **777 K**, this reaction is spontaneous.

What is the ΔG for the following reaction at standard conditions? Is it spontaneous?

$$Ca\,(s) + 2HCl\,(aq) \longrightarrow CaCl_2\,(s) + H_2\,(g)$$

This problem asks us for the ΔG of a reaction at 298 K and 1 atm. This means we can use table 13.5 and equation 13.7 to answer it. Recognizing that the ΔG_f^o's for both Ca (s) and H_2 (g) are 0, Equation 13.7 becomes:

$$\Delta G = (1\,\cancel{\text{mole}} \times -750 \frac{\text{kJ}}{\cancel{\text{mole}}}) - (2\,\cancel{\text{moles}} \times -131 \frac{\text{kJ}}{\cancel{\text{mole}}}) = -490\ \text{kJ}$$

This reaction, then, is **spontaneous** at 298 K.

ON YOUR OWN

13.11 The ΔH of a certain chemical reaction is 20.0 kJ, and the ΔS is 123 J/K. At what temperatures is this reaction spontaneous?

13.12 Is the following synthesis of ethyl alcohol a spontaneous reaction at 298 K?

$$2CH_4\,(g) + O_2\,(g) \longrightarrow C_2H_6O\,(l) + H_2O\,(l)$$

As we have cautioned you in the past, do not get so wrapped up in solving problems that you forget the chemical significance of what you are doing. Remember, ΔG is so important because it takes inventory of all of the entropy of both the chemical reaction and its surroundings. By doing that, we can use it to determine whether or not any chemical equation is consistent with the second law and therefore whether or not it will occur naturally. Finish this module with "On Your Own" questions 13.11–13.12.

SUMMARY OF KEY EQUATIONS AND TABLES IN MODULE 13

Equation 13.1: 1 calorie = 4.184 Joules
Conversion calorie and Joules

Equation 13.2: ΔH = Energy required to break bonds - Energy released when bonds form

Equation 13.3: $\Delta H^\circ = \Sigma\ \Delta H_f^\circ\ (\text{products}) - \Sigma\ \Delta H_f^\circ\ (\text{reactants})$
Standard change in enthalpy for a reaction

Equation 13.4: $\Delta S_{universe} \geq 0$
Second law of thermodynamics in terms of ΔS

Equation 13.5: $\Delta S^\circ = \Sigma\ S^\circ\ (\text{products}) - \Sigma\ S^\circ\ (\text{reactants})$
Standard change in entropy for a reaction

Equation 13.6: $\Delta G = \Delta H - T \cdot \Delta S$ Gibbs free energy equation

Equation 13.7: $\Delta G^\circ = \Sigma\ \Delta G_f^\circ\ (\text{products}) - \Sigma\ \Delta G_f^\circ\ (\text{reactants})$
Standard change in Gibbs free energy for a reaction

Table 13.1: ΔH Relationships in Reactions

Table 13.2: Bond Energies

Table 13.3: Standard Enthalpies of Formation in kJ/mole

Table 13.4: Standard Absolute Entropies in J/(mole·K)

Table 13.5: Standard Gibbs Free Energies of Formation in kJ/mole

ANSWERS TO THE "ON YOUR OWN" QUESTIONS

13.1 (a) The formula for ammonium nitrate is NH_4NO_3. A formation reaction involves using the elements in the compound as reactants and the compound itself as the only product. The unbalanced equation is:

$$N_2 + H_2 + O_2 \rightarrow NH_4NO_3$$

Remember, all reactants are homonuclear diatomics; that's why they each get subscripts of 2. Now we balance the equation and since ΔH is negative, add the energy as a product:

$$2N_2 + 4H_2 + 3O_2 \rightarrow 2NH_4NO_3 + 365.1\ \text{kJ}$$

(b) The formula of manganese (IV) oxide is MnO_2. When a compound decomposes, we write a chemical equation in which the compound is the sole reactant and its constituent elements are the products. The unbalanced chemical reaction is:

$$MnO_2 \rightarrow Mn + O_2$$

Oxygen is a homonuclear diatomic, so it gets a subscript of 2. The equation is already balanced, and since ΔH is positive, we add energy as a reactant:

$$MnO_2 + 519.7\ \text{kJ} \rightarrow Mn + O_2$$

(c) You should recognize H_2CO_3 as an acid and KOH as a base. This is an acid-base reaction that produces salt and water:

$$H_2CO_3 + KOH \rightarrow H_2O + K_2CO_3$$

Now we just balance the equation, and since the reaction releases energy, we add that energy as a product:

$$H_2CO_3 + 2KOH \rightarrow 2H_2O + K_2CO_3 + 21\ \text{kcals}$$

13.2 To solve virtually any problem involving chemical reactions, we first must find the balanced chemical equation. Realizing that chlorine is a homonuclear diatomic, we can say that the unbalanced equation is:

$$C + Cl_2 \rightarrow CCl_4$$

This equation is rather simple to balance:

$$C + 2Cl_2 \rightarrow CCl_4$$

Now we just need to draw the Lewis structures:

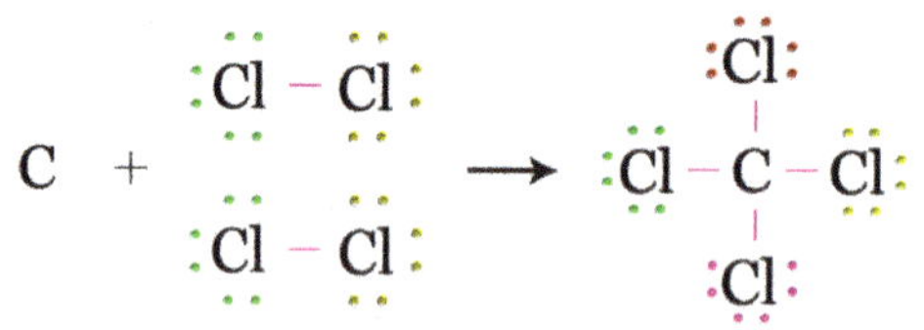

In this reaction, then, 2 Cl-Cl bonds are broken and 4 C-Cl bonds are formed. According to table 13.2, the bond energies for these bonds are 240 kJ/mole and 327 kJ/mole, respectively.

$$\Delta H = (2\ \cancel{\text{mole}} \times 240\ \frac{\text{kJ}}{\cancel{\text{mole}}}) - (4\ \cancel{\text{moles}} \times 327\ \frac{\text{kJ}}{\cancel{\text{mole}}}) = -830\ \text{kJ}$$

The change in enthalpy for this reaction, then, is -830 kJ, indicating that it is an exothermic reaction. Note that we could report our answer only to the tens place in this problem because the bond energy of the Cl-Cl bond is reported to the tens place.

13.3 The reaction in question is a combustion reaction, and by definition, the unbalanced equation would be:

$$C_2H_2 + O_2 \rightarrow CO_2 + H_2O$$

This balances to:

$$2C_2H_2 + 5O_2 \rightarrow 4CO_2 + 2H_2O$$

To use bond energies to calculate ΔH, we need to see the Lewis structures of the chemicals involved:

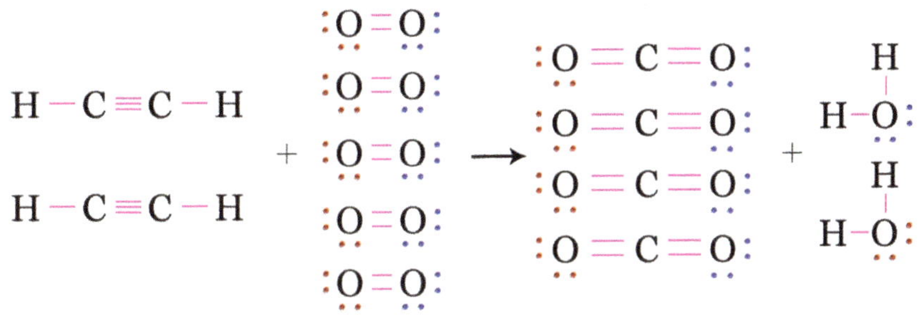

Now we can see exactly what bonds need to be broken and what ones need to be formed to make this reaction work. We need to break 2 C≡C bonds (each of which has an energy of 835 kJ/mole), 4 C-H bonds (each of which has an energy of 411 kJ/mole), and 5 O=O bonds (each of which has an energy of 494 kJ/mole). In the process, 8 C=O bonds (worth 799 kJ/mole each) and 4 H-O bonds (worth 459 kJ/mole each) are formed. Equation 13.2, then, looks like this:

$$\Delta H = [(2\ \cancel{\text{moles}} \times 835\ \frac{\text{kJ}}{\cancel{\text{mole}}}) + (4\ \cancel{\text{moles}} \times 411\ \frac{\text{kJ}}{\cancel{\text{mole}}}) + (5\ \cancel{\text{moles}} \times 494\ \frac{\text{kJ}}{\cancel{\text{mole}}})] -$$

$$[(8\ \cancel{\text{moles}} \times 799\ \frac{\text{kJ}}{\cancel{\text{mole}}}) + (4\ \cancel{\text{moles}} \times 459\ \frac{\text{kJ}}{\cancel{\text{mole}}})]$$

$$\Delta H = -2{,}444\ \text{kJ}$$

The ΔH for this reaction is -2,444 kJ, and it is exothermic.

13.4 Since the problem says that we have to use Hess's law to solve it, we will use Hess's law. First we need a balanced equation:

$$2C_4H_{10}\,(g) + 13O_2\,(g) \longrightarrow 8CO_2\,(g) + 10H_2O\,(g)$$

Now that we have a balanced chemical equation, we must check to see if table 13.3 contains all of the substances in their proper phases. We don't need the ΔH_f° for O_2 (g) since it is the elemental form of oxygen, so table 13.3 contains everything that we need. Now we just have to apply equation 13.3:

$$\Delta H = [(8\,\cancel{\text{moles}} \times -394\,\frac{\text{kJ}}{\cancel{\text{mole}}}) + (10\,\cancel{\text{moles}} \times -242\,\frac{\text{kJ}}{\cancel{\text{mole}}})] -$$

$$(2\,\cancel{\text{moles}} \times -126\,\frac{\text{kJ}}{\cancel{\text{mole}}}) + (13\,\cancel{\text{moles}} \times 0\,\frac{\text{kJ}}{\cancel{\text{mole}}})]$$

$$\Delta H = -5.320 \times 10^3 \text{ kJ}$$

13.5 To solve this problem, we first need to determine the chemical equation. According to the problem, CaO (s) and CO_2 (g) are the reactants because the problem says that they are reacted together. The product is an ionic compound containing the polyatomic ion CO_3^{2-}. Based on our rules for determining the formula of an ionic compound from its name (Modules 4 and 5), we can determine that the product was $CaCO_3$ (s). So our chemical equation is:

$$CaO\,(s) + CO_2\,(g) \longrightarrow CaCO_3\,(s)$$

Table 13.3 contains all of these substances in their proper phase, so we just need to apply equation 13.3:

$$\Delta H^\circ = [(1\,\cancel{\text{mole}} \times -1{,}207\,\frac{\text{kJ}}{\cancel{\text{mole}}})] - [(1\,\cancel{\text{mole}} \times -635\,\frac{\text{kJ}}{\cancel{\text{mole}}}) + (1\,\cancel{\text{mole}} \times -394\,\frac{\text{kJ}}{\cancel{\text{mole}}})] = -178 \text{ kJ}$$

From this ΔH°, we see that the reaction is exothermic. This is one way that shellfish can keep themselves warm if they get too cold. They just produce more shell, and the exothermic reaction that makes the shell warms the shellfish!

13.6 Since this problem asks about energy in relationship to a chemical reaction, the first thing we must do is calculate the ΔH of the reaction. To do that, we need a balanced chemical equation for the combustion of ethane:

$$2C_2H_6\,(g) + 7O_2\,(g) \longrightarrow 4CO_2\,(g) + 6H_2O\,(g)$$

Since all of the substances in this equation that are not elements appear in table 13.3, we can use Hess's law to determine the ΔH°:

$$\Delta H^\circ = [(4\,\cancel{\text{moles}} \times -394\,\frac{\text{kJ}}{\cancel{\text{mole}}}) + (6\,\cancel{\text{moles}} \times -242\,\frac{\text{kJ}}{\cancel{\text{mole}}})] - (2\,\cancel{\text{moles}} \times -84.7\,\frac{\text{kJ}}{\cancel{\text{mole}}}) = -2{,}859 \text{ kJ}$$

This means that energy is a product in the reaction:

$$2C_2H_6\,(g) + 7O_2\,(g) \longrightarrow 4CO_2\,(g) + 6H_2O\,(g) + 2{,}859 \text{ kJ}$$

Now we are finally ready to do stoichiometry. To do this, we first convert to moles of ethane:

$$\frac{500.0 \cancel{\text{ g } C_2H_6}}{1} \times \frac{1 \text{ mole } C_2H_6}{30.1 \cancel{\text{ g } C_2H_6}} = 16.6 \text{ moles } C_2H_6$$

Then we realize that the chemical equation tells us that 2 moles of ethane give us 2,859 kJ of energy:

$$\frac{16.6 \cancel{\text{ moles } C_2H_6}}{1} \times \frac{2{,}859 \text{ kJ}}{2 \cancel{\text{ moles } C_2H_6}} = 2.37 \times 10^4 \text{ kJ}$$

Burning 500.0 g of ethane provides 2.37×10^4 kJ of energy.

13.7 To solve this problem, we first need to calculate ΔH. This can be done with Hess's law since we already have the equation, and everything in the equation is in table 13.3:

$$\Delta H^{o} = (1 \cancel{\text{ mole}} \times -814 \frac{\text{kJ}}{\cancel{\text{mole}}}) - [(1 \cancel{\text{ mole}} \times -242 \frac{\text{kJ}}{\cancel{\text{mole}}}) + (1 \cancel{\text{ mole}} \times -396 \frac{\text{kJ}}{\cancel{\text{mole}}})] = -176 \text{ kJ}$$

This means that energy is a product in the reaction:

$$H_2O\,(g) + SO_3\,(g) \longrightarrow H_2SO_4\,(l) + 176 \text{ kJ}$$

Now we can do stoichiometry:

$$\frac{25.0 \cancel{\text{ g } H_2SO_4}}{1} \times \frac{1 \text{ mole } H_2SO_4}{98.1 \cancel{\text{ g } H_2SO_4}} = 0.255 \text{ moles } H_2SO_4$$

$$\frac{0.255 \cancel{\text{ moles } H_2SO_4}}{1} \times \frac{176 \text{ kJ}}{1 \cancel{\text{ mole } H_2SO_4}} = 44.9 \text{ kJ}$$

When 25.0 grams of H_2SO_4 are made in this way, 44.9 kJ of energy are produced.

13.8 For reaction A, we know 2 things. It is very hard to start. This tells us that the difference in the energy of the reactants and the intermediate state is large. In other words, our energy diagram needs to contain a large hump. Second, we know that the reaction is highly exothermic. This means that the reactants must have an energy that is much larger than the products. Putting that information into an energy diagram, we get:

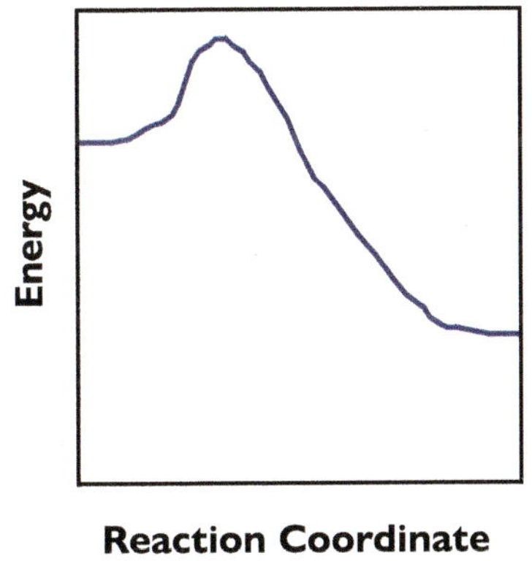

For reaction B, we know that it was easy to start. This means that the difference in energy between the reactants and the intermediate state is small. In other words, this energy diagram has a small hump. Secondly, although the reaction is exothermic, it doesn't release much energy. The reactants will have a higher energy than the products, but not by much:

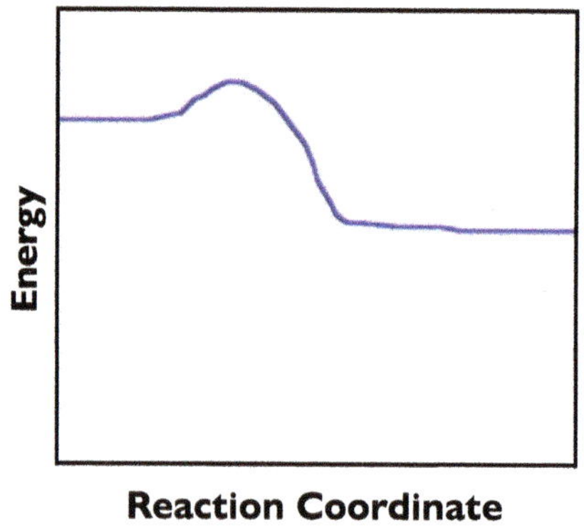

For reaction C, we know it is hard to start. This means that the difference in energy between the reactants and intermediate state is large. In other words, this energy diagram has a large hump. Second, although the reaction is endothermic, it is only slightly endothermic because the container gets only slightly cold during the reaction. This means our enthalpy or the distance between the starting and ending energies should be small. Thus, the reactants have a lower energy than the products by a small distance.

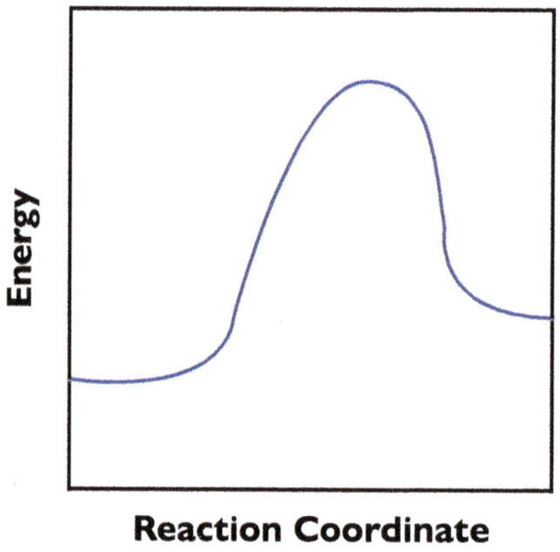

For reaction D, we know it is easy to start. This means the difference in energy between the reactants and intermediate state must be smaller than what we drew for reaction C. In other words, this energy diagram has a small hump. Second, the reaction is very endothermic, which we know because the container was said to get very very cold during the reaction. This means our enthalpy or the distance between the starting and ending energies should be large. Thus, the reactants have a lower energy than the products by a larger distance than what was drawn for reaction C.

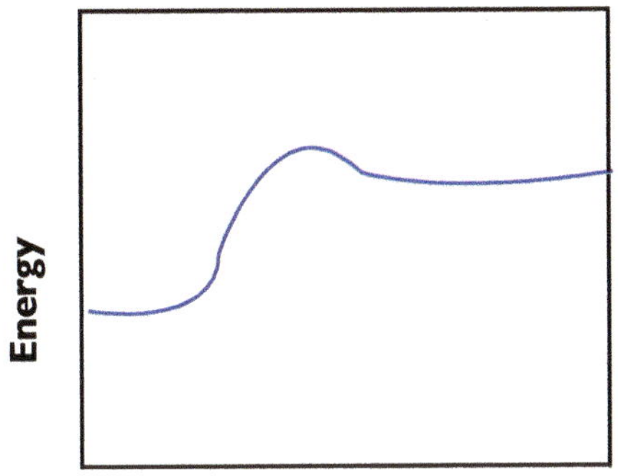

13.9 (a) This reaction has 1 gas molecule on the reactant side of the equation and 0 on the product side. This means that the reactants have more entropy than the products, so **ΔS is negative.**

(b) In this reaction, there are 2 molecules of gas on the product side and 3 on the reactant side. The reactants have higher entropy than the products, so **ΔS is negative.**

(c) This reaction has no molecules or atoms in the gas phase, so we must look at the liquid (or aqueous) phase. There are 3 molecules in the aqueous phase on the product side and only 1 molecule in the aqueous phase on the reactant side. As a result, **ΔS is positive.**

13.10 Table 13.4 has all of the information we need, so we just set this up like a Hess's law problem:

$$\Delta S^\circ = \left(1\,\cancel{\text{mole}} \times 92.9\,\frac{\text{J}}{\cancel{\text{mole}}\cdot\text{K}}\right) - \left[\left(1\,\cancel{\text{mole}} \times 38.2\,\frac{\text{J}}{\cancel{\text{mole}}\cdot\text{K}}\right) + \right.$$

$$\left.(1\,\cancel{\text{mole}}) \times 214\,\frac{\text{J}}{\cancel{\text{mole}}\cdot\text{K}}\right)\Big] = -159\,\frac{\text{J}}{\text{K}}$$

13.11 This problem gives us ΔH and ΔS and asks us at what temperature the reaction is spontaneous. In other words, we need to see for what temperatures ΔG is negative.

$$\Delta H - T\Delta S < 0$$

To use this equation, though, we need to get our units consistent:

$$\frac{123\,\cancel{\text{J}}}{\text{K}} \times \frac{1\text{ kJ}}{1{,}000\,\cancel{\text{J}}} = 0.123\,\frac{\text{kJ}}{\text{K}}$$

Now we can use the equation:

$$20.0\text{ kJ} - T \times 0.123\,\frac{\text{kJ}}{\text{K}} < 0$$

We solve an inequality just like a regular algebraic equation, except when we multiply or divide both sides by a negative value. If that happens, we must switch the direction of the inequality sign:

$$-T \times 0.123 \frac{kJ}{K} < -20.0\ kJ$$

$$T > \frac{-20.0\ \cancel{kJ}}{-0.123 \frac{\cancel{kJ}}{K}}$$

$$T > 163\ K$$

13.12 This question asks us to determine the spontaneity of a chemical reaction. The only way we can do this is to calculate ΔG. Since this reaction is supposed to run at 298 K, we can use table 13.5 and equation 13.7:

$$\Delta G° = [(1\ \cancel{mole} \times -237 \frac{kJ}{\cancel{mole}}) + (1\ \cancel{mole} \times -175 \frac{kJ}{\cancel{mole}})] - (2\ \cancel{mole} \times -50.8 \frac{kJ}{\cancel{mole}}) = -3.10 \times 10^2\ kJ$$

This means that the reaction is spontaneous at 298 K.

NOTE: All Lewis structure illustrations in this module are by Connor Hughes.

STUDY GUIDE FOR MODULE 13

REVIEW QUESTIONS

1. In a chemical reaction, where is the potential energy? Where is the kinetic energy?

2. If a chemical reaction has a positive ΔH, will the beaker that contains the reaction feel hot or cold once the reaction is finished?

3. Why is Hess's law a more exact way of determining ΔH than the technique that uses bond energies?

4. What is a state function? Give 2 examples.

5. Which of the following substances will have a ΔH_f° of 0?

   $NaOH\,(aq)$, $Na^+\,(aq)$, $O_2\,(g)$, $O\,(g)$, $Cl_2\,(g)$, $H_2\,(l)$

6. Which of the following diagrams indicates an endothermic reaction? What is the ΔH of both reactions?

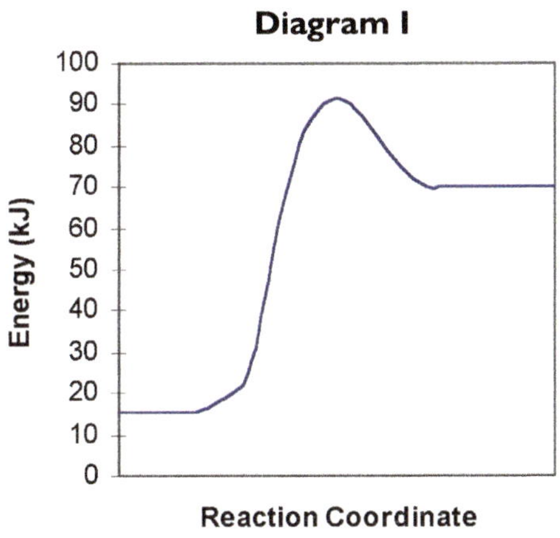

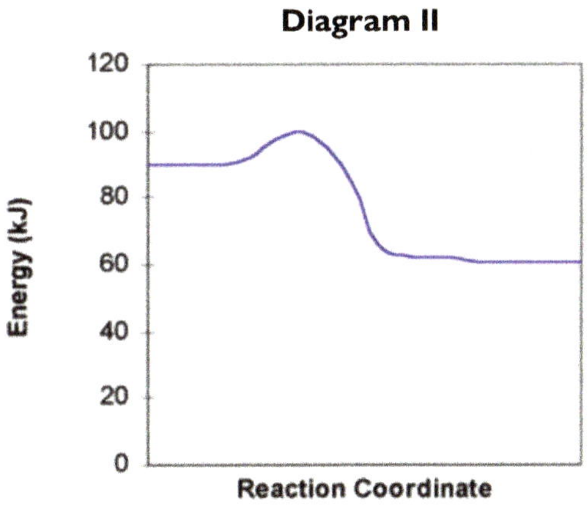

7. Draw an energy diagram for a reaction that has a large activation energy and a ΔH of 0.

8. If you have two 50.0 kg blocks of copper, and one is at a temperature of 50°C while the other is at a temperature of 25°C, which has more entropy?

9. If a chemical reaction is exothermic but has a negative ΔS, what could you do to the temperature to make it possible to run the reaction?

PRACTICE PROBLEMS

(You will need to use tables 13.2–13.5)

1. The chlorine smell of pool water is not, in fact, due to chlorine. When chlorine hits water, some of it reacts with the water in the following way:

$$2Cl_2 + 2H_2O \longrightarrow 4HCl + O_2$$

The smell of a chlorinated pool is often the smell of the HCl produced in this reaction. What is the ΔH of this reaction?

2. Benzene (C_6H_6) is a toxic liquid with a $\Delta H_f°$ of 49.00 kJ/mole. The best method for disposing of benzene is to burn it in the presence of excess oxygen. What is the ΔH° for this reaction? All phases are gases except for benzene, so use Hess's law.

3. When coal is burned in a power plant, it usually contains a small amount of sulfur contamination. The sulfur burns along with the coal to make SO_2. This sulfur dioxide then begins to travel up the smokestack, where it reacts with oxygen as follows:

$$2SO_2\,(g) + O_2\,(g) \longrightarrow 2SO_3\,(g)$$

If the ΔH° of this reaction is -198 kJ, what is the $\Delta H_f°$ of $SO_2(g)$?

4. Using the ΔH you got in problem 2, calculate the number of kJ of energy released when 250.0 grams of benzene are burned in excess oxygen.

5. Suppose you synthesized ammonia with the following reaction:

$$2NO_2(g) + 7H_2(g) \longrightarrow 2NH_3(g) + 4H_2O(g)$$

How much energy will be released if 70.0 g of NH_3 are formed?

6. What is the sign of ΔS for the following reactions?
a. $NaCl\,(aq) + H_2O\,(l) + CO_2\,(g) \longrightarrow NaHCO_3\,(aq) + HCl\,(aq)$
b. $NH_4Cl\,(s) + H_2O\,(l) \longrightarrow HCl\,(g) + NH_4OH\,(g)$
c. $SF_6\,(g) + O_2\,(g) \longrightarrow SO_2\,(g) + 3F_2\,(g)$

7. What is the ΔS° of the following reaction?

$$H_2O\,(l) + CaO\,(s) \longrightarrow Ca(OH)_2\,(s)$$

8. If the ΔH of a certain reaction is -1,023 kJ/mole and the ΔS is -324 J/(mole·K), what is the temperature range for which this reaction is spontaneous?

9. Determine the activation energy and ΔH for the following reaction:

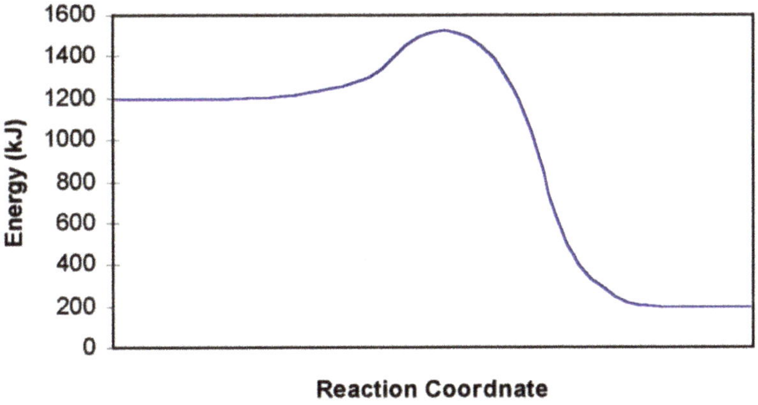

Is this reaction endothermic or exothermic?

10. Is the following reaction spontaneous at 298 K?

$$2C_6H_6\,(l) + 12H_2\,(g) + 3O_2\,(g) \longrightarrow 6C_2H_6O\,(l)$$

Cartoon by Speartoons, Inc.

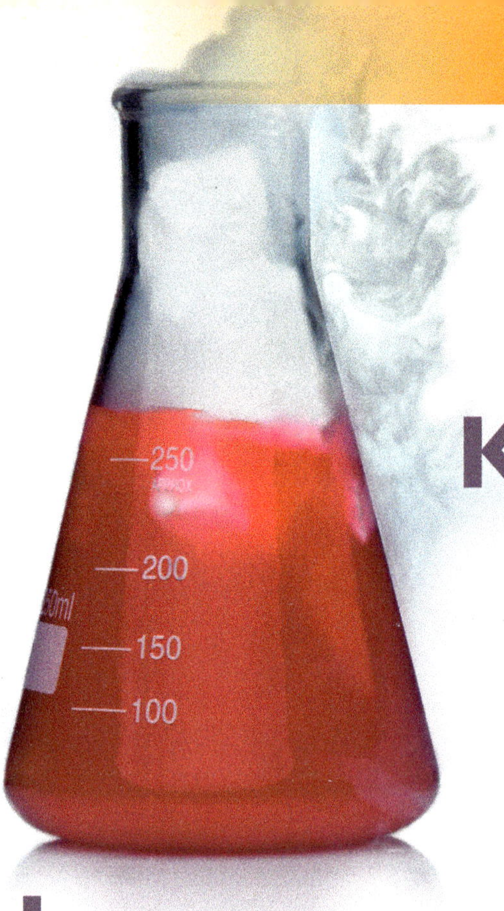

KINETICS

first reactions

Will a chemical reaction need something to start it? How fast will it progress? How long it will last? Sometimes we use equations to help us model real-world reactions. Why will a reaction wait for us to heat the solution? Why did it take a long time? These are some of the questions we'll find answers to in this module.

In the previous module, we spent a great deal of time learning how we could determine whether or not a reaction was spontaneous. We learned that a reaction had to meet certain enthalpy and entropy requirements before it could occur. Since enthalpy and entropy are both concepts in the field of thermodynamics, chemists often call these requirements thermodynamic criteria. We could say that a reaction will not occur spontaneously unless it meets certain thermodynamic criteria.

When studying a chemical reaction, we must consider other criteria as well as thermodynamic criteria. We should also understand how quickly the reaction proceeds. For example, consider the decomposition of hydrogen peroxide:

$$2H_2O_2\ (aq) \longrightarrow 2H_2O\ (l) + O_2\ (g)$$

The standard Gibbs free energy ($\Delta G°$) for this reaction is -190 kJ/mole. *Standard* indicates that it is a reaction at room temperature and normal atmospheric pressure, and the negative sign indicates that the reaction is spontaneous. However, if you watched a sample of aqueous hydrogen peroxide all day, you would never see bubbles of oxygen being produced, as you would expect from the above reaction. In fact, bottles of hydrogen peroxide solution can be left on the shelf for years and will still have some hydrogen peroxide solution in them once they are opened, although the solution will not be as concentrated as it was when it was purchased.

If the decomposition of hydrogen peroxide is spontaneous (as our thermodynamic criteria say it is), why can't we see it decompose? Why does it sit on the shelf without decaying away completely? The answer is quite simple. The reaction occurs very, very slowly. Even though this reaction is thermodynamically spontaneous, it is not necessarily a good way of producing oxygen since it proceeds so slowly.

In studying a chemical reaction, then, we need to consider 2 very important things. First, we need to look at the thermodynamic criteria. After all, studying a reaction that will not occur would be rather dull. Once we have examined the reaction using thermodynamic criteria, we must also look at how fast the reaction can go. Even if the reaction is thermodynamically spontaneous, if it doesn't proceed quickly enough, there is no use running it.

REACTION KINETICS

When we study how fast a reaction proceeds, we call that a study of the reaction kinetics (kih net' iks). In kinetics, the first important definition we come to is the rate of the chemical reaction:

Reaction rate—The rate of change of a product in a chemical reaction.

In mathematical terms, we could say:

$$R = \frac{\Delta\,[\text{product}]}{\Delta t} = -\frac{\Delta\,[\text{reactant}]}{\Delta t} \qquad \text{Equation 14.1}$$

At first, this equation might look quite mysterious, but it isn't really that bad. First, R stands for rate. Second, chemists often abbreviate concentration with square brackets, so [product] means the concentration of a product in the reaction, and [reactant] refers to the concentration of a reactant in the equation. As always, the Δ means change in, and t refers to time.

Equation 14.1 says that if we measure how much the concentration of a reaction product changes and divide it by the time it took to make that change, we will get the rate of the chemical reaction. Think about the units that will result from such a calculation. If we take concentration (usually expressed in molarity) and divide by time (usually expressed in seconds), we will get a unit of molarity/second (pronounced molarity per second), which is usually abbreviated as M/s. This is the unit for the rate of a chemical reaction.

Notice that we can also calculate the rate of a chemical reaction by using the change in concentration of one of the reactants in the equation. If we do that, we need to take the negative of that change to calculate the rate of a chemical reaction. This should make sense. In a chemical reaction, products are made; therefore, throughout the course of a chemical reaction, the concentration of the products increases. As a result, when we take the final concentration and subtract the initial concentration, then the change in concentration of a product is positive. Since reactants are used up in a chemical reaction, their concentration decreases. The change in concentration of the reactants, therefore, is negative. That's why the negative sign is present in front of the reactant quotient in Equation 14.1. Since the change in the concentration of a reactant is negative, we need to take the negative of that change to ensure that both sides of Equation 14.1 are equal.

FACTORS THAT AFFECT THE KINETICS OF A CHEMICAL REACTION

Since we now know that chemical reactions do not occur instantaneously but instead take a certain amount of time depending on the reaction rate, it is important to determine what factors affect the rate of a chemical reaction. For example, if a reaction is slow, is there any way to speed it up? Alternatively, suppose a reaction happens too rapidly. Is there a way to slow it down? Perform Experiment 14.1 to figure that out.

EXPERIMENT 14.1

PURPOSE: To learn how concentration and temperature affect chemical reaction rates.

MATERIALS:

- Liquid toilet bowl cleaner (The Works was used when this experiment was tested, but any liquid that contains hydrochloric acid or hydrogen chloride should work. Stay away from the thickened toilet bowl cleaners that cling, however.)
- Antacid tablets (TUMS works best. In principle, any antacid tablet that has calcium carbonate as its main ingredient should work.)
- Sharp, non-serrated knife
- Spoon
- Stirring rod (something thin enough to stir the contents of the test tubes)
- 4 test tubes (Small glass containers will work.)
- Flame heater (An oven burner will work.)
- Large beaker (A short, fat glass will work.)
- Small beaker (A short, fat glass will work. If you are not using test tubes, make sure that one of your glass containers fits into this.)
- Rubber gloves
- Safety goggles

QUESTION: How does varying concentration and temperature affect the rate of a reaction?

HYPOTHESIS: Write a hypothesis for this experiment.

PROCEDURE:

1. Fill your small beaker ¾ full with tap water and start heating it. The water needs to come to a boil.
2. While you are waiting for the water to boil, put on the gloves.
3. Fill 3 of the test tubes halfway with toilet bowl cleaner. We will call these tube 1, tube 2, and tube 3.
4. Put the 3 tubes in the large beaker (not the one being heated) so that they stand up. The large beaker will act as a test tube holder.
5. To the last test tube (4), add only enough toilet bowl cleaner to fill it ⅛ full. Use tap water to dilute the toilet bowl cleaner until the test tube is ½ full, like the others.
6. Stir the resulting diluted solution.
7. Put that tube in the large beaker as well.
8. Use the knife to cut an antacid tablet into 4 equal pieces.
9. Grind one of those pieces into a fine powder with a spoon. When you are done

grinding, fill the spoon with the powder you just created.

10. Once the water in the beaker is boiling, remove the beaker from the flame so that the boiling stops.
11. Put test tube 3 into the hot water so that the toilet bowl cleaner in it gets hot.
12. Place a piece of antacid tablet into test tubes 1, 3, and 4.
13. Take the powdered antacid in the spoon and pour it into tube 2.
14. Observe what goes on in each test tube.
15. The antacid should be bubbling, indicating the production of a gas. Use your stirring rod to stir the contents of each test tube several times over the course of 5 minutes.
16. After the 5 minutes are up, observe what is left in each of the test tubes.
17. Clean up and return everything to the proper place.

CONCLUSION: Write something about the outcome and your hypothesis.

Let's evaluate what happened in each test tube in the experiment. In test tube 1, there will probably be a small amount of antacid tablet left. However, all of the powder should be gone in test tube 2, and the piece of tablet should be gone from test tube 3. In test tube 4, there should still be antacid tablet present. As a matter of fact, there should be a lot more antacid tablet left in test tube 4 than there was in test tube 1. What explains these results?

The reaction that took place in this experiment comes from the hydrochloric acid in the toilet bowl cleaner reacting with the calcium carbonate in the antacid tablet:

$$\underset{\text{acid in the toilet bowl cleaner}}{2HCl\,(aq)} + \underset{\text{calcium carbonate in the antacid tablet}}{CaCO_3\,(s)} \rightarrow H_2O\,(l) + CaCl_2\,(aq) + CO_2\,(g)$$

The carbon dioxide gas formed in this reaction was responsible for the bubbles you saw.

By observing the amount of antacid tablet left after a certain amount of time, you are essentially learning about the rate of the chemical reaction. Since test tube 4 still had a lot of antacid left after 5 minutes, the reaction in test tube 4 proceeded rather slowly. On the other hand, since little or no antacid was left in test tubes 3 and 2, the reaction proceeded rather quickly in those test tubes. Finally, since there was still antacid left in test tube 1, the reaction proceeded slowly in that test tube. However, since there was less antacid left in test tube 1 than there was in test tube 4, the reaction in test tube 1 was faster than the one in test tube 4.

Now that we see the relative rates of the chemical reactions, we can draw some general conclusions from this experiment. First, the only difference between the reaction in test tube 1 and the one in test tube 3 was the temperature. Test tube 3 (with a reaction that was faster than that of test tube 1) was hotter, indicating that chemical reactions have a faster rate at higher temperatures. In addition, the only difference between the reactions in test tubes 1 and 4 was the concentration of the toilet bowl cleaner. Remember that test tube 4 had diluted toilet bowl cleaner in it. Since the reaction in test tube 1 was faster than the one in test tube 4, we can conclude that dilute reactants lead to lower reaction rates. Finally, the only difference between test tube 1 and test tube 2 was that the antacid in test tube 2 was powdered. Since that led to a reaction with a higher rate, we can conclude that powdered solids react more quickly than non-powdered solids. We will see *why* these factors influence reaction rates in a moment. For now, just understand how this experiment leads us to these conclusions.

Before leaving this experiment, please notice one thing. Throughout the previous paragraph, we compared every reaction to the one that occurred in test tube 1. That's because the reaction in test tube 1 was performed under normal conditions (regular concentration of toilet bowl cleaner, room temperature, uncrushed antacid tablet). In the other test tubes, we changed only one thing in comparison to test tube 1. This is a common experimental technique called a controlled experiment. In such an experiment, one test is done under standard (normal) conditions, and all other tests are compared to those standard conditions. The test done under standard conditions is referred to as the control. Ideally, only one thing should change between the control and any other test. This allows us to make better conclusions. In this experiment, test tube 1 was our control, and each of the other 3 test tubes had only one condition different from the control. This allowed us to make firm conclusions regarding how temperature, reactant concentration, and the surface area of a solid affect the rate of a chemical reaction.

Now that the experiment has shown us how certain factors affect the rate of a chemical reaction, it is time to learn *why* they do. In Module 3, you learned that the valence electrons in an atom are responsible for its chemical behavior. In Module 4, you learned that molecules are formed through the transfer or the sharing of those valence electrons. Ionic compounds are formed when valence electrons are transferred, while covalent compounds are formed when valence electrons are shared.

Based on these principles, it should be easy to see that for a chemical reaction to proceed, the reactants must be able to get close enough to each other so that their electrons can either be transferred or rearranged. To do this, the reactants must essentially collide into one another. That's the only way that they can get close enough to do what must be done.

To react, chemicals must collide with one another so that their electrons can be either transferred or rearranged.

This is the guiding principle through which we can understand why certain factors influence the rate of a chemical reaction.

For example, consider the effect of temperature. We learned through our experiment that when temperature increases, the rate of a chemical reaction increases. If we think about this result in terms of the principle we just learned, the reason for it should become obvious. After all, if the reactants must collide together for the chemical reaction to proceed, then the faster each one is moving, the more likely it is for a collision to occur. Based on the kinetic theory of matter, when temperature increases, the motion of the molecules or atoms in a substance increases.

Higher temperature increases chemical reaction rate because the reactant molecules (or atoms) move faster, increasing the chance for a collision.

This is why increasing temperature increases the rate of a chemical reaction. It is a fact worth remembering.

Next, consider the effect of concentration. In the experiment, we learned that as the concentration of a reactant decreases, the chemical reaction rate decreases as well. Once again, this can be understood completely in terms of collisions between reactant molecules. Remember, the concentration of a substance tells us how many molecules (or atoms) of that substance exist within a certain volume. When concentration decreases, there are fewer molecules (or atoms) in the same amount of volume. In other words, the reaction vessel is less crowded, so the chance for collisions between reactant molecules (or atoms) decreases.

Decreasing concentration decreases the chemical reaction rate because there are fewer reactant molecules (or atoms) to collide with one another.

Of course, if we want to slow a reaction down even more, we can decrease *all* reactant concentrations rather than just the concentration of one reactant. After all, the fewer reactants there are in a certain volume, the less crowded it will be and the lower the reaction rate will be.

Finally, we need to see why crushing the antacid tablet increased the rate of our chemical reaction. When a solid object is crushed, the molecules spread out over a larger area. Chemists generally call this an increase in the surface area of a reactant, which will increase the rate of the reaction. Once again, we can understand this effect in terms of collisions. When we crush a solid, it can be more evenly distributed through the reaction vessel. As a result, the chance for a collision between this reactant molecule (or atom) and another reactant molecule (or atom) goes up. After all, a collision is more likely when the molecules (or atoms) involved mingle closely together. When a solid exists as a chunk, its molecules (or atoms) cannot easily mingle with the other substances in the reactant vessel. If we crush the solid, its molecules (or atoms) can travel throughout the vessel, making them much more likely to collide with something else.

Increasing the surface area of a reactant increases the reaction rate because the molecules (or atoms) of the reactant can more easily mingle with the molecules (or atoms) of the other reactants.

In the end, then, we can see that all the major factors affecting the rate of a chemical reaction can be understood in terms of the collisions that must take place between reactants in a chemical reaction. It is important for you to know these 3 factors, how they affect rate, and why. Many times chemists must make decisions about how to change the rate of a chemical reaction. These facts help them to make such decisions. See if you understand these principles by answering "On Your Own" questions 14.1–14.2.

ON YOUR OWN

14.1 Two containers are filled with the same mass of both nitrogen and hydrogen gas. If the first container is significantly larger than the second, in which container will the reaction be fastest?

14.2 The muscles in an animal's body are run by chemical reactions. When these reactions occur, they contract the muscle, making it move. A biologist is studying a lizard and notices that when the lizard is in a cold place, it moves very slowly, but when it is in a warm place, it moves very quickly. Explain this observation in terms of chemical kinetics.

THE RATE EQUATION

Now that we know the factors that affect the rate of a chemical reaction, it is time to make our study of chemical kinetics a little more detailed. Do you remember the equation we gave you at the beginning of this module for calculating the rate of a reaction? If you don't, turn back a few pages and review Equation 14.1. This equation is very limited in its result of calculating the rate, R. The rate calculated using Equation 14.1 is good only as an average over the time interval for which ΔT was calculated. So if we calculate R after 30 seconds, then that R from Equation 14.1 is only good at that 30-second mark.

Why is that? What is happening to the concentrations of the reactants as time increases? Our experiment results showed that R gets slower as concentrations get smaller. So we should expect that an R calculated at 45 seconds would be smaller than an R calculated at 30 seconds.

Is there a way to figure out what the rate would be at any time during a chemical reaction? Yes. Chemists can write mathematical equations to represent the rate of a chemical reaction at any time. These mathematical equations are based on the concentration of the reactants in the chemical equation since the concentration of the reactants in a chemical reaction has a profound influence on the rate of the chemical reaction.

We will use the following generic chemical equation to represent all chemical reactions.

$$aA + bB \longrightarrow cC + dD$$

This expression represents a general chemical equation that has reactants A and B, which react to form the products C and D. The lowercase letters (a, b, c, and d) represent the stoichiometric coefficients for each of the substances (A, B, C, and D) in the chemical equation.

For this generalized chemical reaction, we can develop a mathematical equation that we call the rate equation:

$$R = k[A]^x[B]^y \qquad \text{Equation 14.2}$$

In this equation, R represents the rate of the chemical reaction, [A] represents the concentration of reactant A, [B] stands for the concentration of reactant B, and k is called the rate constant for the chemical reaction. The letters x and y represent exponents for the concentrations of the reactants.

If all of this seems rather confusing, don't worry. We first have to pick the rate equation apart bit by bit to understand each of its components. Let's start with the 2 concentration terms. If we look at Equation 14.2, what happens to the reaction rate (R) when the concentration of either reactant ([A] or [B]) increases? Based on the equation, we would expect R to increase. So this equation accurately predicts a fact that we have already learned: When reactant concentration increases, the rate of the chemical reaction increases. Remember that when using reactants to calculate the reaction rate (Equation 14.1), we used a negative sign to indicate the loss of reactants over time. The rate equation (Equation 14.2) addresses only the overall speed of the reaction; we don't add a negative sign when using reactants.

Of course, chemists are detail-oriented people. We not only want to know that reaction rate increases when reactant concentration increases, but we also want to know how much it increases. For example, if we double the concentration of a reactant in a chemical equation, will the reaction rate double? Will it triple? Will it quadruple? In fact, all of those things are possible. For some chemical reactions, doubling the concentration of a reactant doubles the reaction rate. For others, doubling the concentration of a reactant quadruples the reaction rate. For other reactions, doubling the concentration of a reactant will increase the reaction rate by a factor of 8! This is where x and y come in. These exponents allow us to determine exactly how much reaction rate changes when reactant concentration changes. These exponents are called the order of the chemical reaction with respect to the reactant they affect; x is the order of the chemical reaction with respect to reactant A, while y is the order of the reaction with respect to reactant B. If we add the 2 together (x + y), we get the overall order of the chemical reaction.

Finally, k represents a number that is unique for each chemical reaction. Its value is totally independent of the concentration of any reactant, and that is why it is called the rate constant. The important thing to remember is that there is only one way to determine x, y, and k: by experiment. To figure out the rate equation for any chemical reaction, we must do some carefully controlled experiments. You will learn how to evaluate those experiments in this course.

Before we jump into a study of these experiments, let's clear up any remaining confusion by applying what we have learned to Example 14.1.

EXAMPLE 14.1

Write the rate equation for the following chemical reaction:

$$N_2 + 3H_2 \longrightarrow 2NH_3$$

According to Equation 14.2, we get the rate equation by just taking the concentration of each reactant and raising it to the power of x or y:

$$R = k[N_2]^x[H_2]^y$$

Of course, that wasn't very hard, but it also wasn't very useful. This is, indeed, the rate equation for the reaction written above, but it tells us virtually nothing. To make the rate equation useful, we need to figure out what x, y, and k are. That's where the experiments come into play.

USING EXPERIMENTS TO DETERMINE THE DETAILS OF THE RATE EQUATION

As we learned in the section above, the rate equation is pretty useless unless we can figure out the values of k, x, and y. Once we have those, the rate equation becomes very powerful because it allows us to predict the rate of the chemical reaction regardless of the concentrations of either reactant. Do you see how figuring out these 3 terms is a pretty important task? There is a standard experimental technique we can perform on any chemical reaction that allows us to calculate the values of k, x, and y.

To perform such an experiment, the first thing we need to do is run the reaction and experimentally measure its rate. That's the tricky part. Generally, to measure the rate of a chemical reaction, we need some pretty fancy equipment. As a result, very few reactions can be studied this way in a high school chemistry course. Even though we cannot perform very many experiments like this, however, we can still learn how to analyze the data from such experiments so that we can determine the values of k, x, and y in the rate equation. Study Example 14.2. It will show you how these experiments are constructed and how to determine k, x, and y from the resulting data.

EXAMPLE 14.2

A chemist runs an experiment designed to determine the rate equation for the following reaction:

$$2ICl + H_2 \rightarrow I_2 + 2HCl$$

To do this, she runs the reaction 3 times, varying the concentration of the reactants:

Trial	Initial Concentration of ICl (M)	Initial Concentration of H_2 (M)	Instantaneous Reaction Rate (M/s)
1	0.25	0.25	0.0102
2	0.25	0.50	0.0204
3	0.50	0.50	0.0408

Use this information to determine the rate equation for this reaction.

Before we begin to solve the problem, let's look at the table and see what it tells us. The columns of the table are labeled to indicate what values they contain. The first column just tells which trial the data come from. The chemist ran the reaction a total of 3 times. Each time she ran it, she called it a trial run of the reaction. The next 2 columns contain the initial concentrations of the 2 reactants. In other words, these are the concentrations that the chemist began the experiment with. The (M) just tells us that these concentrations are expressed in molarity. The last column of the table tells us the instantaneous rate the chemist measured when she mixed the reactants together. If you don't understand why the term *instantaneous* is used, don't worry. We will explain that later in the module. For now, just realize that this column represents the experimentally measured rate of the chemical reaction. The (M/s) just tells us that the units are molarity per second, which, as you should recall, is the standard unit for chemical reaction rate.

Now that we know what the data table is all about, let's start to solve the problem. First, let's write down the rate equation to the best of our abilities. Since ICl and H_2 are the reactants, we know that the basic rate equation should look something like this:

$$R = k[ICl]^x[H_2]^y$$

As always, we have no idea what k, x, and y are, so as it stands, this rate equation is pretty useless. However, we can figure out these terms in the equation if we look at the data from the experiment.

In the first row of data in the table, we see that the chemist started out with equal concentrations of both reactants. She then measured the reaction rate to be 0.0102 M/s. The next row of data tells us that the second time she ran the experiment, she doubled the concentration of H_2, but she left the concentration of ICl alone. These concentrations were chosen very carefully. If the concentration of ICl stayed the same between these 2 trials, what does that tell us about any rate changes that we see between the 2 trials? Since the concentration of ICl was not changed, any difference between trial 1 and trial 2 must be due solely to the influence of the concentration of H_2. Between trial 1 and trial 2, the concentration of H_2 doubled. What happened to the rate? It doubled as well. This information is all we need to calculate the value for y.

Think about it. If the concentration of H_2 doubled and the rate also doubled, what does that tell us about the value for y? It tells us that y = 1. After all, if doubling the concentration of H_2 doubles the rate of the reaction, then rate and H_2 concentration are linearly proportional to each other, which means y = 1. Do you get the idea here? To determine the exponent associated with a reactant's concentration, all we have to do is hold everything else constant and see what happens to the rate when the concentration of only that reactant changes. This will tell us what the exponent must be.

Now that we have determined y, we can work on figuring out the value for x. To do this, we must look at 2 trials that keep the concentration of H_2 the same and vary only the concentration of ICl. Which trials do that? In trials 2 and 3, the concentration of H_2 stays the same (0.50 M), but the concentration of ICl doubles. Any change in rate between trials 2 and 3 is due solely to the influence of ICl. What happened? The concentration of ICl doubled, and the rate doubled as well. This tells us that x = 1, and our rate equation has now become:

$$R = k[ICl][H_2]$$

We are almost done now. All we have to do is figure out a value for k. How do we do that? We just wrote down the rate equation that now contains only one unknown: k. We also have 3 trials of data that tell us the values of every other variable in the equation. To calculate the value of k, all we have to do is choose *any one* of the 3 trials and put that data into the equation. The only unknown in the equation will be k, and we can then solve for it. As we said, we can choose any one of the 3 trials, so we will just use the first one. In that trial, [ICl] = 0.25 M, $[H_2]$ = 0.25 M, and R = 0.0102 M/s. Putting those numbers in the rate equation gives us:

$$R = k\,[ICl][H_2]$$

$$0.0102\,\frac{M}{s} = k \times 0.25M \times 0.25M$$

Solving for k gives us:

$$k = \frac{0.0102\,\frac{\cancel{M}}{s}}{0.25\cancel{M} \times 0.25M} = 0.16\,\frac{1}{M \cdot s}$$

Since k is constant regardless of the concentration of the reactants, this k is the same for all 3 trials, and for any other trials that the chemist might run later on at the same temperature. If we had used the data in either trial 2 or trial 3 to calculate k, we would have gotten the exact same answer. Notice the units on k. When we took the rate in M/s and divided by the first concentration, the molarity unit canceled. However, when we divided by the second concentration, there was nothing to cancel that molarity unit. Also, nothing canceled the second's unit. That's why k has such a strange unit.

In the end, then, the final rate equation for this reaction is:

$$R = 0.16 \frac{1}{M \cdot s} [ICl][H_2]$$

Now that we have the final rate equation, we can predict the rate of this chemical reaction for any concentration of reactants. If the chemist wished to run a fourth trial with [ICl] = 1.0 M and $[H_2]$ = 0.75 M, we could determine the rate of the reaction under these conditions by just plugging those numbers into our rate equation:

$$R = (0.16 \frac{1}{\cancel{M} \cdot s}) \times (1.0 \cancel{M}) \times (0.75\ M) = 0.12 \frac{M}{s}$$

That's why the rate equation is so important. Once we have figured it out, it can be used to determine how fast the reaction goes no matter what the concentration of the reactants.

If this long and complicated example seemed a little confusing, don't worry about it. This is an important subject, so we will do 2 more examples in a moment to help you become more confident in solving problems like the one you just read. Before we do that, it is helpful to sit back and review exactly how problems like these are approached and why they give us the information that they do.

The first thing that we must do in analyzing an experiment such as the one given in Example 14.2 is to look at the data that results from each trial. We need to find 2 trials in which the concentration of one reactant didn't change, but the concentration of the other did change. We can then say that any difference in the rate between the 2 trials is due solely to the reactant whose concentration did change. Then, by comparing the change in the reaction rate to the change in concentration of the reactant, we can determine the value of the exponent attached to the concentration of that reactant. We can then use the same procedure to determine the value of the exponent attached to the other reactant. Finally, after the exponents are determined, we can determine the value for the rate constant by using the data from any one trial, plugging it into the rate equation, and solving for k. Let's see how this all pans out in Example 14.3.

EXAMPLE 14.3

A chemist studies the following reaction:

$$2NO\ (g) + Cl_2\ (g) \longrightarrow 2NOCl\ (g)$$

To determine the rate equation, he runs the reaction 3 different times and comes up with the following data:

Trial	Initial Concentration of NO (M)	Initial Concentration of Cl_2 (M)	Instantaneous Reaction Rate (M/s)
1	0.50	0.50	0.019
2	1.00	0.50	0.076
3	1.00	1.00	0.152

What is the rate equation for this reaction?

Once again, we can start solving this problem by first writing down as much of the rate equation as we already know:

$$R = k[NO]^x[Cl_2]^y$$

Now we need to figure out the values of x and y. To do this, we need to find 2 trials in which the concentration of one reactant changes while the concentration of the other reactant does not. In trials 1 and 2, the concentration of NO doubles, while Cl_2 remains constant. The data from these 2 trials will allow us to calculate the value for x, the exponent for the concentration of NO. The concentration of NO doubled from trial 1 to trial 2, but the instantaneous reaction rate quadrupled. What does x need to be to explain this data? The NO concentration increased by a factor of 2, but R increased by a factor of 4. What do we need to do to turn a factor of 2 into a factor of 4? We need to square it, so x = 2. That way, when the concentration of NO doubles, the rate will go up by a factor of 4.

Now that we have determined x, we need to determine y. To do this, we need to find 2 trials in which the concentration of NO stayed the same but the concentration of Cl_2 changed. Trials 2 and 3 fit this bill. In these trials, the concentration of NO stayed the same, but the concentration of Cl_2 doubled. This will give us the information we need to determine y. The concentration of Cl_2 doubled, and the rate doubled as well. This means y = 1 since rate and concentration of Cl_2 are directly proportional to one another. Therefore, our rate equation becomes:

$$R = k\,[NO]^2[Cl_2]$$

Notice that the exponents in a rate equation do not have to be the same. This particular rate equation tells us that to increase the rate of the chemical reaction, it is more effective to increase the concentration of NO than it is to increase the concentration of Cl_2. After all, the rate depends on the square of the concentration of NO, and it only depends on the concentration of Cl_2 to the first power. Any increase in the concentration of NO will increase the rate much more than the same increase in the concentration of Cl_2.

Now the only thing left to do is to determine the value for k. To do this, we choose any trial and plug the data from that trial into our rate equation. The only unknown will be k, so we can solve for it:

$$R = k\,[NO]^2[Cl_2]$$

$$0.019\,\frac{M}{s} = k \times (0.50\text{ M})^2 \times 0.50\text{ M}$$

$$k = \frac{0.019\,\frac{\cancel{M}}{s}}{0.25\text{ M}^2 \times 0.50\,\cancel{M}} = 0.15\,\frac{1}{M^2 \cdot s}$$

Notice the units on the rate constant this time. Since the rate equation has a square in it, we had to square the 0.50 M concentration of NO. When we squared that concentration, we got 0.25 M^2. This resulted in an M^2 term that could not cancel, making the units on the rate constant look strange indeed. The units for the rate constant depend on the rate equation.

As a result, we cannot know the units for the rate constant until we have determined the exponents in the rate equation. This is a very important point and must be remembered:

The units for the rate constant vary.
They depend on the exponents of the rate equation.

Now that we have the value for the rate constant, we can write down the final rate equation:

$$R = 0.15 \frac{1}{M^2 \cdot s} [NO]^2[Cl_2]$$

To determine the rate equation for the following reaction:

$$C_4H_9Br + OH^- \longrightarrow C_4H_9OH + Br^-$$

a chemist performs a rate study and gets the following results:

Trial	Initial Concentration of C_4H_9Br (M)	Initial Concentration of OH^- (M)	Instantaneous Reaction Rate (M/s)
1	0.10	0.10	0.0010
2	0.20	0.10	0.0020
3	0.10	0.20	0.0010

What is the rate equation for this reaction?

Right now, we know the rate equation will take this form:

$$R = k[C_4H_9Br]^x[OH^-]^y$$

To determine x, we need to find 2 trials in which the concentration of OH^- didn't change but the concentration of C_4H_9Br did. Trials 1 and 2 fit the bill. In these trials, the concentration of C_4H_9Br doubled, and the reaction rate doubled. This tells us that x =1. To determine y, we look for 2 trials in which the concentration of C_4H_9Br didn't change but the concentration of OH^- did. Those would be trials 1 and 3. In those trials, the concentration of OH^- doubled, but the rate stayed the same. What does this tell us? It tells us that the reaction rate is *independent* of the concentration of OH^-. What value does y need to be for the rate equation to be independent of the concentration of OH^-? It needs to equal 0 because any number raised to the 0 power is 1. The rate equation is:

$$R = k[C_4H_9Br][OH^-]^0$$

Since any number raised to the 0 power is equal to 1, we usually just drop the $[OH^-]$ term since it always equals 1. Another way to state this is to say that since the concentration of OH^- has no influence on R, then $[OH^-]$ does not deserve a place in the R equation. Only variables that affect R should be in the equation. So the rate equation can also be expressed as

$$R = k[C_4H_9Br]$$

Now we just need to solve for k by using any of the 3 trials:

$$R = k[C_4H_9Br]$$

$$0.0010\,\frac{M}{s} = k \times 0.10\ M$$

$$k = \frac{0.0010\,\frac{\cancel{M}}{s}}{0.10\ \cancel{M}} = 0.010\,\frac{1}{s}$$

The final rate equation is:

$$\mathbf{R = 0.010\,\frac{1}{s}\,[C_4H_9Br]}$$

In the end, then, we see that one of our earlier statements was not really correct. Initially, we said that increasing the concentration of the reactants in a chemical equation will increase the rate of the reaction. We see that this is not necessarily true for all reactants in a chemical equation. In this particular reaction, we can increase the concentration of OH^- all we want, and we will never, ever, affect the rate of the reaction. We must keep this in mind:

Sometimes, the rate of a chemical reaction will be unaffected by the concentration of one of its reactants. In this case, the exponent of the rate equation is 0 for that reactant.

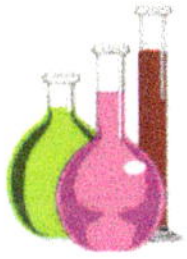

If you purchased the MicroChem kit mentioned in the Student Notes, you can perform Experiment 15 in that kit. It shows you a similar experimental method you can use to determine the rate equation.

Now that you have seen some examples of how to determine the rate equation for a chemical reaction, it is time to cement your newfound skills with "On Your Own" questions 14.3–14.4.

ON YOUR OWN

14.3 A chemist does a rate analysis on the following chemical reaction:

$$H_2(g) + 2NO(g) \longrightarrow N_2O(g) + H_2O(l)$$

He comes up with the following data:

Trial	Initial Concentration of H_2(M)	Initial Concentration of NO (M)	Instantaneous Reaction Rate (M/s)
1	0.70	0.30	0.005670
2	0.70	0.60	0.02268
3	2.10	0.60	0.06804

What is the rate equation for this reaction?

14.4 Ozone can be destroyed by certain pollutants, one being the nitrogen dioxide that is expelled from the exhaust of cars and planes. The reaction that governs this process is:

$$NO_2(g) + O_3(g) \longrightarrow NO_3(g) + O_2(g)$$

A chemist does the following experiment to determine how fast this ozone destruction occurs:

Trial	Initial Concentration of NO_2(M)	Initial Concentration of O_3(M)	Instantaneous Reaction Rate (M/s)
1	0.000100	0.000010	0.088
2	0.000050	0.000020	0.044
3	0.000100	0.000020	0.088

What is the rate equation for this reaction?

RATE ORDERS

Now that you know how to use experimental data to determine rate equations, we need to review some terminology. Remember that the exponents in a rate equation are referred to as the order of the chemical reaction. A chemist would look at the following rate equation:

$$R = k[HNO_3][Cu(OH)_2]^2$$

and say that this reaction is first order with respect to HNO_3 (because the exponent for HNO_3 is 1) and second order with respect to $Cu(OH)_2$ (because the exponent for $Cu(OH)_2$ is 2). In addition, a chemist would say that this rate equation has an overall order of 3 because the sum of the exponents is equal to 3. Chemists often communicate the rate equation in terms of reaction order. You need to be familiar with this kind of communication, so study Example 14.4 and answer "On Your Own" questions 14.5–14.6 to make sure you understand it.

EXAMPLE 14.4

A chemist does a rate analysis of the following reaction:

$$HCl + NaOH \longrightarrow H_2O + NaCl$$

The experiment reveals that this reaction is first order with respect to each of its reactants. What is the rate equation and the overall order of the reaction?

If the reaction is first order with respect to each reactant, then the exponent attached to each reactant is 1. The rate equation, then, is:

$$R = k[HCl][NaOH]$$

Since we have no data to calculate k, we just have to leave it as an unknown in the equation. The overall reaction order is just the sum of all exponents in the rate equation. Since there are 2 exponents and each of them is 1, the overall order of the reaction is 1 + 1 = 2.

ON YOUR OWN

14.5 The following reaction:

$$2PO + Br_2 \longrightarrow 2POBr$$

is determined by experiment to be second order with respect to PO and third order with respect to Br_2. What is the rate equation? What is the overall order of the reaction?

14.6 The following reaction:

$$C_7H_{15}Cl + NaF \longrightarrow C_7H_{15}F + NaCl$$

is determined to be fourth order with respect to $C_7H_{15}Cl$ and 0 order with respect to NaF. What is the rate equation and what is the overall order of the reaction?

You might have already noticed that in some of the rate equations we have developed so far in this module, the orders of the rate equation are often the same as the stoichiometric coefficients in the chemical equation. For example, in the first problem of Example 14.3, the stoichiometric coefficient for each reactant in the chemical equation was the same as the exponent attached to that coefficient in the rate equation. This was also the case in "On Your Own" questions 14.3 and 14.5. It is often tempting to simply use the stoichiometric coefficients in a chemical equation as exponents in the rate equation, but that is not correct! You must always remember:

The reaction rate orders for a chemical reaction must be determined experimentally; they are NOT equal to the stoichiometric coefficients in the chemical equation.

If you take a second year of high school chemistry or when you take chemistry in college, you might learn a way to determine the reaction rate orders from the stoichiometric coefficients in a chemical equation. Such a topic is beyond the scope of this particular course because it is quite complicated.

USING RATE EQUATIONS

Now that you have a thorough understanding of rate equations, we need to spend some time using them to learn something more about how chemical reactions proceed. For example, consider the following hydrogen iodide formation reaction:

$$H_2\,(g) + I_2\,(g) \longrightarrow 2HI\,(g)$$

Experimentally, we can determine that the rate equation is:

$$R = 0.012\,\frac{1}{M^2 \cdot s} \times [H_2][I_2]^2$$

If we decided that we wanted to run this reaction, we could calculate the rate of the chemical reaction for any set of reactant concentrations.

Suppose that we wanted to run this reaction with initial reactant concentrations of $[H_2] = 0.50$ M and $[I_2] = 0.50$ M. We could calculate the rate of this reaction to be:

$$R = 0.012\,\frac{1}{\cancel{M^2} \cdot s} \times 0.50\text{ M} \times (0.50\,\cancel{M})^2 = 0.0015\,\frac{M}{s}$$

But what does this rate tell us? According to the definition of reaction rate, this number tells us that every second, the concentration of the reactants will decrease by 0.0015 M. But if the reactant concentration decreases, what will happen to the rate? It will go down! When the reaction is halfway finished, half of the reactants will be gone and only half will be left. This means that the concentrations of the reactants will only be 0.25 M each. At that point, the rate will be:

$$R = 0.012\,\frac{1}{\cancel{M^2} \cdot s} \times 0.25\text{ M} \times (0.25\,\cancel{M})^2 = 0.00019\,\frac{M}{s}$$

This is almost a factor of 10 lower than the initial rate of the chemical reaction!

So we see that as soon as a reaction begins, the concentrations of the reactants decrease, which causes the reaction rate to decrease as well. The rate that we calculate from the rate equation is not the reaction rate throughout the entire reaction. In fact, the reaction rate will go down immediately because the reactant concentrations keep going down. As a result, we must be sure to make this distinction when we use the rate equation. Therefore, we call the rate calculated from rate equation an instantaneous rate. It tells us that at the instant the reactants have the concentration we used, the rate will be what we calculate. In the next instant, the concentration of the reactants will be lower, reducing the rate from what we had just calculated.

To really bring this fact home, examine Figure 14.1.

FIGURE 14.1
How the Rate of a Reaction Changes with Time

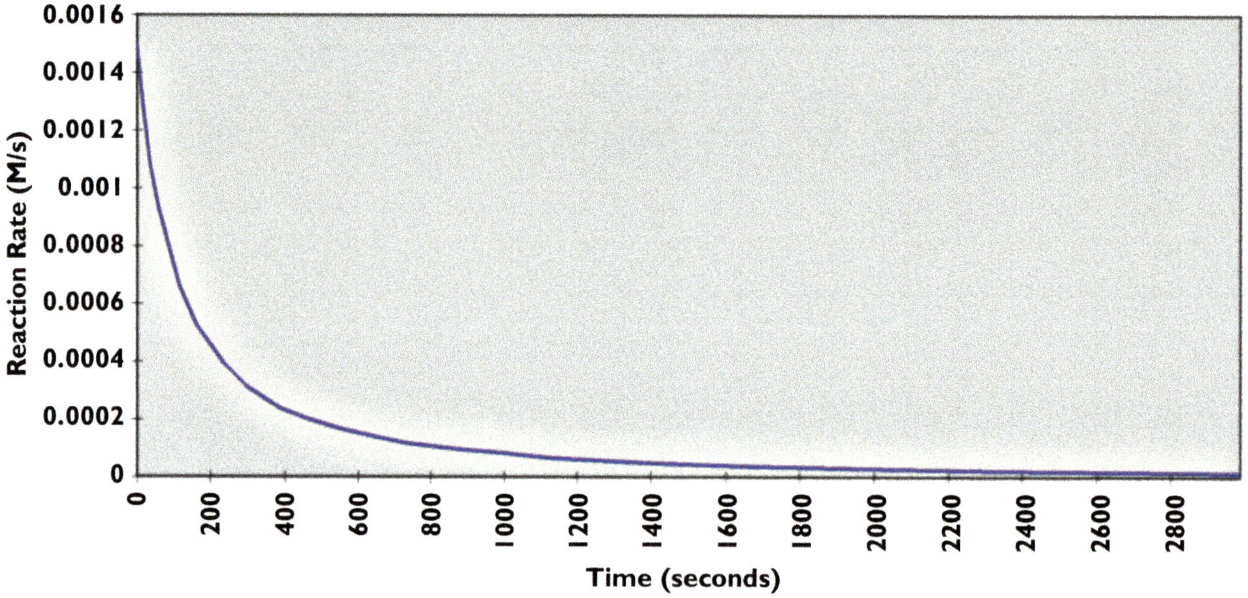

In this figure, the rate of the reaction we have been discussing is plotted versus the time we allow the reaction to run. Notice that when we first start the reaction (at time = 0), the rate is what we calculated initially: 0.0015 M/s. However, the reaction rate lowers right away because as soon as the reaction is allowed to proceed, the concentration of the reactants lowers, reducing the rate of the reaction. After a few thousand seconds, the rate of the reaction has lowered so much that is it very close to 0.

This graph tells us something very important about how chemical reactions proceed. When a chemical reaction begins, its rate is the largest that it will ever be. In the first few minutes of the reaction, the rate slows down enormously. Notice that for this reaction, in only 10 minutes (600 seconds), the rate of the reaction drops from 0.0015 M/s to about 0.00015 M/s, a factor of 10 less than the original rate. After those first few minutes, the rate becomes very slow and does not change much. Notice that for this reaction, in the next 10 minutes, the reaction rate drops from about 0.00015 M/s to about 0.00006 M/s, which represents only a 60% change. In general, then, reactions occur with a flurry of early activity, and then the reaction rate drops off so quickly that the reaction slows down to a crawl.

In Figure 14.2, the concentrations of the reactants in this chemical reaction are plotted for the same time period.

FIGURE 14.2
How the Concentration of a Reactant Changes with Time

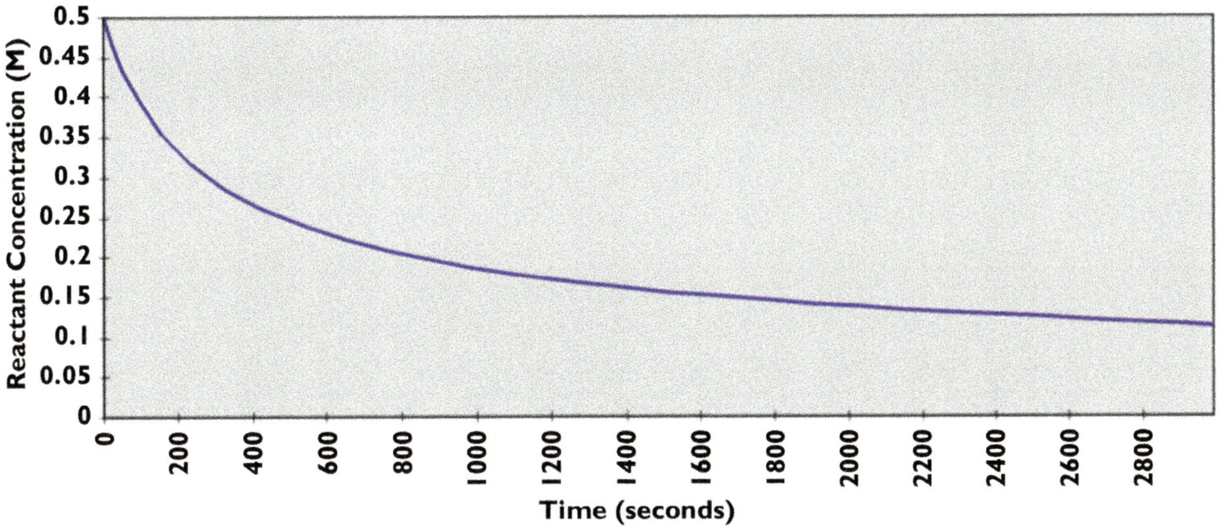

Notice that the concentration of the reactant first drops very quickly. This is a reflection of the fact that at first, the reaction is proceeding quickly. However, as time goes on, the reactant concentration starts changing quite slowly because the reaction rate has dropped so much that the reaction is proceeding at a crawl.

The most important thing to notice about this figure is that it looks like the reactant concentration will never really reach 0. If we extrapolate the curve in Figure 14.2 out to longer times, it will just get flatter and flatter, but it will never actually reach 0! This demonstrates that the reactants will never really go completely away in a chemical reaction.

Chemical reactions never actually finish. They just keep getting slower and slower until they proceed imperceptibly slowly. This means that for any chemical reaction, no matter how long we wait, there will always be some reactants left.

This probably goes against your preconceived notions, but it is nevertheless true.

In all of our stoichiometry problems, we assume that the limiting reactant runs out during the course of a chemical reaction. Does this discussion mean that all of our stoichiometry rules are wrong? No, not really. For the vast majority of chemical reactions, even though the reaction never really finishes, it comes so close to finishing that there is almost none of the limiting reactant left. For purposes of stoichiometry, we can assume that the limiting reactant in a chemical reaction is always used up. Nevertheless, you must remember that this is only an approximation. In reality, when the concentration of the reactants gets very low, the reaction begins to proceed so slowly that it never truly finishes.

TEMPERATURE DEPENDENCE IN THE RATE EQUATION

We must consider one final thing when discussing the rate equation. We learned near the beginning of this module that the rate of a chemical reaction depends on temperature. The higher the temperature, the higher the reaction rate. However, the rate equation has no term that involves temperature. The rate equation accurately portrays the fact that as the concentration of reactants goes up, the rate of the reaction goes up. But there seems to be no temperature dependence in the rate equation. Does this mean that our rate equation is wrong? No, of course not.

Temperature is in the rate equation; it is just hidden so that we cannot really see it. The temperature dependence of reaction rate is taken care of by the rate constant. You see, despite its name, the rate constant is not totally constant. It changes with temperature. Figure 14.3 shows this change by plotting the rate constant for the reaction depicted in the hydrogen iodide formation reaction equation versus temperature.

FIGURE 14.3
How the Rate Constant Depends on Temperature

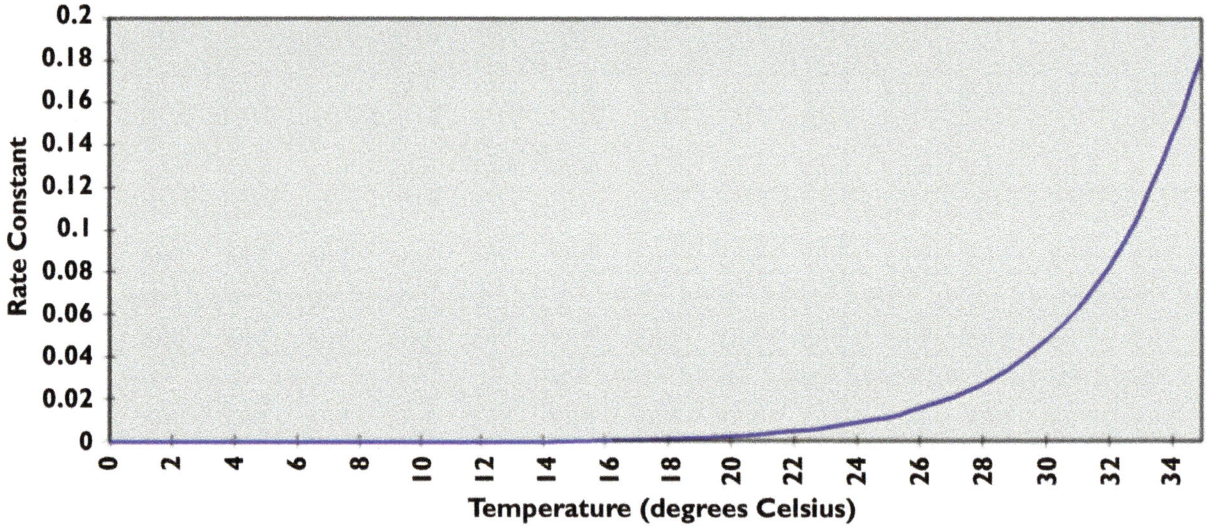

Notice from the figure that for temperatures below 14°C, the rate constant for this reaction is very near 0. This means that at these temperatures, the reaction proceeds rather slowly. After all, since the rate constant multiplies the concentration terms in the rate equation, if the rate constant is near 0, then the rate itself is rather small. Notice also that after 14°C, the rate constant shoots up quickly, indicating that the rate increases dramatically with temperature, as we learned earlier in the module. The fact that the rate constant increases with increasing temperature is important to remember:

The temperature dependence of reaction rate is determined by the variation of the rate constant with temperature. Since the rate constant increases dramatically with increasing temperature, the reaction rate does as well.

There is an equation that relates the rate constant to the temperature at which the reaction is run. It is called the Arrhenius (uh ree' nee us) equation, but the use of that equation is beyond the scope of this course. Instead, we will teach you a general rule of thumb that is applicable for many chemical reactions at the conditions under which you will typically work:

The rate of many chemical reactions doubles for every 10°C increase in temperature.

Please understand that this is just a general rule of thumb that is not applicable in all cases. Nevertheless, it is a good estimate for standard conditions and at least gives you some idea of the temperature dependence of reaction rate. Study Example 14.5 and answer "On Your Own" questions 14.7–14.8 to make sure you remember this rule and can use it.

EXAMPLE 14.5

A chemist performs the following reaction at a temperature of 25°C:

$$2NO\ (g) + Cl_2\ (g) \longrightarrow 2NOCl\ (g)$$

In an earlier example, we determined that its rate equation was as follows:

$$R = k \times [NO]^2[Cl_2]$$

The rate constant for a temperature of 25°C is $0.15\ \frac{1}{M^2 \cdot s}$. What will the instantaneous rate of the reaction be at 25°C if the chemist runs the reaction with both reactants at a concentration of 0.15 M? What will the rate be if he runs the reaction at 45°C instead of 25°C?

To answer the first question in this problem, we merely have to plug all of our numbers into the rate equation. After all, the problem gave us the rate equation, the concentrations of the reactants, and the rate constant. That's all we need to find rate:

$$R = k \times [NO]^2[Cl_2]$$

$$R = 0.15\ \frac{1}{\cancel{M^2} \cdot s} \times (0.15\ \cancel{M})^2 \times 0.15\ M = 0.00051\ \frac{M}{s}$$

We can't simply use the rate equation to answer the second question because the rate constant given is *only* for 25°C, not for 45°C. However, we have the rule of thumb that for every 10°C increase in the temperature, the rate of a chemical reaction doubles. Since 45°C is 20°C above 25°C, the reaction rate doubles (for the first 10°C increase in temperature) and then doubles again (for the second 10°C increase in temperature). We can calculate what the new rate will be by simply multiplying the old rate by 2 and then multiplying by 2 again:

$$\text{At 45°C: } R = 0.00051\ \frac{M}{s} \times 2 \times 2 = 0.0020\ \frac{M}{s}$$

To use our rule of thumb, then, we can find the reaction rate at a new temperature by multiplying the old reaction rate by 2 for each 10°C increase in the temperature. Alternatively, if the temperature is lowered, we simply divide the old rate by 2 for every 10°C that the temperature is lowered.

If you purchased the MicroChem kit mentioned in the Student Notes, you can perform Experiment 16 in that kit, which demonstrates that our general rule of thumb doesn't always work!

ON YOUR OWN

14.7 A chemist runs a chemical reaction at 100°C and determines the instantaneous rate to be 0.24 M/s. She decides to speed up the reaction by running it at 180°C. What is the new reaction rate?

14.8 In "On Your Own" question 14.5, you determined that the rate equation for the following reaction:

$$2PO + Br_2 \longrightarrow 2POBr$$

is:

$$R = k[PO]^2[Br_2]^3$$

a. If a chemist runs this reaction at 30°C with reactant concentrations of 2.0 M each and then runs the reaction again at 0°C with the same reactant concentrations, what will the reaction rates be in both cases? The rate constant for this reaction at 30°C is $1.25 \frac{1}{M^4 \cdot s}$.

b. Why do the units in the rate constant include M^4 in the denominator?

Before we leave this section, you might be asking yourself why the rate constant is called a constant when, in fact, it changes depending on the temperature. Even though the rate constant varies with temperature, it does not vary with the concentration of reactants. So no matter what the concentration of the reactants, the value of the rate constant does not change. Since most reactions are always carried out at the same temperature and the only thing that varies is the reactant concentrations, it is convenient to think about k as a constant. Just remember it this way: The rate constant is, indeed, constant, as long as we are always working at the same temperature.

CATALYSTS AND REACTION RATE

The last thing that we need to consider when it comes to reaction kinetics is the topic of catalysts (kat' uh lists). We have already learned that there are 3 things we can do to increase the rate of a chemical reaction. We can increase the temperature at which the reaction is being run, increase the concentration of one or more of the reactants, and increase the surface area of any solids in the reaction. There is one more thing that we can do: we can add a catalyst. Catalysts can dramatically increase the rate of a chemical reaction. Perform Experiment 14.2 to see how this can happen.

EXPERIMENT 14.2

PURPOSE: To determine the effect of a catalyst on the decomposition of hydrogen peroxide.

MATERIALS:

- 100 mL beaker (A small glass will do.)
- Watch glass (A small saucer that covers the beaker will do.)
- Graduated cylinder (A ⅛ measuring cup will work.)
- ¼ measuring teaspoon
- Hydrogen peroxide (available at supermarkets and drugstores)
- Baker's yeast (Any kind of bread yeast, even bread machine yeast, will do.)
- Safety goggles

QUESTION: Can adding a catalyst speed up the rate of a chemical reaction?

HYPOTHESIS: Write a hypothesis for this experiment.

PROCEDURE:

1. Pour 25 mL (1/8 cup) of hydrogen peroxide into your beaker.
2. Observe the clear liquid for a moment. Not much seems to be happening, does it? In fact, however, something is happening. As mentioned early in this module, the hydrogen peroxide is decomposing into water and oxygen according to the following equation:

$$2H_2O_2\ (aq) \longrightarrow 2H_2O\ (l) + O_2\ (g)$$

   The problem is that this reaction proceeds incredibly slowly. Even though the hydrogen peroxide is decomposing, it is doing so at such an imperceptibly slow rate that you cannot see it happening. We will change this.
3. Measure ¼ teaspoon of yeast and pour it on top of the hydrogen peroxide. You should start to see some bubbles forming.
4. Quickly cover the beaker with the watch glass.
5. Grasp the beaker with your hand and place one finger on the center of the watch glass to hold it in place. Then gently swirl the yeast into the hydrogen peroxide. You should start to see many bubbles forming. In fact, bubbles should form so quickly that they rise up the beaker and smash into the watch glass. That's why the watch glass is there. It stops the bubbles from overflowing out of the beaker.
6. What happened here? The yeast acted as a catalyst and sped up the otherwise incredibly slow decomposition of hydrogen peroxide. What's neat about a catalyst is that although the catalyst speeds up the reaction, *it doesn't actually get used up in the reaction*. It's hard for you to see because the yeast has spread out in the liquid, but there is still ¼ of a teaspoon of yeast in the solution. Not a single milligram of yeast was used up to make this reaction go fast! You can confirm this if you want by adding another 25 mL of hydrogen peroxide. You will see that the bubbles still form. You can add another 25 mL, and the bubbles will still form. In fact, the yeast that you put in the beaker will decompose as much hydrogen peroxide as you want to add because the yeast never gets used up.
7. Clean up and return everything to the proper place.

CONCLUSION: Write something about the outcome and your hypothesis.

The behavior demonstrated by this experiment is important to remember:

Catalysts speed up the reaction rate without actually getting used up in the chemical reaction.

You might be asking yourself how in the world a catalyst can speed up a chemical reaction without being used up in the process. In general, a catalyst speeds up a chemical reaction by reducing the activation energy of that reaction. The lower the activation energy of a chemical reaction, the faster the reaction can proceed. This is an important point:

Decreasing the activation energy of a reaction increases its rate.

If we think about this in terms of an energy diagram, it should make some sense. Look at Figure 14.4.

FIGURE 14.4
The Effect of a Catalyst on Activation Energy
Illustration by David Weiss

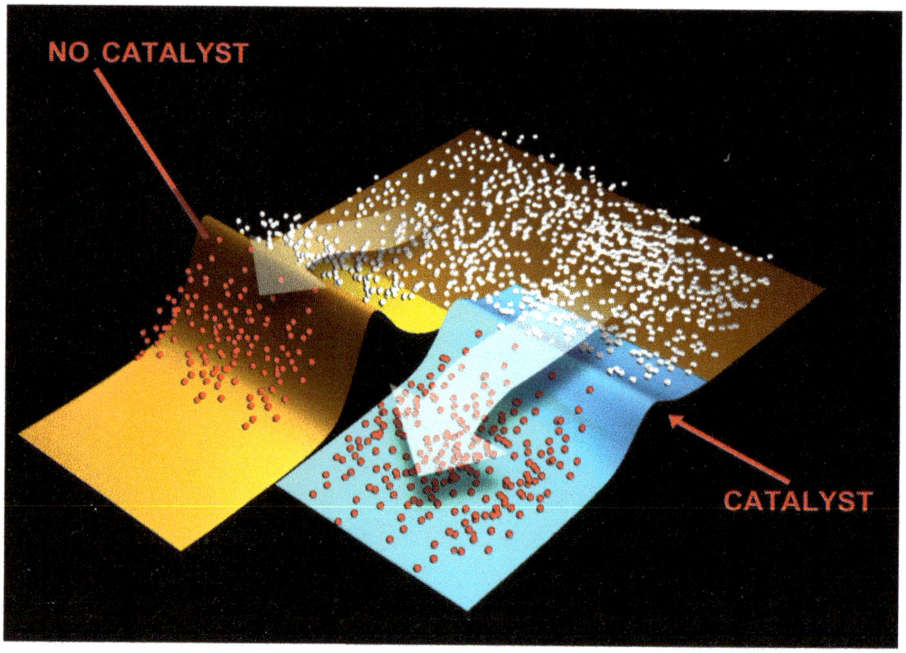

The figure shows a stylized version of an energy diagram. The reactants (the white balls) have a certain energy, and the products (the red balls) have a lower energy. For the reactants to form products, they would normally have to climb the hump in between, which represents the activation energy of the reaction. A catalyst provides a shortcut for the reactants so that they need not climb all of the way up the hill. This speeds up the reaction.

Catalysts speed up reaction rates by lowering the activation energy of the reaction.

For a given reaction, then, there can be more than one energy diagram. If the reaction is done without a catalyst, the hump between the reactants and products will be one height. If the reaction is done with a catalyst, the reactants and products will be in the same place on the diagram, but the hump between them will be much smaller.

You might be asking yourself *how* a catalyst reduces the activation energy of a reaction. There are 2 ways that catalysts do their job. As a result, there are 2 types of catalysts:

heterogeneous catalysts and homogeneous catalysts.

Heterogeneous catalysts—Catalysts that are in a different phase than the reactants.

Homogeneous catalysts—Catalysts that have the same phase as at least one of the reactants.

We will look at examples of each type of catalyst and how they end up reducing the activation energy of the reaction.

In Module 7, we mentioned that catalytic converters were put in automobiles in the late 1970s to reduce the amount of carbon monoxide (a poison) in automobile exhaust. A catalytic converter is an example of a heterogeneous catalyst. You see, carbon monoxide is naturally converted into carbon dioxide according to the equation:

$$2CO(g) + O_2(g) \longrightarrow 2CO_2(g)$$

Although this reaction occurs spontaneously, it is very slow and must be sped up so that the carbon monoxide has been converted into carbon dioxide before it leaves the exhaust pipe. The catalytic converter accomplishes this task.

A catalytic converter is a ceramic structure coated with a metal catalyst, which is usually platinum, rhodium, and/or palladium. The exhaust flows through this structure, exposing the exhaust to the catalyst as well as oxygen. The catalyst works by attracting carbon monoxide molecules and oxygen molecules to its surface. As the molecules get close to the catalyst's surface, they also get close to each other. In fact, they get so close to each other that they begin to exchange electrons and the reaction starts. By attracting carbon monoxide molecules and oxygen molecules to its surface, the metal causes these molecules to get much closer than they would normally get to each other, forcing the reaction to proceed. Figure 14.5 illustrates this process.

FIGURE 14.5
A Heterogeneous Catalyst
Illustration by Megan Whitaker

A heterogeneous catalyst works by attracting molecules to its surface. This brings the molecules much closer together than they would normally be.

Once near the catalyst's surface, the molecules are close enough together to exchange electrons and thus start the reaction.

Another way to understand this process is to think about it in terms of the energy necessary to get molecules close together. Since the electrons in a molecule repel the electrons in another molecule, molecules do not like to get too close to one another. However, since the catalyst in a catalytic converter attracts the molecules to its surface, the catalyst is, in fact, pulling the molecules close together. In the end, this reduces the amount of energy that the molecules themselves need to exert to get close enough to react together.

In contrast to heterogeneous catalysts, homogeneous catalysts work in an entirely different way. Instead of forcing the reactants to get closer together, a homogeneous catalyst actually participates in the chemical reaction, entirely changing the way in which the chemicals interact. For example, the oxidation of sulfur dioxide is a spontaneous reaction:

$$2SO_2\,(g) + O_2\,(g) \longrightarrow 2SO_3\,(g)$$

Although this reaction is spontaneous, it is also very slow. However, nitrogen monoxide gas is known to speed up this reaction immensely, without ever being used up in the reaction. Therefore, NO (g) is a catalyst for the reaction. Since it is in the same phase as the reactants, it must be a homogeneous catalyst.

Here's how it works. Without NO, there must be a collision between sulfur dioxide molecules and an oxygen molecule for the reaction to go. However, in the presence of NO, the reaction completely changes. First, the NO reacts with the oxygen to make NO_2:

$$2NO\,(g) + O_2\,(g) \longrightarrow 2NO_2\,(g)$$

NO and O_2 react much faster than SO_2 and O_2. After that reaction is complete, the NO_2 molecules then react with the SO_2 molecules:

$$2NO_2\,(g) + 2SO_2\,(g) \longrightarrow 2SO_3\,(g) + 2NO\,(g)$$

This ends up making the SO_3 gas that was supposed to be made in the original equation.

Do you see what has happened here? The NO gas reacts in the first step of this process, and in the next step, it is produced again. Even though it participates in the reaction, it never gets used up because each time 2 NO molecules are destroyed in the first step, 2 more NO molecules are made in the second step. As a result, the number of NO molecules present never really changes.

When a single chemical reaction can be written as a series of other chemical reactions, we call the series a reaction mechanism.

Reaction mechanism—A series of chemical equations that tells the step-by-step process by which a chemical reaction occurs.

We could say that the reaction:

$$2SO_2\,(g) + O_2\,(g) \longrightarrow 2SO_3\,(g)$$

will proceed by the following mechanism when catalyzed with NO gas:

$$\text{Step 1:}\quad 2NO\,(g) + O_2\,(g) \longrightarrow 2NO_2\,(g)$$

$$\text{Step 2:}\quad 2NO_2\,(g) + 2SO_2\,(g) \longrightarrow 2SO_3\,(g) + 2NO\,(g)$$

If you don't exactly understand all of this, don't worry. The only thing you need to understand is the difference between homogeneous and heterogeneous catalysts. They both reduce the activation energy of a chemical reaction, but they go about it in different ways. Homogeneous catalysts change the steps that the reaction takes. Heterogeneous catalysts, on the other hand, force the reactants in a chemical equation close together by attracting them to a surface. You also need to know the definitions of heterogeneous and homogeneous catalysts, and you need to be able to look at a reaction mechanism and see which substance in that mechanism is a catalyst. "On Your Own" questions 14.9–14.11 will test your knowledge on these points.

ON YOUR OWN

14.9 To catalyze the following reaction:

$C_2H_4(g) + H_2(g) \longrightarrow C_2H_6(g)$

solid platinum is often used. Is the platinum a heterogeneous or a homogeneous catalyst?

14.10 Two energy diagrams for the same reaction are presented below. Which one represents the reaction with a catalyst?

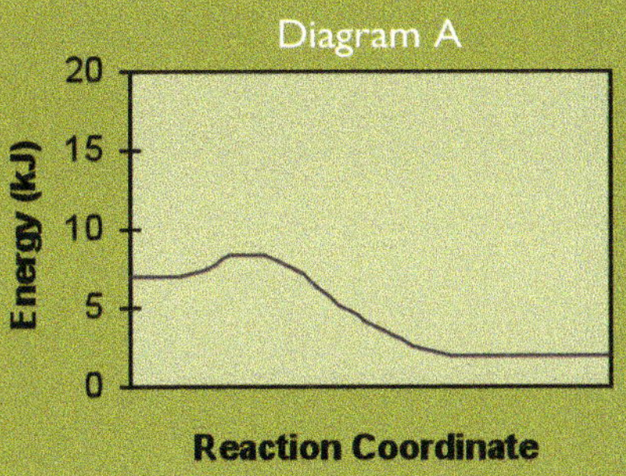

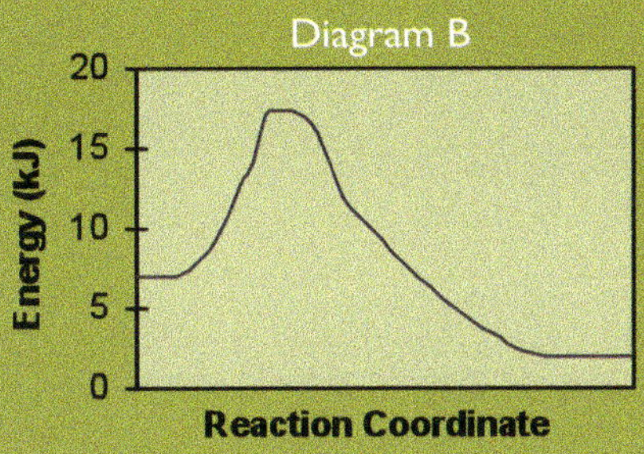

14.11 Identify the catalyst in the following reaction mechanism:

Step 1: $2H_2O_2 + 2I^- \longrightarrow 2H_2O + 2IO^-$

Step 2: $2H_2O + IO^- + IO^- \longrightarrow 2H_2O + O_2 + 2I^-$

SUMMARY OF KEY EQUATIONS IN MODULE 14

Equation 14.1: $R = \frac{\Delta[\text{product}]}{\Delta t} = -\frac{\Delta[\text{reactant}]}{\Delta t}$ Rate of chemical reaction equation

Equation 14.2: $R = k[A]^x[B]^y$ Rate equation

ANSWERS TO THE "ON YOUR OWN" QUESTIONS

14.1 In the smaller of the 2 vessels, the volume is lowest. Since concentration is number of moles divided by volume, the gases in the smaller container will have much larger concentrations. This means that the reaction in the smaller vessel will proceed faster because its reactants have larger concentrations.

14.2 The rate of a chemical reaction depends on the temperature. When the lizard is cold, the chemical reactions that run its muscles proceed very slowly. As a result, the muscles move slowly. When the lizard is warm, the chemical reactions that run its muscles have a faster rate, so the muscles can move faster. Note that the external temperature does not affect the motion of mammals (such as humans, dogs, and cats) because mammals are warm-blooded. This means our body temperatures, and hence the temperatures at which the chemical reactions in our bodies run, are mostly independent of the outside temperature.

14.3 The rate equation will take this form:

$$R = k[H_2]^x[NO]^y$$

To determine x and y, we look at trials where the concentration of one reactant stayed the same and the concentration of the other reactant changed. In trials 1 and 2, the concentration of H_2 remained the same, but the concentration of NO doubled. When that happened, the rate increased by a factor of 4. This means that y = 2. In the same way, between trials 2 and 3, the NO concentration remained constant, but the H_2 concentration tripled. When that happened, the rate tripled. This means x = 1. The rate equation, then, looks like:

$$R = k[H_2][NO]^2$$

To solve for k, we can use the data from any trial and plug it into our rate equation. We can then solve for k:

$$R = k[H_2][NO]^2$$

$$0.005670\,\frac{M}{s} = k \times (0.70\text{ M}) \times (0.30\text{ M})^2$$

$$k = \frac{0.005670\,\frac{\cancel{M}}{s}}{0.090\text{ M}^2 \times 0.70\,\cancel{M}} = 0.090\,\frac{1}{M^2 \times s}$$

The overall rate equation, then is

$$R = 0.090\,\frac{1}{M^2 \cdot s}\,[H_2][NO]^2$$

14.4 The rate equation will take this form:

$$R = k[NO_2]^x[O_3]^y$$

To determine x and y, we look at trials where the concentration of one reactant stayed the same and the concentration of the other reactant changed. In trials 1 and 3, the concentration of NO_2 remained the same, but the concentration of O_3 doubled. When that happened, the rate stayed the same. This means that y = 0 because the rate is not affected by the change in concentration of O_3. In the same way, between trials 2 and 3, the O_3 concentration remained constant, but the NO_2 concentration doubled. When that happened, the rate doubled. This means x = 1. The rate equation, then, looks like:

$$R = k[NO_2][O_3]^0\text{, which is the same as } R = k[NO_2]$$

To solve for k, we can use the data from any trial and plug it into our rate equation. We can then solve for k:

$$R = k[NO_2]$$

$$0.088\ \frac{M}{s} = k \times 0.000100\ M$$

$$k = \frac{0.088\ \frac{\cancel{M}}{s}}{0.000100\ \cancel{M}} = 880\ \frac{1}{s}$$

The overall rate equation, then, is:

$$R = 880\ \frac{1}{s}\ [NO_2]$$

14.5 If the reaction is second order with respect to PO, then the exponent attached to PO in the rate equation is a 2. Third order in Br_2 means that its exponent is 3. The rate equation is:

$$R = k[PO]^2[Br_2]^3$$

The overall order is the sum of the exponents, or 2 + 3 = 5.

14.6 If the reaction is fourth order with respect to $C_7H_{15}Cl$, then its exponent is 4. However, the exponent for NaF is 0 because the reaction is 0 order with respect to NaF. The rate equation is:

$$R = k[C_7H_{15}Cl]^4\ [NaF]^0$$

Usually, we drop exponents of 0 from the rate equation, so the real answer is:

$$R = k[C_7H_{15}Cl]^4$$

This simply means that the concentration of NaF has no effect on the rate of the reaction. The overall reaction order is the sum of the exponents, which is just 4.

14.7 To calculate the rate of a chemical reaction at an elevated temperature, we just multiply the old rate by 2 for every 10°C increase in temperature. When the chemist went from 100°C to 180°C, he increased the temperature by eight 10-degree units. Therefore, we must multiply the old rate by 8 twos to get the new rate:

$$R = 0.24\,\frac{M}{s} \times 2 \times 2 \times 2 \times 2 \times 2 \times 2 \times 2 \times 2 = 61\,\frac{M}{s}$$

14.8 (a) To get the rate of the reaction at 30°C, we just need to plug things into the rate equation:

$$R = k \times [PO]^2[Br_2]^3$$

$$R = 1.25\,\frac{1}{\cancel{M^4}\cdot s} \times (2.0\,\cancel{M})^2 \times (2.0\,\cancel{M})^3 = 4.0 \times 10^1\,\frac{M}{s}$$

To get the rate of the reaction at a lower temperature, we just divide the old rate by 2 for every 10°C decrease. The chemist dropped the temperature by three 10-degree increments:

$$R = 4.0 \times 10^1\,\frac{1}{s} \div 2 \div 2 \div 2 = 5.0\,\frac{M}{s}$$

(b) The rate constant needs to ensure the calculated reaction rate has units in M/s. Since the rate reaction equation yields a concentration in units of M^5, the rate constant needs to cancel out M^4, leaving M in the numerator the calculated reaction rate.

14.9 The platinum must be a heterogeneous catalyst because it has a different phase from the reactants.

14.10 Diagram A has the lowest activation energy, so it must represent the catalyzed reaction.

14.11 A catalyst is not used up in the reaction. So the substance that is both used in the first step of the mechanism and then produced in the last step of the mechanism is the catalyst. I^- is used in step 1 and produced in step 2. This makes it the catalyst for the reaction.

STUDY GUIDE FOR MODULE 14

REVIEW QUESTIONS

1. If 2 molecules want to react together to form new chemicals, what is the first thing that must happen?

2. A chemist runs a reaction at a low temperature and then runs the same reaction again at a high temperature. Which reaction runs faster?

3. Explain in your own words why the rate of a chemical reaction increases with increasing reactant concentration.

4. A chemistry book lists the rate constants for 2 different reactions. The first rate constant is $0.12\frac{1}{s}$ and the second is $34.5\frac{1}{M \cdot s}$. Even though the rate equations for these reactions are not given, an astute student says that the overall order of these 2 chemical reactions must be different from each other. In addition, the student says that the second reaction is faster than the first if the concentrations of the reactants are roughly the same. Why can the student make these statements?

5. The order of a chemical reaction with respect to one of its reactants is 0. If you double the concentration of that reactant, what happens to the rate?

6. A student is studying a chemistry book and reads that for the following reaction:

$$CH_3Cl + NaF \longrightarrow CH_3F + NaCl$$

the rate equation is:

$$R = k[CH_3Cl][NaF]$$

The student looks at the rate equation and says that since temperature does not appear in the rate equation, the rate of this reaction is obviously not affected by temperature. Is the student correct? Why or why not?

7. What do we call a chemical that increases the rate of a chemical reaction without getting used up in the process?

8. A chemist studies 2 different reactions and finds that the activation energy for reaction 1 is significantly higher than that of reaction 2. Assuming that the reactant concentrations are roughly the same in the reactions, which reaction is the fastest?

9. What is the main difference between a heterogeneous catalyst and a homogeneous catalyst?

10. What is the purpose of a catalytic converter in an automobile?

PRACTICE PROBLEMS

1. Milk sours because of a chemical reaction that takes place between some of the chemicals that make up the milk. Why does a carton of milk left out on a countertop sour sooner than a carton of milk left in the refrigerator?

2. A chemist does a reaction rate analysis on the following reaction:

$$2NO\ (g) + O_2\ (g) \longrightarrow 2NO_2\ (g)$$

She collects the following data:

Trial	Initial Concentration of NO (M)	Initial Concentration of O_2(M)	Instantaneous Reaction Rate (M/s)
1	0.0250	0.0253	0.0281
2	0.0500	0.0253	0.112
3	0.0250	0.0506	0.0562

What is the rate equation for this reaction?

3. A chemist does a reaction rate analysis of the following reaction:

$$C_4Cl_9OH + F^- \longrightarrow C_4Cl_9F + OH^-$$

He collects the following data:

Trial	Initial Concentration of C_4Cl_9OH (M)	Initial Concentration of F^-(M)	Instantaneous Reaction Rate (M/s)
1	0.25	0.50	0.0202
2	0.25	1.00	0.0202
3	0.50	1.00	0.0404

What is the rate equation for this reaction?

4. The following reaction:

$$Br_2\ (g) + I_2\ (g) \longrightarrow 2IBr\ (g)$$

is determined by experiment to be second order with respect to I_2 and first order with respect to Br_2. Its rate constant is $1.1 \times 10^{-3}\ 1/(M^2 \cdot s)$. What is the overall order of the reaction? If the reaction was run with both reactants at a concentration of 1.00 M, then what would the instantaneous rate be?

5. A chemist runs a reaction with a rate of 0.0167 M/s at 100°C. If the reaction is run at 40°C, what will its rate be?

6. A chemist runs a chemical reaction at 30°C and decides that it proceeds far too slowly. As a result, he decides that the reaction rate must be increased by a factor of 32. At what temperature should the chemist run the reaction to achieve this goal?

7. Which of the following graphs accurately depicts the dependence of the rate constant on temperature?

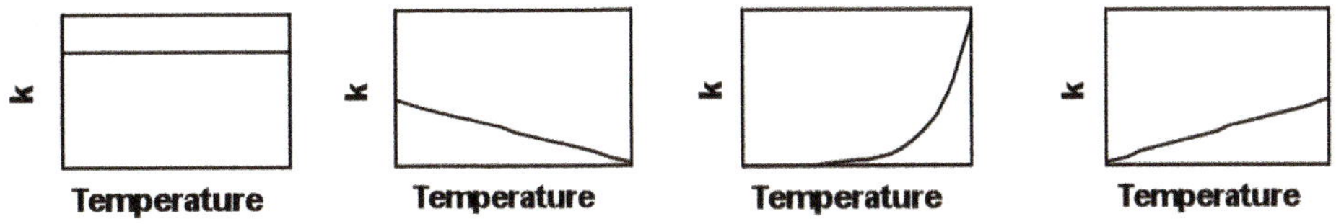

I II III IV

8. A chemist wants to use a Cu (s) catalyst on the following reaction:

$$2H_2\,(g) + O_2\,(g) \longrightarrow 2H_2O\,(g)$$

Would the copper be considered a heterogeneous catalyst or a homogeneous catalyst?

9. Energy diagrams are presented below for the same reaction. One diagram is for the reaction without a catalyst, one is for the same reaction with a catalyst that increases the reaction rate by a factor of 3, and one is for the same reaction with a catalyst that increases the reaction rate by a factor of 10. Identify the diagram for each case. Note: reaction is abbreviated as rxn in the diagrams below.

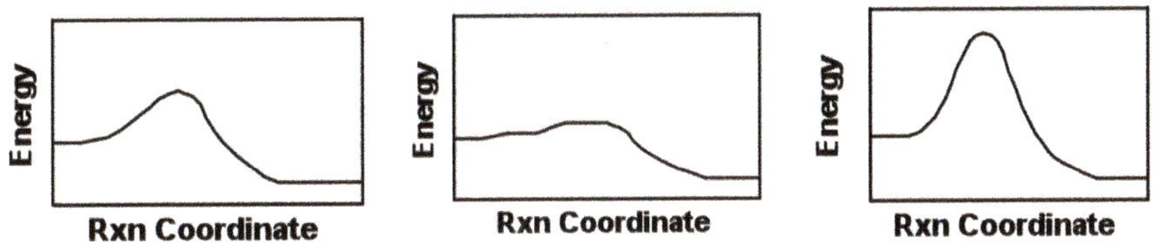

I II III

10. In the reaction mechanism below, indicate which substance is acting like a catalyst.

$$\text{Step 1. } 2Cl\,(g) + 2O_3\,(g) \longrightarrow 2O_2\,(g) + 2OCl\,(g)$$

$$\text{Step 2. } 2OCl\,(g) \longrightarrow O_2\,(g) + 2Cl\,(g)$$

Is this a heterogeneous or a homogeneous catalyst?

CHEMICAL EQUILIBRIUM

first reactions

Not all reactions occur in the directions that we think they should. Chemistry doesn't happen in a certain way just because we drew the equation in that direction. In some reactions, once a certain amount of product is produced, further product will push the reaction backward to produce the original samples again!

Changes constantly occur, but some changes are not permanent. Water freezes and becomes ice; ice can be melted back to water. After running out of energy, some kinds of batteries can be recharged. Up until this point, we have talked about chemical reactions in only one way: reactants forming products. In other words, if we were to present you with a chemical equation like this one:

$$NO_2\,(g) + CO\,(g) \rightarrow NO\,(g) + CO_2\,(g)$$

you would say that the NO_2 and CO react to form NO and CO_2. This is the way you've been taught.

We could look at the equation another way. Instead of assuming that the NO_2 and CO are reactants, we could assume that they are products. In other words, we could write the equation backward:

$$NO_2\,(g) + CO\,(g) \leftarrow NO\,(g) + CO_2\,(g)$$

Since the arrow is now pointing the other way, this equation indicates that the NO and CO_2 react to form NO_2 and CO. Which chemical equation do you think is correct? *Both* are correct.

If we were to start out with pure NO_2 and pure CO and mix them together, at first we would see them react to form NO and CO_2 As time went on, some of the NO and CO_2 that had been formed would collide into each other and turn back into NO_2 and CO! As we watched the reaction for a while, there would eventually come a time where for every NO_2 and CO molecule that react to form NO and CO_2, an NO molecule and a

CO_2 molecule would react to form NO_2 and CO. Eventually, as we watched the reaction, we would see it going *both ways*. This kind of situation is called chemical equilibrium (ee' kwuh lib' ree uhm), and makes up the bulk of what we will cover in this module.

THE DEFINITION OF CHEMICAL EQUILIBRIUM

Since the reaction we have been discussing can actually operate both forward and backward, we often use the following notation in the chemical equation:

$$NO_2\,(g) + CO\,(g) \rightleftharpoons NO\,(g) + CO_2\,(g)$$

The double arrow indicates that this particular reaction can operate in either direction. If NO_2 reacts with CO to make NO and CO_2, we call that the forward reaction. However, if NO reacts with CO_2 to make NO_2 and CO, we call that the reverse reaction.

If this chemical reaction is going forward and backward all of the time, what happens? Does the chemical reaction ever stop? Does it just keep going back and forth and back and forth? The answers to these questions lie in a discussion of the *rate* of both chemical reactions. Suppose we started off with a container of NO_2 and another container of CO. When we mixed these substances together, they would begin to react according to the above equation. Since there would not be any NO or CO_2 initially, the reverse reaction would not go. However, since there would be plenty of NO_2 and CO, the forward reaction would start to go. Since the concentrations of these reactants would be high, the rate of the forward reaction would be high. As time went on, the rate of the forward reaction would decrease because the concentration of the reactants would decrease. Of course, this forward reaction would produce NO and CO_2, allowing the reverse reaction to start running as well.

At first, the rate of the reverse reaction would be low because not much NO and CO_2 would have been made; therefore, the concentration of these reactants would be very low. As time went on, the forward reaction would keep making more NO and CO_2, increasing their concentration. This would keep increasing the rate of the reverse reaction. At the same time, since the forward reaction would be using up NO_2 and CO, their concentrations would go down. This would decrease the rate of the forward reaction.

Do you see what's happening here? When the reaction begins, the rate of the forward reaction is high and the rate of the reverse reaction is 0. As time goes on, the rate of the forward reaction *decreases* and the rate of the reverse reaction *increases*. Eventually, *the 2 rates will equal each other*. At that point, every time one forward reaction occurs, a reverse reaction will occur as well. In other words, every time an NO_2 molecule and a CO molecule are used up to make an NO molecule and a CO_2 molecule, the reverse reaction will also happen. As a result, despite the fact that both the forward and reverse reactions are going like crazy, the average amount of each substance *will not change*. This situation is referred to as chemical equilibrium.

Chemical equilibrium—The point at which both the forward and reverse reactions in a chemical equation have equal reaction rates. When this occurs, the amounts of each substance in the chemical reaction will not change, despite the fact that both reactions still proceed.

If all of this sounds a little confusing so far, don't worry. This is such an important concept that we want to explain it again using Figure 15.1. Even if you think you fully grasped the discussion the first time, carefully study Figure 15.1 and the explanation that follows.

FIGURE 15.1
Chemical Equilibrium

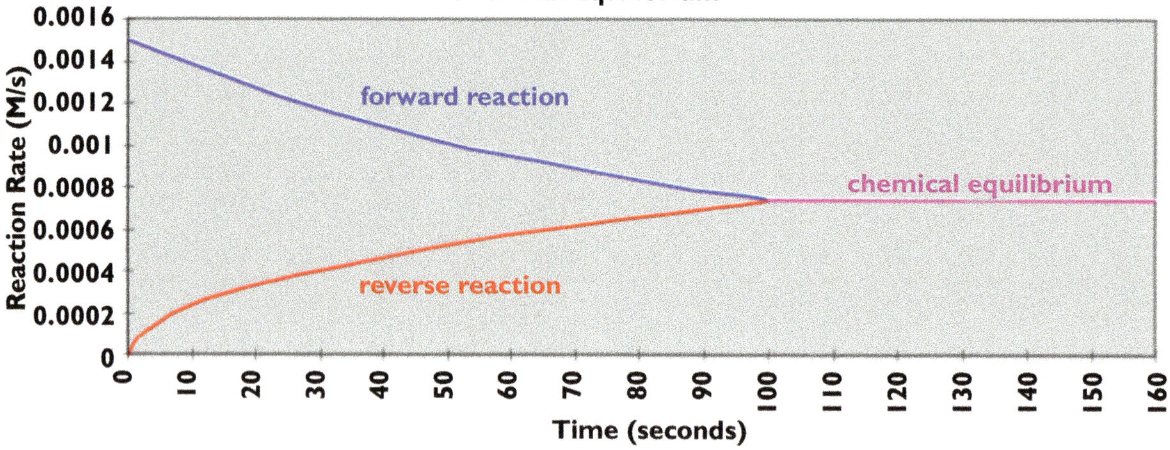

In the figure, the rates of the forward and reverse reactions of this equation are plotted versus time. Notice that when the reaction starts, the rate of the forward reaction is high and the rate of the reverse reaction is 0. This is because (as stated earlier) the reaction started with lots of NO_2 and CO and no CO_2 or NO. Once the forward reaction got started, its rate went down. At the same time, since NO and CO_2 were being formed by the forward reaction, the rate of the reverse reaction began to increase. In about 100 seconds (according to the graph), the forward reaction rate had decreased and the reverse reaction rate had increased enough so that they became equal. At that point, every time a forward reaction occurred, a reverse reaction did as well. As a result, the concentrations of the substances didn't change, and therefore the rate didn't change. No matter how long we watched this chemical reaction after the first 100 seconds, the rates of the forward and reverse reactions would stay equal; therefore, the concentrations of all the substances in the chemical reaction would never change. This is what we refer to as chemical equilibrium.

So chemical equilibrium is a very interesting state. If we were to look only at the concentrations of the substances in the chemical equation, we would notice that their concentrations did not change. As a result, we might think that nothing is really happening in the reaction. However, if we were able to study the individual molecules involved, we would see that they would be reacting together like mad. Despite the fact that a lot of reacting is going on, the amount of each substance does not change because every time a forward reaction occurs, a reverse reaction also occurs.

It's very important to realize that, as illustrated in Figure 15.1, chemical reactions do not *start* at equilibrium. When a chemical reaction begins, the rate of the forward and reverse reactions will not be equal. It takes *time* for a reaction to achieve chemical equilibrium. According to Figure 15.1, the reaction that we have been studying took about 100 seconds to reach equilibrium. Once there, the reaction will stay at equilibrium forever, unless it is disturbed by some outside intervention. Perform Experiment 15.1 to get an idea of how reactions reach and maintain equilibrium.

EXPERIMENT 15.1

PURPOSE: To demonstrate equilibrium.

MATERIALS:
- 3 plastic 2-liter bottles (Plastic milk cartons will work as well.)
- 4 small cups
- 2 bowls that are taller than the small cups
- A serrated knife (like a steak knife)
- A small, Phillips-head screwdriver
- Water
- A person to help you
- A kitchen counter
- Towels
- Safety goggles

QUESTION: How does equilibrium work?

HYPOTHESIS: Write a hypothesis for this experiment.

PROCEDURE:

NOTE: This experiment will get the counter wet. Use the towels to keep the mess to a minimum.

1. Use the knife to cut the tops off 2 of the plastic bottles. ***Be careful—do not cut yourself in the process!*** You want to remove each top completely so that the bottles have large openings at their tops. The openings need to be large enough to easily dump the contents of a cup into the bottle without spilling anything (see the picture after step 4).
2. Use the screwdriver to make a small hole near the bottom of the 2 bottles. ***Be careful—don't poke yourself in the process!*** On one of the bottles, make the hole about 2 inches from the bottom of the bottle. We will call this bottle 1. On the other bottle, make the hole about an inch from the bottom of the bottle. We will call this bottle 2.
3. Put the 2 bowls on the counter upside down.
4. Stand each of the bottles on a bowl with the hole pointing toward you. Place a small cup under each hole, and place the other cups right next to those cups. The setup should look something like this:

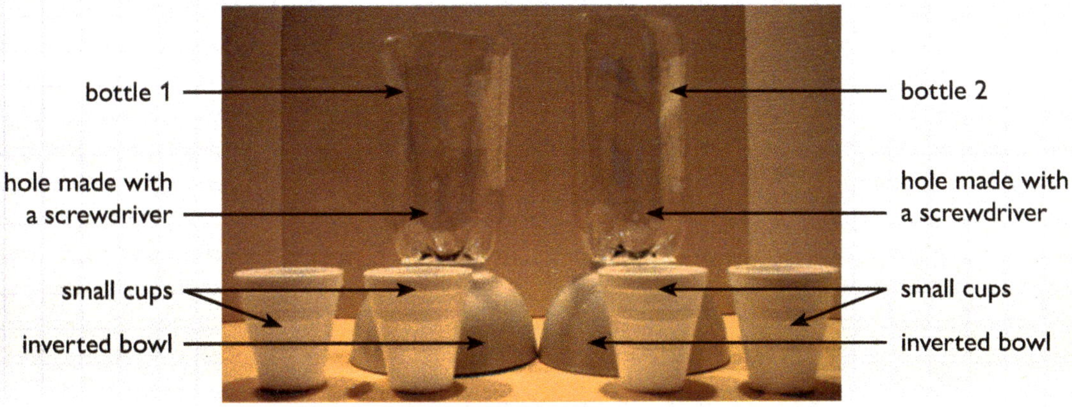

5. Fill the remaining 2-liter bottle (the one you haven't cut up) with water.
6. Have your helper stand in front of bottle 2.
7. You need to stand in front of bottle 1.
8. Have your helper hold bottle 1 as you use the bottle you filled in step 6 to pour water into it. Fill bottle 1 with as much water as possible without risking spillage. As soon as the water level reaches the hole, water will start to come out of the hole. Use your free hand to hold the cup and catch the water as it comes out. As the water level rises, the water will shoot out farther from the hole, so you will need to follow the water stream to keep catching the water.
9. Once bottle 1 is as full as you can get it, put down the bottle you have been using to pour water. Grab the other cup and quickly pull the cup that is being filled with water out from under the stream of water and replace it with the other cup.
10. Tell your helper to let go of bottle 1.
11. Pour the water in the full cup into bottle 2.
12. Once the cup that is currently being filled with water gets pretty full, pull it from under the stream of water and replace it with the cup in your other hand.
13. Once again, take the full cup and pour its contents into bottle 2.
14. Eventually, the water level in bottle 2 will rise above its hole. At that point, water will start coming out of the hole. Have your helper catch the water with one of the cups while he holds the other cup in his free hand. He will lose some water at first because the water will just trickle out of the hole. That's fine; just have him do the best that he can.
15. Once both you and your helper are catching water with your cups, start counting to 20. Each time you reach 20, you should switch the cups so that the cup that has been filling is out from under the water, and the cup in your other hand is filling with water. Then, dump the contents of the full cup into the other person's bottle. You should always dump water into bottle 2, and your helper should always dump water into bottle 1. Once you have each dumped your water, count to 20 again and repeat the process.
16. Keep doing this for several minutes. Once you both have the rhythm down, start looking at the water level in the bottles. After a while, the average water level in each bottle should not change significantly. Despite the fact that water is leaving each bottle through the hole and entering each bottle through the top, the average water level in each bottle should stay constant.
17. Clean up and return everything to the proper place.

CONCLUSION: Write something about the outcome and your hypothesis.

The experiment you just did provides an excellent illustration of equilibrium. Bottle 1 represents the reactants in a chemical equilibrium, and bottle 2 represents the products. At first, there was lots of water in bottle 1 (lots of reactants) and no water in bottle 2 (no products). However, bottle 1 began losing water (the forward reaction began making products). When this happened, bottle 2 began to fill with water (products were forming). Eventually, the water level in bottle 2 rose to the point where it began to lose water (the reverse reaction began running). As a result, some of the contents of bottle 2 started going back into bottle 1.

Since the holes were at different heights in the bottles and since the water levels in the bottles were different, the bottles lost water at different rates. This represents the fact that the forward and reverse reactions in an equilibrium have different rates. Eventually, the rate of water loss in bottle 1 (the rate of the forward reaction) decreased, and the rate of water loss in bottle 2 (the rate of the reverse reaction) increased to the point where the 2 rates were equal. At that point, although water was being continually transferred between the bottles, the average amount of water in the bottles did not change. This is just like chemical equilibrium, where the forward and reverse reactions are both running but the average amounts of the products and reactants do not change.

THE EQUILIBRIUM CONSTANT

Now that we know a little about *how* a chemical reaction reaches equilibrium, it is time to learn about *where* a chemical reaction ends up once it has reached equilibrium.

$$NO_2\,(g) + CO\,(g) \rightleftharpoons NO\,(g) + CO_2\,(g)$$

For example, when the reaction above reached equilibrium, how many NO_2 and CO molecules were left? Before this module, we would have assumed that by the end of the reaction, no CO and NO_2 molecules would be left because they all would have reacted. Now that we know about chemical equilibrium, we know that the reaction reaches some point where the forward reaction is running at the same rate as the reverse reaction; therefore, NO_2 and CO always remain. The question is, how much remains? If we were to start out with 1 mole of each, how much would be left at equilibrium?

To determine this, chemists have developed the equilibrium constant. This constant, abbreviated with a capital K, tells us all we need to know about how much reactant and how much product exist at equilibrium. For the generic chemical equilibrium reaction where A and B are reactants, C and D are products, and a, b, c, and d are their respective stoichiometric coefficients:

$$aA + bB \rightleftharpoons cC + dD$$

we define the equilibrium constant as:

$$K = \frac{[C]_{eq}^{c}\,[D]_{eq}^{d}}{[A]_{eq}^{a}\,[B]_{eq}^{b}} \qquad \text{Equation 15.1}$$

As always, the square brackets simply stand for concentration. The eq subscript on each concentration term reminds us that to calculate the equilibrium constant, we must always use the concentrations of the substances once equilibrium is reached. Before we learn what the equilibrium constant is and what it means, let's make sure we understand the equation by studying Example 15.1 and answering "On Your Own" questions 15.1–15.2.

EXAMPLE 15.1

The following reaction:

$$N_2\,(g) + O_2\,(g) \rightleftharpoons 2NO\,(g)$$

reaches equilibrium when the concentration of N_2 is 0.354 M, the concentration of O_2 is 0.124 M, and the concentration of NO is 4.51 M. What is the equilibrium constant for this reaction?

According to Equation 15.1, we can calculate the equilibrium constant for any reaction by taking the equilibrium concentrations of the products (raised to their stoichiometric coefficients) and dividing them by the equilibrium concentrations of the reactants (raised to their stoichiometric coefficients). For this particular reaction, the equilibrium constant is given by:

$$K = \frac{[NO]^2_{eq}}{[N_2]_{eq}[O_2]_{eq}}$$

Since the concentrations given are the concentrations at equilibrium, we can simply plug them into the formula:

$$K = \frac{(4.51\,\cancel{M})^2}{0.354\,\cancel{M} \times 0.124\,\cancel{M}} = 463$$

Notice that the units of M^2 cancel out in this case, leaving the equilibrium constant with no units. Just like the reaction rate constant k you learned about in Module 14, the units of the equilibrium constant can be different for different reactions, so you must always be careful to keep track of the units while you are calculating K.

A chemist is studying the following reaction:

$$4NH_3\,(g) + 7O_2\,(g) \rightleftharpoons 4NO_2\,(g) + 6H_2O\,(g)$$

If the chemist starts with 1.2 M NH_3, 3.4 M O_2, and no NO_2 or H_2O, he finds that the reaction reaches equilibrium when the concentration of NH_3 reaches 0.80 M and the concentration of O_2 reaches 2.7 M. At that point the concentration of NO_2 is 0.40 M, and the concentration of H_2O is 0.60 M. What is the equilibrium constant for this reaction?

According to Equation 15.1, the equilibrium constant for this particular reaction can be calculated with the following formula:

$$K = \frac{[NO_2]^4_{eq}\,[H_2O]^6_{eq}}{[NH_3]^4_{eq}\,[O_2]^7_{eq}}$$

Remember, only *the concentrations at equilibrium* can be used to calculate K. In this case, the first 2 concentrations are irrelevant in the problem because those are the concentrations that the chemist *started with*. We must use the concentrations at equilibrium, which are given later in the problem. Plugging those numbers into the equation gives us:

$$K = \frac{(0.40\ M)^4 \times (0.60\ M)^6}{(0.80\ M)^4 \times (2.7\ M)^7} = 2.8 \times 10^{-6}\ \frac{1}{M}$$

Notice that this time one of the molarity units in the denominator did not cancel, leaving the equilibrium constant with a unit of $\frac{1}{M}$.

It doesn't do us much good to calculate the equilibrium constant unless we know what it means and how we can use it. So that's what we have to learn now. The first thing that we need to know about the equilibrium constant is that it never changes, as long as the temperature doesn't change. In other words, once we calculate the equilibrium constant for a particular reaction, we never have to calculate it again at that temperature. No matter what the initial concentrations of the substances in the chemical equation are, they will always change so that, at equilibrium, the equilibrium constant is always the same.

The more important thing to remember about the equilibrium constant is that it can be used as a gauge to tell us the relative concentrations of reactants and products when the reaction reaches equilibrium. In other words, the equilibrium constant can tell us *where* the reaction will end up at equilibrium. For example, suppose the equilibrium constant for a chemical reaction is very large. This tells us that at equilibrium, there will be a lot of products and only a few reactants. After all, look at Equation 15.1. The equilibrium constant is calculated with the concentration of the products in the numerator of the fraction and the concentration of the reactants in the denominator. The only way such a calculation will result in a large number is if the number in the numerator is larger than the number in the denominator. That means the concentration of products will be greater than the concentration of the reactants.

ON YOUR OWN

15.1 What is the value of the equilibrium constant for the following reaction?

$N_2\,(g) + 3H_2\,(g) \rightleftharpoons 2NH_3\,(g)$

The equilibrium concentration of N_2 is 0.242 M, the equilibrium concentration of H_2 is 0.468 M, and the equilibrium concentration of NH_3 is 2.28 M.

15.2 A chemist is studying the following reaction:

$2SO_3\,(g) \rightleftharpoons 2SO_2\,(g) + O_2\,(g)$

If the chemist starts with 2.60 M SO_3 and no SO_2 or O_2, he finds that the reaction reaches equilibrium when the concentration of SO_3 reaches 0.40 M. At that point the concentration of SO_2 is 2.20 M, and the concentration of O_2 is 1.10 M. What is the equilibrium constant for this reaction?

On the other hand, if the equilibrium constant is small, it tells us that at equilibrium, there will be a lot of reactants and only a few products. Once again, think about Equation 15.1. The only way that the equation can work out to a small number is when the number in the numerator is smaller than the number in the denominator. This will happen when the concentration of reactants is greater than the concentration of the products. Finally, when the equilibrium constant is close to 1, then there will be approximately equal amounts of reactants and products when the reaction reaches equilibrium.

When an equilibrium reaction has more products than reactants at equilibrium, we say that it is weighted toward the products side of the equation. On the other hand, when there are more reactants than products at equilibrium, we say that the reaction is weighted toward the reactants side of the equation. Finally, when there are roughly the same amounts of products and reactants at equilibrium, we say that the equilibrium is balanced between reactants and products. Therefore, the principles we just learned can be restated as follows:

If [reactant] << [product] then

$$K = \frac{[\text{product}]}{[\text{reactant}]} = \frac{\text{large number}}{\text{small number}} = \text{large value for K}$$

so we use → for the reaction arrow (toward products side)

If [reactant] >> [product] then

$$K = \frac{[\text{product}]}{[\text{reactant}]} = \frac{\text{small number}}{\text{large number}} = \text{small value for K}$$

so we use ← for the reaction arrow

If [reactant] = [product] then

$$K = \frac{[\text{product}]}{[\text{reactant}]} = 1$$

so we use ⇌ for the reaction arrow

ON YOUR OWN

15.3 In Example 15.1, the equilibrium constants of 2 equilibrium reactions were calculated. For each reaction, tell whether the equilibrium is weighted toward reactants, weighted toward products, or balanced between reactants and products.

These are important principles to remember. Answer "On Your Own" question 15.3 to make sure you understand them.

At this point, it's time to let you in on a little secret. *All* chemical reactions are equilibrium reactions. To be perfectly correct, *every* equation that you have ever seen in this course should be written with a double arrow rather than a single arrow. Does this mean that everything we've learned up to this point is wrong? No, of course not. As we just learned, when the equilibrium constant for a reaction is large, the reaction is weighted toward the products side of the equation. When equilibrium is achieved, there are a lot of products and only a few reactants. If the equilibrium constant for a chemical reaction is *very* large, then that equilibrium is *heavily* weighted toward the products side of the equation. At some point, the equilibrium constant becomes so large that once the reaction has hit equilibrium, there are mostly products and virtually no reactants present. At that point, chemists just say that the amount of reactants present for such a reaction is so small that we can just ignore them. In the same way, if the equilibrium constant is very, very small, we can say that the reverse reaction is really what happens because only reactants exist at equilibrium.

Therefore, when the equilibrium constant for a reaction is very large, we replace the double arrow with a single arrow, indicating that, as far as we are concerned, the reaction completely runs out of reactants. When the equilibrium constant is very small, we write the reaction in reverse. From now on, then, when you see a reaction that contains a single arrow, remember that the reaction is, in fact, an equilibrium, but the equilibrium is so heavily weighted toward the products side that we just assume that no reactants are present when the reaction finishes. Answer "On Your Own" question 15.4 to make sure you understand these types of equations.

ON YOUR OWN

15.4 Three chemical equilibria (the plural of *equilibrium*) are written below. Below each equation, the value for the equilibrium constant is given. If any of these equations can be written with a single arrow, then do so, making sure that the arrow is pointing in the proper direction:

a. $2NO_2\ (g) + Cl_2\ (g) \rightleftharpoons 2NOCl\ (g) + O_2\ (g)$
   $K = 4.6 \times 10^4$

b. $2Cl_2\ (g) + 2H_2O\ (g) \rightleftharpoons 4HCl\ (g) + O_2\ (g)$
   $K = 3.2 \times 10^{-14}\ M$

c. $CH_4\ (g) + Cl_2\ (g) \rightleftharpoons CH_3Cl\ (g) + HCl\ (g)$
   $K = 1.2$

A FEW MORE DETAILS ABOUT THE EQUILIBRIUM CONSTANT

Now that you've had a chance to become a bit familiar with the concept of the equilibrium constant and how it is calculated, it is time to throw you a curveball. Equation 15.1 tells us that we calculate K by taking the products of the equation (raised to their stoichiometric coefficients) and dividing them by the reactants of the equation (raised to their stoichiometric coefficients). There are a few things we ignore in the equation for K, although they might appear in the chemical equation.

First, any solids that happen to be in the chemical equation are not used when calculating K. For the following reaction:

$$2NaOH\ (aq) + MgCl_2\ (aq) \rightleftharpoons Mg(OH)_2\ (s) + 2NaCl\ (aq)$$

we would calculate the equilibrium constant with the following equation:

$$K = \frac{[NaCl]^2_{eq}}{[MgCl_2]_{eq}[NaOH]^2_{eq}}$$

Notice that although $Mg(OH)_2$ is a product, it is not included in the equation for the equilibrium constant. This is because $Mg(OH)_2$ is a solid, and no solids are ever included in the equation for the equilibrium constant. You must remember this:

When a solid appears in a chemical equilibrium, it is not included in Equation 15.1.

You might be wondering why solids are not included in the equilibrium constant. After all, if something is a part of the chemical equation, shouldn't it be included? Remember that in the solid state, molecules are very close to one another. The distance between them is governed by the properties of the particular molecule involved, and that distance cannot change for a given temperature. As a result, *the concentration of a solid is constant*. Think about it this way: At a given temperature, the *density* of a solid is always the same. The number of grams per mL is constant. We could take the density of a solid and convert grams to moles by looking at the periodic table. We could also convert mL to L. What would we have then? We would have *moles per liter*, which is the concentration. The concentration of a solid is constant for any given temperature. This means that while the *mass* of a solid might increase or decrease through the course of a chemical reaction, its *concentration* (moles per liter) remains the same.

Along the same lines, when calculating the equilibrium constant, we can ignore liquids that appear in the chemical reaction. The rationale is the same. Since the density (and the concentration) of a pure liquid is constant for a given temperature, liquids should not be included in Equation 15.1. Please note that this applies only to *liquids*, not aqueous substances. For example, in the following reaction:

$$2H_2O_2\,(aq) \rightleftharpoons 2H_2O\,(l) + O_2\,(g)$$

the equilibrium constant would be calculated as follows:

$$K = \frac{[O_2]_{eq}}{[H_2O_2]^2_{eq}}$$

Even though the reaction contains liquid water, we ignore it as a part of the equation for the equilibrium constant. However, we *do not* ignore the aqueous H_2O_2. This, then, is another important principle:

When a liquid appears in a chemical equation, we do not include it in Equation 15.1. This applies to the liquid phase; it does not apply to the aqueous phase.

FIGURE 15.2
Equilibrium in Aqueous Solution
Illustration by David Weiss

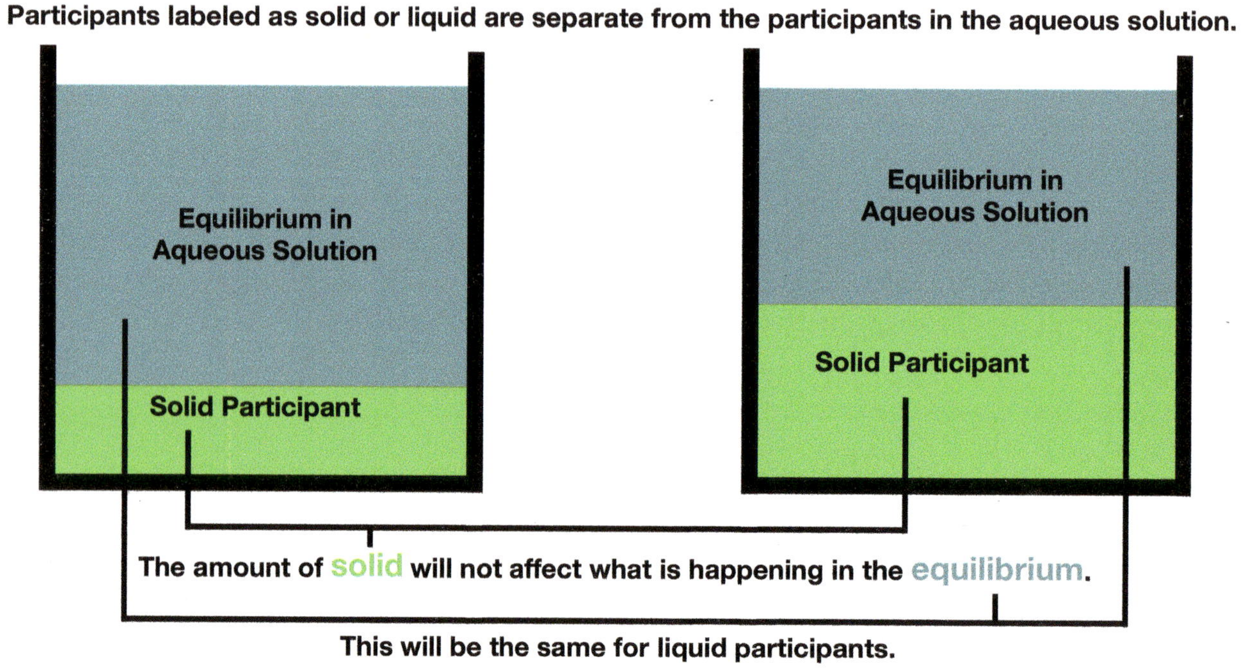

Study Example 15.2 and then answer "On Your Own" questions 15.5–15.6 so that you are sure you know how to apply these new rules.

EXAMPLE 15.2

What is the equation used to calculate the equilibrium constant in the following reaction?

$$NH_4Cl\,(s) \rightleftharpoons NH_3\,(g) + HCl\,(g)$$

To solve this problem, we just use Equation 15.1, recognizing that we ignore the solid in the equation:

$$K = [NH_3][HCl]$$

What is the expression for the equilibrium constant for this reaction?

$$CH_3CO_2H\,(aq) + C_2H_5OH\,(aq) \rightleftharpoons CH_3CO_2C_2H_5\,(aq) + H_2O\,(l)$$

To solve this problem, we again use Equation 15.1, remembering that we ignore the liquid in the equation:

$$K = \frac{[CH_3CO_2C_2H_5]}{[CH_3CO_2H][C_2H_5OH]}$$

ON YOUR OWN

15.5 Give the expression for the equilibrium constant of this reaction:
$2Li(s) + 2H_2O(l) \rightleftharpoons 2LiOH\,(aq) + H_2\,(g)$

15.6 What is the equation for the equilibrium constant for the following reaction?
$Ba(OH)_2H_{16}O_8(s) + 2NH_4SCN(s) \rightleftharpoons$
$2NH_4OH(aq) + Ba(SCN)_2(aq) + 8H_2O(l)$

USING THE EQUILIBRIUM CONSTANT TO PREDICT THE PROGRESS OF A REACTION

Besides telling us the concentration of reactants and products at equilibrium, the equilibrium constant can also be used to predict whether or not a chemical reaction has reached equilibrium yet. To do this, we must first know the value for the equilibrium constant for whatever temperature at which we are running the reaction. Next, we can measure the concentrations of each substance in the chemical reaction at any given time. We can then plug those concentrations into the equation for the equilibrium constant. If the reaction has reached equilibrium, by definition, the answer we get should equal the value of the equilibrium constant. If it does not, we know that the equation is not at equilibrium yet.

This will tell you more than just that. It will allow you to predict what the reaction will do next to reach equilibrium. What we mean is this: If a reaction has not yet reached equilibrium, it must either make more products (which will lower the rate of the forward reaction) or make more reactants (which will lower the rate of the reverse reaction). When a chemical reaction does something like this, we say that the reaction shifts. The equilibrium constant can be used to predict the direction that a chemical reaction will shift if it is not already at equilibrium. Study Example 15.3 to see how that is done.

EXAMPLE 15.3

A chemist is studying the following reaction:

$$N_2\,(g) + O_2\,(g) \rightleftharpoons 2NO\,(g)$$

She knows that the equilibrium constant for this reaction at this temperature is 0.050. Suppose she knows that the concentrations of each substance are as follows: $[N_2]$ = 1.5 M, $[O_2]$ = 1.3 M, and [NO] = 0.020 M. Determine whether or not the reaction has reached equilibrium yet. If it has not, determine the direction it will shift to reach equilibrium.

First, we have to see whether or not the reaction has reached equilibrium. To do this, we just plug the concentration values into Equation 15.1. If the result is equal to 0.050, then we know that the reaction is at equilibrium. If not, then the reaction has not reached equilibrium yet.

$$\frac{[NO]^2_{eq}}{[N_2]_{eq}[O_2]_{eq}} = \frac{(0.020\,\cancel{M})^2}{(1.5\,\cancel{M})(1.3\,\cancel{M})} = 0.00021$$

According to the problem, the correct value of the equilibrium constant is 0.050. Since this is not the answer we got, we are left to conclude that the concentrations we used were not, in fact, equilibrium concentrations. This means that the **reaction is not at equilibrium**.

To determine which way the reaction will shift to reach equilibrium, we just need to compare our calculation to the actual value of K. Our calculation is smaller than the correct value. This means that to reach equilibrium, the concentrations need to change in such a way as to increase the value of the equation we used. How can that happen? In the equation, the concentration of NO is in the numerator of the fraction. If the concentration of NO is increased, the value we calculate from the equation will increase, bringing it closer to the true value for K. How can we get the concentration of NO to increase? The reaction just needs to shift toward the products side of the equation. This will increase the concentration of NO (because more NO will be made) and at the same time decrease the concentrations of N_2 and O_2 (because they will be used up). Both of these things will result in a larger value calculated from Equation 15.1, bringing the reaction closer to equilibrium. The reaction needs to make more products. In chemical terminology, we would say that **the reaction must shift toward the products**.

The same chemist studied the same reaction under different concentrations. Suppose this time she measures the following concentrations: $[N_2]$ = 0.100 M, $[O_2]$ = 0.200 M, [NO] = 1.00 M. Is the reaction at equilibrium under these conditions? If not, how will the reaction need to shift to reach equilibrium?

Once again, we have to see whether or not the reaction has reached equilibrium. To do this, we just plug the concentration values into Equation 15.1. If the result is equal to 0.050, we know that the reaction is at equilibrium. If not, the reaction has not reached equilibrium yet.

$$\frac{[NO]^2_{eq}}{[N_2]_{eq}[O_2]_{eq}} = \frac{(1.00\,\cancel{M})^2}{0.100\,\cancel{M} \times 0.200\,\cancel{M}} = 50.0$$

According to the problem, the correct value of the equilibrium constant is 0.050. Since this is not the answer we got, we are left to conclude that the concentrations we used were not, in fact, equilibrium concentrations. The **reaction is not at equilibrium.**

To determine which way the reaction will shift to reach equilibrium, we just need to compare our calculation with the value of K. Our calculation is larger than the correct value. This means that to reach equilibrium, the concentrations need to change in such a way as to decrease the value of the equation we used. How can that happen? In the equation, the concentration of NO is in the numerator of the fraction. Therefore, if the concentration of NO is decreased, the value we calculate from the equation will decrease, bringing it closer to the value for K. How can we get the concentration of NO to decrease? The reaction just needs to make more reactants. This will decrease the concentration of NO (because NO will be used up) and at the same time increase the concentrations of N_2 and O_2 (because they will be made). Both of these things will result in a smaller value calculated from Equation 15.1, bringing the reaction closer to equilibrium. Therefore, **the reaction must shift toward the reactants**.

So you see that all we need to do to predict which way a reaction will shift to reach equilibrium is take the concentrations at the moment and plug them into Equation 15.1. If the number we get is equal to the equilibrium constant, then the reaction is at equilibrium and will not change. If the number we get is less than the equilibrium constant, the reaction needs to shift toward the products. However, if the number we calculate is greater than K, the reaction must shift toward reactants to reach equilibrium. Make sure that you grasp this concept by answering "On Your Own" question 15.7.

ON YOUR OWN

15.7 A chemist is studying the following reaction:

$$2NH_3\,(g) \rightleftharpoons N_2\,(g) + 3H_2\,(g)$$

The reaction has an equilibrium constant of 17 M^2. The chemist measures the concentrations of each substance to be $[NH_3] = 0.80$ M, $[N_2] = 2.5$ M, $[H_2] = 2.3$ M. Is the reaction at equilibrium? If not, which way will it shift to attain equilibrium?

LE CHATELIER'S PRINCIPLE

Earlier in this module, we said that a reaction that has attained equilibrium will stay at equilibrium forever unless it is disturbed by some outside influence. Now it's time to learn how a chemical equilibrium behaves when such a thing happens. For example, suppose we have the following equilibrium:

$$H_2\,(g) + I_2\,(g) \rightleftharpoons 2HI\,(g)$$

If we start with just H_2 and I_2, the forward reaction will occur at a fast rate, and the reverse reaction won't happen at all. As soon as some HI is formed, the reverse reaction will start to occur as well. As time goes on, the reverse reaction will get faster while the forward reaction gets slower and, eventually, equilibrium will be achieved.

Let's suppose now that equilibrium has been achieved. What would happen if we added some H_2 into the system? Before the H_2 was added, the reaction was at equilibrium. This means that the rate of the forward reaction was equal to the rate of the reverse reaction. If extra H_2 were suddenly added, the rate of the forward reaction would go up because the concentration of H_2 would have increased. This would throw the reaction out of equilibrium. Since the forward reaction would suddenly be faster than the reverse reaction, more HI would be made. This would start to slow down the forward reaction (because H_2 and I_2 would be used up), and it would start to increase the rate of the reverse reaction (because more HI would be made). Eventually, the reaction rates would equal each other again, and a new equilibrium would be established. This process is illustrated in figure 15.3.

FIGURE 15.3
The Effect of Adding More Reactant to an Equilibrium

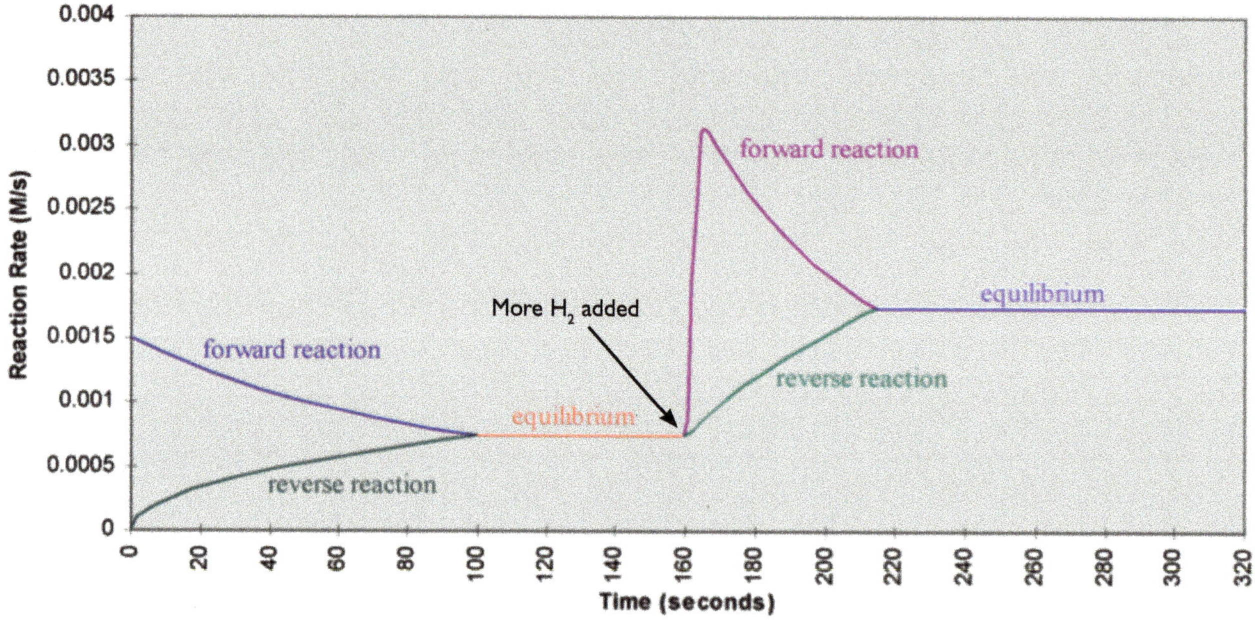

When more H_2 was added to the system, the equilibrium was upset. To restore equilibrium, the reaction simply shifted in such a way as to make the forward and reverse reaction rates equal again. This is a general principle, called Le Châtelier's (luh shat' tee ayes) principle:

> **Le Châtelier's principle**—When a stress (such as a change in concentration, pressure, or temperature) is applied to an equilibrium, the reaction will shift in a way that relieves the stress and restores equilibrium.

Le Chatelier's principle allows us to predict how an equilibrium responds to stress and is therefore a useful principle to apply.

Let's look back to our original example again. We already determined that if our equilibrium is stressed by adding more H_2 to the system, the equilibrium will relieve the stress by making more HI. This will lower the rate of the forward reaction and raise the rate of the reverse reaction, eventually bringing the system back to equilibrium. We know that at the new equilibrium point, the concentration of HI will be larger than it was at the first equilibrium point. In other words, the equilibrium will shift to the products side of the equation to relieve the stress.

What about the concentration of I_2? At the new equilibrium, will the concentration of

I_2 be greater, lower, or the same as the original equilibrium? When the rate of the forward reaction increases, more HI will be made. The only way that the reaction can do this is to use up I_2. So, in the end, the concentration of I_2 will be less at the new equilibrium because when the reaction shifted to restore equilibrium, I_2 was used up. This is the kind of reasoning you must be able to do when it comes to Le Châtelier's principle. Study Example 15.4 to make sure you can reason this way.

EXAMPLE 15.4

A chemist is experimenting with the following equilibrium:

$$4NH_3\,(g) + 5O_2\,(g) \rightleftharpoons 4NO\,(g) + 6H_2O\,(g)$$

She allows the reaction to come to equilibrium and then stresses it in several different ways.

a. What will happen to the concentration of NO and O_2 if the chemist adds extra NH_3 to the system?

When extra NH_3 is added, the rate of the forward reaction will increase. This means that more NO and H_2O will be made. The reaction will shift toward the products side of the equation, increasing the concentration of NO and H_2O. To make more NO and H_2O, the reaction will have to use up some O_2. That will reduce the concentration of O_2. Therefore, **the concentration of O_2 will decrease (because it will be used up), and the concentration of NO will increase (because more of it will be made).**

b. What will happen to the concentration of H_2O and NH_3 if the chemist removes O_2 from the system?

When O_2 is removed from the system, the forward reaction slows down, making the reverse reaction suddenly faster than the forward one. The equilibrium will begin making extra NH_3 and O_2. In other words, the equilibrium will shift toward the reactants side. This will increase the concentration of NH_3. To make extra NH_3 and O_2, the reaction must use up NO and H_2O, which will cause their concentrations to decrease. When the system reaches its new equilibrium, **the concentration of H_2O will be lower (because it will be used up), and the concentration of NH_3 will be greater (because more of it will be made).**

c. What will happen to the concentration of H_2O and NH_3 if the chemist adds extra NO?

When the chemist adds extra NO, the reverse reaction speeds up. This results in more NH_3 and O_2 being made, shifting the equilibrium to the reactants side. To do that, the reaction must use up NO and H_2O. Therefore, when the new equilibrium is achieved, **the concentration of H_2O will be lower (because it will be used up), and the concentration of NH_3 will be greater (because it will be produced).**

Before we go on, let's revisit something mentioned earlier. We told you that all chemical reactions are equilibrium reactions. However, if the equilibrium constant is large, the equilibrium will be so heavily weighted toward the products that we can ignore the few reactants that exist. In that case, we can assume that the reactants react completely to form products. Consider the following reaction:

$$CaCO_3\,(s) \rightleftharpoons CaO\,(s) + CO_2\,(g)$$

The equilibrium constant for this reaction at 25°C is 0.86 M. According to the analysis we did before, we cannot consider this a complete reaction. In fact, this reaction is weighted toward the reactants side of the equation since $K < 1$.

However, Le Chatelier's principle provides us a way to *force* this reaction to go to completion so that no $CaCO_3$ remains. How is that possible? Consider what would happen if the CO_2 were simply allowed to escape the reaction vessel while the reaction took place. What would that do to the equilibrium? The concentration of CO_2 would decrease, which would stress the equilibrium. As a result, the equilibrium would shift to the products side of the equation, making more CaO and CO_2 and using up $CaCO_3$. If the CO_2 were continually allowed to escape, the equilibrium would continually shift to the products side of the equation until no more $CaCO_3$ remained. In this case, although the equilibrium constant for this reaction is less than 1, the reaction would go until all of the reactant is used up. In fact, the only way we could get this reaction to reach equilibrium is to run it in an enclosed vessel so that the carbon dioxide could not escape. Under those conditions, the concentration of CO_2 would not be continually decreasing, so the reaction could reach equilibrium.

Le Chatelier's principle, then, tells us how to force an equilibrium to react so that the limiting reactant is always used up. All we have to do is continually remove a product. If we do that, the equilibrium will continually be stressed and will continually shift to the products side of the equation to relieve the stress. As long as we continually remove one or more products, this will happen until the limiting reactant is completely used up.

This concept is used quite often in the commercial production of chemicals. For example, ammonia is made commercially via its formation reaction:

$$N_2\,(g) + 3H_2\,(g) \rightleftharpoons 2NH_3\,(g)$$

This reaction is run at high temperatures (usually around 400–600°C) to increase the rate of the reaction, but at those temperatures, the equilibrium constant is not very large (less than $10\,\frac{1}{M}$). As a result, the equilibrium has more products than reactants, but it certainly cannot be treated as a reaction that uses up all of its reactants. At equilibrium, significant quantities of nitrogen and hydrogen gas are still present.

People in the business of making ammonia want to produce all of the ammonia they can from the reactants they have. To force this reaction to go to completion, the ammonia is continually removed from the reaction. As a result, the equilibrium is continually stressed so that it continually shifts to the products side of the equation. As a result, all of the nitrogen and hydrogen gas ends up reacting to form ammonia.

So you see that although all chemical reactions are, in fact, equilibria, we can stress an equilibrium so that it shifts to the products side of the equation. If we continue to do this throughout the course of the reaction, the reaction will never be able to reach equilibrium, and it will eventually use up all of its limiting reactant, regardless of the value of the equilibrium constant.

ON YOUR OWN

15.8 Assume that the following reaction has achieved equilibrium.

$CO_2\,(g) + H_2\,(g) \rightleftharpoons CO\,(g) + H_2O\,(g)$

a. What will happen to the concentration of CO and CO_2 if extra H_2O is added to the system?

b. What will happen to the concentration of CO_2 and H_2O if CO is removed from the system?

c. What will happen to the concentration of H_2O and H_2 if extra CO_2 is added to the system?

PRESSURE AND LE CHÂTELIER'S PRINCIPLE

In the definition of Le Châtelier's principle, we mentioned that stress can come in 3 different forms. We can stress an equilibrium by changing concentration, but we can also do it by changing either the pressure or the temperature. In this section, let's explore what happens when we stress an equilibrium by changing the pressure.

Before we can do that, you need to be reminded of something from an earlier part of the module. Remember that when we use Equation 15.1, we are supposed to ignore solids and liquids. The same thing is true with Le Châtelier's principle. If we change the amount of a solid or liquid in a chemical reaction, the equilibrium is not stressed. As a result, nothing happens. This is an important fact to remember:

Le Châtelier's principle ignores solids and liquids as a source of stress to the equilibrium.

As mentioned before, the aqueous phase is not the same as the liquid phase. We cannot ignore aqueous substances in the reaction. We can only ignore substances in the solid or liquid phase.

Now that that's out of the way, we can learn how changes in pressure stress a chemical equilibrium. When we increase pressure, what happens to the volume of any gases present? According to the gas laws that we learned in Module 11, increasing the pressure of a gas will lower its volume. What happens to concentration when volume is lowered? According to the definition of concentration, volume is in the denominator of the fraction. Therefore, when volume is lowered, concentration goes up. That's the effect of increasing the pressure on a chemical reaction: The concentration of the gases in the reaction increases.

What happens to the concentration of solids, liquids, or aqueous substances when the pressure increases? Nothing, really. Only gases are affected by a change in pressure. As a result, if an equilibrium contains no gases, a change in pressure does not cause it any stress. If it does contain gases, a change in pressure does produce a stress. In fact, whichever side of the equation has more gas molecules is more affected by a change in pressure. For example, consider the following equilibrium:

$$CaCO_3\,(s) \rightleftharpoons CaO\,(s) + CO_2\,(g)$$

In this equation, only the products side of the equation has a gas in it. As a result, an increase in pressure will only increase the concentration of CO_2. When that happens, the reverse reaction will speed up, and the equilibrium will shift toward the reactants. If, on the other hand, the pressure decreases, that will decrease the concentration of CO_2, causing the reverse reaction to slow down, making the forward reaction faster than the reverse. Therefore, the equilibrium will shift to the products side of the reaction.

The equilibrium tends to shift based on the side of the equation that has more total gas molecules. If pressure is increased, the reaction will shift away from the side that contains more gas molecules. On the other hand, if the pressure is reduced, the reaction will shift toward the side with the most gas molecules on it. If both sides have the same number of gas molecules on them, then both sides of the reaction are affected equally and, as a result, the reaction will not shift. These are important rules to remember:

When an equilibrium is subjected to an increase in pressure, it will shift away from the side with the largest number of gas molecules. If pressure decreases, the equilibrium will shift toward the side with the largest number of gas molecules. Finally, if there are no gases in the equation or if the number of gas molecules is the same on both sides, nothing will happen.

Make sure you can apply these rules by studying Example 15.5 and answering "On Your Own" question 15.9.

EXAMPLE 15.5

A chemist studies the following reaction, which is already at equilibrium:

$$NH_4Cl\ (s) \rightleftharpoons NH_3\ (g) + HCl\ (g)$$

a. Which way will the reaction shift if pressure is increased?

There are 2 molecules of gas on the products side and none on the reactants. An increase in pressure causes a shift away from the side with the largest number of gas molecules. Therefore, **the reaction will shift toward the reactants side of the equation**.

b. What will happen to the amount of NH_4Cl present if pressure is decreased?

A decrease in pressure causes a shift toward the side with the largest number of gas molecules. Since the products side has the largest number of gas molecules, the reaction will shift that way. When this happens, NH_4Cl will be used up, so **the amount of NH_4Cl present will decrease**. Note that the *concentration* of the NH_4Cl did not change because the concentration of a solid is constant. However, the *amount* did change.

c. Which way will the reaction shift if more NH_4Cl is added?

We were just taught that when solids (or liquids) are added or taken away from an equilibrium, it does not stress the system. Therefore, **the reaction will not shift**. Remember, the amount of a solid (or liquid) can change as a result of the shift in an equilibrium, but the concentration of a solid (or liquid) does not change. Changing the *amount* of a solid (or liquid) in a chemical equilibrium does not change the *concentration* and therefore cannot stress the equilibrium.

TEMPERATURE AND LE CHÂTELIER'S PRINCIPLE

We are still not quite done with Le Châtelier's principle. We still have to figure out how a change in temperature affects an equilibrium. This requires you to remember a fact we learned in Module 13: When a reaction is exothermic, its ΔH is negative and energy can be considered a product of the reaction. Endothermic reactions, on the other hand, have positive ΔHs and are considered to have energy as a reactant. With these facts in mind, let's see how temperature stresses an equilibrium.

Consider the following reaction:

$$2H_2O\,(g) \rightleftharpoons 2H_2\,(g) + O_2\,(g) \qquad \Delta H° = 242\text{ kJ}$$

Because $\Delta H° > 0$, we can think of energy as a reactant in the equation. We can think of the equation this way:

$$2H_2O\,(g) + 242\text{ kJ} \rightleftharpoons 2H_2\,(g) + O_2\,(g)$$

When we raise the temperature of the system, what are we really doing? We are adding energy. For this reaction, when we raise the temperature, we are really adding extra reactant. What did we learn about equilibria that have extra reactants added? The forward reaction increases in rate, making more products and less reactant. Therefore, when the temperature of this reaction is raised, more H_2 and O_2 are made, while H_2O is used up.

In the same way, if we were to lower the temperature of this reaction, it would be like taking a reactant away. When that happens, the rate of the forward reaction slows down, making the reverse reaction faster than the forward reaction. This causes the reaction to shift to the reactants side of the equation, making more H_2O and using up H_2 and O_2. The general trend, then, is:

When temperature is raised, an equilibrium will shift away from the side of the equation that contains energy. When temperature is lowered, the reaction will shift toward the side that contains energy.

Perform Experiment 15.2 to get more experience with Le Châtelier's principle and temperature changes.

ON YOUR OWN

15.9 Assume that the following reaction has already attained equilibrium:

$2SO_2\,(g) + O_2\,(g) \rightleftharpoons 2SO_3\,(g)$

a. Which way will the reaction shift if the pressure is decreased?

b. Which way will the reaction shift if the pressure is increased?

EXPERIMENT 15.2

PURPOSE: To show temperature effects on reactions and Le Châtelier's principle.

MATERIALS:

- 2 test tubes
- 2 eyedroppers
- 2 beakers
- 2 small cups
- Clear ammonia solution (This is sold with the cleaning supplies in most supermarkets. It must be clear. A colored solution will mess up the colors you are supposed to see.)
- White vinegar (It must be white. Colored vinegar will mess up the colors you are supposed to see.)
- Flame or stove
- Pot for boiling water on the stove
- Water
- Ice (preferably crushed)
- A few leaves of red cabbage (It must be red cabbage, not regular cabbage.)
- Safety goggles

QUESTION: How does equilibrium work with respect to temperature and concentration?

HYPOTHESIS: Write a hypothesis for this experiment.

PROCEDURE:

1. Put about 200 mL of water into the pot and add the leaves of red cabbage.
2. Put the pot on the stove and allow the water to boil for about 3 minutes.
3. While you are waiting for the water to boil, pour a small amount of ammonia solution into one of the cups and mark the cup so that you know it contains ammonia.
4. Pour a small amount of the white vinegar into the other cup and mark the cup so that you know it contains vinegar.
5. Fill the larger beaker with a mixture of ice and water. The mixture should have a lot of ice and only a little water because the ice will melt throughout the experiment.
6. Fill the smaller beaker about ¾ full with water.
7. Once the water in the pot has been boiling for 3 minutes, remove it from the stove.
8. Place the smaller beaker (the one with just water in it) on the stove or on a heating flame so that the water will begin to boil.
9. Use an eyedropper to put 20 drops of cabbage water into each of the test tubes. ***Be careful. The cabbage water is hot!***
10. Rinse out the eyedropper.
11. Use that same eyedropper to add 15 drops of vinegar to each test tube.
12. Swirl the test tubes to mix the cabbage water and vinegar. The solution in each test tube should be pink.
13. Use the other eyedropper to add ammonia to one of the test tubes. Add the ammonia

drop by drop. Swirl the test tube after each drop is added. As soon as the solution turns any shade of green, stop adding ammonia.

14. Repeat step 13 for the other test tube. When you are done, each solution should be approximately the same shade of green.
15. Place both test tubes in the solution of ice water in the big beaker.
16. Allow them to sit in the ice water for at least 1 minute.
17. Pull them both out of the ice water and compare the color of the solutions in the test tubes. They should be the same.
18. Put one of the test tubes back into the ice water.
19. Once the water in the smaller beaker is boiling, place the other test tube in the boiling water. ***Once again, be careful.*** As long as you hold the test tube by the very top, you will not get burned.
20. Allow the test tube to sit in the boiling water for 30 seconds, and then pull it out.
21. Pull the other test tube out of the ice water and compare the color of the solutions. The solution that was in the boiling water should be noticeably bluer than the one in the ice water. If it is not, try putting it back into the boiling water for another 30 seconds.
22. Put both test tubes back into the ice water.
23. After both test tubes have been in the ice water for a few minutes, pull them out again.
24. Compare the color of the solutions. They should be the same again.
25. If you like, repeat steps 18–24 again. Each time, the solution that gets hot should become bluer than the test tube in the ice water, and each time, the solutions should be roughly the same color once they have been in the ice water for a while. Do not allow the test tube to sit in the hot water for more than 30 seconds at a time because you do not want to drive the ammonia out of solution.
26. Clean up and return everything to the proper place.

CONCLUSION: Write something about the outcome and your hypothesis.

What happened in the experiment? When you mixed the ammonia and vinegar in the test tubes, you set up the following equilibrium:

$$NH_3\,(aq) + C_2H_4O_2\,(aq) \rightleftharpoons NH_4^+\,(aq) + C_2H_3O_2^-\,(aq)$$

This equilibrium is heavily weighted toward the products, but nevertheless, it is an equilibrium, so some of the reactants remain. The cabbage water you added was an indicator, telling you how much acid or base was present. When you added just vinegar (the source of $C_2H_4O_2$), the cabbage water turned pink to indicate the presence of an acid. When you added the ammonia to the solution, it eventually turned a shade of green, indicating that the acid was neutralized by the base.

This reaction is exothermic. Energy is a product in the reaction. When you put one test tube into ice water, you lowered its temperature. This, in effect, removed a product from the reaction, so the reaction shifted toward the products to relieve the stress. That caused the cabbage water to turn a bit more green. When you put the other test tube in the boiling water, you raised its temperature. In effect, this added a product to the equation, causing the equilibrium to shift toward the reactants. Therefore, the amounts of acid and base present in the test tube changed, and the color of the cabbage water changed accordingly.

This experiment is an excellent illustration of how Le Châtelier's principle works with temperature changes. The cabbage water provided an indication of the relative amounts of reactants and products. A pink color indicated more reactants, while a green color indicated more products. When the temperature of the reaction was lowered, the equilibrium shifted to the products because energy (a product) was being removed from the equilibrium. This caused the solution to turn green. When the temperature of the reaction was increased, the reaction shifted toward the reactants because energy (a product) was being added. As a result, the solution turned pink. Make sure you can do this kind of reasoning by studying Example 15.6 and answering "On Your Own" question 15.10.

EXAMPLE 15.6

For the following equation:

$$H_2\,(g) + F_2\,(g) \rightleftharpoons 2HF\,(g)$$

$\Delta H° = -273$ kJ. If the reaction has already reached equilibrium:

a. What will happen to the concentrations of each substance when the temperature is raised?

When the temperature is raised, equilibria shift away from the side that contains energy. Since the $\Delta H°$ for this reaction is less than 0, it is exothermic, making energy a product. The reaction will shift toward the reactants. Therefore, **the HF concentration will decrease while the H_2 and F_2 concentrations will increase.**

b. What will happen to the concentrations of each substance if the temperature is lowered?

When the temperature is lowered, equilibria shift toward the side that contains energy. Since the ΔH for this reaction is less than 0, it is exothermic, making energy a product. The reaction will shift toward the products. Therefore, **the HF concentration will increase, while the H_2 and F_2 concentrations will decrease.**

ON YOUR OWN

15.10 Consider the following reaction that has already attained equilibrium:

$CaO(s) + CO_2(g) \rightleftharpoons CaCO_3(s)$

The $\Delta H = -132$ kJ.

a. What will happen to the amount of $CaCO_3$ if the temperature is raised?

b. What will happen to the concentration of CO_2 if the temperature is lowered?

ACID-BASE EQUILIBRIA

Now that we've learned the basics of chemical equilibrium, it is time to examine a specific application of this interesting phenomenon. In Module 9, we introduced you to the subject of acid-base chemistry. To understand more, you need to understand chemical equilibrium. With the basic material of this module under your belt, however, you can now learn about acids and bases in a bit more depth.

Remember that we defined an acid as any molecule that donates an H^+ ion. Similarly, a base was defined as anything that accepts an H^+ ion. The molecule HCl (hydrochloric acid) acts as an acid in the following reaction:

$$HCl\,(aq) + OH^-\,(aq) \rightarrow H_2O\,(l) + Cl^-\,(aq)$$

because it donates an H^+ ion to the OH^- ion. Similarly, the hydroxide ion acts as a base because it accepts an H^+ ion. This reaction is actually a 2-step process. In the first step, the H^+ ion must first pull away from the HCl molecule as follows:

$$\text{Step 1: } HCl\,(aq) \rightarrow H^+\,(aq) + Cl^-\,(aq)$$

Only then can the H^+ ion be donated to the OH^- ion:

$$\text{Step 2: } H^+\,(aq) + OH^-\,(aq) \rightarrow H_2O\,(l)$$

From what you learned in the previous module, you should recognize that the step 1 and step 2 reaction equations represent the reaction mechanism of the first equation.

So you see that if an acid is going to donate an H^+ ion, it must first go through a reaction like the step 1 reaction equation where the H^+ ion actually separates from the molecule. Without this step, the acid cannot do its job. Reactions like that equation are called acid ionization reactions.

Acid ionization reaction—The reaction in which an H^+ ion separates from an acid molecule so that it can be donated in another reaction.

Since this step is the most critical in allowing an acid to do its job, we need to look at it in some detail and see how well this reaction works for different acids.

For example, the acid we have been using is HCl. This is considered a strong acid, which means its acid ionization reaction nearly goes to completion. When a solution of HCl is made, virtually all of its molecules break apart into H^+ and Cl^-. As a result, we never really bother to write the reaction as an equilibrium because this particular reaction has such a huge equilibrium constant that we can assume there are no reactants left once the reaction finishes.

Other acids are not nearly as good at ionization. Acetic acid ($C_2H_4O_2$) is an example of such an acid. This acid's ionization reaction is represented by the following equation:

$$C_2H_4O_2\,(aq) \rightleftharpoons H^+\,(aq) + CH_3CO_2^-\,(aq)$$

This equation is written as an equilibrium because the H^+ ion does not separate well from the acetic acid molecule. As a result, this reaction reaches equilibrium after only a few products are made. In other words, this is an equilibrium that is heavily weighted toward the reactants side of the equation.

Remember, the first acid we looked at (HCl) was so good at separating its H^+ ion from the rest of the molecule that its ionization reaction wasn't even written as an equilibrium. As a result, we call HCl a strong acid. However, the acid we are considering now (acetic acid) does not ionize very well. As a result, the equilibrium constant for its ionization is only 1.8×10^{-5} M. Since that equilibrium constant is small, we know that this reaction produces few products. When this reaction comes to equilibrium, then, it has mostly reactants present with only a few products present. Since acetic acid's ionization reaction does not produce many H^+ ions, we can say that acetic acid is a weak acid.

It's important to be able to tell the difference between weak and strong acids. Luckily, that's not very difficult. All we have to do is look at the value of the equilibrium constant for the acid's ionization reaction. The larger the equilibrium constant, the stronger the acid. As a rule of thumb, an acid is considered weak if the equilibrium constant is less than 1. As soon as the equilibrium constant gets larger than that, we consider the acid to be a strong acid.

Since the equilibrium constant of an acid's ionization reaction is so important in determining the strength of an acid, we give it a special name. We call it the acid ionization constant, which is usually abbreviated as K_a.

Acid ionization constant—The equilibrium constant for an acid's ionization reaction.

Since the acid ionization constant is simply the equilibrium constant for an acid's ionization reaction, we can say that the equation for the ionization constant for acetic acid looks like this:

$$K_a = \frac{[H^+][CH_3CO_2^-]}{[C_2H_4O_2]}$$

If we are given the value of this ionization constant, or if we are given the equilibrium concentrations with which to calculate its value, we can determine the strength of an acid.

Thus for any acid we can substitute our acid for HA:

$$aHA\ (aq) \rightleftharpoons bH^+\ (aq) + cA^-\ (aq)$$

$$K_a = \frac{[H^+]^b_{eq}[A^-]^c_{eq}}{[HA]^a_{eq}} \qquad \text{Equation 15.2}$$

If the ionization constant of an acid is large, the acid is considered a strong acid. If the ionization constant is not large, the acid is considered a weak acid. In general, the larger the ionization constant, the stronger the acid.

Not surprisingly, just as there can be strong and weak acids, there can also be strong and weak bases. However, since a base's job is quite different from an acid's job, the reactions that tell us whether a base is strong or weak are quite different. To determine the strength of a base, we have to look at how well it *accepts an H^+ ion from water*. For example, the reaction that determines the strength of the base NH_3 (ammonia) is:

$$NH_3\ (aq) + H_2O\ (l) \rightleftharpoons NH_4^+\ (aq) + OH^-\ (aq)$$

The base ionization constant, then, is given by the equilibrium constant for this reaction:

$$K_b = \frac{[NH_4^+][OH^-]}{[NH_3]}$$

Once again, the larger the K_b, the stronger the base. Also, don't forget to drop the H_2O (l) from the K_b equation.

This equation can also be generalized for all bases

$$bB\,(aq) + H_2O\,(l) \rightleftharpoons cHB^+\,(aq) + dOH^-\,(aq)$$

Which gives a base equilibrium constant of:

$$K_b = \frac{[HB^+]^c_{eq}[OH^-]^d_{eq}}{[B]^b_{eq}} \qquad \text{Equation 15.3}$$

Let's see how this works with Example 15.7 and "On Your Own" questions 15.11–15.12.

EXAMPLE 15.7

Nitrous acid (HNO_2) has an ionization constant equal to 4.5 x 10^{-4} M. Give the equation for this ionization constant. Is this acid stronger or weaker than acetic acid (K_a = 1.8 x 10^{-5} M)?

The ionization constant is the equilibrium constant for the acid ionization reaction. To determine the ionization reaction, we simply take the acid in its aqueous phase and remove an H^+ ion When you remove an H^+ ion from HNO_2, you are left with NO_2^-. If that doesn't make sense to you, review Module 9, where we first learned to reason this way. In the end, then, the aqueous acid is the reactant, and the H^+ and NO_2^- (both in aqueous phase) will be the products:

$$HNO_2\,(aq) \rightleftharpoons H^+\,(aq) + NO_2^-\,(aq)$$

The equilibrium constant for this reaction is the ionization constant, K_a:

$$K_a = \frac{[H^+][NO_2^-]}{[HNO_2]}$$

That is the equation for the ionization constant. Since the value given for this ionization constant is an order of magnitude larger than acetic acid, **this is a stronger acid than acetic acid.**

Methyl amine, CH_3NH_2, is a base with a K_b of 4.4 x 10^{-5}. Give the equation for its ionization constant. Would this be considered a strong or weak base?

The base ionization reaction involves having the base accept an H^+ ion from water. If CH_3NH_2 accepts an H^+ ion, it becomes $CH_3NH_3^+$. When water gives up that H^+, it becomes OH^-. All of this takes place in water, so the water phase is liquid and everything else is aqueous:

$$CH_3NH_2\,(aq) + H_2O\,(l) \rightleftharpoons CH_3NH_3^+\,(aq) + OH^-\,(aq)$$

The ionization constant, then, is just the equilibrium constant of this reaction. Remember that we must ignore water here since it is in its liquid phase:

$$K_b = \frac{[OH^-][CH_3NH_3^+]}{[CH_3NH_2]}$$

Since the value given for this ionization constant is less than 1, **this would be considered a weak base.**

ON YOUR OWN

15.11 What is the equation for the acid ionization constant of $HClO_2$?

15.12 What is the equation for the base ionization constant of PH_3?

THE pH SCALE

Since the strength of an acid is really dependent on how many H^+ ions it can make in an aqueous solution, chemists have developed a special scale to help us classify acids and bases. This scale is called the pH scale. The pH scale was developed by the Danish biochemist S. P. L. Sørenson in 1909 and is an abbreviation in French for *pouvier d'hydrogene* or in English, power of hydrogen.

The pH scale ranges from 0 to 14 and gives an idea of the concentration of the hydrogen ion. Having a scale of 0 to 14 is a lot easier to work with than a range of actual concentration of 1 to 10^{-14} M. The pH scale is a convenient way to describe the concentration of hydrogen (H^+) and hydroxide ions (OH^-). When a solution has a pH of 0, we consider it a very strongly acidic solution. Therefore, it has a lot of H^+ ions. When a solution has a pH of 14, it is strongly basic and has a lot of OH^- ions. When a solution is purely neutral (equal acid-base properties), it has a pH of 7.

Generally speaking, the lower the pH number, the more acidic a solution is. As a pH of 7 indicates a neutral solution, a pH of 6 (weak acid) to pH of 3 is increasing in acidity. Solutions with a pH 2-0 would be considered strongly acidic. Again with a pH of 7 as neutral, the higher the pH number the more basic a solution is. A pH of 8 (weak base) to a pH of 11 is increasing in alkalinity. Solutions with a pH 12-14 would be considered strongly basic.

In other words, the lower the pH, the more acidic a solution is; the higher the pH, the less acidic (and therefore more basic) the solution is.

If you purchased the MicroChem kit mentioned in the Student Notes, you can perform Experiment 11 to get more experience with the pH scale and pH indicators.

It's important for you to understand that pH is a measure of the *overall* acidity of a solution. For example, the pH of a 1.0 M solution of acetic acid is about 2.3. As mentioned before, acetic acid is a weak acid. However, the pH of a 0.0050 M solution of HCl is *also* 2.3, despite the fact that HCl is a strong acid. Why does the solution made with a weak acid have the same pH as a solution made with a strong acid? Look at the concentrations. The weak acid has a concentration of 1.0 M, while the strong acid has a concentration that is 200 times smaller. That's what's nice about pH. It is a measure of the acidity of a *solution*, which takes both the strength of the acid and the concentration of the acid into account. Remember, the chemistry of any substance depends on its concentration. A weak acid can still make a strongly acidic solution if its concentration is large enough.

The other nice thing about the pH scale is that it is very flexible. For every one unit change of pH, there is a tenfold change in the acidity of the solution. In other words, a solution with a pH of 3 is 10 times more acidic than a solution with a pH of 4, and it is 100 times more acidic than a solution with a pH of 5.

FIGURE 15.4
pH Scale
Illustration: Piercetheorganist

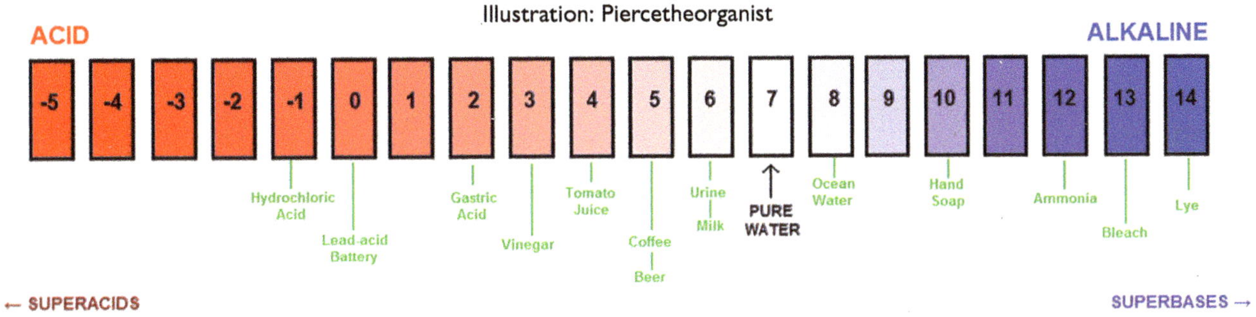

You will find the pH scale on a lot of products, especially things that come into contact with hair or skin. Our bodies prefer everything they come into contact with to be neutral. The best thing for our skin, hair, etc. is something that has a pH of 7, which is often called pH balanced. You may have heard or read that in advertisements before. It simply means that the product is has equal acid-base properties. Answer "On Your Own" question 15.13 to test your understanding of pH.

ON YOUR OWN

15.13 Equal concentration aqueous solutions are made with the following 4 chemicals. The pH of each solution is written next to the chemical formula:

C_6H_6OH (pH=3.2), KOH (pH=12.0), C_2H_7N (pH=9.1), and HBr (pH=0.50)

Identify the acids and the bases. In addition, indicate their relative strengths.

ACID RAIN

Now that we are back onto the subject of acids and bases, it is time to bring up an environmental subject again. You have probably heard something about acid rain. To truly understand this problem, you need to understand a bit about where acid rain comes from.

The first thing to learn is that rain is naturally acidic. When most people hear that, they are usually shocked. Isn't acid dangerous? Not necessarily. As you learned in Module 9, many of the foods we eat and drink are acidic. The acids in those foods provide a pleasant sour taste. The key is that to be safe, acids have to be weak. The acid in normal rain is carbonic acid, the same acid we drink in soda pop. The carbonic acid gets into the rain because the atmosphere has carbon dioxide in it. When that carbon dioxide mixes with the water in a raindrop, the following reaction occurs:

$$CO_2\ (g) + H_2O\ (l) \rightleftharpoons H_2CO_3\ (aq)$$

The pH of the resulting solution is usually around 5.3.

In a way, then, all rain is acid rain. The problem is that pollutants in the air can mix with the water in a raindrop just like CO_2 can. For example, coal that is burned in power plants is often contaminated with sulfur. The sulfur burns right along with the coal, making sulfur trioxide. When this mixes with the water in a raindrop, it makes sulfuric acid:

$$SO_3\ (g) + H_2O\ (l) \rightleftharpoons H_2SO_4\ (aq)$$

This puts more acid in the raindrop, making it more acidic.

In addition, a major source of pollutants on this planet is the automobile. When an automobile burns gasoline, it also burns some of the nitrogen in the surrounding air. This makes nitrogen dioxide. Like carbon dioxide and sulfur trioxide, this can react with the water in a raindrop to form acid as well:

$$2NO_2\ (g) + H_2O\ (l) \rightleftharpoons HNO_3\ (aq) + HNO_2\ (aq)$$

Once again, this adds more acid to the raindrop, making it more acidic.

With all of this extra acid in the rain, the pH of the rain goes down. There have been instances when the pH of rain was measured to be 3.2. Remember, every 1 pH unit represents a 10-fold change in acidity. This pH represents a raindrop that is more than 100 times more acidic than it should be! At such pHs, rain can be hazardous to living organisms.

Therefore, acid rain is a problem today. However, many people do not realize that acid rain is *less* of a problem today than it was 40 years ago. The air we are breathing today is *significantly cleaner* than the air we were breathing 40 years ago. Government regulations in many countries have succeeded in cutting the total amount of pollutants in the atmosphere significantly. This means there is less sulfur trioxide and nitrogen dioxide in the air, which means acid rain is less of a problem. As stewards of God's creation, we must continually address problems of pollution, like acid rain.

SUMMARY OF KEY EQUATIONS IN MODULE 15

Equation 15.1:	$K = \frac{[C]^c_{eq}[D]^d_{eq}}{[A]^a_{eq}[B]^b_{eq}}$	Equilibrium constant for the reaction $aA + bB \rightleftharpoons cC + dD$
Equation 15.2:	$K_a = \frac{[H^+]^b_{eq}[A^-]^c_{eq}}{[HA]^a_{eq}}$	Acid ionization constant
Equation 15.3:	$K_b = \frac{[HB^+]^c_{eq}[OH^-]^d_{eq}}{[B]^b_{eq}}$	Base ionization constant

ANSWERS TO THE "ON YOUR OWN" QUESTIONS

15.1 For this reaction, the equilibrium constant is given by:

$$K = \frac{[NH_3]^2_{eq}}{[N_2]_{eq}[H_2]^3_{eq}}$$

Since the concentrations given are, indeed, the equilibrium concentrations, we can simply plug them into our equation:

$$K = \frac{(2.28\ \cancel{M})^2}{(0.242\ M)(0.468\ \cancel{M})^3} = 2.10 \times 10^2\ \frac{1}{M^2}$$

15.2 According to Equation 15.1, the equilibrium constant for this reaction can be calculated with the following formula:

$$K = \frac{[SO_2]^2_{eq}[O_2]_{eq}}{[SO_3]^2_{eq}}$$

Remember, *only the concentrations at equilibrium* can be used to calculate K. The first concentration is irrelevant in the problem because it is the concentration that the chemist *started with*. We must use the concentrations at equilibrium, which are given in the next part of the problem:

$$K = \frac{(2.20\ \cancel{M})^2(1.10\ M)}{(0.40\ \cancel{M})^2} = 33\ M$$

15.3 In the first reaction of Example 15.1, the equilibrium constant was calculated to be 463. Since this is a number much larger than 1, this reaction is weighted toward the products side of the equation.

In the second reaction, the equilibrium constant was calculated to be 2.8×10^{-6} 1/M. Since K is a small number, the reaction is weighted toward the reactants side of the equation.

15.4 (a) If the equilibrium constant of a reaction is very large, the double arrow can be replaced by a single arrow, indicating that the reaction leads to all products and no reactants. Since this reaction has a very large equilibrium constant, it qualifies:

$$2NO_2\,(g) + Cl_2\,(g) \rightarrow 2NOCl\,(g) + O_2\,(g)$$

(b) This reaction has a very small equilibrium constant. That means there are virtually no products. The double arrow in this reaction can be replaced by a single arrow pointing toward the reactants:

$$2Cl_2\,(g) + 2H_2O\,(g) \leftarrow 4HCl\,(g) + O_2\,(g)$$

Since this looks awkward, we usually turn the reaction around so that it is written in the direction that it actually goes:

$$4HCl\,(g) + O_2\,(g) \rightarrow 2Cl_2\,(g) + 2H_2O\,(g)$$

(c) In this case, the equilibrium constant is close to 1. This means that the equilibrium is well balanced; we cannot remove the double arrow in this case.

15.5 To solve this problem, we simply use Equation 15.1, recognizing that we ignore Li since it is a solid. We also ignore H_2O because it is in its liquid state. The equilibrium constant is given by the following equation:

$$K = [LiOH]^2[H_2]$$

15.6 In this equation, there are 2 solids to ignore. We can also ignore the water since it is in its liquid state. Equation 15.1 becomes:

$$K = [NH_4OH]^2[Ba(SCN)_2]$$

15.7 We first have to see whether or not the reaction has reached equilibrium. To do this, we just plug the concentration values into Equation 15.1. If the result is equal to 17 M^2, then we know that the reaction is at equilibrium. If not, then the reaction has not reached equilibrium yet.

$$\frac{[N_2]_{eq}[H_2]^3_{eq}}{[NH_3]^2_{eq}} = \frac{(2.5\text{ M})(2.3\cancel{\text{ M}})^3}{(0.80\cancel{\text{ M}})^2} = 48\text{ M}^2$$

According to the problem, the correct value of the equilibrium constant is 17 M^2. Since this is not the answer we got, we must conclude that the concentrations we used were not equilibrium concentrations. This means that the reaction is not at equilibrium. To determine which way the reaction will shift to reach equilibrium, we just need to compare our calculation to the value of K. Our calculation is larger than the correct value. This means that to reach equilibrium, the concentrations need to change in such a way as to decrease the value of the equation we used. The reaction must shift toward the reactants.

15.8 (a) If extra H_2O is added to the system, the rate of the reverse reaction will increase. This will make more CO_2 and H_2. To do this, the reaction will use up CO and H_2O. At the new equilibrium, the concentration of CO will be lower (because it will be used up), and the concentration of CO_2 will be higher (because it will be made).

(b) If CO is removed from the system, the rate of the reverse reaction will decrease, making the forward reaction suddenly faster than the reverse. This will make more CO and H_2O. To do this, the reaction will use up CO_2 and H_2. Therefore, at the new equilibrium, the concentration of CO_2 will be lower (because it will be used up), and the concentration of H_2O will be higher (because it will be made).

(c) If extra CO_2 is added to the system, the rate of the forward reaction will increase. This will make more CO and H_2O. To do this, the reaction will use up CO_2 and H_2. At the new equilibrium, the concentration of H_2 will be lower (because it will be used up), and the concentration of H_2O will be higher (because it will be made).

15.9 (a) When pressure is lowered, an equilibrium will shift toward the side with the most gas molecules on it. This means the reaction will shift toward the reactants side of the equation.

(b) When the pressure is increased, an equilibrium will shift away from the side with the most gas molecules. There are 3 gas molecules on the reactants side (2 + 1), and only 2 on the products side. Therefore, the reaction will shift away from the reactants side of the equation, which is toward the products side of the equation.

15.10 (a) The ΔH tells us that this equation is exothermic, making energy a product. Therefore, when temperature is raised, the equilibrium will shift as if extra products had been added. This means that the equilibrium will shift toward the reactant side of the equation. Since that means $CaCO_3$ will be used up, the amount of $CaCO_3$ present will decrease.

(b) When temperature is lowered, energy is taken away. Since energy is a product in this reaction, lowering the temperature is like removing a product. Therefore, the equilibrium will shift toward products. This will cause CO_2 to be used up, and the concentration of CO_2 will go down.

15.11 The ionization constant is simply the equilibrium constant for the acid ionization reaction. To determine the ionization reaction, we simply take the acid in its aqueous phase and remove an H^+ ion When we remove an H^+ ion from $HClO_2$, we are left with ClO_2^-. In the end, then, the aqueous acid is the reactant, and the H^+ and ClO_2^- ions (both in aqueous phase) will be the products:

$$HClO_2\,(aq) \rightleftharpoons H^+\,(aq) + ClO_2^-\,(aq)$$

The equilibrium constant for this reaction is the ionization constant, K_a:

$$K_a = \frac{[H^+][ClO_2^-]}{[HClO_2]}$$

15.12 The base ionization reaction involves having the base accept an H^+ ion from water. If PH_3 accepts an H^+ ion, it becomes PH_4^+. When water gives up that H^+ ion, it becomes OH^-. All of this takes place in water, so the water phase is liquid and everything else is aqueous:

$$PH_3\,(aq) + H_2O\,(l) \rightleftharpoons PH_4^+\,(aq) + OH^-\,(aq)$$

The ionization constant, then, is just the equilibrium constant of this reaction:

$$K_b = \frac{[OH^-][PH_4^+]}{[PH_3]}$$

15.13 When pH is 7, the solution is neutral. Less than 7 means acid, and the lower the pH, the more acidic. More than 7 means base, and the higher the pH, the more basic. The HBr solution is strongly acidic; the C_6H_6OH solution is weakly acidic; the C_2H_7N solution is weakly basic; and the KOH solution is strongly basic.

STUDY GUIDE FOR MODULE 15

REVIEW QUESTIONS

1. What is the definition of chemical equilibrium?

2. Does a chemical reaction actually stop once equilibrium is reached?

3. A chemist is studying 2 chemical equilibria. Reaction A has an equilibrium constant equal to 1,213 M, while Reaction B has an equilibrium constant equal to 0.344 M^2. Which reaction will make more products?

4. After taking a cursory glance through a chemistry book, a student says that sometimes a chemical reaction is written with a single arrow and sometimes it is written with a double arrow. The student says that this means some chemical reactions are equilibrium reactions and some are not. What is wrong with the student's statement?

5. Why do we ignore solids in the equilibrium constant and when using Le Chatelier's principle?

6. Why are acid ionization reactions important in chemistry?

7. What is the range of the pH scale?

8. Three solutions have the following pH levels:

   Solution A: pH = 1

   Solution B: pH = 10

   Solution C: pH = 7

   Which is (are) the acidic solution(s)?

9. Three acid solutions of equal concentration have the following pH levels:

   Solution A: pH = 1

   Solution B: pH = 4

   Solution C: pH = 6

   Which solution is made with the acid that has the *lowest* ionization constant?

10. Where does acid rain come from?

PRACTICE PROBLEMS

1. For the following reaction:

$$2NH_3\,(g) \rightleftharpoons N_2\,(g) + 3H_2\,(g)$$

The equilibrium concentrations are: $[NH_3] = 0.48$ M, $[N_2] = 0.90$ M, and $[H_2] =$ 1.26 M. What is the value of the equilibrium constant for this reaction?

2. A chemist is studying the following equilibrium:

$$2Pb(NO_3)_2\,(s) \rightleftharpoons 2PbO\,(s) + 4NO_2\,(g) + O_2\,(g)$$

He starts out with 20 g of $Pb(NO_3)_2$ and, at equilibrium, has 4.04 g of PbO. The concentrations of NO_2 and O_2 at equilibrium are 0.36 M and 0.090 M, respectively. What is the value of the equilibrium constant?

3. Three chemical equilibria are written below, along with their equilibrium constants. If any of these equations can be written with a single arrow, do so.

a. $2NOCl\,(g) + O_2\,(g) \rightleftharpoons 2NO_2\,(g) + Cl_2\,(g)$ $\qquad K = 2.2 \times 10^{-5}$

b. $2SO_3\,(g) \rightleftharpoons 2SO_2\,(g) + O_2\,(g)$ $\qquad K = 1.6$ M

c. $2NO_2 \rightleftharpoons N_2O_4$ $\qquad K = 9.3 \times 10^{4}\,\frac{1}{M}$

4. A chemist is studying the following reaction:

$$H_2CO_3\,(aq) \rightleftharpoons CO_2(g) + H_2O\,(l) \qquad K = 2.3 \times 10^{4}$$

She measures the concentrations of H_2CO_3 and CO_2 when there is 1.23×10^{3} kg of water present. If their concentrations are 4.6 M and 7.0 M, respectively, is the reaction at equilibrium? If not, which way must the reaction shift to attain equilibrium?

5. The following reaction:

$$2SO_2\,(g) + O_2\,(g) \rightleftharpoons 2SO_3\,(g)$$

has an equilibrium constant equal to $4.4\,\frac{1}{M}$. If the following concentrations are present: $[SO_2] = 0.170$ M, $[O_2] = 0.087$ M, $[SO_3] = 0.560$ M, is the reaction at equilibrium? If not, which way must it shift to reach equilibrium?

6. Consider the following reaction that has reached equilibrium:

$$CaCO_3\,(s) + H_2O\,(l) \rightleftharpoons H_2CO_3\,(aq) + CaO\,(aq)$$

a. What will happen to the concentration of H_2CO_3, H_2O, and CaO if extra $CaCO_3$ is added?

b. What will happen to the concentration of $CaCO_3$, H_2CO_3, and CaO if more H_2O is added?

c. What will happen to the amount of $CaCO_3$, H_2CO_3, and CaO if H_2CO_3 is removed?

7. Consider the following reaction that has reached equilibrium:

$$2HF\,(g) \rightleftharpoons H_2\,(g) + F_2\,(g) \qquad \Delta H = -541\ kJ$$

a. What will happen to the concentration of H_2, F_2 and HF if the temperature is raised?

b. What will happen to the concentration of H_2, F_2 , and HF if the temperature is lowered?

8. Consider the following reaction that has reached equilibrium:

$$2NH_3\,(g) \rightleftharpoons N_2\,(g) + 3H_2\,(g) \qquad \Delta H = 92.2\ kJ$$

a. What will happen to the concentration of N_2, H_2, and NH_3 if the temperature is raised?

b. What will happen to the concentration of N_2, H_2, and NH_3 if the pressure is raised?

c. What will happen to the concentration of N_2, H_2, and NH_3 if the pressure is lowered?

9. What is the equation for the acid ionization constant of $HC_2H_3O_2$?

10. What is the equation for the base ionization constant of PO_4^{3-}?

REDUCTION-OXIDATION REACTIONS

Have you ever wondered how a battery works? You stick a D-cell into a flashlight, turn on the switch, and the light comes on. We know that for the light to work, it must have electricity. How is the battery supplying this electricity? Does a battery store electricity and then just release it when the switch is turned on? If a battery does *store* electricity, how does it get there in the first place? What is electricity, anyway? These questions will be answered in this module. Of course, to be able to answer these questions, we need to cover a lot of background material, so we'll begin with that.

first reactions

We've been on a journey of discovering the science of creation together. "For since the creation of the world God's invisible qualities—his eternal power and divine nature—have been clearly seen, being understood from what has been made, so that people are without excuse" (Romans 1:20, NIV). As we begin our final module, we hope that you are able to say with confidence: "I know that everything God does will endure forever; nothing can be added to it and nothing taken from it" (Ecclesiastes 3:14, NIV).

OXIDATION NUMBERS

To understand the processes that govern batteries, we first need to learn about oxidation numbers. All along we've talked about 2 fundamentally different types of compounds: ionic and covalent. The big difference between them is that ionic compounds are made when electrons are given up by one atom and taken by another, and covalent compounds are made when atoms share their electrons. Oxidation numbers provide a way to bridge the gap between these 2 types of compounds.

As we learned in Module 5, even though atoms share electrons in a covalent molecule, some atoms in the molecule have a bigger share of those electrons than others. We learned that the more electronegative an atom is, the bigger the share of electrons it receives. In other words, when 2 atoms share electrons, the more electronegative of the 2 ends up with the bigger portion of this electron pair. In many covalent molecules, electrons are not shared equally. As a result, there are fractional charges in these molecules, and we call them polar covalent molecules. This is all review, but it's important that you see where we're going with this.

Since the more electronegative atoms get a larger portion of the electrons they share, it might be instructive to see what charges *would* develop on the atoms of a molecule, *if* the more electronegative atoms got *all* of the electrons they share. When we do this, we are calculating the oxidation number of the atoms.

Oxidation number—The charge that an atom in a molecule would develop if the most electronegative atoms in the molecule took the shared electrons from the less electronegative atoms.

For example, the HCl molecule is made up of a hydrogen atom and a chlorine atom sharing one pair of electrons, as shown in the following Lewis structure:

Even though these atoms share a pair of electrons, the Cl atom is much more electronegative than the H atom. As a result, it gets more than its fair share of that electron pair. When we calculate oxidation numbers, we are asking what charges would develop if the Cl atom just took that electron pair and stopped sharing it. If that happened, the Cl atom would suddenly have 8 electrons, and the H atom would have none. Since a Cl atom usually has 7 electrons, the fact that it would now have 8 electrons indicates that it would have a -1 charge. In the same way, an H atom usually has 1 electron. If the H atom were left with 0 electrons, it would have a +1 charge. The oxidation number of hydrogen in this molecule is +1, while the oxidation number of chlorine is -1.

Do you see now why we say that oxidation numbers bridge the gap between ionic and covalent compounds? By calculating oxidation numbers, we are calculating *possible* charges on the atoms in a molecule *assuming* that the electrons were not shared but were transferred from one atom to another. Oxidation numbers treat all compounds as ionic, attempting to assign charges to all atoms that make up the compound. Remember, oxidation numbers *aren't real*. In our example above, H *does not* have a charge of +1 and Cl *does not* have a charge of -1. Oxidation numbers are a way of saying "what if?" They don't have any significance in the real world.

If oxidation numbers don't have any real physical or chemical meaning, why are we bothering to learn them? Oxidation numbers are an excellent bookkeeping technique that will allow us to determine where electrons travel during a chemical reaction. As we will see later, knowing where electrons go during a chemical reaction is a major key to understanding the reactions we will study in this module. Oxidation numbers are a powerful *tool* that we can use in analyzing chemical equations, so we need to learn how to calculate and use them.

The first rule to learn in calculating oxidation numbers is that in any compound, the sum of all oxidation numbers (times the number of atoms that have that oxidation number) must equal the electrical charge on the compound. For example, we determined the oxidation numbers for H and Cl in the HCl molecule. The H atom has an oxidation number of +1, and the Cl atom has an oxidation number of -1. They add to 0, which is the overall charge of the molecule.

The sum of all oxidation numbers in a molecule must equal the charge of that molecule.

This simple rule can make all the difference in calculating the oxidation numbers of atoms. Example 16.1 will help you understand oxidation numbers.

EXAMPLE 16.1

What is the oxidation number of iron in Fe?

The sum of all oxidation numbers must add up to the total charge of the substance. Since Fe has no charge, and since there is only 1 atom here, the oxidation number must be **0**.

What is the oxidation number of iron in Fe^{3+}?

Once again, the sum of all oxidation numbers must equal the charge. Since the charge is +3, and since there is only 1 atom in the substance, the oxidation number of Fe must be **+3**.

Suppose you know that the oxidation number of Mg in $MgBr_2$ is +2. What is the oxidation number of Br in that same compound?

The total charge of $MgBr_2$ is 0. This means that all oxidation numbers must add up to 0. Since the oxidation number of Mg is +2, this means that the sum of all Br's oxidation numbers must be -2. Since there are 2 Br atoms, each must have an oxidation number of **-1** to make this happen.

Suppose you know that the oxidation number of H is +1 and O is -2 in the compound $HBrO_3$. What is Br's oxidation number?

Since H has an oxidation number of +1 and there are 3 O atoms, each with an oxidation number of -2, the sum of the oxidation numbers so far is 1 + 3 x (-2) = -5. For all of the oxidation numbers to add up to the total charge (which is 0), the Br must have an oxidation number of **+5**.

This example brings up a good point. Notice that Br had an oxidation number of -1 in $MgBr_2$ and an oxidation number of +5 in $HBrO_3$. The same atom can have different oxidation numbers depending on the molecule it is in. Remember, oxidation number tells us the charge that Br would have if the more electronegative atoms got all of the shared electrons. In $MgBr_2$, bromine is the most electronegative atom. Therefore, it gets all of the electrons, making it negative. In $HBrO_3$, the oxygen is more electronegative. Therefore, bromine loses its shared electrons and becomes positive.

ON YOUR OWN

16.1 What is the oxidation number of phosphorus in each of the following substances given O has an oxidation number of -2?

a. P

b. P^{3-}

c. P_2O_5

Answer "On Your Own" question 16.1 to make sure you understand this part of calculating oxidation numbers.

DETERMINING OXIDATION NUMBERS

Of course, you need to be able to calculate the oxidation numbers of *all* atoms in a molecule. We really shouldn't have to tell you any of them. We just had you answer "On Your Own" question 16.1 as a warm-up to the real thing, which is often a bit difficult. You see, the best way to calculate the oxidation numbers of all atoms in a molecule is to draw a Lewis structure of the molecule and then give the shared electrons to the most electronegative atoms. Then we can count the electrons on each atom and determine the charge they would have under those conditions. This is what we did in the previous section to show you the oxidation numbers of H and Cl in HCl.

The problems with figuring out oxidation numbers this way are pretty enormous. First, many of the compounds that participate in the reactions we are going to study have extremely difficult Lewis structures, so first-time chemists often get very frustrated figuring them out. Also, even though you learned in Module 4 how to tell which atoms are more electronegative than others, the rules you learned are often broken, making it very difficult to determine which atoms get the shared electrons. In the end, the Lewis structure method for determining oxidation numbers cannot really be taught to chemistry students at either the high school or college level.

Are we stuck then? Not really. Chemists have developed rules of thumb that allow us to determine the oxidation numbers of certain atoms. Once you learn those rules, you can usually use those oxidation numbers to begin your calculations, which will then help you calculate the oxidation numbers of other atoms. The first 4 rules we will use in this text are:

1. When a substance has only 1 type of atom in it (F_2, O_3, Ca^{2+}, or Mg for example), the oxidation number for that atom is equal to the charge of the substance divided by the number of atoms present.

2. Group 1A metals (Li, Na, K, Rb, Cs, and Fr) always have oxidation numbers of +1 in molecules that contain more than 1 type of atom.

3. Group 2A metals (Be, Mg, Ca, Sr, Ba, and Ra) always have oxidation numbers of +2 in molecules that contain more than 1 type of atom.

4. Fluorine always has a -1 oxidation number in molecules that contain more than 1 type of atom.

What's nice about these 4 rules is that they are *always true*. The problem is that after these 4 rules, things get a little fuzzy because there are a few exceptions to the next rules. In general, though, the next rules *usually* apply:

5. When grouped with just 1 other atom that happens to be a metal, H has an oxidation number of -1. In *all other cases* in which it is grouped with other atoms, H has an oxidation number of +1.

6. Oxygen has an oxidation number of -2 in molecules that contain more than 1 type of atom.

There are exceptions to both rule 5 and rule 6. However, you will be happy to know that for this course, we will ignore those exceptions. As far as you are concerned, rules 1–6 always are true. After that, though, things get really fuzzy. If you are really stuck and none of these rules apply, you can follow this general (and often not true) guideline:

7. If all else fails, assume that the atom's oxidation number is the same as what it would take on in an ionic compound. The atoms that are most likely to follow this rule are in groups 3A, 6A, and 7A.

This means that you might be able to assume that chlorine has a -1 oxidation number because it is in group 7A, and those atoms take on a -1 charge in ionic compounds. Similarly, it might be a good guess that S has a -2 oxidation number because all group 6A atoms take a -2 charge in ionic compounds. This general principle has so many exceptions that it should be used only as a last resort!

In the end, we have 6 rules that we can assume will always work, and 1 general principle that has many exceptions but is a good guess when all else fails. How do we use these rules to determine oxidation numbers? In general, we end up trying to determine the oxidation numbers of all atoms using rules 1–6. If that doesn't work, before we resort to rule 7, we see if rules 1–6 apply to all but 1 of the atoms in the molecule. If they do, then we can determine the oxidation number of the last atom the same way we did in the previous section, by using the condition that all oxidation numbers have to add up to the charge on the molecule. Therefore, we use rule 7 *only* when rules 1–6 do not apply to either all or all but 1 of the atoms in the molecule. This may sound really complex right now, but some practice with Example 16.2 will make you a virtual master at this.

EXAMPLE 16.2

What is the oxidation number of each atom in the following molecules?

a. H_2

According to rule 1, H's oxidation number is 0 divided by 2, which is **0**.

b. $Ca(OH)_2$

In this compound, we can apply our rules to every atom. Ca is a group 2A metal, so its oxidation number is +2 (rule 3). Rule 6 tells us that oxygen has an oxidation number of -2. Finally, since H is not grouped with just 1 other atom, its oxidation number is +1 (rule 5). As a check, we need to make sure that these numbers add up to the total charge (0) of the molecule. There are 1 Ca atom, 2 O atoms, and 2 H atoms in the molecule. Adding the oxidation numbers:

$$(+2) + 2 \times (-2) + 2 \times (+1) = 0$$

Therefore, the oxidation numbers are: **Ca: +2, O: -2, H: +1**.

c. F_2CO

Rule 4 tells us that F is always -1. Rule 6 says that oxygen is -2. We have no rule for C, but that's okay. Since we know all but 1 of the atoms, we can figure out C by making sure that the oxidation numbers add up to the overall charge of the molecule (0). There are 2 F atoms and 1 O atom, so the total oxidation number so far is 2 x (-1) + (-2) = -4. To make all of the numbers add up to 0, the C must have an oxidation number of +4. Therefore, the oxidation numbers are: **F: -1, C: +4, O: -2**.

d. $NaSO_4^-$

Rule 2 tells us that Na's oxidation number is +1 since it is a group 1A metal. Rule 6 says that O must have a -2 oxidation number. Once again, we have no rule for S, but we can figure out its oxidation number because the sum of all oxidation numbers must equal the charge of the molecule, which is -1. Since Na is +1 and O is -2, we have (+1) + 4 x (-2) = -7. For the sum of oxidation numbers to be -1, S must have an oxidation number of +6. Therefore, the oxidation numbers are **Na: +1, S: +6, O: -2**.

e. C_2H_4

According to rule 5, when H is grouped with just 1 other atom that happens to be a metal, its oxidation number is -1; otherwise it is +1. In this molecule, H is grouped with just 1 other atom, but that atom is not a metal. H's oxidation number is +1. We have no rules for C, but we know that the sum of oxidation numbers must equal the overall charge (0). Since H is +1, we have a +4 sum so far. For them to add to 0, the 2 C atoms must account for a sum of -4. Each C atom must have a -2 oxidation number. In the end, then, the oxidation numbers are **C: -2, H: +1**.

f. NCl_3

This is the first time that rules 1–6 do not apply. As a result, we must fall back on our general principle (7). According to this principle, groups 3A, 6A, and 7A are the most likely to follow the rule. This molecule contains a group 7A atom, so we can apply the principle to it. This means that Cl will have an oxidation number of -1. If you don't understand how we got this, you need to review Module 4, where we learned how to predict the charge of an atom in an ionic compound. Since Cl has an oxidation number of -1, and since there are 3 of them, N must have an oxidation number of +3 for the oxidation numbers to add up to the total charge (0). Therefore, the oxidation numbers are: **N: +3, Cl: -1**.

g. $MgCoBr_4$

In this compound, we know that Mg is +2 because it is a group 2A metal (rule 3). For the other 2, rules 1–6 do not apply. We are stuck with rule 7. We will apply this to Br for 2 reasons. First, the rule states that group 7A is a group that likely follows the rule, and Br is in group 7A. Also, Co is not in an A group, and our rules for determining charge apply only to groups 1A–8A. Br is the only atom we can apply this rule to. Since Br is in group 7A, it develops a -1 charge when in ionic compounds. That leaves us to determine Co's oxidation number by making use of the fact that all oxidation numbers add up to the overall charge (0). Therefore, the oxidation numbers are **Mg: +2, Co: +2, Br: -1**.

Try this yourself in "On Your Own" question 16.2 to make sure you can calculate oxidation numbers.

Before you go on in this module, make sure that you understand how to calculate the oxidation numbers of all atoms in a molecule. The rest of this module depends on knowing this important skill. If you are still having trouble at this point, please find another chemistry book and see how it explains this important topic. You must be able to do this to go on!

ON YOUR OWN

16.2 What are the oxidation numbers of all atoms in the following compounds?

a. CO_2 b. S_8 c. BF_4^-
d. N_2H_2 e. ClNO f. $LiNH_2$
g. PO_4^{3-} h. $Ca(NO_2)_2$

OXIDATION AND REDUCTION

Now that you have learned how to calculate oxidation numbers, we can go on to learn about reduction-oxidation reactions, which are often called redox reactions. The first thing we must do to accomplish this is to learn the definitions of these terms:

Oxidation—The process by which an atom loses electrons.

Reduction—The process by which an atom gains electrons.

It's so important for you to remember these definitions that we will give you a mnemonic to help you. Leo is a name often used to refer to a lion. When lions are angry (when they study chemistry, for example), they growl. We could say:

LEO says GER

LEO stands for Lose Electrons Oxidation, and GER stands for Gain Electrons Reduction. Use this mnemonic to help you keep these definitions straight.

To understand redox reactions, you must be able to recognize when an atom is undergoing oxidation and when it is undergoing reduction. This is why oxidation numbers are so important. Let's say an atom has an oxidation number of +5. If it undergoes a chemical reaction, it will end up in a different molecule, and its oxidation number might change. Let's suppose that after a chemical reaction, its oxidation number changed to +3. What happened? To go from +5 to +3, the number of electrons owned by this atom changed. Since electrons are negative, the atom must have *gained* 2 electrons. That's the only way to change the oxidation number from +5 to +3. Since the atom gained electrons (GER), we say that it underwent reduction. When an atom undergoes reduction, we often say that it has been reduced. Similarly, when an atom undergoes oxidation, we say that it has been oxidized.

To determine whether an atom is oxidized or reduced, we must look at how its oxidation number changes. We then determine whether negatives had to be added to or subtracted from the original oxidation number to get the new one. If negatives must be added, electrons were gained, and therefore the atom was reduced. If negatives must be subtracted, the atom lost electrons and therefore was oxidized. Study Example 16.3 and answer "On Your Own" questions 16.3–16.4 to cement these ideas in your head.

EXAMPLE 16.3

A sulfur atom goes from a -2 oxidation number to a +6 oxidation number. Has it been oxidized or reduced? How many electrons participated in this change?

Remember, to determine oxidation or reduction, we must think in terms of electrons. We have to ask ourselves whether we must add negative numbers to or subtract negative numbers from the original oxidation number to get the final one. To go from -2 to +6, negative numbers would have to be subtracted. This means that the atom lost electrons, therefore it was **oxidized**.

The second question simply asks how many electrons it took to make the oxidation happen. To go from -2 to +6, we must subtract -8 because:

$$-2 - (-8) = +6$$

Therefore, the sulfur must have lost **8 electrons**.

A carbon atom goes from a -2 oxidation number to a -4 oxidation number. Was it oxidized or reduced? How many electrons participated in the change?

In terms of electrons, there is only one way to go from -2 to -4. The atom must gain negatives and must gain electrons. Therefore, the atom was **reduced**. To see how many electrons participated, we just figure out how many negatives must be gained to go from -2 to -4:

$$-2 + (-2) = -4$$

Therefore, **2 electrons** participated in the change.

An atom goes from a +5 oxidation number to +3. Was it oxidized or reduced? How many electrons participated in the change?

To go from +5 to +3, we must add 2 negatives. Since negatives were added, the atom was **reduced**. Since it took 2 negatives, **2 electrons** participated in the change.

An atom goes from a -6 oxidation number to -1. Was it oxidized or reduced? How many electrons participated in the change?

To go from -6 to -1, 5 negatives must be subtracted. Since negatives must be subtracted, the atom lost electrons. Therefore, it was **oxidized**. Since 5 negatives must be subtracted, **5 electrons** participated in the change.

ON YOUR OWN

16.3 An atom changes its oxidation number from +7 to -1. Was it oxidized or reduced? How many electrons participated in this change?

16.4 An atom goes from an oxidation number of +2 to +6. Was it oxidized or reduced? How many electrons participated in this change?

RECOGNIZING REDUCTION-OXIDATION REACTIONS

We are now finally ready to actually study redox reactions. We call a chemical reaction a redox reaction if, during the process of the chemical reaction, an atom's oxidation number changes. For example, in the following reaction:

$$Mg\,(s) + 2HCl\,(aq) \longrightarrow H_2\,(g) + MgCl_2\,(aq)$$

the oxidation numbers of both Mg and H changed. On the reactants side of the equation, Mg has an oxidation number of 0, H is +1, and Cl is -1. On the products side of the equation, H has an oxidation number of 0. In addition, Mg has an oxidation number of +2, while Cl still has an oxidation number of -1. Let's rewrite this reaction, putting the oxidation number above each atom so that you can see this clearly:

$$\overset{0}{Mg}\,(s) + 2\overset{+1}{H}\overset{-1}{Cl}\,(aq) \longrightarrow \overset{0}{H_2}\,(g) + \overset{+2}{Mg}\overset{-1}{Cl_2}\,(aq)$$

Mg went from an oxidation number of 0 to +2, while H went from an oxidation number of +1 to 0. We could therefore say that Mg was oxidized and H was reduced. This makes the reaction a redox reaction. Notice that the oxidation number of chlorine did not change. This just means that Cl was neither oxidized nor reduced. Not every atom in a redox reaction needs to change oxidation numbers. If *any* atoms change oxidation numbers in a reaction, it is a redox reaction.

Compare that reaction to the following one:

$$NaOH\,(aq) + HCl\,(aq) \longrightarrow H_2O\,(l) + NaCl\,(aq)$$

In this reaction, no atom changed oxidation number. On the reactants side, the oxidation numbers are: Na: +1, O: -2, H: +1 (in both molecules), and Cl: -1. On the products side, the oxidation numbers are still: Na: +1, O: -2, H: +1, and Cl: -1.

$$\overset{+1}{Na}\,\overset{-2}{O}\,\overset{+1}{H}\,(aq) + \overset{+1}{H}\overset{-1}{Cl}\,(aq) \longrightarrow \overset{+1}{H_2}\overset{-2}{O}\,(l) + \overset{+1}{Na}\overset{-1}{Cl}\,(aq)$$

In this reaction, no atom changed its oxidation number, so this is not a redox reaction. Make sure that you can identify redox reactions by studying Example 16.4 and answering "On Your Own" question 16.5.

EXAMPLE 16.4

For each of the following, determine whether or not it is a redox reaction. If it is a redox reaction, identify the atoms whose oxidation numbers changed and tell whether they were oxidized or reduced.

a. $Ca\,(s) + Cl_2\,(g) \longrightarrow CaCl_2\,(s)$

On the reactants side, the oxidation numbers are as follows: Ca: 0, Cl: 0. On the products side, the oxidation numbers are: Ca: +2, Cl: -1. The oxidation numbers did change, so **this is a redox reaction**. Since Ca went from 0 to +2, negatives were lost. This means that **Ca was oxidized**. On the other hand, Cl went from 0 to -1, indicating that negatives were gained. This means that **Cl was reduced**.

b. Ag^{+} (aq) + OH^{-} (aq) $\longrightarrow$ AgOH (s)

On the reactants side, the oxidation numbers are: Ag: +1, O: -2, H: +1. On the products side, Ag: +1, O: -2, H: +1. Therefore, **this is not a redox reaction.**

c. $3IF_5$ (aq) + 2Fe (s) $\longrightarrow$ $2FeF_3$ (aq) + $3IF_3$ (aq)

On the reactants side, the oxidation numbers are: F: -1, I: +5, Fe: 0. On the products side, the oxidation numbers are Fe: +3, F: -1 (in both molecules), I: +3. Therefore, **this is a redox reaction** because Fe and I changed oxidation numbers. Since Fe went from 0 to +3, it lost 3 electrons. This means that **Fe was oxidized.** On the other hand, I went from +5 to +3, indicating that it gained 2 electrons. Therefore, **I was reduced.**

ON YOUR OWN

16.5 For each of the following, determine whether or not it is a redox reaction. If it is a redox reaction, identify the atoms whose oxidation numbers changed and tell whether they were oxidized or reduced.

a. Cu^{2+}(aq) + Zn (s) $\longrightarrow$ Zn^{2+}(aq) + Cu (s)

b. $2H_2O \longrightarrow 2H_2 + O_2$

c. NaCl + AgF $\longrightarrow$ AgCl + NaF

d. Ca(s) + $2HNO_3$(aq) $\longrightarrow$ $Ca(NO_3)_2$(aq) + H_2(g)

If you purchased the MicroChem kit mentioned in the Student Notes, you can perform Experiment 6 in that kit. It will give you some experience working with redox reactions.

AN IMPORTANT CHARACTERISTIC OF REDUCTION-OXIDATION REACTIONS

Hopefully you noticed something while studying the example and answering the "On Your Own" questions. In each one of the problems, if the reaction was a redox reaction, there were always 2 atoms that changed their oxidation numbers. In addition, one of those atoms was always oxidized and the other was always reduced. This is the way it has to happen, and that's why we call them reduction-oxidation reactions:

When a redox reaction occurs, there will always be at least 1 atom that is oxidized and at least 1 atom that is reduced.

Just like acids can work only if there is a base (and we therefore call those reactions acid-base reactions), reduction can only occur if oxidation occurs as well (and we therefore call those reactions reduction-oxidation reactions).

If you think about this, it should make sense. After all, an oxidation number tells us what charge an atom would take if it got full control of the electrons for which it already owns the largest share. If the oxidation number of an atom changes, it has, in effect, lost or gained electrons. Since electrons cannot simply be created out of thin air, if one atom gains electrons, another atom has to lose them. Therefore, for an atom to be reduced (gain electrons), another atom must be oxidized (lose electrons).

The fact that atoms gain and lose electrons in a redox reaction forms the basis of how a battery works. Therefore, to understand batteries, you need to understand this point. Whereas many chemical equations involve simply the *rearrangement* of electrons, redox reactions involve the *transfer* of electrons. In a redox reaction, one atom gains electrons and another loses. In the next section, you will see how, under the proper circumstances, this ends up making a battery. Before you learn that, do Experiment 16.1, which utilizes a reduction-oxidation reaction.

EXPERIMENT 16.1

PURPOSE: To demonstrate an oxidation-reduction reaction.

MATERIALS:

- Large bowl
- Blank notebook paper without lines
- Scissors
- Cotton swab or small paintbrush
- Lemon juice (This must be *real* lemon juice, either fresh or from concentrate.)
- Iodine solution (This is available in large drugstores. The pharmacy department within your supermarket will probably not have this, and neither will a small corner drugstore. A major chain will have it, however. Please note that *iodide* will not work. Iodide, as you should know, is I^-, while iodine is I_2.)
- Water
- Measuring cups
- Medicine dropper
- Stirring rod (A **plastic** spoon or knife will work as well. Do not use any metal flatware around the iodine solution because iodine will rust metal.)
- Safety goggles

QUESTION: How do oxidation and reduction happen?

HYPOTHESIS: Write a hypothesis for this experiment.

PROCEDURE:

1. Cut the notebook paper into a square that will fit in the bottom of the bowl.
2. Wet the cotton swab in the lemon juice and begin using it to write your first name on the paper. The lemon juice will appear clear as you write with it, but you will be able to see what you are writing because the cotton swab will leave a wet trail. Make sure the trail left by the cotton swab is wet enough so that you can easily see what you are writing. You will probably have to dip your cotton swab in the lemon juice several times. Once you have finished writing your name in this way, allow the paper to dry completely, so that your name has essentially vanished.

3. While you are waiting for the paper to dry, fill the bowl with 1 cup of water.
4. Add to that water 40-50 drops of iodine solution. Mix the water and iodine together thoroughly with the stirring rod.
5. Once the paper is dry, examine it. The name that you wrote on the paper should be pretty much invisible.
6. Put the paper (face up) into the iodine/water solution. Be sure to fully submerge the paper into the solution.
7. Observe what happens. You should see the paper slowly start to turn blue because all paper has starch in it. Starch and iodine react to form a very complex molecule that happens to be blue. As you wait, you should see that the name you wrote with lemon juice will not turn blue. As a result, the paper will be blue except where you wrote your name. Your name, therefore, will be visible.
8. Clean up and return everything to the proper place.

CONCLUSION: Write something about the outcome and your hypothesis.

What happened in the experiment? The blue iodine-starch compound could form only where there was no lemon juice because to form this compound, the iodine must have an oxidation number of 0. In an iodine solution, the I_2 molecule is present, which has iodine with an oxidation number of 0. However, when ascorbic acid (found in lemon juice) reacts with iodine, it *oxidizes* it, producing a color change. As a result, wherever the ascorbic acid is, there is no iodine with a 0 oxidation number. Therefore, the blue iodine-starch compound could not form where the ascorbic acid was, leaving your name in white.

HOW BATTERIES WORK

Now that we have learned how to analyze redox reactions, we are finally able to see how a battery works.

Remember from the last section that redox reactions occur because electrons are transferred from one chemical compound to another. For example, consider the following simple equation:

$$Cu^{2+}(aq) + Zn(s) \longrightarrow Zn^{2+}(aq) + Cu(s)$$

This is clearly a redox reaction because the reactants side has Cu with an oxidation number of +2 and Zn with an oxidation number of 0. On the products side, the Cu now has a 0 oxidation number, while Zn has the +2 oxidation number. In this reaction, Cu is reduced and Zn is oxidized.

This reaction happens because Zn gives 2 electrons to the copper. Electrons are transferred between zinc and copper. If we were to put solid zinc into a solution of Cu^{2+} ions, we would slowly see the solid zinc disappear because it would begin to turn into aqueous Zn^{2+}. Remember, *aqueous* means dissolved in water, so once the zinc turns into Zn^{2+}, it dissolves into the water. To an observer, it would look like the zinc was disappearing. At the same time, we would start to see solid copper begin to appear. Before the reaction, the copper was dissolved as Cu^{2+}. However, when the reaction begins, it forms solid copper, which would settle down at the bottom of the container. If we mixed these 2 chemicals together, we could observe this reaction.

Rather than mixing the 2 chemicals, what if we just put them in 2 separate beakers and placed the beakers next to each other?

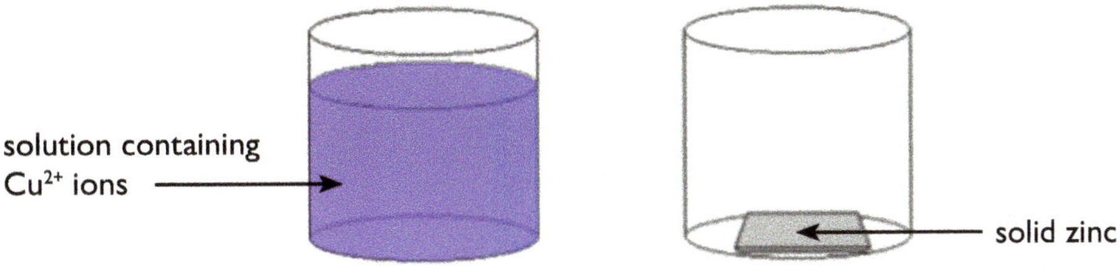

Would anything happen? Of course not! For the reaction to occur, the reactants would have to get close enough together to exchange electrons. Since they can't even touch each other, there's no way to get the electrons transferred. Or is there?

Suppose we were to put the 2 chemicals in contact, but not directly. Instead of mixing the chemicals together, we connect them with a wire called a conductor that can transport the electrons.

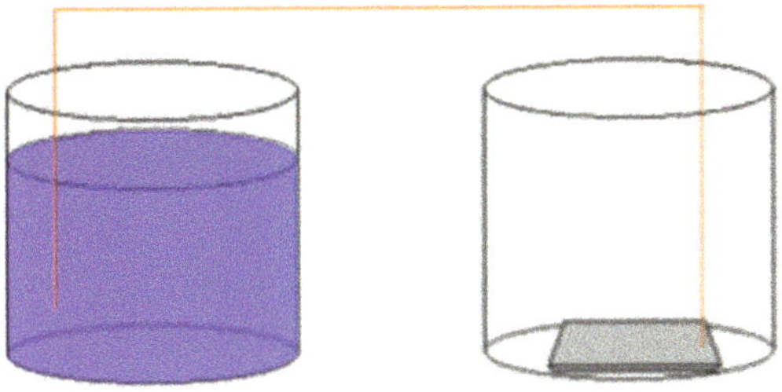

All that is left to do now is to include the products. As zinc loses it electrons, Zn^{2+} is produced, but Zn^{2+} cannot exist by itself. Zinc ions can exist in water, so we need to fill the zinc beaker with water. Also, the copper ions will be turning into solid copper. This process can be helped out by putting a piece of solid copper in the beaker.

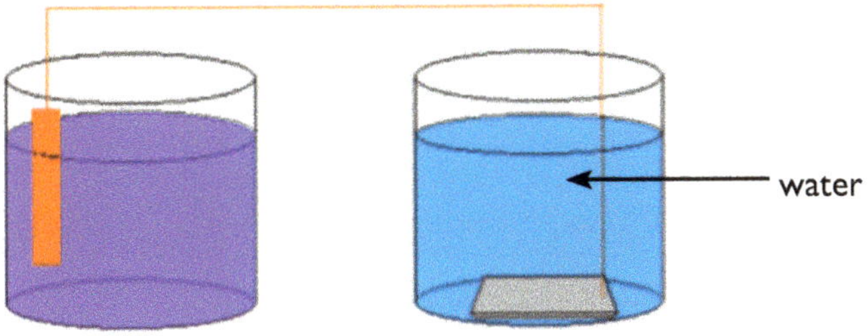

Believe it or not, under these circumstances, *the reaction would proceed*! To get this reaction to work, all we need to do is get electrons from the zinc to the Cu^{2+}. The wire allows this to happen because it conducts electrons. Electrons will flow out of the zinc electrode, through the wire, into the copper strip, and into the Cu^{2+} solution. When they leave the zinc, the zinc becomes Zn^{2+}. As the electrons arrive in the Cu^{2+} solution, the Cu^{2+} turns into Cu. The reaction occurs even though the reactants are in different containers! If we watched this setup for a while, we would notice that the zinc strip would slowly disappear as it turned into Zn^{2+} and dissolved into solution. We would also notice that the reaction would stop.

When the first few electrons leave the solid zinc, it turns into Zn^{2+}. As a result, the entire solution in the beaker containing zinc develops a positive charge. As the solution in the beaker becomes more and more positive, electrons will be less and less likely to leave it because that positive charge will attract the electrons, keeping them from leaving. In the same way, as electrons reach the solution containing Cu^{2+}, that solution will start becoming negative. This will repel electrons traveling down the wire, making them less likely to get there. In the end, then, after a few electrons were transferred in this way, the battery would stop working because electrons could no longer travel across the wire. A salt bridge keeps this from happening.

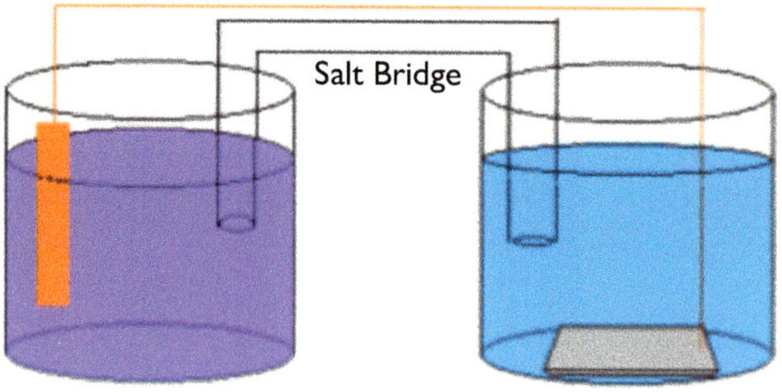

Remember, *salt* is just another term for an ionic compound. A salt bridge is just made up of an aqueous ionic compound. When ionic compounds are aqueous, their ions break apart and are able to move independently. When the beaker that holds the zinc begins to become positively charged, the negative ions in the salt bridge travel to that solution. This cancels out the positive charge that is developing and, as a result, more electrons are free to travel down the wire. In the same way, as the solution that contains the Cu^{2+} becomes more negative, the positive ions in the salt bridge are attracted to that beaker. This cancels out the negative charge that is developing, making sure that electrons are not hampered as they move toward the Cu^{2+}. If you don't quite understand all of this, don't worry. Just keep in mind that a salt bridge is something that contains an aqueous ionic compound and that for the battery to work, the salt bridge must be there to balance out the charges that otherwise would develop in the beakers.

FIGURE 16.1
A Simple Battery
Illustration by Megan Whitaker

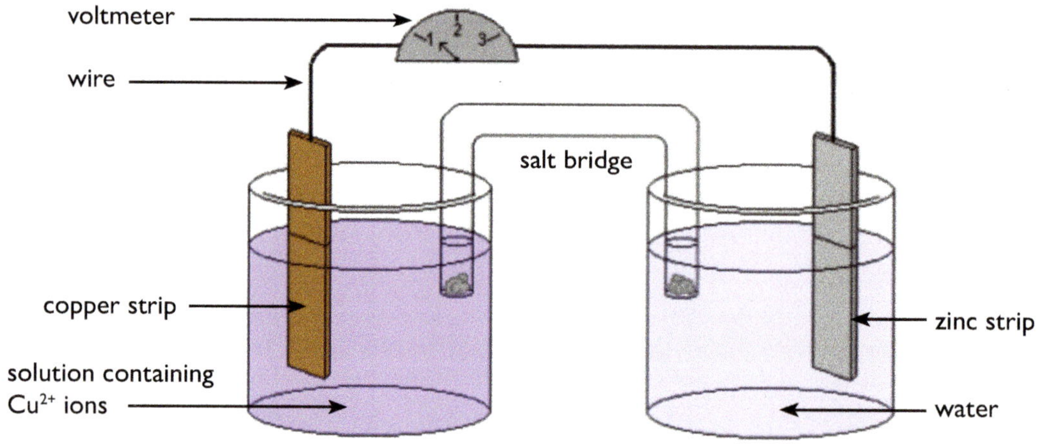

Make sure you understand what's going on here. The electrons leave the zinc. We know this happens because according to the chemical equation presented earlier, zinc's oxidation number is changing from 0 to +2. This means that the zinc is losing electrons. We know that the electrons are traveling to the Cu^{2+} solution because the equation tells us that the copper's oxidation number is changing from +2 to 0. This means that the copper is being reduced, which tells us that it is gaining electrons. This is why we know that electrons flow from the zinc to the copper.

Electricity, of course, is just electrons traveling through a wire. Since this chemical reaction causes electrons to run through a wire, it is making electricity. This setup can be used to run an electrical device. For example, Figure 16.1 shows a voltmeter connected to the wire. A voltmeter measures the potential difference between 2 points. For example, if we hook a voltmeter to the ends of a D-cell battery, the voltmeter will read about 1.5 volts. This indicates the voltage of the battery. In Figure 16.1, notice that the voltmeter is reading just slightly more than 1 volt. Therefore, this chemical reaction is supplying almost the same voltage as a D-cell battery.

FIGURE 16.2
Household Batteries
Photo: Lead holder (CC BY-SA 3.0)

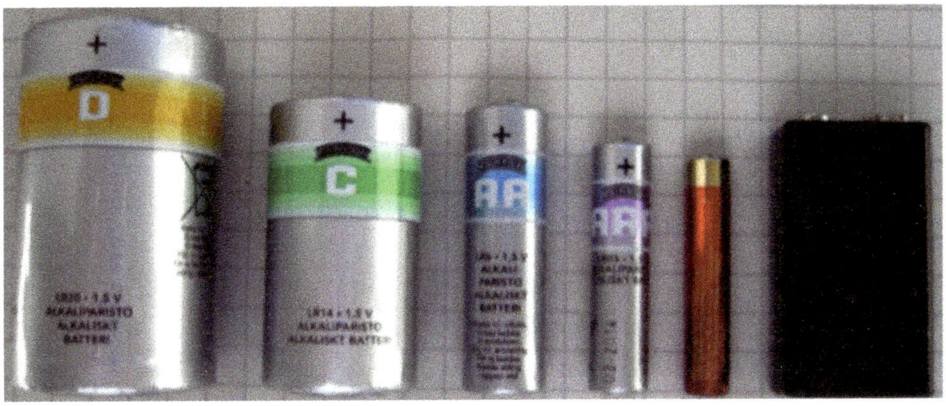

Remember, Figure 16.1 is a very simple picture of a battery. If you look at a D, C, AA, or AAA battery, you will see that the ends are labeled as positive (+) and negative (-). In the battery pictured in Figure 16.1, which beaker would be the positive side and which would be the negative side? Electrons are flowing away from the zinc. Since electrons are negative, they are repelled by other things that are negative. This means that the container of zinc can be thought of as negatively charged. In the same way, since electrons travel toward the container of Cu^{2+}, it must be positive because electrons travel toward positive charges.

Before we can continue, we need to nail down some terminology. Chemists call the negative side of the battery the anode (an' ohd). This is because it is a source of negative ions, which are usually called anions (an' eye uns). The positive side of the battery is called the cathode (kath' ohd) because it is the source of positive charges, which are usually called cations (kat' eye uns). Also, the term *battery* is a bit too specific for chemists because there are many different types of chemical reactions like this. Chemists call this kind of setup a Galvanic cell, after Italian physiologist Luigi Galvani, who first discovered that electricity can flow between 2 metals. While dissecting a frog, he noticed that when his steel scalpel touched a brass hook, the frog's leg would twitch. He realized that this meant he was somehow making electricity. Of course, now we know that a redox reaction occurred between the 2 metals, causing electricity to flow through the frog's nerves, activating a response in the frog's muscles. Finally, the metal strips are usually called electrodes (ee lek' trohds), which means that they form the electrical contact between the wire and the contents of the beakers.

In the end, then, we can label our picture as is done in Figure 16.3.

FIGURE 16.3
Electron Flow in a Galvanic Cell
Illustration by Megan Whitaker

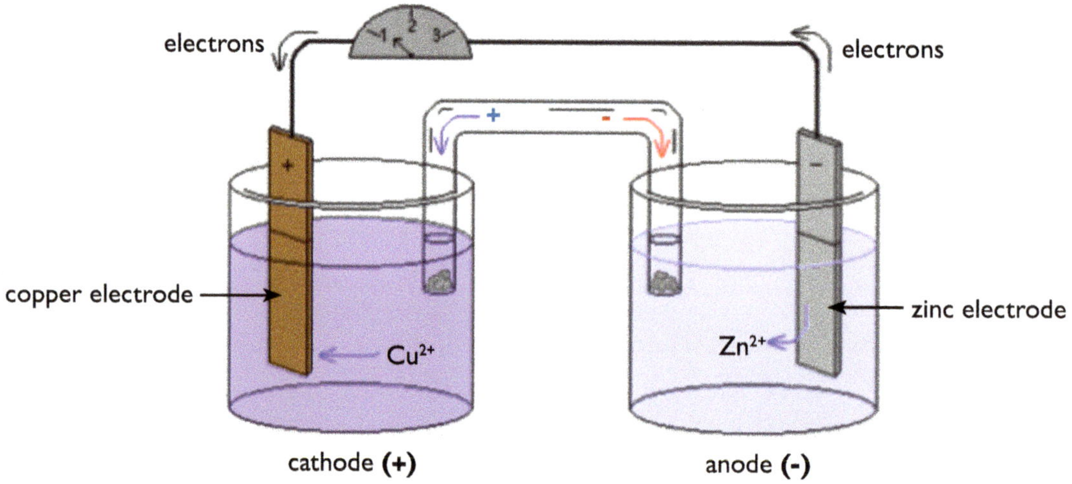

From your own experience, you know that batteries eventually stop working. Hopefully, you can see why. A battery produces electricity because electrons flow from one reactant in the battery's chemical reaction to another reactant. As long as the reaction can proceed, the battery will produce electricity. However, once one of the reactants is completely used up, the chemical reaction stops. This stops the flow of the electrons and causes the battery to go dead. A battery does not store electricity; it stores chemicals. The electricity a battery produces is made through the reaction of those chemicals. When the chemicals are used up, the battery stops working.

What about rechargeable batteries? When a battery uses up its chemicals, it is dead. How does a battery recharger work? The details of this are a bit too complicated to get into here, but we can provide an abbreviated explanation. A battery charger produces an electrical force that causes the battery's chemical reaction to reverse. When a battery is dead, it has all products and no reactants. A battery recharger forces those products to react with each other and reform the reactants. Once that happens, the battery is ready to go again. This doesn't work for all batteries, however. That's why many batteries are not rechargeable.

A real battery is a little more complicated than the simple picture we have discussed so far, but all of the principles that exist in this simple battery are the same as those that exist in a real one. In a moment, you will see some examples of real batteries and how they work. First you need to be comfortable with analyzing a Galvanic cell. Given the chemical equation, you need to be able to draw a figure similar to Figure 16.3, label the anode and cathode, and identify the electron flow. This will take a bit of practice, so study Example 16.5, perform Experiment 16.2, and answer "On Your Own" questions 16.6–16.7.

EXAMPLE 16.5

A Galvanic cell runs on the following chemical reaction:

$$2Ag^{+}(aq) + Mg(s) \longrightarrow 2Ag(s) + Mg^{2+}(aq)$$

Draw a diagram of the Galvanic cell, labeling all reactants and products, the electron flow, the anode, the cathode, and the positive side and negative side of the Galvanic cell.

The first thing that we have to do is look at the reactants and see which is being oxidized and which is being reduced. Since Ag^+ turns into Ag, it must be gaining electrons, so it is reduced. This means that the container that holds the Ag^+ will have the electrons flowing into it. The Mg, on the other hand, loses electrons because it goes from an oxidation number of 0 to an oxidation number of +2. This means that the Mg is oxidized and the container that holds it will have electrons leaving it. Since the electrons are flowing from the Mg to the Ag^+, the container that holds Mg is negative (electrons are repelled by negative charge) and is therefore the anode. The container that holds the Ag^+ is positive (electrons are attracted to the positive charge), and is therefore the cathode. In the equation, Ag^+ is aqueous. This means that the Ag^+ is dissolved, so the solution that holds it will be clear. The Mg, on the other hand, is solid, so there should be a magnesium electrode in its beaker. That will be the source of the solid magnesium. As always, there must be a salt bridge as well. In the end, then, the picture should look something like this:

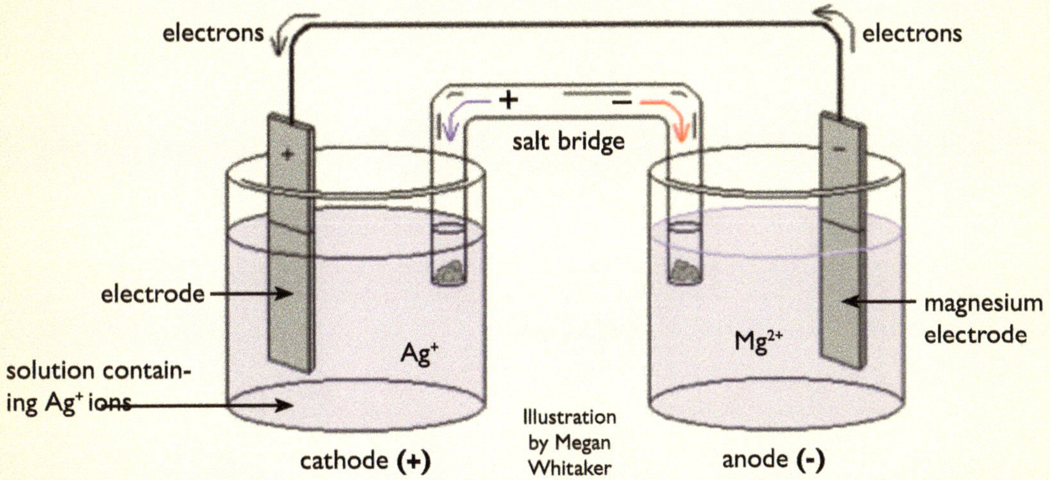

Illustration by Megan Whitaker

We did not label the electrode that is in the Ag^+ solution because it does not need to be any specific metal. It just needs to conduct electricity. The electrode on the right must be magnesium because solid magnesium is a reactant in the equation. The presence of solid magnesium helps the Mg^{2+} ions turn into Mg solid. Since the only other reactant needed is Ag^+, and since Ag^+ is in the solution, there is no need for the electrode in that solution to be silver. It can be any metal. Usually, we do use the metal version of the ion in that ion's beaker, but since silver is expensive, a chemist would probably not use silver as the electrode in this case. When you are drawing Galvanic cells, you need not describe what an electrode is made of unless the electrode *must* be made of something. For example, in this problem you would be required to label the magnesium electrode on the right because that electrode must be made of magnesium.

Finally, notice that we also did not include the voltmeter in the diagram. After all, the voltmeter is not a necessary part of the Galvanic cell. Electrons will flow through the wire with or without the voltmeter being there. We just used it in figures 16.1 and 16.2 to emphasize that electricity will flow through the wires. It would be hard for you to draw the voltmeter because you would need to determine the voltage, which is different for every Galvanic cell. There is a way to do that, but we don't want to teach it here. If you take advanced chemistry later on, you will learn how to calculate Galvanic cell voltages.

A Galvanic cell runs on the following chemical reaction:

$$I_2\,(aq) + Fe\,(s) \rightarrow 2I^-\,(aq) + Fe^{2+}\,(aq)$$

Draw a diagram of the Galvanic cell, indicating the electron flow, labeling the anode and cathode, and indicating the positive and negative sides of the Galvanic cell.

In this reaction, iodine is going from an oxidation number of 0 to an oxidation number of -1. This indicates that it is gaining electrons. The solution holding aqueous iodine will have electrons flowing into it. This makes that container positive (electrons are attracted to a positive charge), and it will be the cathode. The Fe is going from an oxidation number of 0 to an oxidation number of +2. This means it loses electrons. Since it is losing electrons, the electrons are flowing away from the container holding the solid Fe. This makes that container the negative side of the Galvanic cell, and it is therefore the anode. The picture, then, looks like this:

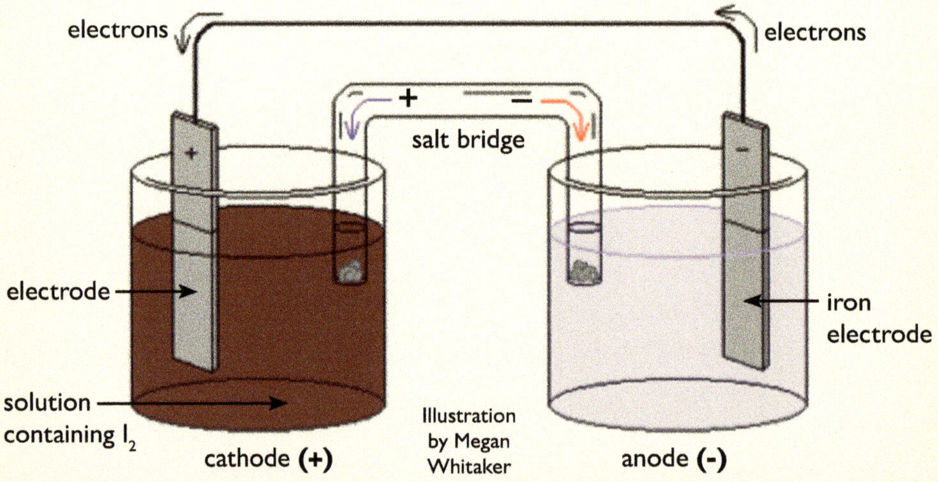

Notice once again that the metal that makes up the electrode on the left is not mentioned. That's because it could be any metal since the reactant on that side of the Galvanic cell is I_2, which is in solution. The iron electrode is labeled because that electrode must be iron since iron is a reactant in the equation. Also, notice the color of the solution that contains I_2. We used that color because it is roughly the correct color for such a solution. We do not expect you to know that, so you need not use colors in your diagrams.

If you purchased the MicroChem kit mentioned in the Student Notes, you can perform Experiment 17, which will give you some experience with making simple Galvanic cells.

EXPERIMENT 16.2

PURPOSE: To create a Galvanic cell from lemons.

MATERIALS:

- 4 juicy lemons
- 4 copper pennies (Pennies from before 1982 have higher copper content. You can also use copper wire instead of a penny.)
- 4 two-inch zinc coated (galvanized) nails
- 5 small wires, ideally with alligator clips
- Volt meter (optional)
- LED (Light Emitting Diode) bulb
- Safety goggles

QUESTION: Can you create a Galvanic cell battery from common household items that can light an LED bulb?

HYPOTHESIS: Write a hypothesis for this experiment.

PROCEDURE:

1. Squeeze the lemon gently with your hands or roll it on a table with some pressure to release the lemon juice. Don't rupture the lemon's skin.
2. Make a small penny-sized cut in the lemon and insert the copper penny into it with half of the penny sticking out.
3. Insert the galvanized nail into the other side of the lemon, ensuring that the nail does not touch the penny.
4. This is a single cell of a battery. The nail and the penny are the electrodes, and the lemon juice is the electrolyte.
5. If you have a volt meter, connect one wire of the volt meter to the nail and the other to the penny. Note what it reads. Now try to connect one wire from the nail to the LED wire and another wire from the penny to the other LED wire. What do you see? This current is not enough to light an LED light. To have enough current, you will need to add more lemons.
6. Create 3 more of the same lemon batteries.
7. Connect 5 wires in this order:
   Wire 1: LED wire to penny in lemon 1
   Wire 2: galvanized nail in lemon 1 to penny in lemon 2
   Wire 3: galvanized nail in lemon 2 to penny in lemon 3

Wire 4: galvanized nail in lemon 3 to penny in lemon 4
Wire 5: galvanized nail in lemon 4 to LED wire

8. What do you see? Can you determine which is the cathode, and which is the anode in this experiment based on Figure 16.3?
9. Clean up and return everything to the proper place.

CONCLUSION: Write something about the outcome and your hypothesis.

ON YOUR OWN

16.6 A Galvanic cell runs on the following chemical equation:

$2H^+(aq) + Mn(s) \longrightarrow H_2(g) + Mn^{2+}(aq)$

Draw a diagram for this Galvanic cell, labeling the electron flow, the anode and cathode, and the positive and negative sides of the Galvanic cell.

16.7 A Galvanic cell runs on the following chemical equation:

$3Cl_2(aq)+2Al(s) \longrightarrow 2Al^{3+}(aq)+6Cl^-(aq)$

Draw a diagram for this Galvanic cell, labeling the electron flow, the anode and cathode, and the positive and negative sides of the Galvanic cell.

In this experiment, we didn't separate the 2 solutions of a Galvanic cell. The copper penny is the cathode and the zinc nail is the anode. The lemon acted as both the salt bridge and Galvanic cell solutions for the reaction.

REAL BATTERIES

As already noted, batteries in real life are a bit more complicated than those we have discussed here. The reactions that run batteries are a bit more complex than those that we have studied so far. Also, since we try to make batteries as small as possible, their architecture is significantly more complicated. Nevertheless, batteries are such an important part of our everyday life that we need to have at least a rudimentary understanding of why they work.

Let's start our discussion with the battery used in automobiles. This battery is called a lead-acid battery, and it runs on the following lead-acid battery reaction:

$$PbO_2\,(s) + Pb\,(s) + 2H_2SO_4\,(aq) \longrightarrow 2PbSO_4\,(s) + 2H_2O\,(l)$$

Notice that this reaction is much more complicated than the ones we have discussed so far. Therefore, we need to analyze it carefully. Let's first see what is being oxidized and what is being reduced. Here are the oxidation numbers for the atoms in the reaction:

$$\overset{+4\,-2}{PbO_2}\,(s) + \overset{0}{Pb}\,(s) + 2\overset{+1+6-2}{H_2SO_4}\,(aq) \longrightarrow 2\overset{+2+6-2}{PbSO_4}\,(s) + 2\overset{+1\,-2}{H_2O}\,(l)$$

Before we get into the analysis of what happens in this reaction, note that while you could have come up with all of the oxidation numbers on the reactants side of the equation, it would have been hard for you to come up with the oxidation numbers in $PbSO_4$. You would have had to recognize that it is an ionic compound made with the sulfate ion (SO_4^{2-}) and a Pb^{2+} ion. Using those ions, you could have come up with the proper oxidation numbers. Don't worry about that, however. We do not expect you to analyze an equation that is this difficult.

Let's look at what happened to the oxidation numbers in this reaction. First, notice that the oxidation numbers of H, O, and S did not change. They are not being oxidized or reduced. What atoms are being oxidized and reduced? The *lead* atoms are. Notice that the Pb in PbO_2 has a +4 oxidation number. On the products side of the equation, the only lead atom is in $PbSO_4$, and in that compound, Pb has an oxidation number of +2. The Pb in PbO_2 went from an oxidation number of +4 to one of +2 in $PbSO_4$. Therefore, the Pb in PbO_2 gained electrons and was reduced. On the other hand, the Pb in Pb (s) has an oxidation number of 0, and the only lead atom on the products side of the equation (the Pb in $PbSO_4$) has an oxidation number of +2. Therefore, the Pb (s) went from an oxidation number of 0 to an oxidation number of +2. It therefore lost electrons and was oxidized.

Do you think it is strange that 2 different lead atoms started out with 2 different oxidation numbers on the reactants side of the equation and ended up with the same oxidation number on the products side of the equation? You shouldn't. After all, one lead atom (the one in PbO_2) was reduced, and another lead atom (the one in solid Pb) was oxidized. They just ended up making 2 lead atoms that happened to have the same oxidation number!

Before we discuss the battery itself, notice that H_2SO_4 is a reactant. That's why this kind of battery is called a lead-acid battery. It uses 2 forms of lead [solid lead and lead(IV) oxide] as well as an acid (sulfuric acid). This acid is always present in an automobile battery, and that's why you must use it with caution. If you were to tip an automobile battery over and spill its contents on yourself, the acid would begin eating away at your clothes and skin. Therefore, you must always handle automobile batteries with extreme care.

Let's look at the way a lead-acid battery is put together. Figure 16.4 contains a simplified illustration of a lead-acid battery.

FIGURE 16.4
A Lead-Acid Battery
Illustration by Michael Dailey

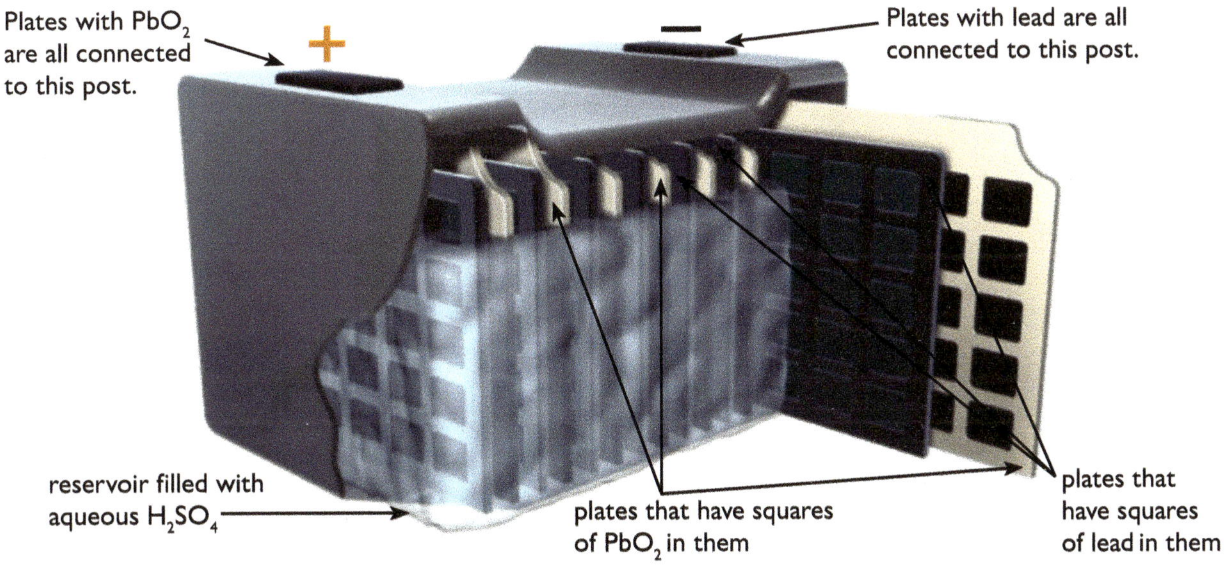

Notice from the figure that there are 2 different kinds of plates in a lead-acid battery: ones that contain squares of lead and others that contain squares of lead(IV) oxide. Those are 2 of the reactants in the lead-acid battery equation. The third reactant, sulfuric acid, is in the reservoir, so the plates are constantly bathed in it. The plates containing solid lead are connected to the negative post of the battery because solid lead is being oxidized in the reaction and is thus losing electrons. This means that electrons flow away from the solid lead, and therefore the solid lead makes up the anode of the battery. This means (of course) that the plates containing lead(IV) oxide make up the cathode, and they are all connected to the positive post of the battery. This should make sense since the lead(IV) oxide is being reduced. Therefore, electrons flow to the lead(IV) oxide, making it the positively charged cathode.

Notice one more thing from the figure: There are several plates containing lead and several plates containing lead(IV) oxide. Why are there several plates? The reaction represented by the lead-acid battery equation produces only about 2 volts. Most automobile batteries run at higher voltages. Therefore, to get a higher voltage, the individual plates that make up the reaction are hooked together. If done in the right way, this allows the voltages to add. Seven plates of lead hooked up to the negative side of the battery (the anode) and 7 plates of lead(IV) oxide hooked up to the positive side of the battery (the cathode) will give the battery about 14 volts instead of just 2 volts.

As you probably know, a lead-acid battery is rechargeable. If you hook up a battery charger to a lead-acid battery, it will force the reaction to run in reverse, converting products to reactants so that the battery can be used again. In automobiles, once the battery is used to start the engine, it can be recharged by the alternator or generator in the engine.

Before we leave the discussion of the lead-acid battery, you might be wondering where the salt bridge is since a battery must have a salt bridge to balance out the charges that develop during the reaction. Remember that the plates are bathed in aqueous sulfuric acid. What happens when sulfuric acid is added to water? The following reaction occurs:

$$H_2SO_4\,(aq) + H_2O\,(l) \longrightarrow H_3O^+\,(aq) + HSO_4^-$$

Notice that this reaction produces 2 ions: H_3O^+ and HSO_4^-. Those ions do the job of the salt bridge in a lead-acid battery. They move in response to the charges that develop during the chemical reaction, making sure that the anode and cathode do not build up excess charge that would keep the electrons from flowing.

Let's continue our discussion of real batteries by briefly discussing a typical AA battery. One major difference between an AA battery and a lead-acid battery is that an AA battery does not have any aqueous solutions in it. Since aqueous solutions take up far too much room, we could never make a reasonably sized battery that way. Because there are no aqueous solutions in this kind of battery, it is called a dry cell.

There are many, many different kinds of dry cells, so chemists must further distinguish between them. The dry cell that we use as an AA battery is called an alkaline cell because the salt bridge it uses is made from KOH, a base. The alkaline cell uses solid zinc as the anode and MnO_2 as the cathode. As the battery begins to produce electricity, the MnO_2 turns into Mn_2O_3, and the zinc turns into Zn^{2+}. When this happens, it breaks down the inner construction of the battery. This is why a typical alkaline battery, unlike a lead-acid battery, cannot be recharged. Even if you hooked a standard alkaline battery up to a battery recharger, it would not recharge because the inner construction of the battery would be unable to reform.

CORROSION

Lest you think that all redox reactions are useful reactions, you should realize that corrosion, one of the most destructive chemical processes in the world, is also a redox reaction. For example, when iron is exposed to water and oxygen, the following reaction occurs:

$$2Fe\,(s) + 2H_2O\,(l) + O_2\,(g) \longrightarrow 2Fe(OH)_2\,(s)$$

FIGURE 16.5
Corrosion
Photo: Vsolymossy (CC BY-SA 3.0)

In this reaction, the iron goes from an oxidation number of 0 to +2, while the oxygen goes from an oxidation number of 0 to -2. Iron is oxidized and oxygen is reduced, making this a redox reaction. The $Fe(OH)_2$ is further oxidized by the oxygen in the air to form Fe_2O_3, which is what we typically call rust. This process turns strong, useful iron into brittle, useless rust. It has been estimated that replacing rusted iron and trying to prevent or slow down the corrosion process costs the United States economy as much as 276 *billion* dollars every year! Therefore, although redox reactions are very useful in making batteries, they also can be harmful.

SOME FINAL WORDS

Believe it or not, once you do the problems and take the test at the end of this module, you will be finished with your first year of chemistry. If you sit back and think about all that you've learned, you should be rather impressed with yourself. After all, you can do quite a bit more chemistry now than you could when you started this course.

If you were able to maintain an A or a B while working through these modules, you obviously have a talent for chemistry and should consider a career in it. If you earned a C, you should not be discouraged from a career in chemistry or some other natural science because this course is designed so that a student getting a C or better will be very well prepared for college chemistry.

Regardless of your grade, you should have gained a lot from this course. You should be able to appreciate how *difficult* natural sciences like chemistry are. You should also have more appreciation for the amazingly intricate earth. Besides that, you should have a much deeper respect for the wisdom and power of our God who created it. Whatever you intend to do with the remainder of your academic career, continue exploring creation, and continue learning more about the marvelous Creator who gave it to us!

salt and light

William Phillips. Photo: Markus Pössel (CC BY-SA 3.0)

William Phillips, while working at the U.S. National Institute of Standards and Technology, was corecipient of the 1997 Nobel Prize in Physics for development of methods to cool and trap atoms with laser light. He commented on his Nobel Prize: "There are many people I want to thank, and I'd also like to thank God for giving us such a wonderful and interesting universe to explore" (*Washington Times*, 2001). The church-going physicist told a conference at Harvard University: "Being an ordinary scientist and an ordinary Christian seems perfectly natural to me. . . . For others, it may appear strange, even astonishing, that someone could be serious about science and about faith." "I'm strongly of the conviction that God is personal, and this is the foundation of my faith" (Phillips 2001).

ANSWERS TO THE "ON YOUR OWN" QUESTIONS

16.1 (a) Since the overall charge is 0 and there is only 1 atom, the oxidation number is 0.

(b) Since the overall charge is -3 and there is only one atom, the oxidation number is -3.

(c) The overall charge is 0, so all oxidation numbers must add up to that. Since O is -2 and there are 5 O atoms, the total so far is -10. For the sum of oxidation numbers to be 0, P must be +5.

16.2 (a) Rule 6 tells us that O has a -2 oxidation number. Since the total charge is 0, C must have a +4 oxidation number. C: +4, O: -2.

(b) Rule 1 says that when a molecule is made up of only 1 type of atom, the oxidation number is the charge (0) divided by the number of atoms (8). S: 0.

(c) Rule 4 tells us that F is always -1. For the oxidation numbers to add up to the total charge (-1), that means B must be +3. B: +3, F: -1.

(d) Rule 5 tells us that since N is not a metal, H's oxidation number is +1. Therefore, to make all oxidation numbers add up to the total charge (0), N must be -1. The oxidation numbers, then, are N: -1, H: +1.

(e) Rule 6 tells us that O is -2. We have no ironclad rules for N or Cl, so now we must resort to general principle 7. This works best on groups 3A, 6A, and 7A. Since Cl is in 7A, we will use it on Cl. Therefore, Cl has an oxidation number of -1. To get all oxidation numbers to add up to the overall charge (0), N must be +3. Cl: -1, N: +3, O: -2.

(f) Rule 2 says Li has an oxidation number of +1. Rule 5 tells us that in this molecule, H has an oxidation number of +1. We have no rule for N, but we can figure it out because all oxidation numbers need to add up to the total charge (0). The oxidation numbers from Li and H sum up to +3. Therefore, N must be -3. The oxidation numbers, then, are: Li: +1, N: -3, H: +1.

(g) Rule 6 tells us that O is -2. Since the oxidation numbers must add up to the total charge (-3), P must be +5. P: +5, O: -2.

(h) Rule 3 says that Ca is +2. Rule 6 tells us that O is -2. There is 1 Ca atom and 4 O atoms, so the total oxidation number so far is (+2) + 4 x (-2) = -6. There are 2 N atoms, and they must combine to make +6 for the oxidation numbers to add up to the total charge (0). Therefore, each N must have a +3. The answers, then, are: Ca: +2, N: +3, O: -2.

16.3 To go from +7 to -1, 8 negatives must be added. This means that electrons were gained, which means the atom was reduced. Since 8 negatives were added, 8 electrons participated.

16.4 To go from +2 to +6 , 4 negatives must be subtracted. This means that electrons were lost, which means the atom was oxidized. Since 4 negatives were subtracted, 4 **electrons** participated.

16.5 (a) On the reactants side of the equation, the oxidation numbers are: Cu: +2, Zn: 0. On the products side, Cu: 0, Zn: +2. This means that both Cu and Zn changed oxidation numbers, telling us that this is a redox reaction. Since Cu went from +2 to 0, it gained electrons; therefore, Cu was reduced. Since Zn went from 0 to +2, it lost electrons; therefore, Zn was oxidized.

(b) On the reactants side of the equation, the oxidation numbers are: H: +1, O: -2. On the products side, H: 0, O: 0. This means that both H and O changed oxidation numbers, telling us that this is a redox reaction. Since H went from +1 to 0, it gained electrons; therefore, H was reduced. Since O went from -2 to 0, it lost electrons; therefore, O was oxidized.

(c) On the reactants side of the equation, the oxidation numbers are: Na: +1, Cl: -1, Ag: +1, F: -1. On the products side, Na: +1, Cl: -1, Ag: +1, F: -1. This means that no atom changed oxidation number, telling us that this is not a redox reaction.

(d) On the reactants side of the equation, the oxidation numbers are: Ca: 0, H: +1, N: +5, O: -2. On the products side, Ca: +2, H: 0, N: +5, O: -2. This means that both H and Ca changed oxidation numbers, telling us that this is a redox reaction. Since H went from +1 to 0, it gained electrons; therefore, H was reduced. Since Ca went from 0 to +2, it lost electrons; therefore, Ca was oxidized.

16.6 In this reaction, H^+ is going from an oxidation number of +1 to an oxidation number of 0. This indicates that it is gaining electrons. The solution holding aqueous H^+ will have electrons flowing into it. Since electrons are attracted to a positive charge, this must be the positive side and therefore the cathode. The Mn is going from an oxidation number of 0 to an oxidation number of +2. This means it loses electrons. Since it is losing electrons, the electrons are flowing away from the container holding the solid Mn. This makes that container the negative side of the battery, and it is therefore the anode. Since solid Mn is a reactant, the electrode on the anode must be Mn. Since the other reactant is aqueous H^+, there is no need to identify the electrode on the cathode. As always, there must be a salt bridge. The picture, then, looks like this:

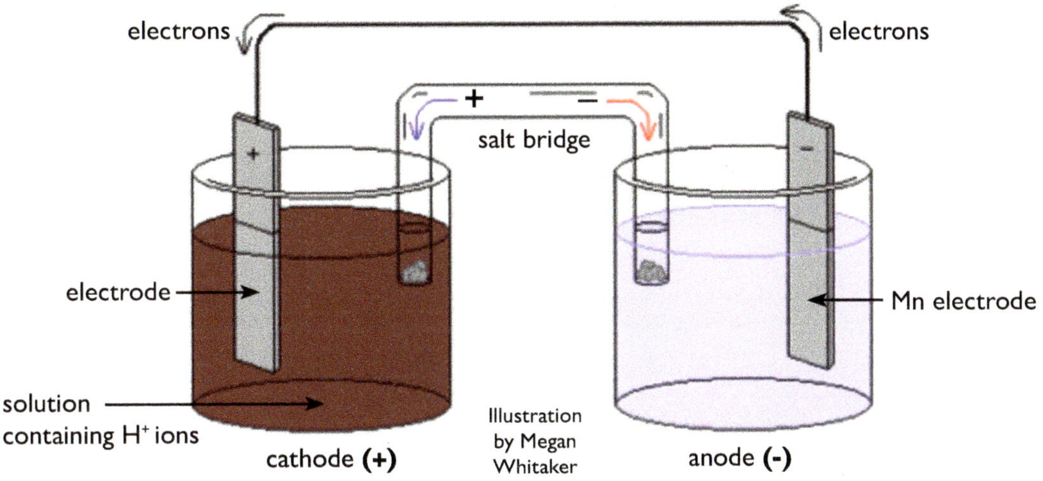

16.7 The first thing that we have to do is look at the reactants and see which is being oxidized and which is being reduced. Since chlorine goes from an oxidation number of 0 to an oxidation number of -1, it must be gaining electrons, so it is reduced. This means that the container that holds the chlorine will have the electrons flowing into it, so it must be positive. The Al, on the other hand, loses electrons because it goes from an oxidation number of 0 to an oxidation number of +3. This means that the Al is oxidized, and the container that holds it will have electrons leaving it; therefore, it must be negative. Since the electrons are flowing from the Al to the Cl_2, the container that holds Al is negative and is therefore the anode, while the container that holds the Cl_2 is positive and is the cathode. In the equation, Cl_2 is aqueous. This means that the chlorine is dissolved in water. The Al, on the other hand, is solid, so it must be used as the electrode in its container. As always, there must be a salt bridge as well. In the end, then, the picture should look something like this:

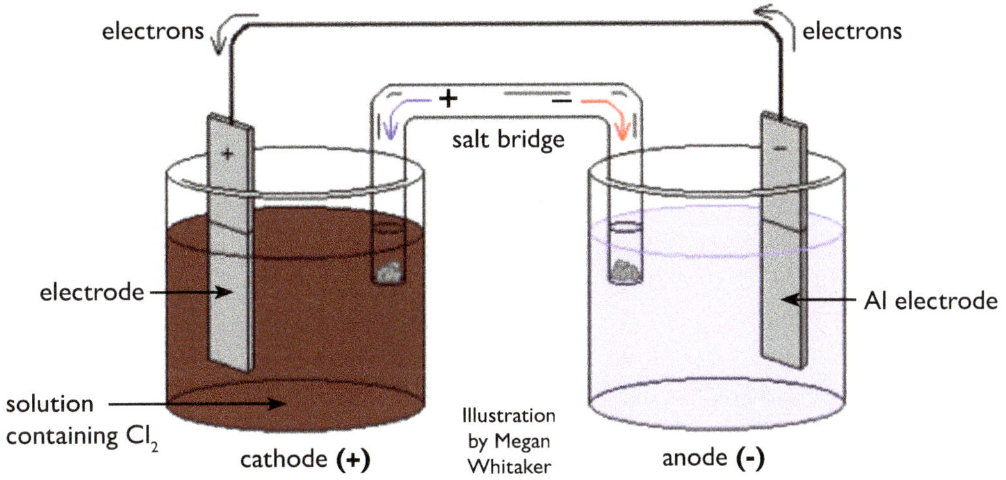

STUDY GUIDE FOR MODULE 16

REVIEW QUESTIONS

1. Define oxidation number.

2. An atom has an oxidation number of -3 in one molecule, but it has an oxidation number of +3 in a second molecule. What can we say about the electronegativities of the other atoms that make up the 2 molecules?

3. What must the oxidation numbers of all atoms in a molecule add up to?

4. Define oxidation and reduction.

5. Would there ever be a case in which oxidation could occur without reduction?

6. Is the anode of a battery positive or negative? What about the cathode?

7. When figuring out oxidation numbers, what rule do you use only as a last resort? Why is it only a last resort?

8. What are the 3 reactants needed for a lead-acid battery?

9. Why are some batteries rechargeable and others not?

10. Name 2 differences between an alkaline dry cell and a lead-acid battery.

PRACTICE PROBLEMS

1. Give the oxidation numbers of all atoms in the following substances:

   a. Cl_2 b. O_3 c. S^{2-} d. Fe e. Fe^{2+}

2. Give the oxidation numbers of all atoms in the following compounds:

   a. $MgCl_2$ b. VS c. SF_6 d. CO_3^{2-}
   e. MnO_2 f. KNO_3 g. H_2SO_4 h. $IrCl_6^{3-}$

3. An atom changes its oxidation number from +5 to +1. Is it oxidized or reduced? How many electrons did it take to do this?

4. An atom changes its oxidation number from 0 to +2. Is it oxidized or reduced? How many electrons did it take to do this?

5. An atom changes its oxidation number from -3 to 0. Is it oxidized or reduced? How many electrons did it take to do this?

6. An atom changes its oxidation number from -2 to -4. Is it oxidized or reduced? How many electrons did it take to do this?

7. Determine whether or not each of the following is a redox reaction. If it is, determine which atom is being reduced and which is being oxidized.

   a. $2Na\,(s) + Cl_2\,(g) \rightarrow 2NaCl\,(s)$
   b. $NiO\,(s) + Cd\,(s) \rightarrow CdO\,(s) + Ni\,(s)$
   c. $Mg(NO_3)_2\,(aq) + 2NaOH\,(aq) \rightarrow Mg(OH)_2\,(s) + 2NaNO_3\,(aq)$
   d. $H_2SO_4\,(aq) + 2NH_3\,(aq) \rightarrow 2NH_4^+\,(aq) + SO_4^{2-}\,(aq)$
   e. $2VO_3^-\,(aq) + Zn\,(s) + 8H^+\,(aq) \rightarrow 2VO^{2+}\,(aq) + Zn^{2+}\,(aq) + 4H_2O\,(l)$

8. A Galvanic cell runs on the following reaction:

$$Co\,(s) + Cu^{2+}\,(aq) \rightarrow Co^{2+}\,(aq) + Cu\,(s)$$

   Draw a diagram for this Galvanic cell, labeling the electron flow, the anode and cathode, and the positive and negative sides of the Galvanic cell.

9. A Galvanic cell runs on the following reaction:

$$Al\,(s) + Fe^{3+}\,(aq) \rightarrow Al^{3+}\,(aq) + Fe\,(s)$$

   Draw a diagram for this Galvanic cell, labeling the electron flow, the anode and cathode, and the positive and negative sides of the Galvanic cell.

GLOSSARY

Absolute temperature scale—The Kelvin temperature scale. It is not possible to reach or go below 0 Kelvin.

Accuracy—An indication of how close a measurement is to the true value.

Acid—A molecule that donates protons or H^+ ions.

Acid ionization constant—The equilibrium constant for an acid's ionization reaction.

Acid ionization reaction—The reaction in which an H^+ ion separates from an acid molecule so that it can be donated in another reaction.

Activated complex—Another name for the intermediate state of a chemical reaction.

Activation energy—The energy necessary to start a chemical reaction.

Alkaline dry cell—A dry cell made with a potassium hydroxide salt bridge.

Amphiprotic compounds—Compounds that can act as either an acid or a base, depending on the situation.

Amphoteric compounds—Another name for amphiprotic compounds.

Amplitude—A measure of the height of the crests or the depths of the troughs on a wave.

Anion—A negatively charged ion.

Anode—The electrode at which oxidation occurs.

Atom—The smallest chemical unit of matter.

Atomic mass—The number on the periodic table representing the average mass of the atom's isotopes in atomic mass units.

Atomic mass units—The mass unit used on the periodic table: 1.00 amu = 1.66×10^{-24} g.

Atomic number—The number on the periodic table that represents the number of protons in the nucleus of the atom.

Atomic radius—The average radius of an atom.

Avogadro's number—The number of molecules or atoms in a mole: 6.02×10^{23}.

Base—A molecule that accepts protons or H^+ ions.

Bent—A molecular shape in which a central atom is bonded to 2 other atoms and the bond angle is 105 degrees.

Boiling—The process by which a substance changes from its liquid phase to its gas phase.

Boiling point—The temperature at which the vapor pressure of a liquid is equal to normal atmospheric pressure.

Boiling-point elevation—When a solute is dissolved in a solvent, the boiling point of the solution will be higher than that of the pure solvent.

Bond energy—The energy required to break a bond.

Boyle's law—At constant temperature, the pressure and volume of a gas are inversely related.

Calibration—The process of using certain physical measurements to define the scale of a measuring device.

Calorie (cal)—The amount of heat necessary to warm 1 gram of water 1 degree Celsius.

Calorimeter—An experimental device that measures the heat released or absorbed during a chemical or physical change.

Calorimetry—An experimental process that measures the heat released or absorbed during a chemical or physical change.

Catalyst—A substance that alters the rate of a chemical reaction without changing concentration throughout the course of the reaction.

Catalytic converter—A device on an automobile that converts gaseous carbon monoxide produced by the engine into gaseous carbon dioxide

Cathode—The electrode at which reduction occurs.

Cathode ray tube—Another name for a Crookes tube.

Cation—A positively charged ion.

Celsius—A temperature scale defined so that water freezes at 0° and boils at 100°.

Change in enthalpy (ΔH)—The energy change that accompanies a chemical reaction.

Charles's law—At constant pressure, the temperature and volume of a gas are linearly proportional.

Chemical change—A change that affects the type of molecules or atoms in a substance.

Chemical equation—A representation of a chemical reaction.

Chemical equilibrium—The point at which both the forward and reverse reactions in a chemical equation have equal reaction rates. When this occurs, the amounts of each substance in the chemical reaction will not change, despite the fact that both reactions still proceed.

Chemical formula—A notation that indicates how many of each atom make up a particular molecule.

Chemical reaction—A process by which one or more substances change into one or more different substances.

Chemical symbol—An abbreviation for an element.

Combustion—A reaction where a substance reacts with oxygen gas.

Complete combustion reaction—A reaction in which oxygen is added to a compound containing carbon and hydrogen, producing CO_2 and H_2O.

Compound—A substance that can be decomposed into elements by chemical means.

Condensing—The process by which a substance changes from its gas phase to its liquid phase.

Continuous theory of matter—The idea that substances are composed of long, unbroken blobs of matter.

Covalent bond—A shared pair of valence electrons that holds atoms together in covalent compounds.

Covalent compound—A compound formed by atoms that share electrons.

Crookes tube—An experimental apparatus developed by William Crookes. It consists of a glass tube filled with a small amount of gas. Electrodes in the tube allow for the passage of electricity through the gas.

Dalton's law of partial pressures—When 2 or more ideal gases are mixed together, the total pressure of the mixture is equal to the sum of the pressures of each individual gas.

Decomposition—The process by which a substance is broken down into its constituent elements.

Decomposition reaction—A reaction that changes a compound into its constituent elements.

Density—An object's mass divided by the volume that the object occupies.

Derived unit—A unit formed by the multiplication and/or division of other units.

Dilution—Adding water to a solution in order to decrease the concentration.

Dimensionless quantity—A quantity with no units.

Diprotic acid — An acid that can donate 2 H^+ ions.

Discontinuous theory of matter—The idea that substances are composed of tiny, individual particles like grains of sand.

Double bond—A total of 4 electrons shared between atoms.

Dry cell—A battery made with no aqueous solutions.

Electromagnetic radiation—Another term for light, including all wavelengths, both visible and not visible.

Electromagnetic spectrum—The total range of wavelengths of light that come from the sun.

Electron—One of the 3 particles that make up the atom. It is negatively charged and occupies orbitals outside the nucleus of the atom.

Electron configuration—A notation that lists the number of electrons that occupy each orbital in an atom.

Electron-dot diagram—Another name for a Lewis structure.

Electronegativity—A measure of how strongly an atom attracts extra electrons to itself.

Element—Any substance that cannot be broken down (decomposed) into simpler components.

Empirical formula—A chemical formula that provides a simple, whole-number ratio for the atoms in a molecule.

Endothermic process—A process that absorbs heat.

Endpoint of the titration—The point in a titration where the acid and base have completely reacted with one another. It is usually accompanied by an indicator's change in color.

Energy—The ability to do work.

Energy diagram—A diagram that shows the energy of the reactants, intermediate state, and products in a chemical reaction.

Enthalpy of formation (ΔH_f)—The ΔH of a formation reaction.

Entropy—A measure of the disorder that exists in any system.

Equilibrium constant—A value that describes the concentrations at which a reaction comes to equilibrium. The equilibrium constant varies with temperature.

Excess reactant—The reactant or reactants that are left over at the end of a chemical reaction.

Exothermic process—A process that releases heat.

Extrapolation—Following an established trend in the data even though no data are available for that region.

Factor-label method—A method of conversion using the multiplication of fractions.

Fahrenheit—A temperature scale defined so that water freezes at 32° and boils at 212°.

First law of thermodynamics—Energy cannot be created or destroyed. It can only change form.

Food calorie (Cal)—1,000 chemistry calories (cal).

Formation reaction—A reaction that starts with 2 or more elements and produces 1 compound.

Freezing—The process by which a substance changes from its liquid phase to its solid phase.

Freezing-point depression—When a solute is dissolved in a solvent, the freezing point of the solution will be lower than that of the pure solvent.

Frequency—The number of wave crests (or troughs) that pass a given point each second.

Galvanic cell—An electricity-producing cell that runs on a redox reaction.

Gay-Lussac's law—The stoichiometric coefficients in a chemical equation relate the volumes of gases in the equation as well as the number of moles of substances in the equation.

Graduated cylinder—An instrument used for measuring the volume of liquids.

Gram—The metric unit of mass.

Ground state—The lowest possible energy state for a given substance.

Heat—Energy that is transferred as a consequence of temperature differences.

Hess's law—Enthalpy is a state function.

Heterogeneous catalysts—Catalysts that are in a different phase than the reactants.

Heterogeneous mixture—A mixture with a composition that is different depending on what part of the sample you are observing.

Homogeneous catalysts—Catalysts that have the same phase as at least one of the reactants.

Homogeneous mixture—A mixture with a composition that is always the same no matter what part of the sample you are observing.

Homonuclear diatomic—A molecule composed of 2 identical atoms.

Hypothesis—An educated guess that attempts to explain observations.

Ideal gas—A gas whose molecules (or atoms) are very small compared to the total volume of the gas, make elastic collisions with one another as well as the walls of the container, and have no attraction between one another.

Incomplete combustion reaction—A reaction in which oxygen is added to a compound containing carbon and hydrogen, producing either CO or C, along with H_2O.

Indicator—A substance that turns one color in the presence of acids and another color in the presence of bases.

Insoluble—Unable to dissolve.

Intermediate state—A complex formed in the process of a chemical reaction when the reactants and products are interacting.

Ion—An atom that has gained or lost electrons and thus has become electrically charged.

Ionic compound—A compound formed by ions.

Ionization—The process by which an atom turns into an ion by gaining or losing electrons.

Ionization energy—The amount of energy needed to take an electron away from an atom.

Isotopes—Atoms with the same number of protons but different numbers of neutrons.

Isotopic enrichment—The process by which the abundance of one isotope in an element is increased. This is typically used to make the fuel for nuclear bombs.

Joule—The metric unit for energy.

Kelvin—The absolute temperature scale. It is not possible to reach or go below 0 Kelvin.

Kinetic energy—Energy that is in motion.

Kinetic theory of matter—The theory that the atoms or molecules which make up a substance are in constant motion, and the higher the temperature, the greater their speed.

Law of definite proportions—The proportion of elements in any compound is always the same.

Law of mass conservation—Matter cannot be created or destroyed; it can only change forms.

Law of multiple proportions—If 2 elements combine to form different compounds, the ratio of masses of the second element that react with a fixed mass of the first element will be a simple whole-number ratio.

Le Châtelier's principle—When a stress (such as a change in concentration, pressure, or temperature) is applied to an equilibrium, the reaction will shift in a way that relieves the stress and restores equilibrium.

Lead-acid battery—A battery that runs on the oxidation of lead and the reduction of lead (IV) oxide in the presence of sulfuric acid.

Lewis structure—A schematic representation of the valence electrons in an atom or molecule.

Limiting reactant—The reactant that runs out first in a chemical reaction. It determines the amount of products made.

Liter—A metric unit of volume.

Mass—A measure of the amount of matter in an object.

Mass number—The total number of neutrons and protons in an atom.

Matter—Anything that has mass and takes up space.

Melting—The process by which a substance changes from its solid phase to its liquid phase.

Melting point—The temperature at which a substance changes from its solid phase to its liquid phase.

Meniscus—The curved surface of a liquid, typically in a glass container.

Metal—An element that tends to give up its electrons. Metals are found on the left side of the jagged line on the periodic table of elements (with the exception of hydrogen, which is a nonmetal).

Meter—The metric unit of length.

Mixture—A substance that contains different compounds and/or elements.

Model—A constructed image of something that we cannot see with our eyes.

Molality (m)—The number of moles of solute per kilogram of solvent.

Molar mass—The mass of 1 mole of a given compound.

Molarity (M)—A concentration unit that tells how many moles of a substance are in a liter of solution. It is determined by dividing the number of moles of a substance by the number of liters of solution.

Mole—A group of atoms or molecules that number 6.02×10^{23}.

Mole fraction—The number of moles of a given component in a mixture divided by the total number of moles in the mixture.

Molecular formula—A chemical formula that provides the number of each type of atom in a molecule.

Molecular mass—The mass of a single molecule.

Molecule—More than one atom bonded together to form a compound.

Neutralization reaction—A reaction between an acid and base that neutralizes both, typically forming salt and water.

Neutron—One of the 3 particles that make up the atom. It is electrically neutral and is in the nucleus of the atom.

Newton—The metric unit of force.

Nonmetal—An element that tends to take electrons from other elements. Nonmetals are found on the right side of the jagged line on the periodic table of elements. Hydrogen is also a nonmetal, even though it is on the left side of the jagged line.

Nonpolar bond—A covalent bond in which the electrons are shared equally between the atoms involved.

Nonpolar molecule—A covalent molecule in which the electrons are evenly distributed throughout the whole molecule.

Nucleus—The center of the atom that contains the neutrons and protons.

Octet rule—Most atoms strive to attain 8 valence electrons.

Orbital—A specific shape that confines the position of an electron relative to the nucleus.

Oxidation—The process by which an atom loses electrons.

Oxidation number—The charge that an atom in a molecule would develop if the most electronegative atoms in the molecule took the shared electrons from the less electronegative atoms.

Partial pressure—The pressure exerted by a single gas in a mixture of gases.

Particle/wave duality theory—The theory that light sometimes behaves as a particle and sometimes behaves as a wave.

Periodic property—A characteristic of atoms that varies regularly across the periodic chart.

pH scale—A scale that measures the acidity of a solution. A pH below 7 indicates an acidic solution. A pH above 7 indicates a basic solution. A pH of 7 indicates a neutral solution.

Phase—One of 3 states of matter: solid, liquid, or gas.

Phase change—The process by which a substance changes from one phase (solid, liquid, or gas) to another phase (solid, liquid, or gas).

Photon—A "particle" of light.

Physical change—A change in which the atoms or molecules in a substance stay the same.

Physical constant—A measurable quantity in nature that does not change.

Planck's constant—The physical constant that relates the energy of light to its frequency: 6.63×10^{-34} J/Hz.

Planetary model—Another name for the Rutherford model of the atom.

Plum pudding model—A model of the atom developed by J. J. Thomson in which the negatively charged particles (the plums) exist in an environment of positive charge (the pudding).

Polar bond—A covalent bond in which the electrons are shared unequally between the atoms involved.

Polar molecule—A covalent molecule that has an imbalance in the distribution of electrons throughout the whole molecule.

Polyatomic ions—Ions that are formed when a group of atoms gains or loses electrons.

Polyprotic acid—An acid that can donate more than one H^+ ion.

Potential energy—Energy that is stored.

Precipitation—The process by which a solid product falls out of a solution.

Precision—An indication of the scale on the measuring device that was used.

Pressure—The force per unit area exerted on an object.

Products—The substances on the right side of a chemical equation.

Proton—One of the 3 particles that make up the atom. It is positively charged and is in the nucleus of the atom.

Proton acceptor—Another name for a base. Bases accept H^+ ions, which are protons.

Proton donor—Another name for an acid. Acids donate H^+ ions, which are protons.

Pure substance—A substance that contains only one element or compound.

Pyramidal—A molecular shape in which a central atom is surrounded by 3 bonds with a bond angle of 107 degrees.

Quantum assumption—The assumption that a physical quantity (such as energy) is restricted to discrete values.

Quantum mechanical model—The modern-day model of the atom in which electrons are found outside the nucleus in orbitals.

Reactants—The substances on the left side of a chemical equation.

Reaction mechanism—A series of chemical equations that indicates the step-by-step process by which a chemical reaction occurs.

Reaction rate—The rate of change of a product in a chemical reaction.

Reduction—The process by which an atom gains electrons.

Rutherford model—Ernest Rutherford's model that said the atom is made of a dense, positively charged nucleus with electrons orbiting the nucleus.

Saturated solution—A solution in which the maximum amount of solute has been dissolved.

Scientific law—A description of the natural world that has been confirmed by an enormous amount of data.

Second law of thermodynamics—The entropy of the universe must always either increase or remain the same. It can never decrease.

Significant figure—A digit in a measurement that is either nonzero, a zero that is between 2 significant figures, or a zero at the end of the number and to the right of the decimal.

Slug—The English unit of mass.

Solubility—The maximum amount of solute that can dissolve in a given amount of solvent.

Solute—The substance being dissolved to make a solution.

Solution—The result of one or more solutes being dissolved in a solvent.

Solvent—The substance in which the solute is being dissolved to make a solution.

Specific heat—The amount of heat necessary to raise the temperature of 1 gram of a substance by 1 degree Celsius.

Spectrometer—A device that analyzes light emitted or absorbed by a substance.

Spectroscopy—The process by which individual wavelengths of light emitted by a substance are analyzed. This process can be used to identify the elements in a substance.

Standard enthalpy of formation—The ΔH of a formation reaction at $T = 298$ K and $P = 1.00$ atm.

Standard temperature and pressure (STP)—A temperature of 273 K and a pressure of 1.00 atm.

State function—Any quantity that depends solely on the final destination, not on the way you get to that destination.

Stoichiometric coefficients—The numbers that appear to the left of the chemical formulas in a chemical equation. They represent the number of moles of each substance.

Stoichiometry—The process by which the amount of one substance in a chemical reaction is related to the amount of another substance in a chemical reaction.

Sublimation—The process by which a solid turns directly into a gas without going through the liquid phase.

Tetrahedron—A molecular shape in which a central atom is surrounded by 4 bonds with bond angles of 109 degrees.

Theory—A hypothesis that has been confirmed by experimental data.

Titration—The process of slowly reacting a base of unknown concentration with an acid of unknown concentration (or vice versa) until just enough acid has been added to react with all of the base. This process determines the concentration of the unknown base (or acid).

Transition metal—An element in the d-orbital block of the periodic table of elements.

Trigonal—A molecular shape in which a central atom is surrounded by 3 bonds with bond angles of 120 degrees.

Triple bond—A total of 6 electrons shared between atoms.

Triprotic acid—An acid that can donate 3 H^+ ions.

Valence electrons—The electrons that exist farthest from an atom's nucleus. They are the electrons with the highest energy level number.

Vapor pressure—The pressure exerted by the vapor that sits on top of any liquid.

Visible spectrum—The range of light wavelengths that are visible to the human eye.

VSEPR theory—Valence Shell Electron Pair Repulsion theory, which states that molecules will attain whatever shape keeps the valence electrons of the central atom as far apart from one another as possible.

Wavelength—The distance between the crests (or troughs) of a wave.

Work—The force applied to an object times the distance that the object travels parallel to that force.

APPENDIX A

TABLES, LAWS, AND EQUATIONS

TABLES

TABLE 1.1

Physical Quantities and Their Base Units

Physical Quantity	Base Metric Unit	Base English Unit
Mass	gram (g)	slug (sl)
Distance	meter (m)	foot (ft)
Volume	liter (L)	gallon (gal)
Time	second (s)	second (s)

TABLE 1.2

Common Prefixes Used in the Metric System

Prefix	Numerical Meaning
micro (μ)	0.000001
milli (m)	**0.001**
centi (c)	**0.01**
deci (d)	0.1
deca (D)	10
hecta (H)	100
kilo (k)	**1,000**
mega (M)	1,000,000

TABLE 1.3
Relationships between English and Metric Units

Measurement	English/Metric Relationship
Distance	1.00 inch = 2.54 cm
Mass	1.00 slug = 14.59 kg
Volume	1.00 gallon = 3.78 L

TABLE 1.4
Examples of Significant Figure Rules

Number	Example	Number of Significant Figures
Nonzero digits	9,341	4
Zeros in front of first 1–9 digit	0.000564	3
Zero between 2 significant figures	120.043	6
Zero at end of number *and* to right of decimal point	510.0 510	4 2

TABLE 2.1
The *-ide* Atom Names of Nonmetals

Atom	*–ide* Name	Atom	*–ide* Name
H	hydride	P	phosphide
C	carbide	S	sulfide
N	nitride	Cl	chloride
O	oxide	Br	bromide
F	fluoride	I	iodide

TABLE 2.2
The *-ide* Atom Names of Nonmetals

Prefix	Meaning	Prefix	Meaning
mono	1	hexa	6
di	2	hepta	7
tri	3	octa	8
tetra	4	nona	9
penta	5	deca	10

TABLE 2.3

Element Names and Symbols

The list is compiled in the order in which they appear on the periodic table.
The elements in bold are the ones you need to memorize for the test.

Name	Symbol	Name	Symbol	Name	Symbol
Hydrogen	H	Niobium	Nb	Thallium	Tl
Helium	He	Molybdenum	Mo	Lead	Pb
Lithium	Li	Technetium	Tc	Bismuth	Bi
Beryllium	Be	Ruthenium	Ru	Polonium	Po
Boron	B	Rhodium	Rh	Astatine	At
Carbon	C	Palladium	Pd	Radon	Rn
Nitrogen	N	Silver	Ag	Francium	Fr
Oxygen	O	Cadmium	Cd	Radium	Ra
Fluorine	F	Indium	In	Actinium	Ac
Neon	Ne	Tin	Sn	Thorium	Th
Sodium	Na	Antimony	Sb	Protactinium	Pa
Magnesium	Mg	Tellurium	Te	Uranium	U
Aluminum	Al	**Iodine**	I	Neptunium	Np
Silicon	Si	Xenon	Xe	Plutonium	Pu
Phosphorus	P	Cesium	Cs	Americium	Am
Sulfur	S	Barium	Ba	Curium	Cm
Chlorine	Cl	Lanthanum	La	Berkelium	Bk
Argon	Ar	Cerium	Ce	Californium	Cf
Potassium	K	Praseodymium	Pr	Einsteinium	Es
Calcium	Ca	Neodymium	Nd	Fermium	Fm
Scandium	Sc	Promethium	Pm	Mendelevium	Md
Titanium	Ti	Samarium	Sm	Nobelium	No
Vanadium	V	Europium	Eu	Lawrencium	Lr
Chromium	Cr	Gadolinium	Gd	Rutherfordium	Rf
Manganese	Mn	Terbium	Tb	Dubnium	Db
Iron	Fe	Dysprosium	Dy	Seaborgium	Sg
Cobalt	Co	Holmium	Ho	Bohrium	Bh
Nickel	Ni	Erbium	Er	Hassium	Hs
Copper	Cu	Thulium	Tm	Meitnerium	Mt
Zinc	Zn	Ytterbium	Yb	Darmstadium	Ds
Gallium	Ga	Lutetium	Lu	Roentgenium	Rg
Germanium	Ge	Hafnium	Hf	Copernicium	Cn
Arsenic	As	Tantalum	Ta	Nihonium	Nh
Selenium	Se	Tungsten	W	Flerovium	Fl
Bromine	Br	Rhenium	Re	Moscovium	Mc
Krypton	Kr	Osmium	Os	Livermorium	Lv
Rubidium	Rb	Iridium	Ir	Tennessine	Ts
Strontium	Sr	Platinum	Pt	Oganesson	Og
Yttrium	Y	Gold	Au		
Zirconium	Zr	Mercury	Hg		

TABLE 4.1
Common Ions and Their Charges for Groups 1A–8A

Metals Lose Valence Electrons			Nonmetals Gain Valence Electrons				Noble Gases
1A	2A	3A	4A	5A	6A	7A	8A
Li^+							He
Na^+	Mg^{2+}	Al^{3+}	C^{4-}	N^{3-}	O^{2-}	F^-	Ne
K^+	Ca^{2+}		Si^{4-}	P^{3-}	S^{2-}	Cl^-	Ar
Rb^+	Sr^{2+}			As^{3-}	Se^{2-}	Br^-	Kr
Cs^+	Ba^{2+}					I^-	Xe

TABLE 4.2
Common Metals That Form More Than One Positive Ion

Element	Possible Ions	Name of Ions
Chromium	Cr^{2+}, Cr^{3+}	Chromium (II), Chromium (III)
Iron	Fe^{2+}, Fe^{3+}	Iron (II), Iron (III)
Copper	Cu^+, Cu^{2+}	Copper (I), Copper (II)
Gold	Au^+, Au^{2+}	Gold (I), Gold (II)
Tin	Sn^{2+}, Sn^{4+}	Tin (II), Tin (IV)
Lead	Pb^{2+}, Pb^{4+}	Lead (II), Lead (IV)

TABLE 5.1
Important Polyatomic Ions

Ion Name	Formula	Ion Name	Formula
ammonium (uh moh' nee uhm)	NH_4^+	cyanide (sigh' uh nide)	CN^-
hydroxide (hye drox' ide)	OH^-	carbonate (kar' bun ate)	CO_3^{2-}
chlorate (klor' ate)	ClO_3^-	chromate (krohm' ate)	CrO_4^{2-}
chlorite (klor' ite)	ClO_2^-	dichromate (dye krohm' ate)	$Cr_2O_7^{2-}$
nitrate (nye' trate)	NO_3^-	sulfate (suhl' fate)	SO_4^{2-}
nitrite (nye' trite)	NO_2^-	sulfite (suhl' fite)	SO_3^{2-}
acetate (as' uh tate)	$C_2H_3O_2^-$	phosphate (fahs' fate)	PO_4^{3-}

TABLE 5.2
Examples of Shapes of Molecules

Molecule	Lewis Structure	Bonded Pairs	Nonbonded Electron Pairs	Molecular Shape	
CO_2	:Ö=C=Ö:	2	0	linear	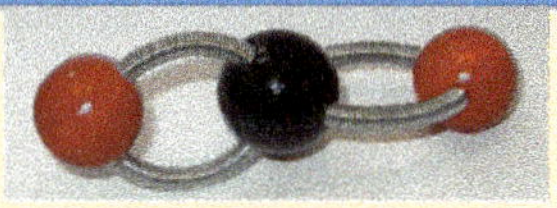
CH_2O	H–C=Ö: with H below C	3	0	trigonal	
CH_4	H above C; H–C–H; H below C	4	0	tetrahedral	
NH_3	H above N; H–N:; H below N	3	1	pyramidal	
H_2O	H above O; H–Ö:	2	2	bent	

TABLE 6.1
Phase Symbols

Phase	Symbol
Solid	(s)
Liquid	(l)
Gas	(g)
Aqueous (dissolved in water)	(aq)

TABLE 9.1
Common Acids

Formula	Name	Formula	Name
HCl	hydrochloric (hy' droh klor' ik) acid	H_2SO_4	sulfuric (sul fyur' ik) acid
HBr	hydrobromic (hy' droh brohm' ik) acid	H_3PO_4	phosphoric (fos for' ik) acid
HF	hydrofluoric (hy' droh flor' ik) acid	H_2CO_3	carbonic (kar bon' ik) acid
HNO_3	nitric (nye' trik) acid	$C_2H_4O_2$	acetic (uh see' tik) acid

TABLE 10.1
Summary of the Methods by Which Solutes Dissolve in Liquid Solvents

Solute Phase	Requirements for Dissolving
Solid	The solvent molecules must be attracted to the solute molecules so strongly that the solvent molecules can get between the solute molecules (or ions) and pull them far apart from each other.
Liquid	The solvent molecules need to be attracted to the solute molecules only a small amount because the solvent does not need to separate the solute molecules very much. The solvent merely needs to get between the solute molecules.
Gas	The solvent molecules must be attracted to the solute molecules enough to pull the solute molecules closer to one another.

TABLE 10.2
The Effect of Temperature on Solubility

Solute Phase	Rules for the Effect of Temperature
Solid	The solubility of solid solutes usually increases with increasing temperature.
Liquid	The solubility of liquid solutes is not affected by temperature.
Gas	The solubility of gases decreases with increasing temperature.

TABLE 11.1
Units of Pressure

Atmosphere (atm)	kiloPascals (kPa)	Millimeters of mercury (mm Hg)	Torr (torr)
1.000 atm	101.3 kPa	760.0 mm Hg	760.0 torr

TABLE 11.2
The Vapor Pressure of Water

Temperature (°C)	Vapor Pressure (torr)	Temperature (°C)	Vapor Pressure (torr)
20	17.5	29	30.0
21	18.7	30	31.8
22	19.8	40	55.3
23	21.1	50	92.5
24	22.4	60	149.4
25	23.8	70	233.7
26	25.2	80	355.1
27	26.7	90	525.8
28	28.3	100	760.0

TABLE 12.1
Specific Heats of Common Substances

Substance	Specific Heat (J / g°C)
Aluminum	0.9000
Copper	0.3851
Glass	0.8372
Gold	0.1260
Iron	0.4521
Silver	0.2330
Water	4.184
Zinc	0.387

TABLE 13.1
ΔH Relationships in Reactions

Reaction is	ΔH is	Heat is	Potential energy stored in the reaction's products is
Endothermic	Positive (+)	Absorbed	Greater than the reactants
Exothermic	Negative (-)	Released	Less than the reactants

TABLE 13.2
Bond Energies

Chemical Bond	Bond Energy (kJ/mole)	Chemical Bond	Bond Energy (kJ/mole)
C-H	411	H-H	432
C-Cl	327	H-O	459
C-C	346	H-Cl	428
C-N	305	O=O	494
C-O	358	N≡N	942
C=C	602	N-H	386
C=O	799	Cl-Cl	240
C≡C	835	Br-Br	190

TABLE 13.3
Standard Enthalpies of Formation in kJ/mole

Substance	ΔH_f°	Substance	ΔH_f°	Substance	ΔH_f°	Substance	ΔH_f°
H_2O (g)	-242	NO_2 (g)	33.2	$Ca(OH)_2$ (s)	-987	C_2H_6O (l)	-278
H_2O (l)	-286	NH_3 (g)	-45.9	$CaCO_3$ (s)	-1207	C_2H_4 (g)	52.5
SO_3 (g)	-396	CO_2 (g)	-394	C_2H_6 (g)	-84.7	C_4H_{10} (g)	-126
NO (g)	90.3	CaO (s)	-635	CH_3OH (l)	-239	H_2SO_4 (l)	-814

TABLE 13.4
Standard Absolute Entropies in J/(mole · K)

Substance	S°	Substance	S°	Substance	S°	Substance	S°
H_2O (g)	189	NO_2 (g)	240	$Ca(OH)_2$ (s)	76.1	C_2H_6O (l)	161
H_2O (l)	69.9	NH_3 (g)	193	$CaCO_3$ (s)	92.9	C_2H_4 (g)	219
SO_3 (g)	257	CO_2 (g)	214	C_2H_6 (g)	230	C_4H_{10} (g)	260
NO (g)	211	CaO (s)	38.2	CH_3OH (l)	127	H_2SO_4 (l)	157

TABLE 13.5
Standard Gibbs Free Energies of Formation in kJ/mole

Substance	ΔG_f°	Substance	ΔG_f°	Substance	ΔG_f°	Substance	ΔG_f°
CH_4 (g)	-50.8	NO_2 (g)	240	$CaSO_4$ (s)	-1,320	C_2H_6O (l)	-175
C_6H_6 (l)	125	H_2O (l)	-237	$CaCO_3$ (s)	-1,129	$CaCl_2$ (s)	-750
CO (g)	-137	H_2O (g)	-229	H_2CO_3 (aq)	-623	HCl (aq)	-131
CO_2 (g)	-394	CaO (s)	38.2	CH_3OH (l)	-166	$Ca(OH)_2$ (s)	-897

SCIENTIFIC LAWS

Boyle's law:

$$PV = \text{constant or}$$

$$P_1V_1 = P_2V_2$$

Charles's law:

$$\frac{V}{T} = \text{constant or}$$

$$\frac{V_1}{T_1} = \frac{V_2}{T_2}$$

Combined gas law:

$$\frac{PV}{T} = \text{constant or}$$

$$\frac{P_1V_1}{T_1} = \frac{P_2V_2}{T_2}$$

Dalton's law of partial pressures:

$$P_T = P_1 + P_2 + P_3 + \ldots \text{ or}$$

$$P_1 = X_1P_T$$

Ideal gas law:

$$PV = nRT\text{, where } R = 0.0821\ \frac{L \cdot atm}{mole \cdot K}$$

EQUATIONS

Density (ρ)	$\rho = \frac{m}{V}$
Temperature Conversions	$°C = \frac{5}{9}(°F - 32)$
Heat (q)	$K = °C + 273.15$
Frequency (f)	$f = \frac{c}{\lambda}$
Energy (E)	$E = h \times f$
Molarity (M)	$M = \frac{\text{\# moles}}{\text{\# liters}}$
Dilution Equation	$M_1V_1 = M_2V_2$
Molality (m)	$m = \frac{\text{\# moles solute}}{\text{\# kg solvent}}$
Freezing-Point Depression	$\Delta T = -i \times K_f \times m$
Boiling-Point Elevation	$\Delta T = i \times K_b \times m$
Pressure (P)	$P = \frac{F}{A}$
Mole Fraction (X)	$x = \frac{\text{\# of moles of compound}}{\text{total \# of moles in the mixture}}$
Change in Enthalpy (ΔH)	$\Delta H° = \Sigma \Delta H_f°\text{ (products)} - \Sigma \Delta H_f°\text{ (reactants)}$
Change in Entropy (ΔS)	$\Delta S° = \Sigma S°\text{ (products)} - \Sigma S°\text{ (reactants)}$
Gibbs Free Energy (ΔG)	$\Delta G = \Delta H - T \cdot \Delta S$ or $\Delta G° = \Sigma \Delta G_f°\text{ (products)} - \Sigma \Delta G_f°\text{ (reactants)}$

Reaction Rate	$R = \frac{\Delta\,[\text{product}]}{\Delta t} = -\frac{\Delta\,[\text{reactant}]}{\Delta t}$
Rate Equation	$R = k[A]^x[B]^y$
Equilibrium Constant (K)	$K = \frac{[C]^c_{eq}[D]^d_{eq}}{[A]^a_{eq}[B]^b_{eq}}$

APPENDIX B

EXTRA PRACTICE PROBLEMS

EXTRA PRACTICE PROBLEMS FOR MODULE 1

1. Two students measure the mass of a kilogram standard known to be exactly 1.000000 kg. The first student uses one scale and reports a mass of 1,002.0 g. The second student uses a different scale and reports a mass of 1,000 g. Which student is more precise? Which is more accurate?

2. How many cg are in 403.1 g?

3. If an object has a volume of 1,355.1 mL, what is its volume in liters?

4. How many inches is 26.2 miles? (1 foot = 12 inches; 1 mile = 5.280×10^3 feet)

5. How many kilometers are equal to 5,000,700 mm?

6. The volume of a box is given by the equation:

   $$V = \text{length} \times \text{width} \times \text{height}$$

   If a box measures 0.12 m by 0.34 m by 0.050 m, what is its volume in liters?

7. A string measures 32.0 inches. How many centimeters is the string?

8. How many hours are in 4,200 seconds?

9. Convert these numbers into scientific notation.
   a. 70,800 b. 0.007840

10. Convert these numbers into decimal notation.
    a. 9.510×10^{-14} b. 9.0050×10^{6}

11. How many significant figures are in these numbers?
    a. 0.007350 b. 7,350 c. 7.005×10^{4}

12. Convert 34.5°C into °F.

13. Convert 350.0°F into K.

EXTRA PRACTICE PROBLEMS FOR MODULE 2

1. How can you experimentally determine whether a compound is ionic or covalent?

2. How can you determine from the periodic chart whether a compound is ionic or covalent?

3. Which of the following molecules are covalent?
   a. SO_2 b. $MgCl_2$ c. NaI d. C_2H_4O

4. Which of the following molecules are ionic?
   a. NO_2 b. $RbNO_3$ c. SiO_2 d. Na_2SO_4

5. In order to burn a 5.13 kg log, oxygen gas has to be added. If a chemist collects 1.04 kg water, 308 g of ash, and 5.05 kg of carbon dioxide, then how many kilograms of oxygen gas are needed?

6. In making calcium chloride, 100.0 g of calcium plus 100.0 g of chlorine makes 156.5 g of product along with some leftover calcium. How much chlorine and calcium should be added together to make 1.500 kg of calcium chloride without any leftovers?

7. A chemist adds 50.0 g of nitrogen to 50.0 grams of hydrogen. This ends up making 60.7 g of ammonia and some leftover hydrogen. What masses of nitrogen and hydrogen should be reacted to make 150.0 g of ammonia with no leftovers?

8. If you react 12.0 grams of carbon with 4.00 grams of hydrogen, you will make methane with no leftovers. If a chemist reacts 150.0 grams of carbon with 150.0 g of hydrogen, how many grams of which element will be left over after the methane is made?

9. In making aluminum sulfide, 54.0 g of aluminum and 100.0 grams of sulfur make 150.3 g of aluminum sulfide and leftover sulfur. How much aluminum and sulfur should be used to make 150.0 g of aluminum sulfide with no leftovers?

10. Give the chemical formula for each of the following compounds:
    a. diphosphorus hexaoxide b. dinitrogen tetrahydride c. nitrogen monoxide

11. Give the names of the following compounds:
    a. SO_2 b. $CaCl_2$ c. OCl_2 d. Na_2S

12. Classify each of the following as an element, compound, homogeneous mixture, or heterogeneous mixture:
    a. A bowl of milk and cereal
    b. A bottle of H_2SO_4
    c. A copper wire
    d. Kool-Aid dissolved in water

EXTRA PRACTICE PROBLEMS FOR MODULE 3

(1 nm = 10^{-9} m; c = 3.0×10^{8} m/sec; h = 6.63×10^{-34} J/Hz)

1. Give the number of protons, electrons, and neutrons in the following atoms:
   a. ^{23}Na b. ^{40}Ar c. ^{55}Fe d. ^{238}Pu

2. If you have a yellow light bulb and a blue one, which emits the largest wavelength, the higher frequency, the higher energy, and the fastest speed?

3. What is the frequency of light that has a wavelength of 425 nm?

4. What is the wavelength of light that has a frequency of 6.15×10^{14} Hz?

5. What is the energy of light that has a frequency of 7.05×10^{14} Hz?

6. What is the energy of light that has a wavelength of 1.95×10^{-7} m?

7. What is the wavelength of light that has an energy of 1.50×10^{-18} J?

8. Give the full electron configurations of the following atoms:
   a. Sc b. Cl c. Ca

9. Give the abbreviated electron configurations for the following atoms:
   a. P b. Mo c. Ba

10. What is wrong with the following electron configurations?
    a. $1s^{2}2s^{6}2p^{2}$
    b. $1s^{2}2s^{2}2p^{6}3s^{2}3p^{6}4s^{2}3d^{8}4p^{6}5s^{1}$
    c. $1s^{2}2s^{2}2p^{6}3s^{2}3p^{6}4s^{2}3d^{10}4p^{6}4d^{10}5s^{2}$

EXTRA PRACTICE PROBLEMS FOR MODULE 4

1. Draw the Lewis structures for the following atoms:
   a. Sr b. Kr c. S d. Bi

2. Give the chemical formulas for the following molecules:
   a. aluminum oxide b. calcium sulfide
   c. beryllium carbide d. potassium nitride

3. Give the chemical formulas for the following molecules:
   a. chromium(III) chloride b. zinc(I) sulfide c. titanium(IV) phosphide

4. Order the following atoms in terms of increasing ionization energy: Sr, Ra, Mg.

5. Which atom has the greatest desire for extra electrons: Ge, Ca, or Br?

6. Order the following in terms of increasing atomic radius: P, Mg, Al, Ar.

7. What is the Lewis structure for NF_3?

8. What is the Lewis structure for CCl_2S?

9. What is the Lewis structure for SiS_2?

10. What is the Lewis structure for CS?

11. What is the Lewis structure for SBr_2?

12. Which of the molecules in problems 6–10 would be the hardest to break, based on their Lewis structures?

EXTRA PRACTICE PROBLEMS FOR MODULE 5

1. Give the chemical formulas for the following compounds:
   a. lithium sulfite b. ammonium phosphate
   c. calcium carbonate d. aluminum dichromate

2. Name the following compounds:
   a. NH_4ClO_2 b. $Mn(NO_2)_2$ c. $Be_3(PO_4)_2$ d. $NiPO_4$

3. Determine the shape of a CS molecule. Give its bond angle and draw a picture of it.

4. Determine the shape of a SiI_4 molecule. Give its bond angle and draw a picture of it.

5. Determine the shape of a NF_3 molecule. Give its bond angle and draw a picture of it.

6. Determine the shape of a CCl_2S molecule. Give its bond angle and draw a picture of it.

7. Determine the shape of a SF_2 molecule. Give its bond angle and draw a picture of it.

8. Identify each of the molecules in problems 3–7 as polar covalent or nonpolar covalent.

9. Classify each of the following molecules as ionic, polar covalent, or nonpolar covalent:
   a. $SiFCl_3$ b. SiS_2 c. AlF_3 d. N_2 e. CCl_4 f. PF_3

10. Which of the substances in problem 9 would you expect to dissolve in water?

EXTRA PRACTICE PROBLEMS FOR MODULE 6

1. If a liquid goes through a phase change and all you know is that the molecules sped up and moved farther apart, what phase did the liquid turn in to?

2. Describe what the inside of a container would look like if all 3 phases of acetone (fingernail polish remover) were present at the same time.

3. Classify the following as physical or chemical changes:
   a. A window is broken.
   b. Eggs, flour, sugar, milk, and yeast are baked into bread.
   c. Water boils.
   d. A wood log is burned in a fireplace.

4. If the mass of a liquid is 23.13 grams and its volume is 35.0 mL, what is its density?

5. The density of a metal is 2.12 grams per mL. What is the volume of 45 g of the metal?

6. What is the mass of 0.67 liters of gold? The density of gold is 19.3 g/mL.

7. Balance the following equation:

$$HCl\,(aq) + Ca\,(s) \longrightarrow CaCl_2\,(aq) + H_2\,(g)$$

8. Balance the following equation:

$$HBr\,(aq) + Na\,(s) \longrightarrow NaBr\,(aq) + H_2\,(g)$$

9. Balance the following equation:

$$AgCl\,(aq) + Zn\,(s) \longrightarrow ZnCl_2\,(aq) + Ag\,(s)$$

10. Balance the following equation:

$$MgCl_2\,(aq) + H_2S\,(aq) \longrightarrow MgS\,(s) + HCl\,(aq)$$

11. Balance the following equation:

$$CO_2\,(g) + H_2O\,(l) \longrightarrow C_{18}H_{32}O_{16}\,(s) + O_2\,(g)$$

12. Balance the following equation:

$$FeCl_3 + Na_2CO_3 \longrightarrow Fe_2C_3O_9 + NaCl$$

13. Balance the following equation:

$$NH_4Cl + Ba(OH)_2 \longrightarrow BaCl_2 + NH_3 + H_2O$$

EXTRA PRACTICE PROBLEMS FOR MODULE 7

1. Balance the following combustion reaction: $CoS_2 + O_2 \rightarrow Co_2O_3 + SO_2$
2. What is the balanced chemical equation for the formation of C_2H_4O?
3. What is the balanced chemical equation for the complete combustion of $C_{16}H_{34}$?
4. What is the mass in kilograms of a mercury atom (Hg)?
5. What is the mass in grams of a lithium carbonate molecule? (1.00 amu = 1.66×10^{-24} g)
6. How many moles of H_2CO_3 are in a 1,250 g sample of the compound?
7. What is the mass of a $CaCl_2$ sample if it contains 0.344 moles of the compound?
8. A sample of potassium chloride has a mass of 25.0 g. How many moles is that?
9. How many grams are contained in a 9.02 mole sample of carbon dioxide?
10. How many atoms are in a 100.0 g sample of copper?
11. 1.78 kg of nickel (IV) bromide contains how many molecules?
12. In the formation of ammonium phosphate, how many moles of N_2 are required to make 15 moles of product?
13. In the complete combustion of C_3H_8, how many moles of CO_2 will be made when 13.2 moles of C_3H_8 are burned?

EXTRA PRACTICE PROBLEMS FOR MODULE 8

1. Nitric acid is manufactured with the following reaction:

$$3NO_2 + H_2O \longrightarrow 2HNO_3 + NO$$

How many grams of nitrogen dioxide are required to produce 2,945 kg HNO_3 in excess water?

2. Tungsten metal, W, is produced by the following reaction:

$$WO_3 + 3H_2 \longrightarrow W + 3H_2O$$

How many grams of tungsten can be made from 2.0×10^6 g of hydrogen in excess WO_3?

3. How many grams of WO_3 would be used in the above reaction under the conditions specified in problem 2?

4. Carbon tetrachloride is made according to the following reaction:

$$CS_2 + 3Cl_2 \longrightarrow CCl_4 + S_2Cl_2$$

How many grams of CS_2 are needed to make 8.0 kg of CCl_4?

5. To make a phosphorus fertilizer ($CaH_4P_2O_8$), agricultural companies use the following reaction:

$$Ca_3P_2O_8 + 2H_2SO_4 + 4H_2O \longrightarrow CaH_4P_2O_8 + 2CaH_4SO_6$$

If 5.000×10^3 grams of $Ca_3P_2O_8$ are reacted with excess H_2SO_4 and H_2O, how many grams of fertilizer can be made?

6. In the following reaction:

$$Mg\ (s) + 2H_2O\ (l) \longrightarrow Mg(OH)_2\ (s) + H_2\ (g)$$

What mass of Mg must you use to make 2.8×10^4 g of $Mg(OH)_2$?

7. Which of the following are empirical formulas?

a. SO_3 b. $K_2S_2O_3$ c. $C_6H_{12}O_6$ d. P_2H_4

8. Decomposition of 200.0 g of a dark green powder yields 79.4 g K, 55.8 g Mn, and an unknown amount of oxygen. What is the empirical formula?

9. The mass of the molecule above is 197.1 amu. What is the molecular formula?

10. A compound has an empirical formula of CH_2O. If its molecular mass is 60.0 amu, what is its molecular formula?

EXTRA PRACTICE PROBLEMS FOR MODULE 9

1. Identify the acid and base in the following reaction:

$$H_2CO_3 + H_2O \longrightarrow H_3O^+ + HCO_3^-$$

2. Identify the acid and base in the following reaction:

$$KOH + C_2H_6O \longrightarrow KC_2H_5O + H_2O$$

3. Identify the acid and base in the following reaction:

$$2HCl + Mg(OH)_2 \longrightarrow 2H_2O + MgCl_2$$

4. What is the reaction between HI and aluminum hydroxide?

5. What is the reaction between H_3PO_4 and sodium hydroxide?

6. What is the reaction between H_3PO_4 and the covalent base CH_5N?

7. Give the concentration (in M) of each of the following solutions:
   a. 2.22 moles of HCl are dissolved in enough water to make 1.5 liters of solution.
   b. 23.1 grams of NaOH are dissolved in enough water to make 1,000.0 mL of solution.
   c. 29.0 grams of H_2CO_3 are dissolved in enough water to make 400.0 mL of solution.

8. A chemist has a stock solution that is 20.34 M HCl. He wants to make 1,500.0 mL of 3.45 M HCl. What should he do?

9. You need to make 100.0 mL of a 9.0 M solution of KCl. All you can find is a 10.22 M solution of KCl. What do you do?

10. A chemist needs to determine the concentration of an unlabeled bottle of HBr. She titrates 75.0 mL of it with 1.14 M NaOH. It takes 135.0 mL of NaOH to reach the endpoint. What is the concentration of the acid?

11. 25.0 mL of an unknown concentration of $Mg(OH)_2$ is titrated with 1.00 M HCl. The endpoint is reached when 1.76 mL of acid are added. What is the concentration of the $Mg(OH)_2$?

EXTRA PRACTICE PROBLEMS FOR MODULE 10

(K_f for water is $1.86\frac{°C}{m}$; K_b for water is $0.512\frac{°C}{m}$)

1. You are having trouble dissolving a gas into a liquid. In terms of temperature, what should you do to increase the solubility of the gas?

2. The following reaction is performed in a lab:

$$3K_2CO_3\,(aq) + 2Al(NO_3)_3\,(aq) \rightarrow Al_2(CO_3)_3\,(s) + 6KNO_3\,(aq)$$

If 382 mL of 1.25 M aluminum nitrate is added to an excess of potassium carbonate, how many grams of aluminum carbonate will be produced?

3. Potassium cyanide can be made in the following way:

$$KOH\,(aq) + HCN\,(aq) \rightarrow H_2O\,(l) + KCN\,(aq)$$

How many mL of a 1.51 M solution of HCN would be required to make 25.0 grams of potassium cyanide? Assume that there is excess KOH.

4. When copper is mixed with nitric acid, the following reaction occurs:

$$3Cu\,(s) + 8HNO_3\,(aq) \rightarrow 3Cu(NO_3)_2\,(aq) + 2NO\,(g) + 4H_2O\,(l)$$

How many mL of a 3.50 M solution of HNO_3 would be required to completely use up 500.0 grams of solid copper in this reaction?

5. What is the molality of a solution in which 100.0 g of $Mg(NO_3)_2$ are dissolved in 500.0 grams of water?

6. How many grams of $CaCl_2$ would you have to add to 1,250 g of water to make a solution with a concentration of 2.0 m?

7. If you wanted to protect water from freezing, which compound would accomplish this best—$Ca_3(PO_4)_2$, $Al(NO_3)_3$, or NaCl? Assume that the molality of the solution is the same in each case.

8. What is the freezing point of a solution that is made by dissolving 20.0 grams of KF in 100.0 grams of water?

9. If you want to lower water's freezing point 10.00°C by adding $CaCl_2$, what must be the molality of the salt solution?

10. What is the boiling point of a solution made by mixing 50.0 g ammonium sulfide with 375.0 grams of water?

EXTRA PRACTICE PROBLEMS FOR MODULE 11

1. A balloon is filled with air so that its volume is 11.2 L at 25.0°C. It is then lowered into a vat of liquid nitrogen so that it reaches a temperature of –196.0°C. Assuming that the pressure stays constant, what is the new volume?

2. A balloon is filled with helium at a temperature of 25.0°C and a pressure of 755 mm Hg so that its volume is 13.3 L. It then rises to an altitude where the temperature is 15.0°C and the pressure is 625 mm Hg. What is its volume at that altitude?

3. A chemist collects 56.7 mL of nitrogen gas at 25°C and 370 torr. What would the volume of the nitrogen be at STP?

4. A mixture of nitrogen and oxygen gases is stored at a temperature of 25.0°C and a pressure of 2.0 atm. If the partial pressure of nitrogen in the mixture is 1.4 atm, what is the partial pressure of oxygen in the mixture?

5. A mixture of gases has a pressure of 15.0 atm. The mixture is analyzed and found to contain 2.00 g of H_2, 40.0 g of CO_2, and 14.5 g of N_2. What is the partial pressure of each gas in the mixture?

6. What is the volume of 24.2 grams of nitrogen gas at STP?

7. What is the pressure of 15.0 grams of nitrogen gas when the temperature is 21.0°C and the gas occupies 1.00 L of space?

8. A chemist performs the following reaction:

$$CS_2\,(s) + 3O_2\,(g) \longrightarrow CO_2\,(g) + 2SO_2\,(g)$$

If the chemist starts with 11.0 grams of CS_2 and an excess of O_2, what volume of sulfur dioxide will be produced at a temperature of 341°C and a pressure of 2.1 atm?

9. It is difficult to make solid titanium, but since it is used in the manufacture of airplane engines and frames, it must be produced. When mined, titanium ore is mostly TiO_2. To make solid titanium, the titanium ore must first be converted to titanium tetrachloride via the following reaction:

$$3TiO_2\,(s) + 4C\,(s) + 6Cl_2\,(g) \longrightarrow 3TiCl_4\,(g) + 2CO_2\,(g) + 2CO\,(g)$$

How many liters of Cl_2 gas at STP would be required to completely convert 1.0000 kilogram of TiO_2 into $TiCl_4$?

10. Heating table salt and liquid sulfuric acid produces gaseous HCl, a strong acid:

$$NaCl\,(s) + H_2SO_4\,(l) \longrightarrow NaHSO_4\,(s) + HCl\,(g)$$

How many liters of HCl (g) will be produced if 25.0g of H_2SO_4 are heated in excess NaCl? Assume the pressure is 1.00 atm and the temperature is 350.0°C.

EXTRA PRACTICE PROBLEMS FOR MODULE 12

1. Which substance takes the least energy to heat up—water or glass? (Refer to table 12.1.)

2. 10.0 g each of aluminum and silver are placed side by side on a burner. After 2.0 minutes, the silver reaches a temperature of 50.0°C. Will the aluminum temperature be higher or lower than 50.0°C? (Refer to table 12.1.)

3. A liquid has a specific heat of 2.40 J/g °C. A 150.0 gram sample of the liquid is initially at 25.0°C. If 3,250 J of energy are removed from the liquid, what is the new temperature?

4. A total of 506 J of heat are added to 50.0 g of water initially at 45.0°C. What is the final temperature of the water?

5. How much heat is necessary to raise the temperature of a 15.0 g metal from 15.0°C to 35.0°C?

$$(c_{metal} = 1.91 \frac{J}{g\ °C})$$

6. A 500.0 g mass of unknown metal at 100.0°C is dropped in a calorimeter initially at 25.0°C. There are 120.0 g of water in the calorimeter. The final temperature of the water, calorimeter, and metal is 30.0°C. What is the c of the metal? You can ignore the calorimeter.

7. A 125.0 g mass of metal at 112.4 °C is dropped in a calorimeter (c = 1.40 J/g °C m = 5.0 g) containing 150.0 g of water. The water and calorimeter are initially at 25.0°C. At the end of the experiment, the temperature of the water, calorimeter, and metal is 31.0°C. What is the c of the metal?

8. A 50.0 g chunk of metal (c = 0.506 J/g °C) is heated to 60.0°C. It is then dropped into a calorimeter (c = 1.40 J/g °C, m = 5.0 g) that contains 200.0 g of an unknown liquid. The temperature of the liquid rises from 15.0°C to 25.0°C. What is the c of the liquid?

9. A 150.0 g piece of metal (c = 0.560 J/g °C) is dropped into a calorimeter that contains 400.0 g of water. The metal was initially at 100.0°C and the calorimeter was initially at 25.0°C. The mass of the calorimeter is 5.0 g. If the temperature of the calorimeter increased to 27.0°C, what is the c of the calorimeter?

10. A 75.0 g sample of copper (c = 0.3851 J/g °C) is initially at 100.0°C. It is dropped into a calorimeter (c = 1.40 J/g °C; m = 5.0 g) that is initially at 25.0°C. The final temperature of the experiment is 30.0°C. What mass of water is in the calorimeter?

EXTRA PRACTICE PROBLEMS FOR MODULE 13

1. Which of the following substances will have a ΔH_f of zero?

$$Ca(OH)_2\ (aq),\ Br^-\ (aq),\ SO_2\ (g),\ Cl\ (g),\ Cl_2\ (g),\ N_2\ (l),\ N_2\ (g)$$

2. Which of the following diagrams indicates an endothermic reaction? What is the ΔH of that reaction?

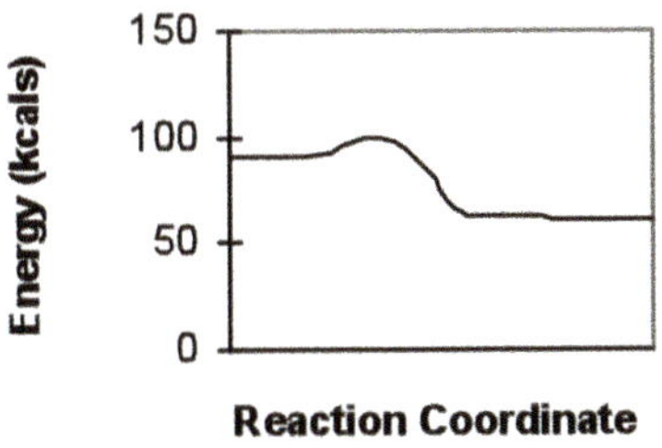

I

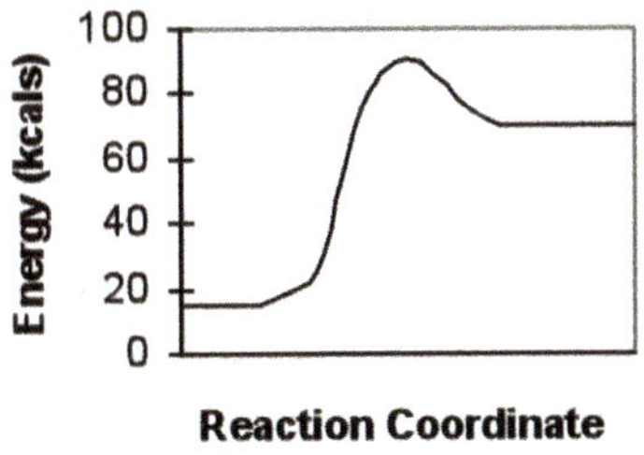

II

3. For the diagrams above, which represents the reaction with the highest activation energy?

4. Is it possible for a reaction to occur if the reaction results in a decrease in the entropy of the chemicals involved in the reaction?

5. What is the sign of ΔS for the following reactions?
   a. $H_2SO_4\ (aq) + 2NH_3\ (aq) \rightarrow (NH_4)_2SO_4\ (aq)$
   b. $4NH_3\ (g) + 5O_2\ (g) \rightarrow 4NO\ (g) + 6H_2O\ (g)$
   c. $NaCl\ (s) + H_2SO_4\ (l) \rightarrow NaHSO_4\ (s) + HCl\ (g)$
   d. $NaCl\ (aq) + H_2O\ (l) + CO_2\ (g) \rightarrow NaHCO_3\ (aq) + HCl\ (aq)$

6. What is the ΔH of the following reaction? (Use bond energies.)

$$CO_2 + 4HCl \rightarrow CCl_4 + 2H_2O$$

7. Use Hess's law to determine the $\Delta H°$ of:

$$4NH_3\ (g) + 5O_2\ (g) \rightarrow 4NO\ (g) + 6H_2O\ (g)$$

8. If the standard absolute entropy of O_2 (g) is 205.0 (J/mole · K), what is the $\Delta S°$ of the reaction in problem 7?

9. If the $\Delta H°$ of a certain reaction is −623 kJ/mole and the $\Delta S°$ is −114 (J/mole · K), what is the temperature range for which this reaction is spontaneous?

EXTRA PRACTICE PROBLEMS FOR MODULE 14

1. A chemistry book lists the rate constant for a reaction as $34.5\frac{1}{M \cdot s}$. What is the overall order of the reaction?

2. The order of a chemical reaction with respect to one of its reactants is 3. If you quadruple the concentration of that reactant, what happens to the rate?

3. A chemist does a reaction rate analysis on the following reaction:

$$2NO\,(g) + Cl_2\,(g) \longrightarrow 2NOCl\,(g)$$

She collects the following data:

Trial	Initial Concentration of NO (M)	Initial Concentration of Cl_2 (M)	Instantaneous Reaction Rate (M/s)
1	0.100	0.050	0.113
2	0.200	0.050	0.452
3	0.200	0.100	0.904

What is the rate equation for this reaction?

4. A chemist does a reaction rate analysis on the following reaction:

$$C_3H_6Br_2 + 3I^- \longrightarrow C_3H_6 + 2Br^- + I_3^-$$

He collects the following data:

Trial	Initial Concentration of $C_3H_6Br_2$ (M)	Initial Concentration of I^- (M)	Instantaneous Reaction Rate (M/s)
1	0.100	0.200	0.234
2	0.200	0.200	0.468
3	0.100	0.400	0.468

What is the rate equation for this reaction?

5. A chemist runs a chemical reaction at 25°C and decides that it proceeds far too slowly. As a result, he decides that the reaction rate must be increased by a factor of 16. At what temperature should the chemist run the reaction to achieve this goal?

6. In the reaction mechanism below, indicate what substance is acting like a catalyst.

Step 1. $2N_2O\,(g) + Cl_2\,(g) \longrightarrow 2N_2\,(g) + 2ClO\,(g)$

Step 2. $2ClO\,(g) \longrightarrow Cl_2\,(g) + O_2\,(g)$

Is this a heterogeneous or a homogeneous catalyst?

7. What does a catalyst do to the size of the hill between reactants and products in an energy diagram?

EXTRA PRACTICE PROBLEMS FOR MODULE 15

1. Three solutions have the following pH levels:

   Solution A: pH = 3   Solution B: pH = 6   Solution C: pH = 12

   Which is (are) the acidic solution(s)?

2. Three acid solutions of equal concentration have the following pH levels:

   Solution A: pH = 4   Solution B: pH = 6   Solution C: pH = 1

   Which solution is made with the acid that has the *largest* ionization constant?

3. A chemist is studying the following equilibrium:

   $$NH_4^+\,(aq) + CO_3^{2-}\,(aq) \rightleftharpoons NH_3\,(aq) + HCO_3^-\,(aq)$$

   He starts out with NH_4^+ at a concentration of 4.0 M and a CO_3^{2-} concentration of 4.0 M. The concentrations of NH_4^+, CO_3^{2-}, NH_3, and HCO_3^- at equilibrium are 1.60 M, 1.60 M, 2.4 M, and 2.4 M, respectively. What is the value of the equilibrium constant?

4. The following reaction:

   $$2SO_2\,(g) + O_2\,(g) \rightleftharpoons 2SO_3\,(g)$$

   has an equilibrium constant equal to 4.4 1/M. If the following concentrations are present: $[SO_2]$ = 0.140 M, $[O_2]$ = 0.147 M, $[SO_3]$ = 0.280 M, is the reaction at equilibrium? If not, which way must it shift to reach equilibrium?

5. Consider the following reaction that has reached equilibrium:

   $$2NH_3\,(g) \rightleftharpoons N_2\,(g) + 3H_2\,(g) \qquad \Delta H = 92.2\text{ kJ}$$

   a. What will happen to the concentration of H_2 if the temperature is lowered?
   b. What will happen to the concentration of NH_3 if the pressure is lowered?
   c. What will happen to the concentration of N_2 if the concentration of NH_3 is raised?

6. Consider the following reaction that has reached equilibrium:

   $$2CO\,(g) \rightleftharpoons CO_2\,(g) + C\,(s) \qquad \Delta H = -172.5\text{ kJ}$$

   a. What will happen to the concentration of CO if the temperature is lowered?
   b. What will happen to the concentration of CO if the pressure is lowered?
   c. What will happen to the concentration of CO if carbon is removed?
   d. What will happen to the amount of carbon if more CO is added?

7. What is the equation for the acid ionization constant of NH_4^+?

EXTRA PRACTICE PROBLEMS FOR MODULE 16

1. Give the oxidation numbers of all atoms in the following substances:
   a. Mg   b. K^+ c. P_4 d. Br_2

2. Give the oxidation numbers of all atoms in the following compounds:
   a. $KMnO_4$ b. $NaCHO_2$ c. $Zn(OH)_2$ d. SCl_2

3. An atom changes its oxidation number from –1 to +2. Is it oxidized or reduced? How many electrons did it take to do this?

4. An atom changes its oxidation number from +4 to 0. Is it oxidized or reduced? How many electrons did it take to do this?

5. Determine whether or not each of the following reactions is a redox reaction.
   a. $2Mg\,(s) + O_2\,(g) + 2H_2O\,(l) \longrightarrow 2Mg^{2+}\,(aq) + 4OH^-\,(aq)$
   b. $2H_2O\,(l) + 2Cl^-\,(aq) \longrightarrow H_2\,(g) + Cl_2\,(g) + 2OH^-\,(aq)$
   c. $H_2CO_3\,(aq) + 2NH_3\,(aq) \longrightarrow 2NH_4^+\,(aq) + CO_3^{2-}\,(aq)$
   d. $Zn(NO_3)_2\,(aq) + 2NaOH\,(aq) \longrightarrow Zn(OH)_2\,(s) + 2NaNO_3\,(aq)$

6. For the redox reactions in problem 5, which substance is oxidized? Which is reduced?

7. For the redox reactions in problem 5, how many electrons are transferred in each reaction?

8. A Galvanic cell runs on the following reaction:

   $$3Mg\,(s) + 2Al^{3+}\,(aq) \longrightarrow 3Mg^{2+}\,(aq) + 2Al\,(s)$$

   Draw a diagram for this Galvanic cell, labeling the electron flow, the anode and cathode, and the positive and negative sides of the Galvanic cell.

9. A Galvanic cell runs on the following reaction:

   $$F_2\,(aq) + Sn\,(s) \longrightarrow 2F^-\,(aq) + Sn^{2+}\,(aq)$$

   Draw a diagram for this Galvanic cell, labeling the electron flow, the anode and cathode, and the positive and negative sides of the Galvanic cell.

10. A Galvanic cell runs on the following reaction:

    $$2Cr\,(s) + 3Cu^{2+}\,(aq) \longrightarrow 2Cr^{3+}\,(aq) + 3Cu\,(s)$$

    Draw a diagram for this Galvanic cell, labeling the electron flow, the anode and cathode, and the positive and negative sides of the Galvanic cell.

APPENDIX C

COMPLETE LIST OF LAB SUPPLIES

Note: *Items in* **boldface blue type** *are found in the laboratory equipment set that is available through Apologia. The other materials are available at grocery stores, hardware stores, or drug stores.*

MODULE 1

- **Safety goggles**
- Meterstick (A yardstick will work as well.)
- 2 eight-inch or larger balloons
- 2 pieces of string long enough to tie the balloons to the meterstick
- Tape
- Tall glass
- Paper towel
- Sink full of water
- Book (not oversized)
- Metric and English rulers

MODULE 2

- **Safety goggles**
- **100 mL beaker** (A small juice glass can be used instead.)
- **250 mL beaker** (A large juice glass can be used instead.)
- **Watch glass** (A small saucer can be used instead. It must cover the mouth of the 100 mL beaker or juice glass listed above.)
- Teaspoon
- Lye (This is commonly sold with the drain cleaners in hardware stores and supermarkets. Make sure that your bottle is labeled 100% lye. If you cannot find lye, you can order it online. It may be listed under its chemical name, sodium hydroxide. *Always use this chemical in a well-ventilated area.*)
- White vinegar
- Several leaves of red (often called purple) cabbage

- Water
- Small pot for boiling water
- Measuring cup
- Heat source such as a stove or alcohol burner
- Mass scale
- Stirring rod (A spoon will work.)
- Rubber cleaning gloves
- Distilled water (Half a gallon is plenty.)
- Baking soda
- Sugar
- 9-volt battery *(Do not use an electrical outlet in place of the battery. The electrical energy in a wall socket will hurt you and can kill you!)*
- 2 pieces of wire (preferably insulated), each of which is at least 15 cm long
- Scissors or wire cutters to strip insulation from wire (if it is insulated)
- Tape (preferably black electrical tape)
- ½ teaspoon
- Sand (Kitty litter is an acceptable substitute, but don't use the kind that clumps.)
- Table salt
- Funnel
- Filter paper (You can cut a circle out of the bottom of a coffee-maker filter.)

MODULE 3

- Safety goggles
- Comb
- Aluminum foil
- Cellophane (Scotch) tape
- 2 plain white sheets of paper (There cannot be lines on them.)
- Bright red marker (A crayon will also work, but a marker is better.)

MODULE 4

There are no experiments in this module.

MODULE 5

- Safety goggles
- Glass of water
- Vegetable oil
- Styrofoam or paper cup
- Comb
- Pen
- 2 test tubes (Thin glasses will work, but they must be transparent.)
- Table salt
- Water

MODULE 6

- Safety goggles
- Plastic graduated cylinder (A rain gauge will work as well. A measuring cup may work, but if you have a frost-free freezer, the opening of whatever you use should be very small.)
- Egg in its shell
- Toilet bowl cleaner (The list of ingredients must include hydrochloric acid or hydrogen chloride. The Works was the brand used when this experiment was tested. You can use vinegar if you cannot find a proper toilet bowl cleaner, but the experiment will have to sit overnight in order for it to work. However, if you do use vinegar, the egg will be much more interesting to observe!)
- Tall glass
- Sink
- Spoon
- Rubber cleaning gloves
- Empty milk jug
- Water
- Small pot
- Stove
- Hot pads
- 2 glass canning jars or peanut butter jars, both the same size
- Food coloring (any color)
- Vegetable oil
- Graduated cylinder or measuring cups
- Maple syrup (Natural syrup does not work as well as something like Mrs. Butterworth's.)
- Large glass
- Mass scale

MODULE 7

- Safety goggles
- Eyedropper
- Stirring rod (or spoon)
- Water
- Large glass (at least 16 ounces)
- Graduated cylinder (Measuring cups and measuring spoons can be used, but they will be less precise.)
- Dishwashing liquid (It must be the kind used for washing dishes by hand.)
- Large bowl (It should have a diameter larger than 10 inches but smaller than 12 inches. If you don't have a bowl in that size range, then use one larger than 12 inches; it will just make it a little harder to use the ruler.)
- Pepper
- Ruler

MODULE 8

- Safety goggles
- Baking soda
- Vinegar
- String or tape measure
- Graduated cylinder (or measuring cups and spoons)
- Ruler
- Round balloon
- Plastic 2-liter bottle (or other large bottle)
- Mass scale
- Funnel or butter knife (Read the experiment to see what is meant.)

MODULE 9

- Safety goggles
- Red litmus paper
- Blue litmus paper
- Apple
- Orange juice or soda pop
- Toilet bowl cleaner (Both The Works and Lime-Away have been tested, but any toilet bowl cleaner designed to combat lime should work.)
- Bar soap (Make sure it isn't labeled "pH balanced.")
- All-purpose cleaner (Windex and 409 have been tested, but any spray cleaner not specifically designed for toilets should work.)
- Powdered drain cleaner (like Red Devil Lye) or scouring powder (like Comet)
- 4 test tubes (Small cups will work.)
- Watch glass (A small saucer will work.)
- Stirring rod (or spoon)
- Rubber gloves (Wear gloves whenever you use powdered drain cleaners and toilet bowl cleaners because these chemicals are caustic.)
- Eyedropper
- Mass scale
- Distilled water (available at any grocery store)
- Sheet of unlined white paper
- A few leaves of red cabbage (It must be red cabbage, not regular cabbage.)
- 2 beakers (If you don't have beakers, use a short, fat glass that is transparent and a small pot to boil water in.)
- Alcohol burner or stove for heating
- Graduated cylinder (Measuring cups and spoons will work, but the experiment will be much harder.)
- Clear ammonia solution (This is sold with the cleaning supplies in most supermarkets. It must be clear. A colored solution will mess up the endpoint.)
- Clear vinegar (Once again, colored vinegar will mess up the endpoint.)

MODULE 10

- Safety goggles
- 250 mL beaker (A short, fat glass or canning jar might work, but **be careful.** *Glasses tend to crack when subjected to the temperature extremes of experiment 10.1*. If you don't have a beaker, you may want to just read this experiment, unless you aren't afraid of losing a glass or 2.)
- 100 mL beaker (Another short, fat glass or canning jar might work.)
- Beaker (A ¼ measuring cup will work.)
- Alcohol burner or stove for heating
- Stirring rod (A spoon will work.)
- Thermometer
- Mass scale
- Table salt
- Filter paper (You can cut a circle out of the bottom of a coffee filter.)
- Funnel
- Oven mitt
- 2 beakers (Two saucepans will work.)
- Test tube (A tall, thin glass will work, but it must fit easily in the saucepans.)
- Ice
- Cold soda pop (Pepsi, Coke, Sprite, etc. It must be carbonated.)
- Lye (This is commonly sold with the drain cleaners in hardware stores and supermarkets. Make sure that your bottle is labeled 100% lye. If you cannot find lye, you can order it online. It may be listed under its chemical name, sodium hydroxide. *Always use this chemical in a well-ventilated area.*)
- Rubber gloves
- Water
- Sink
- Tablespoon
- ¼ measuring teaspoon

MODULE 11

- Safety goggles
- Kitchen sink or large bucket
- Towels or napkins
- Beaker
- Butane lighter (A cigarette lighter with a clear case will work best.)
- Mass scale
- Thermometer
- Weather report that contains the atmospheric (sometimes called barometric) pressure for the day
- Plastic 2-liter bottle
- Round balloon with an 8-inch diameter
- Vinegar
- Baking soda
- Seamstress's tape measure (A piece of string and a ruler will work as well.)

MODULE 12

- Safety goggles
- Thermometer
- 250 mL beaker (A small glass container will do, but make sure it is safe for boiling water.)
- Water (preferably distilled water, which can be purchased at any supermarket)
- Ice (preferably crushed)
- Graph paper or computer with spreadsheet program
- Alcohol burner or stove for heating
- Mass scale
- 2 Styrofoam cups
- A chunk of metal that has mass of at least 30 grams (a lead sinker or a very large steel nut, for example)
- Boiling water (either in a pot or a beaker)
- Kitchen tongs

MODULE 13

- Safety goggles
- 2 Styrofoam coffee cups
- Thermometer
- Vinegar
- Mass scale
- Measuring tablespoon and ½ teaspoon
- Lye (This is commonly sold with the drain cleaners in hardware stores and supermarkets. Make sure that your bottle is labeled 100% lye. If you cannot find lye, you can order it online. It may be listed under its chemical name, sodium hydroxide. *Always use this chemical in a well-ventilated area.*)

MODULE 14

- Safety goggles
- Liquid toilet bowl cleaner (The Works was used when this experiment was tested, but any liquid that contains hydrochloric acid or hydrogen chloride should work. Stay away from the thickened toilet bowl cleaners that cling, however.)
- Antacid tablets (TUMS works best. In principle, any antacid tablet that has calcium carbonate as its main ingredient should work.)
- Sharp, non-serrated knife
- Spoon
- Stirring rod (Something thin enough to stir the contents of the test tubes.)
- 4 test tubes (Small glass containers will work.)
- Alcohol burner or stove for heating
- Large beaker (A short, fat glass will work.)
- Small beaker (A short, fat glass will work. If you are not using test tubes, make sure that one of your glass containers fits into this.)
- Rubber gloves

- Watch glass (A small saucer that covers the beaker will do.)
- Graduated cylinder (A ⅛ measuring cup will work.)
- ¼ measuring teaspoon
- Hydrogen peroxide (available at supermarkets and drug stores)
- Baker's yeast (Any kind of bread yeast, even bread machine yeast, will do.)

MODULE 15

- Safety goggles
- 3 plastic 2-liter bottles (Plastic milk cartons will work as well.)
- 4 small cups
- 2 bowls that are taller than the small cups
- Serrated knife (like a steak knife)
- Small Phillips-head screwdriver
- Water
- Person to help you
- Kitchen counter
- Towels
- 2 test tubes
- 2 eyedroppers
- 2 beakers
- 2 small cups
- Clear ammonia solution (This is sold with the cleaning supplies in most supermarkets. It must be clear. A colored solution will mess up the colors you are supposed to see.)
- White vinegar (It must be white. Colored vinegar will mess up the colors you are supposed to see.)
- Alcohol burner or stove for heating
- Pot for boiling water on the stove
- Ice (preferably crushed)
- A few leaves of red cabbage (It must be red cabbage, not regular cabbage.)

MODULE 16

- Safety goggles
- Large bowl
- Blank notebook paper without lines
- Scissors
- Cotton swab or small paintbrush
- Lemon juice (This must be *real* lemon juice, either fresh or from concentrate.)
- Iodine solution (This is available in large drugstores. The pharmacy department within your supermarket will probably not have this, and neither will a small corner drugstore. A major chain will have it, however. Please note that *iodide* will not work. Iodide, as you should know, is I^-, while iodine is I_2.)
- Water
- Measuring cups

- Medicine dropper
- Stirring rod (A **plastic** spoon or knife will work as well. Do not use any flatware around the iodine solution, as iodine will rust metal.)
- 4 juicy lemons
- 4 copper pennies (Pennies from before 1982 have higher copper content. You can also use copper wire instead of a penny.)
- 4 two-inch zinc coated (galvanized) nails
- 5 small wires, ideally with alligator clips
- Volt meter (optional)
- LED (Light Emitting Diode) bulb

REFERENCES

Ahrens, Richard. 1977. "William Prout (1785–1850) A Biographical Sketch." *The Journal of Nutrition* 107:15–23.

Corey, Michael Anthony. 2002. *The God Hypothesis: Discovering Design in Our "Just Right" Goldilocks Universe*. Lanham, MD: Rowman & Littlefield Publishers.

Encyclopædia Britannica Online. 2014. S. v. "William Thomson, Baron Kelvin." Accessed June 26. http://www.britannica.com/EBchecked/topic/314541/William-Thomson-Baron-Kelvin.

Graves, Dan. 1996. *Scientists of Faith: Forty-Eight Biographies of Historic Scientists and Their Christian Faith*. Grand Rapids, MI: Kregel Resources.

Kelvin Library. 2008. "Science Affirms Creative Power." Updated December 14, 2008. http://zapatopi.net/kelvin/papers/science_affirms_creative_power.html.

Lewis, Neil. 2013. "Robert Grosseteste." In *The Stanford Encyclopedia of Philosophy*, edited by Edward N. Zalta (summer 2013 edition). http://plato.stanford.edu/archives/sum2013/entries/grosseteste/.

MacIntosh, J. J. and Anstey, Peter. 2010. "Robert Boyle." In *The Stanford Encyclopedia of Philosophy*, edited by Edward N. Zalta (fall 2010 edition). http://plato.stanford.edu/archives/fall2010/entries/boyle/.

Morris, Henry M. 1982. *Men of Science, Men of God: Great Scientists Who Believed the Bible*. San Diego, CA: Creation-Life.

Pagel, Walter. 1944. *The Religious and Philosophical Aspects of van Helmont's Science and Medicine*. Baltimore: Johns Hopkins.

Phillips, William D. 2006. "Ordinary Faith, Ordinary Science." In *Science and the Search for Meaning: Perspectives from International Scientists*, edited by Jean Staune, 193–208. West Conshohocken, PA: Templeton Foundation Press.

Stevenson, Francis Seymour. 1899. *Robert Grosseteste, Bishop of Lincoln: A Contribution to the Religious, Political and Intellectual History of the Thirteenth Century*. London: Macmillan.

Stokes, George G. 2010. *Memoir and Scientific Correspondence of the Late Sir George Gabriel Stokes, Bart*, vol. 1. Edited by Joseph Larmor. Cambridge: Cambridge University Press.

The Washington Times. 2001. "Putting Faith in Science." *The Washington Times*, October 22. http://www.washingtontimes.com/news/2001/oct/22/20011022-030258-4265r/.

INDEX

A

B

D

E

F

G

CONVERSION FACTORS

1.00 inch = 2.54 cm

1.000 slug = 14.59 kg

1.00 gallon = 3.78 L

1.00 amu = 1.66×10^{-24} g

1.000 calorie = 4.184 Joules

1 food calorie (Cal) = 1,000 calories (cal)

1.000 atm = 760.0 torr

1.000 atm = 760.0 mm Hg

1.000 atm = 101.3 kPa

1.000 atm = 760.0 torr = 760.0 mm Hg = 101.3 kPa

PHYSICAL CONSTANTS

Specific heat of water: c = cal $\frac{cal}{g \cdot °C}$ or 4.184 $\frac{J}{g \cdot °C}$

Avogadro's number: $6.02 \times 10^{23} \frac{\text{particles}}{\text{mole}}$

Speed of light (in vacuum): $3.0 \times 10^{8} \frac{m}{s}$

Planck's constant: $h = 6.63 \times 10^{-34} \frac{J}{Hz}$

Freezing point depression constant of water: $K_f = 1.86 \frac{°C}{m}$

Boiling point elevation constant of water: $K_b = 0.51 \frac{°C}{m}$

Ideal gas constant: $R = 0.0821 \frac{L \cdot atm}{mole \cdot K}$

Standard temperature and pressure (STP): T = 273 K, P = 1.00 atm

	Mass	Charge
Proton	1.673×10^{-27} kg	+1
Neutron	1.675×10^{-27} kg	0
Electron	9.109×10^{-31} kg	–1

THE PERIODIC TABLE OF ELEMENTS

1A	2A	3B	4B	5B	6B	7B	8B	8B	8B	1B	2B	3A	4A	5A	6A	7A	8A
1 **H** 1.01																	2 **He** 4.00
3 **Li** 6.94	4 **Be** 9.01											5 **B** 10.8	6 **C** 12.0	7 **N** 14.0	8 **O** 16.0	9 **F** 19.0	10 **Ne** 20.2
11 **Na** 23.0	12 **Mg** 24.3											13 **Al** 27.0	14 **Si** 28.1	15 **P** 31.0	16 **S** 32.1	17 **Cl** 35.5	18 **Ar** 39.9
19 **K** 39.1	20 **Ca** 40.1	21 **Sc** 45.0	22 **Ti** 47.9	23 **V** 50.9	24 **Cr** 52.0	25 **Mn** 54.9	26 **Fe** 55.8	27 **Co** 58.9	28 **Ni** 58.7	29 **Cu** 63.5	30 **Zn** 65.4	31 **Ga** 69.7	32 **Ge** 72.6	33 **As** 74.9	34 **Se** 79.0	35 **Br** 79.9	36 **Kr** 83.8
37 **Rb** 85.5	38 **Sr** 87.6	39 **Y** 88.9	40 **Zr** 91.2	41 **Nb** 92.9	42 **Mo** 96.0	43 **Tc** (98)	44 **Ru** 101.1	45 **Rh** 102.9	46 **Pd** 106.4	47 **Ag** 107.9	48 **Cd** 112.4	49 **In** 114.8	50 **Sn** 118.7	51 **Sb** 121.8	52 **Te** 127.6	53 **I** 126.9	54 **Xe** 131.3
55 **Cs** 132.9	56 **Ba** 137.3	57 **La** 138.9	72 **Hf** 178.5	73 **Ta** 181.0	74 **W** 183.8	75 **Re** 186.2	76 **Os** 190.2	77 **Ir** 192.2	78 **Pt** 195.1	79 **Au** 197.0	80 **Hg** 200.6	81 **Tl** 204.4	82 **Pb** 207.2	83 **Bi** 209.0	84 **Po** (209)	85 **At** (210)	86 **Rn** (222)
87 **Fr** (223)	88 **Ra** (226.0)	89 **Ac** (227)	104 **Rf** (265)	105 **Db** (268)	106 **Sg** (271)	107 **Bh** (270)	108 **Hs** (277)	109 **Mt** (276)	110 **Ds** (281)	111 **Rg** (280)	112 **Cn** (285)	113 **Nh** (284)	114 **Fl** (289)	115 **Mc** (288)	116 **Lv** (293)	117 **Ts** (294)	118 **Og** (294)

58 **Ce** 140.1	59 **Pr** 140.9	60 **Nd** 144.2	61 **Pm** (145)	62 **Sm** 150.4	63 **Eu** 152.0	64 **Gd** 157.2	65 **Tb** 158.9	66 **Dy** 162.5	67 **Ho** 164.9	68 **Er** 167.3	69 **Tm** 168.9	70 **Yb** 173.0	71 **Lu** 175.0
90 **Th** 232.0	91 **Pa** 231.0	92 **U** 238.0	93 **Np** (237)	94 **Pu** (244)	95 **Am** (243)	96 **Cm** (247)	97 **Bk** (247)	98 **Cf** (251)	99 **Es** (252)	100 **Fm** (257)	101 **Md** (258)	102 **No** (259)	103 **Lr** (262)

Nonmetals

Standard metals

Transition metals

Inner transition metals